Introduction to
CHEMISTRY

E. Russell Hardwick

University of California, Los Angeles

Burgess Publishing Company
Minneapolis, Minnesota

Acquisitions Editor: Gerhard Brahms, Jeff Holtmeier
Assistant Editor: Elizabeth Weinstein
Development Editor: Janilyn Richardson
Copy Editor: Sandra Chizinsky
Production Coordinator: Melinda Radtke
Production: Pat Barnes, Morris Lundin, Judy Vicars
Typesetter: TriStar Graphics

Cover: A paper chromatograph separation of a dye mixture. Developed by 10% acetone/water.

© 1984 by Burgess Publishing Company
Printed in the United States of America

Burgess Publishing Company
7108 Ohms Lane
Minneapolis, Minnesota 55435

Library of Congress Cataloging in Publication Data

Hardwick, E. Russell.
  Introduction to chemistry.

  Includes index.
  1. Chemistry. I. Title.
QD31.2.H377 1984   540   83-23146
ISBN 0-8087-4740-1

J I H G F E D C B A

Contents

PREFACE   xiii

INTRODUCTION   1

CHAPTER 1   MEASUREMENT AND PROBLEM SOLVING IN CHEMISTRY   5

1.1   Measurement and Uncertainty   6
1.2   Significant Figures   8
1.3   Solving Problems in Chemistry   9
1.4   Using Unit Analysis to Solve Problems   13
1.5   Summary   18
       Questions   18
       Problems   19

CHAPTER 2   SYSTEMS OF MEASUREMENT   21

2.1   Metric System of Units and SI   22
2.2   Measurement of Length   23
2.3   Units of Volume   24
2.4   Mass and Weight   26
2.5   Energy, Heat, and Temperature   28
2.6   Measurement of Temperature   31
2.7   Density   33
2.8   Laboratory Instruments of Measurement   35
2.9   Summary   37
       Questions   37
       Problems   38

CHAPTER 3   MATTER AND ENERGY   43

    3.1   **Matter**   **44**
    3.2   **Energy**   **45**
    3.3   **Conservation of Mass and Energy**   **48**
    3.4   **Homogeneity and Purity in Samples of Matter**   **49**
    3.5   **Separating and Identifying Substances**   **51**
    3.6   **Summary**   **54**
           Questions   **56**

CHAPTER 4   THE ATOMIC THEORY   58

    4.1   **Early Views of the Nature of Matter**   **59**
    4.2   **Chemical Analysis**   **59**
    4.3   **The Laws of Chemical Combination**   **62**
    4.4   **Dalton's Atomic Theory**   **63**
    4.5   **The Law of Multiple Proportions**   **66**
    4.6   **Deficiencies of Dalton's Theory**   **67**
    4.7   **Modern Atomic Theory**   **68**
    4.8   **Atomic Mass: Isotopes**   **70**
    4.9   **The Atomic Weight Scale**   **71**
    4.10   **Isotopic Abundance and Atomic Weight**   **73**
    4.11   **Summary**   **74**
           Questions   **74**
           Problems   **76**

CHAPTER 5   FORMULAS, REACTION EQUATIONS, AND THE MOLE   78

    5.1   **The Elements**   **79**
    5.2   **Compounds and Their Formulas**   **82**
    5.3   **Molecular Weight**   **84**
    5.4   **Names of Compounds**   **86**
    5.5   **Chemical Reactions and Their Equations**   **87**
    5.6   **Energy Changes in Chemical Reaction**   **92**
    5.7   **Reading Reaction Equations in Units of Moles**   **93**
    5.8   **The Atomic Weight Scale and the Mole**   **94**
    5.9   **Gram-Molecular Weight**   **97**
    5.10   **Summary**   **97**
           Questions   **98**
           Problems   **100**

CHAPTER 6   MASS RELATIONS IN COMPOUNDS   103

    6.1   **Review of Some Important Ideas**   **104**
    6.2   **Laboratory Data and the Numbers of Moles in Samples**   **105**
    6.3   **Formulas of Compounds and the Mole**   **105**

6.4   **Mass Calculations Using the Mole   107**
6.5   **Percent Composition of Compounds   109**
6.6   **Quantitative Calculations Involving Laboratory Data,
          Compound Formula, and Gram-Atomic Weight   111**
6.7   **Calculating the Actual Molecular Formula From the Empirical Formula   117**
6.8   **Summary   119**
          Questions   120
          Problems   121

CHAPTER 7   CALCULATIONS USING INFORMATION
                      FROM REACTION EQUATIONS   125

7.1   **The Mole Ratio Method of Making Calculations From Reaction Equations   126**
7.2   **Mole Ratios as Conversion Factors   128**
7.3   **Limiting Reactant Problems   130**
7.4   **Mole Ratio Problems Involving Mass Units   133**
7.5   **Limiting Reactant Problems Using Mass Units   136**
7.6   **Stoichiometric Problems Involving Density Data   138**
7.7   **Calculating the Energies of Reactions   140**
7.8   **Summary   142**
          Questions   142
          Problems   143

CHAPTER 8   THE NUCLEAR ATOM   148

8.1   **Electric Charge and Electricity   149**
8.2   **The Electron   149**
8.3   **Positive and Negative Charges in Atoms: The Proton   151**
8.4   **Mass of the Proton: Mass Spectroscopy   151**
8.5   **The Nuclear Atom   152**
8.6   **Atomic Number, Mass Number, and Neutrons   155**
8.7   **The Numbers of Neutrons in Nuclei   157**
8.8   **Summary   159**
          Questions   159
          Problems   160

CHAPTER 9   ELECTRON SHELL STRUCTURE IN ATOMS   162

9.1   **Electron Shell Structure   163**
9.2   **Theory of Electron Behavior in Atoms   164**
9.3   **Quantum Numbers and Electrons   166**
9.4   **Details of the Quantum Number System: Shells, Subshells, Orbitals   168**
9.5   **Energies of Shells and Subshells: The Order of Shell Filling   171**
9.6   **Electron Configurations of Atoms   173**
9.7   **The Octet Rule   175**
9.8   **Electron Dot Diagrams of Atoms   175**

9.9   **Hund's Rule for Subshell Filling   178**
9.10   **Electron Energy Level and Electron Location   181**
9.11   **Summary   182**
    Questions   184
    Problems   185

CHAPTER 10   FAMILIES OF ELEMENTS AND THE PERIODIC TABLE   188

10.1   **Historical Development of the Periodic Table   189**
10.2   **Form of the Periodic Table   191**
10.3   **Electron Shell Structure and the Periodic Table   192**
10.4   **Examples of Periodicity and Characteristics of Families of Elements   197**
10.5   **Changes in Properties Within Elemental Families   198**
10.6   **Formulas of Compounds and the Periodic Table   200**
10.7   **The Special Nature of Hydrogen   201**
10.8   **Transition Elements   201**
10.9   **Summary   201**
    Questions   202
    Problems   204

CHAPTER 11   HOW ELEMENTS FORM COMPOUNDS   206

11.1   **Introduction   207**
11.2   **Forces on Electrons in Atoms   208**
11.3   **Ionization Energy and Electron Affinity   210**
11.4   **The Periodic Table and Forces on Electrons   210**
11.5   **Formation of Ions by Electron Transfer Reactions   212**
11.6   **Ionic Compounds   215**
11.7   **Two New Terms: Oxidation and Reduction   218**
11.8   **Formulas of Ionic Compounds   220**
11.9   **Covalent Bonding   222**
11.10   **Drawing Covalent Molecules   223**
11.11   **Atoms Forming More Than One Covalent Bond   224**
11.12   **Multiple Bonds and Coordinate Covalent Bonds   225**
11.13   **Lewis Structures of Larger Molecules   226**
11.14   **Polyatomic Ions   230**
11.15   **Shapes of Molecules   232**
11.16   **Motions of Atoms in Molecules   234**
11.17   **Writing Simplified Bond Structures   234**
11.18   **Isomerism   236**
11.19   **Summary   237**
    Questions   237
    Problems   239

CHAPTER 12   PROPERTIES AND CHEMICAL REACTIONS OF GASES   242

12.1   **Characteristics of Gases   243**
12.2   **Kinetic Molecular Theory of Matter   246**

12.3   Measuring Gas Pressure   249
12.4   Pressure Volume Relations in Gases: Boyle's Law   252
12.5   Relation of Volume to Amount of Sample: Avogadro's Hypothesis   260
12.6   Relation of the Volume of a Gas to Its Temperature: Charles's Law   262
12.7   Relation Between Pressure and Temperature: Gay-Lussac's Law   265
12.8   The Ideal Gas Law   266
12.9   Calculations Using the Ideal Gas Law   267
12.10   Standard Temperature and Pressure   271
12.11   Gas Densities and Molecular Weight   273
12.12   Dalton's Law of Partial Pressures   275
12.13   Stoichiometric Calculations Involving Gases   277
12.14   Nonideality in Gases   280
12.15   Summary   281
        Questions   282
        Problems   283

CHAPTER 13   THE CONDENSED STATES:
              THEIR PROPERTIES AND ENERGETICS   289

13.1   Condensed States   291
13.2   Specific Heat   293
13.3   Changes of Phase   295
13.4   Energetics of Melting-Freezing
        and Vaporization-Condensation   297
13.5   Vapor Pressure   298
13.6   Phase Equilibrium   300
13.7   Boiling Point   302
13.8   Interparticle Forces   304
13.9   Crystalline Solids   310
13.10   Some Properties of Liquids   318
13.11   Summary   318
        Questions   319
        Problems   321

CHAPTER 14   WATER AND ITS ELEMENTS   323

14.1   Properties and History of Oxygen   324
14.2   Preparation and Commercial Use of Oxygen   325
14.3   Ozone   325
14.4   Properties of Hydrogen   326
14.5   Preparation and Use of H_2 326
14.6   Uniqueness of Hydrogen   327
14.7   Hydrogen Peroxide   328
14.8   Water   328
14.9   Our Endangered Water Supplies   329
14.10   Hydrogen as a Fuel   331
        Questions   331

CHAPTER 15   CHEMICAL REACTIONS IN SOLUTION   333

15.1   Language of Solutions   334
15.2   Kinds of Substances That Dissolve in One Another   336
15.3   Solubility and Saturated Solutions   337
15.4   Ionic Solutions   337
15.5   Solubility Rules   341
15.6   Acids and Bases   342
15.7   Aqueous Solutions of Acids and Bases   343
15.8   Acidic and Basic Oxides   344
15.9   Neutralization and the Formation of Salts   345
15.10   Reactions of Ions in Solution   345
15.11   Concentration   347
15.12   Dilution   352
15.13   Calculations From Reactions in Solution   354
15.14   Titration   356
15.15   Ionization of Water   357
15.16   The pH Scale   359
15.17   Industrial Preparation of H_2SO_2 and NaOH   363
15.18   Summary   365
        Questions   366
        Problems   367

CHAPTER 16   REACTION RATE AND CHEMICAL EQUILIBRIUM   372

16.1   Completion in Chemical Reaction   373
16.2   Dynamic Equilibrium Revisited   374
16.3   Collision Theory and Activation Energy   375
16.4   External Factors That Affect Reaction Rates   376
16.5   Catalysts   380
16.6   The Equilibrium Constant and the Equilibrium Equation   380
16.7   Finding the Values of Equilibrium Constants   384
16.8   Interpreting the Values of Equilibrium Constants   386
16.9   Driving a Chemical Reaction   387
16.10   Le Chatelier's Principle   390
16.11   The Ionization Constant   391
16.12   Buffer Systems   395
16.13   Solubility Product   397
16.14   Summary   399
        Questions   400
        Problems   403

CHAPTER 17   OXIDATION-REDUCTION REACTIONS   407

17.1   Review of Oxidation-Reduction Reactions   408
17.2   Oxidation Number System   408
17.3   Balancing Redox Equations   413

17.4   **Some Important Redox Reactions   417**
17.5   **Names Indicating Relative Oxidation States   419**
17.6   **Electrolytic Reactions   421**
17.7   **Batteries   423**
17.8   **Activity Series   425**
17.9   **Summary   429**
        Questions   429
        Problems   430

CHAPTER 18   CHEMISTRY OF SEVERAL COMMON ELEMENTS   432

18.1   **Characteristics of the Chemistry of Nonmetals   433**
18.2   **Noble Gases   435**
18.3   **Halogen Family   435**
18.4   **Heavier Members of the Oxygen Family   441**
18.5   **Nitrogen Family   442**
18.6   **Inorganic Carbon; Silicon   445**
18.7   **Some Generalizations About Nonmetals   447**
18.8   **Characteristics of the Chemistry of Metals   448**
18.9   **Alkali Metal Family   448**
18.10   **Alkaline Earth Metals   449**
18.11   **Aluminum   451**
18.12   **Transition Metals   452**
18.13   **Complex Ions and the Transition Elements   455**
18.14   **Metallurgy of Iron   457**
18.15   **Coinage Metals   460**
18.16   **Summary   461**
        Questions   461

CHAPTER 19   CARBON AND THE COMPOUNDS OF CARBON   464

19.1   **Carbon   465**
19.2   **Hydrocarbons   469**
19.3   **Unsaturated Hydrocarbons   473**
19.4   **Ring Hydrocarbons   473**
19.5   **Functional Groups   476**
19.6   **Naming Organic Compounds   477**
19.7   **Properties and Reactions of Organic Compounds   483**
19.8   **Reactions of Unsaturated Molecules   484**
19.9   **Reactions of Aromatic Compounds   488**
19.10   **Alcohols and Their Oxidation Reactions   489**
19.11   **Aldehydes and Ketones   490**
19.12   **Amines   491**
19.13   **Carboxylic Acids, Esters, and Amides   492**
19.14   **Condensation Polymers   494**
19.15   **Unusual Structures   497**

19.16   Petroleum   497
19.17   Summary   499
        Questions   501
        Problems   503

CHAPTER 20   CHEMISTRY OF LIVING ORGANISMS   508

20.1   Chemistry in the Laboratory and in Living Organisms   509
20.2   The Living Cell   510
20.3   Biological Energetics: Photosynthesis, Respiration,
        and the ADP-ATP System   511
20.4   Proteins   514
20.5   Structure of Proteins   516
20.6   Nucleic Acids   521
20.7   Enzymes   526
20.8   Hormones   529
20.9   Food and Digestion   530
20.10   Summary   536
        Questions   536

CHAPTER 21   CHEMISTRY AND THE ATOMIC NUCLEUS   538

21.1   Early Developments in Nuclear Science   539
21.2   How Radioactivity Occurs   542
21.3   Half-Life   543
21.4   Radiocarbon Dating   543
21.5   Measurement of Radioactivity   544
21.6   Writing Nuclear Equations   546
21.7   Induced Transmutation and Artificial Radioactivity   550
21.8   Energy From Nuclear Reactions   552
21.9   Nuclear Fission   554
21.10   Nuclear Fusion   556
21.11   Nuclear Power   557
21.12   Summary   559
        Questions   559
        Problems   561

APPENDIX A   MATHEMATICS REVIEW   562

A.1   Arithmetic   563
A.2   Percent   570
A.3   Working with Positive and Negative Numbers   572
A.4   Algebra   574
A.5   Exponents   576
A.6   Scientific Notation   577
A.7   Calculations Using Exponential Notation   578

A.8   **Logarithms**   **580**
A.9   **Significant Figures**   **582**
            Problems   **584**

APPENDIX B   RELATIVE PRECISION IN SCIENTIFIC MEASUREMENT   589
            Problems   **592**

APPENDIX C   TEMPERATURE SCALES AND CONVERSIONS   595
            Problems   **598**

APPENDIX D   VALENCE SHELL ELECTRON PAIR REPULSION THEORY   599
            Problems   **602**

APPENDIX E   NAMING IONIC COMPOUNDS   603

APPENDIX F   VAPOR PRESSURE OF WATER AT VARIOUS TEMPERATURES   606

GLOSSARY   607

ANSWERS TO EXERCISES AND SELECTED PROBLEMS   618

INDEX   655

Preface

TO THE INSTRUCTOR

This book is designed to give students a thorough introduction to beginning chemistry. Fundamental concepts are discussed in detail, and the emphasis is on quantitative thinking. A course using this book would ordinarily be followed either by a standard, comprehensive, first-year course in college chemistry for those going on in the sciences or by a terminal course in organic and biological chemistry for those intending to pursue careers in the health sciences or similar fields. The book would also be appropriate for a terminal course for nonscience majors.

Because the book is introductory in nature, I have chosen to treat a few subjects thoroughly and in detail rather than to cover all the topics usually included in a first course in college chemistry. The objective is to provide a firm foundation, particularly in problem solving, that can be applied to further work. The text does not, therefore, contain sophisticated interpretations of what is observed in the laboratory, nor does it discuss advanced theory. The organic chemistry chapter is included primarily for the practice it offers in understanding molecular structure. The chapters on biochemistry and nuclear chemistry are intended mostly to acquaint the student with those topics. In those course sequences that introduce organic, biological, and nuclear chemistry at a later stage, the chapters treating these three topics can be ignored without affecting the presentation of the first 18 chapters.

The book is designed to be used flexibly; certain sets of chapters can be interchanged to fit individual preferences. In particular, Chapters 8, 9, and 10 on atomic structure and families of elements can be studied earlier, following Chapter 5. Or, if your course includes a laboratory, those chapters could be delayed until calculations from reaction equations have been covered in Chapter 11.

Although the book is intended to be used as the text for a one-semester course, there is more material than could reasonably be covered in such a course. This gives you wide latitude in choosing the topics you wish to teach as well as the sequence in which you wish to take them, all depending on the particular purposes of your course.

During three decades of teaching beginning chemistry, I have found that students are far better motivated and achieve far greater understanding if they are told in detail *why* certain assertions are so and *how* problem-solving methods reach desired results. In the attempt to attain brevity, many beginning texts often take the tacit approach, "believe this statement because I say it is so," and the student may indeed believe it. But the information will not be retained long nor be of maximum use if it is merely memorized and not thoroughly understood. So many students go on to more advanced courses only to find that they arrive there with only a set of memorized data and a number of highly structured rules with which to solve certain kinds of problems, that is, with little understanding of what chemistry is all about. For this reason, I have attempted as often as possible to make detailed, comfortable, and understandable explanations of the background of the great concepts of chemistry and of problem-solving algorithms. The book is characterized by this approach: First understand thoroughly, then apply.

The book is written in informal, first-and-second-person colloquial language. It is cast this way not only for easier reading but also to be as little threatening as possible. The analogies are meant to be as familiar as possible without doing violence to the underlying facts and theories, thus we find experiments with tire pumps, ping-pong balls, coins, toy boats, and other ordinary objects.

In working problems, which are the *sine qua non* of beginning chemistry, I believe that the student should make every possible concerted effort to achieve a result before looking up the solution in the back of the book. Even so, some students will be tempted to take the easy way out. For this reason, I have reserved the majority of the solutions for your own use, to release or not to release as you choose. In the book there are solutions to all of the exercises and to selected problems, generally one of each identifiable kind. Solutions to the other problems and answers to selected questions appear in the Instructor's Manual.

Supplementary Materials

Study Guide. The *Study Guide to E. Russell Hardwick's Introduction to Chemistry* was written by Joan Bouillon, Mt. San Antonio College. The chapters in the study guide correspond exactly to the chapters in the textbook. Each chapter includes a chapter review, example problems with solutions, questions and problems for the students to work, and answers to the questions and problems.

Instructor's Guide. The *Instructor's Guide to Introduction to Chemistry,* written by Joan Bouillon and me, is available to those who adopt the text. The guide contains chapter overviews, suggestions for teaching, and answers to selected questions in the text and to all problems not answered in the text.

Acknowledgments

Writing a chemistry textbook is something that no one person can do single-handedly; because so many people contribute, it is often impossible to name and thank everyone individually. In my particular case, the acknowledgments that follow probably omit as many names as they include, but I do want to give my special thanks to certain individuals who have contributed in outstanding ways.

First, my wife, family, and friends deserve thanks and sympathy for having to put up with me during the five years that this book was in gestation. In particular, my wife Pat midwifed the book by acting as secretary, critic, and handholder. I cannot thank her enough.

My coauthor in the Instructor's Manual and the author of the Study Guide, Joan Bouillon, has also made enormous and essential contributions to the text by inventing a large proportion of the problems and exercises and by reading proof again and again. Any errors that remain can be only a miniscule proportion of those that were caught by her eagle eye. Outside my own family, I have never before worked with a person so dedicated, meticulous, and unwavering as Joan. How lucky I was to have found her.

Among the reviewers, whose names appear below, I owe particular thanks to Professor Oren Anderson, who went over parts of the discussion again and again, each time making excellent suggestions. I was dismayed (but very relieved) when Dr. Anderson caught me out in one or two particularly egregious gaucheries, and I hasten to say that other mistakes that may appear are my own and were probably made after his last review.

The text was reviewed at various stages by faculty members who teach courses at the level for which the book is intended. Without exception, these excellent and careful individuals made useful and important comments, and the book would be of far less value without their efforts. While I will not single each of them out for individual thanks, I do gratefully acknowledge their contributions. Listed alphabetically are the reviewers who lent their expertise and experience to the book:

Oren P. Anderson
Colorado State University

Stanley Ashbaugh
Orange Coast College

Robert H. Becker
Mankato State University

Ralph A. Burns
St. Louis Community College at Meramec

Bill W. Callaway
Oscar Rose Junior College

Andrew C. Dachauer, S.J.
University of San Francisco

Jerry A. Driscoll
University of Utah

Ronald Garber
Long Beach City College

David E. Goldberg
Brooklyn College

Robert Graham
Mankato State University

Julius C. Hastings
San Joaquin Delta College

Russell H. Johnson
Florida State University

Dwight Kinzer
Berry College

J. Thomas Knudtson
Northern Illinois University

Jean P. McNeal
McLean, VA

R. S. Monson
California State University—Hayward

Wayne Moxley
San Diego Community College

Martin G. Ondrus
University of Wisconsin—Stout

William Owen
University of Wisconsin—Stout

David Smith
Pennsylvania State University—Hazleton

E. E. Sorman
San Diego Mesa College

David L. Wilson
Valencia Community College

TO THE STUDENT

This book is designed for you. You will notice that I talk with you in idiomatic and personal language. When I say "we," I am referring to you and me. I have written the book this way to make our communication relaxed and friendly so that you will feel comfortable with chemistry.

The format of each chapter is designed to guide you through the chapter and to help you learn. The basic structure is described in the following paragraphs.

Chapter Introductions

Each chapter begins with an introduction containing three kinds of information. First, a brief preview introduces the subject to be covered. Read the preview carefully so that you will know what the chapter is about and how the sections fit together. Following the preview is a statement of objectives designed to tell you what you should expect to learn from the chapter. Study the objectives and keep them in mind as you read. Finally, there is a list of new terms that should be memorized and often a list of terms with which you should become familiar. Repeat the terms to yourself a few times so that you will be alerted to them when you see them in the text.

The introductions describe the framework around which the book is organized. They tell you not only what to watch for and learn but how the new information in each chapter fits into what you already know.

Vocabulary

Any detailed discussion of science requires special words and phrases. When terms are first defined in the text, they are printed in **bold letters,** a signal to you that a significant new term is being introduced. Many of these boldfaced words also appear in the chapter introductions. In other places you will see words or sentences printed in *italics* to indicate that something particularly important is being said.

Glossary

At the back of the book is a glossary. In it you will find definitions of many of the words printed in boldface type in the text. You may want to place a tab of tape at the glossary so that you can turn to it quickly; it is easier to look up words there than to dig back through the text for them. Whenever you are uncertain about a word, look it up right away. Do not allow your understanding of a passage to suffer because you have forgotten the meaning of a key word.

Exercises, Questions, and Problems

To understand chemistry, it is necessary to understand quantitative calculations (those using information expressed in numerical form). *Problem solving cannot be given too much emphasis.* The best way to learn chemistry is to work problems, work problems, and work more problems.

Whenever appropriate, the discussion of a new concept is accompanied by one or more quantitative **examples,** problems worked out in detail. Most examples are followed immediately by similar **exercises** for you to do before you go on. (Keep your calculator and a pad and pencil handy as you study.) The answer to an exercise can be obtained by using the method of attack shown in the preceding example.

At the end of each chapter is a set of **questions** with which you can test your understanding of the chapter. This is followed by a set of **problems** with which you can practice and reinforce your understanding. Plan to spend as much or more time on the problem sets than you did on the chapters themselves. The problems given in the book should be considered a minimum. Do not shortchange yourself by neglecting the problem sets.

Once you are sure that you've exhausted all your resources in attempting to solve an example or an end-of-chapter problem, you may want to look at the printed solution. Solutions for the exercises and for selected problems are found at the end of the book, just past the Glossary. Asterisks mark the problems for which there are worked-out solutions. If a problem has lettered parts, only the parts with asterisks are answered in the back of the book.

If you begin your study of each chapter by spending a few minutes with the chapter introduction, you will find that the chapter itself will be easier and more interesting.

Good luck in your study of chemistry. I hope that the book will not only give you a good start but will also help you feel the satisfaction gained by understanding something of how the universe is put together.

Introduction

The next few pages are an introduction to science in general and to chemistry in particular. You will be shown how science differs from other kinds of activity. You will get a close-up look at the scientific method and the nature of experiment. The nature of chemistry and its place in the sciences are explained, and the importance of chemistry to civilization is discussed.

AFTER STUDYING THIS CHAPTER, YOU WILL BE ABLE TO

- explain how the scientific method works
- explain the relations between observations, natural laws, and theories
- explain how chemistry relates to other sciences
- explain how chemistry relates to society

TERMS TO KNOW

biology	molecular biology	reproducibility
chemistry	natural law	scientific method
experiment	physics	theory

SCIENTIFIC METHOD

Chemistry, along with other sciences, employs what is often called the scientific method. Because chemistry is at the very heart of science, we can start our study of it by learning how to use the scientific method.

Let us make up a story.

You introduce yourself to the scientific method on the day that you go out to ride your bike to class and find that it has a flat tire. Still half asleep, you attach your little tire pump and begin pumping away. Then you begin to notice the way the pump behaves. Your curiosity wakes you up (curiosity is one of the most important characteristics of a scientist), and you begin to pay careful attention to what is going on. At the beginning of a stroke on the pump it is easy to press the plunger down, but as you near the bottom it takes much more force. Funny. Why is that? You have made an observation; now you theorize a bit and devise a little experiment. Maybe the effect is caused by the elasticity in the tire. You pinch the air tube of the pump so that no air can get out, and you push the plunger again. The effect is now even more pronounced. So much for that theory. You finish filling the tire and go to class, but you think about this all day (persistence is also a valuable quality for scientists).

Late that afternoon, you make a careful experiment. You fasten the pump upright and build a little platform on the top of the plunger. Now you pinch the tube again to trap air in the pump and then start putting bricks on the platform, increasing the downward force on the plunger as you increase the number of bricks (Figure i.1).

As a result of many observations, here is what you find: as you increase the number of bricks on the platform, thus increasing the force on the plunger, the position of the plunger goes farther and farther down, compressing the trapped air into a smaller and smaller volume. By repeating the experiment and remembering what you have observed about tires, pumps, and similar equipment, you are able to write a general statement of what you have found out:

> The volume of air trapped in a tire pump gets smaller as the force on the
> plunger gets larger.

You have just stated a natural law. True, it is limited to air in pumps; it stops short of saying exactly how much force is needed to produce various changes in volume. However, as far as you can tell, the law is valid within its limitations. Moreover, its validity is soon reinforced by a friend who makes similar experiments. She uses a different kind of pump but gets the same results. Your law has been tested and verified.

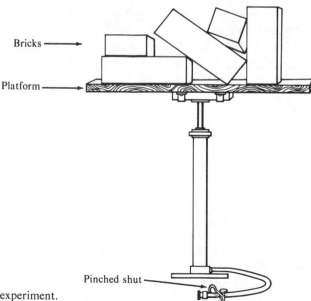

Figure i.1   A simple experiment.

What's next? Well, you might continue your investigation in two ways. First, you will want to accumulate more facts about the phenomena you have observed. Exactly *how* does the volume of the air change with the number of bricks on the platform? What happens if you let some of the air out? On a hot day, you notice that more bricks are needed to get the plunger to go down a certain distance; what about that? In addition to gathering physical facts from your experiment, you will eventually want an explanation for them. **Why** does the plunger behave that way? You now need to try to explain what you have seen by making some guesses about things you cannot see: that is, you make a theory. Perhaps a tiny little man in the pump pushes back on the plunger as you squeeze him. (The history of science includes such little men and similar creatures, but when you start inventing little creatures, you approach superstition. It is best to seek your explanations in physical things you know to exist.) The little man seems like a fruitless idea, so you look further. Aha! Maybe there is a spring inside the pump. You already know that a spring works in a way that is similar to the effects you have observed. So you experiment: you open the pump and look inside. No spring (no little man, either)! Back to the drawing board.

What kinds of forces are there that you already know about and that might be responsible? Gravitational, perhaps. True, the bricks are attracted by gravity, but you can't see how something pushing *up* on the plunger works by gravity. How about electric and magnetic forces? You try some simple experiments and decide that such forces are not involved. One evening on television, however, you see pictures of a hurricane and notice that moving air can exert tremendous force. Can something be moving and colliding with the plunger even though there is no wind inside the pump? Coming up with a theory is going to require a lot of thought and effort.

Although we cannot pursue our story much further now without going into considerable detail, we will later spend a great deal of time on just this subject. However, the story *has* demonstrated the **scientific method.** First you made **observations** of a physical phenomenon. Then you organized and generalized your findings into a statement describing what you had seen: you proposed a **natural law.** You, and others, used **experiments** to test the law. You then tried to make a **theory,** a reasonable explanation of the natural law in terms of things that you were not able to observe. You made further observations and experiments to check the theories, perhaps finding more about the phenomena themselves in the process. You might have gone even further, discovering more laws and making more theories to explain them.

An essential feature of the scientific method is *reproducibility;* if no one else can observe what you claim to have observed, few scientists are likely to take your experiments seriously. Reproducibility is an important difference between the work of scientists and that of fakes and charlatans.

This, then, is the way science works: observations, laws, theories, experiments to test the theories, modifications of the theories, more experiments, and on and on. As you continue through the chapters that follow, you will see the process played out again and again.

THE PLACE OF CHEMISTRY IN THE SCIENCES

If we were to arrange some of the sciences according to the degree to which they are general and apply to all of nature, the order would be physics (the most general of all), chemistry, biochemistry, the life sciences, and the social sciences. Let us briefly review the subject matter of several of these fields.

If a characteristic statement can be made about **physics,** it is that physics is the science that seeks to develop laws and theories that apply to samples of any kind of matter. For example, physics gives us the laws that govern falling bodies in gravitational fields. Those laws apply to all kinds of falling bodies and are valid in gravities of all strengths, including those of the moon and the earth.

It is the goal of **chemistry** to explain the properties of specific kinds of matter and to unravel the complexities of the ways in which various kinds of matter interact. The systems that chemistry seeks to explain are more complex than those analyzed by physics. Chemistry is concerned with the differences between kinds of matter and the reasons for those differences, as much as with generalizations that hold for any kind of matter.

Biology, the study of living tissues and organisms, seeks eventually to describe in intimate detail how living tissues behave, and, more important, to explain why they behave as they do. To do this completely, the biologist must apply sophisticated knowledge derived from chemistry and physics to the analysis of incredibly complex systems. (How, for example, does the brain work?)

A relatively new field of scientific research is **molecular biology.** Here the powerful tools of physics and chemistry are being applied to the simplest biological units such as viruses and single cells, in the hope of gaining fundamental understanding that can ultimately be applied to the functioning of larger organisms. Although this kind of research has been carried on for a long time, only now, as we enter the final fifth of the twentieth century, have chemical theory and instrumentation become sufficiently refined to allow their intense and organized application to biology. Great advances have been made in recent years, and there is the promise of more in the near future.

Serious and fruitful inquiry into the physical nature of our world began with quantitative, reproducible experiments at the simplest level. First were the measurements of the early astronomers. Later, Galileo dropped weights from the leaning tower of Pisa to show that the rate of descent did not depend on the size of the weight. Newton made essential studies of motion and gravitation. As the base of scientific knowledge grew, more complex studies became possible: these days the physicist probes the interior of the atomic nucleus and the organic chemist confidently designs totally new kinds of matter. Now, science finally seems to be on the threshold of unlocking the secrets of life itself.

Chapter 1

Measurement and Problem Solving in Chemistry

Observation is an essential part of the scientific method. As science advanced from its early days, when dropping weights from a tower or floating in the bathtub could produce significant new insights, the task of observing physical systems grew more complex and demanding. It became increasingly important to record numerical data. Before an observation could be of much use, it was necessary to record not only *what* was seen but also *how much* was seen. Precise and well-defined systems of measurement began to be developed. Techniques of error analysis became more refined. As connections were made among many different kinds of physical phenomena, the mathematical method of analyzing data began to be more widely used. Now we live in the world in which a large part of science is quantitative, and information from scientific experiments is analyzed using sophisticated mathematical techniques, often with the help of computers.

Chapter 1 begins by helping us see how much information we can expect to gain from the measurements we make. We will learn how to take account of experimental error in reporting the results of our measurements. We will tackle problem solving: what to do, what not to do, and how to use unit analysis, a helpful method of setting up problems.

AFTER STUDYING THIS CHAPTER, YOU WILL BE ABLE TO

- explain the difference between precision and accuracy
- use an organized thought process to approach chemistry problems
- use the unit analysis method to set up mathematical solutions to problems
- analyze your answers to problems in terms of reasonableness, precision, and the units used

TERMS TO KNOW

accuracy	error	precision
conversion factor	experimental error	random error

rounding   systematic error   units

scale   unit analysis   values

significant figures

1.1 MEASUREMENT AND UNCERTAINTY

Chemistry is a physical science; it deals with the quantitative measurement of the properties of matter. To measure something is simply to find the numerical value of one of its properties, perhaps its length or its mass or its temperature. The results of quantitative measurements are called **values.** A value consists of a numerical quantity and the set of units in which the measurement was made. **Units** are the quantities that have been defined for making measurements. In the metric system, for example, the meter is a unit of length. If the speed of a car is measured, the units of the result could be miles/hour (that is, miles per hour) or meters/second, or any of several other sets. Sometimes the units in which a value is expressed are called the *dimensions* of that value.

It is meaningless to say that the distance to a friend's house is 60. How far is 60? It is too far to walk if it is 60 miles, but if it is 60 inches, you can almost reach out and touch the house. When giving the result of a measurement, it is *always* necessary to state the *units* as well as the *numerical quantity*. The importance of conscious and careful attention to units cannot be overemphasized. It is necessary for communication and is a tremendous aid to problem solving. In Section 1.4, we will explore in detail the technique of using units as a guide to working problems.

That science is exact, that scientific experiments are so precise as to be free of error, are popular misconceptions. Even the simplest results contain an element of uncertainty. Ask ten people to measure the length of a steel bar as accurately as possible, and you will get ten different answers. You cannot be certain which answer, if any, is correct; the best you can do with your information is to report an average value. Where there is uncertainty, there is likely to be error. Even if an experiment is designed to obtain a simple yes or no answer, the chance for error still exists.

Because error may always be present, we must be prepared to assess its effect. Three terms help us describe uncertainty in measurement:

Accuracy The closeness with which an experimental result approaches the true value.

Precision The closeness with which several measurements of the same quantity agree.

Error The difference between the true value and the result of the experimental measurement.

You can see from Figure 1.1 that an experiment may be quite precise while still containing large errors. The precision of a result merely indicates that the measurements agree, not that the result was close to the true value. For instance, if we use a bathroom scale and weigh someone several times, we might see a number very close to 60 again and again. We would confidently report the weight to be 60 pounds with high precision, not ever noticing that the scale was marked in *kilograms* (1 kilogram = 2.2 pounds) and that the weight of the person in *pounds* was close to 132. This particular error is obvious, once detected, but others may not always be so.

Errors that occur again and again in the same way because of a flaw in experimental machinery or technique are called **systematic errors.** Such an error would result, for example,

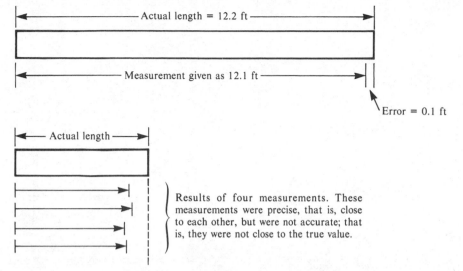

Figure 1.1   Accuracy, error, and precision.

if the bathroom scale had a weak spring so that it always indicated a weight that was 2% too high. Systematic errors are often undetectable and go unnoticed in reported results.

Along with systematic errors, **random errors** can affect results. Some inescapable uncertainty is always present in measurements themselves, regardless of what the true values are. Suppose, for example, that you are asked to measure the length of a book. You are given a rod, or scale, on which there are several marks exactly 1 inch apart. (The word **scale** is applied to weighing instruments, but it is also used to mean anything that is marked in units and is used for measurement.) You place your scale next to the book and estimate that the length is 9.7 inches. Making the same measurement again, you might estimate the length to be 9.6 or 9.8 inches, while another person might make any of these estimates. Everyone would agree on the 9, but there would be uncertainty about the number of tenths of an inch (Figure 1.2). But suppose that the scale had been marked with inches and tenths. Then you might report that the length of the book is 9.73 inches or 9.74 inches, estimating the last hundredth of an inch (Figure 1.3). If the scale were marked with hundredths of an inch, you probably could not do any better, because the roundness of the edge of the book and the way in which you positioned the scale would be causing more than a hundredth of an inch of uncertainty. The improvement in the scale would not improve your measurement.

No matter how finely a scale is marked, there will always be some uncertainty in a measurement made with it. There is no such thing as an *exact* measurement except when you are counting small numbers of discrete units such as coins, people, or automobiles.

In most cases, then, we cannot know an exact value. Because we do not know the exact value, we cannot know exactly how much the measurement deviates from it. The best we can do is to measure as precisely as possible, and attempt to eliminate systematic errors by using many different methods of measurement and improving our instruments and techniques.

The two main things to remember about experimental error are: it is always present, and we must be careful not to believe that our experiments are more precise than they actually are.

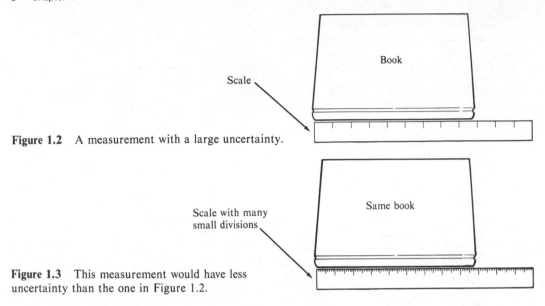

Figure 1.2   A measurement with a large uncertainty.

Figure 1.3   This measurement would have less uncertainty than the one in Figure 1.2.

1.2 SIGNIFICANT FIGURES

When we report the result of a measurement, it is important to indicate how precise the measurement is: how much uncertainty there is in it. We can indicate the uncertainty by actually giving a numerical statement of the amount of probable error (as explained in Appendix B) or by employing what is known as the system of **significant figures** (Appendix A.9). We will use the latter system in this book.

A frequently used method of indicating the precision of a numerical measurement is to give the number of figures or digits that are said to be *significant,* that is, to give *all the digits that are known to be accurate, along with one last digit that is uncertain.* Let us return to your measurement of the book. When you reported the result as 9.7 inches, the last digit, 7, was an estimate, that is, uncertain. With the better scale, you reported 9.73 inches, and the uncertain digit was the 3. In the first case, you used two significant figures; in the second case, three.

In reporting the result of a single measurement, it is not too difficult to decide how many significant figures to use. When you make calculations using several numerical values, however, the number of digits in your answer is not likely to be the number appropriate to the precision of your measurements—unless you are careful to make adjustments. The number of digits often increases considerably during long calculations, even when none of the measurements is especially precise. For this reason, you must examine the results of calculations and be sure that the precision they indicate is not unrealistic. The number of digits will usually have to be decreased by a process called **rounding.** The use of significant figures, the determination of the proper number of digits to use in an answer, and the techniques for handling significant figures in calculations are the subject of Appendix A.9. Unless you are already familiar with the use of significant figures, you will need to study carefully the examples, exercises, and problems in the appendix. You should have these techniques in your mental toolbox before using the results of chemical measurements to make calculations.

1.3 SOLVING PROBLEMS IN CHEMISTRY

1.3.1 General Principles

Nearly everyone who has been through first-year college chemistry feels that solving numerical problems is the toughest part. By attacking problems in an organized way, however, you can make the task of problem solving considerably easier.

This section offers a general approach to problem solving, showing what to do and what not to do. Once you understand the basic principles, you will be ready for the next section, which describes a detailed method, called **unit analysis,** of finding and setting up the relations that will lead to desired answers.

In chemistry, measured values are expressed as numbers, words, equations, or in other ways. Chemistry problems require that the values be put together mathematically to produce a wanted result. To achieve this, several steps are necessary:

1. Read the problem carefully. Be sure you understand it.
2. Put the individual statements into mathematical form whenever possible.
3. Make a mental or written plan for working the problem (a "solution map").
4. Make sensible approximations when you can.
5. Set up and make your calculation, being sure to include the units.
6. Check your answer—both the numerical quantity and the units.

1.3.2 How to Begin

1. *Read the problem and be sure you understand it.* Start by reading slowly. Pause a few seconds to think about what you read; take care to see and understand the meaning of every word and symbol. Be sure that you know what is being asked for and what the units of the answer should be.

2. *Write mathematical versions of the values given.* It is often difficult to decide how a problem should be set up until the data statements are organized in mathematical form, complete with units. Make a table of the relations, similar to those shown later in this step. It's worth the time it takes, just to avoid confusion.

Suppose you are told that there are seven oranges in a box, that a car can run 21 miles on a gallon of gasoline, and that three eggs, two golf balls, a potato, and a bottle of beer make 4½ pints of soup. Here is the first of those relations:

$$\frac{7 \text{ oranges}}{1 \text{ box}}$$

Ordinarily, when exactly one of something appears in the denominator of an equation, the numeral 1 is not shown. We would write the first relation as

$$\frac{7 \text{ oranges}}{\text{box}}$$

This statement is read as *7 oranges per box,* the horizontal line meaning *per* and indicating division. (For example, the number of oranges per box could have been determined by counting all the oranges from several boxes and dividing by the number of boxes. You know the old joke: To count the number of sheep in a field, count the number of legs and divide by four.)

The rest of the relations are:

$$\frac{21 \text{ mi}}{\text{gal gasoline}} \qquad \frac{3 \text{ eggs}}{4\frac{1}{2} \text{ pt soup}} \qquad \frac{2 \text{ golf balls}}{4\frac{1}{2} \text{ pt soup}}$$

$$\frac{1 \text{ potato}}{4\frac{1}{2} \text{ pt soup}} \qquad \frac{1 \text{ bottle beer}}{4\frac{1}{2} \text{ pt soup}}$$

The same mathematical relations may be written in the inverse form if necessary. If there are 7 oranges in every 1 box, it is surely true also that there is 1 box for every 7 oranges, or,

$$\frac{1 \text{ box}}{7 \text{ oranges}}$$

The form you will use in your mathematical statements will depend on the nature of the problem and the way in which you plan to work it. Once the statements are written in mathematical form, you may invert them later if you wish. In any case, write down all the relations you find in the problem; it is best to have them where you can easily refer to them.

3. *Make a mental or written plan for working the problem.* Begin by deciding which chemical principles or concepts apply and the order in which they must be used. Plan the steps you will use in getting from the initial data to the final answer. Write the plan by making what we will call a "solution map." (Using the units as a guide, as explained in Section 1.4, will help you in this step.) Next, check to see that you have all the data necessary to work the problem; sometimes you are expected to supply some of the information yourself. Do not begin to use formulas or make calculations until your solution map, or "roadmap," to the answer is complete. (When instructors grade papers, they often give substantial credit for a step-by-step plan for working a problem even if no answer is given.) Now you are ready to work out a mental or scratch-paper estimate of what the numerical part of the answer will be. This exercise helps check your solution map and will also be valuable when you finish the problem. Write down your estimated answer.

4. *Make sensible approximations when you can.* You can often spare yourself much work and confusion by leaving out part of a calculation or by making an easy but inexact calculation when the exact one would be difficult and unnecessary. Properly done, such approximations will not affect the answer. Shortcuts like this are especially appropriate when small numbers are to be added to or subtracted from large numbers of low precision.

EXAMPLE 1.1.   What is the total number of pounds of cake mix that is made from 33 pounds of flour, 17 pounds of sugar, 4 pounds of butter, and 0.01 pounds of salt?

Solution:

The flour, sugar, and butter together weigh 54 pounds. Because it is possible to have an error of a pound or so in each of these quantities, the 54-pound total could have an error of as much as 3 or 4 pounds. Compared with this, the 0.01 pound of salt is negligible; that is,

$$
\begin{array}{r}
54. \text{ lb} \\
+ \ \ 0.01 \text{ lb} \\
\hline
+ \ 54.01 \text{ lb}
\end{array}
$$

But this answer must be rounded to 54 pounds and the 0.01 is lost (see Appendix A.9). As an approximation, simply leave the salt out from the beginning (Figure 1.4). Although this is a particularly simple example, later we will encounter problems in which making an approximation at the right time can save an enormous amount of work.

5. *Set up and make the calculation.* After deciding what steps are necessary to get from data to answer, write down the data in the mathematical form that will lead to the answer. Put the data into your solution map.

6. *Check your answer* (a most important step). Are the units correct? (Presumably so; you have used them in setting up.) Is the number of significant figures appropriate? (If not, round off.) Finally, is the answer reasonable? Students working first-year chemistry problems sometimes devise a correct setup but make a mistake elsewhere that leads to an absolutely impossible, ridiculous answer. Here is an example of the way that you can go wrong by not checking your answer.

> **Question:** How many gallons of coffee can be made with 2.35 pounds of ground coffee and as much water as needed, if each pound of ground coffee will make 3.41 gallons?

> **Answer:** 801.35, right? Wrong.

The answer is wrong on three counts. First, it should carry its units, gallons. Second, even assuming that "each pound" means exactly 1 pound, there still should be only three significant figures in the answer. Finally, it should be clear that the answer ought to have been nearer 8 gallons than 800. (See Figure 1.5.) There is no way to get more than a few gallons of coffee out of less than 4 pounds of ground coffee. The answer should have been 8.01 gallons.

Now is the time to look back at your estimated answer. How does the calculated answer compare? If it is close, then everything is probably all right. If it is way off, then what went wrong? Don't go on until you are satisfied.

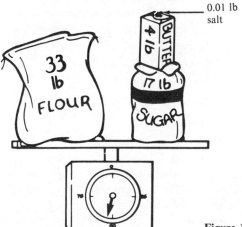

Figure 1.4   The salt will make no difference in this weighing.

Figure 1.5   It is advisable to check your answers to see that they are reasonable. If a pound of coffee makes 3.41 gallons, it is not reasonable that 2.35 pounds will make 801.35 gallons—that would be very weak coffee!

Review your assumptions. Are you taking anything for granted? If so, you will have to be sure your assumptions are correct. Here is another example.

> **Problem:** Jennifer can run at a speed of 7 miles per hour. She starts at her front door and runs for 23 minutes. Now she turns around and walks back at 3.5 miles per hour. How long will it take her to get back to her starting point?

> **Answer:** 46 minutes, right? Wrong. Jennifer was running around the outside of her house. When she finished running, she was only 12 feet away from her front door.

Even when you think you have all the information, it is possible to go wrong by assuming too much. You may have been tricked by this problem because nothing was said about where Jennifer ran. In working the problem, you may have assumed that Jennifer ran to a particular place and then walked back, but you had no more reason to make that assumption than to decide that she was running around her house. The mere fact that nothing was said about where she ran does not entitle you to assume whatever you choose. To work the problem properly, you would need either to obtain more information, or to state your assumption in your answer. For example:

> **Answer:** Assuming that she retraced her path exactly, it would take Jennifer 46 minutes to return.

1.3.3. What to Avoid

There are a few problem-solving techniques that can allow you, after some practice, to work certain kinds of simple chemistry problems without really thinking about what the problems mean. One such method requires that you write chemical equations, put particular kinds of data in particular positions on your paper, then transform the resulting pattern into a ratio equation and solve it. This approach can yield an answer, without requiring any understanding whatever of the science involved. It can also produce errors without your even being aware that the answers are wrong. It is essential that you try to understand a problem before you begin to solve it; any other approach only hinders learning. Be sure you know the fundamental principles that the problem is designed to teach, then use problem-solving techniques only to help devise the setup and to check your answer. (If you are taking a test, and you are unable to grasp a problem, leave it and work another. Come back and make a fresh attempt later.)

Students often attempt to understand chemistry by memorizing countless equations and planning to use them without really knowing what they mean. Little good comes of memorizing equations unless the concepts that underlie them have been understood. Remember that your goal is to achieve a better understanding of chemistry: the equations are merely tools to allow you to employ that understanding.

1.4 USING UNIT ANALYSIS TO SOLVE PROBLEMS

In our discussion of problem solving, we touched on what is probably the most critical and at the same time the hardest part: putting the fragments of the problem into a mathematical setup that will lead to a solution. This section describes a method that can guide you in setting up many kinds of problems. The method, called **unit analysis** (sometimes called **dimensional analysis**), is based on the fact that arithmetic and algebra work on units in exactly the same way that they work on numerical values. Before you begin to study the method itself, you may want to get a running start by reading Appendix A.1.2, which reviews some of the properties of fractions.

At some point, most chemical problems involve the conversion of values having one kind of units to values having another kind. Such calculations can often be set up simply by doing the arithmetic on the units themselves and letting the numbers follow along. Unit analysis is the method of approaching problems this way. (But unit analysis will not substitute for understanding!)

Conversions from one set of units to another can be accomplished using **conversion factors.** (Conversion factors are also sometimes called **unit factors,** for a reason that we will soon see.) A conversion factor can be written in either of two ways, each of which is the reciprocal of the other. Although the relation is the same regardless of which way the factor is written, it is convenient to choose in advance the form you plan to use. For example, there are 12 inches in 1 foot. This relation is written in abbreviated form as

$$\frac{12 \text{ in.}}{1 \text{ ft}} \qquad \text{or as the reciprocal} \qquad \frac{1 \text{ ft}}{12 \text{ in.}}$$

Notice that periods are not used after abbreviations for units except in the case of the abbreviation for *inches,* which is written as *in.* to distinguish it from the preposition *in.* From this point, we will begin extensive use of the abbreviations of units.

These two conversion factors contain the same information; but by means of multiplication, one converts feet to inches and the other converts inches to feet. Because this conversion is a direct statement of the definition relating inches and feet, it contains as many significant figures as you wish it to have. That is, it can be ignored when you are determining how many significant figures to keep in an answer to any calculation that includes the conversion.

On the other hand, it is possible that we might have been given the same conversion in the form 1 in. = 0.0833 ft, for which we would write the factors as

$$\frac{1 \text{ in.}}{0.0833 \text{ ft}} \quad \text{or} \quad \frac{0.0833 \text{ ft}}{1 \text{ in.}}$$

In this case, the 0.0833 was obtained by dividing 1 by 12 and rounding to three digits. It is not exact. If you make a calculation using the conversion in this form, you must take notice of its four significant figures when you round your answer.

As we will see somewhat later, exact numbers arise out of chemical reaction equations as well as from definitions.

To see why conversion factors are called unit factors, look at the conversion factor

$$\frac{1 \text{ in.}}{0.0833 \text{ ft}}$$

Because the length of 1 in. is stated to be the same as the length of 0.0833 ft, we can substitute 0.0833 ft for the 1 in. of that equation and obtain a result of unity; that is,

$$\frac{0.0833 \text{ ft}}{0.0833 \text{ ft}} = 1 \text{ (unity)}$$

True conversion factors all have the numerical value of one. Because this is so, we can multiply or divide anything whatever by a conversion factor without changing its actual value, although we will indeed change its units. Watch the change of units happen as we work the problems that follow.

EXAMPLE 1.2.   A vase is 0.442 ft high. How high is it in inches?

Solution:

Notice that our answer is required to be in inches. Of the two possible relations between foot and inch, which form should we use? One gives an answer in the correct units, and one leads to nonsense. Let's look at both.

$$0.442 \text{ ft} \times \frac{1 \text{ ft}}{12 \text{ in.}} = 0.0368 \frac{\text{ft}^2}{\text{in.}}$$

$$0.442 \text{ ft} \times \frac{12 \text{ in.}}{\text{ft}} = 5.30 \text{ in.}$$

After checking our second answer, 5.30 in., we can accept it because it is in the correct units—inches—and is reasonable. The first answer cannot be correct because it has the wrong units, feet^2 per inch. Our check of the units in the first calculation tells us to go back and set up the problem correctly.

Obtaining a solution to a problem often requires more than a single conversion step. Here is a problem that uses a conversion in the metric system, with which we will soon become familiar.

EXAMPLE 1.3.   A room is 15.3 ft across. How many centimeters (cm) is it across? One inch is the same as 2.54 cm.

Solution map: feet $\longrightarrow$ inches $\longrightarrow$ centimeters

Solution:

Write the initial information with its units, then multiply it by the proper conversion factors to obtain the answer in the desired units. Although we already know how to convert feet to inches, we must also use the conversion from inches to centimeters.

$$15.3 \text{ ft} \times \frac{12.0 \text{ in.}}{\text{ft}} \times \frac{2.54 \text{ cm}}{\text{in.}} = 466 \text{ cm}$$

We check and accept the answer. If either conversion factor had been used incorrectly, the numerical answer would have been incorrect and its units would not have been centimeters, as we required them to be.

EXERCISE 1.1.   Four teaspoons of cocoa are needed for 1 cup of hot chocolate. How many tablespoons do you need for 5 cups if 1 tablespoon equals 3 teaspoons?

EXERCISE 1.2.   A mile is 5,280 ft. How many yards are in 0.242 mi?

Although it is essential that you become familiar with the units of measurement used in chemistry, you can use conversion factors whether you know the meaning of the units or not.

EXAMPLE 1.4.   We will solve a problem using imaginary units from the planet Xanthu. No one on earth knows the meaning of these units, but anyone can do the problem using unit analysis.
There are 3.21 borks to every neb. How many borks are there in 7.2 nebs?

Solution:

$$7.2 \text{ nebs} \times \frac{3.21 \text{ bork}}{\text{neb}} = 23 \text{ borks}$$

Conversion factors can be used more than once in a single calculation; that is, it may be necessary to use the square or cube of a factor.

EXAMPLE 1.5.   How many cubic inches are there in a box that measures 2.5 ft × 1.5 ft × 3.7 ft?

Solution:

To convert the volume of the box to cubic inches, we could multiply each of the sides by the conversion factor from feet to inches.

$$\left(2.5 \text{ ft} \times \frac{12 \text{ in.}}{\text{ft}}\right) \times \left(1.5 \text{ ft} \times \frac{12 \text{ in.}}{\text{ft}}\right) \times \left(3.7 \text{ ft} \times \frac{12 \text{ in.}}{\text{ft}}\right) = 2.4 \times 10^4 \text{ in.}^3$$

But this is the same as

$$2.5 \text{ ft} \times 1.5 \text{ ft} \times 3.7 \text{ ft} \times \left(\frac{12 \text{ in.}}{\text{ft}}\right)^3 = 2.4 \times 10^4 \text{ in.}^3$$

To convert a value in which some of the units appear raised to a power, you must raise the entire conversion factor for that unit. And, of course, *raising the factor to a power means raising both the numerical part and the units as well.* Here is a somewhat more difficult problem.

EXAMPLE 1.6. A plot of ground contains 12,555 square rods. If one rod contains 16.5 ft and one acre contains 43,560 square ft, how many acres are there in the plot? (The words *one* or *a* are often used in problems. Like the digit "1" in defined terms, or like any number that is written as a word, they are taken to mean exactly that amount. With such defined quantities, you can use as many significant figures as you wish.)

Solution map: rods² ⟶ feet² ⟶ acres

Solution:

Step 1. Square the factor converting rods to feet.

$$\frac{16.5 \text{ ft}}{\text{rod}} \times \frac{16.5 \text{ ft}}{\text{rod}} = \frac{272.25 \text{ ft}^2}{\text{rod}^2}$$

Step 2. Make the setup and do the arithmetic.

$$12{,}555 \text{ rods}^2 \times \frac{272.25 \text{ ft}^2}{\text{rod}^2} \times \frac{1 \text{ acre}}{43{,}560 \text{ ft}^2} = 78.46875 \text{ acres}$$

Step 3. Notice how the number of digits increased as we did the calculation. We round to 78.5 acres. Then, having checked our answer to see if it is reasonable and in the correct units, we accept it.

EXERCISE 1.3. A rectangular tank measures 1.32 yd by 3.55 yd. If the tank is filled with water to a depth of 0.23 yd, how many gallons of water are in it? A gallon contains 231 cubic in.

EXERCISE 1.4. You are going to wallpaper a wall that is 10.5 ft by 8.0 ft. If one roll of wallpaper covers 4.0 square yd, how may rolls do you need?

The next problem requires that you make several conversions, and that one of the conversion factors be raised to the third power if the units are to cancel properly. Because the problem is somewhat more complex than the preceding ones, we will approach it with the full problem-solving technique discussed in the last section.

EXAMPLE 1.7.   Given that there are exactly 36 in. in 1 yd, that a cubic inch of water weighs 0.554 oz, and that a pound contains 16 oz, how many pounds does a cubic yard, yd^3, of water weigh?

Solution:

Step 1.   Write down the data mathematically and look at them.

$$\frac{36 \text{ in.}}{yd} \qquad \frac{0.554 \text{ oz water}}{\text{in.}^3} \qquad \frac{1 \text{ lb}}{16 \text{ oz}}$$

Step 2.   Although we may not have seen some of these units before, we have read and understood the question. There do not seem to be any approximations we can make, so we will guess at an answer. A cubic yard might be a box about as big as a card table. Imagine such a box filled with water; lift it. You can't. It must weigh hundreds of pounds. That's our guess: hundreds of pounds.

Step 3.   We set up our solution map. We are to calculate the number of pounds in a cubic yard of water. (Units are sometimes stated in plural form. You need not worry about that. Whether we say pound or pounds makes no difference in the calculation.) Here is what the solution map looks like, starting with cubic yards and going to pounds.

Solution map: $yards^3 \longrightarrow inches^3 \longrightarrow ounces \longrightarrow pounds$

Because it would be inconvenient to include the conversion factor 36 in./yd three times in the calculation, we will raise it to the third power, $(36 \text{ in./yd})^3 = 46,656 \text{ in.}^3/yd^3$, before using it. What this operation means physically is that if we take a cube that measures 36 in. on a side, it will have 46,656 in.^{3} in its volume.

Step 4.   We make the setup and calculate the answer.

$$1 \text{ yd}^3 \times \frac{46,656 \text{ in.}^3}{yd^3} \times \frac{0.554 \text{ oz}}{\text{in.}^3} \times \frac{1 \text{ lb}}{16 \text{ oz}} = 1615.464 \text{ lb/yd}^3$$

Step 5.   Now to check our work. The units are correct. The answer is in the same ballpark as our guess, many hundreds of pounds. Significant figures? All the values were defined except the 0.544 oz/yd^3, so that is our limiting value. After rounding to 1.62×10^3 lb of water, we accept the answer.

EXERCISE 1.5.   If a cubic inch of metal weighs 4.55 oz, how much will a cubic foot of the metal weigh in pounds? (See Example 1.2 for conversions.)

EXERCISE 1.6.   A bag contains 0.050 yd^3 of topsoil. If the soil is spread 0.25 in. thick, how many square feet will a bag of it cover?

Although unit analysis is a powerful tool to help us set up problems, it falls short of being magic. It cannot think for us. We can use the method as a valuable guide—especially to check answers—but we must not yield to the temptation of letting it do the work for us. Our objective is to learn chemistry, not merely a technique for working problems automatically.

In the chapters that follow, we will encounter numerous problems, many of which will be worked-out examples. Occasionally, the examples will use the entire problem-solving sequence explained in Section 1.3. As we become increasingly familiar with the technique, only parts of the sequence will appear. When you do your *own* problems, however, you should continue to use the entire method until you are thoroughly at home with problem solving and have developed your own methods of approach.

1.5 SUMMARY

Quantitative measurement is essential to science, but measurements are meaningless unless results can be reliably communicated. Because there is no such thing as a totally accurate measurement, reports of results must contain some indication of the precision of the measurements themselves. The precision of a set of data is often expressed numerically through the use of significant figures.

Part of the language of scientific communication is the system of units by which numerical values are related to one another. Science uses the metric system of units, which contains the subset called SI. Chemists use some of the SI units, along with others from the metric system. Measured values must always be accompanied by the units in which the measurements were made.

In chemistry, as in many other sciences, it is necessary to make mathematical calculations using values from experimental measurements. Such calculations can be approached in an organized way. It is possible to use the units of the measurements as an aid in setting up and checking the calculations.

A good method of approaching calculations is to follow these steps:

1. Read and understand the problem.
2. Write mathematical versions of the statements given.
3. Make a plan, or solution map, for working the problem.
4. Make sensible approximations when appropriate.
5. Set up the calculation and do the arithmetic.
6. Check your answer thoroughly.

QUESTIONS

Section 1.1

1. Without looking them up, define accuracy, precision, and error.
2. Is it ever possible to make a measurement without error? Is it ever possible to make a measurement without error and know that there is no error?
3. Does making a measurement with high precision (small deviations from an average) necessarily guarantee that error will also be small? Why or why not?

Section 1.3

4. List the six steps used in the approach to problem solving.
5. What are the mathematical relations (all the possible unit factors) given in Problem 14?

PROBLEMS

Section 1.3

*1. How many pounds of chocolate-chip cookie dough will you obtain from 5.0 lb of flour, 3.0 lb of white sugar, 3.0 lb of brown sugar, 8.0 lb of butter, 10.0 lb of chocolate chips, 0.03 lb of baking powder, and 0.02 lb of salt?

2. Assuming the volumes are additive, how many gallons of solvent will result if you mix together the following liquids: 2.0 qt of heptane, 6.5 gal of hexane, 3.5 qt of pentane, and 7 drops of methanol?

3. You are wrapping gifts for five children. If a package requires 8.0 ft of ribbon, a bow requires 2.0 ft, and the card is attached with 2.0 in. of ribbon, how many rolls of ribbon do you need for five packages? Each roll contains 54 yd.

4. A recipe calls for ¾ cup of vinegar, but the cook has only a single measuring spoon, which holds 2 teaspoons. How many of these spoonfuls of vinegar will be required if a cup holds 16 tablespoons and a tablespoon holds 3 teaspoons?

Section 1.4

5. If one box of grass seed covers 250 ft², how many boxes will you need to plant one acre? One acre contains 43,560 ft².

6. A German shepherd eats five cups of dry food a day. A pound of food contains 10.5 cups. How much does it cost to feed the dog each day if a 50.0-lb bag costs $22.50?

*7. An American college student is in Paris and has 550 German marks that she now wants to change into French francs. A franc equals 0.185 dollars and 0.49 dollars equals 1.41 marks. Does she have enough money to stay two days in a hotel that costs 145 francs per day?

8. If a stopped car takes up 13.5 ft on a freeway, how many cars will there be in a 12.6-mi traffic jam on a four-lane freeway?

9. A mason can carry 12 bricks at one time. If four bricks cover a wall area of 0.22 yd², how many trips up the ladder must the mason make to cover a panel that is 1.7 yd by 3.25 yd?

10. Here is another nonsense problem. Besides illustrating that unit analysis will work even on units with which you are not familiar, it provides practice in inverting conversion factors and raising them to powers.

  You have a fream which is 21 nuds wide × 75 nuds long. A greep has 41 sec, there are 17 glors to every nud, every square glor contains 22 snaff, but a snaff can be accomplished in 0.16 sec. How many greeps are used in this fream?

  You can use unit analysis to do this apparently complex problem. Start by writing down the conversion factors. Here are two of them:

$$\frac{17 \text{ glor}}{\text{nud}} \qquad \frac{22 \text{ snaff}}{\text{glor}^2}$$

  (See how the words and phrases *to every, can be accomplished in, has,* and *contains* all become simple conversion factors.) Now take it from here.

*11. Lines on an interstate highway are 2.5 ft long and 4.0 in. wide. One quart of paint covers 43 ft². How many lines can be painted with 15 gal of paint?

12. A contractor is constructing a concrete sidewalk that is to be exactly 4 ft wide, 30 yd long, and 4.0 in. thick. Each cubic yard of concrete contains 6.6 sacks of cement. How much cement will be required for the sidewalk?

13. How many quarts of paint will you have to buy to paint all four walls and the ceiling of a room that is 24 ft 9.25 in. long × 19 ft 4.75 in. wide and has a ceiling height of 8.00 ft? A quart of paint covers 43 ft² of surface. Paint is sold only by the quart.

14. A rectangular park measures 0.22 mi by 0.074 mi. How many acres are there in the park if an acre contains 43,560 ft²?

15. A brewery fills 8.2 million 12-oz cans of beer per can line per day. If there are ten can lines, how many gallons of beer will be canned in one day?

16. If approximately 26 million cupcakes are produced each year, and if one cupcake has 0.70 oz of cake and 1.30 oz of cream filling, how many pounds of sugar are used per year for the cupcakes? Each pound of cake contains 4.0 oz of sugar and each pound of filling contains 12.0 oz of sugar.

MORE DIFFICULT PROBLEMS

17. On the planet Xanthu a neb contains 3.21 borks, a bip equals a square neb, 4.576 curps have 8.43 bips, and 2.2 curps make a square blump. A space explorer discovers that a blump equals 0.42 mi on earth. How many inches are there in 5.3 borks?

18. A pet-shop owner is planning to buy three tanks of goldfish, which he will keep for 31 days before selling. Each tank contains 144 fish. Each fish will eat exactly an ounce of fish-food mixture every 83 days (that is, 1 oz food per fish per 83 days or 1 oz food/83 day fish). To each ounce of fish food is added 0.000355 oz of an antibiotic called Xot to keep the fish healthy. How much antibiotic will the owner need?

> **Solution map:** tanks ⟶ fish ⟶ days of food needed ⟶ ounces of food ⟶ antibiotic

*19. The state of Wyoming is 400 mi × 500 mi in size. If an average snowflake is 0.0000223 in. thick and 0.250 in.² in area, how deep would the snow be (in feet) if 6.0×10^{23} snowflakes covered all of Wyoming? Would the snow be deep enough to ski on?

20. If each man will drink 7 oz of punch and each woman 5 oz, how many couples may be served by a punch made with the juice of three dozen oranges mixed with 1.10 gal of ginger ale, 0.5 lb of sugar, and 1 teaspoon of mint? A dozen oranges will make 1.7 pt of juice. A pound of sugar adds 0.33 pt of volume. One cup = 48 teaspoons, 1.00 pt = 16.0 oz, 1.00 gal = 128 oz, and 1.00 cup = 8.00 oz. Assume that 2% of the punch will be lost in mixing and pouring.

Chapter 2

Systems of Measurement

We have learned that for the quantitative result of a measurement to have meaning, it must be expressed in terms of a unit of measurement. It follows that if the result is to be used by others, the unit must have meaning to all who use it. An ideal system of units would have the same quantitative meaning to everyone. Although the world is progressing toward that ideal, it has not yet reached it. The set that is most widely accepted, the metric system, has been used in most of the world for generations.

The metric system is the language used by scientists worldwide and by the general populations of many countries to communicate the numerical results of their measurements, from the intensities of cosmic rays to the weights of fish offered for sale. (The United States is now slowly adopting the metric system.) A subset of the metric system is called the *Système International d'Unités,* or SI. Although we will familiarize ourselves with SI, we will usually use the more convenient units of the larger metric system.

After describing the characteristics of the metric system and SI, the chapter covers the subject of temperature and ways to make certain conversions between temperature scales. Mastering the information contained in Chapters 1 and 2 will give us the necessary tools to begin the serious study of chemistry.

AFTER STUDYING THIS CHAPTER, YOU WILL BE ABLE TO

- convert measurements from the English system to the metric system and the reverse
- convert units to other units within the metric system
- convert temperatures from one temperature scale to another
- use density measurements to solve problems involving mass and volume

TERMS TO KNOW

calorie	Fahrenheit scale	liter
Celsius scale	gram	mass
centimeter	heat	meter
cubic centimeter	intensive property	metric system
cubic decimeter	Kelvin scale	milligram
cubic meter	kilocalorie	milliliter
deciliter	kilogram	millimeter
density	kilojoule	*Système International (SI)*
energy	kilometer	temperature
extensive property	kinetic energy	weight

2.1 METRIC SYSTEM OF UNITS AND SI

The **metric system** of units originated in France and was officially adopted there in 1799. It was generally based on two units, the meter and the kilogram. (The meter was at first intended to be a measurement equal to 10^{-8} times the distance from either pole of the earth to the equator.) As we will see, the definitions of the various units of the system have changed over the years. In 1960, the General Conference on Weights and Measures, an international organization that defines the units of the metric system, set up the system called **SI,** in which certain of the metric units were to be used in preference to others whenever convenient. SI is now used generally by most of the countries of the world and by many scientists. In the United States it is used by the Bureau of Standards and the military. SI defines seven independent basic units, five of which we will become familiar with, the **meter** (m), the **kilogram** (kg), the **second** (s), the **kelvin** (K), and the **mole** (mol). (The remaining two units are the **candela** (cd) for measuring luminous intensity, and the **ampere** (A) for measuring electric current.) The SI units form the core of the larger metric system, in which the other units of measurement are derived primarily from the basic SI units.

Unlike the English system, which is based on a series of units (foot, pound, quart, and so on) having no organized relation to one another, the metric system relates the units of length, volume, and mass in an easily understood way. Moreover, the different sizes of units for each kind of measurement, for instance, length, are related to each other by simple powers of ten. Its structure and organization makes the metric system much easier to use than the English system.

Because in your study of science it is essential that you be acquainted with metric units and because the United States has traditionally used British units, you should consider memorizing at least one English-metric conversion factor for each category of measurement: length, volume, and mass. In the chapters that follow, we will have little occasion to make conversions between the English and metric systems; however, problems involving such conversions do appear in the present chapter. The problems are included to familiarize you with the sizes of metric units and to provide practice in making unit conversions.

To express larger or smaller quantities of any of the basic units, the metric system uses a series of prefixes that are added to the name of the unit. Because each prefix of the metric system represents multiplication or division by a factor of ten, the system is entirely decimal in nature. Table 2.1 shows the names, symbols, and numerical values of the most commonly used metric prefixes.

Table 2.1
Metric Prefixes

Prefix	Symbol	Numerical Value
mega-	M	1,000,000 or 10^6
kilo-	k	1000 or 10^3
deca-	da	10 or 10^1
deci-	d	0.1 or 10^{-1}
centi-	c	0.01 or 10^{-2}
milli-	m	0.001 or 10^{-3}
micro-	μ	0.000001 or 10^{-6}

Examples of the use of prefixes in the metric system:

    1 milliliter = 0.001 liter
    1 kilogram = 1000 grams
    1 decimeter = 0.1 meter

In the strict use of SI, the basic units themselves, or units derived directly from those basic units, are to be used. The prefixes *milli-* (0.001) and *kilo-* (1000) are to be used whenever possible. In chemistry, however, some of the basic SI units are inconvenient; for example, there is wide use of the gram rather than the kilogram, and of the centimeter rather than the meter.

2.2 MEASUREMENT OF LENGTH

The standard unit of length in the metric system is the **meter.** Until recently, the meter was defined as the distance between two marks placed on a platinum bar stored in Sevres, France (that distance supposedly being 0.00000001 of the distance from the pole to the equator). The old definitions of many of the metric units, however, have been replaced with more sophisticated and precise ones, and the new definition of the meter is based on the measurement of the wavelength of light emitted by atoms of the element krypton.

A meter is the same as 39.37 in., which is, of course, a little more than a yard in the English system. A **kilometer** is about 0.62 mi. The **centimeter,** 0.01 m, is a convenient unit for working in the laboratory. One inch is 2.54 cm.

Other units frequently used in chemistry are the **millimeter,** mm, which is 0.001 m or 0.1 cm, and the **Angstrom,** Å, 10^{-8} cm or 10^{-10} m. A small atom is about 1 Å in diameter.

Conversion factors among the metric length units are:

$$\frac{1000 \text{ m}}{\text{km}} \qquad \frac{100 \text{ cm}}{\text{m}} \qquad \frac{10 \text{ mm}}{\text{cm}} \qquad \frac{10^8 \text{ Å}}{\text{cm}}$$

Since we are already familiar with the conversions within the English system (inches to feet to yards to miles), we will practice some English-metric conversions and some conversions within the metric system itself. A useful English-metric conversion factor for length is

$$\frac{2.54 \text{ cm}}{\text{in.}}$$

EXAMPLE 2.1.   We are traveling in Europe and decide to buy some clothes. One of us has a neck measurement of 15½ in., but the shirt collar sizes are given in centimeters. What size collar shall we ask for?

Solution:

$$15.5 \text{ in.} \times \frac{2.54 \text{ cm}}{\text{in.}} = 39.4 \text{ cm}$$

We'll try a 39-cm collar and see if it will fit.

EXAMPLE 2.2.   A room is 3.219 m across. What is this in feet and inches?

Solution:

$$3.219 \text{ m} \times \frac{100 \text{ cm}}{\text{m}} \times \frac{\text{in.}}{2.54 \text{ cm}} = 126.77 \text{ in.}$$

We round to 127 in. and then convert the inches to feet and inches. Because 120 in. is exactly 10 ft, we have (127 − 120) in. = 7 in. more than 10 ft. Our answer is 10 ft 7 in.

EXERCISE 2.1.   You are buying a cut-glass wine bottle that measures 290 mm high. Will it fit on your bookshelf at home, where the shelves are spaced exactly 1 ft apart?

EXERCISE 2.2.   The earth is approximately 24,000 mi in circumference. How many meters is this?

2.3 UNITS OF VOLUME

In the metric system, there is a direct relation between the units of length and those of volume. This is in contrast to the English system, in which units of length (e.g., feet) and volume (e.g., quarts) have no logical relation whatever.

Although the **cubic meter,** m^3, is the standard unit of volume in SI, the most widely used and most convenient volume unit is the **liter,** L, which is almost exactly 1000 **cubic centimeters** (cm^3) or 1 **cubic decimeter** (dm^3). This is very roughly equivalent to the English quart; 1 qt = 946 cm^3 = 0.946 L. For English-metric conversions, we will use the conversion factors

$$\frac{0.946 \text{ L}}{\text{qt}} \qquad \frac{29.6 \text{ mL}}{\text{fl oz}}$$

EXAMPLE 2.3.   How may liters are there in 1.000 m^3?

Solution map: cubic meters ⟶ cubic centimeters ⟶ liters

(m^3 ⟶ cm^3 ⟶ L)

Solution:

$$1.000 \text{ m}^3 \times \left(\frac{100 \text{ cm}}{\text{m}}\right)^3 \times \frac{1 \text{ L}}{1000 \text{ cm}^3} = 1000 \text{ L}$$

Notice that it was necessary to raise the conversion cm/m to the third power.

The **milliliter,** mL, which is almost exactly the same as 1 cm³, is frequently used in chemistry. In common practice, these two units are used interchangeably. The **deciliter** (deci- = 0.1; 1 dL = 0.1 L) is widely used as a commercial unit in Europe; it is about 3 oz, or somewhat more than ⅓ cup (Figure 2.1).

The metric-to-metric conversion factors for volume are:

$$\frac{1000 \text{ cm}^3}{\text{L}} \qquad \frac{1000 \text{ mL}}{\text{L}} \qquad \frac{10 \text{ dL}}{\text{L}}$$

EXAMPLE 2.4.   You are camping your way across Europe, carrying your Sierra Club cup on your belt. It holds 0.331 qt (about a cup and a half). At a sidewalk cafe, the price of beer is 1 mark for 3 dL. Will your cup hold a mark's worth?

Solution:

$$0.331 \text{ qt} \times \frac{0.946 \text{ L}}{\text{qt}} \times \frac{10 \text{ dL}}{\text{L}} = 3.1326 \text{ dL}$$

Round to 3.13 dL, then go ahead and buy.

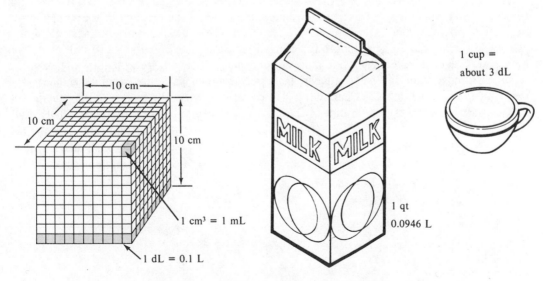

Figure 2.1 Comparisons of volume measurements.

EXAMPLE 2.5.   Your friend has a tin box that contained cookies. The box is 3.3 in. × 2.5 in. × 4.0 in. How many deciliters will it hold?

Solution map: inches³ $\longrightarrow$ centimeters³ $\longrightarrow$ liters $\longrightarrow$ deciliters

Solution:

$$3.3 \text{ in.} \times 2.5 \text{ in.} \times 4.0 \text{ in.} \times \left(\frac{2.54 \text{ cm}}{\text{in.}}\right)^3 \times \frac{1 \text{ L}}{1000 \text{ cm}^3} \times \frac{10 \text{ dL}}{\text{L}} = 5.40773 \text{ dL}$$

Round to 5.4 dL.

EXERCISE 2.3.   Because you were unable to pay for the beer, the proprietor has put you to work. You are required to fill a large tank with water by carrying it in your 1-gal jug. The tank holds 0.75 m³. How many trips must you make with the jug?

EXERCISE 2.4.   In Germany, a soft-drink costs about 1.10 German marks for 5.0 dL. How many German marks would you pay for a 12.0-oz can of the drink, if there are 32.00 oz in 1 qt?

2.4 MASS AND WEIGHT

Although we will be discussing the nature of matter as such in the next chapter, we first need to learn about the property of matter called mass, and its relation to weight and the measurement of weight. The amount of matter in a sample is usually measured by taking its weight.

Mass is that property of matter that gives the matter its inertia, or resistance to change in velocity. The mass of a sample of matter does not depend on the location of the sample or on its relation to other bodies of matter. To get an intuitive feel for mass and its relation to change of velocity, imagine a smooth tabletop with a ping-pong ball and a golf ball on it. You blow gently on each ball. Which ball moves most rapidly? The ping-pong ball, of course; it has much less mass than the golf ball and is therefore much easier to get into motion. The experiment you have just done has nothing to do with gravity. Without weighing them, you have found that there is a difference in the mass of the two balls (Figure 2.2).

Returning a tennis ball

Returning a lead ball

Figure 2.2   A lead ball is hard to move.

Because of its mass, a sample of matter in motion tends to remain in motion, and a sample of matter at rest tends to stay at rest. The greater the mass of the sample, the more difficult it is to bring about a change in its motion. Ocean liners, moving at only a fast walk, have crunched into docks and adjoining buildings, slowly but irresistibly destroying them before finally coming to a stop. The huge mass of a large ship has an equally huge tendency to remain in motion.

In addition to its effect on motion, mass is what makes matter subject to gravitational force. The **weight** of a sample depends both on the amount of its mass and on the strength of gravitational attraction where the sample is. The weight of a body of matter at the surface of the earth, for instance, is simply the force by which the body is attacted by the earth's gravity. The greater the mass of the body, the greater the weight. Although the mass of a sample is independent of location, its weight is not. For instance, an astronaut weighs less on the surface of the moon than on earth, even though his mass is the same in both places (Figure 2.3).

Because the amount of matter in a sample is measured by the amount of mass, many measurements in chemistry are measurements of mass. In some cases, mass is measured directly, as in experiments using the same approach that you used with the ping-pong and golf balls. For example, the experiments discussed in the next chapter (which concerns the first measurements of the masses of atoms) involve changes in the motion of atoms and thus measure mass directly. Ordinarily, however, the mass of a sample that can be seen and handled in the laboratory is measured indirectly, using gravitational attraction.

Balances, which are widely used in science, measure the relation between the *weight* of a sample and that of an object whose mass is known. Because the objects are both on the balance, they are subject to the same gravitational field; comparing their weights therefore also compares their masses. A balance measures mass, then, by comparing weight. (See Section 2.8 for a discussion of balances.)

Even though mass and weight are not the same thing, weights at the earth's surface are usually quoted in mass units instead of force units, and sample weights are usually given in kilograms or grams. If a sample has a mass of 1 g in the metric system, it is also said to weigh

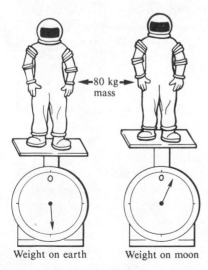

80 kg mass

Weight on earth   Weight on moon

Figure 2.3 The mass of an object is constant; its weight depends on gravitational force.

1 g. Of the measurements of sample mass that we will be examining, most will have been performed by weighing. Because weighing, as we have seen, can be used as a technique to measure mass itself, it is appropriate that we use mass units to report our weighings.

The **kilogram** is defined by SI as the mass of a prototype kilogram sample of metal stored in France. The **gram** is then defined, according to the system of prefixes, as 1/1000 kg. (The gram was formerly defined as the mass of a sample of pure water having a volume of 1 cm³ at a temperature of 4 °C.) A useful unit for small samples is the **milligram,** mg, which is 1/1000 g.

The English-metric conversion that we will find most useful is

$$\frac{453.6 \text{ g}}{\text{lb}}$$

EXAMPLE 2.6.   After riding a bicycle over a hundred miles of French roads, you step on a scale to weigh yourself. The number you read is 47.7, but at home, you weighed 112 lb. Could you possibly have lost that much?

Solution:

The French scale is in kilograms. Your weight in pounds is

$$47.7 \text{ kg} \times \frac{1000 \text{ g}}{\text{kg}} \times \frac{\text{lb}}{453.6 \text{ g}} = 105 \text{ lb}$$

You have lost only a little.

EXERCISE 2.5.   At home you usually eat about half a pound of meat. When you go to the local French market, the meat is sold in units of hundreds of grams. About how much should you buy?

EXERCISE 2.6.   The corrosive compound sulfuric acid is sometimes sold in large, carefully protected jugs called carboys. A particular carboy holds 167.5 lb of sulfuric acid. How many grams is this? How many kilograms?

2.5 ENERGY, HEAT, AND TEMPERATURE

Energy is often defined as work or the ability to do work. When we discuss energy in more detail we will broaden that definition, but it is sufficient for our present purposes.

A body of matter in motion possesses what is called **kinetic energy,** or energy of movement. As we will discover, all matter is composed of microscopic individual particles (atoms) so small as to be invisible. In any sample of matter, these particles constantly move in relation to one another; but that motion is not visible to the eye, nor does it cause the sample itself to move. The **temperature** of a sample is a measure of the average kinetic energy of the particles of that sample, and what we call **heat** is the transfer of that energy from the particles of one sample to those of another. Temperature is what determines the direction in which heat will flow. As you know, heat flows from areas of high temperature to areas of low temperature (Figure 2.4). Let's examine temperature, heat, and heat flow in an ordinary situation.

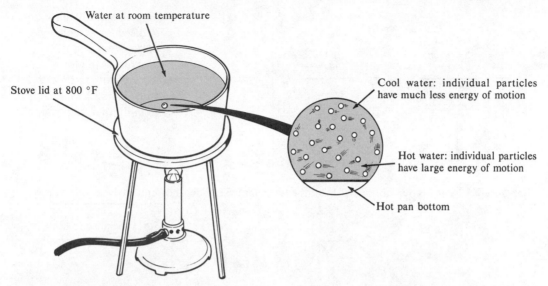

Water at room temperature

Stove lid at 800 °F

Cool water: individual particles
have much less energy of motion

Hot water: individual particles
have large energy of motion

Hot pan bottom

Figure 2.4   Heat flows from areas of higher temperature to areas of lower temperature.

Suppose you have a pan, a stove, and some water. You put 1 L of water in the pan and place it on the burner. Heat flows from the stove to the water in the pan. The water becomes hot; that is, its temperature rises. Beginning again, you put 2 L of water in the pan and heat it to the same temperature as you did the first time. Except for a small effect caused by the pan itself, it will take twice as long to heat the 2 L of water as it did the single liter, simply because twice as much heat is required. Why two times as much heat? Because 2 L of water has two times as many particles as 1 L. Although the larger sample has received twice as much heat as the smaller, its temperature will be the same as that of the smaller. Temperature measures the average kinetic energy of the individual particles of a sample, and that is the same in both samples. If we were to take a 10-L sample of water at this same temperature, all its particles together would have ten times as much total kinetic energy as those of the single liter. At any temperature, the amount of kinetic energy of the particles in a sample depends on the size of the sample taken. (See Figure 2.5.) In Chapter 13, we will consider the relation of heat, temperature, sample size, and nature of the sample in more detail.

We can now define two useful words. Quantities, such as total particle energy, that depend on the size of the sample taken, are called **extensive** quantities. Quantities like temperature, that are independent of sample size, are said to be **intensive** quantities. The total amount of kinetic energy of the individual particles of a sample is extensive, but temperature is intensive. What about volume? Is it an extensive or intensive quantity? The amount of volume of a sample clearly depends on the amount of sample taken, as measured by its mass. Two kilograms of water have twice the volume of 1 kg. Volume, then, is extensive.

The **calorie** is a measurement of heat that is easy to understand. A calorie is the amount of heat required to raise the temperature of 1 g of water from a temperature of 15.5 °C to 16.5 °C (Figure 2.6). (Temperature scales are discussed in the next section.) The **kilocalorie** is 1000 cal. What nutritionists call a **Calorie** (with a capital C) is really a *kilocalorie;* if your chocolate bar contains 250 Cal, it really has 250,000 cal (small c), or enough energy to raise

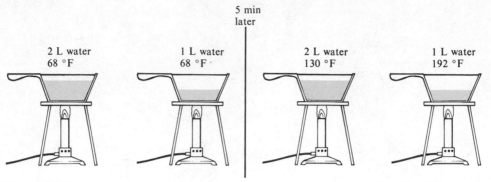

Figure 2.5   For a given amount of heat, the temperature rise depends on the size of the sample.

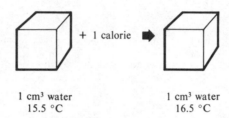

Figure 2.6   How the calorie is defined.

the temperature of 250,000 g of water (~250 L) 1 °C. That is enough energy to bring approximately 3 L of water at room temperature to the boiling point. Although the calorie was first defined in terms of heat energy, it is used to measure other kinds of energy as well, an example being food energy.

The energy unit appropriate to SI is the **kilojoule,** kJ; 1 kJ is 239 cal. There is no need for us to discuss the basis for the definition of the kilojoule at this point, but it might help us to know that about 300 kJ of energy will heat 1 L of water from room temperature to the boiling point.

Our conversion factor from calories to kilojoules is

$$\frac{kJ}{239 \text{ cal}}$$

Even though the calorie is the more familiar unit, we will use the kilojoule because eventually the calorie will probably fall into disuse in scientific work.

EXAMPLE 2.7.   One bite of a chocolate bar contains 27 Cal. How many calories is this? How many kilojoules?

Solution:

$$27 \text{ Cal} \times \frac{1000 \text{ cal}}{\text{Cal}} = 2.7 \times 10^4 \text{ cal}$$

$$(2.7 \times 10^4 \text{ cal}) \times \frac{\text{kJ}}{239 \text{ cal}} = 1.1 \times 10^2 \text{ kJ}$$

EXERCISE 2.7.   How many calories are needed to heat 1 L of water from room temperature, 20 °C, to the boiling point?

EXERCISE 2.8.   How many kilojoules will be necessary to raise 25.0 g of water from 15.5 °C to 16.5 °C? From 22 °C to 63 °C?

2.6 MEASUREMENT OF TEMPERATURE

As we have seen, temperature is a measure of the amount of motion in the microscopic particles of a sample. In the English system, the standard for measuring temperature is the **Fahrenheit** scale. In our study of chemistry, however, we will use two scales, the **Celsius** scale, and the **Kelvin,** or **absolute,** scale. Temperature scales are usually counted in degrees, but the degrees of different scales may be of different sizes. (The term *kelvins* instead of *degrees Kelvin* is becoming more common, and we will use it.)

The abbreviations for these units are

°C = degrees Celsius
 K = kelvins
°F = degrees Fahrenheit

To define a temperature scale, it is first necessary to find an actual temperature that can be easily reproduced by anyone. The temperature at which pure water freezes is one place to start. In the Celsius scale, the freezing point of water under carefully defined conditions is at 0°. But designating only a zero point is not enough. To state any temperature other than the freezing point, we must say how far and in which direction that temperature lies from the freezing point. A second known temperature must be selected for this purpose; then the interval between the two temperatures is divided into the units usually called degrees. Any temperature can then be designated if we say how many degrees and in which direction that temperature is from the zero point. In the Celsius scale, the temperature at the boiling point of water is 100° at sea level. The interval between the freezing and boiling points of water, then, contains 100 degrees. A temperature that is 20/100 of the total difference between freezing and boiling, for instance, is 20 °C (20 °C is approximately room temperature). Water that is so hot that you can barely put your hand in it is about 50 °C. In the Celsius scale, temperatures below freezing are designated in the same way as those above freezing, but they are given negative signs. The Fahrenheit scale, which is used in the English system, is also arbitrarily defined, but with different values for the freezing and boiling points. Water freezes at 32 °F and boils at 212 °F. Thus, while there are 100 Celsius degrees between the freezing and boiling points of water, there are 212 − 32, or 180, Fahrenheit degrees between the same two points. The individual Celsius degrees are clearly larger than the Fahrenheit degrees, since fewer of the Celsius degrees cover the same range. In fact, 100 Celsius degrees are the same as 180 Fahrenheit degrees: a Celsius degree is 9/5 of a Fahrenheit degree.

   The Kelvin scale is defined in a more fundamental way, which we will learn about in Chapter 12. Although kelvins are the same size as Celsius degrees, the zero point on the Kelvin scale is placed at −273 °C. The freezing temperature of water is therefore 273 K and the boiling temperature 373 K. The three scales are shown in Figure 2.7.

   If a temperature value is known in one scale, a conversion can be made to another scale as long as the relation between the two scales is known. The most common temperature conversion you will encounter in beginning chemistry is the change from Celsius to Kelvin and vice versa. The conversion is quite easy, because kelvins are the same size as Celsius degrees. To go from Celsius to Kelvin, just add 273. To go from Kelvin to Celsius, merely subtract 273.

$$°C = K − 273$$
$$K = °C + 273$$

EXAMPLE 2.8.   What is the Kelvin equivalent of a temperature of 37 °C?

Solution:

$$37 °C + 273 = 310 K$$

EXAMPLE 2.9.   What is the Celsius equivalent of a temperature of 258 K?

Solution:

$$258 K − 273 = −15 °C$$

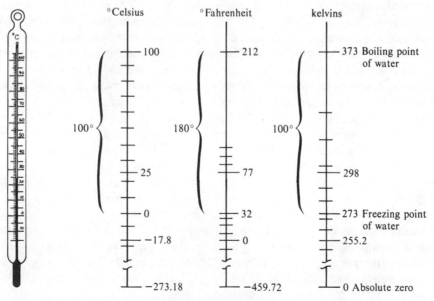

Figure 2.7   Celsius, Fahrenheit, and Kelvin temperature scales.

EXERCISE 2.9.   What are the Kelvin equivalents of 32 °C and 178 °C?

EXERCISE 2.10.   What are the Celsius equivalents of 212 K and 298 K?

Because Fahrenheit temperatures are rarely used in science, Celsius-Fahrenheit conversions are useful to us primarily as exercises. Appendix C explains the general approach to making temperature conversions and, in particular, demonstrates Celsius-to-Fahrenheit and Fahrenheit-to-Celsius conversions.

As we have seen, temperature is an intensive property; that is, it is a property the value of which is independent of sample size. If you have 1 L of water at a temperature of 40 °C and you divide the sample into two parts, each part will still have a temperature of 40 °C.

2.7 DENSITY

The ratio of the mass of a sample to its volume is called the **density,** d, of that sample. Density is one of the characteristic properties of matter and is one of the means of identifying particular samples.

If two substances have different densities and if we take samples of equal volume from the two substances, the samples will have different masses. The density of a substance depends on both the mass of the individual particles of the substance and the closeness with which those particles are packed together. Let us consider both effects. Suppose you have some shoeboxes, all of the same volume. Into one pair of boxes you place some golf balls, each carefully wrapped in cotton (which, we will assume for our example, has no mass). One box will hold 15 golf balls. Into the other box you pack golf balls as tightly as you can, and manage to put 45 golf balls into it. The second box will weigh three times as much as the first. Because the boxes have equal volumes, the ratio of mass to volume for the second box (its density) will be three times greater than that of the first. That the balls are packed more closely into the second box causes the difference. Suppose you now take a third box and place 45 ping-pong balls in it. The third box will weigh far less than the second, even though it has the same number of balls. The density of the third box is less than that of the second because each ball in the third box has less mass than each ball in the second box.

A block of gold measuring 10 cm on a side will weigh about 20 times as much as a block of ice of exactly the same size. The density of gold is about 20 times as great as that of ice. The individual particles of the gold are packed about twice as tightly as those of water, and each particle of gold has about ten times the mass of a particle of water. A pint of water weighs 1 lb, but a pint-sized sample of the element platinum would weigh more than 21 lb. Platinum is one of the most dense substances.

The most commonly used density units for liquids and solids, the ones we will use, are grams per cubic centimeter. (The SI units of kilograms and cubic meters are inconvenient.) Because milliliters and cubic centimeters are almost exactly the same, sometimes density is written in units of grams per milliliter. Because density changes with temperature, it is important that you specify the temperature at which the measurements were taken when quoting values of density.

The density of water at 4 °C is 1 g/cm^3.

$$d = \frac{1 \text{ g}}{1 \text{ cm}^3}$$

(In writing density values, a subscript is used to identify the substance and a superscript gives the Celsius temperature at which the measurement was made, thus

$$d_{H_2O}^{4 \text{ °C}}$$

denotes the density of water, H_2O, at 4°.)

At other temperatures, water has somewhat different densities. For example,

$$d_{H_2O}^{25 \text{ °C}} = 0.999 \text{ g/cm}^3$$

$$d_{H_2O}^{95 \text{ °C}} = 0.962 \text{ g/cm}^3$$

The density of liquid mercury at 20 °C is 13.6 g/cm³, a value indicating that mercury is more than 13 times as dense as water. If we were to place some mercury in a beaker and put some water in with it, the water would float on the surface of the mercury, just as oil, which is less dense than water, floats on water.

In Table 2.2 are values for the densities of some substances. Notice that limestone, one of the more plentiful components of the earth's crust, has a density not quite three times that of water, whereas iron is about eight times as dense. The only density to memorize from this table is that of mercury, which we will use from time to time. Other density values can be looked up when necessary.

It is useful also to measure the densities of gases. Gases are not nearly as dense as liquids or solids. Because 1 or 2 g will ordinarily occupy a liter or so of volume, it is convenient to use the units grams per liter. Densities of gases depend on temperature as well as pressure (a subject that we will examine in detail somewhat later), so it is necessary when reporting gas densities to specify both the temperature and the pressure. Table 2.3 gives some sample gas densities, all taken at 0 °C and 1 standard sea level atmospheric pressure.

Density is an intensive quantity. Metallic iron has a density of 7.87 g/cm³. A sample of iron as large as a peach will weigh about a kilogram, and its density will be 7.87 g/cm³. An

Table 2.2
Densities of Some
Liquids and Solids at 20 °C

Substance	Density (g/cm³)
Mercury	13.55
Aluminum	2.70
Iron	7.87
Silica sand	2.65
Salt	2.16
Limestone	2.8
Lead	11.3
Gold	19.6

Note: The average density of the earth is about 5.5 g/cm³.

Table 2.3
Densities of Some
Gases at 1 atm, 0 °C

Gas	Density (g/L)
Air	1.29
Helium	0.17
Oxygen	1.43
Freon	6.11
Hydrogen	0.090
Methane	0.714
Carbon dioxide	1.963
Chlorine	3.17

ingot of iron at a foundry will weigh several metric tons, but its density will nevertheless be 7.87 g/cm³.

Density calculations often appear in chemistry problem solving. Although many laboratory operations depend on measurement of volume, the calculations for those operations involve masses. It is thus often necessary to convert arithmetically from mass to volume and the reverse. Density is the key to making such conversions.

EXAMPLE 2.10.   You wish to buy enough honey to fill your honey jar, which holds 460 mL. Honey has a density of 1.44 g/cm³ and costs $2.25 per kg. How much will it cost to fill your jar?

Solution:

$$460 \text{ cm}^3 \times \frac{1.44 \text{ g}}{\text{cm}^3} \times \frac{1 \text{ kg}}{1000 \text{ g}} \times \frac{\$2.25}{\text{kg}} = \$1.49$$

EXERCISE 2.11.   Your salt box contains 3.31 kg of salt, the density of which is 2.16 g/cm³. What is the volume of the salt?

EXERCISE 2.12.   You are studying the way in which the density of water changes with temperature. You find that a sample of water weighing 32.7 g has a volume of 33.1 mL at 50 °C. What is the density of water at 50 °C? Is this more or less than the density of water at 4 °C?

2.8 LABORATORY INSTRUMENTS OF MEASUREMENT

In beginning chemistry laboratory work, the quantities most often measured are mass, liquid volume, and temperature.

Mass is usually measured by comparing weights. Weighings are made on balances, machines that allow two weights to be matched exactly. Depending on sample size and the degree of precision required, one of several kinds of balances can be used. For large samples or for weighings not requiring high precision, a *triple-beam pan balance* is often employed (Figure 2.8). The object to be weighed is placed on the pan, and the moveable weights are positioned by hand until the machine balances. The amounts of the weights can then be read. Triple-beam balances are ordinarily precise to about 0.01 g.

For small samples or for weighings of greater precision, several kinds of balances are available, but the single-pan analytic balance is used most often (Figure 2.9). To use this instrument, the operator places the sample on the pan and controls the placing of the internal weights by means of exterior knobs. The balance point is indicated by a pointer on a scale. Balances like this are generally used for weighings in which the precision must be in the milligram range or slightly better. The machines are delicate and expensive and do not have the capacity to weigh large amounts of sample.

Liquids are measured by volume in what is called volumetric glassware (Figure 2.10). Graduated cylinders are used for rough measurements. Pipets and volumetric flasks are used for more precise work. They are designed so as to minimize error. Volumetric glassware is made in various sizes, pipets being used for small amounts and volumetric flasks for volumes

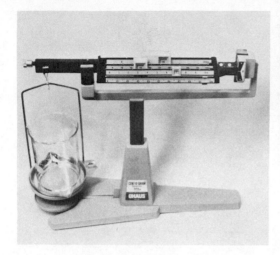

Figure 2.8 Triple-beam balance. (Photo by author.)

Figure 2.9 Automatic, single-pan balance. (Photo courtesy of Mettler Instrument Corporation.)

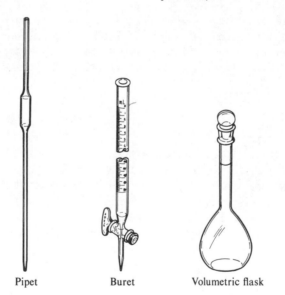

Pipet   Buret   Volumetric flask

Figure 2.10 Samples of volumetric glassware.

ranging usually from 100 mL to 1 or 2 L. A buret is a long, graduated tube with a valve at the bottom. It is used to measure and dispense arbitrary amounts of liquid.

As we have seen, temperature can be measured with a thermometer. Glass thermometers with mercury columns are most often used in early courses in chemistry, although other kinds of thermometers using alcohol or other liquids are also common.

2.9 SUMMARY

In science, the standard set of units of measurement is the Système International, SI, which is a subset of the larger metric system. The units that we will most often encounter are shown in the table that follows.

Property	Unit
Volume	Liter or milliliter (or cubic deciliter or cubic centimeter)
Mass	Gram
Length	Centimeter
Temperature	Kelvins or degrees Celsius

Metric units exist in several sizes that are uniformly related by factors of ten and can be used to measure any quantity. For example, three units of mass are the milligram, gram, and kilogram:

$$1 \text{ kg} = 10^3 \text{ g} = 10^6 \text{ mg}$$

Conversion factors can be used to change English units into metric units; for example, 1 in. = 2.54 cm.

The practice of chemistry rests on measurements of the mass of samples of matter. Mass is a property of matter that causes matter to be subject to gravitational force and to have inertia. The mass of a sample is usually measured by weighing, that is, by comparing the gravitational force on the sample with that on a known mass, using an instrument called a balance.

Energy is usually expressed in units of calories or kilojoules; 1 kJ = 239 cal. Temperature, which is a measure of the average kinetic energy of the atoms of a sample, is usually reported in units of degrees Celsius or kelvins. Celsius temperatures can be converted to Kelvin temperatures by adding 273 to the value; the reverse is accomplished by subtracting 273. Temperature is an intensive property.

Density, also an intensive property of matter, is defined as the ratio of the mass of a sample to its volume (example: grams/liters). Calculations in chemistry often involve density.

QUESTIONS

Section 2.1

1. What is the relationship of the Système International, SI, to the metric system?
2. Name six of the basic SI units. Which of these has an unusual characteristic that makes it different in kind from the others?
3. Without looking them up, name the metric prefixes that indicate multiplication and division by powers of ten, as listed in Table 2.1.

Section 2.3

4. Explain the relation between the length unit, meter, and the volume unit, liter. Does such a defined relation exist in the units of the English system?

Section 2.4

5. Explain how mass differs from weight. If the two are different, how can it be that we measure mass by weighing? Design a simple experiment by which you could measure mass in an environment free of gravity—in a manned satellite, for example.

Section 2.5

6. What is the difference between heat and temperature?
7. Explain the difference between intensive and extensive properties. Try to list some of each. Which kind of property is density?

Section 2.6

8. Explain why it is necessary to designate two actual fixed temperatures in defining a temperature scale.

Section 2.7

9. Explain the relation between the volume unit, cubic meter (m^3), and the mass unit, kilogram, in the metric system. Why must temperature be involved in this relation?

PROBLEMS

Section 2.2

1. In the table shown, numbers on the same horizontal line are to have the same values expressed in the units shown at the heads of the columns. The second number in the first line is given as an example. Fill in the empty spaces.

Meters	Centimeters	Millimeters	Kilometers	Inches
____	153.7	1537	____	____
2.25	____	____	____	____
____	____	339	____	____
____	____	____	____	17
____	____	____	33.876	____

*2. For each of the following, convert the given value into the units requested. Show your calculations for each in a unit analysis form.

*a.	2.3 km	=	mi	e.	4.11 m	=	in.	i.	9.9 ft	=	m
*b.	4 ft 2.2 in.	=	cm	f.	55.66 cm	=	ft	j.	32.113 mi	=	km
*c.	223 Å	=	mm	g.	44 mm	=	in.	k.	32 mm	=	ft
*d.	1.433 cm	=	Å	h.	3.2 yd	=	m	l.	22 cm	=	in.

3. Your waist is 32.5 in., and you want to buy a pair of pants in Berlin. What size in centimeters should you try on?
*4. The distance between Los Angeles and Denver is 1182 mi. How many kilometers is it?

5. The distance between the city of Munich and the village of Holzkirchen is 12.5 km. How many miles is it?

6. An airline is offering a round trip ticket to Hawaii from Los Angeles for $100, if you buy another round trip ticket for a distance greater than 550 km each way. Can you get the Hawaii ticket if you fly from Los Angeles to Fresno, California, and back? The distance between these cities is 382 mi.

7. A chemist is 5 ft 6.5 in. tall. What is her height in centimeters? Write your own height in feet and inches, then convert it to centimeters.

*8. Visible light, as well as ultraviolet, infrared, X ray, and other radiation, is characterized by what is called *wavelength*. The wavelength of certain infrared light is 30 μm. How many meters is this? How many millimeters?

9. Visible light has wavelengths between 3.75×10^{-9} m and 8.0×10^{-9} m. Can we see light with a wavelength of 5.5×10^{-2} μm?

10. X rays have wavelengths between 180 Å and 500 Å. What are these wavelengths in centimeters and in millimeters?

11. A molecule is approximately 1.00 Å in length. How many molecules will fit end-to-end in a row 1.00 m long?

Section 2.3

12. In the table shown, numbers on the same horizontal line are to have the same values expressed in the units shown at the heads of the columns. Fill in the empty spaces.

Cubic meters	Cubic Centimeters	Liters	Quarts
_____	435	_____	_____
0.75331	_____	_____	_____
_____	_____	1.005	_____
_____	_____	_____	2.99

13. For each of the following, convert the given value into the units requested. Show your calculations for each in a unit analysis form.

a. 22.4 L = qt
b. 8 fl oz = mL
c. 432 mL = pt
d. 2 dL = fl oz
e. 448 mL = qt
f. 91 gal = m³

14. How many milliliters are in 4.76×10^{-2} L?

*15. How many cubic centimeters are in 5.89×10^2 L?

16. How many liters equals 432 mL of solution?

17. A gas has a volume of 250 mL. What is its volume in liters?

18. A recipe requires 2.5 cups of wine. Wine is usually sold in 750-mL bottles. Will one bottle be enough wine for the recipe?

19. Will a hollow cube with sides equal to 0.500 m hold 8.5 L of solution?

*20. A student health center will vaccinate 1555 students. Each vaccination requires 10.0 cm³ of serum. Will 1.75 L of serum be enough to vaccinate all the students?

21. A sip of wine is about 2.0 mL. If a 1965 bottle of Lafite Rothschild is $235 and contains 24.5 fl oz, how much does just one sip cost?

22. In Germany you can buy a liter of beer for 2.00 German marks. In the United States you can buy 1.5 cups of a soft drink for $0.40. Which drink costs more per quart, if $1.00 is 1.90 German marks?

Section 2.4

23. In the table shown, numbers on the same horizontal line are to have the same values expressed in the units shown at the heads of the columns. Fill in the empty spaces.

Grams	Milligrams	Kilograms	Pounds
_____	4.32×10^6	_____	_____
0.0059	_____	_____	_____
_____	_____	7.52	_____
_____	_____	_____	2.643

*24. For each of the following, convert the given value into the units requested. Show your calculations for each in a unit analysis form.

   *a. 155 lb = kg   c. 3.10 lb = g   e. 14 g = oz
   *b. 44 kg = lb   d. 0.065 lb = mg   f. 12 oz = g

25. How many milligrams of copper are in a copper sample weighing 2.75×10^{-3} kg?
26. You need 5.75×10^{-2} kg of the compound heptane. How many milligrams of heptane do you need?
27. For a particular chemical equation you need to calculate the mass of a gas in kilograms. What is 4.52×10^3 mg of gas in units of kilograms?
28. A chemist wishes to know how many grams of the compound sodium carbonate are contained in a 55-lb drum. Make the calculation.
29. You need 300.0 g of lead chloride for a chemical reaction. Will a 12.0-oz jar of lead chloride be sufficient?
30. You have a French friend who weighs 65 kg. What is her weight in pounds? What is your weight in kilograms?
31. A mechanic uses a lift to raise automobiles so that he can work on them. The lift has a capacity of no more than 3000 lb. A man drives in a German limousine that needs some work. His car might weigh more than the lift will raise. The owners' manual says that the car weighs 1568 kg. If 1 kg contains 2.205 lb, will the lift work?
32. A pharmacist needs to fill five prescriptions of a drug. Each prescription is for 40 capsules, each containing 50.0 mg of drug. He has 2.5 g of the drug. Does he have enough?
*33. Rice in Japan costs 12 yen per kilogram. In the United States, Japanese-style rice costs about $0.32 per pound. If 185 yen equals $1.00, where is the rice more expensive?
34. In France, gold is sold by the gram. How much does a gram of gold cost in French francs if there are 4.25 francs in a dollar and if gold costs $400 per ounce.

Section 2.5

35. Calculate the number of calories and the number of kilojoules in 39 Cal.
36. How many calories are there in 1.2×10^{-2} kJ?
*37. How many kilojoules will heat 250 g of water from room temperature to the boiling point?
38. How many calories will be necessary to raise 2.45 kg of water from 15.5 °C to 16.5 °C?
39. A certain reaction produces enough heat to raise 55 g of water from 15.5 °C to 16.5 °C. Calculate the number of kilojoules produced by the reaction.

40. How many milligrams of water will rise in temperature from 15.5 °C to 16.5 °C, on receiving 723 kJ of heat?

41. How many deciliters of water can you heat from room temperature to the boiling point with 4.5 × 10^4 cal?

42. Your friend finishes the honey (mentioned in Problem 49) and decides to clean the box with hot water. He fills it with cold water at 18 °C and heats the water to 85 °C. Making the approximation that 1 cal will raise the temperature of 1 mL of water 1 °C at any temperature (neglecting the heat required by the box itself), calculate how many calories would be required to heat the water. Calculate this value in kilojoules.

Section 2.6

43. Calculate the following temperature conversions:

   a. 331 °C =   K   c. −221 °C =   K   e. 520 K =   °C
   b. 285 K =   °C   d. 11.4 °C =   K   f. 33 K =   °C

44. A gas at 328 K has what Celsius temperature?
45. Would an object at 280 K feel cold?

Section 2.7

46. Complete the table by performing the following density conversions.

g/cm³	kg/m³	lbs/gal	g/L
1.000	——	——	——
——	0.331	——	——
——	——	6.002	——
——	——	——	2.55

47. A block of metal measures 3.3 cm × 0.51 cm × 19 cm. Its mass is 0.198662 kg. What is its density?

48. The density of pure sulfur is 2.07 g/cm³. What is the volume of a 1.44-kg sample of sulfur?

49. Your friend in Example 2.5 now wants to fill his cookie box (dimensions are 8.4 cm × 6.4 cm × 10.2 cm) with honey, the density of which is 1.44 g/cm³ and which costs 5.00 marks per kg. How much must he pay for his honey?

50. An alloy for making tire weights is prepared by melting and mixing 500 g of lead, density 11.4 g/cm³, with 45 g of antimony, density 6.69 g/cm³, and 39 g of tin, density 7.31 g/cm³. Assuming that there is no loss in the volume of the metals, calculate the volume of the finished alloy and its density.

51. An empty graduated cylinder has a mass of 44.113 g. Into it is put 19.6 cm³ of a liquid. The mass of the cylinder with the liquid in it is 91.552 g. Give the density of the liquid, using the appropriate number of significant figures and the correct units.

52. The graduated cylinder of Problem 51 is cleaned and into it is put 16.3 cm³ of mercury. What is the mass of the cylinder with the mercury?

53. Does the saying "A pint's a pound the world around," refer to a pint of any particular substance? Must it do so? How many pounds would a pint of mercury weigh?

MORE DIFFICULT PROBLEMS

*54. In a chemical reaction, 3.45 L of one solution reacts exactly with 8.56 L of another solution. If you have 150 mL of the first solution, how many milliliters of the second solution do you need for an exact reaction?

55. You need twice the weight of hydrochloric acid as of sodium hydroxide for an experiment. You have 4.75×10^{-2} g of acid. How many milligrams of sodium hydroxide do you need?

56. You are going to run a chemical reaction in which you need 16 g of oxygen gas for every 7.0 g of nitrogen gas that will be used. If you have 0.554 kg of oxygen, how many milligrams of nitrogen do you need?

57. If 36 g of water yields 4.0 g of hydrogen gas after reaction, how many kilograms of hydrogen will you obtain from 4.35×10^4 mg of water?

*58. A tall glass vessel with a rectangular cross section has inside dimensions of 15.2 cm by 22.9 cm and is about a meter tall. Water is put into the vessel until it stands 55.0 cm high in it. A jagged piece of rock weighing 5.21 kg is placed in the vessel; it sinks to the bottom and the water level rises to 58.3 cm. What is the density of the rock?

59. A zeppelin, which rises by floating on air, contains 2.4×10^4 kg of helium gas in its gasbags. Why would it not be better to carry that same weight in lead and save space? How many liters of volume would be occupied by a piece of lead weighing 2.4×10^4 kg? How many liters are occupied by the helium? How many cubic feet? (See Table 2.3 in Section 2.7, and the conversions earlier in the chapter.)

Chapter 3

Matter and Energy

Chemistry is the study of matter and of the interactions, called chemical reactions, between samples of matter. Because chemical reactions usually involve energy changes, both matter and energy are subjects that we will need to study.

We are about to learn what energy is and about two kinds of energy, stored energy (potential energy) and energy of motion (kinetic energy). We see in this chapter how energy can be changed from one form to another.

We will learn a system by which matter can be classified, as well as the definitions of two important terms, element and compound. We study some of the ways in which samples of matter can be identified and find out how it is possible to determine the purity of a sample of matter.

In the chapter that follows we meet for the first time the natural law that says that neither matter nor energy can be created or destroyed, although each can be converted to the other.

AFTER STUDYING THIS CHAPTER, YOU WILL BE ABLE TO

- give examples of the different states of matter
- describe the relations between force, mass, and acceleration
- state the law of conservation of mass and the law of conservation of energy
- give examples of the conversion of energy from potential to kinetic and vice versa
- give examples of heterogeneous and homogeneous mixtures
- give examples of compounds and elements and explain how to tell the difference between these two classes of matter
- name and describe some common methods of separating mixtures and suggest appropriate purification methods for particular kinds of mixtures
- explain how chemical reactions differ from physical processes and give examples of each

TERMS TO KNOW

acceleration	heterogeneous	phase
atom	homogeneous	physical property
chemical property	kinetic energy	potential energy
chemical reaction	law of conservation of	product
compound	energy	pure substance
condensed state	law of conservation of	reactant
crystal	mass	solid
element	liquid	solution
endothermic	macroscopic	vapor
energy	matter	vaporization
exothermic	melting point	velocity
gas	melting temperature	work
glass	microscopic	

TERMS WITH WHICH TO BE FAMILIAR

distillation
radiant energy
sublimation

3.1 MATTER

Matter is anything that occupies space and possesses mass. Mass, as we learned in Section 2.4, is that property of matter that causes it to respond to the force of gravity (to have weight) and to resist changes in motion (to have inertia). The amount of matter in a sample is often determined by weighing.

Matter can exist in any of the three familiar states: solid, liquid, or gas. A **solid** has definite volume and shape. A **liquid** has definite volume, but takes the shape of its container. Solid and liquid are called **condensed states.**

A **gas** has neither definite shape nor volume; it will fill its container and press everywhere against the walls of the container. It is easy to think about solids and liquids occupying space and having momentum, but seeing a gas that way is more difficult. Consider, however, the fact that moving air—wind—can pick up an entire house. Try to think what keeps a liquid from collapsing into the bubbles under its surface: it is the gas in the bubbles, occupying its own space. If you drop a 5-lb piece of dry ice on your toe, you will surely notice that it has mass. If, however, that chunk vaporizes into carbon dioxide gas, the gas will retain all the mass it had as a solid. Gases can be compressed into small volumes or expanded into large ones. Liquids and solids, however, are practically incompressible.

Most kinds of matter can exist in any of the three states, although we are usually familiar only with one, or possibly two, forms of most matter we encounter every day. (See Table 3.1.) The state of a particular kind of matter depends on its pressure and temperature. We know, for instance, that below a particular temperature, water exists as ice, and above another, it exists as the gas we call steam.

Table 3.1 Common Solids, Liquids, and Gases		
Solids	**Liquids**	**Gases**
Ice	Molasses	Methane
Sugar	Mercury	Air
Iron	Gasoline	Carbon dioxide
Aluminum	Alcohol	Steam
Carbon	Water	Oxygen

Table 3.2 Examples of Crystalline and Glassy Substances	
Crystalline	**Glassy**
Salt	Window glass
Sugar	Hard taffy
Diamond	Plastic
Ice	Rubber

Although it appears to be continuous, matter is made of tiny particles called **atoms** (which are discussed in the next chapter). Atoms are so small that they cannot be seen individually, even with microscopes. Chemistry is the study of atoms and the ways in which they give rise to various characteristics of matter. Anything that is small enough to fall in the realm of the atom is said by chemists to be **microscopic.** Larger things (even if they must be viewed with a microscope) that can be manipulated individually, using instruments of any kind, are said to be **macroscopic.** (According to these definitions, even bacteria are macroscopic, although they are visible only through a microscope; the definition of microscopic differs somewhat from one scientific field to another.) The chemist's work deals with particles that are smaller than those studied by most other scientists.

The atoms in solids are held close together in fixed locations by strong forces, causing the solids to have definite shapes. The individual particles in liquids are also close together, but can move in relation to one another; that is why liquids can flow. The particles of gases are far apart and have no fixed positions relative to one another, as is illustrated by the ability of gas to expand, filling whatever space is available.

A solid can be either crystalline or glassy. (See Table 3.2.) In a **crystal,** the atoms are arranged in an orderly way according to a pattern, whereas in a **glass** they are not. The melting of a crystal into a liquid occurs at a definite, characteristic temperature called the **melting point,** or **melting temperature.** When crystal melts there occurs a definite, sharp transition in which the organized pattern of the crystal is lost. The change from a glass to a liquid, however, is gradual; there is no pattern to be lost. As glass is heated, it becomes softer and softer, until it flows. Because it has no definite melting point, a glassy solid may be simply thought of as a stiff liquid. Glass (even window glass) will change its shape (flow) over a long time. Crystal does not do this.

We will learn more about the three states of matter after we have studied the nature of the atoms of which they are composed.

3.2 ENERGY

In Section 2.5, we described **energy** as work or the ability to do work. Let us now examine energy and work in more detail. To understand work in the technical sense, and thereby to know what energy is, we first must consider the concept of force.

Force and work have to do with bodies of matter in motion. When something is moving, it has **velocity,** which refers to how fast an object is changing its position in any given direction.

For the velocity of a sample of matter to be changed, the object must be accelerated. Although you may ordinarily think of accelerating something as causing it to move faster, the term **acceleration** can be used in three senses: to increase velocity (positive acceleration), to decrease it (negative acceleration), or to change the direction of the motion. Mass can be accelerated only if force is exerted on it. Force is something that produces acceleration, either positive or negative.

We must be careful to note that acceleration can be produced only by *unbalanced* forces. Although nearly all objects at rest are constantly being acted on by forces, no acceleration occurs because forces in one direction are exactly balanced by forces in the opposite direction. Consider a mountain climber hanging on a rope. He experiences the force of gravity, but the rope he is holding exerts an equal upward force. Because the opposing forces are balanced, he is not accelerated either downward or upward. An inept mountain climber, however, may experience unbalanced forces (Figure 3.1).

The amount of force needed to produce a given amount of acceleration depends on the mass of the object being accelerated. The larger the mass, the greater the force required. (Remember blowing on the golf ball and the ping-pong ball?) The relation between force and mass can be expressed in mathematical language:

force = mass × acceleration

When a force is exerted on a body of matter, which then moves over a distance with or against that force, work is done. **Work** is simply the product of the amount of force used and the distance over which the force is exerted. This is stated mathematically as

work = force × distance

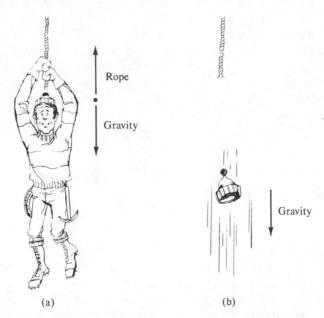

(a)   (b)

Figure 3.1   Unbalanced forces cause acceleration. (a) No net force.
(b) With net force—undergoing acceleration.

Because a force acting through a distance can produce acceleration, there must be some relation between work and acceleration. Although we do not need to see this relation mathematically, a bit of understanding can help us tie all the facts together. We must do work to accelerate any object. What happens to that work, to the energy that we expend? Since it is used to accelerate the object, perhaps it goes into the object itself. This is in fact what happens. The work we do to accelerate an object is then possessed by that object as **kinetic energy,** the energy of motion. We do work on objects to speed them up; they do work on us when we reverse the process and slow them down.

A moving body, then, has kinetic energy. This energy is the sum of the work that was done to give it the velocity it has. **Potential energy** is another kind of energy that is related to the work done on an object. Potential energy is energy of position in relation to a force or forces. Suppose that you lift a brick from the floor and place it on a table. To lift it, you had to exert an upward force somewhat greater than the force of gravity on the brick. Since you applied continuous force to move the brick through the distance, you did work on the brick. That work is then stored in the brick, because of its new position relative to the earth. Lifting the brick increased its potential energy. Energy that is given to a system and does not appear as kinetic energy is stored in the system as potential energy and can be recovered. As we will see, chemical energy, the energy released by a chemical reaction, is a form of potential energy, as is nuclear energy.

Kinetic energy can be transformed into potential energy and vice versa. Look at the slingshot and rock in Figure 3.2. In the slingshot, potential energy is stored in the rubber bands as they are stretched; when they are let go, they accelerate the rock upward, thus turning some of their potential energy into the kinetic energy of the rock in its vertical flight. As the rock rises, it continuously converts its kinetic energy into the potential energy of its elevated position. When the rock stops at the top of its flight, it has no kinetic energy left, but it soon gains some as the force of gravity accelerates it toward the earth's surface. The potential energy of the rock is continuously converted to kinetic energy during its fall, as the rock continues to accelerate downward until it strikes the earth.

Although kinetic energy and the many kinds of potential energy are all associated with matter, there is also a form of pure energy that can exist in a vacuum. This is called **radiant energy,** one form of which is familiar to us as light.

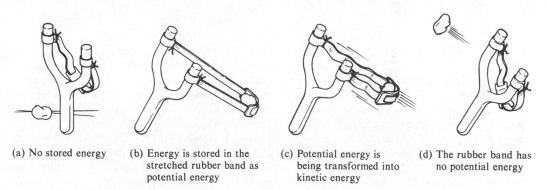

(a) No stored energy

(b) Energy is stored in the stretched rubber band as potential energy

(c) Potential energy is being transformed into kinetic energy

(d) The rubber band has no potential energy

Figure 3.2 Conversion of potential energy into kinetic energy.

3.3 CONSERVATION OF MASS AND ENERGY

Two of the most important and widely known laws of science are the **law of conservation of mass** and the **law of conservation of energy.** We can combine these laws into one statement:

Neither mass nor energy can be created or destroyed.

Although under certain circumstances either mass or energy can be converted to the other and back, there are otherwise no known exceptions to the laws. Chemistry, as such, does not include processes in which mass and energy are interconverted, thus we will usually apply the laws as just stated. In particular, to understand chemical reaction, we need to be aware that during this process neither any matter nor any energy is created or destroyed. In the history of chemistry, the reaction process was more or less a mystery until it was realized that matter is conserved during reaction (Figure 3.3).

Our slingshot-and-rock system offers an example of the conservation of energy. Our muscles provided energy to start the process. That energy next went into the rubber band, then into the flying rock. When the rock hit the earth, it may have seemed as though the energy had been lost, but this is not so. The kinetic energy of the falling rock was converted mostly to heat, both in the rock and in the earth near where it struck. Heat energy is discussed in a later chapter.

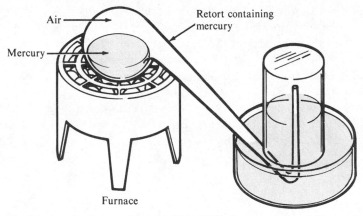

Figure 3.3 Lavoisier's experiments (about 1774–1785). In one of the earliest examples of quantitative chemistry, Antoine Lavoisier (often called the father of modern chemistry) made a series of experiments demonstrating the law of conservation of mass and determining the proportion of oxygen in the air. In several instances, he sealed samples of tin or lead and air into glass vessels, weighed the vessels carefully, heated them strongly until the metal was oxidized, and weighed them again. He showed that no weight gain or loss accompanied the chemical reaction. When he then opened the vessels, air rushed in, and the resulting weight gain was exactly the same as the weight increase of the metals when they turned to "calc" (their oxides). This result showed that the metals combined with something in the air (what we now know as oxygen). In an apparatus like that shown above, he heated mercury until it oxidized, causing the water level to rise in the air vessel as some of the air was used in the reaction. In this way, he was able to show that about a fifth of the air could support combustion while four fifths was inert. (We now know that air is $1/5$ O_2 and $4/5$ N_2, which is inert in these conditions).

In everyday life, and in the laboratory as well, it often seems that energy or mass has disappeared—but appearances are deceiving. A car moving down the street uses a large amount of energy, obtained from burning gasoline. Is that energy lost? No; most of it is transformed into heat; a little is probably converted to various kinds of potential energy; none is destroyed. A massive block of ice sitting on the sidewalk will melt, and eventually the water will evaporate. But the ice has not been destroyed. It now exists in the form of a gas mixed with the air. It may someday be part of a rainstorm.

The laws of conservation of mass and energy will form an important part of our studies to come. We will rely on them implicitly in nearly every chemical statement we make.

3.4 HOMOGENEITY AND PURITY IN SAMPLES OF MATTER

Figure 3.4 shows a useful way to classify matter according to the properties of the sample.

Heterogeneous samples are composed of different substances divided into distinct regions with definite boundaries. An example is a glass of ice chips mixed with rock salt. In addition to the air between the pieces and the glass itself, the system contain two substances, ice and salt, and the boundaries of each piece are distinct.

A heterogeneous mixture is composed of **phases,** each of which is a homogeneous region with distinct boundaries between it and the other phases. Although the sample has many chips of ice and of salt, it is said to have only two phases, the ice phase and the salt phase. (If the glass and air are to be included in what we define as the system, there are two more phases to consider.) Since phases are defined by homogeneity and sets of properties, a sample of ice and liquid water consists of two phases, even though both are water.

If the ice in our system is allowed to melt and the mixture is stirred to dissolve the salt, the boundaries between the salt and the ice (now water), disappear, and the resulting mixture is said to be **homogeneous,** or everywhere the same. A homogeneous mixture is a **solution.**

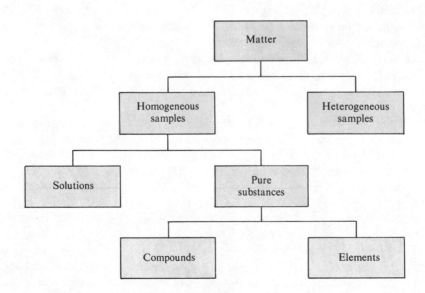

Figure 3.4   Classification of matter.

It is sometimes difficult to say whether a particular sample is heterogeneous or homogeneous. Milk, for instance, looks homogeneous, but actually consists of two phases, one that is mostly water and one that is tiny globules of butterfat. The boundaries of the phases are distinct, but the fat particles are too small to be perceived individually by the unaided eye.

Heterogeneous and most homogeneous mixtures can be separated into simpler substances by physical changes alone. Pure salt and pure water can be obtained from a solution of salt and water simply by boiling the solution until all the water vaporizes. Only salt will be left, and the water vapor can be recondensed to pure water. Both the salt and the water will then be homogeneous samples. In a *physical change,* matter may alter its shape, appearance, or state, but no substance disappears nor is any new substance created.

Homogeneous samples that cannot be separated by physical changes are called **pure substances** (or sometimes merely **substances**). Pure substances have definite composition, in contrast, for example, to the salt solution, which could have had various proportions of water and salt. Although no macroscopic sample of a substance can be said to be totally without impurities, samples subjected to the physical processes of purification to such an extent that continued purification produces no further change are said to be "pure." An example of a pure substance in everyday life is ordinary table sugar, sucrose, which usually contains no more than about one part of impurity (something not sugar) in a thousand, or about 0.1%. If you try to purify table sugar further, you will get little result for much effort.

Physical processes can be used to separate pure substances from one another. Another process, known as **chemical reaction,** can change substances to other substances. The operational definition of chemical reaction is that *clearly identifiable substances* (called **reactants)** *disappear, and new, different substances* (called **products**) *appear in their place.* Energy is often released in chemical reaction, usually in the form of heat. A chemical reaction that releases energy is said to be **exothermic** (or, more precisely, exoenergetic). A reaction that absorbs energy is called **endothermic.**

Because *substances* are converted to other *substances* during chemical reaction, we must not imagine that any *matter* has been created or destroyed; it has merely changed its form. Moreover, because energy is released during reaction, we must not think that the energy was created; it was there all the time in the reactants as potential energy.

An example of change by means of physical process is the purification of sugar; the crystals of sugar are grown from a syrup. An example of a chemical reaction is the burning of the substance gasoline with oxygen in the air, producing the new substances water and carbon dioxide and releasing heat.

Some pure substances can be separated into simpler substances by chemical reaction. A pure substance that can be separated in this way is called a **compound.** A pure substance that cannot be in any way separated into simpler substances is called an **element.** Chemical reactions can separate compounds into their elements, unite elements into compounds, or change compounds into other compounds. Table 3.3 gives examples of elements, compounds, and mixtures.

Our definitions are not yet quite satisfactory, because we have not thoroughly distinguished physical processes from chemical reactions. We have not made the distinction because it is often difficult to tell whether a substance has disappeared, or a new one has been created. If sugar is dissolved in water, is the sugar still there? It would seem to be in this case; the sweetness of sugar helps identify it. But how can we be sure? A chemical reaction might

Table 3.3
Examples of the Classes of Matter

Heterogeneous Mixtures	Solutions	Compounds	Elements
Salt and pepper	Gasoline	Water	Gold
Concrete	Beer	Sugar	Aluminum
Milk	Tea	Salt	Sulfur
This book	Seawater	Alcohol	Oxygen
Mud	Air	Methane	Iron

have occurred that produced some other sweet substance. What might look like a chemical reaction may be only the physical process of dissolving.

The release of energy that usually accompanies chemical reaction can be of some help in deciding whether reaction has occurred, but the surest way to decide is to separate the components in the new system and identify them. Let us look further into the identification of pure substances and the separation of the components of mixtures.

3.5 SEPARATING AND IDENTIFYING SUBSTANCES

Any pure substance can be identified by its *properties,* or characteristics. Every substance has a set of properties that differs from the set of properties of every other substance. Certain kinds of properties, called **physical properties,** can be measured without chemically destroying any of the sample. Other kinds of properties, those that are displayed only in chemical reaction and that change some or all of the sample into another substance, are called **chemical properties.** Some familiar physical properties are color, hardness, and melting and boiling temperatures. Other physical properties may not be as easy to perceive, but are nonetheless characteristic. Compressibility, electrical conductivity, and the way a substance refracts light are examples. Whether a substance will burn, or whether the products of its combustion will etch metals, are examples of chemical properties.

Different pure substances may have some properties in common, but not all; each has its own total set. The properties of mixtures are usually combinations of the properties of the pure substances in them. To characterize any particular sample, you must first determine whether it is a mixture or pure substance, and, if it is a mixture, separate the components. Table 3.4 gives some examples of the properties by which some common substances can be identified. As your study of chemistry progresses, such properties will become familiar to you.

Physical methods can be used to distinguish pure substances from solutions. Pure substances will go through complete changes of state at constant temperature; few solutions will do so. A sample of pure water, once boiling, will boil entirely away without changing temperature. Once it has started freezing, it will freeze completely at the same temperature. The temperature of our sample of salt and water, however, will rise continuously as the water vaporizes during the boiling process. During freezing, the temperature will continuously decrease, as illustrated in Figure 3.5. Although there exist a few solutions that can maintain a constant temperature during boiling, even these change temperature while freezing.

Apart from simple mechanical methods such as sifting, the separation of pure substances from one another (that is, their purification) is usually accomplished by means of a change of

Table 3.4
Physical and Chemical Properties of Some Substances

Substance	Physical Properties	Chemical Properties
Iron	Hard, high melting point,* conducts electricity and heat, shiny	Liberates hydrogen from acids, rusts, forms colored ions
Sulfur	Soft, low melting point, nonconductor, dull yellow color	Burns readily, forms gaseous acidic oxides
Oxygen	Gaseous, odorless, tasteless, colorless	Supports combustion, combines with most other elements
Water	Colorless liquid, low melting point, good solvent	Can be decomposed to oxygen and hydrogen
Table salt	Colorless solid, high melting point, nonconductor as solid (but liquid salt conducts), brittle	Fairly inert, can be decomposed to metallic sodium and gaseous chlorine

*For the purposes of our discussion, we take *low melting point* as referring to melting temperatures at most only 100 or 200 °C above room temperature. Many substances with *high melting points* must be heated to incandescent temperatures before they melt.

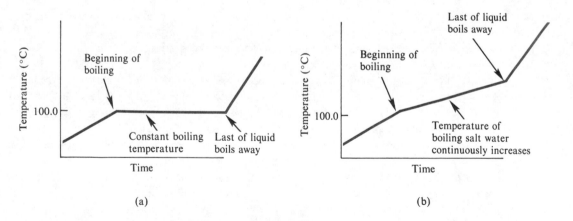

(a)

(b)

Figure 3.5 Effect of dissolved substances on boiling temperature. In the two graphs shown, heat is steadily added to a sample of pure water (a) and to a sample of water in which a substantial amount of salt is dissolved (b). Notice that even though heat continues to be added, the temperature of the pure water remains at 100.0°, while the liquid boils. When all the water is gone, the temperature of the empty container rises. In (b), showing water with salt, the temperature continues to rise *during* boiling.

state. The components of liquid solutions can usually be separated by means of vaporization in a process called **distillation.** (**Vaporization** simply changes a liquid into a gas; gases in contact with the liquids from which they came are often called **vapors.**) Distillation takes advantage of the fact that at the distillation temperature, one of the substances in the solution vaporizes more readily than the others, thus bringing about a separation.

A classic example of distillation is the separation of alcohol from other substances in the preparation of certain beverages (Figure 3.6). If a solution of alcohol in water is boiled, the first samples of the vapor will be richer in alcohol than was the original solution, because alcohol vaporizes more readily than water. The vapor is then captured and recondensed into a liquid richer in alcohol than the original liquid. The process can be repeated until alcohol of 95% purity is obtained (190 proof), but few beverage makers go beyond about 50% (100 proof).

Distillation is used on a large scale to remove salt from seawater. Making this process sufficiently economical to provide water for domestic use and irrigation in desert countries has required considerable engineering, but large plants are presently in operation in several parts of the world. Small-scale, sun-powered distillation apparatuses can now be found in lifeboats, where they can supply drinking water to people who might otherwise perish (Figure 3.7).

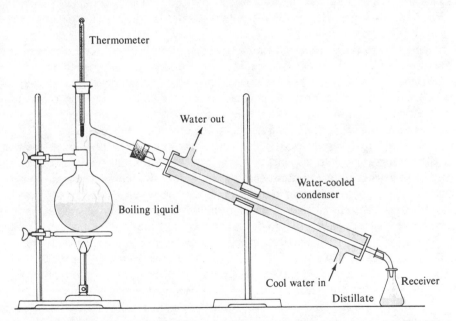

Figure 3.6 Simple distillation apparatus. The solution to be distilled is placed in the left-hand flask and boiled gently. The vapor thus formed passes through the side arm in the neck of the flask and again becomes a liquid in the long, water-cooled tube called a condenser. The new liquid then drips into a delivery flask. If a volatile liquid is being separated from a nonvolatile substance (for example, water from sugar), the vapor contains only the volatile substance, and the separation is complete after one careful operation. If two volatile liquids are to be separated, the process is more complicated.

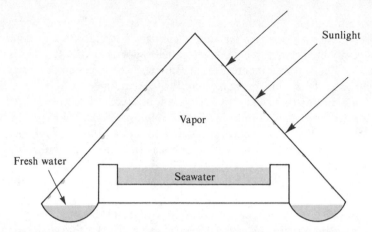

Figure 3.7 Small distillation apparatus used to obtain pure water from seawater. When sunlight heats the interior, the water vaporizes. The pure water then condenses on the walls and collects at the bottom.

Sublimation is similar to distillation, except that the change is from solid to vapor rather than from liquid to vapor. Several industrial substances, sulfur and iodine for instance, are purified by sublimation (Figure 3.8).

Commercially, sugar is separated from water and other impurities by means of **fractional crystallization,** in which the sugar fraction is separated from the other substances. The phase change is from sugar in liquid solution to crystalline sugar. A hot, concentrated sugar solution is allowed to cool slowly; crystals of pure sugar are formed, leaving the other substances behind. The preparation of colorless table sugar from ordinary brown sugar is a simple crystallization experiment often performed in beginning chemistry laboratories.

In other separation procedures, the change of state may be less obvious. A method called **solution chromatography** takes advantage of the fact that different materials, such as paper, glass, and clay, attract different substances with different degrees of strength. When a solution is poured through the chromatography material, the various solution components move through at different rates, and come out separately at the other side. A simple demonstration of this method may be done even without a laboratory. Ordinary black ink usually consists of several dyes of different colors mixed together. You can separate the various dyes if you place a drop of ink in the center of a piece of filter paper (perhaps from a coffee maker) or blotting paper, then use a dropper to feed water to the spot. As the wet portion grows, drawing the ink to the outside edge of the paper, the different dye components move at different rates, causing them to appear as separate bands of color. Figure 3.9 illustrates the process (see also the color plates and the cover). Laboratory applications of chromatography use not only paper and coated glass plates, but also glass columns filled with various materials such as aluminum oxide or even powdered brick. Chromatography is now used routinely to purify many substances once thought impossible to separate.

3.6 SUMMARY

Chemistry is concerned with matter; matter has mass and occupies space. Mass is what causes matter to experience gravitational attraction and to resist changes in motion. Matter

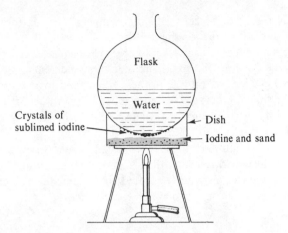

Figure 3.8   Sublimation arrangement for demonstration. In this simple apparatus, a mixture of sand and crystals of the element iodine is placed in the large dish and then heated with a burner. The iodine sublimes, that is, turns to vapor directly from the solid state, while the sand remains unaffected. The iodine vapor returns to solid form when it encounters the cold surface of the flask filled with cold water. After the experiment has run awhile, it is easy to see large flakes of fairly pure iodine that have grown on the flask.

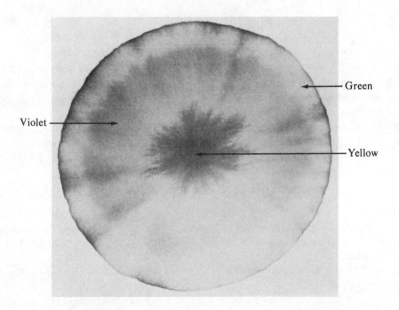

Figure 3.9   Example of paper chromatography. In this ink chromatograph, circles of color form as the expanding spot of moisture leaves behind the dyes in an order that reflects the degree to which they are adsorbed on the paper fibers. See color plates for same photograph in color.

can be accelerated by means of force; when this happens, the matter gains or loses kinetic energy. Energy is the term for work or the ability to do work.

Matter can possess potential energy (stored energy) because of its relation to other matter and the forces between the two. Although neither mass nor energy can be created or destroyed, each can be transformed into the other. Such transformation does not occur to a noticeable extent in chemical processes; what is more likely to occur in chemical reaction is the transformation of energy from one form to another, or the transformation of matter from one state to another.

Matter can be classified into pure substances called elements and compounds, or mixtures of these. Mixtures can be separated into their component substances by physical processes such as distillation; compounds can be separated into elements by chemical reaction. Elements, however, cannot be separated into simpler substances.

QUESTIONS

Section 3.1

1. According to the way in which chemists use the words microscopic and macroscopic, why is it that a germ, which can be seen only with a powerful microscope, would be considered macroscopic by chemists?
2. Define liquid and solid. It is said that in certain old buildings, there are windowpanes that are slightly thicker at the bottom than at the top, although they were presumably uniform when installed. Is window glass a solid or a liquid? How could you find out?
3. Why can gases be compressed easily, whereas liquids and solids cannot?

Section 3.2

4. A rock is thrown into the air, falls back down, strikes a flat roof, and stays there. Discuss the fate of the energy involved in the process from the moment the energy initially leaves the muscle.
5. A train makes an emergency stop and melts part of its wheels. Where did the energy to melt the metal come from?
6. The combination of gasoline and air has stored energy (a subject we will study in detail somewhat later). A car is put into motion using that energy. It goes up a hill, then the brakes are applied and the car stops. Has the energy been destroyed? If not, describe in detail what happened to it from the time it ran the engine until the car was stopped.

Section 3.4

7. Expand Table 3.3 by giving three more examples of each of the four classes listed.
8. Remembering that milk is not a solution, name the phases present in a system consisting of a glass of milk with ice in it.
9. How many phases are in a glass of iced tea? A cloud that contains hail? A steel cylinder that is filled with steel nuts and bolts, which are then covered with oil?
10. What are the differences between a solution and a pure substance? Between an element and a compound?
11. In Lavoisier's experiment with mercuric oxide, shown in Figure 3.3, were there phase changes? Explain. Was there chemical reaction or merely a physical process? How do you know?

12. Matter cannot be created or destroyed. Can phases? What happens to the number of phases when the ice melts in a glass of ice water?

Section 3.5

13. Explain the difference between a physical and a chemical property.
14. Expand Table 3.4 by giving physical properties of three more substances. If you can, do this for the chemical properties as well.
15. A glass of water is left on a table and the water evaporates. Has the water been destroyed? If not, exactly what has happened to it? Could it be recovered somehow? Give two examples of similar disappearances and discuss the possibility of recovery.
16. If water and alcohol are mixed and then distilled, the alcohol may be obtained with less and less water in it until the solution contains only 5% water. Further distillation will make no further separation. Is the alcohol now a pure substance? How might you go about finding out?
17. From everyday life or industry, give three examples of separations of heterogeneous mixtures and three examples of separations of solutions.
18. In the process of making the alcoholic beverages called spirits or liquors, two methods of separation are used. Try to name them.
19. What method of separation would you use to obtain pure antifreeze (ethylene glycol) from the dilute solution in a car radiator?
20. What method of separation does an automobile engine use to remove particles of dirt from motor oil? What kind of a mixture is the dirty oil?

Chapter 4

The Atomic Theory

The assumption that all matter is made of microscopic individual particles is called the atomic theory. This theory is probably the single most important concept in all science; certainly it is essential to thinking about chemistry in an organized way. The atomic theory maintains its status as a theory because no one has ever directly seen an individual atom. Modern instruments, however, have so convincingly demonstrated the particle nature of matter that chemists unhesitatingly use the theory as the cornerstone of their studies.

If we think of matter in terms of atoms and the combinations of atoms called molecules, we will be able to improve our definitions of elements and compounds. We can then use these definitions to make a clear distinction between physical process and chemical reaction.

Chapter 4 gives us a first glance at some of the earliest speculation about the nature of matter. To obtain the background we need to understand the development of modern atomic theory, we will examine the methods by which simple laboratory analyses can be made and review the laws of chemical combination that were discovered through just such analyses nearly two centuries ago. Thus prepared, we will learn about Dalton's atomic theory, its modern descendants, and how the masses of atoms of different elements are related in what is commonly called the atomic weight scale.

AFTER STUDYING THIS CHAPTER, YOU WILL BE ABLE TO

- state, in order, the individual steps that could be taken in a simple laboratory experiment to measure the masses of the elements in a compound
- calculate, from such laboratory data, the relative masses of the atoms of the elements in that compound
- state the law of definite proportions and explain its relation to laboratory data
- describe the features of Dalton's atomic theory and those of the modern theory, and point out and explain the differences between the two
- calculate the atomic weights of geonormal mixtures of elements from isotopic abundance data

TERMS TO KNOW

alchemy	compound	law of constant
atom	element	composition
atomic mass scale	geonormal	law of definite proportions
atomic mass unit	ion	law of multiple proportions
atomic theory	isotope	molecule
atomic weight	isotopic abundance	qualitative analysis
chemical analysis		quantitative analysis

4.1 EARLY VIEWS OF THE NATURE OF MATTER

Speculation about the ultimate basis of matter dates back to several Greek philosophers five centuries before the time of Christ. It was at this time that the word **atom** had its beginning.

The early approach was a simple one, directed to the question of the continuity of matter. If, for example, you were to take a block of gold and divide it in half, both the halves would be gold. If one piece were to be halved again, pieces of gold would still result. If the process were to continue even until the fragments were not visible to the unaided eye, the fragments would still be gold. The Greeks speculated, however, that the process of repeated division would finally lead to a particle of gold that would be indestructible. To this smallest particle they applied the Greek word *atomos,* meaning "uncut."

These ideas were surprisingly correct as far as they went. However, for lack of experimental evidence, they remained for centuries a subject of speculation. Interest in natural philosophy declined during the Middle Ages, and was not again revived until Newton, Boyle, and Bernoulli, scientists of the late seventeenth and early eighteenth centuries, used it in describing the behavior of gases. Later still, and only after the appearance of quantitative laboratory chemistry as we know it today, was the atomic theory established and widely accepted.

The close of the eighteenth century saw the end of the old practice of **alchemy,** which was a somewhat superstitious, unorganized inquiry into experimental chemistry, motivated primarily by the desire to change lead into gold (Figure 4.1). By this time (1770-1780), various experiments of Priestley, Cavendish, and Lavoisier on the nature of combustion and the reactions of metals and their oxides had resulted in the discovery of oxygen and the recognition of the need for painstaking controls on experimental systems. This careful work established quantitative chemistry which, in turn, made possible the discovery of several natural laws of chemical combination. The search for an explanation of these laws motivated the creation of the atomic theory.

4.2 CHEMICAL ANALYSIS

Historically, the development of the atomic theory followed the sequence that we now have learned to think of as the scientific method. The method begins with observation and the gathering of information. The data that motivated the framing of the atomic theory came from studies of the relations of the masses of elements as they reacted to form compounds and from analyses of compounds.

Chemical analysis in the laboratory is the process of separating and identifying the components of samples. If the sample happens to be a mixture of compounds, the compounds are

Figure 4.1 Alchemy preceded science. (Photo courtesy of Aldrich Chemical Co., Milwaukee.)

separated. Each compound is then studied by means of physical tests and chemical reactions. Often, the compounds are themselves broken down into their elements or into simpler compounds that can be easily identified.

There are two kinds of chemical analysis, **qualitative** and **quantitative.** In the first, the objective is to find out what elements or compounds are present in the sample. In the second, the point is to find out not only what substances are present, but also the relative amounts of each. Qualitative work came long before quantitative in the history of chemical analysis.

An essential assumption of quantitative analysis is that no detectable mass can be created or destroyed in chemical reaction. The conclusions based on laboratory data would be impossible without this idea. If a piece of the element magnesium is burned in air, for example, it is discovered that the resulting substance has more mass than did the original sample. From this finding it is assumed that the magnesium must have joined in reaction with some component of the air and formed a new compound heavier than the magnesium itself, but it is not thought that any new mass has been created. In the late eighteenth century, by isolating a sample of tin and oxygen in a sealed flask and showing that the whole system had the same mass both before and after the reaction to make the oxides of tin, Lavoisier demonstrated that mass was conserved in chemical reaction (see Figure 3.3).

To appreciate the laws of chemical combination, let us now examine a popular instructional experiment. The experiment will help us to understand that there is nothing mysterious about this kind of work, nor is there any special talent needed to perform it. We will follow the technique of measuring the relative masses of the elements copper and oxygen in a sample of their black-colored compound that has the chemical formula CuO and the name cupric oxide. With simple equipment, you could easily do this experiment yourself.

An empty test tube is carefully weighed, a sample of CuO is put into it, then the test tube and CuO are weighed together. The mass of the test tube with CuO, minus the mass of the test tube alone, gives the mass of the CuO itself. The test tube with its sample is now heated strongly, and gaseous hydrogen is passed over the sample. The hydrogen reacts with the oxygen in the CuO to form water. (See Figure 4.2.) When the water passes off as steam, it leaves behind pure copper metal with its familiar reddish color.

The hydrogen flow is continued until no more emission of steam or change of color occurs. The tube and copper are then cooled and weighed once more. The difference between the second and third weighings gives the mass of oxygen taken away in the water. By subtracting this amount from the calculated mass of the CuO, the mass of the copper alone can be found. (Or, the copper mass can be calculated from the difference between the first and third masses.) Here is a set of numbers that came from an actual experiment. (Look at the numbers carefully; the whole point of the discussion is here.)

mass of test tube and CuO	9.70 g
−mass of empty test tube	−8.50 g
mass of CuO sample	1.20 g
mass of test tube and CuO	9.70 g
−mass of test tube and Cu metal	−9.46 g
mass of oxygen lost	0.24 g
mass of CuO sample	1.20 g
−mass of oxygen	−0.24 g
mass of Cu metal	0.96 g

$$\text{ratio:} \quad \frac{\text{mass of oxygen}}{\text{mass of copper}} = \frac{0.24 \text{ g}}{0.96 \text{ g}} = \frac{1.0}{4.0}$$

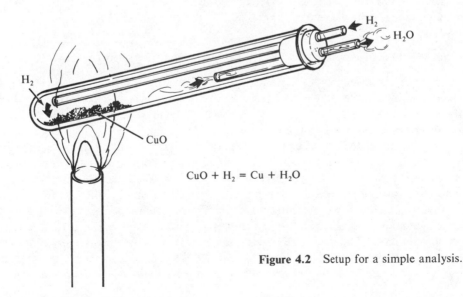

$$CuO + H_2 = Cu + H_2O$$

Figure 4.2 Setup for a simple analysis.

In this compound, our analysis found 4.0 g of copper present for every 1.0 g of oxygen.

EXERCISE 4.1.   What are the relative masses of manganese and oxygen in a sample if the results in an experiment similar to the one just described are: sample mass, 2.10 g, mass of oxygen lost, 1.05 g?

EXERCISE 4.2.   A test tube containing a sample of silver oxide has a mass of 14.50 g. After reaction with hydrogen, the test tube with only silver metal has a mass of 14.09 g. What is the mass ratio of silver to oxygen in the sample, if the empty test tube has a mass of 8.15 g?

4.3 THE LAWS OF CHEMICAL COMBINATION

When elements combine chemically to form compounds, they do not necessarily do so in the proportions in which they happen to be present, but according to specific natural laws, three of which were discovered near the beginning of the nineteenth century. Two of these laws motivated John Dalton, an English schoolteacher, to propose the atomic theory. The third law was implied by the theory and later discovered by Dalton himself. The atomic theory became firmly established after a few years because of the simple and logical explanation it offered for all three laws.

The **law of definite proportions,** the first of the three to become known, was discovered experimentally by Proust in 1797. This simple statement is sometimes called the **law of constant composition:** *a pure sample of any compound, no matter what its origin, is found always to contain its elements in a fixed proportion by mass.* Consider our analysis of cupric oxide. We found that the ratio of the mass of oxygen to that of copper was 1.0 to 4.0. According to the law of definite proportions, every sample of pure cupric oxide, no matter what its history, will contain that same ratio of oxygen to copper. Silver iodide, another compound, is made from the elements silver and iodine. On careful analysis, it is found that silver iodide contains its elements in the ratio

$$\frac{1.17648 \text{ g iodine}}{1.00000 \text{ g silver}}$$

This means that in silver iodide there are 1.17648 g of iodine for every 1.00000 g of silver. Now, the law of definite proportions tells us that all samples of silver iodide will exhibit exactly this same mass ratio. Similarly, every other compound contains its elements in a particular, characteristic, fixed proportion by mass.

The constant composition of compounds is the inevitable result of chemical combination, so much so that we can use it to refine our definition of the term **compound.** *A compound is a chemical combination of two or more elements in fixed proportions by mass.* (We will be able to define the nature of compounds even better after we have studied the atomic theory.)

At first glance, the law of definite proportions might seem obvious or trivial because it is so simple. But stop a moment and think. Samples of silver and iodine, for instance, are merely lifeless, unintelligent chunks of matter. How could they measure themselves out so precisely every time they react chemically? *How can that be?* That is an important step in science: *to ask how can it be and then to try to create a theory that will give the answer.* Nearly two

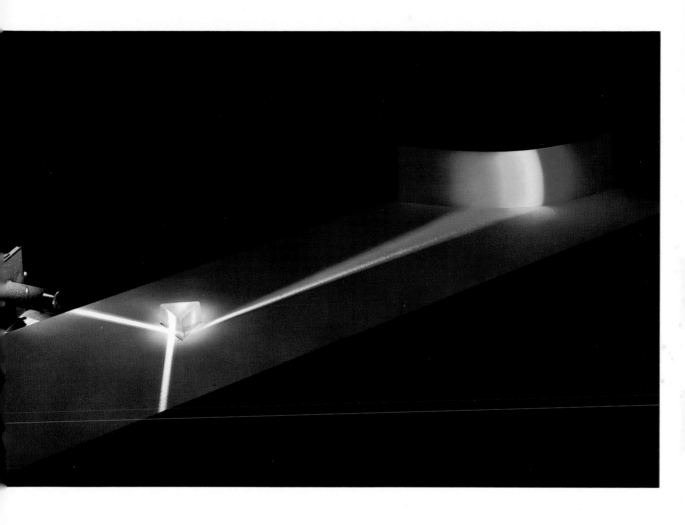

Action of a prism in separating white light into its components.

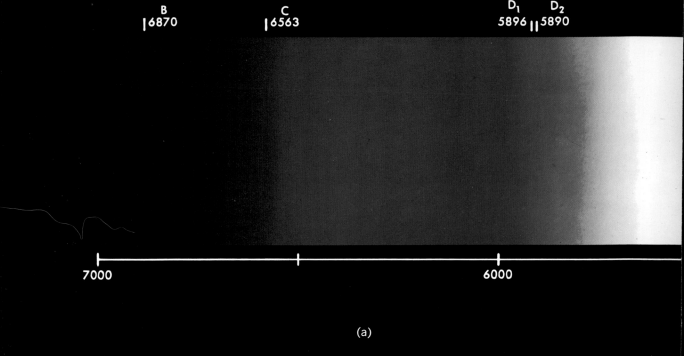

B
|6870

C
|6563

D₁   D₂
5896 || 5890

7000   6000

(a)

(a) The visible spectrum. (b) Characteristic emission lines for hydrogen, helium, and mercury atoms.

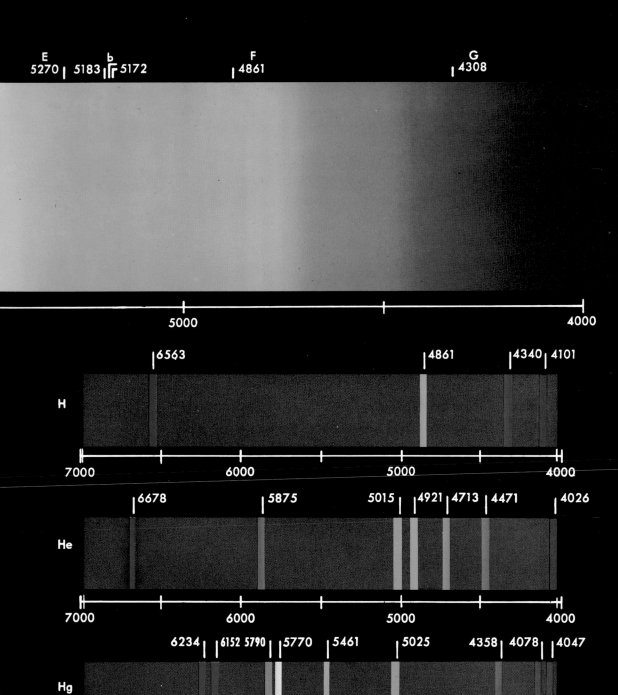

E 5270 | 5183 b | 5172 | F 4861 | G 4308

5000 4000

| 6563 | 4861 | 4340 | 4101

H

7000 6000 5000 4000

| 6678 | 5875 | 5015 | 4921 | 4713 | 4471 | 4026

He

7000 6000 5000 4000

6234 | 6152 5790 | 5770 | 5461 | 5025 | 4358 | 4078 | 4047

Hg

7000 6000 5000 4000

(b)

Example of paper chromatography. In this ink chromatograph, circles of color are formed as the expanding spot of wetness leaves behind the dyes in an order that reflects the degree to which they are adsorbed on the paper fibers.

hundred years ago, Dalton knew about the newly discovered law of definite proportions as well as a similar pattern of elemental behavior, now known as the law of equivalent proportions. Dalton knew about these phenomena and, in fact, checked them in his own primitive laboratory. He became curious. He wanted to know why these things happened. When he asked, "How can it be?," his answer was the **atomic theory.**

Let us put ourselves in Dalton's place and attempt to explain the law of definite proportions by means of a theory.

4.4 DALTON'S ATOMIC THEORY

As we did in our study of the air in the tire pump, we can propose possible causes for what has been observed. Perhaps the tiny man is back, this time with a balance on which he weighs the elements. Perhaps individual elements can flow together at different speeds, and part of the slower element gets left behind. Although we may continue to think of other possibilities, any theory we advanced would have to do two things: (1) explain the known facts and (2) withstand the test of new information derived from further experiments designed to test our theory. Dalton's theory did fit the facts known at that time. The theory has since been tested by thousands of different kinds of experiments. The details have been changed, but the main features of the theory have survived and are now stronger than in Dalton's time. Let us consider what Dalton proposed, and then see how his theory explains the physical facts.

Dalton theorized that the matter in samples of elements existed in individual packages (atoms) that had characteristic masses. He believed that the packages somehow attached themselves to one another to form the larger packages of which compounds were made. Dalton called the larger packages "compound atoms"; we now call them **molecules.** The characteristic masses with which the elements reacted to form compounds, then, simply reflect the masses of the atoms of the elements. In Dalton's view, a sample of a compound was just a large number of molecules of that compound, all alike.

Let us make an analogy. Suppose you are going into the hardware business. You buy a supply of nuts and bolts, all of the same kind (your sample of compound). These come to you assembled, one nut on every bolt. Each bolt happens to weigh 26 g and each nut 13 g.

Consider one nut-bolt pair. The mass of the bolt (26 g) is twice that of the nut (13 g). Now pick up two pairs. The mass of the two bolts (52 g) is twice that of the two nuts (26 g). Take a sample containing any number of pairs, ten, a thousand, a hundred million. The mass of all the bolts in the pile is still twice the mass of all the nuts because of two physical facts: (1) one bolt weighs twice as much as one nut and (2) there is always one bolt for every nut (Figure 4.3).

Dalton's reasoning was similar to that used in our example, except that the nuts and bolts (atoms of different kinds) were so small as to be invisible, and the nut-bolt pairs were what we now call molecules. In the compound hydrogen fluoride, the molecules (the nut-bolt pairs) each contain one hydrogen and one fluorine atom (one nut and one bolt). Because of this ratio and because one fluorine atom weighs 19 times as much as one hydrogen atom, the fluorine in one molecule of hydrogen fluoride weighs 19 times as much as the hydrogen. This mass relation is true for a sample containing two, a hundred, or any number of hydrogen fluoride molecules. Hydrogen fluoride obeys the law of definite proportions: the ratio of the masses of hydrogen and fluorine is the same in all samples of the compound.

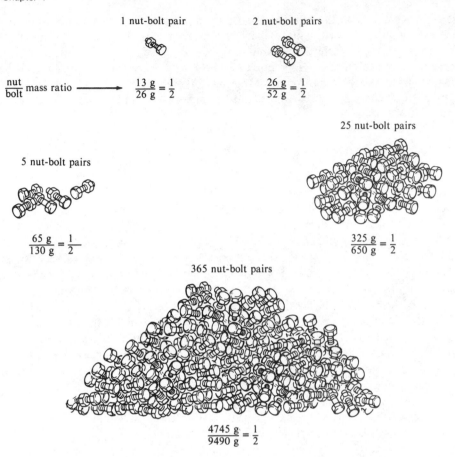

Figure 4.3 Mass ratios of nuts and bolts. As long as there is one nut to every one bolt, the ratio of the mass of nuts to that of bolts is 1/2 regardless of the number of pairs taken. Notice that the units (grams) cancel out in each case.

Here is Dalton's theory in detail, most of which is still accepted:

1. All matter is made of indestructible, discrete particles. (Dalton called these atoms.)
2. All atoms of a given element are exactly alike, particularly in mass and size.
3. Atoms of different elements have different masses and sizes.
4. Chemical reaction is simply the joining or separation of atoms, none of which is destroyed in the process.
5. Compounds are composed of "compound atoms," which are combinations of atoms joined to one another.
6. Different compounds of the same two elements can result from combinations of the atoms in different ratios, but such ratios are always in the form of small whole numbers, 1/1, 1/2, 2/3, and so on.
7. The identity and properties of a compound are determined by the numbers and kinds of atoms composing it.

Using Dalton's reasoning and remembering our nut-and-bolt analogy, let us examine another slightly more complicated example. The compound water happens to have two hydrogen atoms and one oxygen atom in each molecule (like having two nuts on one bolt). In round numbers, one oxygen atom weighs 16 times as much as one hydrogen atom. But in one molecule of water, all the oxygen (one atom) weighs only eight times as much as all the hydrogen. Why? Because there are two hydrogen atoms, and the mass ratio in the molecule is

$$\frac{16}{(1 + 1)} = \frac{16}{2} = 8$$

That mass ratio holds for two molecules of water,

$$\frac{(2 \times 16)}{2(1 + 1)} = \frac{32}{4} = 8$$

or for any number of water molecules taken as a sample (Figure 4.4).

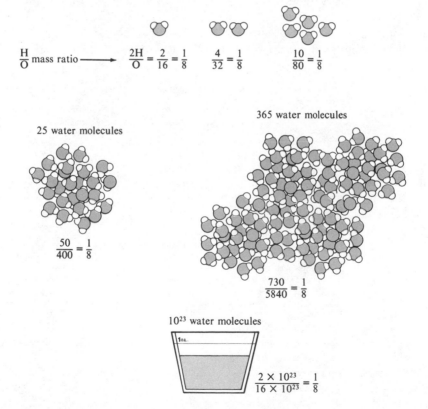

Figure 4.4 Mass ratios of H and O in water molecules. As long as there are two H atoms for one O atom, the mass ratio H/O is 1/8, no matter how many molecules are in the sample. Notice that it is not necessary to specify the units in which the mass is measured, because the units cancel in every case.

By considering the nuts and bolts and going through the example with water, you can see that the atomic theory does indeed explain why compounds always have their elements in constant proportion by mass. If the atoms of any given kind always have the same mass, and if the molecules all have the same number of atoms of each kind, the mass ratios have to be constant. In a similar way, the theory also explains the other laws of chemical combination. We will study one such law next, because it will be useful when we begin to investigate how to tell the number of atoms of each kind that are contained in a molecule.

4.5 THE LAW OF MULTIPLE PROPORTIONS

A third series of mass relations, now called the **law of multiple proportions,** was predicted by the atomic theory and demonstrated by Dalton in 1804. A given pair or set of elements will often combine to make two or more distinct compounds, each with its own fixed elemental mass ratio. As we might by now expect, there are definite relations between those mass ratios. To demonstrate this, we select five compounds of nitrogen and oxygen, and for each we analyze a sample exactly large enough to contain 1.00 g of nitrogen. The results of the analyses are shown in Table 4.1.

Glancing at the table, we find that the masses of oxygen combining with 1 g of nitrogen to form the successive compounds are, in order, 1×0.57, 2×0.57, . . . 5×0.57; that is, the ratios are related by small whole numbers. Now look at Table 4.2. Here are the masses compared with the number of atoms in the different molecules.

Table 4.1
An Example of the Law of Multiple Proportions

Compound	Grams of Nitrogen Found	Grams of Oxygen Found	Grams of Oxygen Found (Rewritten)
Nitrous oxide	1.00	0.57	1×0.57
Nitric oxide	1.00	1.14	2×0.57
Dinitrogen trioxide	1.00	1.71	3×0.57
Nitrogen dioxide	1.00	2.28	4×0.57
Dinitrogen pentoxide	1.00	2.85	5×0.57

Table 4.2
Molecular Structure and Multiple Proportions

Name of Compound	Number of Atoms in Molecule	Number of Oxygen Atoms for Every Nitrogen Atom	Grams of Oxygen for Every Gram of Nitrogen
Nitrous oxide	Two nitrogen and one oxygen (2N + 1O)	One half	One $\times$ 0.57
Nitric oxide	One nitrogen and one oxygen (1N + 1O)	Two halves	Two $\times$ 0.57
Dinitrogen trioxide	Two nitrogen and three oxygen (2N + 3O)	Three halves	Three $\times$ 0.57
Nitrogen dioxide	One nitrogen and two oxygen (1N + 2O)	Four halves	Four $\times$ 0.57
Dinitrogen pentoxide	Two nitrogen and five oxygen (2N + 5O)	Five halves	Five $\times$ 0.57

Notice that as the number of oxygen atoms for every nitrogen atom goes from one half to two halves and on to five halves, the number of grams of oxygen for every gram of nitrogen increases in exactly the same way, a plain illustration that the mass ratios themselves depend directly on the relative numbers of atoms in the compound (Figure 4.5). Of course, there is no such thing as half an oxygen atom; the table is written that way merely for clarity. The actual numbers of atoms in the molecules are shown in the second column. Don't leave this section without understanding the tables and being sure that you have the main idea, namely, that *mass ratios in compounds depend both on the masses of the individual atoms and on the number of atoms of each kind in the molecules*. We'll be using these data again soon.

4.6 DEFICIENCIES OF DALTON'S THEORY

The fundamental ideas of Dalton's theory are still valid today, but several of the details have since been shown to be incorrect. Most important of the changes is the discovery that atoms of the same element do not necessarily all have the same mass. At first glance, this seems ridiculous; the entire theory was constructed around the concept that the atoms of any element were indeed characterized by having the same mass. This assumption is nevertheless incorrect, and we must find out how the atomic theory can survive under this great change.

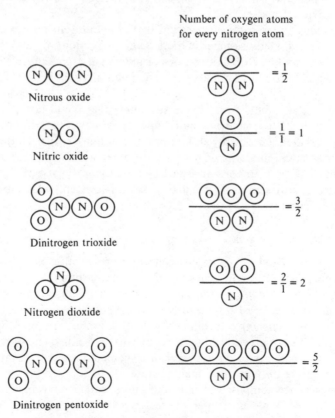

Figure 4.5 Atom ratios in the oxides of nitrogen.

Consider the nuts and bolts again. Now we'll do the experiment in reverse. You have a large number of new nuts and bolts. You want to find out how much more the bolts weigh than the nuts (since it's not the same batch, you're not sure). You can't find a scale that will weigh a single bolt or single nut, but you can weigh a large number of either. You take the bolts and nuts apart and stack them in separate piles. Now you weigh each pile. (How about that! You've done a quantitative chemical analysis.) The bolts weigh exactly twice as much as the nuts. You repeat the experiment several times, always with the same result. From your finding, you conclude that one bolt weighs exactly twice as much as one nut; after all, you weighed equal numbers of each every time.

Your conclusion could be correct—but it may not be. Suppose that some of the bolts were slightly longer and heavier and others shorter and lighter, and that in any large number of bolts you took, there would always be five of the larger and five of the smaller bolts for every thousand of the others. Then in every large-number sample of bolts you weighed, the average mass of a bolt would be constant, and an average bolt compared with an average nut would always yield a mass ratio of two to one. Although you would have a constant nut-bolt mass ratio, not all the bolts would necessarily weigh the same, nor, for that matter, would all the nuts. Atoms can follow the pattern of our example. The atoms of an element do not necessarily have the same mass, but in an ordinary sample of any element, the atoms have the same *average* weight. When you perform a chemical analysis, you find the average masses of the atoms, not the individual masses.

Atoms that belong to the same element but that have different masses are called **isotopes** of that element. (We will study isotopes in more detail in Sections 4.9 and 8.6.) At this point, you may wonder how we know that the masses of individual atoms of an element are not the same. It is now possible to find the masses of individual atoms, but Dalton did not have the tools to do that.

Dalton believed that atoms were simple structures that could not be subdivided in any way, but atoms are neither simple nor indivisible. It is, however, still considered true that atoms are neither created nor destroyed in chemical reaction—although their complex structures may exchange small amounts of matter with other atoms in processes called electron transfer reactions. The most important aspect of the atomic theory, the idea that matter is composed of atoms, is still with us. It is unlikely that this concept, which is crucial to science, will ever be proved incorrect.

4.7 MODERN ATOMIC THEORY

Taking advantage of today's enormous collection of scientific knowledge, we define the atom in a way somewhat different from that of Dalton. *An **atom** is the smallest particle of matter that can carry the distinguishing characteristics of an element; it is the smallest particle of any element that can enter into chemical reaction.*

Atoms of nearly all elements can combine with other atoms, either of the same or of other elements, to form stable structures held together by what are called chemical bonds. There are only a few elements whose atoms do not enter into such combinations. Many other elements, however, occur in nature *only* in the combined state.

Some compounds are composed of individual units called molecules. With some important exceptions, the word **molecule** is used to mean *a particle composed of two or more atoms* *held together by a chemical bond or bonds and capable of independent existence.* A molecule can have atoms of the same element; nitrogen molecules are each made up of two nitrogen atoms. Or a molecule can have two or more elements in it; the molecules of ethyl alcohol each contain six hydrogen atoms, two carbon atoms, and one oxygen atom. Some inert elements form almost no chemical bonds and exist in the elemental form as molecules containing only a single atom (what are often called monatomic molecules, one of the exceptions noted earlier). Helium is an example of such an element (Figure 4.6).

Not all compounds, however, are necessarily molecular. A sample of sodium chloride (table salt), for instance, contains no individually existing particles at all—no molecules. Sodium chloride pairs have strong attractions for other sodiums and chlorides, so that a single sodium chloride "molecule" can exist *independently* only under unusual conditions and for a very short time. In any single piece of sodium chloride, large numbers of what are called the **ions** of sodium and chloride are stuck together in an organized pattern forming the solid compound. An ion can be an atom or a group of atoms, but it has electrical properties that ordinary atoms and molecules do not have. We will study ions in detail in Section 11.5.

We should be sure to remember that compounds, whether they are molecular or not, always contain two or more elements in fixed proportion by mass.

Using what we know of atomic theory, we can now broaden our definition of the word *element.* Although it is still important to remember that an element cannot be separated into simpler substances, it is also useful to think of the word **element** as referring to *a sample of matter in which all the atoms are alike.* (In what way they are alike is a question we must leave open for awhile; as we already know, they do not all necessarily have the same mass.)

Here is a brief list of the teachings of modern atomic theory. If you compare it with the list of Dalton's teachings and observe the differences we have discussed, you will see that the main ideas have survived.

1. All matter is made up of discrete particles called atoms. Atoms are neither created nor destroyed during chemical reaction.
2. All atoms of a given element are alike in a particular respect called the atomic number. (We will soon study the atomic number.) Atoms of the same element do not necessarily have the same mass.

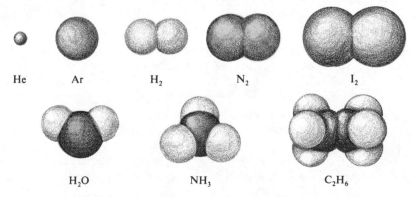

He   Ar   H$_2$ N$_2$ I$_2$

H$_2$O   NH$_3$ C$_2$H$_6$

Figure 4.6 Scale models of some molecules.

3. Atoms of different elements usually have different masses and sizes.
4. Chemical reaction is the joining or separation of atoms; no atoms are destroyed in the process.
5. Compounds are composed of atoms or ions, held together by chemical bonds. Compounds may or may not be molecular in structure.
6. Different compounds of the same two elements can result from combinations of the atoms in different ratios. The relations between such ratios are often small whole numbers.
7. The identity and properties of a compound are determined by the numbers and kinds of atoms that compose it and by the way in which those atoms are combined.

4.8 ATOMIC MASS

We have learned that quantitative chemistry depends on knowing the mass relations of elements in compounds and in chemical reaction. Because such mass relations arise from the relative masses and relative numbers of the individual atoms that react to form compounds, it is important to learn about the masses of individual atoms.

Modern instruments are capable of determining the absolute masses of individual atoms, and the masses of the atoms of most elements are well known. The instrument that measures atomic mass is called a mass spectrometer. We will examine the theory and operation of this device in Section 8.4.

4.8.1 Isotopes

As you know, not all the atoms of any particular element necessarily have the same mass; samples of atoms belonging to the same element and having different *masses* are said to be different **isotopes** of that element. Hydrogen, for example, has three species of atoms, that is, three isotopes. The most common is the lightest; the atoms of the second weigh about twice as much as those of the lightest, and the atoms of the third weigh about three times as much. Nearly all the elements have two or more isotopes.

For reasons we will discuss in more detail in Section 8.6, the isotopes of an element are identified by numerical superscripts determined by the masses of the atoms of those isotopes. For example, the three isotopes of hydrogen are identified as ^1H, ^2H, and ^3H, in order of increasing mass. The two isotopes of oxygen are ^{16}O and ^{18}O, the two of helium ^3He and ^4He, and so on.

A natural sample of an element is a mixture of the different isotopes of that element. The proportions of the mixture are referred to as the **isotopic abundance.** With a few exceptions, the history of the elements on earth is such that the natural isotopic abundance found in any sample of an element is independent of the origin of the sample. There are, however, a few elements in which the isotopic abundance is not constant. For these elements, measurements of average atomic mass are inconsistent. A sample containing the natural abundance of the isotopes of an element is called a **geonormal** sample. Because the isotopes of an element can now be separated from one another in the laboratory, it is possible to obtain pure samples of single isotopes of many of the elements. Such samples, of course, are not geonormal.

4.8.2 Isotopes and the Laws of Chemical Combination

In most ordinary processes, such as chemical reactions or purification procedures, the behavior of atoms is not influenced by their relative masses: a mixture of isotopes of an element behaves the same way as a pure sample of any particular isotope. In other words, the laws of chemical combination work as well for geonormal samples as they would for isotopically pure samples. As a consequence, the existence of isotopes was not even suspected until more than a hundred years after the establishment of the atomic theory, and the laws of chemical composition were thus written without that information. In the light of our present knowledge, the law of definite proportions might now be written in this way: A compound, samples of whose elements have constant isotopic distribution, always contains its elements in fixed proportions by mass. (Because early workers had *only* samples of constant isotopic distribution, they would never have guessed that there could be samples of another kind or that such a modification would have been required in their language.)

4.9 THE ATOMIC WEIGHT SCALE

As an example of the masses of individual atoms, let us consider the elements hydrogen and oxygen. The average mass of the atoms in a sample of geonormal hydrogen happens to be 1.67366×10^{-24} g, and in a sample of geonormal oxygen atoms it is 26.5660×10^{-24} g. We could use these values in making calculations, but for many uses scientists have devised a different, more convenient approach.

Several decades ago, an international committee of scientists adopted a scale that lists the relative masses of atoms of different kinds. A somewhat modified version of that scale is now in universal use. In the sense we are using it here, let us examine what is meant by a *relative scale*.

Referring to individual isotopes, we could state the relation between the mass of an atom of ^1H and one of ^{16}O by saying that the mass of the oxygen atom is 15.87 times greater than that of the hydrogen atom. Similarly, we would say that an atom of ^{35}Cl has a mass 2.1862 times greater than that of the ^{16}O atom. Although we could use a series of statements like these to show the relations between the masses of the atoms of all the elements, this would be clumsy. It would be better to assign one particular kind of atom an arbitrary mass designation and relate the other atoms to that. For example, we could make a list, or scale, in which we would take the mass of the ^1H atom as being exactly 1. In that list, the mass of the ^{16}O atom would then be 15.87, that of the ^{35}Cl would be 34.70, and so on. Notice that it is not necessary to designate any particular mass unit in making our scale; the numbers relate directly to one another.

For reasons that will not concern us, it has recently been decided to base the scale of atomic masses not on H as exactly 1, but on the most common isotope of carbon, ^{12}C, with a mass of exactly 12. According to this scale, the mass of an atom of ^1H comes out to be 1.0078, that of an atom of ^{16}O 15.995. The scale of the relative masses of the atoms of individual isotopes is called the **atomic mass scale.**

Although values on the atomic mass scale can be expressed without using mass units of any kind, it has been found useful to define what is called an **atomic mass unit, amu,** such that the mass of a single ^{12}C atom contains exactly 12 amu. Then the mass of a single atom of ^{1}H is 1.0078 amu and that of a single atom of ^{16}O 15.995 amu. In laboratory units, the mass of 1 amu is 1.6604×10^{-24} g. The atomic mass scale can be read in a purely relative way, or in units of amu per atom. The mass of ^{40}Ca, for example, is 39.96 amu/atom. In a later section, we will find that the scale can be used with other units as well.

At this point, we encounter an inconsistency in the language, but one that is so widespread that we will use it. The terms *atomic mass, atomic mass scale,* and *atomic mass unit* are ordinarily used only in reference to the masses of *individual* atoms. Most chemistry, however, is done with geonormal mixtures of atoms, and the masses used are the masses of those atoms taken as an average. For historical reasons, and to distinguish average masses from the masses of real individual atoms, *the mass of an "average" atom on the relative scale is commonly called the **atomic weight**.* The list of relative atomic weights is called the *atomic weight table*. For example, the *atomic weight* of chlorine is given as 35.453, although there are no chlorine atoms that have an atomic mass of 35.453. A sample of geonormal chlorine is made up of 75.5% ^{35}Cl atoms and 24.5% ^{37}Cl atoms. The *average* mass of all these atoms is 35.453 on the atomic mass scale. *The term atomic weight actually refers to average atomic masses and is not a weight designation at all.* The term is nonetheless used by almost all chemists, and we will use it—remembering that it is somewhat inappropriate.

A table of atomic weights on the ^{12}C scale appears inside the back cover of the book, and atomic weights also appear below the symbols for the elements in the periodic table inside the front cover. Remember that for the atoms in a geonormal sample, each atomic weight value gives the *average* mass on the scale that includes ^{12}C as exactly 12. Although there are tables that give masses individually for each isotope, we will have little need for such information because we will work almost entirely with geonormal mixtures. Although there is no such thing as an "average" atom (just as there is no such thing as an average family that has two and a half children), the average mass of a large number of atoms is a perfectly workable concept and yields useful information.

To understand the atomic weight table better, we will work some simple problems.

EXAMPLE 4.1.   What is the ratio of the atomic weight of O to that of He?

Solution:

First, look up the atomic weights in the table, then form the ratio.

$$\frac{15.9994}{4.0026} = 3.997$$

Oxygen atoms are about four times as massive as those of helium.

EXERCISE 4.3.   What is the ratio of the atomic weight of Be to that of Sc?

EXAMPLE 4.2. What is the mass in grams of an "average" chlorine atom?

Solution:

Multiply the atomic weight (the average atomic mass in amu) of Cl by the unit conversion of amu to grams.

$$\frac{35.453 \text{ amu}}{\text{atom}} \times \frac{1.660 \times 10^{-24}\text{g}}{\text{amu}} = 5.885 \times 10^{-23} \text{ g/atom}$$

EXERCISE 4.4. What is the mass in grams of an "average" geonormal iodine atom? Of an "average" geonormal boron atom?

4.10 ISOTOPIC ABUNDANCE AND ATOMIC WEIGHT

As we found in Section 4.8.1, different samples of any element will ordinarily have the same atomic weight, because the samples will all contain the same proportions of the isotopes of that element. It is possible to calculate the atomic weight of geonormal samples of any element, using the atomic masses of the isotopes of that element and its isotopic abundances.

Isotopic abundances are often given as percentages. If the percent abundance of each isotope in the geonormal mixture is converted to a decimal fraction and multiplied by the atomic mass of that isotope, the sum of the results will give the atomic weight of the mixture.

EXAMPLE 4.3. Samples of geonormal copper, Cu, contain 69.1% of ^{63}Cu and 30.9% of ^{65}Cu. The atomic mass of ^{63}Cu is 62.9 and that of ^{65}Cu is 64.9. What is the atomic weight of copper?

Solution:

Step 1. Multiply the atomic mass of each isotope by the percentage of that isotope taken as a decimal fraction.

Step 2. Add the results.

$$
\begin{array}{lc}
^{63}Cu & 62.9 \times 0.691 = 43.5 \\
^{65}Cu & 64.9 \times 0.309 = \underline{20.1} \\
\text{Atomic weight of Cu} & 63.6
\end{array}
$$

EXERCISE 4.5. What is the atomic weight of geonormal magnesium, Mg? The isotopic abundances of geonormal Mg are as follows:

78.7% ^{24}Mg, atomic mass = 24.0
10.1% ^{25}Mg, atomic mass = 25.0
11.2% ^{26}Mg, atomic mass = 26.0

4.11 SUMMARY

The atomic theory, which is a cornerstone of science, was first proposed to explain certain laws of chemical combination. Those laws, in turn, represented the combined results of quantitative analytical measurements.

One such law, the law of definite proportions, states that any compound contains its elements in fixed proportions by mass. We have now learned to use this law as one of the criteria by which we can distinguish compounds from mixtures. (But we now also know that the law applies only if the elements in the compounds are geonormal, a concept that was not understood in the early days of chemistry.) The laws of chemical composition can be explained only on the basis of the atomic theory. The theory states that matter is composed of atoms and that atoms of different elements combine in fixed number ratios to give compounds. The weight relations in compounds arise from atom number ratios and characteristic atomic weights.

Although there were certain faults in the original atomic theory, the fundamental concept is still considered valid, and our understanding of the nature of matter is based on it.

The masses of individual atoms can be measured. An atomic mass scale has been established that relates these masses to one another. Not all the atoms of a given element have the same mass. Atoms belonging to the same element and having different masses are said to belong to different isotopes of that element. The atomic *mass* scale, which is based on the isotope 12 of carbon as having a relative mass of 12, compares the masses of atoms of individual isotopes of all the elements. The atomic *weight* of an element, on the other hand, compares the *average* mass of the atoms in a natural (geonormal) sample of that element in the same scale. Chemists deal mostly with atomic weights instead of atomic masses.

If the natural abundances and the atomic masses of an element's isotopes are known, it is easy to calculate its atomic weight.

QUESTIONS

Section 4.2

1. Explain the processes of chemical analysis, both qualitative and quantitative.
2. Explain why chemical analysis depends on the assumption that no significant amount of matter is created or destroyed during chemical reaction. Find a book on the history of science and look up the theory of phlogiston. Compare that approach with our present reliance on the idea of the conservation of mass.
3. Explain why theories about the atomic nature of matter were only guesses or speculation until the beginning of quantitative chemical analysis.
4. In a careful experiment, samples of wood and of iron are burned in oxygen, and the resulting solid products are weighed. In the case of the wood, the ashes weigh less than did the wood itself, but the solid produced by the burning of the iron weighs more than the iron did. Do either of these processes violate the law of conservation of mass? Explain the difference in the results.
5. In one compound of fluorine and oxygen, the mass ratio of fluorine to oxygen is 19.00 g to 8.00 g. In a compound of fluorine with sulfur, the mass ratio of fluorine to sulfur is also 19.00 g to 8.00 g. Since sulfur atoms are much heavier than oxygen atoms, how can this be possible? Explain.

Section 4.3

6. Explain the law of definite proportions.
7. How is the law of definite proportions related to the experiment using CuO described in Section 4.2?
8. The law of definite proportions holds for compounds. Does it also hold for solutions? Explain.

Section 4.4

9. Give a brief description of Dalton's atomic theory.
10. What does the term *compound atoms* mean to you?
11. Show how the atomic theory explains the law of definite proportions.

Section 4.5

12. Explain how the law of multiple proportions applies to the compounds listed in Table 4.2.
13. Water is not the only compound formed by hydrogen and oxygen. Write another formula for a possible combination of these two elements and give its mass ratio.
14. How is the law of multiple proportions related to Dalton's atomic theory?
15. How are the laws of definite proportions and multiple proportions related? Can one be true without the other? Explain.
16. Dalton proposed that all atoms of a given element had to be alike, especially in respect to size and mass. Is there anything in the laws of chemical combination that has to do with atomic size? Can the laws be explained without the assumption that atoms of an element all have the same mass? How?

Section 4.6

17. List the deficiencies of Dalton's atomic theory.
18. Explain how it is possible to describe the average mass of the atoms of an element as being constant.
19. Give an example from technology showing that Dalton was incorrect in assuming that atoms could not be subdivided.

Section 4.7

20. How are atoms, elements, and compounds related to each other?
21. Under what circumstances can an atom be classified as a molecule?
22. How is it possible for a substance to be a compound and not be composed of molecules?

Section 4.8

23. Explain the relationships between an atom, an element, and an isotope.
24. Does the existence of isotopes change the laws of chemical combination? Explain fully.

Section 4.9

25. How would you go about developing a relative scale of any kind?
26. Explain the basis for the atomic mass scale used by chemists. Why is ^{12}C used for the standard? How can a single isotope be used as a standard for geonormal mixtures?

27. Explain the difference between atomic mass and atomic weight. For what is each term used?
28. How does the existence of isotopes affect the atomic weight scale? The atomic mass scale?
29. Explain why it is not possible to find an atom of carbon with a mass equal to 12.011 amu.

PROBLEMS

Section 4.2

*1. In an experiment similar to that described in Section 4.2, an analysis is made of the mass ratio of zinc to oxygen in zinc oxide. The empty test tube weighs 10.43 g. The test tube with the sample of zinc oxide weighs 12.11 g. After the experiment, the test tube with only zinc in it weighs 11.78 g. What is the mass ratio of zinc to oxygen in zinc oxide?

2. A porcelain crucible is found to weigh 10.371 g empty and 11.711 g when it contains a sample of calcium metal. The crucible is strongly heated, and the calcium reacts with the oxygen in the air to form calcium oxide. After cooling, the crucible and the compound are found to weigh 12.376 g. What is the mass ratio of Ca to O in the oxide?

3.

$$\begin{array}{cc} 8\ g & 1\ g \\ O \longleftrightarrow H \\ \times \\ Ca \longleftrightarrow F \\ 20\ g & 19\ g \end{array}$$

According to the diagram, 8 g of oxygen reacts with 1 g of hydrogen, 8 g of oxygen reacts with 19 g of fluorine, and so on.
a. How much F reacts with 1 g of H?
b. How much F reacts with 1 g of Ca?
c. How much H reacts with 5 g of Ca?
d. Which early law of chemical combination is represented by this problem?

4. In a large test tube weighing 34.6 g is placed 10.0 g of CuO. The mass ratio of copper to oxygen in this compound is 4.0 to 1.0. All the oxygen is driven off as steam by heating the CuO in the presence of hydrogen. The tube, now containing only copper, is cooled. What is the mass of the test tube and copper?

5. A crucible weighs 45.3 g when empty and 51.7 g when it contains a sample of zinc oxide. The mass ratio of zinc to oxygen is 4.1 to 1.0. What will be the mass of the crucible containing only zinc metal if all the oxygen is removed from the zinc oxide?

*6. A sample of copper wire weighing 1.588 g is placed in a solution of silver nitrate. A reaction occurs. The copper disappears and is replaced by small flakes of silver. For every atom of copper originally present, two atoms of silver appear. (The chemistry of this experiment is discussed in a later chapter, but no further information is needed to solve this problem.) The silver metal is washed and dried and found to weigh 5.348 g. What is the ratio of the mass of an average silver atom to that of an average copper atom?

7. When hydrogen and oxygen combine to form water, the mass ratio is 8 g of oxygen to 1 g of hydrogen. In a closed reaction vessel, 8 g of oxygen is mixed with 8 g of hydrogen and the reaction is run. What substances are present in the vessel when the reaction finishes? How many grams are there of each?

8. If 15.5 g of nitrogen is mixed with 5.40 g of hydrogen and made to react, forming a compound with a mass ratio of 4.63 nitrogen to 1.00 hydrogen, will one of the reactants be left over? Which one?

9. A sample containing 12 g of iron is placed in a vessel with 15 g of oxygen. Reaction occurs, forming a compound that has a mass ratio of 2.32 iron/1.00 oxygen. What substances are present in the vessel after reaction?

10. In the compound ammonia, 14 g of nitrogen is combined with 3 g of hydrogen. In nitrogen iodide, 14 g of nitrogen is combined with 381 g of iodine. What is the probable mass ratio in hydrogen iodide? According to these data, could there be other ratios in this compound? Explain:

*11. Find the ratios of the masses of "average" geonormal atoms of the following pairs of elements.

  *a. Al, O   e.   B, H
   b. Ar, S   f.   Rb, N
   c. C, Cl   g.   Sc, Br
   d. Mn, O   h.   Sn, F

*12. Calculate the atomic weight ratios for the following pairs of elements:

  *a. B, H   c.   Sc, Br
  *b. Rb, N   d.   Sn, F

13. What is the mass in grams of an "average" geonormal atom of each of the following elements: Ar, Ba, Ca, C, Cr, Co, Au?

Section 4.10

*14. The isotopic abundances and isotopic atomic masses of three elements are listed in the table. What are the atomic weights of these elements?

		Percent abundance	Isotopic atomic mass
*a.	Silicon	92.2% ^{28}Si	27.98
		4.7% ^{29}Si	28.98
		3.1% ^{30}Si	29.97
b.	Silver	51.82% ^{107}Ag	106.9
		48.18% ^{109}Ag	108.9
c.	Gallium	60.4% ^{69}Ga	68.9
		39.6% ^{71}Ga	70.9

15. What is the atomic mass of ^{65}Cu? The atomic weight of Cu is 63.546, and the other naturally occurring isotope is ^{63}Cu, which has an abundance equal to 69.09% and atomic mass equal to 62.9298.

*16. Calculate the atomic mass of ^{79}Br if the atomic weight of Br is 79.904. The only other naturally occurring isotope, ^{81}Br, has an atomic mass equal to 80.9163 and an abundance of 49.46%.

MORE DIFFICULT PROBLEMS

*17. If the atomic mass scale were based on the definition of ^{1}H exactly equal to 1, what would be the atomic mass of ^{40}Ca, which is 39.96 amu/atom on the present scale?

18. On the planet Xanthu, the relative atomic weight scale is based on the isotope 12 of carbon, just as on Earth. Instead of assigning carbon a relative atomic mass of 12.0000 . . ., the scientists of Xanthu have chosen a relative mass of 1.0000 . . ., for carbon. On the Xanthu scale, what is the atomic weight of Ni?

Chapter 5

Formulas, Reaction Equations, and the Mole

Now that we know the characteristics of elements and compounds, we will take time in this important chapter to acquaint ourselves with several of the elements and with the elemental classifications of metals and nonmetals. We next investigate compounds and learn the names and formulas of some compounds. We will learn how reaction equations can be used to describe chemical reactions.

At this point, we are about to encounter one of the most important concepts in our early study of chemistry: the mole. We will begin interpreting chemical formulas and chemical reaction equations in terms of the units of moles, rather than individually as atoms and molecules. A mole, to a chemist, is like a dozen to a seller of eggs; it is a number that is itself a unit of measure. As merchants often think in dozens, chemists usually think in moles, but the mole is a number much larger than 12. With a full understanding of how chemists use the mole, we will be ready to tackle chemistry quantitatively and to make useful chemical calculations.

AFTER STUDYING THIS CHAPTER, YOU WILL BE ABLE TO

- give the names and symbols of many common elements
- list typical properties of metals
- list typical properties of nonmetals
- tell whether an element is a metal or nonmetal
- make estimates concerning the properties of many elements
- list the elements that are diatomic in their natural state
- name some common compounds and give their formulas
- balance chemical equations
- distinguish between an exothermic and an endothermic reaction by reading the value of the enthalpy change
- define and explain the term *mole*
- read reaction equations in terms of moles

- calculate the gram-atomic weights of the elements
- calculate the molecular weights and gram-molecular weights of compounds, given their formulas

TERMS TO KNOW

Avogadro's number	exothermic reaction	molecular weight
balanced equation	formula	noble gas
binary compound	gram-atomic weight	nonmetal
coefficient	gram-molecular weight	periodic table of the
diatomic molecule	metal	elements
empirical formula	metalloid	product
endothermic reaction	mole	reactant
enthalpy	molecular formula	reaction equation

5.1 THE ELEMENTS

All substances in the universe, including those that make up the earth, the seas, the air, living creatures, the sun, and the most distant stars, are composed of combinations of the few substances we call **elements,** or are elements themselves. In the entire history of science, only a few more than a hundred elements have been discovered; it is unlikely that many more will be found. The last few "discovered" were actually created in the laboratory. Because the atoms of such man-made elements are unstable, they spontaneously disintegrate only fractions of a second after they have been prepared. For instance, no samples of fermium, mendelevium, or nobelium exist on earth. Even plutonium, now a familiar substance, did not exist naturally on earth until the 1940s.

More than 99% of the earth and its contents is made from fewer than 20 elements, the other 85 being present either in small amounts or not at all. By far the most abundant element on earth is oxygen, constituting nearly half the mass of the earth's crust. The atmosphere is about 20% oxygen, but this is a tiny fraction of the total amount of oxygen on the earth. Most oxygen exists in combination with other elements, in the rocks that form the earth's crust and in the water of the seas. (See Table 5.1.)

Table 5.1
Abundances of Elements in the Earth's Crust
(Percentages by Mass)

Element	Percentage	Element	Percentage
Oxygen	49.20	Chlorine	0.19
Silicon	25.67	Phosphorus	0.11
Aluminum	7.50	Manganese	0.09
Iron	4.71	Carbon	0.09
Calcium	3.39	Sulfur	0.06
Sodium	2.63	Barium	0.04
Potassium	2.40	Fluorine	0.03
Magnesium	1.93	Nitrogen	0.03
Hydrogen	0.87		
Titanium	0.58	All others	0.47

The elements were discovered one by one in the course of history. Many were in common use before the idea of an element had even been thought of, and because of their history, no system for naming the elements was ever established. A list of their names is a colorful mix of Latin, Greek, German, English, and other languages; each name is meant to give some information about the element. Each element also has its own symbol, an abbreviation used to save time. Sometimes the abbreviations fit the names: for example, C for carbon, H for hydrogen, Cl for chlorine. Other symbols refer to the ancient Latin names and bear no resemblance to names used today; examples are Na for sodium (from natrium), Fe for iron (from ferrum), and Ag for silver (from argentum). Both the names and the symbols are used constantly. It is important that you begin now to memorize the names and symbols of elements most commonly encountered in chemistry. The list given in Table 5.2 is a good place to start.

EXERCISE 5.1.   Write the full names for the following elements: Mg, Mo, Mn, Ta, Ti, K, Na, Ag, Fe. (See the table inside the back cover.)

EXERCISE 5.2.   What are the symbols for barium, fluorine, silicon, and arsenic?

For reasons having to do with their atomic structure and their physical and chemical properties, the elements are often listed in an arrangement called the **periodic table of the elements,** shown in Figure 5.1. (See also the inside front cover.)

Although we will later be studying the periodic table in depth, we can now begin to learn a few things about the elements and their properties. The table can be divided into several areas, each of which contains elements of similar properties. On the left are the **metals,** distinguished by their shiny appearance and their ability to conduct electricity and to be bent and hammered into different shapes without breaking. Metals have little tendency to combine with one another but are often found in compounds with the nonmetals, particularly oxygen and sulfur. Much of the earth's crust is made of compounds called silicates, combinations of metals with silicon and oxygen. The so-called active metals, such as sodium and calcium, are

Table 5.2
Names and Symbols
of Some Important Elements

Metals		Nonmetals	
Name	Symbol	Name	Symbol
Aluminum	Al	Bromine	Br
Calcium	Ca	Carbon	C
Chromium	Cr	Chlorine	Cl
Copper	Cu	Helium	He
Gold	Au	Hydrogen	H
Iron	Fe	Iodine	I
Lead	Pb	Neon	Ne
Magnesium	Mg	Nitrogen	N
Potassium	K	Oxygen	O
Sodium	Na	Phosphorus	P
Silver	Ag	Sulfur	S

Legend:

6	— Atomic number
C	— Symbol
12.0	— Atomic weight

IA	IIA	IIIB	IVB	VB	VIB	VIIB	VIII	VIII	VIII	IB	IIB	IIIA	IVA	VA	VIA	VIIA	0
1 **H** 1.00																	2 **He** 4.00
3 **Li** 6.94	4 **Be** 9.01											5 **B** 10.8	6 **C** 12.0	7 **N** 14.0	8 **O** 16.0	9 **F** 19.0	10 **Ne** 20.2
11 **Na** 23.0	12 **Mg** 24.3											13 **Al** 27.0	14 **Si** 28.1	15 **P** 31.0	16 **S** 32.1	17 **Cl** 35.5	18 **Ar** 39.9
19 **K** 39.1	20 **Ca** 40.1	21 **Sc** 45.0	22 **Ti** 47.9	23 **V** 50.9	24 **Cr** 52.0	25 **Mn** 54.9	26 **Fe** 55.8	27 **Co** 58.9	28 **Ni** 58.7	29 **Cu** 63.5	30 **Zn** 65.4	31 **Ga** 69.7	32 **Ge** 72.6	33 **As** 74.9	34 **Se** 79.0	35 **Br** 79.9	36 **Kr** 83.8
37 **Rb** 85.5	38 **Sr** 87.6	39 **Y** 88.9	40 **Zr** 91.2	41 **Nb** 92.9	42 **Mo** 95.9	43 **Tc** (98)‡	44 **Ru** 101.1	45 **Rh** 102.9	46 **Pd** 106.4	47 **Ag** 107.9	48 **Cd** 112.4	49 **In** 114.8	50 **Sn** 118.7	51 **Sb** 121.8	52 **Te** 127.6	53 **I** 126.9	54 **Xe** 131.3
55 **Cs** 132.9	56 **Ba** 137.3	57 **La*** 138.9	72 **Hf** 178.5	73 **Ta** 180.9	74 **W** 183.9	75 **Re** 186.2	76 **Os** 190.2	77 **Ir** 192.2	78 **Pt** 195.1	79 **Au** 197.0	80 **Hg** 200.6	81 **Tl** 204.4	82 **Pb** 207.2	83 **Bi** 209.0	84 **Po** (210)	85 **At** (210)	86 **Rn** (222)
87 **Fr** (223)	88 **Ra** (226)	89 **Ac**† (227)	104 – § (260?)	105 – § (260?)	106						112?						

* Lanthanides:

58 **Ce** 140.1	59 **Pr** 140.9	60 **Nd** 144.2	61 **Pm** (147)	62 **Sm** 150.4	63 **Eu** 152.0	64 **Gd** 157.3	65 **Tb** 158.9	66 **Dy** 162.5	67 **Ho** 164.9	68 **Er** 167.3	69 **Tm** 168.9	70 **Yb** 173.0	71 **Lu** 175.0

† Actinides:

90 **Th** 232.0	91 **Pa** (231)	92 **U** 238.0	93 **Np** (237)	94 **Pu** (242)	95 **Am** (243)	96 **Cm** (247)	97 **Bk** (247)	98 **Cf** (249)	99 **Es** (254)	100 **Fm** (253)	101 **Md** (256)	102 **No** (254)	103 **Lr** (256)

‡ Parentheses around atomic weight indicate that weight given is that of the most stable known isotope.

§ Both Russian and American scientists have claimed the discovery of elements 104 and 105. Official names have not been adopted yet.

Figure 5.1 Periodic table of the elements.

found naturally only in the combined state, but the coinage metals, gold, silver, and copper, as well as a few others, do sometimes occur naturally in their elemental form.

The **nonmetals,** such as fluorine, chlorine, oxygen, nitrogen, and carbon, are shown toward the upper right of the periodic table. The properties of nonmetals are distinctly different from those of metals. Nonmetals do not conduct electricity, shatter rather than bend, and are dull in appearance or are gases or liquids. Most metals have high melting and boiling temperatures, whereas most nonmetals have relatively low melting and boiling temperatures. The nonmetals combine chemically with one another to form molecular substances. Examples are water, carbon dioxide, hydrogen chloride, methane, and ammonia.

Lying between metals and nonmetals is the group of elements called the **metalloids,** which combine many of the properties of both. Finally, almost unique in their characteristics, are the **noble gases,** the family of elements in the vertical column on the far right. The noble gases have almost no chemistry whatever.

Not all metals are equally metallic, nor are all nonmetals alike. The extent to which an element displays the properties of its type depends roughly on its distance in the periodic table from fluorine, for the nonmetals, and from francium, for the metals. If we begin at the lower left of the table, where the most metallic elements are, we find that the elements become generally less metallic as we scan the table, moving toward the upper right corner.

EXERCISE 5.3.   Classify the following elements as metals or nonmetals: Mg, S, Ne, Cs, V, C, P, I.

Except for mercury, metals are solids at room temperature, although a few are liquid at the temperature of boiling water. Only one other element is liquid under ordinary conditions: bromine, a brown, corrosive, poisonous substance. Eleven of the elements are gases at room temperature. These are listed in Table 5.3.

EXERCISE 5.4.   Classify the following elements according to whether each is solid, liquid, or gas in the natural state: Na, Sc, He, Kr, Br, I, Hg, H.

As can be seen in Table 5.4, several elements exist in their natural state as **diatomic molecules,** molecules having two atoms. Except for two, I_2, a solid, and Br_2, a liquid, the diatomic elements are all gases. A few other elements, such as S_8 and P_4, are also molecular in their natural state. The noble gases are monatomic.

5.2 COMPOUNDS AND THEIR FORMULAS

We have learned that there are only about a hundred elements, and that only a third of those are at all plentiful. Even so, we see as we look around us enormous numbers of different kinds of substances, far more than a hundred, or even a thousand, different kinds. Because not all of these can be elements, it is obvious that most of the matter in the world consists of compounds. There are in fact more than half a million different compounds known, and there is no limit indicated on the number yet to be discovered.

In the language of chemistry, compounds are distinguished by their formulas. The **formula** of a compound is a combination of symbols showing (1) which elements appear in the

Table 5.3	
The Gaseous Elements	
Element	**Symbol**
Hydrogen	H
Nitrogen	N
Oxygen	O
Fluorine	F
Chlorine	Cl
Helium	He
Neon	Ne
Argon	Ar
Krypton	Kr
Xenon	Xe
Radon	Rn

Table 5.4		
The Diatomic Elements		
Element	**Molecular Formula**	**Natural State**
Hydrogen	H_2	Colorless gas
Nitrogen	N_2	Colorless gas
Oxygen	O_2	Colorless gas
Fluorine	F_2	Pale yellow gas
Chlorine	Cl_2	Pale green gas
Bromine	Br_2	Red-brown liquid
Iodine	I_2	Purple-black solid

compound and (2) the relative numbers of atoms of each element in a sample of the compound. Formulas can also be used to show how many atoms of each element there are in an individual molecule of a compound. Here is a set of rules that will help you write formulas.

1.   The formula includes the symbols of all the elements that appear in the substance.

   Example: Water is made of hydrogen and oxygen. The symbols for both those elements appear in the formula for water, H_2O.

2.   The number of atoms of each kind is indicated by subscripts following the symbols for those atoms. If the formula contains only one atom of an element, no subscript is used.

   Example: The formula for ammonia, NH_3, tells us that there are three atoms of hydrogen and one of nitrogen in the molecule.

3.   Sometimes atoms appear in groups. If more than one such group appears in a formula, the group is shown in parentheses and the number of times it appears is shown as a subscript to the right of the closing parenthesis.

   Example: The formula for the compound $Ba(NO_3)_2$ contains two of the groups known as a nitrate ion. (As we continue, we will gradually learn the names and characteristics of the various groups.)

EXAMPLE 5.1.   Name the elements and the number of atoms of each element in the formula $(NH_4)_3PO_4$.

Solution:

No subscript appears after P, so only one phosphorus is present. The subscript 4 says that there are four atoms of oxygen in the formula. The subscript 3 beside the parenthesis means that we must multiply all the numbers within the parentheses by three. We know, then, that there are three nitrogen atoms and 12 hydrogen atoms in the formula.

EXERCISE 5.5.   For the following formulas, name the elements and say how many atoms of each appear: MgO_2, Na_2SO_4, $Al_2(SO_4)_3$.

Let us investigate formulas a little further. As we know, some compounds form substances that have no individual molecules. The formula for such a compound takes no notice of that; it merely tells us the relative numbers of the different kinds of atoms. For example, a sample of barium chloride will contain twice as many atoms of chlorine as of barium. That ratio is reflected in the formula $BaCl_2$. There are, however, no individual molecules of $BaCl_2$, nor need there be for the formula to be written.

If molecules exist in a compound, formulas are sometimes written to indicate only the *relative* numbers of atoms present, not the actual numbers of atoms in the molecules. Such formulas are called **empirical,** or **simplest, formulas.**

Many formulas, however, give more information. If a compound happens to be made up of molecules, its formula can tell us how many atoms of each element are actually in each molecule. For example, CH is the empirical formula of a compound called acetylene, but the acetylene molecule actually contains two carbon atoms and two hydrogen atoms; its molecular formula is C_2H_2. The **molecular formula** is always a multiple of the empirical formula, the multiplying factor being one, two, three, or another whole number, usually small.

EXERCISE 5.6.   Give the empirical formula for the compounds with the following molecular formulas: H_2O_2, C_3H_6, $C_4H_6O_2$.

5.3 MOLECULAR WEIGHT

Just as an individual atom has a characteristic mass, so has an individual molecule, whether it consists only of a single atom or of several atoms bound together. The mass of a single molecule is the sum of the masses of the atoms it contains.

Let us consider a water molecule, H_2O, in which the two hydrogen atoms are ^1H and the oxygen atom is ^{16}O. The mass of the molecule is the sum of the masses of the atoms.

$$
\begin{array}{rl}
\text{one } ^1H \text{ atom} = & 1.008 \text{ amu} \\
\text{one } ^1H \text{ atom} = & 1.008 \text{ amu} \\
\underline{\text{one } ^{16}O \text{ atom} =} & \underline{15.995 \text{ amu}} \\
\text{mass of molecule} = & 18.011 \text{ amu}
\end{array}
$$

We calculated the molecular mass in amu, but if we had wished, we could have made the calculation without units. The result (numerically the same) would have been the relative mass of the molecule on the atomic mass scale.

A sample of molecules made of geonormal mixtures of elements may contain molecules of several different masses. For example, it is possible to have six varieties of water molecules made of ^1H, ^2H, ^{16}O, and ^{18}O.

EXAMPLE 5.2.   A molecule of water contains one oxygen atom and two hydrogen atoms. Calculate, in amu, the mass of each of the water molecules listed. The mass of an atom of ^1H is 1.008

amu, that of an atom of ^2H is 2.014 amu, that of an atom of ^{16}O is 15.995 amu, and that of an atom of ^{18}O is 17.999 amu.

a. One atom of ^1H + one atom of ^2H + one atom of ^{16}O.

Solution:

$$1.008 \text{ amu} + 2.014 \text{ amu} + 15.995 \text{ amu} = 19.017 \text{ amu}$$

b. Two atoms of ^2H + one atom of ^{18}O.

Solution:

$$(2 \times 2.014 \text{ amu}) + 17.999 \text{ amu} = 22.027 \text{ amu}$$

c. Two atoms of ^1H + one atom of ^{16}O. (The most common kind of water molecule and the one we discussed in the preceding paragraph.)

Solution:

$$(2 \times 1.008 \text{ amu}) + 15.995 \text{ amu} = 18.011 \text{ amu}$$

EXERCISE 5.7.   A molecule of ammonia, NH_3, contains one nitrogen atom and three hydrogen atoms. Calculate the mass of each of the different kinds of NH_3 molecules listed on the left. Isotopic masses are given on the right. (Because of its instability, the atomic mass of ^3H is not precisely known.)

a. $^{14}N + 2^1H + 1^3H$ ^{14}N 14.00307
                                  ^2H 2.0141
b. $^{15}N + 1^2H + 2^3H$ ^{15}N 15.00011
                                  ^3H 3.0
c. $^{14}N + 1^1H + 2^2H$ ^1H 1.00797

EXERCISE 5.8.   Give a combination of atoms for the molecule NH_3 different from those given in Exercise 5.7. Calculate the mass of the molecule.

Since a sample containing a large number of molecules will have its elements distributed geonormally, we can speak of "average" molecular masses just as we do of "average" atomic masses. To distinguish between the mass of an individual molecule and that of an "average" molecule, the term *molecular weight* is used. It refers to the mass of an "average" molecule. The **molecular weight** of a compound is the sum of the atomic weights of the elements in the compound, allowing for the fact that more than one atom of an element may be present in the molecule.

EXAMPLE 5.3.   Calculate the molecular weight of $Ca(ClO_4)_2$.

Solution:

Working as we did in Example 5.1, we find that there are two chlorine atoms, eight oxygen atoms, and one calcium atom shown in the formula for $Ca(ClO_4)_2$. We look up the atomic weights of the elements (we are dealing in geonormal mixtures now, not individual isotopes) and multiply each by the number of atoms of that element that appear in the formula. Then we add the results.

$$
\begin{array}{rcl}
2 \times 35.453 & = & 70.906 \\
8 \times 15.995 & = & 127.960 \\
1 \times 40.08 & = & \underline{40.08} \\
& & 238.95
\end{array}
$$

EXERCISE 5.9.   Calculate the molecular weights of the following: BaF_2, $Fe(Cr_2O_7)_3$, and $KBrO_3$.

When we begin to make calculations using molecular weights, we will find it possible to use the idea of molecular weight even with compounds that have no actual molecules. $Ca(ClO_4)_2$ in Example 5.3 is such a compound.

5.4 NAMES OF COMPOUNDS

There exists an internationally established system for naming all compounds in a way that indicates their formulas or establishes relationships among the names of families of compounds. Other, less formal systems have been in common use for decades. Finally, many compounds are called by common names that arose historically and have nothing to do with formulas. (For example, hardly anyone, even a chemist, calls water by its descriptive name, dihydrogen oxide.) The language of chemistry, then, is a combination of official and common names. We will begin to deal with this array of names by looking at a few simple rules and learning the names of some compounds. We will investigate more of the naming systems as the need arises, but we will neither need nor be interested in the more detailed or complex rules for naming exotic compounds. A partial list of compound-naming rules appears in Appendix E, so that you can easily find a rule without searching through the text for it. There are also special rules for naming organic compounds, those built around carbon atoms. These rules appear in Chapter 19.

Consider the names *sodium chloride* and *mercuric oxide*. These names were created according to a rule requiring that **binary compounds,** compounds having only two elements, be given names ending in *-ide*. In such a compound, the name of the more metallic element appears first, followed by the name of the second element written with the ending *-ide*. Thus, in hydrogen chloride, hydrogen is the more metallic element and is followed by chlorine, the name of which is changed to end in *-ide*. Although several "-ide" compounds do contain more than two elements, some of the elements in these compounds are considered as a group: examples are ammonium iodide, NH_4I, and calcium hydroxide, $Ca(OH)_2$. Notice that in writing names and formulas, the symbols of the elements always begin with capital letters, but the names of the elements and compounds are not capitalized.

EXAMPLE 5.4.   Name the compound LiCl.

Solution:

LiCl is a binary compound, hence it has a name ending in *-ide*. Lithium is the more metallic element, so its name comes first. The *chlorine* becomes *chloride*. The name is *lithium chloride*.

EXERCISE 5.10.   Name these three compounds: BaF_2, MoB_2, $Al(OH)_3$.

When two elements can form more than one compound, it is necessary that the names distinguish between those compounds. One method, used for certain compounds, is to include prefixes that indicate the numbers of atoms in the formula. A familiar example is provided us by carbon monoxide, CO, and carbon dioxide, CO_2. *Mono* indicates that *one* oxygen atom appears in the formula and *di* signals that there are *two*. In this system, *tri* means *three* (as in nitrogen triiodide, NI_3), and *tetra* means *four*. A second method of distinguishing between different compounds of the same elements relies on different word endings; for example, CuO and Cu_2O are cupric oxide and cuprous oxide. And finally, a third system employs Roman numerals. CuO and Cu_2O would then be copper (II) oxide and copper (I) oxide. Section 17.5 describes these naming systems in more detail.

Two important series of compounds are the oxyacids and the compounds derived from them. Examples of oxyacids are H_2SO_4 and HNO_3, sulfuric and nitric acids (but see the footnote to Table 5.5). Sodium sulfate, Na_2SO_4, can be considered to be derived from sulfuric acid, and potassium nitrate, KNO_3, from nitric acid. The naming system for the oxyacids and their derivatives is also presented in Section 17.5.

Table 5.5 gives the names and formulas of a few important compounds; memorizing these will give you a start on developing your background knowledge of chemical compounds.

5.5 CHEMICAL REACTIONS AND THEIR EQUATIONS

Chemical reaction is characterized by the disappearance of some substances and the appearance of new substances. It is possible to describe a reaction by writing a simple statement in words; for example:

Hydrogen and oxygen combine to form water.

Table 5.5
A Few Important
Names and Formulas

Name	Formula
Water	H_2O
Sodium chloride	NaCl
Sodium hydroxide	NaOH
Hydrochloric acid*	HCl
Sulfuric acid*	H_2SO_4
Nitric acid*	HNO_3
Ammonia	NH_3
Carbon dioxide	CO_2

*Hydrochloric acid, sulfuric acid, and nitric acid are names given to water solutions of the compounds hydrogen chloride, hydrogen sulfate, and hydrogen nitrate. The actual names of these compounds are not often seen because the compounds are ordinarily used in solution.

Such statements, however, leave out much information and are clumsy to write. What is called a **reaction equation** does the job more completely and easily. Let us see how to write reaction equations.

By agreement, the symbols for the initial substances, called **reactants,** are written on the left side of the equation. The symbols for the substances formed, the **products,** are written on the right side. Any of several signs is used to join the two parts of the equation; you will usually see an arrow, but at times a double arrow or an equal sign is more appropriate. Here is an unfinished equation for the formation of water:

$$H + O \longrightarrow H_2O$$

Several things are missing from this equation. First, hydrogen and oxygen do not naturally occur as single atoms but as molecules containing two atoms, H_2 and O_2. Because those are the starting substances, we use them to write the equation.

$$H_2 + O_2 \longrightarrow H_2O$$

Something important is still missing, however. No atoms are created or destroyed in a chemical reaction, but our statement says that two atoms of hydrogen have combined with two atoms of oxygen to produce a molecule of water that contains only one atom of oxygen. What happened to the other atom of oxygen? We have not finished until we have a **balanced equation.** We must write the equation in such a way that the numbers of atoms of each kind will be the same on both sides. (Later, we will learn that balancing also applies to the particles of which atoms are made.) *We must not attempt to balance the equation by changing the subscripts of the atoms in any of the reactants or products, because this would change the chemical species itself.* To make the number of each kind of atom the same on both sides, we must change the number of times the substances themselves appear. We do this by writing the proper integral numbers (called **coefficients**) in front of the symbols for reactants and products. In the water equation, we place 2's in front of the symbols for hydrogen and water, leaving oxygen as it is. If no coefficient is written, the number is assumed to be unity (one).

$$2 H_2 + O_2 \longrightarrow 2 H_2O$$

In this balanced equation, there are four atoms of hydrogen on each side and two atoms of oxygen on each side (Figure 5.2).

To make certain kinds of calculations, or to predict how some kinds of reactions will run, we need to know the states in which reactants and products occur. We use the letters *g* (for gas), *s* (for solid), *l* (for liquid), and *aq* (for substances in water solution). When it is necessary, the appropriate abbreviation is written just to the right of each formula in the reaction equation, as in the example of solid sodium reacting with liquid water to produce hydrogen gas and a solution of sodium hydroxide.

$$2 Na\ (s) + 2 H_2O\ (l) \longrightarrow H_2\ (g) + 2 NaOH\ (aq)$$

(When we discuss reactions involving ionic compounds such as NaOH, we will learn another way to show solutions of these compounds.)

In most of the examples in this chapter, we will use the abbreviations that indicate the states of the reactants and products. You should do the same when you work the examples and end-of-chapter problems. In later chapters, however, we will indicate the states of chemical species only when necessary.

$$2\text{ H}_2 + \text{O}_2 \longrightarrow 2\text{ H}_2\text{O}$$

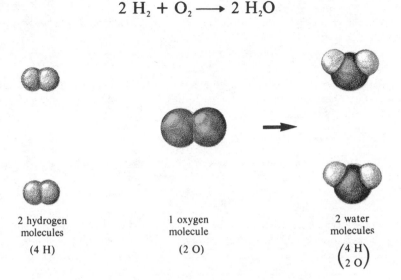

2 hydrogen	1 oxygen	2 water
molecules	molecule	molecules
(4 H)	(2 O)	$\begin{pmatrix} 4\text{ H} \\ 2\text{ O} \end{pmatrix}$

Figure 5.2   The reaction of hydrogen and oxygen.

Using what we have learned so far, we can now write reaction equations in an organized way:

Step 1.   Write the formulas for all the reactants on the left and those for all the products on the right. Put in the arrow or other sign in its proper place.

Step 2.   When it is appropriate for the purpose of the equation, write the states of the reactants and products.

Step 3.   Balance the equation by writing the proper coefficient for each formula. (If a coefficient turns out to be unity, it need not be written.) In balancing the equation, do not alter any formula or add or take away any product or reactant. Work only with coefficients.

As you continue through the chapters that follow, you will often find that you need to furnish parts of reaction equations from your own knowledge. In some cases this depends on your knowing the kind of reaction or the formulas of named compounds. The examples and exercises that follow show us some characteristic kinds of reactions and some of the techniques of balancing reaction equations.

Except for certain kinds of equations for which specific balancing procedures have been developed, equations can usually best be balanced by inspection and trial and error. Inspection means studying the equation to find out which substances are in excess and which will have to be increased. Trial and error means experimenting with some coefficients to see what the results are. If your first attempt doesn't work, try again. It becomes easier with practice.

EXAMPLE 5.5.   When melted, the compound NaCl can be decomposed into its elements by electricity. The metallic sodium appears in liquid form. Write a balanced equation for this process. (Decomposition of compounds into elements is commonplace.)

Solution:

Step 1. Write the formulas for the reactant and the products in equation form. Remember that chlorine is a diatomic gas, as shown in Table 5.4.

$$NaCl \text{ (l)} \longrightarrow Na \text{ (l)} + Cl_2 \text{ (g)}$$

Step 2. Balance the equation. Notice that there are at least two chlorine atoms on the product side; therefore, there must be two on the reactant side. If we place the coefficient 2 in front of the NaCl, we get two each of Na and Cl.

$$2 NaCl \text{ (l)} \longrightarrow Na \text{ (l)} + Cl_2 \text{ (g)}$$

But the equation is still not balanced: there are two Na's on the left, but only one on the right. We fix this by placing a 2 in front of Na on the product side.

$$2 NaCl \text{ (l)} \longrightarrow 2 Na \text{ (l)} + Cl_2 \text{ (g)}$$

Now we check the entire equation, element by element. Left side, two Na's; right side, two Na's. OK. Left side, two Cl's; right side, two Cl's. OK. The equation is now balanced.

EXAMPLE 5.6. Write a balanced equation for the production of ammonia from its elements.

Solution:

Ammonia is one of the compounds with which you will become familiar. It is a gas with the formula NH_3. The N and H begin as diatomic gases, N_2 and H_2.

Step 1. Write the reactants and products.

$$N_2 \text{ (g)} + H_2 \text{ (g)} \longrightarrow NH_3 \text{ (g)}$$

Step 2. Balance the equation. Notice that there must be an even number of N's on the right side, because there can only be an even number of N's on the left; the nitrogen is given to us as N_2. Let's try putting a 2 in front of NH_3. That gives us six H's on the right side (2×3), so we must put six H's on the left. We place a 3 in front of H_2. Checking, we find two N's and six H's on each side.

$$N_2 \text{ (g)} + 3 H_2 \text{ (g)} \longrightarrow 2 NH_3 \text{ (g)}$$

EXAMPLE 5.7. Combustion, or burning, in oxygen is a familiar process. We use combustion to heat houses, run cars, produce electricity, and in numberless other ways. If a substance to be burned contains only C and H, or perhaps C, H, and O, the products are usually CO_2 and H_2O, unless the supply of oxygen is restricted. Write a balanced reaction equation for the complete combustion of liquid heptane, C_7H_{16}.

Solution:

When the words *complete combustion* are used, assume that the products are CO_2 and H_2O. The H_2O usually appears as a gas, steam.

Step 1. $C_7H_{16} \text{ (l)} + O_2 \text{ (g)} \longrightarrow CO_2 \text{ (g)} + H_2O \text{ (g)}$

Step 2. Begin with the carbon. Place a 7 in front of CO_2. Now the hydrogen: placing an 8 before

H_2O gives the necessary 16 H atoms on the right.

$$C_7H_{16} \text{ (l)} + O_2 \text{ (g)} \longrightarrow 7\ CO_2 \text{ (g)} + 8\ H_2O \text{ (g)}$$

Now balance the oxygen. The total number of O atoms on the right is $(7 \times 2) + (8 \times 1) = 22$. Write 11 in front of the O_2.

$$C_7H_{16} \text{ (l)} + 11\ O_2 \text{ (g)} \longrightarrow 7\ CO_2 \text{ (g)} + 8\ H_2O \text{ (g)}$$

Check. Seven carbons each side, 16 hydrogens each side, 22 oxygens each side.

EXAMPLE 5.8. Gaseous methane, CH_4, is burned in a restricted oxygen supply. The products are CO and H_2O. Write the balanced equation.

Solution:

Step 1. $CH_4 \text{ (g)} + O_2 \text{ (g)} \longrightarrow CO \text{ (g)} + H_2O \text{ (g)}$

Step 2. First, carbon and hydrogen:

$$CH_4 \text{ (g)} + O_2 \text{ (g)} \longrightarrow CO \text{ (g)} + 2\ H_2O \text{ (g)}$$

Now, oxygen:

$$CH_4 \text{ (g)} + O_2 \text{ (g)} \longrightarrow CO \text{ (g)} + 2\ H_2O \text{ (g)}$$

Oops. Three oxygens on the right side, but only even numbers on the left side. What to do?

Answer: Simply multiply everything we've done so far (CH_4, CO, H_2O) by two, which makes all the numbers even. Then go on with oxygen.

$$2\ CH_4 \text{ (g)} + O_2 \text{ (g)} \longrightarrow 2\ CO \text{ (g)} + 4\ H_2O \text{ (g)}$$

Now we have six oxygen atoms on the right. Simply put a coefficient of 3 in front of O_2, and the equation will balance.

$$2\ CH_4 \text{ (g)} + 3\ O_2 \text{ (g)} \longrightarrow 2\ CO \text{ (g)} + 4\ H_2O \text{ (g)}$$

Check: two carbons, eight hydrogens, six oxygens on each side.

EXAMPLE 5.9. Metallic aluminum reacts with oxygen to give the solid aluminum oxide, Al_2O_3. Write a balanced equation for the reaction.

Solution:

Step 1. $Al \text{ (s)} + O_2 \text{ (g)} \longrightarrow Al_2O_3 \text{ (s)}$

Step 2. This equation is similar to that in Example 5.8. There are two Al's on the right. But multiplying Al on the left by two is not enough, because it leaves an odd number of oxygens on the right. Since there must be an even number of O's on the right, try putting a 2 in front of Al_2O_3. Now we can make the left side balance by putting a 3 in front of O_2 and a 4 in front of Al. A final check shows us that the equation is balanced.

$$4\ Al \text{ (s)} + 3\ O_2 \text{ (g)} \longrightarrow 2\ Al_2O_3 \text{ (s)}$$

EXAMPLE 5.10. Balance $H_2SO_4 + AlCl_3 \longrightarrow Al_2(SO_4)_3 + HCl$. In this case, the states of the compounds are not specified. We will write our finished equation the same way.

Solution:

To begin, notice that S and O appear as SO_4 on both sides of the equation. Because they do, we will treat SO_4 as we would a single symbol, keeping it always together. This equation is easier than it looks. We first arrange to get enough SO_4 and Al on the left side to balance the right side, then we can take care of the H's and Cl's. We place a 3 in front of H_2SO_4 and a 2 in front of $AlCl_3$. That gives six H's and six Cl's on the left. We balance those by putting a 6 in front of HCl. We then check to be sure everything is balanced.

$$3\ H_2SO_4 + 2\ AlCl_3 \longrightarrow Al_2(SO_4)_3 + 6\ HCl$$

EXERCISE 5.11. Balance $NaCl + NaBrO \longrightarrow NaClO_3 + NaBr$.

EXERCISE 5.12. Balance $C_7H_8 + O_2 \longrightarrow CO_2 + H_2O$.

After we have learned more about ions and their compounds, we will have more tools to help us balance the difficult equations. Meanwhile, the equations we will be seeing can be balanced by inspection and trial and error.

5.6 ENERGY CHANGES IN CHEMICAL REACTION

One of the factors that characterizes our twentieth century industrialized society is the lavish, widespread use of energy. The state of advance of various societies throughout the world has often been assessed simply by determining the per-capita consumption of energy. And much of the energy that powers our modern world is chemical energy, the energy emitted when chemical reaction occurs.

Many chemical reactions spontaneously release energy, often in the form of heat. Such reactions are said to be exoenergetic, or **exothermic.** The combustion of hydrocarbons is an example of an exothermic reaction. Other reactions consume energy, and can run only as long as they are supplied with energy. Such reactions are said to be endoenergetic, or **endothermic.** The decomposition of H_2O into H_2 and O_2 is an example of an endothermic reaction.

Because energy cannot be created or destroyed, *the products of an exothermic reaction contain less energy than did the reactants; the products of an endothermic reaction contain more energy than did the reactants.* Scientists use the term **enthalpy,** often called heat content, to refer to the energy contained by a substance. In an exothermic reaction, the enthalpy of the products is less than the enthalpy of the reactants; the reverse is the case in an endothermic reaction.

The symbol **H** represents the amount of enthalpy of a system, and $\triangle H$ represents the *change* of enthalpy during a reaction or other process ($\triangle$ is often used to mean *a change of*). (See Figure 5.3.) If a reaction is exothermic, $\triangle H$ is *negative* because in changing from reactants to products, the system *loses* energy. If the reaction is endothermic, $\triangle H$ is *positive:*

$-\triangle H$ means the reaction is exothermic
$+\triangle H$ means the reaction is endothermic

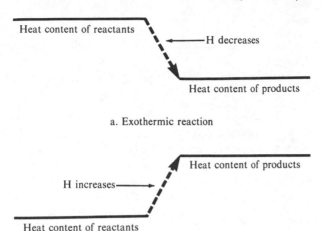

a. Exothermic reaction

b. Endothermic reaction

Figure 5.3   Schematic diagram of energy change during reaction. Part (a) shows the energy change in an exothermic reaction: energy is released and $\triangle H$ is negative. In part (b), which shows the energy change in an endothermic reaction, energy is absorbed and $\triangle H$ is positive.

If it is desired to show the energy of the reaction, it is customary to follow the reaction equation with an indication of the enthalpy change. Here are two examples.

The combustion of butane, C_4H_{10} (sometimes called LPG, or liquified petroleum gas), is exothermic.

$$2\ C_4H_{10}\ (g) + 13\ O_2\ (g) \longrightarrow 8\ CO_2\ (g) + 10\ H_2O\ (g) \qquad \triangle H = -330\ kJ$$

Electric power is used to decompose water. The reaction is endothermic.

$$2\ H_2O\ (l) \longrightarrow 2\ H_2\ (g) + O_2\ (g) \qquad \triangle H = +572\ kJ$$

The amount of enthalpy shown for a reaction is that amount of energy lost or gained by the amounts of reactants and products shown in the reaction equation. To use such values numerically, we will need the additional information contained in sections to follow. We will return to the energies of reactions in Section 7.7.

5.7 READING REACTION EQUATIONS IN UNITS OF MOLES

Although this section is partly devoted to reading and interpreting reaction equations, it is also our introduction to an important new term, the mole—without which our study of chemistry could hardly continue. The mole is one of the most important concepts you will encounter in this book. The idea itself is not complicated, but you should be sure you understand it thoroughly before going on to the next section.

Here is the equation for the formation of water from its elements. Let us wring all the information we can from it.

$$2 \; H_2 + O_2 \longrightarrow 2 \; H_2O$$

Two molecules of hydrogen, each containing two hydrogen atoms, combine with one molecule of oxygen, which contains two oxygen atoms, to produce two molecules of water, each of which contains two hydrogen atoms and one oxygen atom.

People who deal in numbers of individual items often prefer to count in units, rather than by single items. Grocery shoppers rarely buy single eggs; the unit of dozens is more convenient. Egg *dealers* can't be bothered with dozens; they count in grosses (12 dozens).

Because atoms are extremely small, it is inconvenient to count them individually. Let's read our reaction equation in terms of dozens. In your imagination, simply add the word *dozen* to each term in the equation. Here is what the result would look like, written out.

Two dozen molecules of hydrogen combine with one dozen molecules of oxygen to produce two dozen molecules of water.

Even so, the unit of dozen is not large enough to be of much use chemically; a dozen molecules is just as invisible to the eye as a single molecule. We need a much larger unit.

To meet that need, chemists have defined a quantity called the **mole,** whose abbreviation in calculations is mol. The mole is used just as pair, dozen and gross are used—simply for convenience. Although the mole can be applied to any individual countable item, such as tennis balls or people, the number of items is so large that a mole of tennis balls would occupy a huge warehouse, one with the volume of about one fifth of the entire earth. The mole, however, is just right for working with atoms and molecules, and we will soon begin to use it constantly.

For a reason that we will learn, the actual number of individual units in a mole turns out to be 6.02252×10^{23}, or approximately 6 with 23 zeroes after it. This number is called **Avogadro's number,** after an early Italian chemist. *A mole of something is Avogadro's number of the individual units of that thing.* Memorize that definition. Repeat it to yourself until it is a part of you. (Now look back at Problem 19 in Chapter 1 for another example of the size of Avogadro's number.)

Let's now read our H_2O equation once more, this time in moles rather than dozens:

Two moles of hydrogen molecules combine with one mole of oxygen molecules to produce two moles of water molecules.

We have learned that many chemical compounds do not form individual molecules (Section 4.7). In such cases, the concept of the mole—even though there are no individual molecules to be counted—still works quite well. The following equation can be read in mole terms, although NaCl is not molecular.

$$2 \; Na + Cl_2 \longrightarrow 2 \; NaCl$$

5.8 THE ATOMIC WEIGHT SCALE AND THE MOLE

In our study of chemistry, we will usually think in terms of moles of atoms or moles of molecules, rather than in terms of individual particles. For convenience in the sections that follow, we will often carry only three significant figures and will consider a mole to contain 6.02×10^{23} particles. But since the mole is an arbitrary number, just as the dozen is an arbitrary number, why was it not chosen more conveniently? Why not exactly 6×10^{23} or 1×10^{23}?

The answer to that question is that the value of Avogadro's number arose as a result of the choices made when the scale of relative atomic masses was set up. *Avogadro's number, to which we will give the symbol N_A, is defined as the number of atoms of ^{12}C in a sample containing exactly 12 g of ^{12}C.* The definition of the mole simply applies that number: a mole of any kind of particles is Avogadro's number of those particles. We must examine the concept of the mole closely in its relation to the atomic weight scale, because our ability to work chemical problems depends on our understanding and being comfortable with it.

The mole allows us to use laboratory measurements to make calculations involving the mass relations between individual atoms and molecules. It allows us to count numbers of individual particles by weighing samples in the laboratory.

EXAMPLE 5.11.   You want to buy 100,000 beans. How will you and the seller count out the beans?

Solution:

You will certainly not count the beans physically, one by one. Most likely, you and the seller will agree that a small error in the number of beans is acceptable to both of you. Then you will count out perhaps 100 beans, weigh them, and buy 1000 times that weight of beans. You have counted the 100,000 beans by weighing them. When you count atoms by weighing, you will also make an error in the count, but that error is not likely to be proportionately greater than the error in the count of the beans.

An example will show us how the mole concept works. Looking at the scale of atomic weights, we find that the ratio of the mass of a single "average" aluminum atom, Al, to that of an "average" boron atom, B, is $26.98/9.0122 = 2.994$, or, in round numbers, 3. This means that if we start with the mass of a single B atom and multiply it by three, we will have the mass of a single Al atom. Still rounding for simplicity, suppose we begin again with the mass of *two* B atoms, $9 \times 2 = 18$, and multiply that value by three. That will give us the mass of two Al atoms. That is, $18 \times 3 = 54$, just as $27 \times 2 = 54$. There is an easier way to express relations like this for one of each atom, two of each, three of each, or n of each, where n is any number we please.

$$\frac{27}{9} = \frac{2 \times 27}{2 \times 9} = \frac{3 \times 27}{3 \times 9} = \frac{n \times 27}{n \times 9} = 3$$

We can also put the relations into words: the ratio of the mass of one atom of aluminum to one atom of boron is the same as the ratio of the masses of two atoms of each, of three atoms of each, or of any number of atoms of each, *as long as the same number of atoms of each element is taken.* In fact, the ratio of the mass of N_A atoms of Al to N_A atoms of B is the same as the ratio of the masses of the individual atoms, namely 3. Similar relations exist between the masses of atoms of all elements.

Using units of grams, we can measure the mass of N_A atoms of any element, a mole of those atoms. We find that *the relations between the masses of moles of atoms of the various elements are the same as the relations of the masses of the individual atoms as seen in the atomic mass scale.* If the mass of a mole of ^{12}C atoms is 12 g (and it is, by definition), then the mass of a mole of Al atoms must be 26.982 g and that of a mole of O atoms 15.995 g. The general statement of this relation is:

> For any element, a sample will contain exactly one mole of the atoms of that element if it has a mass in grams numerically equal to the value of the atomic weight of that element.

The mass of a mole of atoms of an element is called the **gram-atomic weight** of that element. To measure out a mole of an element, all you need do is weigh out a sample whose weight in grams is the same as the atomic weight.

The mass of a mole of He atoms is 4.0026 g, of Cl atoms 35.453 g, of Na atoms 22.9898 g, of Ag atoms 107.870 g, and so on for all the elements. In the practice of chemistry, the mass of a mole of atoms is given various names, such as molar mass, gram-atomic mass, and gram-atomic weight. We will follow the most general custom and use the term *gram-atomic weight* to mean the mass of a mole of the geonormal mixture of atoms of an element. (The term has the same ambiguity as does *atomic weight;* it really refers to a mass rather than a weight.)

EXAMPLE 5.12.   What is the gram-atomic weight of Ca?

Solution:

Step 1.   Look up the atomic weight of Ca. It is 40.08.

Step 2.   Express this value in the units of grams per mole. The gram-atomic weight of Ca is 40.08 g/mol.

EXAMPLE 5.13.   What is a mole of Ca?

Solution:

A mole of Ca is a sample of Ca weighing 40.08 g and containing N_A atoms of Ca.

EXERCISE 5.13.   What is the gram-atomic weight of P? Of Xe?

The term *atomic weight* is often used when *gram-atomic weight* is meant. This usage is not harmful, as long as the difference between the two terms is understood. The *atomic weight* of an element is the average mass of the atoms in a geonormal sample of that element, taken either in amu or taken without units on the atomic mass scale. The *gram-atomic weight* is the actual mass of a *mole* of a geonormal element. For example, the atomic weight of hydrogen is 1.008 (no units) on the atomic weight scale; that is, the mass of an "average" H atom is 1.008 amu. The gram-atomic weight of hydrogen is 1.008 g, meaning that a mole of geonormal H atoms has a mass of 1.008 g.

5.9 GRAM-MOLECULAR WEIGHT

The reasoning we have just followed to find the gram-atomic weight of an element can also be applied to molecules, and even to compounds that have no individual molecules.

The molecular weight of water, as we have seen, is the sum of the atomic weight of oxygen and twice the atomic weight of hydrogen. An individual molecule of H_2O has a relative mass of 18.01 on the ^{12}C scale. We could, if we wished, write the ratio of the mass of a molecule of H_2O to that of an atom of ^{12}C, just as we wrote similar ratios relating the mass of He atoms or O atoms to that of ^{12}C atoms. In an entirely parallel way, we can calculate that the mass of a mole of water, that is, of Avogadro's number of H_2O molecules, is 18.01 g. The gram-molecular weight of water is the same as the molecular weight of water taken in grams.

*The **gram-molecular weight** of any compound is the same as the molecular weight of that compound taken in grams.*

EXERCISE 5.14.   Calculate the gram-molecular weight of NO_2 and of BF_3.

You might object that many compounds have no individual molecules that can be counted, and therefore can have no Avogadro's number of molecules and no mole of those compounds. Although it is true that some compounds have no countable molecules, the concept of the mole can still be applied as though there were molecules. Mass calculations using the mole are independent of whether there are individual molecules—as long as a formula can be written for the compound and the formula treated as though it represented a molecule. NaCl, for example, has no molecules, but a mole of the ions of Na and a mole of the ions of Cl form a sample of NaCl which, for all purposes of calculation, contains a mole of NaCl. Instead of *mole,* some chemists find it useful to use the term *formula weight* when referring to compounds that have no individual molecules, but we will use gram-molecular weight in two ways: (1) for the mass of a mole of molecules or (2) for the mass of a mole of the imaginary molecules of compounds that form no molecules.

If we are asked the mass of a mole of nitrogen, we cannot know whether the question refers to nitrogen atoms or nitrogen molecules, N_2. The mass in grams of a mole of nitrogen *atoms* is 14 g, whereas that of a mole of nitrogen *molecules* is 2 × 14 g, or 28 g. We do not know what is meant unless we are told. If we are asked for the gram-atomic weight of nitrogen, we can confidently say that it is 14 g. If we are asked for the gram-molecular weight of nitrogen, we can say it is 28 g. When only the word *mole* is used, it is important to specify the species, as in "a mole of nitrogen atoms," "a mole of nitrogen molecules," or "a mole of N_2."

5.10 SUMMARY

Matter is composed either of elements or of combinations of elements called compounds. Elements are arranged in the periodic table, and can be classified as metals, nonmetals, and metalloids. A compound is characterized by its formula, which gives the relative numbers of the atoms of the different elements in the compound. Compounds are named according to a number of systems. Some names indicate the formulas of the compounds; many, such as *ammonia,* do not.

Chemical reactions are described by reaction equations giving the formulas of reactants and products. A reaction equation must be balanced to be complete; that is, there must be the same total number of atoms of each element shown on both sides of the equation. An indication of the enthalpy change is often used to show the energy change that accompanies a reaction. An exothermic reaction is one in which the enthalpy change is negative, that is, *exothermic reaction systems lose enthalpy,* while *endothermic reactions gain enthalpy.*

In the language of chemistry are many terms, some of which sound much alike. Be sure you know how to distinguish among the similar terms that we have just learned. Here is a collected set of definitions for your review.

Atomic mass	On the ^{12}C scale, the mass of an individual atom of any isotope of any element. Atomic mass can be read in a relative way or in amu.
Atomic weight	Actually a term that refers to mass. On the ^{12}C scale, the mass of an "average" atom in a geonormal sample of an element. Atomic weight can be read in a relative way or in amu.
Molecular weight	A term that refers to mass. On the ^{12}C scale, the mass of a molecule composed of geonormally distributed elements. Can also be used to refer to the mass of an imaginary molecule in compounds that form no individual molecules. (In this case, molecular weight is sometimes called formula weight.) Molecular weight can be read in a relative way or in amu.
Gram-atomic weight	A term that refers to mass in the units of grams. The mass of a mole of atoms of any element.
Gram-molecular weight	A term that refers to mass in the units of grams. The mass of a mole of molecules of any compound. The mass of a mole of the imaginary molecules of a compound that forms no individual molecules.
Mole	For any element, a sample that has a mass in grams numerically equal to the atomic weight of the element. A sample containing N_A atoms of an element.
	For any compound, a sample that has a mass in grams numerically equal to the molecular weight of the compound, whether the compound forms individual molecules or not. A sample containing N_A molecules, real or imaginary, of a compound.

Ordinarily, when the term *mole* is used, it refers to a specific species represented by a symbol. If symbols happen not to be used, as in speech, confusion can result.

QUESTIONS

Section 5.1

1. Use the list of elements inside the back cover of the book to make three lists: in the first group, include those elements whose symbols start with the first letters of their names; in the second, those

whose symbols are the first *two* letters of their names; in the third, those whose symbols bear little resemblance to their names. In a source book such as the *Handbook of Chemistry and Physics,* look up the background of the symbols belonging to the third group.

2. Although oxygen is the most abundant element on earth, the atmosphere, which is 20% oxygen, contains only a small fraction of the earth's oxygen. Where and in what form is most of the oxygen on earth?

3. What is a silicate? Where are many silicates found?

4. List the properties of metals and nonmetals.

5. List the elements that exist as liquids and those that exist as gases.

6. Write the names and formulas of the elements that exist as diatomic molecules.

7. Which of the following are diatomic: CO, He, H_2O, N_2O_5, Ar, HBr, $BrCl$, P_2O_5?

Section 5.2

8. What is the primary purpose of a chemical formula? What is the difference between a symbol and a formula? Why are formulas sometimes used for elements, as well as for compounds?

9. Explain the difference between an empirical formula and a molecular formula. Can a formula be both?

Section 5.3

10. Explain the difference between molecular mass and molecular weight.

11. Must molecular mass have units? Why or why not?

Section 5.4

12. What is a binary compound? How are binary compounds named? Give one example of a compound so named that is not truly binary. Why is this not inconsistent with the naming system?

13. How can their names distinguish between two compounds made from the same two elements?

14. A student takes from the shelf a bottle marked "Copper Oxide," and with a sample of the compound runs an analysis exactly like that described in Section 4.2. The mass ratio of copper to oxygen comes out to be 2.0 to 1.0 rather than 4.0 to 1.0. The student is sure that the experiment was run properly. Can you explain what might have caused this result?

15. Acids are an important class of chemicals. How do some binary compounds form acids when put into aqueous solution?

Section 5.5

16. What is a reaction equation? What symbols are used to join the two sides of a reaction equation? Why must an equation be balanced?

17. When reaction equations are balanced, only the coefficients may be rewritten. Why not change the subscripts?

18. Describe how to balance a reaction equation by inspection and trial and error.

Section 5.6

19. What is enthalpy? Explain why an exothermic reaction system loses enthalpy.

20. Why is $\triangle H$ positive for an endothermic reaction?

21. Do you think that the enthalpy loss of these two reactions would be the same or different? If different, by how much? Why? (The quantitative aspects of enthalpy change are discussed in Section 7.7.)

$$H_2 + \tfrac{1}{2} O_2 \longrightarrow H_2O \qquad\qquad 2\,H_2 + O_2 \longrightarrow 2\,H_2O$$

Section 5.7

22. In words, write two different interpretations of the second equation of Question 21.
23. What is a mole? Why is it useful to interpret reaction equations in terms of moles?

Section 5.8

24. Explain how N_A came to be 6.02×10^{23}.
25. Distinguish between atomic mass, atomic weight, and gram-atomic weight.
26. How would you go about measuring out a sample containing a gram-atomic weight of an element?

Section 5.9

27. Distinguish between molecular weight and gram-molecular weight.
28. How do you calculate the gram-molecular weight of a compound?
29. The terms *molecular weight* and *gram-molecular weight* are often used for compounds that do not form molecules. Explain why this practice works.

PROBLEMS

Section 5.1

*1. Give the names of elements whose symbols follow: Mn, Ta, Ti, K.
 2. Give the names of elements whose symbols follow: P, Ba, Pd, Sn.
*3. Write the symbols of the following elements: strontium, rhodium, phosphorus, cerium, cesium.
 4. Write the symbols of the following elements: gold, silver, iron, lead, cobalt.
*5. Of the elements that follow, which are metals and which are nonmetals: Cr, Ba, Te, B, O, Ar, Kr, K, As, P?
*6. Classify the following elements according to whether each is solid, liquid, or gas in the natural state: Na, Ti, B, C, Br, P, Rh, Cs, Cl, I, Si, As.

Section 5.2

*7. Name the elements present in each of the following compounds, and say how many atoms of each element appear in the formula for the compound: H_2SO_4, $NaNO_3$, $CaCl_2$, Cl_2CO, $AlBr_3$.
 8. Name the elements present in each of the following compounds, and say how many atoms of each element appear in the formula for the compound: $(NH_4)_2SO_4$, CS_2, Cs_2S, $(NH_4)_3Fe(CN)_6$, C_2H_7SN.
 9. Give the empirical formulas for the compounds that have the following molecular formulas: P_4O_{10}, N_2H_4, $C_4H_{12}N_2$, Hg_2Cl_2, $C_6H_{12}O_6$.

Section 5.3

*10. Calculate the mass of one molecule of CO consisting of one ^{12}C atom and one ^{18}O atom. (The atomic mass of ^{18}O is 17.999.)
 11. Calculate the mass of one molecule of $HClO_2$ that contains only ^2H (atomic mass 2.0141), ^{35}Cl (atomic mass 34.969), and ^{18}O (atomic mass 17.999).

*12. Use the atomic masses given to calculate the mass of each of the following molecules.

Atomic masses

*a. $^{14}N + 2\ ^{16}O$ $^{14}N = 14.00307$
b. $^{15}N + 1\ ^{16}O + 1\ ^{18}O$ $^{15}N = 15.00011$
c. $^{14}N + 2\ ^{18}O$ $^{16}O = 15.995$
d. $^{15}N + 2\ ^{16}O$ $^{18}O = 17.999$

13. Use the atomic masses given to calculate the mass of each of the following molecules.

Atomic masses

a. $^{12}C + 2\ ^{35}Cl + 2\ ^{37}Cl$ $^{12}C = 12.00000$
b. $^{13}C + 3\ ^{35}Cl + 1\ ^{37}Cl$ $^{13}C = 13.00335$
c. $^{12}C + 4\ ^{35}Cl$ $^{35}Cl = 34.96885$
d. $^{13}C + 1\ ^{35}Cl + 3\ ^{37}Cl$ $^{37}Cl = 36.9474$

*14. Calculate the molecular weights of the compounds listed in Problem 17 (answer given for $CaCl_2$ only).

15. Calculate the molecular weights of the compounds listed in Problem 18.

Section 5.4

16. Write all the possible binary formulas that can be made from combinations of the following elements: Ca, I, N, O, H. (The point of this question is to know which symbol to write first in each formula.)

*17. Name the following compounds: $CaCl_2$, NO_2, $MgBr_2$, KF, AsI_2, BaS_3, Pb_3O_4.

18. Name the following compounds: AgCl, H_2S, Ca_3N_2, BeF_2, MnO_2, Al_2O_3, MgS.

*19. Give formulas for the following binary compounds:
 *a. potassium chloride   *c. boron nitride   e. manganese phosphide
 *b. beryllium sulfide   d. lead oxide   f. sodium hydride

*20. Write formulas for the following compounds:
 *a. bromine pentafluoride   c. iodine dioxide
 *b. carbon tetrachloride   d. sulfur trioxide

Section 5.5

*21. Balance the following equations, using the formulas as they are written (all answered):
 a. $H_2 + O_2 \longrightarrow H_2O$
 b. $N_2 + I_2 \longrightarrow NI_3$
 c. $HCl + NaOH \longrightarrow NaCl + H_2O$
 d. $KClO_3 \longrightarrow KCl + O_2$
 e. $CuCl_2 + H_2S \longrightarrow CuS + HCl$
 f. $Ca(NO_3)_2 + Pb \longrightarrow Ca(NO_2)_2 + PbO$
 g. $H_2SO_4 + KOH \longrightarrow K_2SO_4 + H_2O$
 h. $C_6H_{14}N + O_2 \longrightarrow CO_2 + NO_2 + H_2O$
 i. $BaBr_2 + Cs_2CO_3 \longrightarrow BaCO_3 + CsBr$
 j. $CuO + NH_3 \longrightarrow Cu + N_2 + H_2O$

22. Balance the following equations:
 a. $Na + Cl_2 \longrightarrow NaCl$
 b. $SiO_2 + C \longrightarrow SiC + CO$

c. $CaCO_3 \longrightarrow CaO + CO_2$
d. $NH_3 + O_2 \longrightarrow NO + H_2O$
e. $MnO_2 + HCl \longrightarrow MnCl_2 + Cl_2 + H_2O$
f. $Al + Fe_3O_4 \longrightarrow Al_2O_3 + Fe$
g. $H_3PO_4 + NaOH \longrightarrow Na_3PO_4 + H_2O$
h. $Cu + H_2SO_4 \longrightarrow CuSO_4 + SO_2 + H_2O$
i. $I_2 + HNO_3 \longrightarrow HIO_3 + NO_2 + H_2O$
j. $Ag + HNO_3 \longrightarrow AgNO_3 + NO + H_2O$

Section 5.7

*23. How many atoms of silver are in 3.0 mol of silver?
24. Calculate the number of molecules in 4.4 mol of molecules.
*25. If a mole of ^{135}Ba has a mass of 134.9056 g, what is the mass of one atom of ^{135}Ba in grams?
26. A sample that contains 3.01×10^{23} atoms of carbon contains how many moles of carbon?
*27. If a typewriter prints 75 letters per line and 27 lines per page, how many pages of typing will have a mole of letters?
28. If 1.00 mm³ of sand contains 88 grains, how long would a beach be if it contained a mole of sand grains and was 100 cm deep and 10 m wide?
29. If 52 tadpoles can live in 1.00 m³ of water, how large would a tadpole pond be if it could accommodate Avogadro's number of tadpoles? How does the volume of the tadpole pond compare with the volume of the earth, which is approximately 1.1×10^{21} m³?
30. If a certain kind of molecule has a length of 2.70×10^{-8} cm, how many times will a chain containing 1 mol of such molecules linked end-to-end stretch around the earth? The circumference of the earth is 24,180 mi. One inch equals 2.54 cm, 1 ft contains 12 in., and a mile is 5280 ft.

Section 5.8

31. What are the gram-atomic weights of the elements listed in Problem 1?
*32. Give the gram-atomic weights for the elements in Problem 2.
33. Give the names and symbols of the elements that have the following gram-atomic weights: 26.98 g/mol, 10.81 g/mol, 55.85 g/mol, 207.2 g/mol, 196.97 g/ml.

Section 5.9

*34. What are the gram-molecular weights of the compounds listed in Problem 9?
35. Give the gram-molecular weights of the compounds listed in Problem 18.

MORE DIFFICULT PROBLEMS

*36. If the atomic mass of ^{12}C were exactly 15.000 and a mole were defined as the number of atoms in exactly 15.000 kg of carbon 12, what would be N_A?
37. On the planet Xanthu, the relative atomic weight scale is based on carbon, as it is on Earth. Instead of assigning the relative atomic mass of 12 to carbon, as we do, the scientists of Xanthu have assigned it a relative mass of 7. On Xanthu, then, a mole of carbon is seven nuds of carbon; a nud is the same as 20.5 of our grams. What is the Xanthu equivalent of Avogadro's number?

Chapter 6

Mass Relations in Compounds

The preceding chapters have made us sufficiently familiar with the fundamental language of chemistry to allow us to begin making quantitative calculations with chemical data. Chemistry, like many other sciences, owes much of its present state of advancement to the careful numerical interpretation of quantitative data. To appreciate chemistry as a science, and to be able to perform certain kinds of significant experiments, we need to make such interpretations.

Chapter 6 offers us our first opportunity to make chemical calculations. In it, we will explore the quantitative relations between atomic weight, chemical formulas, and the mass data derived from the analyses of compounds. Chapter 6 is the point of departure for our further advance into the field of chemical calculation.

AFTER STUDYING THIS CHAPTER, YOU WILL BE ABLE TO

- calculate from any two of the following the third: laboratory mass data, gram-atomic weights, and compound formulas
- convert values in mass units to values in moles (and vice versa) for atoms and compounds
- convert from numbers of individual particles to numbers of moles of particles, and vice versa
- calculate the percent composition of compounds from their formulas
- calculate the percent composition of compounds from laboratory mass data given in grams
- determine actual molecular formulas from empirical formulas, as well as from either gram-molecular weights or other laboratory information

TERMS TO KNOW

empirical formula
percent composition

6.1 REVIEW OF SOME IMPORTANT IDEAS

Before going on, let us review and strengthen our knowledge of the fundamental definitions we will use in nearly all our chemical calculations. We must be especially sure that we are fully familiar with the mole and its relation to the units of mass.

We remember that an *atom* is an individual particle of matter, the smallest particle that can exhibit the characteristics of an element. A *molecule* is a discrete group of atoms held together by chemical bonds. (The word *molecule* is often also used for individually existing atoms such as He.) Finally, to our list of individual particles, we add *ions,* the particles from which certain kinds of compounds are formed. Although we first encountered ions in Section 4.7, we will not study their characteristics in detail until Section 11.5.

The mole is not an individual particle, but can be thought of as a unit of measurement, a concept, or a defined number of particles. A mole of some countable individual units is Avogadro's number of those units, where Avogadro's number, N_A, is 6.02×10^{23}. A mole of oxygen atoms contains N_A individual oxygen atoms, and a mole of water molecules, H_2O, contains N_A water molecules.

We can apply the word *mole* to a substance, even though particles of that substance are not individually independent. The term mole is in fact commonly used to refer to ionic substances in which there are no individual particles of the compound at all. We would speak, for instance, of a mole of NaCl, although NaCl is an ionic compound in which there are no molecules.

When we work with samples of ionic compounds, it is often useful to consider the number of moles of the individual ions themselves. If we are dealing with samples of molecular compounds, we can think of the numbers of moles of the different kinds of atoms present. We will explore these alternatives in Section 6.3.

Each individual atom of any element has an atomic mass, and the numerical values of atomic masses of geonormally distributed elements are related to one another by the atomic weight scale. Molecules, ions, and other particles also have individual masses, and those masses are also related to one another and to the entire atomic weight scale. The molecular weight of any particular molecule or of any particular ion is the sum of the atomic weights of all the atoms in that molecule or ion.

Remember that because the atomic weight scale is a relative set of numbers, neither atomic weights nor molecular weights need be expressed in any particular unit. That H has an atomic weight of 1.0 and He an atomic weight of 4.0 on the scale simply means that the He atom weighs four times as much as an H atom. The atomic weight scale does not give the actual numerical mass of an individual atom; however, the mass of an atom *can* be given in grams, pounds, atomic mass units, or any other unit that is convenient.

Although the masses of individual particles, or even of dozens or hundreds of particles, are too small to be widely used in elementary chemistry, the masses of moles of substances are used constantly. Avogadro's number is defined in such a way that the mass of a mole of atoms, the gram-atomic weight, is equivalent to the atomic weight taken in grams. (If you have any doubt about your understanding of this statement and why it is so, I recommend that you review Sections 5.7, 5.8, and 5.9, in which the mole is discussed in detail.) The mass of a mole of real or imaginary molecules is called the gram-molecular weight; it is the molecular weight taken in grams. For example, the molecular weight of water is $16.0 + (2 \times 1.0) = 18.0$; the gram-molecular weight therefore is 18.0 g/mol. The gram-molecular weight of potassium bromide, KBr, is the molecular weight (39.1 + 79.9 = 119.0) taken in grams: 119.0 g/mol.

Once you understand the ideas and definitions of this section, the sections that follow will not be difficult—and might be fun. Do be sure that you are ready before you go on. Spend as much time as you need in this section and in the sections it reviews. The time you take will be paid back several times as you go on.

6.2 LABORATORY DATA AND THE NUMBERS OF MOLES IN SAMPLES

In Chapter 4, we studied a simple analysis of a chemical compound, copper oxide. We found the relative masses of the elements oxygen and copper in copper oxide (that is, the relative amounts of those elements in any sample of the compound) to be 1.0 to 4.0. At that time, we were primarily interested in developing a theory to explain how it could be that copper and oxygen could so consistently and accurately measure out their relative amounts in the compound. Although we learned about the atomic mass scale, we did not ask how the relative masses of the atoms themselves were determined. Now we are ready to ask that question.

There are now many reliable, direct methods of finding atomic masses. The atomic masses of a number of elements, however, had already been accurately measured at the beginning of this century, long before our modern methods were developed. The initial determinations were derived from calculations based on two things: (1) analytical information and (2) knowledge about the formulas of the compounds analyzed.

The three kinds of information, *compound formula, atomic weights,* and *the actual masses of the elements in samples of compounds* are physically related to one another such that any two can be used to calculate the third.

1. If you know the formula of a compound and the masses of the elements in a sample of that compound as measured in the laboratory, you can calculate the relative masses of the individual atoms. Such results are usually placed in relation to the atomic weight scale.
2. If you have the actual masses of the elements in a sample and a table of atomic weights, you can calculate the formula of the compound.
3. If you have the formula of the compound and a table of atomic weights, you can calculate the relations among the masses of the elements in the compound, or the actual mass of each element in any sample of the compound.

We will consider these situations one by one and practice making some calculations. First, however, let us examine the relations between compound formulas and the numbers of moles of the elements present in samples of compounds.

6.3 FORMULAS OF COMPOUNDS AND THE MOLE

In its simplest version, the formula of a compound tells us the relative numbers of the atoms of the different elements in that compound. The formula H_2O says that in this compound there are two H atoms for every O atom, or 2 H/1 O. This particular formula also happens to say more, but we will discuss that in a moment. The formula $KAl(SO_4)_2$ says that in this compound, commonly called alum, there is one potassium atom and one aluminum atom for every two sulfur atoms and every eight oxygen atoms. Formulas may be read in terms of individual particles, as we just did, in terms of dozens of particles, or in moles of particles. The formula for nitric acid, HNO_3, can be read to mean that a mole of nitric acid contains 1 mol of H, 1 mole of N, and 3 mol of O. This gives the conversions 1 mol H/mol HNO_3, 1 mol N/mol HNO_3, and 3 mol O/mol HNO_3. When we make calculations based on laboratory mass data, we must read formulas in terms of moles; laboratory measurements (usually in grams) are made with large numbers of particles, not with individual particles. In the calculations that follow, and for that matter in most of the calculations we will perform in the future, we will work with relative numbers of moles rather than with numbers of individual atoms.

EXAMPLE 6.1. In 1 mol of H_3PO_4, how many moles of H are there? Of P and O?

Solution:

According to the formula, there are 3 mol of H, 1 mol of P, and 4 mol of O for every mole of H_3PO_4. Although it is not necessary in this case, we could have used conversion factors. For example:

$$1 \text{ mol } H_3PO_4 \times \frac{3 \text{ mol H}}{1 \text{ mol } H_3PO_4} = 3 \text{ mol H}$$

EXAMPLE 6.2. In 0.271 mol of sucrose, $C_{12}H_{22}O_{11}$, how many moles of C are there?

Solution:

$$0.271 \text{ mol } C_{12}H_{22}O_{11} \times \frac{12 \text{ mol C}}{1 \text{ mol } C_{12}H_{22}O_{11}} = 3.25 \text{ mol C}$$

EXERCISE 6.1. How many moles of O are there in 1.31 mol of nitrobenzoic acid, $C_6H_5O_4N$?

In Section 5.2, we learned that a compound formula can tell us not only the ratio of the numbers of atoms in a compound, but also the actual number of atoms in a molecule of the compound, if the compound happens to be molecular. An example is water; the formula H_2O gives the ratio of H atoms to O atoms, as well as the number of atoms of each element in a water molecule. We must remember, however, that the number of atoms in *some* molecules is a multiple of a simpler ratio. A hydrogen peroxide molecule, for example, contains two hydrogen atoms and two oxygen atoms, giving a molecular formula of H_2O_2. The *ratio* of the numbers of atoms in the compound, however, is one hydrogen for every one oxygen. The simplest formula for hydrogen peroxide is HO, an expression that gives us the relative number of atoms in the compound but not a description of the molecule. A simplest formula is also called an **empirical formula.** Calculations made from atomic masses and mass measurements from the laboratory lead only to empirical formulas, not to molecular formulas, unless, as in the case of H_2O, the empirical and molecular formulas happen to be identical.

6.4 MASS CALCULATIONS USING THE MOLE

If atomic weights are known, the results of laboratory mass measurements can easily be converted to numbers of moles and vice versa. We simply need to use the correct unit conversion factor, namely the gram-atomic weight of the element.

EXAMPLE 6.3.   How many moles of argon atoms are there in 34.0 g of argon?

Solution:

First, look up the atomic weight of argon and write it down as a conversion factor in moles per gram (why not grams per mole?). Then make the conversion.

$$34.0 \text{ g Ar} \times \frac{1 \text{ mol Ar}}{40.0 \text{ g Ar}} = 0.850 \text{ mol Ar}$$

EXAMPLE 6.4.   How many grams of sodium are there in a sample containing 0.226 mol of Na?

Solution:

$$0.226 \text{ mol Na} \times \frac{22.99 \text{ g Na}}{1 \text{ mol Na}} = 5.20 \text{ g Na}$$

EXERCISE 6.2.   Calculate the number of moles of helium atoms in 10.5 g of helium.

EXERCISE 6.3.   How many grams of silver are in 0.0562 mol of silver?

Similar calculations can be made for amounts of compounds, or the amounts of elements in compounds, if the compound formula is known.

EXAMPLE 6.5.   How many moles of H_2S are there in a sample weighing 92.5 g?

Solution:

We solve this problem exactly as we did the previous one, except that the unit conversion factor comes from the gram-molecular weight of H_2S.
    Molecular weight of H_2S:

$$
\begin{aligned}
2 \times \text{atomic weight of H} &= 2 \times 1.008 = 2.016 \\
1 \times \text{atomic weight of S} &= 1 \times 32.06 = \underline{32.06} \\
\text{molecular weight} & = 34.08 \\
\text{gram-molecular weight} & = 34.08 \text{ g } H_2S/\text{mol } H_2S
\end{aligned}
$$

$$92.5 \text{ g } H_2S \times \frac{1 \text{ mol } H_2S}{34.08 \text{ g } H_2S} = 2.71 \text{ mol } H_2S$$

EXERCISE 6.4.   How many moles of K_2SO_4 are in a sample weighing 176.2 g?

EXERCISE 6.5.   How many grams of CO are there in 1.55 mol of CO?

EXAMPLE 6.6.   How many moles of H are there in a sample of water containing 11.0 g of H_2O?

Solution map: grams of H_2O $\longrightarrow$ moles of H_2O $\longrightarrow$ moles of H

Solution:

The gram-molecular weight of H_2O is 18.02 g H_2O/mol H_2O, obtained by the same method as used for H_2S. The second unit conversion factor comes from the formula H_2O.

$$11.0 \text{ g } H_2O \times \frac{1 \text{ mol } H_2O}{18.02 \text{ g } H_2O} \times \frac{2 \text{ mol H}}{1 \text{ mol } H_2O} = 1.22 \text{ mol H}$$

EXERCISE 6.6.   How many moles of O are there in 320.5 g of Al_2O_3?

EXERCISE 6.7.   Calculate the number of moles of O in 156 g of CO_2.

In many kinds of calculations, it is necessary to convert laboratory data to moles, make a conversion from moles to moles, then convert back to laboratory data.

EXAMPLE 6.7.   How many grams of carbon are there in a sample of C_2H_6 that weighs 35.8 g?

Solution map: grams of C_2H_6 $\longrightarrow$ moles of C_2H_6 $\longrightarrow$ moles of C $\longrightarrow$ grams of C

Solution:

Conversion:

$$35.8 \text{ g } C_2H_6 \times \frac{1 \text{ mol } C_2H_6}{30.07 \text{ g } C_2H_6} \times \frac{2 \text{ mol C}}{1 \text{ mol } C_2H_6} \times \frac{12.01 \text{ g C}}{1 \text{ mol C}} = 28.6 \text{ g C}$$

The first conversion factor is just the gram-molecular weight of C_2H_6 used in inverted form. The second factor (2 mol C/1 mol C_2H_6) can be read right off the compound formula, which says that every molecule of C_2H_6 contains two atoms of carbon and, consequently, that every mole of C_2H_6 contains 2 mol of carbon. The third factor is simply the gram-atomic weight of carbon.

EXERCISE 6.8.   How many grams of O are in 82 g of $Ca(NO_3)_2$?

EXERCISE 6.9.   Calculate the number of grams of H in 46.3 g of H_2S.

In the last kind of problem, one in which a formula is used to find the relation between the numbers of moles of elements in a compound, it helps to remember that the formula is written in the language of nature—in terms of individual atoms. We can use the concept of the mole to translate from our laboratory language of grams into the language of atoms. We then use the information in the formula to return to laboratory language.

1. Translate data given in grams into information given in moles.
2. Interpret the formula in terms of moles.
3. Translate the new information back into the language of the laboratory (grams).

We will later discover that exactly the same technique can be used to make calculations from reaction equations.

6.5 PERCENT COMPOSITION OF COMPOUNDS

In our study of CuO in Section 4.2, the analysis of a 1.20-g sample yielded 0.24 g of O and 0.96 g of Cu, or an oxygen-copper mass ratio of 1/4. We could have reported these results in several ways: by giving the actual masses of the two elements in the sample; by giving only the ratio of the masses; or by showing our results as percent values, the method usually employed. In sections to come, we will make many calculations in terms of percent and we will often be given data in terms of percent.

To give the results of an analysis in percent, we do the following:

Step 1. Find the actual mass of each element in the sample and the total mass of the sample.

Step 2. For each element, write a fraction in which the mass of the element is the numerator and the mass of the entire sample is the denominator.

Step 3. For each element, multiply that fraction by 100%. The result is the percent of that element in the sample. A report of an analysis of all the elements in a compound, as expressed in percent, is the **percent composition.**

EXAMPLE 6.8. Give the percent of O and of Cu present in the oxide of copper just mentioned.

Solution:

Step 1. The mass of O was 0.24 g, that of Cu 0.96 g, and the total mass 1.20 g.

Step 2. The fraction of O in the sample is 0.24/1.20, and that of Cu is 0.96/1.20.

Step 3. The percent of O is

$$\frac{0.24}{1.20} \times 100\% = 20\%$$

The percent of Cu is

$$\frac{0.96}{1.20} \times 100\% = 80\%$$

EXERCISE 6.10. What is the percent of oxygen in a sample of a compound weighing 4.40 g and containing 3.20 g of oxygen?

EXAMPLE 6.9.   A sample of a compound contains 51.10 g of Na, 35.55 g of S, and 71.10 g of O. What is its percent composition?

Solution:

Step 1.   Given the masses of the individual elements, the total mass is

$$51.10 \text{ g} + 35.55 \text{ g} + 71.10 \text{ g} = 157.75 \text{ g}$$

Step 2.   The fraction of each element in the compound is

$$\frac{51.10 \text{ g Na}}{157.75 \text{ g}} \qquad \frac{35.55 \text{ g S}}{157.75 \text{ g}} \qquad \frac{71.10 \text{ g O}}{157.75 \text{ g}}$$

Step 3.   The percent composition is

$$\frac{51.10 \text{ g Na}}{157.75 \text{ g}} \times 100\% = 32.39\% \text{ Na}$$

$$\frac{35.55 \text{ g S}}{157.75 \text{ g}} \times 100\% = 22.54\% \text{ S}$$

$$\frac{71.10 \text{ g O}}{157.75 \text{ g}} \times 100\% = 45.07\% \text{ O}$$

(The sum of the percent values must always equal 100, except for small differences that will sometimes occur because of rounding.)

EXERCISE 6.11.   A sample of compound contains 1.20 g C, 0.10 g H, and 0.70 g N. What is its percent composition?

EXERCISE 6.12.   Calculate the percent composition of a sample of an acid that contains 0.403 g H, 6.412 g S, and 12.800 g O.

EXAMPLE 6.10.   A compound contains only C, H, and Cl. A sample of this compound weighing 112.55 g is found to contain 72.06 g of C and 5.04 g of H, the rest being Cl. What is its percent composition?

Solution:

Step 1.   We have the mass of H and of C, but we must calculate the mass of Cl. If the compound contains only C, H, and Cl, the total mass of the sample is the sum of the masses of those three elements.

$$\text{mass of C} + \text{mass of H} + \text{mass of Cl} = \text{mass of sample}$$

$$72.06 \text{ g C} + 5.04 \text{ g H} + \text{mass of Cl} = 112.55 \text{ g}$$

$$\text{mass of Cl} = 112.55 \text{ g} - 72.06 \text{ g} - 5.04 \text{ g} = 35.45 \text{ g}$$

We can now perform Steps 2 and 3; the answer is 64.02% C, 4.48% H, and 31.50% Cl.

> EXERCISE 6.13.   A sample of a compound containing only Na, Br, and O weighs 118.9 g. What is its percent composition if there are 23.0 g of Na and 16.0 g of O?

6.6 QUANTITATIVE CALCULATIONS INVOLVING LABORATORY DATA, COMPOUND FORMULA, AND GRAM-ATOMIC WEIGHT

6.6.1 Empirical Formulas Calculated From Gram-Atomic Weights and Relative Mass Data From the Laboratory

The empirical formula for a compound can be calculated (1) if the relations of the masses of the elements in the compound are known and (2) if the atomic weights of the elements are available. Because the formula of a compound is a statement, made in whole numbers, of the *ratios of the numbers of atoms of the elements in the compound,* it is therefore also a statement of the *ratios of the numbers of moles of the elements in a mole of the compound.* For instance, the formula for sulfuric acid, H_2SO_4, tells us that a molecule of this compound contains two H atoms and four O atoms for every one S atom. The formula also says that a mole of H_2SO_4 contains 2 mol of H and 4 mol of O for every 1 mol of S.

To find a formula of a compound, we need only find the number of moles of each element in any sample of the compound, then transform this information into a whole-number ratio. Here are the steps:

Step 1.   From the laboratory data, write down the number of grams of each element in a sample of the compound.

Step 2.   Convert the number of grams of each element to the number of moles, using the gram-atomic weight as a unit factor. This will give the actual number of moles of each element in the sample. Except for the fact that these are not whole numbers, they could be used in a molecular formula. Write down this preliminary formula so that you can look at it.

Step 3.   To reduce the numbers of moles in the sample to the simplest whole-number ratio, divide all the numbers by the smallest of those numbers, then round off the answers to whole numbers. A little further adjustment may occasionally be needed to obtain whole numbers.

EXAMPLE 6.11.   A sample of sulfuric acid contains 1.43 g of H, 22.72 g of S, and 45.44 g of O. What is the empirical formula for sulfuric acid?

Solution:

Step 1.   We have been given the number of grams of each element in the compound, so we can go to Step 2. In some problems, it is first necessary to find the numbers of grams from other data.

Step 2.   We convert the mass of each element to the number of moles. Look up the gram-atomic mass in a table and use it as a unit factor. (Careful with the units!)

$$1.43 \text{ g H} \times \frac{1 \text{ mol H}}{1.008 \text{ g H}} = 1.42 \text{ mol H}$$

$$45.44 \text{ g O} \times \frac{1 \text{ mol O}}{16.00 \text{ g O}} = 2.84 \text{ mol O}$$

$$22.72 \text{ g S} \times \frac{1 \text{ mol S}}{32.00 \text{ g S}} = 0.710 \text{ mol S}$$

Step 3. We now divide each of the numbers of moles by the smallest number, namely 0.710.

$$\frac{1.42 \text{ mol H}}{0.710} = 2.00 \text{ mol H}$$

$$\frac{2.84 \text{ mol O}}{0.710} = 4.00 \text{ mol O}$$

$$\frac{0.710 \text{ mol S}}{0.710} = 1.00 \text{ mol S}$$

The formula, then, is $H_{2.00}S_{1.00}O_{4.00}$ or H_2SO_4.

EXERCISE 6.14.   A sample of a compound contains 3.0 g of C, 1.0 g of H, and 4.0 g of O. What is its empirical formula?

EXERCISE 6.15.   Sodium bicarbonate is used in almost every household. What is its empirical formula if a sample contains 5.06 g Na, 2.64 g C, 10.56 g O, and 2.22×10^{-1} g H?

The numbers for the formula will not always come out exactly as whole numbers. As in the next example, some rounding will occasionally be necessary. Notice that in the example that follows, the laboratory data are given as percent composition, necessitating a slightly different approach to the calculation.

EXAMPLE 6.12.   Potassium carbonate is 56.5% K, 8.7% C, and 34.8% O. What is the empirical formula for potassium carbonate?

Solution:

Step 1. To get the actual mass in grams of each element in a sample, we must imagine that we have an actual sample. It is most convenient to imagine a sample of 100.0-g mass, because the percent numbers for each element then convert directly into grams. Remember that 56.7% K means, for example, that 100.0 g of compound contains 56.7 g of K. If our imaginary sample of potassium carbonate weighed 100.0 g, then it would contain the following:

$$100.0 \text{ g of sample} \times \frac{56.5 \text{ g K}}{100.0 \text{ g sample}} = 56.5 \text{ g K in the sample}$$

$$100.0 \text{ g of sample} \times \frac{8.7 \text{ g C}}{100.0 \text{ g sample}} = 8.7 \text{ g C}$$

$$100.0 \text{ g of sample} \times \frac{34.8 \text{ g O}}{100.0 \text{ g sample}} = 34.8 \text{ g O}$$

Step 2. Convert each mass to numbers of moles.

$$56.5 \text{ g K} \times \frac{1 \text{ mol K}}{39.10 \text{ g K}} = 1.45 \text{ mol K}$$

$$34.8 \text{ g O} \times \frac{1 \text{ mol O}}{16.00 \text{ g O}} = 2.18 \text{ mol O}$$

$$8.7 \text{ g C} \times \frac{1 \text{ mol C}}{12.01 \text{ g C}} = 0.724 \text{ mol C}$$

Step 3. As it stands, the formula is $K_{1.44}C_{0.725}O_{2.18}$. To reduce these values to whole numbers, divide each by the smallest.

$$\frac{1.45}{0.724} = 2.00 \qquad \frac{0.724}{0.724} = 1.00 \qquad \frac{2.18}{0.724} = 3.01$$

This operation yields $K_{2.00}C_{1.00}O_{3.01}$, still not a whole-number formula, but one that we can round to whole numbers. We look at the original data, and find that the least accurate value, that for C, was given to us in two significant figures. Because the value for C, 8.7%, could be inaccurate by as much as 1 or 2 parts in about 90, we are justified in rounding our final results by that much. We round 3.01 to 3, giving the final empirical formula as K_2CO_3.

EXERCISE 6.16. A compound contains 52.0% Zn, 9.6% C, and 38.4% O. What is its empirical formula?

In Examples 6.10 and 6.11, the formulas for the compounds each contained an element that appeared with a subscript of unity, namely, S in $H_2S_1O_4$ and C in $K_2C_1O_3$. (Ordinarily, subscripts of unity are not shown. We use them here for emphasis.) In some formulas, however, there is no such element, an example being Fe_2O_3. In such cases, Step 3 leads not to integer numbers but to integer fractions. We must then obtain an integer formula by multiplying all of the subscripts by an appropriate integer.

EXAMPLE 6.13. A compound of phosphorus and oxygen contains 43.7% P and 56.3% O. What is its simplest formula?

Solution:

Step 1. A 100.0-g sample of the compound contains 43.7 g of P and 56.3 g of O.

Step 2.
$$43.7 \text{ g P} \times \frac{1 \text{ mol P}}{31.0 \text{ g P}} = 1.41 \text{ mol P}$$

Step 3.
$$\frac{1.41 \text{ mol P}}{1.41 \text{ mol}} = 1.00 \text{ mol P}$$

$$56.3 \text{ g O} \times \frac{1 \text{ mol O}}{16.0 \text{ g O}} = 3.52 \text{ mol O}$$

$$\frac{3.52 \text{ mol O}}{1.41 \text{ mol}} = 2.50 \text{ mol O}$$

Our formula, then, is $P_{1.00}O_{2.50}$. The atoms appear in the ratio of 1 to 2.5 or 2/2 to 5/2: $P_{2/2}O_{5/2}$. If we multiply both values by 2, we have P_2O_5.

EXERCISE 6.17. A hydrocarbon contains 11.25 g of C and 2.5 g of H. What is its empirical formula?

EXERCISE 6.18. An oxide of iron contains 72.36% Fe and 27.64% O. What is its simplest formula?

6.6.2 Calculating Percent Composition From Formula and Gram-Atomic Weight

It is often desirable to calculate the percent composition of a compound—a kind of analysis by calculation. You can perform such calculations if you know the compound formula and if atomic weights are available: simply read the formula as if it applied to 1 mol of the compound. Because the subscripts in a formula give the number of atoms of each element in a molecule of the compound *or* the number of moles of each element in a mole of compound, the number of moles of each element in your 1-mol sample can be read directly from the formula. You then calculate the number of grams of each element and write your analysis in the required way, either as a simple ratio or in terms of percent composition. Here are the steps:

Step 1.   From the formula of the compound, list the number of moles of each element in 1 mol of the compound.

Step 2.   Convert the number of moles of each element in the list to the number of grams of that element, using the gram-atomic weight as a unit factor. This step yields the number of grams of each element in a sample consisting of 1 mol.

Step 3.   To convert numbers of grams to percent composition, divide the number of grams of each element by the gram-molecular weight of the compound; this gives the number of grams of each element in 1 g of compound. Now multiply each of these answers by 100; this gives the number of grams of each element in 100 g of compound, which is the same as the percent composition.

EXAMPLE 6.14.   What is the percent composition of nitric acid, HNO_3?

Solution:

Step 1.   In 1 mol of compound, there are 1 mol H, 1 mol N, and 3 mol O.

Step 2.   The number of grams of each element is

$$\frac{1 \text{ mol H}}{\text{mol compound}} \times \frac{1.008 \text{ g H}}{\text{mol H}} = \frac{1.008 \text{ g H}}{\text{mol compound}}$$

$$\frac{1 \text{ mol N}}{\text{mol compound}} \times \frac{14.01 \text{ g N}}{\text{mol N}} = \frac{14.01 \text{ g N}}{\text{mol compound}}$$

$$\frac{3 \text{ mol O}}{\text{mol compound}} \times \frac{16.00 \text{ g O}}{\text{mol O}} = \frac{48.00 \text{ g O}}{\text{mol compound}}$$

Step 3. Find the number of grams of each element in 1 g of compound. We use the conversion factor

$$\frac{1 \text{ mol compound}}{63.02 \text{ g compound}}$$

$$\frac{1.008 \text{ g H}}{1 \text{ mol compound}} \times \frac{1 \text{ mol compound}}{63.02 \text{ g compound}} = \frac{0.01599 \text{ g H}}{\text{g compound}}$$

$$\frac{14.01 \text{ g N}}{1 \text{ mol compound}} \times \frac{1 \text{ mol compound}}{63.02 \text{ g compound}} = \frac{0.2223 \text{ g N}}{\text{g compound}}$$

$$\frac{48.00 \text{ g O}}{1 \text{ mol compound}} \times \frac{1 \text{ mol compound}}{63.02 \text{ g compound}} = \frac{0.7617 \text{ g O}}{\text{g compound}}$$

Multiplying each of these values by 100 gives the number of grams of each element in 100 g of compound, or the percent composition: 1.60% H, 22.23% N, 76.17% O.

After you become accustomed to the method, you may find it more convenient to combine Steps 1, 2, and 3 into a unit equation for each of the elements.

$$\frac{1 \text{ mol H}}{\text{mol HNO}_3} \times \frac{1.008 \text{ g H}}{\text{mol H}} \times \frac{1 \text{ mol HNO}_3}{63.02 \text{ g HNO}_3} \times 100\% = 1.599\% \text{ H}$$

Similar calculations can be written for the other two elements. Notice that the units on percent composition are (grams of element)/(grams of compound).

EXERCISE 6.19. What is the percent composition of NaOH?

EXERCISE 6.20. What is the percent composition of calcium chlorite, $Ca(ClO_2)_2$?

EXAMPLE 6.15. What is the ratio of the mass of sulfur to that of copper in cuprous sulfide, Cu_2S? (If, as in this example, only a ratio of the masses of the elements is required, there is no need to do Step 3.)

Solution:

Step 1. In 1 mol of Cu_2S there are 2 mol of Cu and 1 mol of S.

Step 2. To get the number of grams of each element in a 1-mol sample of the compound, multiply the number of moles of each element by its atomic mass.

$$2 \text{ mol Cu} \times \frac{63.55 \text{ g Cu}}{1 \text{ mol Cu}} = 127.1 \text{ g Cu}$$

$$1 \text{ mol S} \times \frac{32.06 \text{ g S}}{1 \text{ mol S}} = 32.06 \text{ g S}$$

The ratio of these numbers of grams is 127.1 g Cu to 32.06 g S, or 3.96 g Cu to 1.00 g S.

EXERCISE 6.21.   What is the mass ratio of cesium to that of nitrogen in cesium nitride, Cs_3N?

6.6.3 Calculating Atomic Weight From Formula and Laboratory Data

The atomic weights of most of the elements were originally calculated from formulas of compounds that had been analyzed in the laboratory. The calculation is not difficult, although it is necessary to know in advance the actual atomic weight of at least one of the elements in the compound before the atomic weights of the other elements can be calculated. Otherwise, the relative masses of the elements can be determined, but the values obtained will not necessarily relate to the atomic weight scale. Because this kind of calculation is somewhat more difficult than others we have performed, we will use an example as we look at the method.

EXAMPLE 6.16.   The compound A_2O is 74.2% A and 25.8% O. What is the gram-atomic weight of A?

Solution:

Step 1.   Begin with an imaginary 100-g sample. From the laboratory mass data, write down the mass of each element in the sample. Our imaginary 100-g sample of A_2O must contain 74.2 g of A and 25.8 g of O.

Step 2.   By using the gram-atomic weight of the *known* element as a unit conversion, find the number of moles of the *known* element in the sample. The known element in this problem is O. We convert to moles of O in the sample.

$$25.8 \text{ g O} \times \frac{1 \text{ mol O}}{16.00 \text{ g O}} = 1.61 \text{ mol O}$$

Step 3.   Find the number of moles of each of the other elements in the sample by converting from the number of moles of the known element. Use as a conversion factor the ratio of the number of atoms of the two elements in the formula. To find the number of moles of A in the sample, we use the conversion factor

$$\frac{2 \text{ mol A}}{1 \text{ mol O}}$$

which we can read right off the formula.

$$1.61 \text{ mol O} \times \frac{2 \text{ mol A}}{1 \text{ mol O}} = 3.22 \text{ mol A}$$

Step 4.   Divide the number of grams of each of the other elements by its number of moles, to obtain the atomic weight. We divide the weight of A in the sample by the number of moles of A in the sample and find the number of grams in each mole of A; that is, the gram-atomic weight of A.

$$\frac{74.2 \text{ g A}}{3.22 \text{ mol A}} = \frac{23.0 \text{ g A}}{\text{mol A}}$$

The element A is Na.

To review what we did, let us look at a map of the whole process:

Solution map: imaginary sample $\longrightarrow$ mass of known element in sample
$\longrightarrow$ moles of known element in sample
$\longrightarrow$ moles of each of the other elements in sample
$\longrightarrow$ grams of each of the other elements in sample
$\longrightarrow$ gram-atomic weights

Because we have worked in the unit of grams, our answer is expressed in the unit of grams per mole, the gram-atomic weight. This value is numerically the same as the atomic weight. To read our answer as atomic weight, we simply delete the units: the atomic weight of A in Example 6.16 is 23.0.

EXAMPLE 6.17.   A sample of K_2DE_4 is 44.9% K, 18.4% D, and 36.7% E. What are the gram-atomic weights of D and E?

Solution:

Step 1.   Start with 100 g of K_2DE_4. The 100 g contains 44.9 g of K, 18.4 g of D, and 36.7 g of E.

Step 2. $44.9 \text{ g K} \times \dfrac{1 \text{ mol K}}{39.1 \text{ g K}} = 1.148 \text{ mol K in the sample}$

Step 3. $1.148 \text{ mol K} \times \dfrac{1 \text{ mol D}}{2 \text{ mol K}} = 0.574 \text{ mol D}$

$1.148 \text{ mol K} \times \dfrac{4 \text{ mol E}}{2 \text{ mol K}} = 2.30 \text{ mol E}$

Step 4. $\text{atomic weight of D} = \dfrac{18.4 \text{ g D}}{0.574 \text{ mol D}} = \dfrac{32.1 \text{ g D}}{\text{mol D}}$

$\text{atomic weight of E} = \dfrac{36.7 \text{ g E}}{2.30 \text{ mol E}} = \dfrac{16.0 \text{ g E}}{\text{mol E}}$

D is sulfur, atomic weight 32.06; E is oxygen, atomic weight 16.00. K_2DE_4 is K_2SO_4.

EXERCISE 6.22.   When 9.63 g of a metal M reacts with 2.63 g of fluorine, MF_3 is formed. What is the gram-atomic weight of M?

EXERCISE 6.23.   A sample of XY_2O_4 contains 60.94% X, 10.66% Y, and 28.40% O. What are the gram-atomic weights of X and Y?

6.7 CALCULATING THE ACTUAL MOLECULAR FORMULA FROM THE EMPIRICAL FORMULA

In Section 6.6.1, we used laboratory mass data and gram-atomic weight to calculate empirical formulas, the simplest whole-number ratios of the numbers of atoms of different elements in compounds. In the case of compounds that form molecules, the empirical formula does not necessarily tell us how many atoms of each kind there are in a molecule. The actual numbers of atoms in the molecule may be a multiple of the empirical formula, as we saw in the example of H_2O_2. It is often the case that two entirely different compounds have the same empirical formula. Consider the formula CH_2, which applies to several compounds, two of which are ethylene and cyclohexane.

Ethylene = $(CH_2)_2 = C_2H_4$
Cyclohexane = $(CH_2)_6 = C_6H_{12}$

Ethylene is a gas from which the familiar plastic, polyethylene, is made. Cyclohexane is a clear liquid with entirely different properties. Yet both have the same empirical formula. Notice that when two compounds have the same empirical formula, they also have the same percent composition. Ethylene and cyclohexane, for instance, both contain 85.6% C and 14.4% H.

Whenever possible, it is useful to know the actual molecular formula of a compound rather than just the empirical formula. You can easily calculate the molecular formula from the empirical formula if you have an approximate molecular weight—perhaps from a laboratory experiment.

Step 1.   Find the "molecular weight" for the empirical formula. We will pretend that the formula represents a molecule and proceed as though we were finding an ordinary molecular weight (see Section 5.9).

Step 2.   Divide the approximate molecular weight by the calculated "molecular weight" and round off the result to the nearest whole number.

Step 3.   Multiply the empirical formula by the result of Step 2.

EXAMPLE 6.18.   The empirical formula for hydrazine is NH_2, and the molecular weight of hydrazine is approximately 30. What is the molecular formula for hydrazine?

Solution:

Step 1.   The calculated "molecular weight" for the empirical formula is

(1 × atomic weight of N) + (2 × atomic weight of H)

$$= \frac{14.007 \text{ g}}{\text{mol}} + \left(2 \times \frac{1.008 \text{ g}}{\text{mol}}\right) = \frac{14.007 \text{ g}}{\text{mol}} + \frac{2.016 \text{ g}}{\text{mol}} = \frac{16.023 \text{ g}}{\text{mol}}$$

Step 2.   Divide the empirical "molecular weight" into the measured approximate molecular weight.

$$\frac{30 \text{ g/mol}}{16 \text{ g/mol}} = 1.9$$

The result, 1.9, is rounded to 2. The molecular formula for hydrazine, then, is

$$2 \times NH_2 = N_2H_4$$

An accurate, actual molecular weight can be calculated from the molecular formula. The actual molecular weight of N_2H_4 is

$$(2 \times 14.007) + (4 \times 1.008) = 32.046$$

It is also possible to obtain an accurate molecular weight by multiplying the empirical "molecular weight" by the multiplying factor you use to obtain the actual formula. In this case, $2 \times 16.023 = 32.046$.

EXERCISE 6.24. The empirical formula of radiator coolant, ethylene glycol, is CH_3O; its estimated molecular weight is 60. What is its molecular formula and exact molecular weight?

EXERCISE 6.25. The empirical formula for a compound of C, H, O, and Cl is C_4H_3OCl, and its molecular weight is approximately 200. What is the molecular formula and exact molecular weight of this compound?

EXAMPLE 6.19. The compound xylene consists only of carbon and hydrogen. It is 90.49% C, and the rest is H. Its gram-molecular weight is approximately 105. What is the molecular formula and the exact gram-molecular weight of xylene?

Solution:

First, determine the empirical formula. In a 100.00-g sample, there are 90.49 g of C and (100.00 g − 90.49 g) = 9.51 g of H. Converting these masses to moles of the two elements, we have 9.43 mol of H and 7.53 mol of C. If both these numbers are divided by 7.53, we find the ratio of moles of C to moles of H to be 1.00 to 1.25, and we write our preliminary empirical formula as $C_{1.00}H_{1.25}$.

To convert the subscripts to whole numbers, we try multiplying the formula by small whole numbers until both subscripts are changed to whole numbers. The number 4 will do this, so we multiply by 4, obtaining $C_{1.00}H_{1.25} \times 4 = C_4H_5$.

But this is only the empirical formula. The "molecular weight" for this formula is 53.08. This will divide into the approximate molecular mass of 105 about two times, so we take the actual molecular formula to be twice the empirical formula and write C_8H_{10}, for which the gram-molecular weight is $2(53.08) = 106.16$.

EXERCISE 6.26. Citric acid contains C, H, and O. It is 37.5% C, 4.20% H, and 58.37% O, and has a molecular weight of about 190. What is its molecular formula and its exact molecular weight?

6.8 SUMMARY

In Chapter 6, we made quantitative calculations using laboratory mass data (often given in the form of percent composition), atomic weights, and the formulas of compounds. We learned how to use the concept of the mole in calculating any one of these if the other two are known.

Formula calculations made with laboratory mass data and atomic weights yield only the simplest formula, the empirical formula. Actual molecular formulas can be determined if empirical formulas and other data, such as approximate molecular weights, are available.

QUESTIONS

Section 6.1

1. Sometimes the values for atomic and molecular weights are written in terms of grams, and sometimes without any mass unit whatever. How can it be correct to leave the units off these values? By what name are the values often called when the unit of grams is attached to them?

Section 6.2

2. How can it be possible to know the relative masses of the atoms of two elements and still not know their atomic weights?

Section 6.3

3. Explain why a chemical formula can be read in terms of individual atoms and also in terms of moles.
4. What is the relation between an empirical formula and a molecular formula? Is there such a thing as a molecular formula for compounds that do not form molecules? In such cases, what is the relation between the empirical formula for the compound and the actual formula?

Section 6.4

5. What conversion factor or factors are used in a mole-to-mass conversion? In a mole of compound-to-mole of atoms conversion? In a grams of compound-to-grams of atoms conversion?
6. Give solution maps for Problems 17 and 21a.
7. Why is it necessary to change grams into moles and moles back into grams when calculating numbers of grams of atoms from numbers of grams of molecules?

Section 6.5

8. Outline the procedure for calculating percent composition from laboratory mass data.
9. What is the relation between the fractional mass of an element in a compound and the percent mass of that element in the compound?

Section 6.6

10. Describe how you can obtain the empirical formula of a compound from the numbers of grams of the individual elements in a sample of the compound.
11. If you are given the percent composition of a compound (rather than the number of grams of each element), what assumption can you make as a starting point for calculating the empirical formula of the compound?
12. Describe how to calculate the percent composition of a compound from its formula.
13. Write solution maps for Problems 25 and 32.

Section 6.7

14. Explain how it is possible for two compounds to have the same empirical formula and completely different physical properties.

15. To determine a molecular formula, you need the empirical formula and an estimated molecular weight. Can you think of some other kind of information that could substitute for the estimated molecular weight and that would still enable you to find the molecular formula? How would such information be used in the calculation?

PROBLEMS

Section 6.3

1. Calculate the number of moles of N atoms in 7.32 mol of N_2H_4.
*2. How many moles of H atoms are in 4.8 mol of CH_4?
3. How many moles of H_2CO_3 contain 2.25 mol of O atoms?
4. How many moles of H_3PO_4 contain 4.37 mol of H atoms?
*5. How many moles of $C_6H_{12}F_2$ contain as much hydrogen as 6 mol of H_2O?
6. The formula for pentane is C_5H_{12}. How many atoms of carbon are there in five molecules of pentane? How many moles of H are there in 3.7 mol of pentane? What is the gram-molecular weight of pentane? How many atoms of carbon are there in 2.2 mol of pentane?

Section 6.4

*7. How many moles of atoms are there in each of the following samples?
    *a. 72.8 g of Al   d. 418 g of Si
    b. 33.9 g of Ne   e. 2.4×10^{-3} g of Au
    c. 0.22 g of Li
8. Calculate the number of moles of atoms in each of the following samples:
    a. 40.0 g of Mg   d. 4.3×10^{-2} g of He
    b. 3.4×10^2 g of Pb   e. 16.8 g of S
    c. 79.6 g of Mn
*9. How many moles of compound are in each of the following samples?
    *a. 53.2 g of Fe_2O_3 d. 409 g of $Fe(H_2PO_2)_3$
    b. 48.7 g of Cl_2 e. 188 g of Al_2S_3
    c. 2.16 g of LiH
10. Calculate the number of moles of compound in each of the following samples:
    a. 125 g of H_2SO_4 d. 8.543×10^{-1} g of CCl_4
    b. 252 g of NaOH   e. 5.24 g of MnO_2
    c. 7.9×10^{-2} g of K_2CrO_4
*11. Calculate the number of grams in each of the following samples:
    *a. 6.30 mol of N_2 c. 4.32×10^{-3} mol of Cr
    b. 9.86×10^{-8} mol of Hg   d. 7.88 mol of H_2
12. What is the mass, in grams, of each of the following samples?
    a. 0.0222 mol of He   c. 2.894 mol of O_2
    b. 4.56×10^{-4} mol of S   d. 1.27 mol of Ag
13. Calculate the number of grams of compound in each of the following samples:
    a. 54.2 mol of LiOH   c. 9.314×10^{-4} mol of $CaCl_2$
    b. 0.0421 mol of H_3PO_4 d. 27.2 mol of CO_2
14. What is the mass in grams of each of the following samples?
    a. 9.27×10^{-3} mol of $K_2Cr_2O_7$ d. 0.0486 mol of $BaBr_2$
    b. 4.20 mol of HCl   e. 3.481×10^{-5} mol of Al_2O_3
    c. 1.1×10^4 mol of CO

*15. How many moles of oxygen atoms are there in each of the following samples?
   *a. 66.6 g of NaOH   d. 9.53×10^{48} molecules of H_2O_2
    b. 18.0 g of $C_{12}H_{16}O_4$ e. 1.21 g of H_2O_2
    c. 122 g of SO_2
16. How many moles of atoms are there in each of the samples a, b, d, and e of Problem 15?
*17. How many grams of S are there in 0.532 mol of $Na_2S_2O_3$?
18. A sample of pentane, C_5H_{12}, weighs 22.7 g. How many grams of carbon are there in that sample?
19. A sample of barium hydroxide, $Ba(OH)_2$, has a mass of 52.11 g. What is the mass of the Ba in the sample?
20. A sample of potassium chromate, K_2CrO_4, has a mass of 541 g. How many grams of potassium are there in the sample? How many grams of oxygen?
21. How many grams of oxygen are contained in each of the following samples?
    a. 8.46 g of NO_3F c. 3.03 mol of H_2O_2
    b. 202 g of $NaClO_3$ d. 26.2 mol of $C_6H_{12}O_6$
22. How many grams of lithium are contained in each of the following samples?
    a. 3.62 g of $Li_2B_4O_7$ d. 2.23 mol of $LiAlH_4$
    b. 5.87 g of $LiSO_3F$ e. 4.01×10^{18} molecules of Li_3N
    c. 4.01 mol of Li_3PO_4 f. 1.74×10^3 mol of Li_4SiO_4

Section 6.5

*23. A 24.4-g sample of sodium carbonate, Na_2CO_3, contains 10.6 g of Na, 2.8 g of C, and 11.1 g of O. What is its percent composition?
24. Calculate the percent composition of a sample of compound that contains 9.20 g of Na, 0.40 g of H, and 6.40 g of O.
25. What is the percent composition of a sample of compound containing 1.56 g of K, 2.20 g of Mn, and 2.56 g of O?
26. A 32.7-g sample of calcium carbonate, $CaCO_3$, contains 3.92 g of C, 13.09 g of Ca, and the rest is O. What is the percent composition of $CaCO_3$?
*27. A sample of the compound $C_7H_6O_4$ weighs 17.6 g and contains 9.60 g of C, 7.31 g of O, and the rest is H. What is its percent composition?
28. A 67.9-g sample of a compound contains 45.8 g of C, 8.68 g of H, and the rest is N. Calculate the percent composition of this compound.
29. A sample of pyridine weighs 9.20 g and contains 6.98 g of C, 1.63 g of N, and the rest is H. What is the percent composition of pyridine?
30. Calculate the percent composition of each of the following samples:
    a. A 33.29-g sample of $CaCl_2$ containing 21.27 g of Cl and 12.02 g of Ca
    b. A sample of Na_2S containing 3.68 g of Na and 2.56 g of S
    c. A sample of $MgBr_2$ containing 9.72 g of Mg and 63.92 g of Br
    d. A 23.95-g sample of LiOH containing 6.94 g of Li and 1.01 g of H
31. A sample of a pure acid has a volume of 45.9 mL and a density of 1.50 g/mL. It contains 1.10 g of H, 15.3 g of N, and the rest is O. What is its percent composition?

Section 6.6

*32. A 13.21-g sample of magnesium will react with 40.71 g of arsenic to form a compound. What is the empirical formula of the compound?
33. Ethanol is found in alcoholic beverages. If a 27.64-g sample of ethanol contains 3.63 g of H, 14.41 g of C, and the rest is O, what is the empirical formula of ethanol?

34. A sample of ascorbic acid contains 5.76 g of C, 0.6464 g of H, and 7.68 g of O. Calculate the empirical formula of ascorbic acid.

*35. Codeine was formerly used in cough syrups; now it is used primarily as a pain-killer. What is its empirical formula if a sample weighing 23.950 g contains 17.294 g of C, 1.697 g of H, 1.121 g of N, and the rest is O?

36. Find the empirical formula of each of the compounds, the percent compositions of which follow.
   a. 31.8% N, 13.6% H, 54.6% C
   b. 71.04% Ag, 7.90% C, 21.07% O
   c. 83.01% K, 16.99% O
   d. 39.69% K, 27.92% Mn, 32.49% O
   e. 42.86% C, 7.14% H, 50.01% N

*37. A certain oxide of chromium contains 23.5% oxygen. What is the empirical formula of this compound?

38. Nitrogen forms six different oxides. Find the empirical formulas (which happen also to be the molecular formulas) for each, given the following:
   a. 46.67% N   c. 22.58% N   e. 36.84% N
   b. 63.64% N   d. 30.43% N   f. 25.93% N

39. What is the percent of oxygen in each of the following compounds?
   a. $AgIO_3$ c. C_6H_7O e. $BaSO_4$ g. PtO_2
   b. $KMnO_4$ d. N_2O_5 f. $MgMoO_4$ h. PON

*40. Calculate the percent of lithium in each of the compounds listed in Problem 22 (answer given for (a) only).

*41. List the following sulfides in order of increasing percent of S: Co_3S_4, Cs_2S_2, Al_2S_3, K_2S_2, SiS_2.

42. Which has the highest percent of
   a. sulfur, SeS_2 or SeS?
   b. sodium, Na_2O or Na_2O_2?
   c. chlorine, S_2Cl_2 or SCl_4?
   d. carbon, C_6H_7O or C_6H_5NO?

*43. *a. What is the percent of Cl in chlorohydroxybiphenyl, $C_{12}H_9OCl$?
   *b. What fraction of all the atoms in this compound are carbon atoms?
   *c. What is the weight in grams of one molecule of $C_{12}H_9OCl$?
   *d. At 25 °C, chlorohydroxybiphenyl is a liquid with a density of 1.25 g/mL. How many moles of the compound are contained in a volume of 250 mL?

*44. Calculate the percent composition of each of the following:
   *a. MgB_6 c. C_6H_5Cl
   *b. $NaHSO_4$ d. ClO_4F

45. Calculate the percent composition of each of the following: $Ca(NO_3)_2$, NH_3, Fe_2O_3, MgC_2O_4.

*46. Give the mass ratio of calcium to boron in CaB_6.

47. Calculate the mass ratio of Li to H in LiH.

48. What is the mass ratio of nitrogen to oxygen in each of the following compounds: NO, NO_3, N_2O, N_2O_3, NO_2, N_2O_5?

49. Calculate the mass ratio of carbon to hydrogen for each of the following hydrocarbons: CH_4, C_5H_6, C_2H_6, C_4H_6, C_6H_6, $C_{12}H_8$.

*50. The ratio of the mass of K to that of Cl, (g K)/(g Cl), in potassium chloride is 1.10 to 1. What is the percent composition of potassium chloride? (Hint: take an imaginary sample weighing 2.10 g. In that sample, the weight of K is 1.10 and that of Cl is what?)

51. What is the percent composition of iron nitride if the mass ratio of iron to nitrogen is 19.93? If it is 7.97?

52. What is the percent composition of silver oxide if the mass ratio of silver to oxygen is 13.48? If it is 6.75?

*53. Find the gram-atomic weight of X, if the compound XN is 65.82% X and 34.18% N. What is X?

54. The compound ACl_2 is 25.53% A and 74.47% Cl. What is the gram-atomic weight of A? What is A?
55. X_3C is 93.20% X. Find the gram-atomic weight of X.
56. The compound PbZ_5 contains 42.77% Z. Calculate the gram-atomic weight of Z. What is Z?
57. NaAB is 57.5% Na, 40.00% A, and 2.52% B. Find the gram-atomic weights of A and B.*
58. A sample of the compound Ag_2XZ_4 weighs 16.59 g and contains 2.60 g of X and 19.29% Z. What are the atomic weights of X and Z?
59. Let X and Y represent two unknown elements that form a compound with oxygen. A 49.00-g sample of this compound contains 18.8 g of X and 9.3×10^{23} atoms of oxygen. If the formula of the compound is $X_3Y_2O_6$, what are the gram-atomic weights of X and Y? What elements are they?
60. A sample of the compound AB_2Cl_2 weighs 11.70 g. The sample is 48.24% A. To form the compound, A and B react with 1.30 L of Cl_2. Cl_2 has a density of 3.214 g/L at the reaction temperature. What are the gram-atomic weights of A and B?*

Section 6.7

61. Vinegar is essentially an aqueous solution of acetic acid, the empirical formula of which is CH_2O. If the estimated molecular weight of acetic acid is 60, what is its molecular formula?
*62. The simplest formula of mercurous iodide is HgI. What is its molecular formula if its estimated molecular weight is 650?
63. The empirical formula for a yellow dye used for wool is C_3H_2O. A mole of the dye is known to have 9.03×10^{24} atoms of C. What is the molecular formula of the dye?
64. A compound is known to contain C, H, N, and O. A sample of the compound weighing 27.4 g is 73.69% C, 7.98% H, 8.58% N, and the rest is O. Calculate the weights of each element present in the sample and find the empirical formula of the compound. The molecular weight of the compound is less than 400. What is the actual molecular weight of the compound?
*65. An 8.00-g sample of a hydrocarbon contained enough hydrogen and carbon to form 12.0 g of H_2O. What is the molecular formula of the hydrocarbon if its molecular weight is 144?

MORE DIFFICULT PROBLEMS

66. The mass ratio of silicon to fluorine in a particular compound is 0.370. What is the empirical formula of this compound?
*67. The mass ratio of carbon to hydrogen is 5.96 in a compound with a molecular weight of 112. What is the molecular formula of the compound?
*68. The empirical formula for octane (a component of gasoline) is C_4H_9. If 6.5 g of octane is burned, 2.50 g of CO_2 results. What is the molecular formula of octane?
69. During the American Revolution, the substance called saltpeter (used to make gunpowder) was often in short supply. The formula for saltpeter is ABO_3 and contains 38.67% A. The mass ratio of oxygen to B is 3.43. What are gram-atomic weights of A and B, and what is the chemical formula for saltpeter?*
70. Aspirin contains 60% carbon and twice as many hydrogen atoms as oxygen atoms. If these are the only elements present in aspirin, what is the empirical formula of aspirin? Find the molecular formula if 1 mol of aspirin molecules contains 4.82×10^{24} atoms of H.

---

*B is not boron in Problems 57, 60, and 69.

Chapter 7

Calculations Using Information From Reaction Equations

When we make calculations using chemical data, we are most likely trying to discover how much product can be made from a certain amount of reacting substance or vice versa. The key to solving a problem of this kind is the reaction equation, a quantitative statement in nature's language of the relation between products and reactants. To solve the problem, we simply translate the laboratory language of measured masses of substances into nature's language of individual reacting particles. The concept of the mole is used to make that conversion.

Chapter 7 is an explanation of the "mole ratio method" of making calculations from reaction equations. Such calculations are an easy and logical next step from the kind of calculations we practiced in the preceding chapter—those made from formulas.

AFTER STUDYING THIS CHAPTER, YOU WILL BE ABLE TO

- calculate the numbers of moles of products, given the numbers of moles of reactants and a reaction equation
- calculate the numbers of moles of reactants, given the numbers of moles of products and a reaction equation
- determine which reactant is the limiting reactant in a reaction
- calculate the amount of product formed in a limited reaction
- combine the mole ratio method with conversions to and from mass units
- combine density measurements with the mole ratio method to make reaction calculations

TERMS TO KNOW

limiting reactant (limiting reagent)
mole ratio
stoichiometric calculation

7.1 THE MOLE RATIO METHOD OF MAKING CALCULATIONS FROM REACTION EQUATIONS

In chemistry we often need to find the relation between the number of grams of a product of a chemical reaction and the number of grams of the reactant or reactants that produced it. Although problems of this kind can appear complex at first, all of them have one central step in common: the interpretation of a balanced reaction equation in terms of moles, rather than in terms of individual molecules. Calculations concerning the relation between amounts of reactants and products are called **stoichiometric calculations.** When you perform stoichiometric calculations (as well as all other chemical calculations) be sure to check your answers for significant figures, units, and reasonableness. Although the answers to the examples that follow have already been checked, we will occasionally go through the routine again as a reminder that we should look carefully at our answers.

Let us recall what a reaction equation says. Take the reaction by which ammonia, NH_3, is made into nitric oxide, NO, in a well-known process known as the Ostwald oxidation. This reaction is an essential part of the preparation of the industrially important compound nitric acid, HNO_3.

$$4\,NH_3 + 5\,O_2 \longrightarrow 4\,NO + 6\,H_2O$$

Interpreted in one way, the equation says that four molecules of NH_3 react with five molecules of O_2 (molecular oxygen) to produce four molecules of NO and six molecules of H_2O. Figure 7.1 shows the equation, using molecular models. The same reaction equation can, as we have seen, be read equally well for sets of dozens of molecules, as in Figure 7.2. And, it can also be read in terms of moles (Figure 7.3).

In this chapter, and in many of those that follow, we will interpret reaction equations in terms of moles. Let us look at the ammonia reaction just once more in moles; after this, we will imagine the mole label to be there when we need it.

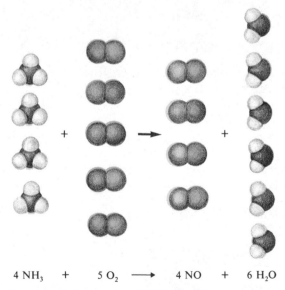

$$4\,NH_3 \quad + \quad 5\,O_2 \quad \longrightarrow \quad 4\,NO \quad + \quad 6\,H_2O$$

Figure 7.1   Representation of the reaction of individual molecules of ammonia and oxygen.

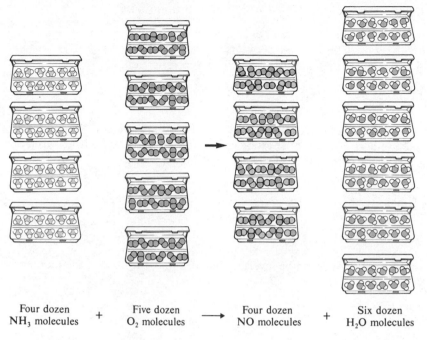

Four dozen   +   Five dozen   ⟶   Four dozen   +   Six dozen
NH₃ molecules   O₂ molecules   NO molecules   H₂O molecules

Figure 7.2 If molecules came in egg cartons, we could write the reaction equation in terms of dozens.

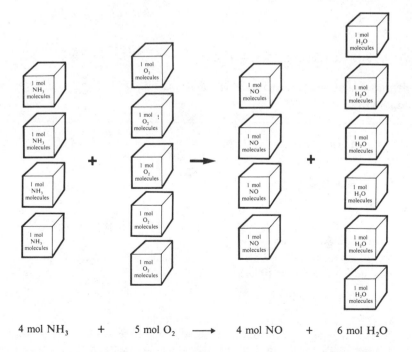

4 mol NH₃   +   5 mol O₂   ⟶   4 mol NO   +   6 mol H₂O

Figure 7.3 Measuring molecules in moles lets us write our equations in moles.

$$4 \text{ mol NH}_3 + 5 \text{ mol O}_2 \longrightarrow 4 \text{ mol NO} + 6 \text{ mol H}_2\text{O}$$

The information in this equation, or in any reaction equation, can be written in a series of mathematical ratios. The relation between the number of moles of NH_3 reacting and the number of moles of NO produced can be written in ratio form as

$$\frac{4 \text{ mol NH}_3}{4 \text{ mol NO}}$$

which says that there are 4 mol of NH_3 reacting for every 4 mol of NO produced. Any ratio between the numbers of moles of substances in a reaction is called a **mole ratio.** Several other mole ratios can be written for the Ostwald reaction:

$$\frac{4 \text{ mol NH}_3}{5 \text{ mol O}_2} \qquad \frac{4 \text{ mol NH}_3}{6 \text{ mol H}_2\text{O}} \qquad \frac{5 \text{ mol O}_2}{6 \text{ mol H}_2\text{O}}$$

$$\frac{6 \text{ mol H}_2\text{O}}{4 \text{ mol NO}} \qquad \frac{5 \text{ mol O}_2}{4 \text{ mol NO}}$$

Any of these six mole ratios can be inverted. For example, the ratio of the numbers of moles of O_2 and NO can be written as

$$\frac{4 \text{ mol NO}}{5 \text{ mol O}_2}$$

If we count the inverted forms, there are 12 possible ratios that can be written from this one reaction equation, and any of the 12 can be useful in a calculation. Depending on the number of reactants and products, different reaction equations can have more or fewer mole ratios. *A mole ratio from a reaction equation can be used as a unit conversion factor in calculations.*

7.2 MOLE RATIOS AS CONVERSION FACTORS

To learn how mole ratios from reaction equations can be used in calculations, we can look at a simple reaction.

EXAMPLE 7.1.   How many moles of N_2O_4 can be prepared from 3.27 mol of NO_2? The equation for the preparation is

$$2 \text{ NO}_2 \longrightarrow N_2O_4$$

Solution:

Step 1.   Be sure that you have a balanced reaction equation for the process involved. The equation is balanced, so we go on.

Step 2.   Begin with the data given and set up your solution map, using the units as a guide. In this case the solution map is simple:

$$\text{moles NO}_2 \longrightarrow \text{moles N}_2\text{O}_4$$

Step 3.   Write down the appropriate mole ratio to use in the calculation. If you don't know at first which is appropriate, write down all the ratios and select the correct one when you set up the calculation. Two mole ratios can be written for this reaction:

$$\frac{2 \text{ mol NO}_2}{1 \text{ mol N}_2\text{O}_4} \qquad \frac{1 \text{ mol N}_2\text{O}_4}{2 \text{ mol NO}_2}$$

The setup is

$$3.72 \text{ mol NO}_2 \times \frac{1 \text{ mol N}_2\text{O}_4}{2 \text{ mol NO}_2} = 1.86 \text{ mol N}_2\text{O}_4$$

The answer is in the correct units and is reasonable. If we had seen 186 mol of N_2O_4 from only about 3 mol of NO_2, for example, we would have known something was wrong. The answer has three significant figures, which is the the proper number. (Since there are three significant figures in the value of the number of moles of NO_2 given, and the coefficients in the reaction equation can be taken as exact, there should be three significant figures in the answer.)

EXERCISE 7.1. The water-gas reaction, the equation for which follows, is an important industrial process. In this reaction, how many moles of H_2 can be formed from 1.60 mol of CO?

$$H_2O + CO \longrightarrow H_2 + CO_2$$

EXERCISE 7.2. In the following reaction, how many moles of H_2 can be formed from 5.45 mol of NaOH?

$$2 \text{ NaOH} + \text{Zn} \longrightarrow \text{Na}_2\text{ZnO}_2 + \text{H}_2$$

EXAMPLE 7.2. How many moles of O_2 are required to prepare 9.11 mol of NO in the Ostwald oxidation? (Use the balanced reaction equation given in Section 7.1.)

Solution:

Step 1. From the mole ratios given by the reaction equation, we select the one that relates moles of O_2 and moles of NO, and is written so that the number of moles of O_2 appears in the numerator. Using this ratio will given the answer in the correct units.

$$\frac{5 \text{ mol O}_2}{4 \text{ mol NO}}$$

The solution map is

moles of NO $\longrightarrow$ moles of O_2

Step 2. Set up and solve:

$$9.11 \text{ mol NO} \times \frac{5 \text{ mol O}_2}{4 \text{ mol NO}} = 13.7 \text{ mol O}_2$$

EXERCISE 7.3. How many moles of NaO_2 are required to produce 8.72 mol of NaCl in the following reaction?

$$2 \text{ HCl} + \text{Na}_2\text{O} \longrightarrow 2 \text{ NaCl} + \text{H}_2\text{O}$$

EXERCISE 7.4.   In the reaction shown in Exercise 7.3, how many moles of HCl would be required to form 3.40 mol of H_2O?

7.3 LIMITING REACTANT PROBLEMS

Several kinds of problems concern the relation between amounts of reactants and products. One such problem requires that we find out in advance whether any of the reactants will be used up, thereby stopping the reaction. Once this has been determined, a mole ratio calculation is based on the reactant that is completely consumed, which is called the **limiting reagent** or **limiting reactant.** Once you understand them, you will not find limiting reactant problems difficult.

Before continuing, we should refresh and enlarge our knowledge of one of the basic facts of chemical reaction. Recall the law of definite proportions: a compound contains its elements in fixed proportions. If we look at it in terms of the numbers of moles of the elements in a compound, the law is another way of saying that compounds have definite formulas. The formula NH_3, for instance, tells us that the compound contains 1 mol of N atoms for every 3 mol of H atoms. Therefore, when N_2 and H_2 react to form NH_3,

$$N_2 + 3 H_2 \longrightarrow 2 NH_3$$

they do so in exactly the mole proportions given in the balanced reaction equation—nothing new so far. The important consequence of all this, however, is that *even if N_2 and H_2 are mixed in different proportions than those required by the reaction, they will react in the correct proportions,* and will continue to do so until one of them is used up. Suppose we mix 3 mol of N_2 and 3 mol of H_2 and cause them to react as much as possible. The reaction will proceed in the proportions of 1 mol of N_2 to 3 mol of H_2, until the H_2 is all used up. At this point, there will still be 2 mol of N_2 left.

EXAMPLE 7.3.   How many moles of N_2 are required to react completely with 2.33 mol of H_2, according to the reaction shown?

$$N_2 + 3 H_2 \longrightarrow 2 NH_3$$

Solution:

Step 1.   From the mole ratios given by the reaction equation, we select the one that relates the number of moles of N_2 to the number of moles of H_2.

$$\frac{1 \text{ mol } N_2}{3 \text{ mol } H_2}$$

The solution map is

moles of $H_2 \longrightarrow$ moles of N_2

Step 2.   Set up and solve:

$$2.33 \text{ mol } H_2 \times \frac{1 \text{ mol } N_2}{3 \text{ mol } H_2} = 0.777 \text{ mol } N_2$$

EXERCISE 7.5. How many moles of C_2H_6 are required to react completely with 3.42 mol of O_2 according to the following reaction?

$$2 \, C_2H_6 + 7 \, O_2 \longrightarrow 4 \, CO_2 + 6 \, H_2O$$

Every reaction proceeds in its own proper proportions regardless of the relative amounts of the reactants present, even though one or more of the reactants may be present in excess amounts. Because the reactant that runs out causes the reaction to stop, it is called the limiting reactant.

EXAMPLE 7.4. How many moles of H_2O are formed in the Ostwald oxidation reaction when 4 mol of NH_3 and 4 mol of O_2 are mixed and caused to react until one of the reactants is used up?

$$4 \, NH_3 + 5 \, O_2 \longrightarrow 4 \, NO + 6 \, H_2O$$

Solution:

Step 1. Determine which reactant will run out first (the limiting reactant) by calculating how many moles of H_2O would be formed by the amount present of *each* of the reactants. The reactant that forms the smallest number of moles of H_2O is clearly the one that will run out first, and is therefore the limiting reactant.

$$4.0 \text{ mol NH}_3 \times \frac{6 \text{ mol H}_2\text{O}}{4 \text{ mol NH}_3} = 6.0 \text{ mol H}_2\text{O}$$

$$4.0 \text{ mol O}_2 \times \frac{6 \text{ mol H}_2\text{O}}{5 \text{ mol O}_2} = 4.8 \text{ mol H}_2\text{O}$$

In this case, we see that the 4 mol of NH_3 would produce more H_2O than 4 mol of O_2 would. There is enough NH_3 to make 6 mol of H_2O, but enough O_2 to make only 4.8 mol of H_2O. When 4.8 mol of H_2O has been formed, the reaction will cease, and some NH_3 will be left over. By comparing the number of moles of product that can be made by each of the reactants, we have found that O_2 is the limiting reactant.

Step 2. Base the answer or any further calculations on the limiting reactant. In this case, our answer must be based on the amount of O_2 present and not on the amount of NH_3.

Our answer is 4.8 mol of H_2O.

EXAMPLE 7.5. Given the reaction between sulfuric acid, H_2SO_4, and aluminum, Al,

$$3 \, H_2SO_4 + 2 \, Al \longrightarrow Al_2(SO_4)_3 + 3 \, H_2$$

how many moles of H_2 will be produced if 1.7 mol of sulfuric acid is mixed with 2.2 mol of aluminum and allowed to react? Which is the limiting reactant?

Solution:

Step 1. Calculate separately how many moles of H_2 will be produced by 1.7 mol of H_2SO_4 and by 2.2 mol of Al.

$$1.7 \text{ mol } H_2SO_4 \times \frac{3 \text{ mol } H_2}{3 \text{ mol } H_2SO_4} = 1.7 \text{ mol } H_2$$

$$2.2 \text{ mol Al} \times \frac{3 \text{ mol } H_2}{2 \text{ mol Al}} = 3.3 \text{ mol } H_2$$

Step 2. Choose as your answer the smaller number of moles, that is, the number made by the limiting reactant: 1.7 mol of H_2.

The reaction will cease when 1.7 mol of H_2 has been made and the H_2SO_4 is gone. The H_2SO_4 is limiting, and the number of moles of H_2 formed is 1.7. Al will be left over.

EXAMPLE 7.6. The NO that comes from the Ostwald process is reacted with oxygen to give NO_2; then the NO_2 is reacted with water to give the desired nitric acid. The balanced equation for the second reaction is

$$3 \text{ NO}_2 + H_2O \longrightarrow 2 \text{ HNO}_3 + \text{NO}$$

How many moles of HNO_3 can be made from 2.2 mol of NO_2 and 0.56 mol of H_2O if these react together until one is used up? Which is the limiting reactant?

Solution:

Step 1. Calculate answers for the amounts of both reactants.

$$2.2 \text{ mol } NO_2 \times \frac{2 \text{ mol } HNO_3}{3 \text{ mol } NO_2} = 1.5 \text{ mol } HNO_3$$

$$0.56 \text{ mol } H_2O \times \frac{2 \text{ mol } HNO_3}{1 \text{ mol } H_2O} = 1.1 \text{ mol } HNO_3$$

Step 2. Choose the answer that is smaller.

The answer is 1.1 mol HNO_3; H_2O is limiting.

EXERCISE 7.6. In the reaction that follows, Ag reacts with HNO_3 to form $AgNO_3$. If 2.5 mol of Ag is mixed with 2.0 mol of HNO_3, how many moles of $AgNO_3$ can be produced?

$$3 \text{ Ag} + 4 \text{ HNO}_3 \longrightarrow 3 \text{ AgNO}_3 + \text{NO} + 2 \text{ H}_2O$$

EXERCISE 7.7. If the reaction of Exercise 7.6 begins with 0.245 mol of Ag and 0.320 mol of HNO_3, how many moles of NO can be produced?

EXAMPLE 7.7. The following reaction was run in the laboratory, starting with 3.3 mol of HNO_3, 2.9 mol of As, and 1.5 mol of H_2O. How many moles of H_3AsO_4 were produced? What reactants were left over?

$$5 \text{ HNO}_3 + 3 \text{ As} + 2 \text{ H}_2O \longrightarrow 5 \text{ NO} + 3 \text{ H}_3AsO_4$$

Solution:

This problem is solved in exactly the same way as the preceding ones except that three reagents are involved instead of two.

$$3.3 \text{ mol HNO}_3 \times \frac{3 \text{ mol H}_3\text{AsO}_4}{5 \text{ mol HNO}_3} = 2.0 \text{ mol H}_3\text{AsO}_4$$

$$2.9 \text{ mol As} \times \frac{3 \text{ mol H}_3\text{AsO}_4}{3 \text{ mol As}} = 2.9 \text{ mol H}_3\text{AsO}_4$$

$$1.5 \text{ mol H}_2\text{O} \times \frac{3 \text{ mol H}_3\text{AsO}_4}{2 \text{ mol H}_2\text{O}} = 2.2 \text{ mol H}_3\text{AsO}_4$$

HNO_3 is limiting; 2.0 mol of H_3AsO_4 is produced; As and H_2O are left over.

EXERCISE 7.8.   Ethanol, C_2H_5OH, can be produced from graphite, C, water, and hydrogen, according to the following reaction:

$$2 \text{ C} + 2 \text{ H}_2 + \text{H}_2\text{O} \longrightarrow \text{C}_2\text{H}_5\text{OH}$$

How many moles of ethanol can be produced from 3.38 mol of C, 3.47 mol of H_2, and 1.79 mol of H_2O?

EXERCISE 7.9.   The compound $Cr_2(SO_4)_3$ is to be produced by the following reaction:

$$\text{K}_2\text{Cr}_2\text{O}_7 + 3 \text{ H}_2\text{S} + 4 \text{ H}_2\text{SO}_4 \longrightarrow \text{K}_2\text{SO}_4 + \text{Cr}_2(\text{SO}_4)_3 + 3 \text{ S} + 7 \text{ H}_2\text{O}$$

We are given 0.72 mol of $K_2Cr_2O_7$, 2.0 mol of H_2S, and 3.0 mol of H_2SO_4 to run the reaction. How many moles of $Cr_2(SO_4)_3$ can be produced from these starting materials? Which reactants will be left over?

7.4 MOLE RATIO PROBLEMS INVOLVING MASS UNITS

Although you will occasionally encounter the kind of problem that we saw in the preceding sections, such problems usually require further steps: the laboratory mass measurements must be converted into moles before the mole ratio calculation is made, then the result of the calculation must be converted back into mass units. We learned how to make conversions of this kind in Chapter 6.

EXAMPLE 7.8.   The reaction of methane, CH_4, and oxygen is used in industry as well as for heating and cooking in homes. The products of the reaction are water and carbon dioxide, CO_2.

$$\text{CH}_4 + \text{O}_2 \longrightarrow \text{CO}_2 + \text{H}_2\text{O}$$

How many moles of water can be made from 6.8 g of CH_4 and excess O_2?

Solution:

Step 1.   First, as always, be sure that the reaction equation is balanced. Since this one is not, we must balance it.

$$CH_4 + 2\,O_2 \longrightarrow CO_2 + 2\,H_2O$$

We have been told that O_2 is present in excess, and that there is only a given amount of CH_4. We base our calculation on the amount of CH_4.

Step 2. Calculate the number of moles of CH_4 present.

$$6.8 \text{ g } CH_4 \times \frac{1 \text{ mol } CH_4}{16 \text{ g } CH_4} = 0.425 \text{ mol } CH_4$$

Step 3. Do the mole ratio calculation using the proper mole ratio obtained from the reaction equation.

$$0.425 \text{ mol } CH_4 \times \frac{2 \text{ mol } H_2O}{1 \text{ mol } CH_4} = 0.85 \text{ mol } H_2O$$

After you have had some practice, you may want to combine Steps 2 and 3 in a single calculation.

$$6.8 \text{ g } CH_4 \times \frac{1 \text{ mol } CH_4}{16 \text{ g } CH_4} \times \frac{2 \text{ mol } H_2O}{1 \text{ mol } CH_4} = 0.85 \text{ mol } H_2O$$

EXERCISE 7.10.   How many moles of O_2 can be formed from 16.4 g of $KClO_3$?

$$KClO_3 \longrightarrow KCl + O_2$$

EXERCISE 7.11.   If 24.3 g of HgO reacted to give Hg and O_2, how many moles of HG resulted?

$$2\,HgO \longrightarrow 2\,Hg + O_2$$

Another similar kind of problem requires that you calculate the number of grams of a reactant necessary to make a required amount of a product.

EXAMPLE 7.9.   You wish to prepare 6.6 mol of H_2O by burning CH_4 in O_2. If enough CH_4 is available to run the reaction, how many grams of O_2 will be needed to prepare the water? Write a balanced reaction equation. (It is the same as that used in Example 7.8.)

Solution:

Step 1.   Using the mole ratio method, find out how many moles of O_2 will be needed. (Because we are told that there is as much CH_4 as can be used, we know that O_2 limits the reaction.)

$$6.6 \text{ mol } H_2O \times \frac{2 \text{ mol } O_2}{2 \text{ mol } H_2O} = 6.6 \text{ mol } O_2$$

Step 2.   Convert the number of moles of O_2 to grams of O_2.

$$6.6 \text{ mol } O_2 \times \frac{32 \text{ g } O_2}{1 \text{ mol } O_2} = 211 \text{ g } O_2 = 2.1 \times 10^2 \text{ g } O_2$$

EXERCISE 7.12.   How many grams of Ag are required to produce 10.3 mol of $AgNO_3$? (Write a balanced reaction equation. Check it against that in Exercise 7.6.)

EXERCISE 7.13.   How many grams of HgO are needed to produce 8.50 mol of pure mercury, Hg? (Write the reaction equation. See Exercise 7.11 to check it.)

Of the problems involving reaction equations, the most common are those in which the data are given in laboratory units (mass units or others), and the answer is also required in laboratory units. Chemistry is, after all, a laboratory science, and problems are derived from measurements in the laboratory. The general method of working such problems is simply to convert the laboratory units into moles, make the mole ratio conversion, then convert back into laboratory units.

EXAMPLE 7.10.   Metallic calcium, Ca, like many other metals, can be burned in oxygen to yield its oxide.

$$2 \text{ Ca} + O_2 \longrightarrow 2 \text{ CaO}$$

How many grams of CaO can be made from 31.02 g of Ca and as much O_2 as is needed for the reaction?

Solution:

Step 1.   Use the gram-atomic weight to find the number of moles of Ca metal that we have.

$$31.02 \text{ g Ca} \times \frac{1 \text{ mol Ca}}{40.08 \text{ g Ca}} = 0.7740 \text{ mol Ca}$$

Step 2.   Use the correct mole ratio from the balanced equation to find the number of moles of CaO produced.

$$0.7740 \text{ mol Ca} \times \frac{2 \text{ mol CaO}}{2 \text{ mol Ca}} = 0.7740 \text{ mol CaO}$$

Step 3.   Convert the number of moles of CaO into grams. Use the gram-molecular weight of CaO, the numerical value of which is 40.08 + 16.00 = 56.08 g/mol.

$$0.7740 \text{ mol CaO} \times \frac{56.08 \text{ g CaO}}{\text{mol CaO}} = 43.40 \text{ g CaO}$$

You will save time and be less likely to make errors if you do problems like the preceding one in a single calculation that combines the three steps.

EXAMPLE 7.11.   In the human body, glucose, $C_6H_{12}O_6$, reacts with O_2 and releases the energy needed for living. The balanced equation for the overall reaction is

$$C_6H_{12}O_6 + 6 O_2 \longrightarrow 6 CO_2 + 6 H_2O$$

How many grams of H_2O result from the reaction of 19.3 g of glucose with excess oxygen?

Solution map: grams of glucose $\longrightarrow$ moles of glucose $\longrightarrow$ moles of H_2O $\longrightarrow$ grams of H_2O

Solution:

We solve the problem in one calculation that combines the three steps of Example 7.10. Large brackets outline the three distinct steps.

$$19.3 \text{ g glucose} \times \frac{1 \text{ mol glucose}}{180 \text{ g glucose}} \times \frac{6 \text{ mol } H_2O}{1 \text{ mol glucose}} \times \frac{18.02 \text{ g } H_2O}{1 \text{ mol } H_2O} = 11.6 \text{ g } H_2O$$

EXERCISE 7.14.   How many grams of Na_2ZnO_2 can be produced if 76.8 g of NaOH react with an excess of zinc? (The reaction equation, if you want to check it, appears in Exercise 7.2.)

EXERCISE 7.15.   How many grams of $AlBr_3$ will be produced if 18.6 g of Al react with an excess of Br_2?

$$Al + Br_2 \longrightarrow AlBr_3$$

When you see a problem in chemistry, especially one that looks something like those we have just seen, stop and ask yourself if the solution depends in some way on the information in a chemical formula or a reaction equation. If it does (and it often will), ask yourself how you can convert the given data into mole language, so that you can use the formula or the equation. You can ordinarily use the steps you have just learned to convert laboratory data to moles; next, use the mole ratio method; finally, convert back to lab data. Easy.

7.5 LIMITING REACTANT PROBLEMS USING MASS UNITS

Just as ordinary mole ratio problems often begin with laboratory units, so do problems involving a limiting reactant. Such problems require one more step than do ordinary mass-mass problems, but they are not harder to understand.

A limiting reactant problem given in mass units can be worked in three steps, all of which we have seen before:

Step 1.   Convert the values given in mass units to numbers of moles.

Step 2.   Using the methods of Section 7.3, find out which of the reactants limits the reaction, that is, runs out first. By doing this step, you automatically find out how many moles of product are produced.

Step 3.   When you have the number of moles of the product that are made, convert that value back into mass units.

EXAMPLE 7.12.   In an experiment, 4.0 g of Na metal was placed in 7.7 g of water. If the reaction ran according to the balanced equation that follows, how many grams of NaOH were produced?

$$2 \text{ Na} + 2 \text{ H}_2O \longrightarrow H_2 + 2 \text{ NaOH}$$

Solution:

Step 1. The number of moles of Na and of H_2O present are

$$4.0 \text{ g Na} \times \frac{1 \text{ mol Na}}{23 \text{ g Na}} = 0.17 \text{ mol Na}$$

$$7.7 \text{ g } H_2O \times \frac{1 \text{ mol } H_2O}{18.02 \text{ g } H_2} = 0.43 \text{ mol } H_2O$$

Step 2. Find the number of moles of product produced by 0.17 mol of Na and the number produced by 0.43 mol of H_2O.

$$0.17 \text{ mol Na} \times \frac{2 \text{ mol NaOH}}{2 \text{ mol Na}} = 0.17 \text{ mol NaOH}$$

$$0.43 \text{ mol } H_2O \times \frac{2 \text{ mol NaOH}}{2 \text{ mol } H_2O} = 0.43 \text{ mol NaOH}$$

From this we find that Na is the limiting reactant and that 0.17 mol of NaOH are produced in the reaction.

Step 3. Convert the number of moles of NaOH into grams, using the gram-molecular weight of NaOH.

$$0.17 \text{ mol NaOH} \times \frac{40.0 \text{ g NaOH}}{\text{mol NaOH}} = 6.8 \text{ g NaOH}$$

EXAMPLE 7.13.   Sulfuric acid reacts with ammonia according to the following balanced reaction equation:

$$2 \text{ } NH_3 + H_2SO_4 \longrightarrow (NH_4)_2SO_4$$

For this reaction, 22.4 g of sulfuric acid was added to 22.4 g of ammonia. How many grams of $(NH_4)_2SO_4$ were produced?

Solution:

Step 1. Convert to numbers of moles of the reactants.

$$22.4 \text{ g } H_2SO_4 \times \frac{1 \text{ mol } H_2SO_4}{98.0 \text{ g } H_2SO_4} = 0.229 \text{ mol } H_2SO_4$$

$$22.4 \text{ g } NH_3 \times \frac{1 \text{ mol } NH_3}{17.0 \text{ g } NH_3} = 1.32 \text{ mol } NH_3$$

Step 2. Find the limiting reactant and the number of moles of product.

$$0.229 \text{ mol } H_2SO_4 \times \frac{1 \text{ mol } (NH_4)_2SO_4}{1 \text{ mol } H_2SO_4} = 0.229 \text{ mol } (NH_4)_2SO_4$$

$$1.32 \text{ mol } NH_3 \times \frac{1 \text{ mol } (NH_4)_2SO_4}{2 \text{ mol } NH_3} = 0.660 \text{ mol } (NH_4)_2SO_4$$

H_2SO_4 is limiting.

Step 3. Calculate the number of grams of $(NH_4)_2SO_4$ formed.

$$0.229 \text{ mol } (NH_4)_2SO_4 \times \frac{132.0 \text{ g } (NH_4)_2SO_4}{\text{mol } (NH_4)_2SO_4} = 30.2 \text{ g } (NH_4)_2SO_4$$

EXERCISE 7.16. Using the reaction equation in Exercise 7.15, calculate how many grams of $AlBr_3$ would result from the reaction of 22.6 g of Al and 38.5 g of Br_2.

EXERCISE 7.17. How many grams of $Ca_3(PO_4)_2$ will be produced if 76.5 g of $Ca(OH)_2$ react with 100.0 g of H_3PO_4?

$$Ca(OH)_2 + H_3PO_4 \longrightarrow Ca_3(PO_4)_2 + H_2O$$

7.6 STOICHIOMETRIC PROBLEMS INVOLVING DENSITY DATA

Laboratory data based on the reactions of liquids or gases are often expressed in units of volume. Measuring the volume of a liquid is usually easier than measuring its mass. The relation between the volume of a sample and its mass is, of course, its density. (As you recall, density is defined as the mass of a sample divided by its volume.) Stoichiometric problems in which some of the data are given in volumes are no more difficult than the problems we have just finished; we simply need to use another conversion step. Here are the steps we will follow when we solve problems in which the reactant data are given in units of volume.

Step 1. Convert reactant volume data into numbers of grams. (If you have to find the limiting reactant, you will need to do this step and the next two steps for more than one reactant.)

Step 2. Convert the gram values into numbers of moles.

Step 3. Use the mole ratio calculation to find out how many moles of product is formed.

Step 4. Convert the numbers of moles of product into other units as necessary.

EXAMPLE 7.14. The density of pure H_2SO_4 is 1.85 g/mL. To an excess of Na_2CO_3 is added 43.2 mL of pure H_2SO_4, and the reaction runs until all the H_2SO_4 is used up. The balanced reaction equation is

$$H_2SO_4 + Na_2CO_3 \longrightarrow Na_2SO_4 + CO_2 + H_2O$$

How many grams of CO_2 are formed in this process?

Solution:

Step 1. We are told that Na_2CO_3 is present in excess, so we will base our calculation on H_2SO_4. We calculate the number of grams of H_2SO_4 present.

$$43.2 \text{ mL } H_2SO_4 \times \frac{1.85 \text{ g } H_2SO_4}{1 \text{ mL } H_2SO_4} = 79.9 \text{ g } H_2SO_4$$

Step 2. We find the number of moles of H_2SO_4.

$$79.9 \text{ g H}_2\text{SO}_4 \times \frac{1 \text{ mol H}_2\text{SO}_4}{98.0 \text{ g H}_2\text{SO}_4} = 0.816 \text{ mol H}_2\text{SO}_4$$

Step 3.   We perform the mole ratio calculation for the product.

$$0.816 \text{ mol H}_2\text{SO}_4 \times \frac{1 \text{ mol CO}_2}{1 \text{ mol H}_2\text{SO}_4} = 0.816 \text{ mol CO}_2$$

Step 4.   We can now convert the number of moles of CO_2 into grams.

$$0.816 \text{ mol CO}_2 \times \frac{44.0 \text{ g CO}_2}{\text{mol CO}_2} = 35.9 \text{ g CO}_2$$

EXAMPLE 7.15.   Octane, C_8H_{18}, is a liquid that burns in air. The density of octane is 0.702 g/mL. How many grams of CO_2 are produced if 1.00 L of octane is burned according to the following balanced reaction equation? Assume that there is sufficient O_2 to use all the octane.

$$2 \text{ C}_8\text{H}_{18} + 25 \text{ O}_2 \longrightarrow 16 \text{ CO}_2 + 18 \text{ H}_2\text{O}$$

Solution:

Step 1.   Find the number of grams of octane.

$$1.00 \text{ L octane} \times \frac{1000 \text{ mL}}{1 \text{ L}} \times \frac{0.702 \text{ g octane}}{\text{mL octane}} = 702 \text{ g octane}$$

Step 2.   Find the number of moles of octane.

$$702 \text{ g octane} \times \frac{1 \text{ mol octane}}{114 \text{ g octane}} = 6.16 \text{ mol octane}$$

Step 3.   Find the number of moles of CO_2.

$$6.16 \text{ mol octane} \times \frac{16 \text{ mol CO}_2}{2 \text{ mol octane}} = 49.3 \text{ mol CO}_2$$

Step 4.   Find the number of grams of CO_2.

$$49.3 \text{ mol CO}_2 \times \frac{44.0 \text{ g CO}_2}{\text{mol CO}_2} = 2169 \text{ g CO}_2 = 2.17 \times 10^3 \text{ g CO}_2$$

Let us now do a problem in which we make all the conversions in a single unit equation.

EXAMPLE 7.16.   Acetic acid is what makes vinegar sour; its formula is H_3CCO_2H. As a pure compound, acetic acid has a density of 1.05 g/mL. It can be made to react with ethyl alcohol to produce a sweet-smelling compound, ethyl acetate, which is used as fingernail polish remover, among other things.

$$\text{H}_3\text{CCO}_2\text{H} + \quad \text{C}_2\text{H}_6\text{O} \longrightarrow \quad \text{C}_4\text{H}_8\text{O}_2 \quad + \text{H}_2\text{O}$$
Acetic acid   Ethyl alcohol   Ethyl acetate

A sample of acetic acid with a volume of 0.733 mL is reacted with excess ethyl alcohol. How many grams of ethyl acetate are produced?

Solution:

$$0.733 \text{ mL H}_3\text{CCO}_2\text{H} \times \frac{1.05 \text{ g H}_3\text{CCO}_2\text{H}}{\text{mL H}_3\text{CCO}_2\text{H}} \times \frac{1 \text{ mol H}_3\text{CCO}_2\text{H}}{60.0 \text{ g H}_3\text{CCO}_2\text{H}}$$

$$\times \frac{1 \text{ mol C}_4\text{H}_8\text{O}_2}{1 \text{ mol H}_3\text{CCO}_2\text{H}} \times \frac{88.1 \text{ g C}_4\text{H}_8\text{O}_2}{1 \text{ mol C}_4\text{H}_8\text{O}_2} = 1.13 \text{ g C}_4\text{H}_8\text{O}_2$$

EXERCISE 7.18.   In the reaction in Exercises 7.15 and 7.16, excess Al was mixed with 12.8 mL of Br_2. How many grams of AlBr_3 were formed? The density of Br_2 is 3.12 g/mL.

EXERCISE 7.19.   A sample of hydrogen chloride has a density of 1.097 g/L. If 43.2 g of Fe_2O_3 was mixed with 2.5 L of HCl, how many grams of FeCl_3 would be obtained?

$$\text{Fe}_2\text{O}_3 + 6 \text{ HCl} \longrightarrow 2 \text{ FeCl}_3 + 3 \text{ H}_2\text{O}$$

7.7 CALCULATING THE ENERGIES OF REACTIONS

In section 5.6, we learned that the energy emitted in an exothermic reaction or absorbed in an endothermic reaction can be shown quantitatively by giving the enthalpy change for the reaction.

Enthalpy changes are stated for the amounts of reactants and products shown in the balanced reaction equation. If the reaction equation is written for different amounts, the amount of enthalpy change will be correspondingly different: the amount of enthalpy change is proportional to the amounts of substances reacting to give products. For example, the energy produced by the combustion of a mole of natural gas, CH_4, is 890 kJ.

$$\text{CH}_4 \text{ (g)} + 2 \text{ O}_2 \text{ (g)} \longrightarrow \text{CO}_2 \text{ (g)} + 2 \text{ H}_2\text{O (g)} \qquad \triangle\text{H} = -890 \text{ kJ}$$

Twice as much energy is produced by the burning of 2 mol of CH_4.

$$2 \text{ CH}_4 \text{ (g)} + 4 \text{ O}_2 \text{ (g)} \longrightarrow 2 \text{ CO}_2 \text{ (g)} + 4 \text{ H}_2\text{O (g)} \qquad \triangle\text{H} = -1780 \text{ kJ}$$

Because the energy of reaction is proportional to the size of the reaction system, the enthalpy of reaction can be used in a conversion factor. For example, for the combustion of a mole of natural gas, we can write four conversion factors, each of which can be inverted, if necessary.

$$\frac{-890 \text{ kJ}}{\text{mol CH}_4} \qquad \frac{-890 \text{ kJ}}{2 \text{ mol O}_2} \qquad \frac{-890 \text{ kJ}}{\text{mol CO}_2} \qquad \frac{-890 \text{ kJ}}{2 \text{ mol H}_2\text{O}}$$

When you use such conversion factors, remember that they pertain to the reaction only as written.

To make calculations, we can use conversion factors based on enthalpy change just as we use mole ratios.

EXAMPLE 7.17. How much energy is produced by the combustion in O_2 of 0.721 mol of CO?

$$2\ CO\ (g) + O_2\ (g) \longrightarrow 2\ CO_2\ (g) \qquad \triangle H = -566\ kJ$$

Solution:

Set up the calculation and solve.

$$\frac{-566\ kJ}{2\ mol\ CO} \times 0.721\ mol\ CO = -204\ kJ$$

Finding that the enthalpy change is −204 kJ for the combustion of 0.721 mol of CO, we report that the amount of energy produced is 204 kJ. Because of the way our answer is worded, we do not use the negative sign with the value of the energy. If we write an answer simply as an *amount* of energy produced or absorbed by a reaction (instead of using enthalpy terms), we always use a positive value and then say whether that energy was produced or absorbed. The use of the negative sign is ordinarily confined to reports of enthalpy decrease when the symbol $\triangle H$ is used.

Was there a limiting reactant in this problem? Notice that the calculation was based on the amount of CO, not of O_2. Because the amount of CO was given and that of O_2 was not, we must assume that there was enough O_2 to burn all the CO.

EXAMPLE 7.18. "Slaked lime," $Ca(OH)_2$, is an important industrial compound. The first step in making $Ca(OH)_2$ is the roasting of limestone, $CaCO_3$, an endothermic reaction.

$$CaCO_3\ (s) \longrightarrow CaO\ (s) + CO_2\ (g) \qquad \triangle H = 176\ kJ$$

How much energy is required to decompose 922 g of $CaCO_3$ by this method?

Solution map: grams of $CaCO_3$ $\longrightarrow$ moles of $CaCO_3$ $\longrightarrow$ energy

Solution:

Set up and solve.

$$922\ g\ CaCO_3 \times \frac{1\ mol\ CaCO_3}{100.09\ g\ CaCO_3} \times \frac{176\ kJ}{mol\ CaCO_3} = 1621\ kJ = 1.62 \times 10^3\ kJ$$

(Because the enthalpy change is positive, the energy is required, not produced. Think about this until you are sure you understand it.)

EXERCISE 7.20. The second step in the preparation of lime is the reaction of "quicklime," CaO, with H_2O, which produces "slaked lime."

$$CaO\ (s) + H_2O\ (l) \longrightarrow Ca(OH)_2 \qquad \triangle H = -65.2\ kJ$$

When 2210 g of CaO is reacted in this way, how much energy is produced?

7.8 SUMMARY

The equations for chemical reactions tell us the numbers of atoms and molecules of reactants and products in those reactions. Reaction equations can also be read in terms of moles, and it is in this language that they are most useful to us. From a balanced reaction equation, it is easy to read the ratios of the numbers of moles of the substances involved. We can then use a mole ratio to calculate the amount of a substance produced or consumed in reaction with a known amount of another substance.

Reactions often stop because one reactant runs out before the others. The reactant used up first is called the limiting reactant, or limiting reagent. Unless we know in advance which reactant is limiting (that is, know that the other reactants are present in excess), we need to make mole ratio calculations for all the reactants present before we can determine how far a reaction will proceed. Once we know the limiting reactant, we can base an ordinary mole ratio calculation on the quantity of that limiting reactant.

If quantities of reacting substances are given in mass units, those quantities must be converted to numbers of moles before we can make mole ratio calculations. We use the gram-atomic weight or gram-molecular weight to make the conversion to moles. It is often necessary to convert the result back into mass units after we have made the mole ratio calculation. If the quantities of the substances are initially given in terms of volume, and if their densities are known, the first conversion is into mass units, followed by a conversion to moles and a mole ratio calculation. The same sequence can be followed in reverse if answers are required in units of volume.

Because the enthalpy change in a chemical reaction is proportional to the amounts of reactants and products that participate in the reaction, enthalpy changes for any amount of reactant or product can be calculated if $\triangle H$ for the reaction is known: simply use the enthalpy change in a conversion factor.

QUESTIONS

Section 7.1

1. Describe what is meant by a stoichiometric calculation.
2. Why do chemists often interpret reaction equation coefficients in terms of numbers of moles rather than numbers of molecules?
3. List all possible mole ratios that can be written for the reaction equation in Problem 40.
4. Why must you have a balanced reaction equation before you can use the mole ratio method?

Section 7.3

5. Define the term *limiting reactant*. At what point in the solution map does the limitation imposed by the limiting reactant appear?
6. If we run a reaction of X and Z starting with 6 mol of X and 4 mol of Z, is it possible for X to be the limiting reactant? Explain.

Section 7.4

7. Write solution maps for Problems 27 and 30.

Section 7.5

8. Outline the procedure for determining which reactant is limiting when you begin with laboratory mass data.
9. Outline the method for determining which reactant is limiting if the laboratory data are given as the volumes and densities of liquids.
10. Write a solution map for Problem 43.

PROBLEMS

Note: Many chemical reactions run only part way to completion, but for the purposes of the problems that follow, assume that reaction proceeds until at least one reactant is entirely used up.

Section 7.2

1. How many moles of I_2 can be produced from 2.31×10^{-2} mol of HI?

$$2\ HI \longrightarrow H_2 + I_2$$

*2. How many moles of product can be formed from 2.80 mol of the *first* reactant in each of the following reactions?
  *a. $4\ Al + 3\ O_2 \longrightarrow 2\ Al_2O_3$ c. $2\ NH_3 + H_2SO_4 \longrightarrow (NH_4)_2SO_4$
  b. $2\ Cu + S \longrightarrow Cu_2S$ d. $Cl_2O_7 + H_2O \longrightarrow 2\ HClO_4$
3. How many moles of product can be formed according to each reaction in Problem 2, if you have 2.54×10^{-2} mol of the *second* reactant?
*4. Iron can be produced according to the following reaction:

$$Fe_3O_4 + 4\ H_2 \longrightarrow 3\ Fe + 4\ H_2O$$

Beginning with 2.74 mol of H_2 and excess Fe_3O_4, how many moles of Fe can be produced according to this reaction?
5. How many moles of H_2 would you need to form 8.69 mol of H_2O?

$$2\ H_2 + O_2 \longrightarrow 2\ H_2O$$

6. How many moles of O_2 would be required to produce 3.85 mol of CO_2 according to the following reaction?

$$2\ CO + O_2 \longrightarrow 2\ CO_2$$

*7. In each of the following reactions, how many moles of the *first* reactant are required to produce 7.22 mol of the *second* product?
  *a. $FeO + C \longrightarrow Fe + CO$
  b. $Al + 2\ HCl \longrightarrow AlCl_2 + H_2$
  c. $Al(NO_3)_3 + 4\ NaOH \longrightarrow NaAlO_2 + 3\ NaNO_3 + 2\ H_2O$
  d. $C_3H_8 + 5\ O_2 \longrightarrow 3\ CO_2 + 4\ H_2O$
8. Given the reactions in Problem 7, how many moles of the *second* reactant will you need to form 6.74 mol of the *first* product?

Section 7.3

*9. For the reaction of CO with O_2 (Problem 6), how many moles of CO would you need to react completely with 4.3 mol of O_2?

10. How many moles of SiO_2 will be required to react completely with 1.60 mol of C, according to this reaction?

$$SiO_2 + 3\ C \longrightarrow SiC + 2\ CO$$

11. Refer to the reaction equations of Problem 7. How many moles of the *first* reactant would be necessary to react with exactly 5.75 mol of the *second* reactant?

*12. The Haber process produces ammonia, a compound manufactured in large quantities in the United States.

$$N_2 + 3\ H_2 \longrightarrow 2\ NH_3$$

If 2.40 mol of N_2 is mixed with 10.0 mol of H_2 and caused to react, how many moles of NH_3 can be produced?

13. If 9.50 mol of O_2 is reacted with 7.65 mol of Fe, how many moles of Fe_2O_3 can result?

$$4\ Fe + 3\ O_2 \longrightarrow 2\ Fe_2O_3$$

14. If 4.30 mol of O_2 is reacted with 6.40 mol of PbS, how many moles of PbO can be produced?

$$2\ PbS + 3\ O_2 \longrightarrow 2\ PbO + 2\ SO_2$$

15. If 1.4 mol of NH_3 reacts with 1.0 mol of O_2, how many moles of N_2 will be produced?

$$4\ NH_3 + 3\ O_2 \longrightarrow 2\ N_2 + 6\ H_2O$$

16. How many moles of iron can be produced from 2.05 mol of Al and 0.57 mol of Fe_3O_4?

$$8\ Al + 3\ Fe_3O_4 \longrightarrow 4\ Al_2O_3 + 9\ Fe$$

17. Beginning with 4.50 mol of Al, 4.00 mol of NaOH, and 3.00 mol of H_2O, how many moles of H_2 can be formed?

$$Al + NaOH + H_2O \longrightarrow NaAlO_2 + H_2$$

18. How many moles of $MnCl_2$ can be produced from 3.65 mol of $H_2C_2O_4$, 4.40 mol of HCl, and 1.50 mol of $KMnO_4$?

$$2\ KMnO_4 + 5\ H_2C_2O_4 + 6\ HCl \longrightarrow 2\ MnCl_2 + 10\ CO_2 + 8\ H_2O + 2\ KCl$$

19. According to the reaction equation shown, how many moles of $CaCO_3$ can be formed from 1.40 mol of $Fe_2(CO_3)_3$, 4.00 mol of $Ca(OH)_2$, and 5.75 mol of HCl?

$$6\ HCl + Fe_2(CO_3)_3 + 3\ Ca(OH)_2 \longrightarrow 2\ FeCl_3 + 3\ CaCO_3 + 6\ H_2O$$

Section 7.4

20. How many moles of NaCl can be produced from 24.3 g of Cl_2 and excess Na?

$$2\ Na + Cl_2 \longrightarrow 2\ NaCl$$

*21. How many moles of N_2 can be prepared from 10.7 g of NH_3 and excess O_2? (Use the reaction equation from Problem 15.)

22. The combustion of butane, C_4H_{10} (one of the fuels sold as bottled gas), results in carbon dioxide and water.

$$C_4H_{10} + O_2 \longrightarrow CO_2 + H_2O$$

How many moles of CO_2 and of H_2O can be obtained from the reaction of 420 g of butane and excess oxygen?

*23. How many moles of H_2 are required to produce 9.00 g of Fe? (Use the reaction equation in Problem 4.)

24. How many moles of HCl are required to produce 23.5 g of $CaCl_2$? How many moles of $Ca(OH)_2$ are required?

$$Ca(OH)_2 + HCl \longrightarrow CaCl_2 + H_2O$$

25. How many moles of Fe_2O_3 and of CO are needed to produce 10.5 g of Fe?

$$Fe_2O_3 + 3\ CO \longrightarrow 2\ Fe + 3\ CO_2$$

26. How many grams of H_2SO_4 can be produced from 27.0 g of SO_3 and excess H_2O?

$$SO_3 + H_2O \longrightarrow H_2SO_4$$

27. If you have 310.5 g of NO and excess O_2, how many grams of NO_2 can you produce?

$$2\ NO + O_2 \longrightarrow 2\ NO_2$$

28. According to the following equation, how many grams of $FeCl_3$ and of H_2 will be formed from 205 g of Fe and excess HCl?

$$Fe + HCl \longrightarrow FeCl_3 + H_2$$

*29. Using the equation in Problem 14, calculate the number of grams of PbO that could be produced starting with 37.0 g of PbS and excess O_2.

30. How many grams of H_2 are required to produce 15.8 g of HCl?

$$H_2 + Cl_2 \longrightarrow 2\ HCl$$

31. Calculate the number of grams of $KClO_3$ that would be needed to produce 11.0 g of KCl. (See Exercise 7.10 for the reaction equation.)

32. How many grams of $MgCO_3$ would you need to produce 8.30 g of MgO? How many grams of $MgCO_3$ would you need to form 50.0 g of CO_2?

$$MgCO_3 \longrightarrow MgO + CO_2$$

33. Silver chloride, AgCl, is quite insoluble and can easily be recovered as a solid from a solution. $CaCl_2$, calcium chloride, reacts with silver nitrate, $AgNO_3$, according to the following reaction:

$$CaCl_2\ (aq) + 2\ AgNO_3\ (aq) \longrightarrow 2\ AgCl\ (s) + Ca(NO_3)_2\ (aq)$$

If 15.4 g of solid AgCl was recovered in an experiment using this reaction, how many grams of $CaCl_2$ were used?

Section 7.5

*34. Find the limiting reactant in the following reactions:
   *a. $CaO + 3\ C \longrightarrow CaC_2 + CO$
      50.0 g   30.5 g

   b. $4\ FeS_2 + 11\ O_2 \longrightarrow 2\ Fe_2O_3 + 8\ SO_2$
      40 g   100.0 g

   c. $2\ C_3H_6O_2 + 7\ O_2 \longrightarrow 6\ CO_2 + 6\ H_2O$
      15.0 g   200.0 g

   d. $2\ Al + 3\ H_2SO_4 \longrightarrow Al_2(SO_4)_3 + 3\ H_2$
      25.0 g   90.0 g

*35. If 75.6 g of $KMnO_4$ reacts with 0.68 g of NO, how many moles of MnO_2 can be produced? How many grams of KNO_3?

$$KMnO_4 + NO \longrightarrow MnO_2 + KNO_3$$

36. If 24 g of mercurous chloride, Hg_2Cl_2, and 2.00 g of H_2O react according to the following equation, how many grams of oxygen will be produced?

$$Hg_2Cl_2 + H_2O \longrightarrow 4\ HCl + O_2 + 4\ Hg$$

How many milliliters of mercury will be formed in this reaction? The density of mercury is 13.6 g/cm^3.

37. Methane, CH_4, can be produced from Al_4C_3 and water. Calculate the number of grams of CH_4 and of $Al(OH)_3$ that can be formed from 41.0 g of Al_4C_3 and 6.50 g of H_2O.

$$Al_4C_3 + 12\ H_2O \longrightarrow 4\ Al(OH)_3 + 3\ CH_4$$

38. NaOH is often prepared commercially from calcium hydroxide, $Ca(OH)_2$, and Na_2CO_3.

$$Na_2CO_3 + Ca(OH)_2 \longrightarrow NaOH + CaCO_3$$

How many grams of NaOH can be produced from 50.0 g of each reactant?

39. The reaction equation for the preparation of phosphorous in an electric furnace is

$$2\ Ca_3(PO_4)_2 + 6\ SiO_2 + 10\ C \longrightarrow 6\ CaSiO_3 + 10\ CO + P_4$$

Calculate the number of grams of P_4 that can be produced from 50.0 g of $Ca_3(PO_4)_2$, 10.0 g of C, and 20.0 g of SiO_2.

40. Ferrous chloride, $FeCl_2$, reacts with hydrochloric acid and potassium dichromate, $K_2Cr_2O_7$, to form ferric chloride, $FeCl_3$, and chromic chloride, $CrCl_3$.

$$6\ FeCl_2 + 14\ HCl + K_2Cr_2O_7 \longrightarrow 6\ FeCl_3 + 2\ KCl + 2\ CrCl_3 + 7\ H_2O$$

If 12 g of *each* reactant are mixed together, how many grams of KCl will result?

Section 7.6

*41. Alcohol is a product of the fermentation of sugar, $C_6H_{12}O_6$.

$$C_6H_{12}O_6 \longrightarrow 2\ C_2H_5OH + 2\ CO_2$$

How many grams of alcohol and of carbon dioxide can be produced from 750 g of sugar? How many milliliters of alcohol result if the density of the alcohol is 0.789 g/mL?

42. Limestone, $CaCO_3$, produces "quicklime," CaO, and carbon dioxide when heated (Example 7.18).

$$CaCO_3\ (s) \longrightarrow CaO\ (s) + CO_2\ (g)$$

If carbon dioxide has a density of 1.98 g/L, how many grams of limestone were originally heated if 6.50 L of carbon dioxide resulted?

43. Calcium carbide, CaC_2, reacts with water to produce acetylene gas, C_2H_2.

$$CaC_2\ (s) + 2\ H_2O\ (l) \longrightarrow C_2H_2\ (g) + Ca(OH)_2\ (aq)$$

This reaction is used in carbide lamps, in which the acetylene is burned to provide illumination. How many grams of water are required to react with 6.4 g of calcium carbide? How many liters of acetylene are produced if the density of the gas is 0.618 g/L?

44. Hematite iron ore, Fe_2O_3, can be reacted with carbon monoxide to form iron.

$$Fe_2O_3 + 3\ CO \longrightarrow 2\ Fe + 3\ CO_2$$

How many grams of hematite are required to produce 1 kg of iron? How many liters of carbon monoxide are required if its density is 1.25 g/L? Calculate the number of liters of carbon dioxide produced if its density is 1.98 g/L.

45. Calculate the number of liters of O_2 at a density of 1.429 g/L that can be produced when you begin with 5.50 L of gaseous H_2O. Under the conditions of this particular reaction, the H_2O has a density of 0.9980 g/mL.

$$2\ H_2O\ (g) \longrightarrow 2\ H_2\ (g) + O_2\ (g)$$

46. How many grams of CO_2 can be produced from 1.00 L of CH_4, measured at a density of 0.992 g/L, and 10.0 g of O_2? Write the reaction and check it against Example 7.8 if you need to.

Section 7.7

*47. Is energy absorbed or released when H_2 reacts with F_2? How much energy is absorbed or released when 3.5 mol of H_2 reacts as follows?

$$H_2 + F_2 \longrightarrow 2\ HF \qquad \triangle H = -539\ kJ$$

48. Living plants produce glucose, $C_6H_{12}O_6$, from CO_2 and water, using energy from sunlight to drive the reaction.

$$6\ CO_2 + 6\ H_2O \longrightarrow C_6H_{12}O_6 + 6\ O_2 \qquad \triangle H = 2816\ kJ$$

How much energy is required to convert 2.2 mol of CO_2 to glucose by this reaction?

49. In producing a mole of slaked lime, $Ca(OH)_2$, is energy consumed or produced, overall? (See Example 7.18 and Exercise 7.20.)

50. The combustion of octane, an ingredient of gasoline, produces energy.

$$2\ C_8H_{18}\ (l) + 25\ O_2\ (g) \longrightarrow 16\ CO_2\ (g) + 18\ H_2O\ (l) \qquad \triangle H = -10,904\ kJ$$

How much energy is produced by the combustion of 1 g of octane?

*51. How much energy is required to produce 1.00 g of glucose? (See the reaction equation in Problem 48.)

52. There is presently considerable interest in substituting other kinds of fuels for gasoline. Among the proposed substitutes are hydrogen gas and methyl alcohol, CH_3OH. Compare the energies obtained by burning 1 g each of H_2, CH_3OH, and gasoline. For gasoline, use the value you calculated for octane in Problem 50.

$$2\ H_2\ (g) + O_2\ (g) \longrightarrow 2\ H_2O\ (l) \qquad \triangle H = -572\ kJ$$

$$2\ CH_3OH\ (l) + 4\ O_2\ (g) \longrightarrow 2\ CO_2\ (g) + 4\ H_2O\ (l) \qquad \triangle H = -1408\ kJ$$

MORE DIFFICULT PROBLEMS

*53. If the metal M reacts with hydrobromic acid, HBr, hydrogen gas is given off.

$$HBr + M \longrightarrow MBr_x + H_2$$

When a sample of HBr was reacted with an excess of the metal, 1.00 g of hydrogen gas and 135 g of MBr_x resulted. Calculate three possible atomic weights of M.

54. When 9.25 g of butane, C_4H_{10}, is burned in oxygen, 456 kJ of energy is released. Write a balanced equation for the reaction, including the enthalpy change.

Chapter 8

The Nuclear Atom

Chemistry seeks to explain the reasons for the properties of different kinds of matter. Although we have learned numerous facts and theories about the structure of matter, nothing we have seen thus far can explain why different kinds of matter behave in different ways. How is it that only slightly more than a hundred elements form hundreds of thousands of different compounds, displaying an enormous range of properties? Limestone is hard, oil is oily, acid corrodes, gasoline burns, sugar is sweet, people live (people are made of compounds). If all substances are made of only a few kinds of atoms, why do the substances differ so? They differ because the atoms themselves have substantial differences. To approach the central concern of chemistry, we must explore the internal secrets of the atom.

In this chapter and the next, we will continue to apply the method used in earlier chapters. Since we cannot view the structure of atoms directly, we will construct a theory about that structure and test the theory against what we observe. Our goal is to explain the differences in chemical and physical properties of individual elements. We want to lay the groundwork for an understanding of how elements form compounds. In the process, we will learn to visualize atoms as something more than tiny, hard spheres.

We begin by learning about electric charge and by following one of the most celebrated experiments of science.

AFTER STUDYING THIS CHAPTER, YOU WILL BE ABLE TO

- name and describe the three major kinds of particles of which atoms are composed
- describe and explain the experiment by which Rutherford discovered the nuclear atom
- calculate the number of neutrons in the nucleus of an atom
- explain the principles of an experiment that can measure the mass of an electron or ion
- show why the atomic number, rather than the atomic mass, is the distinguishing characteristic of an element

TERMS TO KNOW

alpha particle	electron	neutron
atomic number	Faraday	nucleus
electric charge	ion	proton
electric current	mass number	

8.1 ELECTRIC CHARGE AND ELECTRICITY

For more than a hundred years, it has been possible to demonstrate experimentally that electric charge is intimately involved in the inner workings of atoms. Although much study has been devoted to the topic since the early 1800s, the fundamental nature of electric charge is not much better understood than it was then. Because the study of electric charge is the province of physics, our consideration of it will not include the details of the experiments by which the facts were discovered.

An **electric charge** is something that exerts force on other electric charges. Such force is either attractive, tending to pull charges together, or repulsive, tending to separate them. There are two kinds of charge, named positive (+) and negative (−). Two charges of the same kind, either two positive or two negative, repel each other. Charges of different kind attract each other. Forces between electric charges are the fundamental cause of most chemical phenomena.

Electricity is the flow of electric charge, often called **electric current.**

8.2 THE ELECTRON

Not long after Dalton proposed the atomic theory, Michael Faraday, an English chemist, discovered that an electric current passed through certain solutions would cause chemical reaction. He found also that the amount of reaction caused was directly proportional to the amount of current applied. To explain this observation, Faraday developed a theory that electric charge exists in the form of individual units (in much the same way as matter exists in the units called atoms), and that each unit has the same amount of charge. A few decades later, the unit of charge discovered by Faraday was named the **electron.** Electrons have negative charges. The amount of electricity that will bring about the reaction of 1 mol of certain elements is now called a **Faraday** of charge. It consists of 1 mol of electrons.

As you know, metallic wires are used to carry electricity. The electricity that flows through the wires is a flow of electrons. Under proper circumstances, electrons can be made to jump across gaps in the wires. In the late 1800s, J. J. Thomson and others performed several important experiments in which electricity flowed across gaps in wires.

Thomson's experiments revealed that electric charge is always associated with matter, and that the electron is a negatively charged particle of matter. Even more important, the combination of Thomson's results with other data made it possible to calculate the mass of an individual electron. Think again about the experiment with the golf ball and the ping-pong ball (Section 2.4). By blowing with the same force on both, you were able to tell which had the greater mass. In a similar experiment, shown in Figure 8.1, balls of different mass are dropped through the breeze created by an electric fan. Thomson's experiments worked on the same principle. The balls were replaced by electrons jumping a gap, and the force of the moving air was replaced by magnetic and electric forces, as shown in Figure 8.2. Because Thomson knew the amount of electron deflection produced, as well as the amount of magnetic

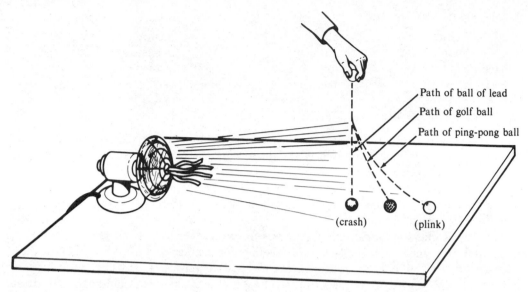

Figure 8.1 Evaluating mass by measuring acceleration. Three balls, all the same size, are dropped through the breeze made by an electric fan. The amount of sideways deflection from a vertical path is greater for the balls that have less mass.

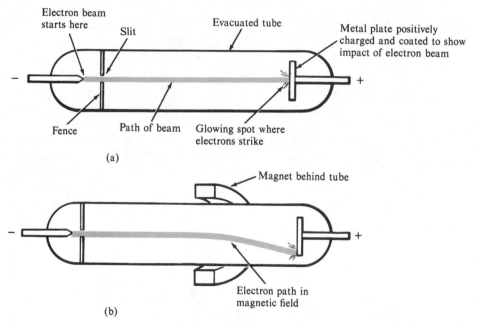

Figure 8.2 Electron deflection experiment in a cathode ray tube. In the apparatus shown in drawing (a), electrons move in a straight line through a space free of magnetic forces. In drawing (b), a magnet is placed near the tube, causing the electron beam to bend toward us just as the path of a ping-pong ball bends in a breeze.

and electric forces applied, he was able to calculate the mass of the electron, a valuable piece of information.*

The mass of the electron is 9.1×10^{-28} g, or about 1/1840 of the mass of a single H atom. Learning how to measure the mass of the electron constituted a breakthrough in the knowledge of atoms and their structures.

8.3 POSITIVE AND NEGATIVE CHARGES IN ATOMS: THE PROTON

A positive charge is repelled by another positive charge and attracted by a negative charge. If, however, we were to put a positive and negative charge together, then bring the combination close to a third charge of either kind, the third charge would feel no overall force. The force exerted by either of the first charges would be balanced by the force of the opposite charge. If a group of charged particles contains as many evenly distributed positive charges as it does negative charges, it will not exert force on other charges. Such a group of charges is said to be electrically neutral. Isolated atoms are ordinarily electrically neutral, although they can be caused to have unbalanced numbers of charges.

Because all atoms contain electrons (which are negative charges), it follows that atoms must contain positive charges as well. Positive charge, like negative charge, has been found to come in units, all of which have the same amount of charge and are associated with particles of matter. Because the size of the positive charge is electrically the same as that of the negative, we will refer to this *amount* of charge as *one unit* of charge. That charges are electrically equal, however, does not necessarily mean that the masses of the particles carrying the charges are equal; the particles come in several sizes.

We will become closely acquainted with a kind of positive particle that is essential to the construction of atoms. That particle, called the **proton,** carries one unit of positive charge. The simplest atom, that of the lightest isotope of hydrogen, consists of one proton and one electron.

8.4 MASS OF THE PROTON: MASS SPECTROSCOPY

During Thomson's time, scientists already knew that individual atoms could lose or gain one or more electrons, thus becoming what are called **ions,** without losing their identity with particular elements. Several experiments entirely parallel to Thomson's experiments were made using streams of positive ions in place of the electrons. Because a hydrogen atom consists of a proton and an electron, removal of the electron leaves only the positively charged proton. In other words, the positive ion of hydrogen is simply a proton. This is usually shown by the symbol H^+.

In a way similar to that used to determine the mass of an electron, the mass of the proton can easily be measured and is found to be about 1840 times that of an electron. In laboratory units, the mass of the proton is 1.67×10^{-24} g, which is only slightly greater than 1 amu. Figure 8.3 gives a schematic version of the apparatus used to measure the mass of a positive ion. In its refined form, the apparatus is called the mass spectrometer.

---

*To calculate the amount of magnetic and electric force exerted, it is actually necessary to know both the strengths of the electric and magnetic fields and the value (in laboratory units) of the charge of the electron. Electron charge is measured by a separate experiment, with which we will not be concerned.

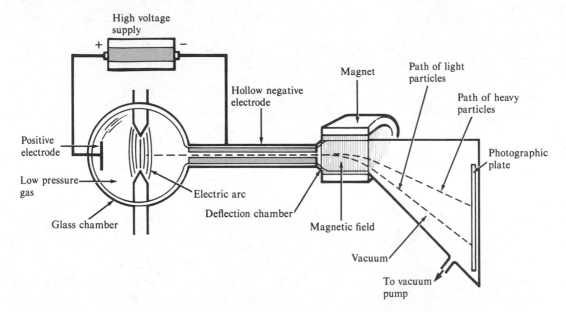

Figure 8.3 Early mass spectrograph. Gas in the glass chamber is bombarded in an electric arc, causing gas molecules and atoms to lose electrons and become positive ions. The ions are accelerated by the high-voltage negative electrode. Most are lost, but a few go through the tiny hole and form a beam. The beam enters a magnetic field in the deflection chamber, and the various particles take up new directions according to their mass. Finally, the separated beams reach the photographic plate, creating images where they strike.

But wait. We have learned that the mass of an entire H atom is about 1840 times that of the electron; now we learn that a single proton has approximately the same mass. Has something gone wrong here? No—compared with an electron, the proton has a huge mass. If you begin with a proton, and form an H atom by adding an electron, you will not have added enough mass to change the mass in a significant way. Only precise measurement can distinguish between the mass of a proton and that of an H atom.

Let us review a few things before we continue. Electrons are extremely light and have negative charges. Protons are much heavier than electrons and have positive charges. Atoms contain protons and electrons in equal numbers, and therefore are electrically neutral. We do not yet know whether other subatomic particles exist, or just how protons and electrons combine to form atoms. Nor do we know, except for H, how many protons and electrons various kinds of atoms contain. We will study these topics as we go on.

8.5 THE NUCLEAR ATOM

Considering what we have learned so far about electric charges and atoms, it seems reasonable to expect that the charged particles in atoms would arrange themselves so that charges of the *same sign* would be as *far apart* as possible, and charges of *different sign* as *close together* as possible. Perhaps the atom has a kind of uniform distribution in which the large protons are imbedded like raisins, surrounded by a pudding of electrons. At the beginning of this century, the logical and attractive "pudding model" was universally accepted.

Shortly after the beginning of the twentieth century, the English scientist Ernest Rutherford was investigating the nature of matter by bombarding thin sheets of metal foil with alpha particles from radioactive substances. We will discuss radioactivity in detail in Chapter 21, but to understand Rutherford's experiments, we need to know a little about alpha particles.

The atoms of a few elements are unstable. In samples of such elements, numbers of atoms are continually disintegrating, and the result of these atomic disintegrations is often the emission of subatomic particles of one sort or another. Many unstable elements emit what are called **alpha particles.** An alpha particle is a helium atom with the electrons stripped off. The alpha particles leave their atoms at extremely high speeds, traveling far faster than is possible for a bullet or any man-made missile.

Rutherford directed alpha particles at a thin metal foil, to see whether their collision with the atoms in the foil would alter the direction of their flight. (Remember that the pudding model of the atom was accepted at that time.)

Rutherford expected his alpha particles to penetrate the foil but to be slowed down as they ploughed through the atoms of the metal. He reasoned that if the atoms of the metal foil were made of uniformly distributed particles, no single particle in the foil would be heavy enough to make the alpha particles deviate much from their paths.

We will set up an analogy to make the experiment more clear. You are given a pile of oranges. In each orange there are embedded ten BBs. You plan to fire .22-caliber rifle bullets through the pile. By observing where the bullets go, you are going to make conclusions about where the BBs are located in the oranges. If the BBs are scattered randomly inside the oranges, no one BB will substantially deflect your bullet, and the bullets will come close to the target beyond the oranges. But suppose that in each orange, all ten BBs are welded together into a solid mass bigger than a .22-caliber bullet, and placed in the center of the orange. Not many bullets will hit these solid chunks—but when one does, it will bounce wildly and land nowhere near the target (Figure 8.4).

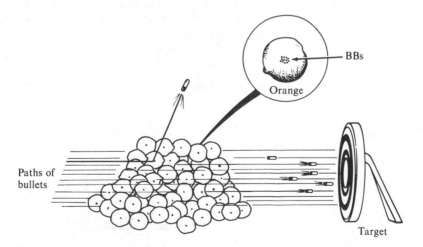

Figure 8.4 Analogy for the Rutherford experiment.

In Rutherford's experiment, most of the alpha particles went through the foil without visible effect, but a very few were greatly deflected. It was clear that the BBs (in this case the heavy protons and similar particles) were closely concentrated in one place in each atom. This was a tremendous surprise to Rutherford, who said that the experiment, planned as a "small research," resulted in "quite the most incredible event that has happened to me in my life . . . almost as incredible as if you fired a 16-inch shell at a piece of tissue paper and it came back and hit you." The pudding model was demolished by Rutherford's results. (See Figure 8.5.)

In 1911, Rutherford published his findings. His calculations from repeated experiments showed that nearly all the mass of the atom was concentrated in one tiny unit, which he named the **nucleus.** He found that the nucleus had a diameter only 1/10,000 that of the atom itself. The electrons, of which there were always enough to balance the number of protons, had negligible mass but occupied most of the volume of the atom.

Many important scientific discoveries have been made by accident; researchers designing experiments for a certain purpose have found totally unexpected results. What is important in scientific work is not that any particular experiment succeed as planned, but that significant experiments be performed by people who understand their experimental systems and have enough knowledge and imagination to interpret their results—regardless of what they are. Rutherford was astonished at what his experiment revealed, yet he continued with it, modified it to display the results better, and brilliantly interpreted what he saw. The result was one of the most famous experimental discoveries in scientific history.

Rutherford's model, which required that the atom consist mostly of empty space and which demanded that the mutually repulsive positive charges stay grouped close together without the gluing effect of negative electrons, was at first regarded as a sophisticated joke. When it became clear, however, that Rutherford was serious, scientists in many laboratories began feverish research to find the characteristics of the nucleus and the outer electrons.

Let us try to visualize a hydrogen atom. Its diameter is about 10,000 times that of its nucleus—the same ratio that exists between the diameter of a basketball and a racetrack 2 mi across. Imagine yourself standing in the center of this racetrack, holding the basketball. The edge of the atom is a mile away in every direction, and all the intervening space is occupied only by an electron, which weighs about as much as a penny (Figure 8.6). Drop an automobile from a building fifty stories high. When it hits the sidewalk, you will think that both the car

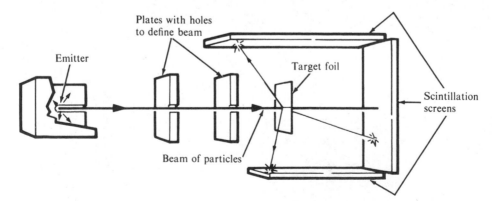

Figure 8.5   Representation of Rutherford's alpha-particle scattering experiment.

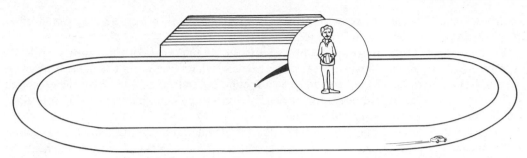

Figure 8.6 The amount of space the nucleus takes up in the atom is proportional to the amount of space a person takes up in the middle of a racetrack.

and the sidewalk are pretty solid, but they are not. Both are made of atoms, and atoms are mostly empty space. Solid matter—truly solid, with no empty space—is unbelievably dense. That basketball you were just holding would have a mass of about 10^{12} lbs, or about a billion tons, if it were "solid" matter.

8.6 ATOMIC NUMBER, MASS NUMBER, AND NEUTRONS

Shortly after the discovery of the nuclear atom, the experiment with alpha particles yielded another important result. From studies of the amount of alpha particle deflection, Rutherford found that (1) he could estimate the number of positive charges in the nuclei of different elements and (2) this number changed in a regular way from one element to the next.

Rutherford's finding suggested what it is that characterizes the atoms of a particular element, what it is that is unique and unchanging about all atoms of that element. When Rutherford discovered that the number of positive charges in the nuclei of atoms changed in a regular way from element to element, he was on the right track.

In the early years of the twentieth century, experiments with ions showed that more than a single electron could be stripped from some kinds of atoms, and it soon became possible to strip all the electrons from the atoms of the lighter elements. In these experiments, it was found that the total number of electrons in an atom, and therefore the number of positively charged protons in its nucleus, was characteristic of the element to which the atom belonged. Hydrogen atoms were found to have one electron, helium two electrons, lithium three, and so on. Such findings agreed with Rutherford's discovery that a specific number of protons was characteristic of a given element. Around 1913, some experiments with X rays, not involving ions at all, confirmed this conclusion, and it is now accepted that each element is characterized by the number of protons in the nuclei of its atoms. This number is called the **atomic number.** An element is identified by its atomic number as much as by its name. The element with atomic number 5, for instance, is boron; that with number 10 is neon; that with number 80 is mercury; and so on. The periodic table shows each element with its symbol and with its atomic number (sometimes denoted by **Z**) printed above the symbol.

As you know, most elements have more than a single isotope. At this point, we may wonder how two atoms can have the same number of protons and the same number of electrons and still have widely different masses, as, for example, the isotopes of hydrogen do. We must conclude that the structure of atoms involves particles other than protons and electrons. Such particles, however, must be electrically neutral; otherwise they would upset the electrical charge balance in the atoms.

The same conclusion was reached by scientists in the first years of this century. In 1932, experiments by Chadwick finally revealed the particle, which was given the name **neutron.** A neutron has approximately the same mass as a proton, but has no charge. With the exception of the lightest isotope of hydrogen, the nucleus of every atom contains one or more neutrons, usually about as many neutrons as protons.

The atomic number of an element identifies the element and gives the number of protons in the nucleus. What is called the **mass number** tells the total number of protons and neutrons in an atom of any isotope of an element. We will consider examples of mass numbers and atomic numbers in the next section.

For any isotope of an element, the mass number is quite close to the value of the atomic mass; that is, to the mass of the atom on the atomic mass scale or as read in amu. Consider why this should be so: atoms consist of protons, neutrons, electrons, and a few light particles that act as "glue" to hold the nucleus together (these are not important to our discussion). Of all these particles, only the protons and neutrons are massive; the others contribute negligible mass. Since each proton or neutron has a mass of approximately unity on the atomic mass scale, the value of the actual mass of an entire atom is nearly the same as the mass number.

EXAMPLE 8.1.   The atomic mass of an isotope of Si is 28.976. What is its mass number?

Solution:

Round the atomic mass to the nearest whole number. The mass number of the Si isotope is 29.

EXERCISE 8.1.   The atomic mass of an isotope of Ar is 35.968. What is its mass number? The atomic mass of an isotope of Ca is 39.963. What is its mass number?

Let us review:

1.   An atom consists of an extremely small, dense nucleus surrounded by one or more electrons.
2.   The two major constituents of the nucleus are protons and neutrons, each of which has a mass of approximately unity on the atomic mass scale. Protons have a charge of +1; neutrons are uncharged.
3.   Electrons have a charge of −1. Their mass is negligible compared with that of protons or neutrons.
4.   The atomic number (the number of protons in the nucleus) determines the element to which the atom belongs.
5.   The mass number (the total number of protons and neutrons) has approximately the same value as that of the atomic mass.
6.   A neutral atom has as many electrons as protons. Ions, which are electrically charged, can have more or fewer electrons.

As we learned in Section 4.8.1, individual isotopes are identified by superscripts, as in ^{12}C for carbon 12, and we can recognize those superscripts as the mass numbers of the isotopes. The atomic number is sometimes shown as a left subscript, an example being ^4_2He for the isotope 4 of helium.

EXERCISE 8.2.   For the Si isotope of Example 8.1, the symbol is ^{29}Si. Give the symbols for the isotopes of Exercise 8.1.

Because an element can be identified either by its symbol or by its atomic number, the practice of writing the atomic number beside the symbol is redundant. For example, chlorine is the element with atomic number 17 and symbol Cl; to write $_{17}Cl$ is repetitious. Moreover, writing the atomic number beside the symbol can be confusing as, for example, in a formula where the diatomic molecule of hydrogen would be written this way:

$$_1^1H_2$$

Although, in certain circumstances, it is useful to show the atomic number, we will not often find it necessary to do so.

8.7 THE NUMBERS OF NEUTRONS IN NUCLEI

Because atomic mass is determined almost entirely by the number of protons and neutrons in an atom, we can find the number of neutrons in the nucleus of an isotope by subtracting the atomic number from the mass number.

EXAMPLE 8.2.   How many neutrons are there in the nucleus of an atom of ^2_1H? Of $^{18}_8O$?

Solution:

The mass number of ^2_1H is 2, and the atomic number is 1. The number of neutrons in each nucleus is $2 - 1 = 1$. (See Figure 8.7.)

The mass number of $^{18}_8O$ is 18, and the atomic number is 8. There are ten neutrons in the nucleus of each atom.

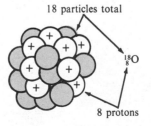

Figure 8.7   The nuclear particles in ^2_1H and $^{18}_8O$.

EXERCISE 8.3. How many neutrons are there in the nucleus of each atom of $^{35}_{17}Cl$? Of $^{14}_{6}C$?

EXERCISE 8.4. How many neutrons are there in the nucleus of an atom of the hydrogen isotope with an atomic mass of 1.014? Of the sodium isotope with an atomic mass of 22.98? (Hint: First find the mass number, then the number of neutrons.)

Table 8.1 gives several more examples.

Notice that the calculation to find the number of neutrons in the nucleus will not work for mixtures containing significant amounts of different isotopes because such mixtures contain atoms with different numbers of neutrons. Doing the calculation for geonormal zinc, for example, leads to a nonsense result.

EXAMPLE 8.3. How many neutrons are there in the nuclei of zinc atoms? The atomic weight of zinc is 65.38, and its atomic number is 30.

Solution:

The question cannot be answered as stated because no particular isotope of zinc is mentioned, nor is a mass number available. If we were to subtract the atomic number from the atomic weight, we would have an answer of 35.38, a fraction. Neutrons are individual particles, and it is not possible for there to be a fraction of a neutron. Unless a particular isotope is specified and a mass number can be found for that isotope, the question is meaningless.

EXAMPLE 8.4. Find the number of neutrons in the atoms of Ba. The atomic weight of Ba is 137.33, and the geonormal mixture is made up primarily of $^{138}_{56}Ba$ and $^{137}_{56}Ba$.

Solution:

You can answer this question for the individual isotopes named, but not for the mixture. There are 82 neutrons in $^{138}_{56}Ba$ and 81 neutrons in $^{137}_{56}Ba$.

In the atoms of the first 15 or so elements, the number of neutrons is about the same as the number of protons, so that the mass number is about twice the atomic number. The common isotope of oxygen, for example, is $^{16}_{8}O$, and that of neon is $^{20}_{10}Ne$.

Table 8.1
Some Calculations of the Numbers of Neutrons in Nuclei

Element	Hydrogen	Carbon	Oxygen	Sodium	Chromium
Symbol	$^{1}_{1}H$	$^{12}_{6}C$	$^{16}_{8}O$	$^{23}_{11}Na$	$^{52}_{24}Cr$
Mass number	1	12	16	23	52
Atomic number	1	6	8	11	24
Number of neutrons	0	6	8	12	28

For the heavier elements, however, the proportional number of neutrons increases with atomic number, growing to about one and a half times the atomic number in the heaviest elements. In the common isotope of gold, for example, $^{197}_{79}\text{Au}$, there are 1.49 neutrons for every proton.

8.8 SUMMARY

Chemistry is concerned with the interior structures of atoms because those structures determine the properties of the atoms. Atoms are made up of subatomic particles, the most important of which to chemistry are the proton, the neutron, and the electron. Each proton carries a positive electric charge, and each electron a negative electric charge. The charge of a proton is the same size as that of an electron. Neutrons are uncharged. A proton or neutron has about 1800 times the mass of an electron.

Atoms are neutral particles that contain subatomic particles with electric charges. For an atom to maintain its neutrality, the number of electrons in the atom must equal the number of protons. Each element is characterized by its atomic number; that is, by the number of protons in the nucleus of each of its atoms. The symbol Z is used to denote atomic number. An element can be identified either by its atomic number or its name.

At the center of each atom is a nucleus, which contains all the protons and neutrons (and therefore all the positive charge and nearly all the mass of the atom). The nucleus is surrounded by the electrons of the atom, which occupy nearly all the volume of the atom. The massive protons and neutrons are crowded into the tiny nucleus, which has a diameter only about 1/10,000 that of the atom itself.

A proton or neutron has a mass of about 1 amu, or about 1.6×10^{-24} g, next to which the mass of an electron is negligible. The mass number of any individual isotope is defined as the integer number nearest the value of its atomic mass on the atomic mass scale; it is also the total number of protons and neutrons in the nuclei of the atoms of that isotope. The number of neutrons in the nuclei of any isotope is the mass number less the atomic number.

QUESTIONS

Section 8.1

1. Describe the physical phenomena associated with electric charge.
2. What is electric current?

Section 8.2

3. What observation led Faraday to conclude that electric charge exists in individual units?
4. What is an electron? What is its mass in grams?
5. Explain in your own language the principle by which the mass of the electron and of other tiny particles can be measured.

Section 8.3

6. Explain why an electric charge that is brought close to a pair of charges, one positive and one negative, feels no overall force.

7. What does it mean to say that an atom is electrically neutral?
8. What is a proton?

Section 8.4

9. Explain the difference between an ion and an atom.
10. What is the symbol for a calcium atom that has lost two electrons?
11. What is the mass of a proton in grams?
12. Explain why the mass of a hydrogen atom is almost the same as that of its proton.

Section 8.5

13. Why was it so easy for scientists before Rutherford to accept the "pudding model" of the atom? What was found to be incorrect about this model?
14. What is an alpha particle?
15. Describe the nuclear model of the atom. What are the relative sizes of an atom and its nucleus?
16. Explain the experiment, including its results, that led Rutherford to conclude that atoms have nuclear structure.

Section 8.6

17. State a second important discovery made possible by Rutherford's experiment.
18. What is an atomic number? How does the atomic number of an element relate to its name? Explain one experimental method of finding the atomic number of one of the lighter elements.
19. Explain how two atoms can belong to the same element and yet have different masses. Define the words *isotope* and *neutron*.
20. What is a mass number? Why is the mass number nearly the same as the value of the atomic weight for some elements, but not for others?
21. In Problem 2, an isotope of Ca is given. Do you think there is likely to be a large natural abundance of this isotope in geonormal Ca? Explain your answer.
22. Is it possible for the atomic number of an element to be greater than its mass number? Explain.

PROBLEMS

Section 8.2

*1. Calculate the mass of the electron in amu.

Section 8.6

*2. Write mass numbers for the following isotopes:

Element	Atomic mass of isotope
*Hg	198.96
*Cs	132.90
Sn	111.90
Sr	83.91
Ca	47.95

3. Give the name and the atomic number of each of the following elements: Na, Co, Au, Li, Po, K, Mg, Hg, Si, Ag, Ti, Sn, F, Fe.

*4. Neon, whose atomic weight is 20.183, has three isotopes, none of which has a mass number smaller than 20. Guess what the mass numbers of the three isotopes might be and write symbols for the isotopes. What might the relative abundances of the three isotopes be in geonormal neon?

5. The atomic weight of Mg is 24.305 and its atomic number is 12. Can you determine a mass number for Mg from these data and if not, why not?

6. Write the complete symbols for the isotopes in Problem 2.

*7. The element carbon has three isotopes whose mass numbers are 12, 13, and 14. Write the symbols for these three isotopes. How many neutrons are there in the nuclei of the atoms of each isotope?

Section 8.7

*8. How many neutrons are there in the nuclei of the atoms of the following isotopes:
 *a. ^{27}Al d. ^{86}Rb
 *b. ^{3}He e. ^{23}Na
 c. ^{24}Mg

9. The following is a list of isotopes, most of which exist. The element symbols are all represented by E.

 $^{16}_{8}E$ $^{14}_{7}E$ $^{18}_{8}E$ $^{13}_{7}E$ $^{15}_{7}E$ $^{15}_{8}E$ $^{14}_{6}E$

 a. Which are isotopes of the same element? Name the elements.
 b. Which have the same mass numbers?
 c. In which are the numbers of neutrons in the nuclei the same?
 d. In which are the numbers of protons in the nuclei the same?
 e. In which are the atomic numbers the same?

Chapter 9

Electron Shell Structure in Atoms

In Chapter 9, we continue our study of the atom. Because the outer portion of an atom is the part that interacts with the rest of the world, it is the structure of that outer portion that is most important for us to understand.

In this chapter, we will learn about the behavior of electrons, those puzzling little particles that refuse to obey the laws we so confidently apply to larger particles of matter. Even though electrons behave in unusual ways, we can learn enough about them to understand how their behavior determines the properties of atoms.

AFTER STUDYING THIS CHAPTER, YOU WILL BE ABLE TO

- state the principal concepts of the Bohr theory of the atom
- explain the basis for the wave treatment of the electron and state its consequences
- describe the differences between the Bohr theory and the wave mechanical theory of the electron
- list, in increasing order of energy, the electron shells and subshells of atoms
- write and diagram the electron configurations of the first several dozen elements
- draw electron dot diagrams of atoms

TERMS TO KNOW

atomic kernel	Hund's rule	quantum mechanics
Bohr atom	mainshell	quantum number
electron configuration	octet rule	spectroscopy
electron dot diagram	orbital	subshell
electron shell	principal quantum number	valence electrons
energy level	quantized energies	wave mechanics
Heisenberg uncertainty principle		

9.1 ELECTRON SHELL STRUCTURE

All the parts of an atom contribute directly or indirectly to the properties of the element to which the atom belongs. The nucleus supplies most of the mass and all of the positive charge. The total charge of the nucleus (the number of protons) determines how many electrons the neutral atom has. *The part of the atom that interacts with the rest of the world is the outside part,* the cloud of electrons surrounding the nucleus. Except for the mass, nearly all the properties that will concern us are directly caused by the electrons and only indirectly by the nucleus. For that reason, we will concentrate our attention on the outer part of the atom and leave the details of nuclear structure to the physicist.

As we have learned, the outer part of the atom consists entirely of electrons and empty space. Electrons are the smallest particles of matter, except possibly for some exotic particles that can sometimes be temporarily coaxed out of the nucleus. Because electrons are so small, they exhibit properties unlike those that characterize larger pieces of matter. The properties of electrons make it difficult to learn much about what they do. Not even a powerful microscope can see the electrons in atoms and allow us to pry into their business. The waves that constitute visible light are not only bigger than electrons, but also bigger than atoms; this makes it impossible to use light to see atoms or electrons. You might be able to learn something about the size of a large ship by looking at the patterns made by the ocean waves that bounce off it, but the same waves would tell you little about a toy sailboat. The sailboat would be so much smaller than the waves that it would make no impression on them at all. The relation between electrons and visible light waves is somewhat similar.

You can find out a little about how electrons behave by shooting fast particles at them and seeing what happens, much as Rutherford did in his experiment with alpha particles and nuclei. But, when you hit an electron with a fast particle, you knock it out of the atom entirely and can say only where it was—not what it was doing. Was it circling the nucleus? Was it vibrating back and forth? Was it standing still? You can't tell. The properties of matter are such that with very small particles, *it is impossible to tell simultaneously where the particles are and what they are doing,* a natural law stated concisely by the **Heisenberg uncertainty principle.**

Fortunately, there are a few ways in which an electron in an atom can make itself known; and from those we can make some good guesses about atomic structure. Electrons have energy, just as any other body of matter can have energy. Because electrons move, they have kinetic energy. They also change position with or against the forces binding them to the nucleus, which gives them energy of position, or potential energy. When an electron in an atom *loses* energy, the lost energy can *leave* the atom as light: pure energy not associated with matter. If the energy of the emitted light is measured, it will reveal the amount of energy change made by the electron. Or light can also *enter* an atom, be totally absorbed, and *increase the energy of an electron* in the atom; such an energy change can also be measured. By bringing about interactions between light and electrons, we can study the energies of light absorbed and emitted by atoms, a study called **spectroscopy.** (See color plates.) Spectroscopy lets us find out the possible energies of the electrons in atoms—not the *locations,* unfortunately, only the energies. Even this much, however, allows us to build a solid theory.

Researchers have found that atoms can absorb or emit light of only certain amounts of energy, not just any amount. This has led to the conclusion that electrons in atoms can *have* only certain amounts of energy. In other words, the energies of electrons are **quantized:**

restricted to certain quantities. Consider, for example, the different, distinct amounts of energy that can be possessed by a brick placed on one or another of the shelves of a bookcase, as shown in Figure 9.1.

Just as you could characterize the shelves on a bookcase by measuring the energies of books placed on various shelves, so can an atom be characterized by the set of energies possible to its electrons. Each atom of an element possesses a set of **energy levels,** one or more of which can be occupied by electrons. No two elements have the same set. Remember that while the picture of the bookshelf shows the *physical location* of the books and only indirectly represents the *energies* of the books, the energy levels of electrons are levels of energy only and have an indefinite relation to electron location. We know an electron not by its position, but by its energy level.

The electron population of an atom occupies the lowest levels available, filling the lowest level first, then succeeding levels until there are no more electrons. The remaining upper levels are ordinarily unoccupied. Because each element has a pattern of energy levels unlike that of any other element, the pattern of light absorption and emission is also unique for each element; this allows the pattern to be used to identify the presence of atoms of a given element in a sample.

The theories about the electron shell structure in atoms are based on spectroscopy and the quantization of electron energies.

9.2 THEORY OF ELECTRON BEHAVIOR IN ATOMS

We should now be accustomed to seeing how theories are devised to explain observed phenomena. This approach was used to explain the known facts of electron behavior. Let us

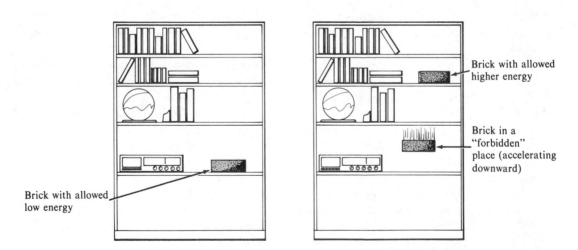

Figure 9.1 Quantized potential energy. The energy of the brick on the shelf at left is less than that of the brick on the shelf at right because work must be done to raise a brick from a lower to a higher shelf. Notice that the brick can have only certain energies, corresponding to its position on any of the shelves. If the brick is placed between shelves, it will fall to a shelf and assume an allowed energy.

list what a useful electron theory should explain, then examine the early approaches to a satisfactory theory.

1. The negatively charged electrons occupy most of the space in an atom, and stay away from the positively charged nucleus. Why doesn't the atom collapse instead?
2. From the experiments of spectroscopy, we know that the energies of electrons are quantized. Why can't an electron have *any* amount of energy, within limits?
3. Although different elements have different physical and chemical properties, the similar sets of properties that occur in many elements let us group elements into "families." Characteristic properties are presumably determined by the electrons in the atoms of the elements. If all electrons are the same, how can they cause such different sets of properties, and how can electron behavior create families of elements?

9.2.1 Bohr's Planetary Theory

What keeps the moon from falling into the earth, or the earth from falling into the sun? It is the balance between gravitational force and centrifugal force. When a rock is whirled around on the end of a string, the string supplies the pull toward the center. In an atom, there is strong attraction between the nucleus and the electrons, but the electrons do not fall into the nucleus. The parallel to a planetary system is obvious, and the first electron theories were planetary.

In 1912, a Danish physicist named Niels Bohr proposed that electrons in atoms occupied orbits around the nucleus, the force of the attraction caused by electric charge being balanced by the centrifugal force. Because Bohr could not explain why the energies of the electrons were quantized, he declared, arbitrarily and without explanation, that the electrons could occupy only certain orbits at particular distances from the nucleus. Bohr explained emission and absorption of light as the result of electrons making transitions from one allowed orbit to another. Apart from the arbitrary features, the **"Bohr atom"** fit the available experimental facts quite well.

Bohr's theory was attractive because it was easy to visualize. It was in serious conflict, however, with other, well-established theories of physics. Moreover, it did not explain the reason for quantization, and it provided no satisfactory explanation of the differences between the properties of elements. What is more, the Heisenberg uncertainty principle was beginning to be understood and seemed to imply that not only could the orbit of an electron not be measured, the electron could not even *have* an orbit.

The Bohr theory was useful primarily because it accustomed scientists to thinking about the quantization of electron energies, and led to efforts to explain that quantization. Although Bohr's theory does not even approach reality, we still see representations of it everywhere (Figure 9.2).

9.2.2 Wave Theory of the Electron

During the second and third decades of the twentieth century, several different kinds of research contributed to the theory of electron behavior. Because of insights furnished by the uncertainty principle, scientists began to suspect that tiny particles like electrons could not be expected to behave like other matter.

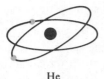
He

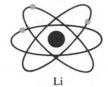
Li

Figure 9.2 Popular representations of atoms.

Researchers already knew that photons of light could behave either like individual particles or like waves. Louis de Broglie used this information to advance the theory that electrons in motion might be considered to do the same; that is, that they might have the character of waves as well as that of particles. In 1924, de Broglie proposed that moving electrons be assigned wavelengths. Until then, wavelength had been used only to describe light and wave phenomena in water, strings, and similar systems. Somewhat later, Ernst Schroedinger took the final step and applied to electrons in atoms a mathematical description similar to that which could be used to describe a wave in a pool of water whose surface had been disturbed. Schroedinger's approach led to a theory that successfully avoided the disadvantages of the Bohr atom. The new theory was given the name **wave mechanics** or **quantum mechanics.**

An understanding of electron behavior that is based on wave mechanics must include the following:

1. Electrons in an atom can in no way be regarded as particles of matter that obey the laws of classical physics.
2. Electrons in atoms must be considered to behave as waves. Just as the wave created by a pebble dropped in a quiet pond exists in all parts of its circular pattern at once, so does the wave representing an electron in an atom exist in many parts of the atom at once (Figure 9.3).
3. No possibility exists of assigning an exact location to an electron in an atom. At any particular moment, we can know only where the electron is most likely to be found, not where it actually is.
4. The only precise thing that can be known about the behavior of an electron in an atom is the energy of that electron.

In return for accepting the idea that electrons in atoms behave as waves rather than particles, we are rewarded by finding that wave mechanics easily explains why the atom does not collapse and why electron energy levels exist. Just as the wave in a plucked guitar string can create only certain tones, depending on which fret it touches, so can the electron in an atom have only certain amounts of energy (exist only in quantized energy levels). (See Figure 9.4.) Wave mechanics answers the most difficult questions satisfactorily, and provides a pattern of electron occupancy in atoms that neatly fits all the known data about the various properties of elements.

9.3 QUANTUM NUMBERS AND ELECTRONS

What we have so far learned about electrons in atoms is that electron behavior is far different from that of flying baseballs or moving planets. Little can be learned directly about

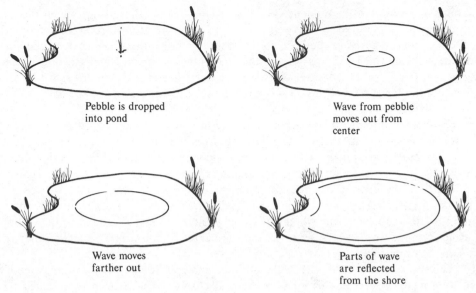

Pebble is dropped
into pond

Wave from pebble
moves out from
center

Wave moves
farther out

Parts of wave
are reflected
from the shore

Figure 9.3   Traveling wave in a pond.

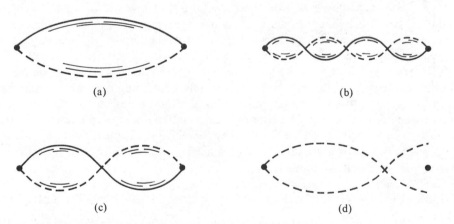

(a)

(b)

(c)

(d)

Figure 9.4   Standing waves in plucked strings. (a), (b), and (c) are allowed wavelengths. (b) and (c) are harmonics of (a). (d) is a non-allowed wavelength: vibrating string cannot have this wavelength if the ends of the string are fixed in place.

exactly what electrons do in the outer shells of atoms. We cannot say what electron sizes are, what paths electrons take from place to place, or even where the electrons *are* in the atoms (although we can say where they *were,* at the expense of knocking them out of their atoms entirely). What we can say with great accuracy and usefulness, however, is what energies electrons in atoms have. This, with the help of wave mechanics, gives enormous insight into why elements have the properties they do.

Every nucleus has a very large number of energy levels that can accommodate electrons. In an atom, some of these levels are occupied by electrons, the number of electrons matching the number of protons in the nucleus. The lowest-lying levels (those from which the largest amount of energy is required to remove the electron from the atom) are preferentially filled. The effect can be compared to the behavior of water in a bowl: the water preferentially lies in the bottom of the bowl. (The lowest part of the bowl is the part from which the largest amount of energy is required to lift the water out.) If more water is added to the bowl, the water level rises, because the lower part of the bowl is already taken. Electrons in atoms behave similarly; they fill the lowest levels first.

Energy levels in atoms are arranged by nature into groups and subgroups that are often called **electron shells** and **subshells.** The terms date from the time when electrons were thought to travel in precise orbits whose paths carved out hollow shells, one within the other like rings in an onion. (Although the terms *energy levels* and *sublevels* are more appropriate than *shells* and *subshells,* we will often use the latter terms for the same reason that we use *atomic weight*—because the terms are in wide use already.) Subshells are further divided into what we might call sub-subshells, but which are called **orbitals.** (The *-als* added to the end of the word avoids the undesirable preciseness of the word *orbit.*) Each orbital has two subdivisions, called spin states or spin levels.

To keep track of the shells, subshells, orbitals, and spin levels, we use a language called the **quantum number** system, derived from wave mechanics. The pattern represented by the system is a direct result of the mathematical treatment of the electron-wave concept of Schroedinger, but it also closely fits the experimental results given us by spectroscopy.

For our purposes, the quantum number system can be compared to the address system in a country. The designation of a certain house in the United States, for instance, proceeds through successively lower levels from state to city to street and finally to street number. When all the levels are given, that house is uniquely identified. The address of an electron is similarly expressed by four quantum numbers, the difference being that the numbers identify the precise energy level of the electron, not its physical location in the atom. Although there is a relation between energy level and location, any information about the location of an electron in an atom is sketchy at best, whereas the energy level information is definite, precise, and useful. Just as no two houses can have the same address, neither can two electrons in an atom have the same set of quantum numbers.

9.4 DETAILS OF THE QUANTUM NUMBER SYSTEM: SHELLS, SUBSHELLS, ORBITALS

We will frequently use quantum numbers in our future discussions, but we will be primarily interested in only two of the four quantum numbers that are needed to specify exactly the energy level of an electron. The first is called the **principal quantum number,** or **main shell** designation, and is given by a number called N, which can be 1, 2, 3, or a greater positive integer. Only seven main shells, given in order of increasing energy as N = 1, N = 2, N = 3, and so on up to N = 7, are needed to accommodate the electrons of all the elements now known.

Within each main shell, the subshells are designated by a second quantum number, represented by a letter. In order of increasing energy, the subshells are designated by the letters *s, p, d,* and *f.* Only these four subshell designations are needed for all atoms now known. The letters *s, p, d,* and *f* represent labels that early research workers attached to their results, and the labels have stuck.

Of the two other quantum numbers, the first designates the sub-subshell, or orbital, and the second gives the spin level. We will be interested in these only in terms of how many of them pertain to any given subshell, not specifically which ones they are or what they represent. Nature's rules specify strictly how many subshells, orbitals, and spin levels can exist in each main shell. *When all the possible subshells, orbitals, and spin levels are filled in any main shell, additional electrons must occupy a higher shell.* Let us look at the rules that electrons follow in filling the energy levels of atoms.

1. Every nucleus possesses the full set of shells, subshells, and orbitals. Some of these may be occupied by electrons, depending on how many electrons are present in the atom.

2. In a normal atom, the shells and subshells are filled in order of increasing energy.

3. The first main shell (N = 1) contains only an *s* subshell.
   The second main shell (N = 2) contains *s* and *p* subshells.
   The third main shell (N = 3) contains *s*, *p*, and *d* subshells.
   The fourth main shell (N = 4) contains *s*, *p*, *d*, and *f* subshells.
   The other main shells (N greater than 4) each contain *s*, *p*, *d*, and *f* subshells. Theoretically, higher subshells exist for these main shells.

Main Shell	Subshells Present
1	*s*
2	*s, p*
3	*s, p, d*
4	*s, p, d, f*
5	*s, p, d, f, –*
6	*s, p, d, f, –, –*
7	*s, p, d, f, –, –, –*

4. The different kinds of subshells contain different numbers of orbitals as follows:

Subshell Type	Number of Orbitals
s	1
p	3
d	5
f	7

5. Each orbital can contain no electrons, one electron, or two electrons, but no more than two. When two electrons are contained in an orbital, their spin levels are shown in diagrams as ↑↓.

Subshell Type	Number of Orbitals	Total Number of Electrons in Subshell
s	1	2
p	3	6
d	5	10
f	7	14

You should memorize these rules because you will use them often. All the rules together allow you to count the number of electrons that can be accommodated in any main shell. Such a count will show that, theoretically, each main shell has enough subshells and orbitals to hold $2N^2$ electrons if each orbital is filled to its capacity of two electrons. No atom of any known element has enough electrons to fill any of the three highest shells entirely, although the shells and their subshells do exist.

Main Shell	Maximum Number of Electrons	$2N^2$
1	2	2×1^2
2	8	2×2^2
3	18	2×3^2
4	32	2×4^2
5	32 (theoretically 50)	2×5^2
6	32 (theoretically 72)	2×6^2
7	32 (theoretically 98)	2×7^2

EXAMPLE 9.1.   What is the electron capacity of the $N = 2$ shell, and in which subshells are the electrons located?

Solution:

The second main shell, $N = 2$, holds eight electrons; $2N^2 = 2 \times 2^2 = 8$. The s subshell has one orbital with two electrons, and the p subshell has three orbitals with six electrons.

EXERCISE 9.1.   What is the electron capacity of the $N = 4$ main shell? In what subshells are the electrons located?

EXERCISE 9.2.   What principal quantum number describes the main shell with 18 electrons?

We will use the shorthand language of science to describe how the energy levels of atoms are occupied by electrons. In this language, the quantum number N that designates the main shell is followed by the letter designating the subshell. For example, $3p$ refers to the p subshell of main shell 3. The number of electrons populating a subshell is given as a superscript to the right of the orbital designation. For example $4d^6$ says that there are six electrons in the d subshell of the fourth main shell. (Although orbitals and spin levels can also be shown, we will often neither need nor use that information.) If no number is shown as a superscript, it is assumed that there is one electron in the subshell.

EXAMPLE 9.2.   (1) Give the shorthand notation for four electrons in the $N = 3$ main shell and the p subshell. (2) If the notation shows $4d$, how many electrons are in the $4d$ subshell?

Solution:

(1) The shell and subshell designation is $3p$: put the four electrons in as a superscript, giving $3p^4$.
(2) Only one electron is present in the subshell if no superscript is given.

EXERCISE 9.3.   What is the shorthand notation for three electrons in the *d* subshell of N = 4?

EXERCISE 9.4.   What is N, and which subshell is occupied by how many electrons if the short-hand notation is $3d^6$?

9.5 ENERGIES OF SHELLS AND SUBSHELLS: THE ORDER OF SHELL FILLING

Although electrons fill the lowest shells and subshells available to them, they do not always fill the main shells in numerical order. In many cases, the difference in energy between the lowest and higher subshells within a main shell is larger than the difference in energy between that main shell and the next higher main shell. The result is that the subshells of some main shells overlap those of the next main shell, an effect that can be more easily understood from a drawing. Figure 9.5 shows both the main shells and their subshells. As an example of overlap, notice that the 3*d* subshell lies higher than the 4*s* subshell, and that both the 4*d* and 4*f* subshells lie higher than the 5*s* subshell. Because the lower levels fill first, *the 4s*

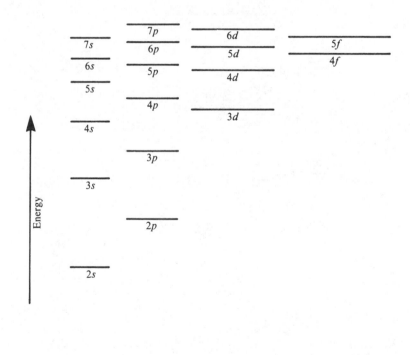

Figure 9.5   Energy levels of shells and subshells through 7*p*.

subshell will fill with electrons before the 3d, even though the 4s subshell lies in a higher main shell. As we will soon see, the effect of this overlap of energies on the filling of energy levels has very important physical and chemical consequences.

We are now ready to put the whole picture into one diagram. Figure 9.6 shows the energies of the main shells and their subshells for the first 80 elements. Also shown are the orbitals in each subshell and the full population of electrons, two electrons in each orbital.

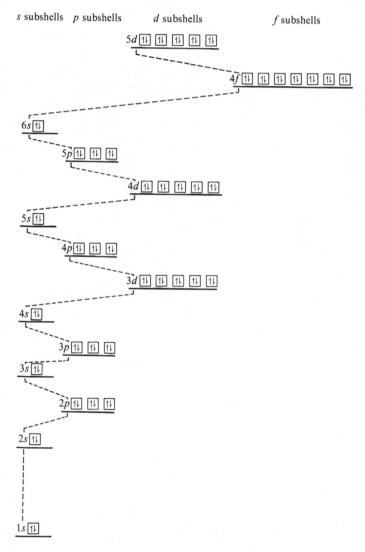

Figure 9.6   Scale drawing of energy levels in the element mercury. The number of orbitals appropriate to each level is indicated by boxes. These are each filled with one electron pair 1↓. The dotted line shows the order in which the subshells fill with electrons, starting with 1s and going through 5d.

Take the time to count the total number of electrons in some of the main shells, and check to see if the number is $2N^2$. For example, in the s and p subshells of the second main shell, are there 2×2^2 electrons?

Although it is essential to remember that the filling of shells and subshells does not follow the simple sequence of increasing principal quantum number, it is fortunately not necessary to *memorize* exactly how the subshells do fill. The periodic table, about which we will learn in the next chapter, has an arrangement that follows the subshell filling sequence and which can easily be interpreted. You can also use a simple device to remember the order of shell filling for any atom. Write the designations of the shells and subshells in order; then, starting at the top right side of the paper, draw a diagonal arrow through 1s, another parallel one through 2s, another through 2p and 3s, and so on through the entire set.

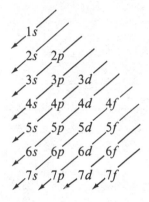

As you draw the arrows one after another, your pencil will pass through the subshells in the order in which they naturally fill, 1s, 2s, 2p, 3s, 3p, 4s, 3d, 4p, 5s, and so on. Write down the order in which the points of the arrows pass through the various subshells, then compare the result with Figure 9.6. It is far easier to remember this pattern than to memorize the entire sequence of energy level overlap.

9.6 ELECTRON CONFIGURATIONS OF ATOMS

We now have all the information we need to find out which electron shells and subshells are populated by the electrons in the atoms of any element. This arrangement, called the **electron configuration,** is easily determined.

1. Find the atomic number of the element in question. Assign to each atom of that element as many electrons as there are protons; that is, the number of electrons in each atom will be the same as the atomic number of the element.
2. Imagine a lone nucleus, then add to it one by one the required number of electrons, beginning with the lowest shell and subshell (1s). Place in each orbital a pair of electrons, listing all the electrons that occupy each main shell before beginning a new main shell.

    Notice that, by custom, the *electron configuration* is listed in the order of increasing N, even though the d and f subshells are not *occupied* in that order. Example 9.4 demonstrates this difference.

3.   When you use this method of successively filling shells (called the *aufbau approach*), remember that atoms are not actually created according to this one-by-one process of adding electrons. The atoms already exist with their electrons. The aufbau method is merely a way to *visualize* the occupation of energy levels by electrons.

Table 9.1 is a list of the electron configurations of all the elements.

EXAMPLE 9.3.   What is the electron configuration of nitrogen?

Solution:

Step 1.   The atomic number of nitrogen is 7. We will put seven electrons in the nitrogen atom.

Step 2.   Adding electrons one by one, we find that the first two electrons go into the $1s$ subshell, that is, $1s^2$. This fills the first main shell. The next two electrons go into the $2s$ level ($2s^2$), filling that subshell but not the entire N = 2 shell. The last three electrons go into $2p$, giving $2p^3$. The entire electron configuration, then, is $1s^2 2s^2 2p^3$. Check your work. The superscripts should add up to seven.

EXAMPLE 9.4.   What is the electron configuration of Se?

Solution:

Selenium is element number 34. Proceeding as before, we place 34 electrons in the atom. The finished configuration is $1s^2 2s^2 2p^6 3s^2 3p^6 3d^{10} 4s^2 4p^4$. Notice that selenium provides us with an example in which the electron configuration ($1s^2 2s^2 2p^6 3s^2 3p^6 3d^{10} 4s^2 4p^4$) is not written in the order of increasing energy ($1s^2 2s^2 2p^6 3s^2 3p^6 4s^2 3d^{10} 4p^4$).

EXERCISE 9.5.   What is the electron configuration of Zr? (The last subshell with electrons in it is $4d^2$.)

EXERCISE 9.6.   What is the electron configuration of V (vanadium)? Of Br?

Although the electron configurations of the elements appear in Table 9.1, it is important that you be able to determine configurations for yourself. Being familiar with electron populations is essential to understanding the properties of the elements. Practice!

There are a few elements in which the electron shells are not populated according to the usual sequence of energy levels. For our purposes, it is not necessary to know which elements these are, or why the exceptions occur; a table of electron configurations is usually available if we need to look up unusual cases. We should remember, however, that some irregularities do exist. As an example, consider the element Cr, which has the configuration $1s^2 2s^2 2p^6 3s^2 3p^6 4s^1 3d^5$. If the usual sequence were followed, the $4s$ subshell would contain two electrons and the $3d$ subshell only four. You can find other exceptions by studying Table 9.1.

9.7 THE OCTET RULE

Armed with our knowledge of electron shell filling order, we are ready to learn what it is about subshell energy overlap that has such great effects on the properties of elements. First, we must briefly consider the connection between electron energy levels and electron geography. (We will go into much greater detail about this subject in the next section.) Generally, electrons in any given main shell spend their time farther from the nucleus than do electrons in a main shell of smaller N. In other words, as the principal quantum number increases, it refers to main shells whose electrons lie increasingly farther from the nucleus. The 2 shell can be said to lie "inside" the 3 shell, the 3 shell "inside" the 4 shell, and so on. The "inside" is in quotation marks because the electrons in any shell easily penetrate the other shells, thus the description applies only to *average* distances of electrons from the nucleus.

Subshell energy overlap causes *new main shells to begin filling after the s and p subshells of the previous shell are full, regardless of whether there is a d subshell in the previous shell.* In any shell, the *s* and *p* subshells hold eight electrons. *After eight electrons go into any shell of an atom, a new shell begins with the next electron.* Any remaining subshells of the previous shell then fill sometime afterward. *No outer shell can hold more than eight electrons.*

As an illustration, let us consider the row of elements K to Kr. The electron configuration of K is $1s^22s^22p^63s^23p^64s$, and the $4s$ electron is the outer electron. Ca adds another electron to the $4s$ subshell for a total of two. Then the $3d$ subshell begins to fill, beginning with Sc ($1s^22s^22p^63s^23p^64s^23d$) and ending with Zn ($1s^22s^22p^63s^23p^64s^23d^{10}$). Twelve electrons have been added since argon, but the outer shell of Zn still contains only the two $4s$ electrons. The ten $3d$ electrons of Zn are not in an outer shell because the N = 3 level electrons all lie "inside" the $4s$ electrons. The $4p$ subshell finishes filling at Kr, but even here there are only eight electrons in the outer (4 level) shell.

When we examine the properties of families of elements, we will see how the so-called rule of eight, or **octet rule,** shapes the sets of properties of many elements.

In Chapter 10, we will learn that the elements with filled outer shells of eight electrons constitute a family called the noble gases. An example is argon, $1s^22s^22p^63s^23p^6$. In writing electron configurations, it is often useful to abbreviate sets of filled shells and subshells by using the symbols of noble gases that have those configurations. For example, we could write the electron configuration of selenium (see Example 9.4) as $\{Ar\}4s^24p^4$.

9.8 ELECTRON DOT DIAGRAMS OF ATOMS

Most properties of the elements result from the action of the electrons in the outer shells, that is, in the shells of highest principal quantum number. Because the outer electrons are by far the most influential ones in the atom, an abbreviated method showing only the outer shell electrons has been devised for writing the symbols of the elements. A symbol written this way is called an **electron dot diagram.** The designation of an atom in electron dot language consists of the symbol for the element surrounded by as many dots as there are electrons in the outer shell, the dots often being shown in pairs.

In an electron dot diagram the symbol of the element stands for the nucleus as well as for all the electrons that are not in the outer shell. This part of the atom is called the **kernel,** or the **atomic kernel.** The outer shell electrons are called the **valence electrons.** The kernel and the valence electrons constitute the entire atom. To write an electron dot diagram for an element:

Table 9.1
Electron Populations of the Elements

Atomic Number	Element	1	2	3	4	5	6	7
		s	$s\ p$	$s\ p\ d$	$s\ p\ d\ f$	$s\ p\ d\ f$	$s\ p\ d\ f$	s
1	H	1						
2	He	2						
3	Li	2	1					
4	Be	2	2					
5	B	2	2 1					
6	C	2	2 2					
7	N	2	2 3					
8	O	2	2 4					
9	F	2	2 5					
10	Ne	2	2 6					
11	Na	2	2 6	1				
12	Mg	2	2 6	2				
13	Al	2	2 6	2 1				
14	Si	2	2 6	2 2				
15	P	2	2 6	2 3				
16	S	2	2 6	2 4				
17	Cl	2	2 6	2 5				
18	Ar	2	2 6	2 6				
19	K	2	2 6	2 6	1			
20	Ca	2	2 6	2 6	2			
21	Sc	2	2 6	2 6 1	2			
22	Ti	2	2 6	2 6 2	2			
23	V	2	2 6	2 6 3	2			
24	Cr	2	2 6	2 6 5	1			
25	Mn	2	2 6	2 6 5	2			
26	Fe	2	2 6	2 6 6	2			
27	Co	2	2 6	2 6 7	2			
28	Ni	2	2 6	2 6 8	2			
29	Cu	2	2 6	2 6 10	1			
30	Zn	2	2 6	2 6 10	2			
31	Ga	2	2 6	2 6 10	2 1			
32	Ge	2	2 6	2 6 10	2 2			
33	As	2	2 6	2 6 10	2 3			
34	Se	2	2 6	2 6 10	2 4			
35	Br	2	2 6	2 6 10	2 5			
36	Kr	2	2 6	2 6 10	2 6			
37	Rb	2	2 6	2 6 10	2 6	1		
38	Sr	2	2 6	2 6 10	2 6	2		
39	Y	2	2 6	2 6 10	2 6 1	2		
40	Zr	2	2 6	2 6 10	2 6 2	2		
41	Nb	2	2 6	2 6 10	2 6 4	1		
42	Mo	2	2 6	2 6 10	2 6 5	1		
43	Tc	2	2 6	2 6 10	2 6 6	1		
44	Ru	2	2 6	2 6 10	2 6 7	1		
45	Rh	2	2 6	2 6 10	2 6 8	1		
46	Pd	2	2 6	2 6 10	2 6 10			
47	Ag	2	2 6	2 6 10	2 6 10	1		
48	Cd	2	2 6	2 6 10	2 6 10	2		
49	In	2	2 6	2 6 10	2 6 10	2 1		
50	Sn	2	2 6	2 6 10	2 6 10	2 2		
51	Sb	2	2 6	2 6 10	2 6 10	2 3		

Table 9.1 (Continued)

Atomic Number	Element	Shells and Subshells Filled						
		1	2	3	4	5	6	7
		s	s p	s p d	s p d f	s p d f	s p d f	s
52	Te	2	2 6	2 6 10	2 6 10	2 4		
53	I	2	2 6	2 6 10	2 6 10	2 5		
54	Xe	2	2 6	2 6 10	2 6 10	2 6		
55	Cs	2	2 6	2 6 10	2 6 10	2 6	1	
56	Ba	2	2 6	2 6 10	2 6 10	2 6	2	
57	La	2	2 6	2 6 10	2 6 10	2 6 1	2	
58	Ce	2	2 6	2 6 10	2 6 10 2	2 6	2	
59	Pr	2	2 6	2 6 10	2 6 10 3	2 6	2	
60	Nd	2	2 6	2 6 10	2 6 10 4	2 6	2	
61	Pm	2	2 6	2 6 10	2 6 10 5	2 6	2	
62	Sm	2	2 6	2 6 10	2 6 10 6	2 6	2	
63	Eu	2	2 6	2 6 10	2 6 10 7	2 6	2	
64	Gd	2	2 6	2 6 10	2 6 10 7	2 6 1	2	
65	Tb	2	2 6	2 6 10	2 6 10 9	2 6	2	
66	Dy	2	2 6	2 6 10	2 6 10 10	2 6	2	
67	Ho	2	2 6	2 6 10	2 6 10 11	2 6	2	
68	Er	2	2 6	2 6 10	2 6 10 12	2 6	2	
69	Tm	2	2 6	2 6 10	2 6 10 13	2 6	2	
70	Yb	2	2 6	2 6 10	2 6 10 14	2 6	2	
71	Lu	2	2 6	2 6 10	2 6 10 14	2 6 1	2	
72	Hf	2	2 6	2 6 10	2 6 10 14	2 6 2	2	
73	Ta	2	2 6	2 6 10	2 6 10 14	2 6 3	2	
74	W	2	2 6	2 6 10	2 6 10 14	2 6 4	2	
75	Re	2	2 6	2 6 10	2 6 10 14	2 6 5	2	
76	Os	2	2 6	2 6 10	2 6 10 14	2 6 6	2	
77	Ir	2	2 6	2 6 10	2 6 10 14	2 6 7	2	
78	Pt	2	2 6	2 6 10	2 6 10 14	2 6 9	1	
79	Au	2	2 6	2 6 10	2 6 10 14	2 6 10	1	
80	Hg	2	2 6	2 6 10	2 6 10 14	2 6 10	2	
81	Tl	2	2 6	2 6 10	2 6 10 14	2 6 10	2 1	
82	Pb	2	2 6	2 6 10	2 6 10 14	2 6 10	2 2	
83	Bi	2	2 6	2 6 10	2 6 10 14	2 6 10	2 3	
84	Po	2	2 6	2 6 10	2 6 10 14	2 6 10	2 4	
85	At	2	2 6	2 6 10	2 6 10 14	2 6 10	2 5	
86	Rn	2	2 6	2 6 10	2 6 10 14	2 6 10	2 6	
87	Fr	2	2 6	2 6 10	2 6 10 14	2 6 10	2 6	1
88	Ra	2	2 6	2 6 10	2 6 10 14	2 6 10	2 6	2
89	Ac	2	2 6	2 6 10	2 6 10 14	2 6 10	2 6 1	2
90	Th	2	2 6	2 6 10	2 6 10 14	2 6 10	2 6 2	2
91	Pa	2	2 6	2 6 10	2 6 10 14	2 6 10 2	2 6 1	2
92	U	2	2 6	2 6 10	2 6 10 14	2 6 10 3	2 6 1	2
93	Np	2	2 6	2 6 10	2 6 10 14	2 6 10 4	2 6 1	2
94	Pu	2	2 6	2 6 10	2 6 10 14	2 6 10 5	2 6 1	2
95	Am	2	2 6	2 6 10	2 6 10 14	2 6 10 7	2 6	2
96	Cm	2	2 6	2 6 10	2 6 10 14	2 6 10 7	2 6 1	2
97	Bk	2	2 6	2 6 10	2 6 10 14	2 6 10 8	2 6 1	2
98	Cf	2	2 6	2 6 10	2 6 10 14	2 6 10 9	2 6 1	2
99	Es	2	2 6	2 6 10	2 6 10 14	2 6 10 10	2 6 1	2
100	Fm	2	2 6	2 6 10	2 6 10 14	2 6 10 11	2 6 1	2
101	Md	2	2 6	2 6 10	2 6 10 14	2 6 10 12	2 6 1	2
102	No	2	2 6	2 6 10	2 6 10 14	2 6 10 13	2 6 1	2

1. Write the symbol for the element.
2. Determine how many valence electrons there are (those in the outer shell). You can do this by writing the electron configuration for the element and seeing how many electrons have the highest principal quantum number. Later, we will learn an easier way that is based on the periodic table.
3. Draw as many dots around the symbol as there are outer shell electrons. It is often convenient, especially when constructing dot pictures of molecules, to draw the electron dots by placing one dot on each of the two sides of the symbol and on the top and bottom. Then add the remaining electrons to make pairs.

EXAMPLE 9.5.   Draw the electron dot diagram for fluorine.

Solution:

Step 1. Write the symbol for fluorine.

$$F$$

Step 2. The electron configuration is $1s^2 2s^2 2p^5$. Therefore, there are seven electrons in the outer shell, which is the N = 2 shell. Draw those electrons.

Add first four electrons $\cdot \overset{\displaystyle \cdot}{\underset{\displaystyle \cdot}{F}} \cdot$

Add the remaining electrons, $\cdot \overset{\displaystyle \cdot\cdot}{\underset{\displaystyle \cdot\cdot}{F}} \cdot$
making pairs when possible

EXERCISE 9.7.   Draw the electron dot diagram of P.

EXERCISE 9.8.   Name three elements that have an electron dot diagram that looks like this one, where X stands for the symbol of the element.

Electron dot diagrams are quite useful for the first 20 elements and for certain others. We will use them often.

9.9 HUND'S RULE FOR SUBSHELL FILLING

In our filling of subshells, we have treated the individual orbitals in every subshell as though they had the same energy, making no indication of which filled first. For example, in placing electrons in the 2*p* subshell in nitrogen, we merely said that there were three electrons; we did not bother to specify which orbitals they were in, nor have we even learned how to specify particular orbitals. For many purposes, merely stating the number of electrons in each subshell is sufficient. To understand some of the finer points of elemental properties, however, we will need to learn a little more about the *sequence* in which the orbitals in a given subshell fill.

The order in which the orbitals of a subshell fill follows what is called **Hund's rule:**

> As successive electrons populate a subshell, they occupy the orbitals of the subshell in such a sequence that each orbital receives a single electron before any orbital receives its full population of two electrons.

The physical basis for the rule is simple:

> Because all electrons have negative charges, they repel one another. When they occupy separate orbitals, electrons can remain farther apart than when they are together in the same orbital. Two electrons will not occupy an orbital as long as there is an empty orbital available in the same subshell.

If we are to give the details of electron distribution in an atom, a simple electron population statement (such $1s^2 2s^2 2p^2$ for carbon) may not be sufficient. We need to specify which orbitals are full, which are half-full, and which are empty. This is most easily done with a diagram that represents the orbitals as sets of boxes grouped together in subshells within main shells. For instance, the set of orbitals through $2p$ would appear as follows:

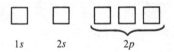

The electrons occupying the orbitals are then shown as vertical arrows, a single arrow for one electron, and a pair of arrows—one pointing up and one down—for a pair of electrons. The following orbital distribution diagram is for carbon. It shows pairs of electrons in the $1s$ and $2s$ orbitals and two electrons in the $2p$ subshell. Notice, however, that the p electrons are not paired: one electron is in each of the two orbitals, while the third orbital is empty. Unless the orbitals were individually labeled, the electron configuration of C would not give this information; it would say that there are two $2p$ electrons, but would not tell us which orbitals they are in.

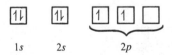

Figure 9.7 shows the orbital distribution diagrams for the first 11 elements. Notice that in the diagram for nitrogen, the orbitals of the $2p$ subshell receive one electron each, and that the first pair in that subshell occurs when we reach oxygen. Because oxygen has four electrons in $2p$, one more than there are orbitals, it is necessary for one orbital to have a pair. (It makes no difference to us now which of the orbitals receives the pair. We have not yet made any distinction among the orbitals in any given subshell.)

Element	$1s$	$2s$	$2p_x$	$2p_y$	$2p_z$	$3s$	Electron configuration
H	↑						$1s^1$
He	↑↓						$1s^2$
Li	↑↓	↑	☐	☐	☐		$1s^2 2s^1$
Be	↑↓	↑↓	☐	☐	☐		$1s^2 2s^2$
B	↑↓	↑↓	↑	☐	☐		$1s^2 2s^2 2p_x^1$
C	↑↓	↑↓	↑	↑	☐		$1s^2 2s^2 2p_x^1 2p_y^1$
N	↑↓	↑↓	↑	↑	↑		$1s^2 2s^2 2p_x^1 2p_y^1 2p_z^1$
O	↑↓	↑↓	↑↓	↑	↑		$1s^2 2s^2 2p_x^2 2p_y^1 2p_z^1$
F	↑↓	↑↓	↑↓	↑↓	↑		$1s^2 2s^2 2p_x^2 2p_y^2 2p_z^1$
Ne	↑↓	↑↓	↑↓	↑↓	↑↓		$1s^2 2s^2 2p_x^2 2p_y^2 2p_z^2$
Na	↑↓	↑↓	↑↓	↑↓	↑↓	↑	$1s^2 2s^2 2p_x^2 2p_y^2 2p_z^2 3s^1$

Figure 9.7 Orbital distribution for first 11 elements.

EXAMPLE 9.6. Diagram the orbital distribution of Fe.

Solution:

Step 1. Write the electron configuration of Fe.

$1s^2 2s^2 2p^6 3s^2 3p^6 3d^6$

Step 2. Draw and label the orbitals that will be needed. Fe will require everything through $4s$, including all of $3d$, although $3d$ will not be entirely filled.

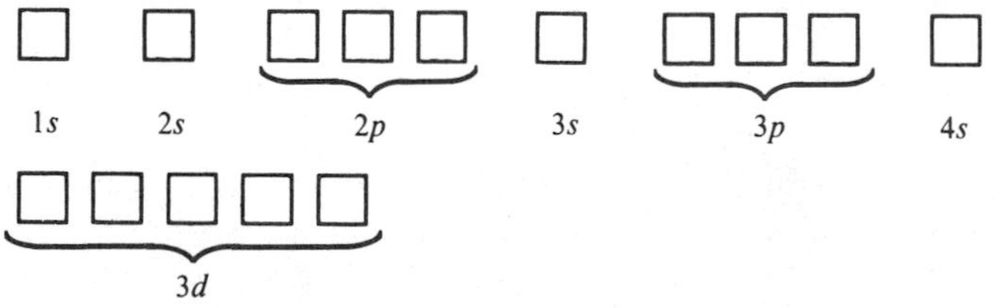

Step 3. Place the electrons in the orbitals. There will be 26 of them because Fe has an atomic number of 26. Use Hund's rule as you go.

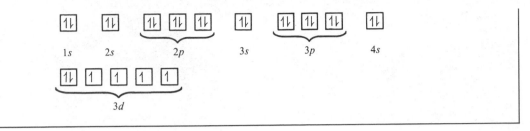

EXERCISE 9.9.   Diagram the orbital distribution of S.

EXERCISE 9.10.   What element has the following orbital distribution?

9.10 ELECTRON ENERGY LEVEL AND ELECTRON LOCATION

According to the Heisenberg uncertainty principle, it is not possible for an electron in an atom to have a definite orbit—neither an orbit that repeats itself, nor even an orbit that gradually changes. The movement of an electron is entirely random, and the paths it takes from place to place in the atom are indefinite and unknowable. All we can discover about the position of an electron in a particular energy level is merely an indication of where that electron spends most of its time. In other words, we can know what the chances are of finding the electron in any particular part of the atom, although we can not say where the electron actually is. An electron is likely to be in certain parts of the atom, but there are other parts to which it almost never goes.

This may seem like little useful information, and it would indeed be better to have more. But even what we know now can help us relate the properties of atoms to their electron shell populations. Data about where electrons spend most of their time can be translated into *plots of probability density,* pictures of the places in atoms where electrons in particular energy levels are most likely to be found. Let us look at an analogy. Suppose you are asked to photograph a baseball game in such a way that it could be determined where each player spent his time on the diamond. One way to do this would be to climb a tall tower and use a long exposure. Although the images of the players would be blurred, and the photo would not reveal exactly where any player had been at any given time, you could say, for example, that the concentrated blur around the pitcher's box meant that you could almost always find the pitcher there. The blur for the center fielder would be in a different place, and would be more smeared out; fielders move around more than pitchers do. By studying the picture, you could come to some valid conclusions about the positions of the players during the game.

If we could, by some magic, take a time exposure that would show an electron in an atom, we would get a smeared blur telling us where the electron spent most of its time, or in what

(a)   (b)

Figure 9.8   Artist's version of the appearance of the 1*s* and 2*s* orbitals (not to a common scale). (a) 1*s* (one orbital, outer view). (b) 1*s* and 2*s* (two orbitals, cross-sectional view).

part of the atom it would most probably be found at any particular time. The electron probability plot, taken mathematically from wave theory, gives us such a picture, and the study of probability plots gives us much useful information about the activities of electrons. Let us see what the plots look like.

The shapes of the plots are determined primarily by the orbitals occupied by the electrons. The *s* orbitals, for example, have the same general shape whether they are 1*s*, 2*s*, 3*s*, or higher. All the *p* orbitals are characterized by a shape different from those of the *s* or *d* orbitals. On the other hand, plots for the same kinds of orbitals in different main shells vary mostly in their average distance from the nucleus. The plot of a 1*s* orbital has the same general shape as the plot of a 2*s* orbital, but the 2*s* is larger. All probability plots have fuzzy boundaries because there is no limit to where the electron can be; the plots show only where it is most likely to be. As they are usually shown, the most heavily shaded parts of the picture represent the regions of the atom within which there is about a 90% chance of finding the electron at any given time. In your baseball photo, you might find the pitcher near first base, but 90% of the time you would find him within about 3 ft of the mound.

As Figure 9.8 shows, the *s* orbitals are spherical.

The *p* orbitals have dumbbell shapes. Under ordinary circumstances, the *p* orbitals in any main shell have the same energy, but the plots for the three separate orbitals are each pointed at right angles to the other two. (See Figure 9.9.)

The plots for the *d* and the *f* orbitals are much more complex in their shapes. Since we will not be much concerned with these orbitals, we will not study their pictures.

In Chapter 11, we will learn that the atoms of individual molecules are arranged in characteristic geometric patterns that contribute to the properties of those molecules. The arrangements of atoms in molecules are, in turn, substantially determined by the shapes of the electron orbitals in the atoms. Orbital patterns thus have significant influence on the properties of many substances.

9.11 SUMMARY

We study the electrons in atoms because they form the part of the atom that interacts most with the world. Nearly all the properties of any element result from the electron shell structure of its atoms.

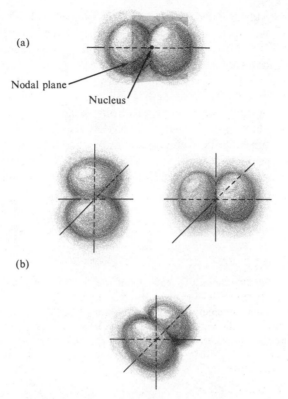

(a)

Nodal plane

Nucleus

(b)

Figure 9.9   (a) Artist's version of the appearance of a 2p orbital. (b) Artist's version of the appearance of the three 2p orbitals, shown separately (nodal plane not shown).

Because electrons are so small, they behave differently from larger particles. Knowing that small particles, such as electrons, cannot simultaneously have a precise location and a precise energy (that is, without any uncertainty whatever) is crucial to our understanding of their behavior. The unavoidable uncertainty means that an electron cannot be said to have a predictable path, or orbit, in an atom.

We can best understand electrons in atoms by treating them as though they were waves instead of tiny particles. Wave theory and spectroscopic measurements of electron energies in atoms have enabled us to learn much about where electrons spend their time.

Electrons in atoms occupy fixed energy levels arranged in a pattern described by the quantum number system. The principal levels, called shells and characterized by a principal quantum number N, are divided into subshells, which are in turn subdivided into orbitals. Orbitals are further subdivided into spin states. To name the principal level (shell), the subshell, the orbital within the subshell, and the spin level of an electron is to identify the electron completely. Quantum numbers are used to give all such designations, although not all quantum numbers are referred to numerically. The quantum numbers of the set of subshells, for instance, are usually designated by the letters *s, p, d,* and *f.*

Each shell and its subdivisions can accommodate only a certain number of electrons; we can therefore characterize an atom by naming the shells and subshells occupied by its electrons. Electrons occupy the lowest set of levels available, and when any shell becomes completely filled, the next starts filling. The sequence of energy level filling creates repeating patterns of outer shell structure (that is, repetition of the number of electrons in the shell of greatest N) as the elements are taken one by one in ascending atomic number. For example, both lithium and sodium have single electrons in the outer shells of their atoms, each element having started a new main shell. Because of the pattern of shell filling, no neutral atom can have more than eight electrons in its outer shell.

Depending on which subshell they occupy, electrons spend most of their time in various characteristic parts of the atom, sweeping out patterns that are sketched in probability density plots. An electron in an *s* subshell, for example, spends most of its time in a spherical shell; *p* electrons sweep out dumbbell-shaped patterns. The electron occupancy patterns of atoms substantially determine the shapes of the molecules made from those atoms.

QUESTIONS

Section 9.1

1. Why is it more important to a chemist to learn about the electrons in atoms than to discover the details of the nucleus, where most of the mass of the atom resides?
2. If it were possible to build a sufficiently powerful microscope, would it then be possible to discover directly the characteristics of the outer shells of atoms? Explain your answer.
3. What is the best source of information about the energies of electrons in atoms?
4. What do we mean when we say that the energies of electrons in atoms are quantized? Use analogies to explain your answer.
5. What is an energy level? How does an electron know which energy level it should occupy?

Section 9.2

6. Name three phenomena that any successful theory of electrons in atoms must explain.
7. Describe the Bohr atom.
8. Use the Heisenberg uncertainty principle to discuss one of the difficulties of the conceptually attractive Bohr theory.
9. Explain what is meant by the de Broglie wavelength.
10. For decades, the electron was regarded only as a particle. Describe the principal characteristic of an electron as viewed in terms of wave mechanics.

Section 9.3

11. Exactly what is an electron shell? What is a subshell? An orbital? Why not just say orbit, instead of orbital?
12. How are electron shells related to the quantization of electron energy?
13. Does a set of four quantum numbers for an electron in an atom give the *location* of the electron? Explain.
14. We say that four quantum numbers can describe an electron in an atom. What information is it that the quantum numbers actually give?

Section 9.5

15. Explain what is meant by the overlap of electron energy levels.
16. Could the N = 4 main shell have more than four subshells? (Use the memory device given in

Section 9.5.) Could the N = 5 main shell have more than four subshells?

17. Why is the nineteenth electron in K located in the fourth main shell, rather than in the third?

Section 9.6

18. What might be the reason for the irregular electron configuration of Cr?
19. Write a brief outline of the aufbau approach.

Section 9.7

20. Explain what is meant by the octet rule.
21. How can the octet rule apply to elements in which N is greater than 2 (that is, when a main shell contains more than eight electrons)?
22. Demonstrate the octet rule by explaining the sequence in which main shells and subshells fill, and which are inner and outer shell electrons. Use the elements Cs to Rn as examples.

Section 9.8

23. What are valence electrons? Atomic kernels?
24. Describe how to write an electron dot diagram of an atom.
25. How does the octet rule relate to electron dot diagrams of atoms?

Section 9.9

26. Explain the pattern of subshell filling known as Hund's rule. What is the physical reason for Hund's rule?
27. Can an electron join an orbital that is already occupied by another electron? If so, under what conditions?
28. In an orbital distribution diagram, what information is given that is not present in an electron configuration statement?
29. When it is said that two electrons are "paired," what does this mean in respect to the sets of quantum numbers of those electrons?

Section 9.10

30. What can we know about the *location* of electrons in atoms? How does the location depend on the quantum number?
31. What is an electron probability plot? Make rough sketches of the probability plots for *s* and for *p* electrons.
32. Would the plot for a 1*s* electron be different than that for a 3*s* electron? Explain.

PROBLEMS

Section 9.4

*1. What main shell, N, would contain a maximum of 50 electrons?
 2. What is the maximum number of electrons that can occupy the N = 6 main shell? The N = 9 main shell?
*3. What is the electron capacity of the N = 1 main shell? Of the N = 2 main shell? In what subshells are the electrons located?

4. What is the electron capacity of the N = 3 main shell? In what subshells are the electrons located?

5. What principal quantum number describes the main shell with eight electrons? With 72 electrons?

*6. Identify the principal quantum number and the subshell, along with the number of electrons in each of the following: $2s^2$, $4p^5$, $7s^1$, $4d^7$.

7. What is N for the electrons described as $4f^8$? How many of these electrons are there? In which subshell are they?

*8. Give the shorthand notation for the following:

   *a. five electrons in the N = 4 main shell and p subshell.

   b. two electrons in the N = 3 main shell and d subshell.

   c. one electron in the N = 2 main shell and s subshell.

   d. ten electrons in the N = 6 main shell and f subshell.

*9. Give the shorthand notation for the following:

N	Number of Electrons	Subshell
*2	2	p
3	1	s
3	4	d
4	10	f
3	8	d

Section 9.5

10. Write in order of increasing energy the designations of the shells and subshells through principal quantum number 6.

11. In Section 9.5 is a memory device for the filling of electron subshells. Complete the memory device, assuming that the g and h subshells have the next higher energy after the f subshells. Could the N = 6 main shell have more than six possible subshells? Explain.

*12. If the next energy subshell after f were g, what would be the maximum number of electrons that the g subshell could contain?

Section 9.6

13. Write the electron configurations for the following elements: Mg, P, B, C.

14. Give the electron configuration for the following elements: Zn, K, Br, O.

15. Three electrons are in the $3d$ subshell of a neutral atom. What element does the atom belong to? How many of these electrons are paired?

16. Another atom has four electrons in the $4p$ subshell. What element does it belong to? How many of these electrons are paired?

17. Which element has $5s^1$ in its electron configuration? Which has $2p^5$? $3d^3$? $4f^2$? $5p^1$?

18. Which element has $3p^4$ in its electron configuration? Which has $4d^6$? $3s^1$? $2p^2$? $4s^1$?

19. What elements have the following outer electron configurations: s^2p^2, s^2d^4, s^1d^5, s^2p^4, s^2d^2?

20. What elements have the following outer electron configurations: $s^2d^{10}p^5$, s^2d^7, s^2p^3, $f^{14}s^2p^6$

21. On Xanthu, the 109th element has been found. How many electrons are in its outermost main shell? What is N for that shell?

Section 9.8

22. Draw the electron dot diagrams of the following elements: N, O, C, Ca.

23. Draw the electron dot diagrams for Cl, Kr, Al, Sn, and As.

*24. Which elements have an electron dot diagram like the following? (*A* stands for the symbol of the element.)

25. Name the elements that have an electron dot diagram like the following. (*X* stands for the symbol of the element.)

Section 9.9

26. Diagram the orbital distributions of Na, C, P, Se, and Ar.
27. Draw the orbital distributions for the following elements: F, Mg, Ga, V.
*28. What element has the following orbital distribution?

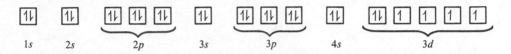

29. Identify the element that has this orbital distribution:

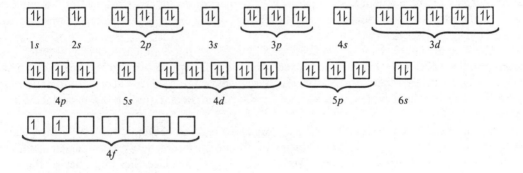

Chapter 10

Families of Elements
and the Periodic Table

In Chapter 9, we learned that atomic electrons exist in fixed energy levels and occupy characteristic patterns in the space around the nucleus. Because electrons form the outer parts of atoms, it is the electrons that interact with the rest of the world. We might reasonably expect, then, that electron shell patterns would cause the occurrence of patterns in the properties of atoms of the elements. Such patterns of properties do indeed exist and are organized and displayed in the periodic table, an arrangement of the elements according to their physical and chemical properties.

The simple classification of elements into families having common sets of properties has been practiced for more than a hundred years, but until fairly recently, no one knew why such relationships existed. The most satisfying achievement of the modern theory of atomic structure was the detailed explanation of these well-known regularities in the behavior of the elements.

In Chapter 10, we review the historic development of the periodic table. We then learn how the form of the table follows changes in atomic structure as we consider elements of successively greater atomic number. We can then begin to relate the properties of the elements to the structures of their atoms, a relation that is central to the study of chemistry.

AFTER STUDYING THIS CHAPTER, YOU WILL BE ABLE TO

- explain how elemental properties were used as the basis for the construction of the first periodic tables
- draw a periodic table in outline, filling in only the atomic numbers of the elements
- describe how the form of the periodic table follows the changes in the electron structures of the elements
- give the positions in the table, the general physical and chemical properties, and the electron shell populations of the elements in the alkali, alkaline earth, halogen, and noble gas families

- describe the relative sizes of atoms within a period and within a group
- predict the formulas of certain simple compounds from the positions of their elements in the table
- list the chemical and physical properties of hydrogen and explain the differences between this element and others with the same outer shell structure
- explain the positions in the table, the electron populations, and the general properties of the transition elements

TERMS TO KNOW

alkali metals	halogens	periodic table
alkaline earth metals	lanthanide and actinide	periods
d block	elements	representative elements
f block	noble gases	s block
groups	p block	transition elements

10.1 HISTORICAL DEVELOPMENT OF THE PERIODIC TABLE

The importance of the concept of elemental substances began to be appreciated at about the time of Dalton. By 1830, more than 50 elements had been discovered and prepared in the pure state. Although chemists had long since begun trying to find systems to classify the substances they studied, little progress was made until the atomic theory had been established and atomic weights measured.

In 1863, the English chemist Newlands made an arrangement of the known elements in order of increasing atomic weight. He found that in many cases, certain sets of characteristics periodically recurred and that the properties of a given element often closely resembled those of the element seven spaces away from it in the list. Newlands noticed, for instance, that lithium, sodium, and potassium (second, ninth, and sixteenth in his list) were similar. At this time, however, atomic weights were considered to be of so little importance that Newlands was ridiculed. Then in 1869, Meyer in Germany and Mendeleev in Russia made similar and somewhat improved listings; Meyer was working with physical properties and Mendeleev mostly with chemical characteristics. Their efforts were warmly received by the scientific world, and the importance of systematic classification of the elements began to be appreciated.

Although several other chemists also deserve credit, the work of Mendeleev shows the greatest insight. Using atomic weights merely as a guide, he arranged a table on the basis of known elemental properties, making vertical family groupings the important factor. Where the order of atomic weights did not agree with his arrangement, he changed the order or left blank spaces, assuming that an element had not yet been discovered (these spaces are shown by dashes in Table 10.1). In the case of tellurium and iodine, for example, he listed the elements out of order, changing the assigned weight of tellurium. Mendeleev's table, unlike that of Newlands, contained no serious exceptions to the similarities of properties. Although Mendeleev believed that the arrangement of the table reflected natural law, his belief received no definite verification for several years.

Table 10.1
Periodic Table as Proposed by Mendeleev, 1869

Series	Group I	Group II	Group III	Group IV	Group V	Group VI	Group VII	Group VIII
1	H 1							
2	Li 7	Be 9.4	B 11	C 12	N 14	O 16	F 19	
3	Na 23	Mg 24	Al 27.3	Si 28	P 31	S 32	Cl 35.5	
4	K 39	Ca 40	—44	Ti 48	V 51	Cr 52	Mn 55	Fe 56, Co 59, Ni 59, Cu 63
5	Cu 63	Zn 65	—68	—72	As 75	Se 78	Br 80	—
6	Rb 85	Sr 87	Yt 88	Zr 90	Nb 94	Mo 96	—100	Ru 104, Rh 104, Pd 106, Ag 108
7	Ag 108	Cd 112	In 113	Sn 118	Sb 112	Te 125	I 127	—
8	Cs 133	Ba 137	Di 138	Ce 140	—	—	—	—
9	—	—	—	—	—	—	—	—
10	—	—	Er 178	La 180	Ta 182	W 184	—	Os 195, Ir 197, Pt 198, Au 199
11	Au 199	Hg 200	Tl 204	Pb 207	Bi 208	—	—	—
12	—	—	—	Th 231	—	U 240	—	—

Note: the dashes represent elements undiscovered in Mendeleev's time.

In a spectacular piece of detective work, Mendeleev published a second paper predicting the properties of three elements that had not yet even been discovered. His predictions are shown in Table 10.2. When these elements (whose names today are gallium, scandium, and germanium) were discovered a few years later and their properties confirmed, Mendeleev's work became widely celebrated. It is ironic that the greatest honors came from England, where the original work of Newlands had been dismissed.

The arrangement begun by Mendeleev and now used by scientists everywhere is called the **periodic table of the elements,** or simply the **periodic table.**

As they are usually shown, elements in the periodic table are each given a square identified by the name of the element and containing several kinds of information.

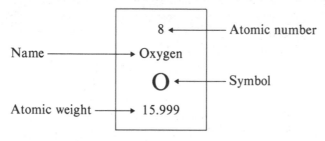

Because of its consistency with known chemical and physical properties, the periodic table was generally accepted by about 1890. But not until another 30 years had passed did scientists begin to understand *why* periodicity (the regular, periodic repetition of similar properties) is characteristic of elemental properties.

Between 1915 and 1930, the concept of atomic number was developed and experimentally confirmed; it then became clear *that the sequence of elements in the periodic table was determined by atomic number,* rather than by atomic weight. Although the arrangement of

Table 10.2
Verification of Mendeleev's Predicted Properties for the Element
Germanium (Ekasilicon, Es)

Property	Mendeleev's Prediction 1871	Observed After Discovery 1885
Atomic weight	72	72.60
Density (g/mL)	5.5	5.47
Gram-atomic volume (mL)	13	13.2
Specific heat capacity (cal/g °C)	0.073	0.076
Formula of oxide	EsO_2	GeO_2
Density of oxide (g/mL)	4.7	4.703
Molar volume of oxide (mL)	22	22.16
Formula of chloride	$EsCl_4$	$GeCl_4$
Boiling point of chloride (°C)	Under 100°	86°
Density of chloride (g/mL)	1.9	1.887
Molar volume of chloride (mL)	113	113.35

the periodic table was verified by the discovery of atomic numbers, the explanation of the causes of elemental properties came only with the arrival of wave theory. By explaining the characteristics of electron shells in atoms, wave theory was able to show why elemental properties depend on nuclear charge. As we study the form of the periodic table, we will learn how that form is determined by the electron shell structures of the atoms of the elements.

10.2 FORM OF THE PERIODIC TABLE

The most common arrangement of the periodic table, called the long form, is shown in Figure 10.1 and inside the front cover of the book.

In this form of the table, what are called the **representative elements** are shown in shaded blocks. Two other sets of elements are shown without shading: the **transition elements** appear as three blocks of ten elements each in the center of the table, the **lanthanide** and **actinide** elements are shown in the two horizontal rows across the bottom of the table. Most of our interest will be focused on the representative elements.

The seven horizontal rows of elements in the table are called **periods** (the lanthanides and actinides belong in the last two periods but are shown separately for convenience.) Each period except the first begins on the left with one of the **alkali metals** (Li–Fr) and ends with one of the **noble gases** (Ne–Rn). Sets of elements with similar properties are called **families** or **groups.** Families appear as vertical rows in the periodic table, thus the alkali metals constitute a family, as do the noble gases, and so on. The elemental families are given group numbers, which are printed above the vertical columns. For the first seven A groups (IA, IIA, and so on) the group number corresponds to the actual number of electrons in the outer shell of each atom of each element in the group. The noble gases are designated as Group 0 (zero) because they have no electrons lying outside closed shells. For the B groups, the numbers are related more to the actual properties of the elements than to electron shell structure.

If the periodic table were to be printed with the lanthanides and actinides in their proper places as members of periods 6 and 7, it would be easy to count the number of elements in

Key: 6 — Atomic number; C — Symbol; 12.0 — Atomic weight

I A	II A	III B	IV B	V B	VI B	VII B	VIII	VIII	VIII	I B	II B	III A	IV A	V A	VI A	VII A	0
1 H 1.00																	2 He 4.00
3 Li 6.94	4 Be 9.01											5 B 10.8	6 C 12.0	7 N 14.0	8 O 16.0	9 F 19.0	10 Ne 20.2
11 Na 23.0	12 Mg 24.3											13 Al 27.0	14 Si 28.1	15 P 31.0	16 S 32.1	17 Cl 35.5	18 Ar 39.9
19 K 39.1	20 Ca 40.1	21 Sc 45.0	22 Ti 47.9	23 V 50.9	24 Cr 52.0	25 Mn 54.9	26 Fe 55.8	27 Co 58.9	28 Ni 58.7	29 Cu 63.5	30 Zn 65.4	31 Ga 69.7	32 Ge 72.6	33 As 74.9	34 Se 79.0	35 Br 79.9	36 Kr 83.8
37 Rb 85.5	38 Sr 87.6	39 Y 88.9	40 Zr 91.2	41 Nb 92.9	42 Mo 95.9	43 Tc (98)‡	44 Ru 101.1	45 Rh 102.9	46 Pd 106.4	47 Ag 107.9	48 Cd 112.4	49 In 114.8	50 Sn 118.7	51 Sb 121.8	52 Te 127.6	53 I 126.9	54 Xe 131.3
55 Cs 132.9	56 Ba 137.3	57 La* 138.9	72 Hf 178.5	73 Ta 180.9	74 W 183.9	75 Re 186.2	76 Os 190.2	77 Ir 192.2	78 Pt 195.1	79 Au 197.0	80 Hg 200.6	81 Tl 204.4	82 Pb 207.2	83 Bi 209.0	84 Po (210)	85 At (210)	86 Rn (222)
87 Fr (223)	88 Ra (226)	89 Ac† (227)	104 —§ (260?)	105 —§ (260?)	106						112?						

*	58 Ce 140.1	59 Pr 140.9	60 Nd 144.2	61 Pm (147)	62 Sm 150.4	63 Eu 152.0	64 Gd 157.3	65 Tb 158.9	66 Dy 162.5	67 Ho 164.9	68 Er 167.3	69 Tm 168.9	70 Yb 173.0	71 Lu 175.0
†	90 Th 232.0	91 Pa (231)	92 U 238.0	93 Np (237)	94 Pu (242)	95 Am (243)	96 Cm (247)	97 Bk (247)	98 Cf (249)	99 Es (254)	100 Fm (253)	101 Md (256)	102 No (254)	103 Lr (256)

‡ Parentheses around atomic weight indicate that weight given is that of the most stable known isotope.

§ Both Russian and American scientists have claimed the discovery of elements 104 and 105. Official names have not been adopted yet.

Figure 10.1 Periodic table of the elements.

each period. In the usual presentation of the table, however, the two series of 14 elements are shown separately. Figure 10.2 is a periodic table with these two series in their proper places. We can get a better understanding of the relation between atomic structure and the periodic table by referring to this version.

> EXERCISE 10.1. Using the periodic table, give the symbol, atomic number, atomic weight, period, and group for silicon.

10.3 ELECTRON SHELL STRUCTURE AND THE PERIODIC TABLE

Let us now see how the form of the periodic table follows the sequence of electron shell filling in atoms. (Remember, however, that the electrons arrange themselves in energy levels, not in fixed layers. The words *shell* and *subshell* are commonly used to mean *energy level* and *sublevel.*) Because the information in this section is important to the understanding of later chapters, you may want to read it several times. Follow the discussion by looking frequently at Figure 10.1 and at the memory device for the order of electron shell filling, which you first met in Chapter 9. The device is printed again here for convenience.

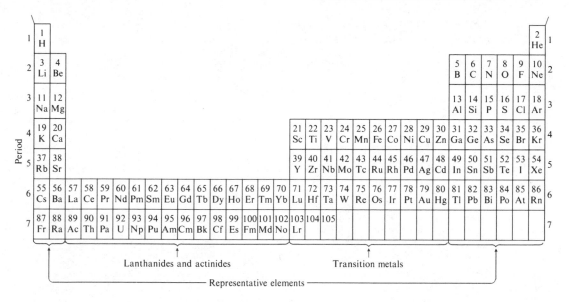

Figure 10.2   The periodic table as it appears when arranged in the natural sequence of increasing atomic number. If you are ever unsure about where the lanthanides and actinides fit, return to this table.

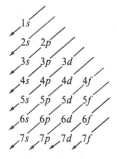

As you study the table, keep in mind that it was arranged according to the *properties* of the elements, but that our knowledge of electron shell structure comes from electron wave *theory*. That the table agrees so well with the theory might make you think that one came from the other, but the two were actually discovered independently. The remarkable agreement between them gives powerful support to the wave theory of the electron.

Here is a simple statement that will help us see in detail the relation between the periodic table and the theory of electron shell structure (as seen in the memory device): *Each of the elements (H, Li, Na, K, Rb, Cs, Fr) that starts a new period also represents the beginning of a new main electron shell.* Let us take some examples by visualizing the first few elements, using the method we learned in Chapter 9. We begin with a bare proton, the nucleus of the H atom. To make this into an atom, we add one electron; it goes into the 1s subshell. Next, we take the second element, He; its nucleus has two protons, and we must add two electrons to make a neutral atom. Both these electrons go into the 1*s* subshell.

H and He are the only two elements in the first period, and we see why as we create Li. Li has a nuclear charge of 3, requiring three electrons for the atom. Of these, two go into the 1*s* subshell, filling it and completing the first main shell. The third electron of Li must then go into the 2*s* subshell. Li has two electrons in the 1 shell and one electron in the 2 shell.

In moving from He to Li, we have started a new main shell and at the same time moved to a new period in the periodic table. To create the next seven elements, Be through Ne, we add electrons to their nuclei, successively filling the N = 1 shell, then the 2*s* and 2*p* subshells—in fact, filling the entire N = 2 main shell as we reach Ne.

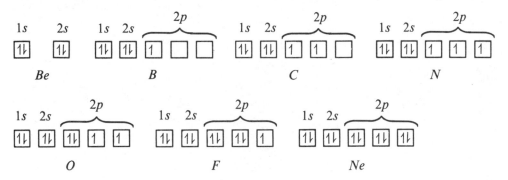

The next element is Na, which stars a new period in the table and opens a new main shell with its single 3*s* electron. Now the 3*s* and 3*p* subshells fill, but the 3*d* must wait until 4*s* is filled (see the memory device). Corresponding to this, the third period of elements that begins with Na has only eight members, the last being Ar. As we saw in Section 9.7, written electron configurations often include abbreviations to stand for filled inner shells of electrons, as in using {Ne} to mean $1s^2 2s^2 2p^6$. Similarly, when drawing electron population diagrams, we can use equivalent abbreviations. For example, we can use the box Ne core to represent the closed shells.

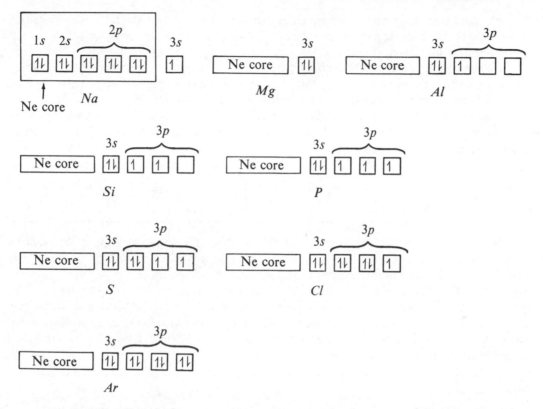

The $4s$ subshell fills next, represented by the two elements K and Ca, and then the $3d$ subshell fills. This process begins with the element Sc and continues through the next ten elements, finishing with Zn. Only now does the $4p$ subshell fill. The final result is that there are 18 elements in the fourth period.

K	Ca	Sc	Ti	V	Cr	Mn	Fe	Co	Ni	Cu	Zn	Ga	Ge	As	Se	Br	Kr

The fourth period

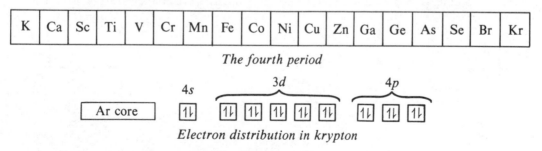

Electron distribution in krypton

Using the memory device and the expanded periodic table in Figure 10.2, you can now follow the form of the table all the way to the end.

The blocks of elements in the periodic table are sometimes given names that correspond to the subshells that are being filled for those elements. The columns headed by H and Be are called the *s* **block;** those headed by B, C, N, O, and F and the elements below He make the *p* **block;** the three horizontal series of ten elements in the center of the table, the transition elements, are the *d* **block;** and the two series of 14 elements (usually shown below the rest of the table) are the *f* **block** (Figure 10.3).

We can now tie all this information together: (1) the periodic table is arranged according to the properties of the elements, (2) the properties of the elements depend primarily on the nature of the outer electron shells of their atoms, and (3) the memory device reflects what wave theory tells us about the structure of electron shells in atoms. The close agreement of experiment (as seen in the periodic table) and theory (as seen in our memory device) is one of the triumphs of modern chemistry.

EXERCISE 10.2. Using the periodic table, give the period, group, block, and electron configuration for magnesium.

In Section 9.8 we learned how to find the number of valence electrons in an atom by writing the electron configuration. Now we can see that it is often possible to say how many valence electrons there are by looking at the position of an element in the periodic table. For the representative elements (the A groups), the number of valence electrons is the same as the group number (examples: one for Rb, four for Si, seven for Br). Although all transition elements formally have two valence electrons, there are many exceptions, and the reactions of these elements often involve inner shell electrons. It is usually best to consider the transition elements individually.

EXERCISE 10.3. What is the number of valence electrons in the atoms of each of the following elements: Al, Se, Xe, F, He, and Ge? Why is He an exception to the rule given earlier?

In the next few sections, we will examine some of the relations among the properties of various elements. Examples of *what* occurs will give us valuable orientation about the periodic table. We will learn in the next chapter *why* the properties are as they are.

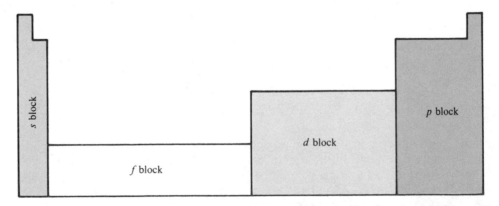

Figure 10.3 Blocks of the periodic table.

10.4 EXAMPLES OF PERIODICITY AND CHARACTERISTICS OF FAMILIES OF ELEMENTS

To find examples of the regular repetition of patterns of elemental properties in the periodic table, we will look first at the elements beginning with Li and ending with K. Li is metallic, but as we follow the row from Be to Ne, the elements become less and less metallic until we reach fluorine, the least metallic of all elements. Following fluorine is the noble gas neon. The next element is Na—which, however, is strongly metallic and resembles Li. As we continue across the row, element by element, we find the metallic character again diminishing until we reach Cl, a strong nonmetal that is followed by the almost inert Ar. A new row starts with the highly metallic K, and the cycle is repeated.

Study Figure 10.4, which shows the relative sizes of the atoms of some of these elements, and see how the pattern of atomic size repeats itself. Each row starts with an element that has large atoms, but the size decreases as we move right. (Because the sizes of the noble gas atoms cannot be measured in the same way as those of other atoms, noble gas atoms are not included in the drawings.) Although there are interruptions for the d and f blocks, the general pattern of decreasing size from left to right repeats itself through the entire table. It is from the periodic repetition of such patterns that the periodic table gets its name.

Of more interest to us, however, are the strong resemblances that elements within the same vertical column, members of the same family, bear to one another. As we have seen (Section 5.1) the table can be divided into sections of metals and nonmetals, but within each section are sets of elements that share many more properties than just their classification (Figure 10.5). The elements of Group VII, the **halogen** family, for example, are all strongly nonmetallic; they form binary compounds in which the halogen atoms combine one-to-one with hydrogen; they all have fairly low melting and boiling temperatures; their solids are all soft. The halogens also share many more properties. Side-by-side with the reactive halogens are the **noble gases,** Group 0, which until recently were known as the inert gases. Because their atoms all have tightly held outer shells of eight electrons, only under the most unusual circumstances can the noble gases react chemically and form bonds. The members of the family, He–Rn, are all monatomic.

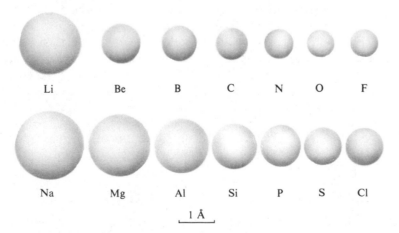

Figure 10.4   Examples of relative atomic diameters.

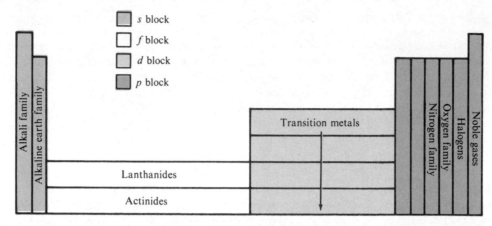

Figure 10.5 How the blocks of the periodic table relate to the subshell structures of the elements.

Table 10.3
Outer Shell Populations of Some Elements

Li	$2s^1$	He	$1s^2$	F	$2s^22p^5$
Na	$3s^1$	Ne	$2s^22p^6$	Cl	$3s^23p^5$
K	$4s^1$	Ar	$2s^23p^6$	Br	$4s^24p^5$
Rb	$5s^1$	Kr	$4s^24p^6$	I	$5s^25p^5$
Cs	$6s^1$	Xe	$5s^25p^6$		

At the other side of the table are the alkali metals, Group I, which share melting and boiling points that are quite low for metals. The alkalis are all soft and malleable, and they all form water-soluble compounds in which the metals appear as singly charged positive ions. Next, the elements of the **alkaline earth** family, Group II, form compounds that are often insoluble in water and contain doubly charged metal ions. Alkaline earths are less reactive, harder, and have higher melting temperatures than do the alkalis. The halogens, the alkali metals, the alkaline earth metals, and the noble gases are the families that exhibit the most distinct family characteristics. The elements in the other vertical columns also display family characteristics, but often to a somewhat lesser extent.

As a review before we go on, let us have another look at the relation between electron shell populations and the properties of elements. The alkali metals, for instance, have strikingly similar chemical and physical properties; they also have only a single outer electron in each of their atoms (see Table 10.3). The noble gases are all quite inert chemically and also have similar physical properties; their atoms all have filled outer shells. Atoms of all the halogens have seven outer electrons, those of the alkaline earth family have two, and so on. The properties of the elements are evidently determined primarily by the outer electrons of their atoms. How and why this occurs is something that we will learn gradually as we go along.

10.5 CHANGES IN PROPERTIES WITHIN ELEMENTAL FAMILIES

Although families of elements are usually characterized by strong resemblances, there are also differences in properties among the elements of most families. Subshell occupancy patterns are the same for all the elements in a vertical column in the periodic table, therefore differences in the properties of such elements must be caused by another factor. That factor is the distance of the valence electrons from the nucleus. Because electrons in shells of larger N spend, on the average, more time farther from the nucleus than do those in shells of smaller N, the properties of the elements change with their vertical position in the table. In the Group V elements, for example, each of the atoms has two s electrons and three p electrons in its outer shell, but those five electrons are in successively larger shells; N = 2 in nitrogen, N = 3 in phosphorus, N = 4 in arsenic, and so on. The size of the atoms and, as we will see, the metallic character within a family generally increases from top to bottom in the column (Figure 10.6).

Changes of properties within families are quite apparent in the center groups of the table, where the columns cross the boundary between the metals and the nonmetals. In Group VA, for instance, the top elements, N and P, are classic nonmetals, but As and Sb display some

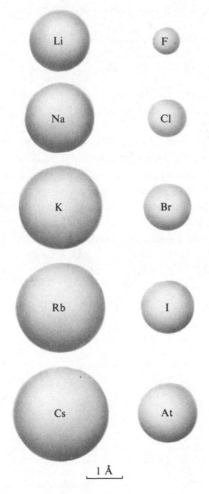

1 Å

Figure 10.6   Relative sizes of the atoms of the alkalis and the halogens.

metallic character, and Sn and Pb must be classed as metals. The increase in metallic character is hardly noticeable, however, either in the strongly nonmetallic elements (because even those at the bottom of the column never really become metallic), or in the metal families (because they are metals anyway). In the alkali metals, for example, the vertical change in properties is not obvious: the melting temperature and hardness of the metals decrease a little from Li to Cs; the tendency to react chemically increases slightly, and there are other small differences.

The step-shaped boundary between the metals and nonmetals reflects the fact that metallic behavior is strongest in the left and lower parts of the table and weakest in the upper and right portions. The few elements lying close to the boundary, the metalloids, display both metallic and nonmetallic properties and cannot properly be included in either class.

EXERCISE 10.4.   Using the periodic table and the known properties of families, describe the element helium in terms of size, reactivity, and metallic character. Do the same for aluminum.

10.6 FORMULAS OF COMPOUNDS AND THE PERIODIC TABLE

A good example of periodicity and family properties is given by the sets of characteristic formulas of compounds formed by the elements of many families. Table 10.4 shows the formulas of the oxides of some of the elements, arranged by family. Notice that the same atom number ratio appears in all the formulas in each vertical column.

Suppose we had been given the formulas only for the oxides of Li, Na, and K, (Li_2, Na_2O, K_2O) and had been asked to make a guess as to what the formula for the oxide of Rb would be. We would undoubtedly have guessed Rb_2O, which is correct. Similarities in formulas occur not only in the oxides of the alkalis, but also in the compounds of many families. In many cases, knowing the formula for one compound of one element of a family will allow you to make an accurate guess about similar compounds formed by other members. The formulas for some of the chlorides of the alkalis are NaCl, RbCl, and CsCl, while those for some of the sodium halides are NaCl, NaBr, and NaF.

Table 10.4
Formulas of Some Oxides
(Arranged According to Periodic Law)

H_2O				
Li_2O	BeO	B_2O_3	CO_2	N_2O_5
Na_2O	MgO	Al_2O_3	SiO_2	P_2O_5
K_2O	CaO			
Rb_2O	SrO			
Cs_2O	BaO			

Note: this list does not necessarily give all of the oxides formed by these elements, only a selected group chosen to illustrate periodicity. The elements of many families, however, form only one type of oxide, BeO and MgO, for example.

EXERCISE 10.5.   What would you guess to be the formula for the compound of Rb and I?

The game is easy with the alkalis, the alkaline earths, and the halogens—and it also works in other places. With experience, you will find the kinds of compounds for which you can correctly make up formulas and you will learn which elements can fool you.

10.7 THE SPECIAL NATURE OF HYDROGEN

Later in the book, we will examine the chemistry of hydrogen in detail. We need to know now, however, that hydrogen is not metallic, although it does lie in the column with the alkali metals. Most of the physical and chemical properties of H are those of a nonmetal, and H has in addition many properties not shared by any other element. There are two reasons for the unusual character of hydrogen: (1) it is the second smallest atom (He is smaller) and (2) its electron lies close to the nucleus, without any inner shell electrons to shield it from the nuclear charge. When H loses its electron to become the ion H^+, only the bare nucleus is left. (Since He is unreactive, H is the only atom that exists in ionic form as only a nucleus.) Because of the unique properties given H by these characteristics, it cannot really be classed in any family but must be studied separately. We will again and again encounter situations in which H displays its unique behavior.

10.8 TRANSITION ELEMENTS

The elements in Groups IB through VIIB and Group VIII are known as the **transition elements** (the d subshell elements). The two groups of 14 elements each that appear at the bottom of the table are called the **lanthanides** and **actinides** (the f subshell elements). These elements occur in periods 4, 5, 6, and 7. In periods 4 and 5, the $3d$ and $4d$ subshells are those that are partially complete (Sc through Cu and Y through Ag). (See Table 9.1.) In period 6, the $4f$ and $5d$ subshells are being filled (La through Lu), while in period 7, the $5f$ subshell fills (Ac through Lr), and $6d$ begins to fill.

Because the transition elements are those in which only the inner electron shells differ, the differences in properties among the members are not as great as those usually seen among elements with *outer* electron arrangements. The three series of ten d-subshell transition elements, all of which are metals, have an interesting and unusual chemistry, which we will study further in a later chapter. In the lanthanides and actinides, the differences in electron shell populations occur only in the f subshells, each of which is buried under subsequent s and p subshells. As a result, the properties of the elements in these two series are almost indistinguishable.

10.9 SUMMARY

Long before the concept of atomic number was developed, chemists had begun to arrange the elements according to atomic weight and to group together elements of similar properties. What is now known as the periodic table of the elements was the result of these efforts. The subsequent discovery of atomic numbers led to improvements in the table and provided a new basis for the arrangement that had previously reflected only observed properties. A full explanation of why the table takes the form it does did not become available, however, until the second quarter of the twentieth century when the wave theory of electrons in atoms was developed.

The elements can conveniently be divided into three sections, the representative elements, the transition elements, and the lanthanide and actinide section. The last two are usually separated from the other elements in the table and shown as two rows of 14 elements each. In the standard form of the periodic table, the representative elements are shown in two blocks (the s block and the p block), separated by the three rows of transition elements (the d block).

The table is further divided into vertical columns called groups and horizontal rows called periods. There are seven periods and 18 groups.

The form of the table follows the electron structures of the atoms of the elements. The first group, for example, the column in the table that is farthest left, contains elements whose atoms have a single electron in the outer shell. If we take the representative elements from left to right of any period, we find that the number of outer shell electrons follows the group number. For example, in the third period, Na to Ar, Na atoms have one outer shell electron, Mg atoms have two, Al atoms have three, and so on.

Because the properties of the elements are determined by the electron shell structures of the atoms, the vertical groups of elements form families whose members exhibit strong resemblances in physical and chemical properties. While there are differences in properties within elemental families, especially in Groups III through VI, it is nonetheless possible to predict many properties of an element just by knowing what group it belongs to.

Although hydrogen falls in the first group of elements with the alkali metals, it does not share the properties of the alkalis but has its own individual characteristics. The distinctive physical and chemical properties of hydrogen arise from its structure: the single outer electron of a hydrogen atom is directly exposed to the nucleus without being shielded by inner electron shells. Moreover, when an atom of H loses an electron in a chemical reaction, it becomes a bare nucleus; no other atom shares that characteristic. (He, remember, is not reactive and does not lose electrons.)

QUESTIONS

Section 10.1

1. What did Newlands use as a basis for his proposed table?
2. What did Mendeleev use as a basis for the periodic table he designed? Explain.
3. Mendeleev published a paper that predicted the properties of then undiscovered elements. Why was this prediction important?
4. The periodic table was originally based on the atomic weights of the elements, rather than on the atomic numbers (which were not then known). Is a periodic table based on atomic weights consistent with the fact that most of the elements have several isotopes? Discuss.

Section 10.2

5. Explain how groups and periods are arranged in the periodic table.
6. What do families of elements have in common besides similar properties?
7. To what period do the actinides belong? The lanthanides?

Section 10.3

8. Explain what is meant by the statement that "each period represents the beginning of a new main electron shell."

9. Why does the first period have only two elements, while periods 2 and 3 each contain eight elements?

10. Distinguish between the terms *s block, p block, d block,* and *f block.* Which of these blocks corresponds to the transition elements? To the lanthanides and actinides?

11. What is the structural factor that relates the elements in any group of the periodic table? In any period? In any block?

Section 10.4

12. In what part of the periodic table are the elements of the most metallic character located? The least metallic?

13. Name some elemental physical and chemical properties that demonstrate periodicity. Give examples in each case to support your answer.

14. List the similarities and differences between Groups IA and IB, and between Groups VIIA and VIIB.

15. In which groups do the atoms of the elements each have four electrons in the outer shell? In which groups do atoms have six?

16. Why are the physical and chemical properties of He, Ne, and Ar similar?

Section 10.5

17. Why does the size of the atoms of the elements in any family increase as the elements are taken from top to bottom in the table?

18. Which of the elements O, S, Se, Te, and Po shows the greatest metallic character? Why?

19. Phosphorus is a nonmetal and Bi is a metal, yet both are in the same family. Explain.

20. The metallic properties of the elements increase from the top to the bottom of the periodic table. Why is the increase in metallic properties not obvious in Group IA?

Section 10.6

21. The compounds NaOH, NaCl, and $MgCl_2$ are well known. What is the formula for the hydroxide of calcium?

22. The following compounds are well known: $MgBr_2$, NaCl, and NH_3. What are likely formulas for the compounds of the following pairs of elements: Ba and I, Rb and Br, As and H?

23. Carbon dioxide has the formula CO_2. Write formulas for the dioxides of the other elements in Group IVA.

24. Each atom of the elements of Group IB has one electron in its outer shell. Atoms of the elements of Group IA also have single-electron outer shells. Knowing that NaCl and KCl exist, would you have predicted that CuCl and AgCl exist? Explain fully.

Section 10.7

25. The alkali metals are the most metallic family known. H lies in the same column as the alkali metals, yet is not considered a metal. Why are the properties of H different from those of the alkali metals?

26. In many periodic charts, H appears twice, once above Li and once above F. Discuss a probable reason for this unique listing.

27. Why doesn't He exhibit the same unusual properties as H?

MORE DIFFICULT QUESTIONS

28. Use the periodic table to make choices in parts a–d. For each answer, base your reason on the structure of the atoms involved, not on the position of the element in the table.
    a. Which has the larger atom size, Na or K? Se or S? O or N?
    b. Which is the more likely compound, PH_2 or PH_3? TeH_2 or TeH?
    c. Which is the better conductor of electricity, P or As?
    d. Which has the greater density, Li or B?
29. The properties of the five elements B through F vary greatly; those of the five elements Ti through Fe are more similar, and the properties of the five elements Nd through Gd are almost identical. Why are there differences in the similarities of properties among these three horizontal series of elements?

PROBLEMS

Section 10.2

1. Which elements are in Group IIA? In period 2?
*2. Give the symbols of the elements that fit these descriptions.
    *a. Period 2, Group VIIA   e. Period 4, representative elements
    *b. Period 4, Group VIB   f. Group VA
    *c. Period 1   g. Group IIB
    *d. Period 6, Group IIA   h. Period 7, Group IIIB
*3. Give the period and group for each of the following elements:
    *a. Ar   c. Sn   e. Sr
    b. As   d. Ir
4. Without looking at another table, make from memory a drawing of the periodic table through element 86. Use atomic numbers to identify the elements. Fill in the symbols for the elements you already know.

Section 10.4

5. Classify the following elements as metals, nonmetals, or metalloids: Cs, C, S, Cl, As, Ne, Si.
*6. In which families would the undiscovered elements with the following atomic numbers belong?
    *a. 107   c. 119
    b. 117   d. 122
*7. Complete the following table:

Atomic Number	Name of Element	Electron Configuration	Class (Metal or Nonmetal)
*12			
25			
14			
9			
19			
34			

*8. Draw the electron dot diagrams for the following elements:

   *a. Rb   *d. I   g. H

   *b. Se   e. As   h. He

   *c. S   f. B   i. Ne

*9. Using the periodic table, give the block, group, period, and outer shell configuration of the following elements. Classify each as a metal or nonmetal.

   *a. magnesium   f. copper

   b. phosphorus   g. americium

   c. potassium   h. vanadium

   d. bromine   i. nitrogen

   e. krypton   j. cesium

10. An element with the atomic number M is a noble gas. What kind of element is the one with atomic number $M - 2$? $M + 2$?

Section 10.5

*11. List the symbols of the elements in the following groups in order of increasing size of the atoms.

   *a. K, As, Ga, F   d. O, Ne, Sn, P

   b. Cs, Rb, Br, Cl   e. Ta, V, U, Cu

   c. Cs, Al, Na, Ar

12. It is said that the element 118 exists on Xanthu. Do you believe that the element could exist anywhere? Why? If it did exist, what would be its period, group, size, block, and outer shell configuration? What other information can you estimate about it?

Chapter 11

How Elements Form Compounds

The process by which elements unite to form compounds is one of the several kinds of chemical reaction. (Other kinds involve compounds that already exist.) Understanding how compounds are created will lead us easily to an understanding of other kinds of reactions. We therefore begin our study of reaction processes with compound formation.

To gain information about the kinds of compounds formed by different kinds of reactions, we examine the distinctive varieties of the chemical reactions of elements. What we learn will eventually enable us to predict, for almost any two elements in the periodic table, the kind of reaction that will be caused and the kind of compound that will be formed.

We will also discover how to predict the formulas of compounds made from particular sets of elements. As we then study the characteristics of molecules, we will learn how to draw their electron dot structures and make some predictions about their geometric shapes.

AFTER STUDYING THIS CHAPTER, YOU WILL BE ABLE TO

- calculate the kernel charge of an atom
- explain the net attractive forces on the valence electrons of individual elements and show how those forces relate to the positions of the elements in the periodic table
- give rough estimates of the ionization energy and electron affinity of several elements and state the relation between those values and the positions of the elements in the periodic table
- tell how many electrons the atoms of any particular element are likely to gain or lose in an electron transfer reaction
- list several properties of ions
- tell whether any pair of elements is likely to react chemically and whether the reaction will be an electron transfer or a sharing of electrons
- recognize the oxidizing and reducing agents in oxidation-reduction reactions

- find the oxidation numbers of the elements in ionic compounds
- predict the formulas of the products of oxidation-reduction reactions on the basis of the oxidation number changes that occur in the process
- make electron dot or bond line drawings of covalent molecules, showing valence electrons and shared pairs
- determine how many covalent bonds the atoms of a particular element are likely to make
- give the formulas for several common polyatomic ions and draw electron dot or bond line structures for them
- write chemical reactions in electron dot language
- draw structural formulas for molecules
- draw diagrams of structural isomers of some molecules
- list the properties of ionic and covalent compounds

TERMS TO KNOW

anion	dissociation	Lewis structure
bond angle	electron affinity	oxidation
bond line drawing	electron transfer	oxidation number
bond structure	ion	oxidizing agent
cation	ionic bond	reducing agent
chemical bond	ionic solid	reduction
coordinate covalent bond	ionization energy	structural isomer
covalent bond	kernel charge	

TERMS WITH WHICH TO BE FAMILIAR

space-filling model	VSEPR theory
tinkertoy model	

11.1 INTRODUCTION

Just as a brick falls to earth in response to the attractive force of gravity, or a compass needle moves under magnetic forces, so atoms and electrons respond to the forces on them by undergoing chemical reaction. By far the most important requirement for an understanding of chemical reaction and its results is an understanding of the forces exerted on electrons before and after reaction and the resulting forces exerted on atoms. The latter forces are called chemical bonds; they are the principal object of most of the study of chemistry.

A **chemical bond** is a strong force that holds two or more atoms together. Chemical bonds create structures that are stable and that can be identified by formulas, either of elements (as in N_2 or S_8) or of compounds (as in H_2O or $NaCl$). Although this definition of a chemical bond is useful for most purposes, it does have some notable exceptions: a sample of diamond, for example, consists of firmly bonded carbon atoms, but no formula can be written for diamond. Many nonstoichiometric compounds are chemically bonded but can be given no formulas; a crystal of $KAl(SO_4)_2$, for example, may contain ions of Co in place of some of the Al ions. Whether or not the H-bond (responsible for the double-helical structure of DNA, among other things) can be called a chemical bond is subject to discussion. Remember that there are such exceptions, although we will ordinarily use the simple definition of the chemical bond.

Although our studies in this chapter will be limited to reactions between elements, we should keep in mind that chemical reaction is not necessarily limited to pairs of elements. The fundamental facts we will learn by examining the reactions of elements, however, also hold for reactions of other species.

11.2 FORCES ON ELECTRONS IN ATOMS

We learned in Section 9.8 about the atomic kernel and the outer shell electrons of atoms (the electrons with the highest principal quantum number, N), also called valence electrons. Because valence electrons are those usually involved in chemical reaction, we must consider these electrons in more detail, particularly in their interactions with the atomic kernel. (You may wish to review Sections 9.8 and 10.3, to refresh your memory about how to find the number of valence electrons in an atom.)

As a rough approximation, let us assume that a valence electron is influenced mostly by the forces exerted on it by the nucleus, and by the electrons of smaller N in the atom. In other words, the force attracting a valence electron to an atom depends primarily on the *net* charge of the atomic kernel, what is called the **kernel charge**: the total positive charge of the nucleus, partially offset by the negative charges of the inner shell electrons. Our approximation is justified by the fact that the inner shell electrons spend their time, on the average, closer to the nucleus than do the valence electrons. The inner shell electrons screen the valence electrons from the full effect of the nuclear charge. (See Figure 11.1.) Later, when we study the finer details of electron energies, we will glance at the effects that outer shell electrons have on each other.

Now, recalling what we learned in Section 10.4 about the electron populations of the outer shells of families of elements, let us examine the kernel charges of the atoms of a few

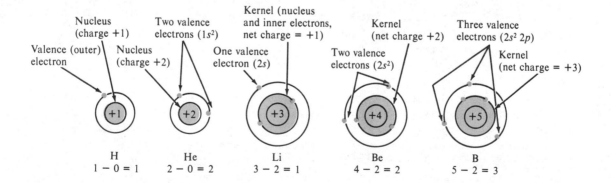

Figure 11.1 Drawings of atoms, showing kernel and valence electron arithmetic.

elements. Hydrogen, of course, has only its single $1s$ electron, which lies close to the bare nucleus. In H, the kernel is the nucleus itself, and the kernel charge is 1. (There is no need to place a + sign in front of a kernel charge because kernel charges are always positive.) Helium is the only other element whose outer electrons are directly exposed to the nucleus, and its kernel charge is 2. With Li, however, an inner shell appears. The $2s$ electron of Li experiences a kernel charge of $(+3 - 2) = 1$, the +3 coming from the nucleus and the −2 from the charges of the two inner-shell $1s$ electrons. Similarly, the valence electron of Na also experiences a kernel charge of 1, that is $(+11 - 10)$. In fact, all the alkali metals have kernel charges of 1 (Figure 11.2).

For Be, we find a kernel charge of $(+4 - 2) = 2$, and for B a kernel charge of $(+5 - 2) = 3$. In neither of these cases do we count outer shell electrons. Valence electrons are not a part of the kernel. Notice that all the electrons in any outer shell experience the same kernel charge. In carbon, for example, the kernel charge of 4 is felt by each of the four outer electrons.

Now, continuing across the first full row of the table, we find that the kernel charge for nitrogen electrons is 5, that for oxygen electrons is 6, that for fluorine electrons is 7, and that for neon electrons is 8 (Figure 11.3). In general, the kernel charge for any outer electron in a neutral atom is the same as the number of valence electrons in that atom.

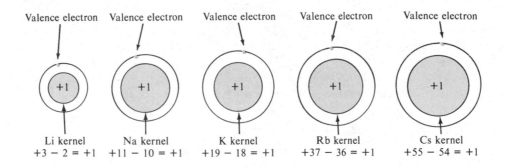

Figure 11.2 Electron structure of alkali metal atoms.

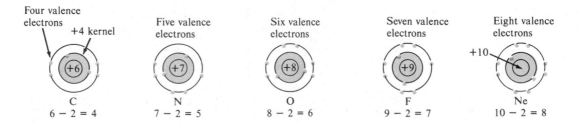

Figure 11.3 Structure of C, N, O, F, and Ne atoms.

EXERCISE 11.1. What is the kernel charge experienced by a valence electron in an atom of aluminum? Of phosphorus?

EXERCISE 11.2. Name five elements whose kernel charge is 4.

11.3 IONIZATION ENERGY AND ELECTRON AFFINITY

As we shall see, chemical reaction involves the loss, gain, or sharing of the outer shell electrons of reacting atoms. For that reason, it is useful to have a way of measuring and referring to the strength of the forces with which atoms attract their own outer electrons as well as extra electrons that might be added to their outer shell. Every atom has substantial attraction for its own electrons, and many have attraction for other electrons as well. The two physical properties primarily involved in these attractions are the **ionization energy** and the **electron affinity.**

Ionization energy is a measure of the amount of work required to remove an electron from a neutral atom and take that electron far away. We see an example of ionization energy when an electron is taken, by one means or another, from a sodium atom.

$$Na\cdot + \text{ionization energy} \longrightarrow \quad Na^+ + \quad e^-$$

Atom   *Ion*   *Electron*
(Separated to a large distance)

Because the removal of an electron requires working against the attractive force of the kernel, the *ionization energy is a measure of the attraction an atom has for the most easily removed outer electron.*

Electron affinity, on the other hand, *is a measure of the attraction that a neutral atom has for an extra electron brought from far away and added to its outer shell.*

Chlorine, for example, readily accepts electrons.

$$Cl + \quad e^- \longrightarrow Cl^- + \quad \text{energy}$$

Atom   *Electron*   *Ion*   *(This energy is called*
(Widely separated)   *the electron affinity)*

A working knowledge of ionization energy and electron affinity is necessary for the understanding of chemical reaction. Knowing the relation between these properties and the position of an element in the periodic table will enable you to make shrewd guesses about what kinds of reactions can take place and what kinds of results will follow for any pair of elements in the table. Let us see how ionization energy and electron affinity values are related to position in the periodic table.

11.4 THE PERIODIC TABLE AND FORCES ON ELECTRONS

As we have seen, the attraction of an atom for its outer electrons is created primarily by the kernel charge. Atoms with a small kernel charge attract their electrons less than do those with greater charge. A second important influence on electron attraction is the average distance of the electron from the nucleus. Other things being equal, electrons that spend their time far from the nucleus are held less firmly than those that lie closer.

Given these two influences, we would expect the ionization energy of the elements in a horizontal row of the table generally to increase as the kernel charge increases. A graph of the ionization energies of the elements from Li to Ne, shown in Figure 11.4, displays that effect. In this series, the kernel charge increases from the 1 charge of Li to the 8 charge of Ne, and the ionization energy rises accordingly.

For most purposes, we can work on the assumption that the ionization energy curve would move smoothly upward as the kernel charge increases from Li to Ne, but the difference in energies of the s and p subshells actually produces a small dip in the curve between beryllium and boron. Although the effects of the dip will not concern us, we can strengthen our understanding of the factors that determine the energies of electrons in atoms by studying Figure 11.5 to see why the dip occurs. A second dip in the ionization energy curve occurs between

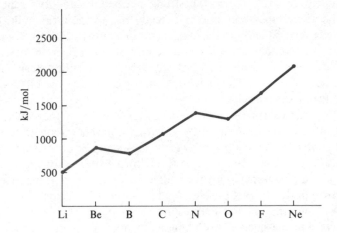

Figure 11.4 Plot of first ionization energies, lithium through neon.

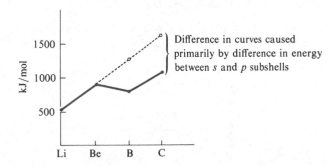

Figure 11.5 Relation of subshell energies to ionization energies for lithium, beryllium, and boron. The dotted line shows what the ionization energies of B and C might be, if being in the p subshell did not help the first electron of these atoms to ionize.

nitrogen and oxygen, caused by the fact that oxygen is the first element in which two electrons must occupy the same *p* orbital. In an orbital with a pair of electrons, the mutual repulsion raises the energy of both electrons, making it easier to remove one of them from the atom. This effect outweighs the increase in kernel charge from nitrogen to oxygen, and oxygen therefore has a slightly smaller ionization energy than that of nitrogen.

To see the effect of increasing the distance between the electron and the nucleus, look at the curve for the ionization energies of the alkali metals, Li to Cs. The single outer 2*s* electron of Li spends most of its time closer to the nucleus than does the 6*s* electron of Cs, because the 6*s* electron lies "outside" five inner shells of electrons (Figure 11.6). As a result of the greater distance, the 6*s* electron of Cs is attracted to the Cs kernel somewhat less than the 2*s* electron of Li is to its kernel, and the ionization energy shows this effect. The ionization energies of the alkali metals become slightly smaller as we take the elements from top to bottom in the family, a general trend that is also found in the other families of the table.

Finally, we should look at the pattern of ionization energies as it relates to the entire periodic table. In Figure 11.7, notice the strong periodicity and the repetition of family characteristics. The curve repeatedly makes a steady rise to the large ionization energy of a rare gas, then plunges dramatically to the low ionization energy of an alkali metal. Notice also that the curve tends to level out when it comes to any of the transition metal series or to the lanthanides or actinides, an effect consistent with the many other close similarities in the properties of these sets of metals. The ionization energy curve gives us one of the best displays of periodicity and familial similarities.

In a general way, the values of electron affinity tend to follow those of ionization energy. If an atom attracts its own electrons strongly, it will also have a strong attraction for an extra electron *provided that there is a place for that electron in the outer shell of the atom.* For example, electron affinity generally increases from Li to F in the first row, but is essentially zero for Ne. The noble gas atoms, with their filled outer shells, have no attraction for extra electrons. An electron in the 3*s* level for Ne would experience no kernel charge. If we look at a vertical column, we find that as the ionization energy decreases from F to I, so does the electron affinity. We will not need to use numerical values of electron affinity and ionization energy; it is sufficient for us to recognize the trends.

11.5 FORMATION OF IONS BY ELECTRON TRANSFER REACTIONS

The atoms of some elements (primarily the nonmetals) hold their valence electrons tightly and exert strong attraction even for electrons not their own. In other words, they have large

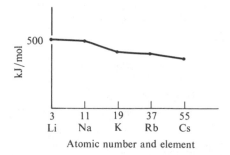

Figure 11.6   Ionization energy curve for alkali metals, lithium through cesium.

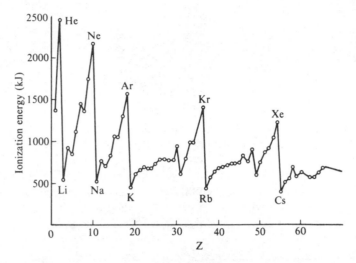

Figure 11.7   Ionization energies of some of the elements. The energy required to remove an outermost electron is plotted against atomic number, Z. Notice the recurrence of peaks at Z = 2, 10, 18, 36, 54, and 86. The differences between these atomic numbers are the same as those between the atomic numbers in Figure 11.6. For example, 18 − 10 = 19 − 11; 54 − 36 = 55 − 37.

ionization energies and electron affinities. Atoms of other elements (the metals) hold their own electrons weakly and have little or no attraction for extra electrons. Considering this information, it seems likely that atoms of nonmetals would take away the loosely held valence electrons of metal atoms and add those electrons to their own valence shells. This process does indeed occur, and is called **electron transfer** or oxidation-reduction. (As we will learn in Section 11.7, a special language is used to refer to electron transfer reactions; the terms *oxidation* and *reduction* are part of that language.)

The transfer of one or more electrons between atoms results in the formation of **ions,** atoms or groups of atoms with electric charges. Ions associate with one another to form compounds. Because electron transfer is one of the principal forms of chemical reaction, we will need to study it in some detail.

We must first take notice of a fact that is often misinterpreted or ignored. Any neutral atom attracts its own valence electrons; thus there is no atom that "wants" to lose electrons. More than that, no atom "wants" an extra electron more than another atom "wants" its own valence electrons. Electron transfer does occur, and frequently, but it usually happens as a part of a process involving several steps, not simply because an atom of one kind steals an electron from an atom of another kind. The chemical reaction that includes electron transfer as a step is driven primarily by the assembly of the ions into a solid compound. To understand electron transfer, we need not consider all the steps in detail, but we do need to know that the reaction includes them. Such a transfer would not occur between two atoms isolated together in space.

Electron transfer reactions, as well as other reactions, often occur in such a way that the resulting ions have outer shells of eight electrons (H being a notable exception). An atom that gains electrons can fill its valence shell with up to eight electrons, but it will then not have attraction for further electrons; any new electrons would encounter only a negative kernel charge and would thus be repelled. (See Figure 11.8.)

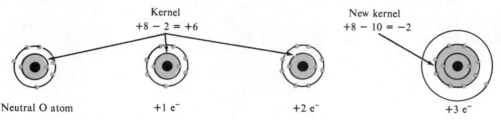

Kernel
+8 − 2 = +6

New kernel
+8 − 10 = −2

Neutral O atom   +1 e⁻   +2 e⁻   +3 e⁻

Figure 11.8   Oxygen atoms can add two electrons to become O^{2-} ions. Notice how the kernel charge changes from +6 to −2 when the second main shell fills.

On the other hand, some kinds of atoms can lose electrons from their valence shells because those electrons experience only a small kernel charge. When all the valence electrons are gone, the resulting ion will then have a new outer shell of eight electrons (formerly the highest inner shell), each of which feels a substantial kernel charge and will not easily be lost. (See Figure 11.9.)

EXERCISE 11.3.   How many electrons is a Cl atom likely to add in an electron transfer reaction? What kernel charge would be experienced by a second or third electron added to a Cl atom? How many electrons would an S atom ordinarily add?

EXERCISE 11.4.   How many electrons is Al likely to lose in an electron transfer reaction? Use electron dot diagrams to show the atom of Al and its ion. After two electrons are lost from Al, what kernel charge does the third electron feel? After three are lost, what kernel charge does the fourth feel?

We will use as our guide the fact that filled outer shells of eight electrons often result from electron transfer. This is a further application of the octet rule (Section 9.7). Using the octet rule, we will soon be able to predict the formulas of ionic compounds.

Since we know that atoms become electrically charged when they lose or gain electrons, we can calculate the amount of charge on the resulting ions. If a neutral atom adds one or more electrons, it will also be adding a corresponding number of extra negative charges, and will become a negative ion. Negative ions are called **anions**. (See Figure 11.10.) When an

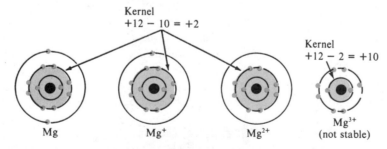

Kernel
+12 − 10 = +2

Kernel
+12 − 2 = +10

Mg   Mg⁺   Mg²⁺   Mg³⁺
(not stable)

Figure 11.9   Loss of electrons by Mg atoms. The kernel charge for the two valence electrons in the Mg atom is 2, but in the Mg^{2+} ion, the kernel charge for the outer electrons (now the $2s$ and $2p$ electrons) is 12 − 2 = 10. This is why there is no stable Mg^{3+} ion.

$$:\ddot{F}: + 1\ e^- = (:\ddot{F}:)^-$$

atom   ion

$$:\ddot{O} + 2\ e^- = (:\ddot{O}:)^{2-}$$

atom   ion

Figure 11.10 Creation of two kinds of negative ions.

$$\text{Na} - 1\ e^- = \text{Na}^+$$

atom   ion

$$\text{Mg} - 2\ e^- = \text{Mg}^{2+}$$

atom   ion

Figure 11.11 Creation of two kinds of positive ions.

atom loses one or more electrons, however, it no longer has enough electrons to equal the number of protons in the nucleus, and the charge balance becomes positive; that is, the atom becomes a positive ion. Positive ions are called **cations**. (See Figure 11.11.) The amount of charge on either a positive or negative ion is exactly the same as the number of electrons gained or lost, since each electron carries one unit of charge. To indicate ions and their charges, use the symbol of the element and a superscript on the right-hand side showing the sign and amount of charge the ion carries. For example, the charge on a Mg ion is +2, written as Mg^{2+}.

EXERCISE 11.5.   Write the symbols, including charge, for the ions of Cl, I, S, Ba, and Al. Use the form just shown.

11.6 IONIC COMPOUNDS

We have now learned some very powerful knowledge; let's see how we can put it to work in a practical way. Because the atoms of many nonmetals are attractive to electrons, and the atoms of many metals lose their electrons relatively easily, electron transfer reactions occur between many metal-nonmetal pairs, sometimes violently. Consider what happens when sodium atoms transfer their electrons to chlorine atoms (Figure 11.12). Positive sodium ions and negative chlorine ions result, and the forces that these ions exert on each other are the chemical bonds in ionic compounds. Notice, however, that the electron transfer process does not create specific bonds between particular pairs of ions. Instead, *all* the positive and negative ions exert forces on one another. Under ordinary circumstances, for example, there are no molecules of the sort

$$\text{Na}^+\ \text{Cl}^-$$

It is too easy for two such sets to join and form

$$\text{Na}^+\ \text{Cl}^-$$
$$\text{Cl}^-\ \text{Na}^+$$

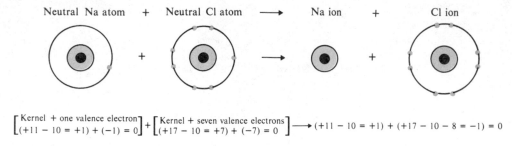

Neutral Na atom + Neutral Cl atom ⟶ Na ion + Cl ion

$$\left[\begin{array}{l}\text{Kernel + one valence electron}\\(+11 - 10 = +1) + (-1) = 0\end{array}\right] + \left[\begin{array}{l}\text{Kernel + seven valence electrons}\\(+17 - 10 = +7) + (-7) = 0\end{array}\right] \longrightarrow (+11 - 10 = +1) + (+17 - 10 - 8 = -1) = 0$$

Figure 11.12 Example of electron transfer arithmetic. Notice that the total charge on each side is zero.

or three to join and form

$$\boxed{\begin{array}{lll}\text{Na}^+ & \text{Cl}^- & \text{Na}^+\\ \text{Cl}^- & \text{Na}^+ & \text{Cl}^-\end{array}}$$

and so on, a process limited only by the number of ions present. In the solids formed by ionic compounds, very large numbers of ions join together so that each ion is surrounded by neighbors of opposite charge, and is equally attracted to all of these. Such substances are called **ionic solids** (Figure 11.13). The ionic solid formed by an electron transfer process is a chemical compound, and differs from the elements that entered the reaction. The forces that hold an ionic solid together are often called **ionic bonds;** they are one kind of chemical bond.

The diagonal separation of any given metal from any nonmetal in the periodic table indicates how likely those two elements are to react chemically in an electron transfer reaction

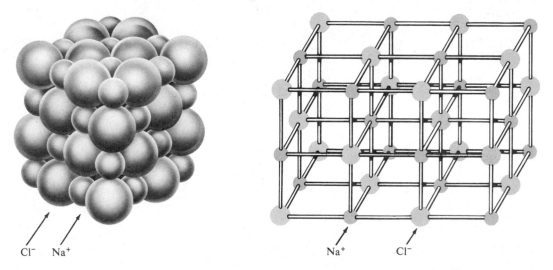

Cl⁻   Na⁺   Na⁺   Cl⁻

Figure 11.13 Models of the sodium chloride crystal. Each diagram represents a small fragment of sodium chloride, which forms cubic crystals. Each sodium ion is surrounded by six chloride ions, and each chloride ion is surrounded by six sodium ions.

and form an ionic solid. (Remember that we ignore the rare gases in these predictions because rare gas atoms cannot ordinarily add extra electrons to their already filled shells.) For example, fluorine and cesium, which lie almost at the corners of the table, will react explosively to give CsF, whereas tellurium and germanium will not react at all, even though one lies to the right of the metal-nonmetal boundary and one to the left (actually, both are so close to the boundary that neither can be classed entirely as a metal or nonmetal). The more metallic an element (low and left in the table), the more likely it is to transfer electrons to nonmetals. The more nonmetallic an element (high and right), the more likely it is to accept electrons.

Remember that since ionic compounds do not form individual molecules, it is incorrect to speak of *molecules* of these compounds, for example, a molecule of NaCl. As we learned in Section 5.9, however, all the kinds of calculations and stoichiometric relations that are so useful in chemistry are unaffected by whether the compounds involved are ionic or not. The molar relations that work so well with individual molecules also work with ionic compounds in which no individual molecules exist, and we often speak of the "molecular weight" of ionic compounds. We will continue to follow this common practice of assigning molecular weights to compounds, whether actual molecules are present or not.

EXAMPLE 11.1.   What are the molecular weights of NH_3 and RbBr?

Solution:

Ammonia, NH_3, is a molecular compound with a real molecular weight of $14 + (3 \times 1) = 17$. Rubidium bromide, however, is ionic. We can nonetheless assign to RbBr the molecular weight of $85.5 + 79.9 = 165.4$.

EXERCISE 11.6.   Assign molecular weights to the following ionic compounds: CsF, Al_2O_3, HgI_2, $Ba_3(PO_4)_2$.

It is important to understand that when atoms lose or accept electrons to become ions, their properties are greatly changed. For example, we think nothing of eating the compound NaCl, table salt, in which the sodium and the chlorine are in ionic form, but either sodium or chlorine in elemental form is extremely harmful to tissues. A piece of Na metal would react almost explosively with the moisture in your mouth, forming poisonous NaOH. Elemental chlorine was used as a poisonous gas in the First World War. Although the ion of chlorine, Cl^-, is not reactive, and Na^+ is almost totally inert (because each has a filled outer shell of eight electrons), the elements themselves are among the most reactive substances known.

Since ions themselves are individual particles, the properties of ionic compounds are combinations of the properties of their ions. Because Cu^{2+} ion is blue, for example, all the compounds containing simple Cu^{2+} ion are blue, unless the blue color is altered by the addition of color from other ions. Because CN^- ions are violently poisonous, the compounds NaCN, KCN, and $Mg(CN)_2$ are poisonous, although Na^+, K^+, and Mg^{2+} are not.

Ionic compounds have another interesting property with which we will become quite familiar. With few exceptions, when ionic compounds dissolve in water, the ions separate, a process called **dissociation.** The solution then contains the individual ions rather than the compound itself. Dissociation in solution creates physical, electrical, and chemical properties that are not exhibited by solutions of nonionic compounds. Sodium chloride gives us an example of dissociation.

$$NaCl \xrightarrow{\textit{In H}_2\textit{O}} Na^+ \text{ (aq)} + Cl^- \text{ (aq)}$$

Ions are structurally different from their atoms. When an Na atom loses its electron to become Na^+, it decreases in size. The outer electrons of Na^+ lie in the 2 shell ($1s^22s^22p^6$), whereas the outer electron of Na ($1s^22s^22p^63s$) lies in the 3 shell (Figure 11.14). When its outer layer of electrons is stripped off, an atom naturally becomes smaller. On the other hand, adding electrons to an atom tends to increase its size, even though the added electrons do not form a new shell. The ion Cl^- is larger than the atom Cl. When an electron is added to the valence shell, the increased repulsion created by the larger number of electrons causes the valence shell to expand a little (Figure 11.15). The increase in size caused by the formation of negative ions is usually not as great as the decrease caused by the loss of electrons in the formation of positive ions.

11.7 TWO NEW TERMS: OXIDATION AND REDUCTION

When an atom gains an electron in an electron transfer reaction, that atom is said to have been reduced. When an atom loses an electron, it is said to have been oxidized. These words are used frequently in chemistry, and we should become familiar with them. Remembering which term is which is easy: when an atom is reduced, its electric charge is also reduced algebraically, that is, made less positive or more negative. In the preparation of sodium metal, for instance, sodium ions, Na^+, are reduced to sodium atoms, Na, and their charge is reduced from +1 to 0. In the reaction of Cl atoms with K atoms, the Cl atoms become Cl^- ions, and their charge is reduced from 0 to −1. Not all cases are as easy—but in general, when something is reduced, its charge is also reduced. (To consider individual Cl atoms ignores the fact

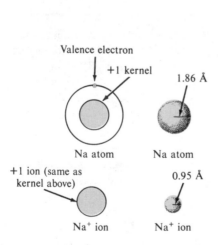

Figure 11.14 Relative sizes of a sodium atom and its ion. Figures at left are schematic; figures at right are to scale.

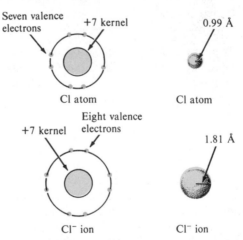

Figure 11.15 Relative sizes of a chlorine atom and its ion. Figures at left are schematic; figures at right are to scale. Eight valence electrons occupy more space than do seven electrons, because of mutual repulsion.

Table 11.1
Changes in Size as the Atoms of a
Few Metals and Nonmetals Are Ionized*

Atomic Radius (Å)		Ionic Radius (Å)		Atomic Radius (Å)		Ionic Radius (Å)	
Li	1.52	Li^+	0.60	F	0.71	F^-	1.36
Na	1.86	Na^+	0.95	Cl	0.99	Cl^-	1.81
K	2.27	K^+	1.33	Br	1.14	Br^-	1.95
Mg	1.60	Mg^{2+}	0.65	O	0.74	O^{2-}	1.40
Al	1.43	Al^{3+}	0.50	S	1.03	S^{2-}	1.84

*The metals lose electrons to become positive ions. The nonmetals gain electrons to become negative ions.

that the element chlorine occurs in the diatomic state as Cl_2. In the reaction of gaseous Cl_2 with K, the overall process also includes breaking the Cl_2 molecules apart and separating the K atoms from the metallic solid. For the purposes of our present discussion, we will assume that those steps have already occurred.)

During **oxidation** (the process of being oxidized) the charge is increased. For instance, potassium can be oxidized by oxygen from K atoms to K^+ ions; each K atom loses an electron and gains an electric charge of +1. See where the word oxidized comes from: K is *oxidized* by *oxygen* when oxygen takes electrons from it. *Anything that takes electrons* from a substance (as oxygen does) *oxidizes* that substance and is called an oxidizer or **oxidizing agent.** *Anything that gives electrons* to a substance *reduces* that substance and is called a **reducing agent** or reducer. The process of being reduced is called **reduction.** Electron transfer reactions are oxidation-reduction reactions because one substance gains electrons and one substance loses them. Now, get ready for this: in an oxidation-reduction reaction, the oxidizer gets reduced and the reducer gets oxidized. Let us look at that again; it seems harder than it is. An oxidizer, such as Cl, takes electrons away from something else, such as Na, and in the process increases the charge of the latter: Na becomes Na^+. The Cl has *oxidized* the Na. But in the process of oxidizing the Na to Na^+, the Cl gains an electron and is reduced to Cl^-; that is, *the oxidizer,* Cl, *is reduced* by the reducer, Na. Now we can turn the situation around and see why the reducer gets oxidized. When the Na atoms reduced Cl atoms to Cl^- ions, the Na atoms were *oxidized* to Na^+ ions: *the reducer,* Na, *was oxidized.* Stay with this paragraph until you have it.

EXERCISE 11.7.   Fill in the blank spaces in the following table.

Reaction	Oxidizer	Reducer	Substance Reduced	Substance Oxidized	Ions Produced
Na + Cl	Cl	Na	Cl	Na	Na^+, Cl^-
Ba + Br	___	___	___	___	Ba^{2+}, Br^-
O + Fe	___	___	___	___	___
Li + Te	___	___	___	___	___

Chemists have found it useful to assign **oxidation numbers** or **oxidation states** to atoms and ions. This is simple for monatomic ions because the oxidation number of the ion is the same as its charge. For example, the oxidation number of Na^+ is $+1$, that of Al^{3+} is $+3$, that of Cl^- is -1, that of S^{2-} is -2, and so on.

EXERCISE 11.8.   Write the symbols for the ions, if any, that would ordinarily be produced from each of these elements: O, Sr, Rb, Ar. Give the oxidation number of each.

Oxidation numbers can also be assigned to atoms that are bonded in ways other than ionic, but because such atoms have no permanent charges, the assignments are less directly related to charge. When the need arises, we will learn to assign oxidation numbers in these more difficult cases (Section 17.2).

One rule to remember about oxidation numbers is: in the formula for any compound, the sum of the oxidation numbers must be zero. Because electrons and protons can neither be created nor destroyed in a chemical reaction, a neutral compound must result from a reaction that began with neutral elements. In other words, the formula for a compound cannot carry a charge. Because oxidation numbers are related to electric charge, oxidation numbers must balance when charges balance. In NaCl, the $+$ charge of the Na^+ exactly cancels the $-$ charge of the Cl^-, and the oxidation numbers ($+1$ for Na and -1 for Cl) cancel similarly. In Al_2O_3, the sum of the positive oxidation numbers is $(+3) + (+3) = (+6)$, and that of the negative oxidation numbers is $(-2) + (-2) + (-2) = (-6)$, all of which add up to 0.

As we have learned, many pairs of elements can form more than one compound. In many cases, this is possible because an element can have different oxidation numbers and can combine with another element in different ways. For example, in CuO, the oxidation number of copper is $+2$, but in Cu_2O it is only $+1$. In cases like this, the resulting compounds must then be named in a way that reflects the oxidation states of the elements. The relation between names and oxidation numbers is discussed in more detail in Section 17.5; that information can easily be reviewed at this time if desired.

11.8 FORMULAS OF IONIC COMPOUNDS

In electron transfer reactions, the atoms involved usually react so as to form ions with filled valence shells, that is, to follow the octet rule. Metals tend to lose all the electrons in their outer shells (but there are many exceptions), and nonmetals tend to fill their outer shells. If we know what the elements in a compound are, if we know the electron populations of the atoms of the element, and if we remember the octet rule, we can make surprisingly accurate predictions of the formulas of many ionic compounds.

Consider the difference in the reactions of K and Mg, each with Br. Remember that all the electrons involved must come from the metal atoms and be transferred to Br atoms. In the reaction of K atoms and Br atoms, the arithmetic is easy: each K atom loses its single valence electron and achieves a filled shell structure, while each Br atom adds a single electron to its seven outer electrons, achieving a filled shell. In other words, one K atom reacts for every Br atom, and the formula for the resulting compound must be KBr, which shows that the compound contains as many K^+ ions as Br^- ions. For the reaction of Mg and Br, however, the proportions are not one to one. In this reaction, each Br atom gains a single electron as before, but because each Mg atom has two electrons in its valence shell, each Mg must lose two electrons to reach a filled shell arrangement. If no electrons are to be left over, there must then be twice as many Br atoms taking part in the reaction as there are Mg atoms, that is, the formula for the compound magnesium bromide must be $MgBr_2$. We can predict the formulas of many ionic compounds by counting the number of electrons lost or gained by the atoms taking part in electron transfer reaction. (See Figure 11.16.)

EXERCISE 11.9.   Using the methods shown in Figure 11.16, work out the electron transfer reactions that would occur between the atoms listed in the following pairs: Li and F, Sr and Cl, Al and O. For each pair, show the number of electrons gained or lost by the atoms of each element and predict the formulas of the resulting compounds.

Because the third pair is a bit more difficult than the other two, here is a hint that will help you find a formula. If you consider only a single Al atom, you find that it loses three electrons, while an O atom gains only two. Using this proportion yields an extra electron, but if you use two O atoms, you are one electron shy. If you start with two Al atoms, you will get six electrons, which can handily be used up by three O atoms. Your proportions must then be two Al^{3+} ions for every three O^{2-} ions. Now, work out the formula.

When oxygen, sulfur, or any of the halogens reacts with members of the alkali or alkaline earth families, you should easily be able to predict the formulas of the resulting compounds. With more experience, you will be able to do the same for several other sets of elements. It is also true, however, that the atoms of some elements react in ways that do not obviously lead to closed shell ions. Copper is an example: a copper atom can lose either one electron or two electrons, forming either the compound Cu_2O or CuO. Many other metal-nonmetal pairs also behave in an unpredictable way. To the extent that it will be useful to us, we will learn about some of these reactions as we go along.

You may have noticed that the information in the last two sections has been confined to the reaction of metals with nonmetals. What happens when nonmetals combine with nonmetals is entirely different; we will study it in Section 11.9. (Since metals do not generally react with other metals, we will have little to say about this subject.)

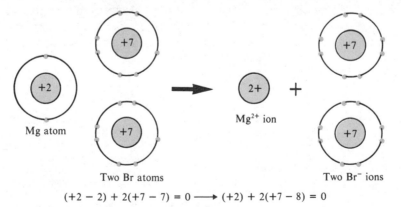

$$(+2 - 2) + 2(+7 - 7) = 0 \longrightarrow (+2) + 2(+7 - 8) = 0$$

Figure 11.16   Electron arithmetic for the reaction of Mg and Br. Two electrons are transferred in the process, but no electrons disappear or are created.

11.9 COVALENT BONDING

Electron transfer reactions are by now familiar to us. If we do a little thinking about some facts we already know, we might guess that there must also be another way in which chemical bonds are formed—even when electrons are not entirely transferred from one atom to another. Turn back to Table 5.4, which gives the formulas for some common gases. For instance, H_2 is a diatomic gas. Elemental hydrogen exists in the form of individual molecules, each of which contains two H atoms. In H_2, the atoms are held together by a chemical bond, but that bond is not ionic. In the first place, the molecules are individual, and we know that ions cannot make individual molecules. Moreover, it is not possible for two H atoms to transfer an electron because the atoms are exactly alike; neither is attractive enough to take an electron from the other. But the atoms of H are held tightly together in the H_2 molecule, and it has been shown experimentally that separating H_2 molecules into individual atoms requires considerable energy.

Examine more closely our two atoms of hydrogen. Each has a single electron in a $1s$ orbital, which is therefore only half filled. Either atom could accept a second electron from a willing giver, but neither can be persuaded to give up its own electron to another hydrogen atom. The need for another electron can be easily satisfied, however, if the atoms join so that their two electrons are shared as a pair. The sharing allows each nucleus to believe that it owns both electrons and has a filled shell of two, and because neither nucleus is willing to forsake its property, the two atoms are forced to remain together. A simple picture showing both electrons in the outer shell of both atoms makes the concept easy to understand (Figure 11.17). Pairs of atoms that share one or more pairs of electrons are said to be united by **covalent bonds.** *A shared pair of electrons is a covalent bond.*

Two chlorine atoms can also share a pair of electrons: each Cl atom has seven electrons in its outer shell, and one of the seven is unpaired. Each Cl atom has a vacancy in its outer shell; one more electron is needed to make a filled shell of eight. Two Cl atoms can achieve a better arrangement by sharing one pair of electrons, and each atom will contribute its single electron to the pair. Sharing the pair holds the atoms together as a molecule. Elemental chlorine is diatomic; its formula is Cl_2. (See Figure 11.18.)

Electron pair sharing can be looked at another way. Suppose we place our two atoms of H near each other. The most attractive place for their two electrons to be is as close as possible to both of the positive-charged nuclei; that is, between the nuclei. In molecules with covalent bonds, the electrons in the bond spend a large part of their time between the nuclei. Since the

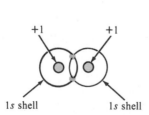

Figure 11.17 Schematic drawing of H_2 molecule. The pair of electrons between the nuclei is shared.

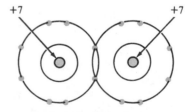

Total of 14 valence electrons

Figure 11.18 Schematic drawing of a Cl_2 (or any halogen) molecule. The pair of electrons between the kernels is shared.

nuclei are attracted to this collection of negative charge, they tend to remain close to it, that is, to be bonded together (Figure 11.19).

Electron sharing often takes place in such a way as to result in filled outer shells of eight electrons, following the octet rule. We saw in the case of the Cl_2 molecule that if the shared electrons are counted as belonging to both atoms, each Cl atom can claim eight electrons in its outer shell. We will find the octet rule quite useful in deciding how molecules are constructed. (With hydrogen, of course, the outer shell contains only two electrons.)

11.10 DRAWING COVALENT MOLECULES

Two kinds of drawings are commonly used to show covalently bonded molecules. As we saw in the preceding section, our electron dot figures can be easily adapted to covalent molecules. We include all the valence electrons of both atoms and show the shared pair or pairs lying between the atoms and the unshared pairs alongside.

$$H : \overset{\cdot\cdot}{\underset{\cdot\cdot}{Br}} : \qquad\qquad : \overset{\cdot\cdot}{\underset{\cdot\cdot}{Br}} : \overset{\cdot\cdot}{\underset{\cdot\cdot}{Cl}} :$$

When you first begin to draw covalent molecules, you may find it helpful to draw circles around the bonded atoms, one circle indicating the valence shell of each atom, and the shared electrons shown as being in both shells. This form does not provide more information, but it makes electron counting easier.

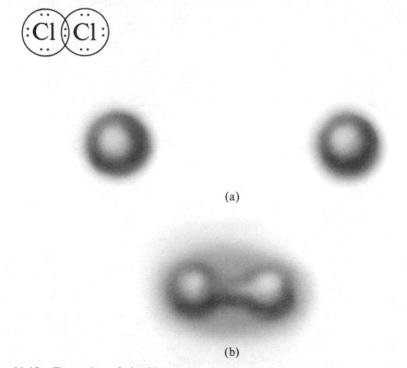

(a)

(b)

Figure 11.19 Formation of the H_2 covalent bond. (*a*) Representation of the electron clouds of two separated hydrogen atoms. (*b*) Electron distribution in the same atoms after bonding.

Although all electrons are alike, it is occasionally convenient to use different symbols for the electrons from different atoms. The shared pair will be shown by one symbol from each atom.

There is a faster method of drawing covalent bonds, but be sure you have had plenty of practice in counting electrons before you begin to use this simple abbreviation. A covalent bond is often shown simply as a line connecting the two atoms.

H—H   Br—Cl

In this notation, the line represents a pair of shared electrons. Using this approach, we can represent *any* pair of electrons with a short line. If we use the line to show unshared as well as shared pairs, we draw the HCl molecule like this:

H—C̅l|

Such a diagram is sometimes called a **bond line drawing.**

In what follows, we will used electron dots (or crosses, x's, or small circles) when we want to keep track of where electrons come from, and we will use the bond line method when we are interested only in the total number of electron pairs. Electron dot drawings and other similar representations of covalent bonding are called **Lewis structures** in honor of G. N. Lewis of the University of California, who, in 1916, first proposed the electron dot formula to explain covalent bonding.

Later, when we are more accustomed to drawing Lewis structures and when it will not be necessary to count electrons, but only to indicate the covalent bonds in molecules, we will sometimes not show the valence electrons that are not actually in bonds. With this further abbreviation, a BrCl molecule looks like this:

Br—Cl

This method is open to more error than is a diagram that uses either lines or dots to show all the valence electrons.

EXERCISE 11.10.   Construct both electron dot and bond line drawings for the following molecules: HF, I_2, IBr. Show all electrons or electron pairs.

11.11 ATOMS FORMING MORE THAN ONE COVALENT BOND

Many kinds of atoms can share more than one pair of electrons, either with one other atom or with more than one other atom. For example, when an oxygen atom combines with two hydrogen atoms to form a molecule of water, it begins the reaction with six valence electrons. The outer shell of oxygen can be completed to a full octet by sharing two more electrons, one from each hydrogen. Each of the new shared pairs becomes a covalent bond, thus bonding the oxygen to both of the hydrogens. Figure 11.20 shows the reaction forming

$$2 \text{ H\texttimes} + \cdot \overset{\cdot \cdot}{\underset{\cdot}{O}} : \longrightarrow \text{H} \overset{\cdot \cdot}{\underset{\underset{H}{\text{\texttimes} \cdot}}{O}} : \qquad\qquad 2 \text{ H} + \text{O} \longrightarrow \underset{\underset{H}{|}}{\text{H}-\text{O}} \qquad\qquad 2 \text{ H} + \text{O} \longrightarrow \text{H}_2\text{O}$$

Figure 11.20 Reaction of H and O to form water. These are three of the ways in which the reaction can be written.

$$3 \text{ H\texttimes} + \cdot \overset{\cdot \cdot}{\underset{\cdot}{N}} \cdot \longrightarrow \text{H} \overset{}{\underset{\underset{H}{\cdot \text{\texttimes}}}{\text{\texttimes} N \text{\texttimes}}} \text{H} \qquad\qquad 3 \text{ H} + \text{N} \longrightarrow \underset{\underset{H}{|}}{\text{H}-\text{N}-\text{H}} \qquad\qquad 3 \text{ H} + \text{N} \longrightarrow \text{NH}_3$$

Figure 11.21 Reaction of H and N to form ammonia.

H_2O. A similar situation occurs in the case of nitrogen, except that the N atom, which has five electrons of its own, can share three pairs with hydrogen atoms (Figure 11.21) forming NH_3. In NH_3, nitrogen counts eight electrons in its outer shell.

EXERCISE 11.11.   In every molecule of phosphorus trichloride, PCl_3, there are three chlorine atoms, each of which is joined by a covalent bond to a central phosphorus atom. Draw both an electron dot diagram and a bond line diagram for PCl_3.

EXERCISE 11.12.   Draw electron dot and bond line diagrams for CH_4. Each of the H atoms is bonded to C.

When you draw more complex molecules that contain atoms forming more than a single bond, decide first which atom is bonded to which. The electron pairs can then be put in and the drawing finished. In the sections that follow, we will learn more about the arrangements of atoms in molecules.

11.12 MULTIPLE BONDS AND COORDINATE COVALENT BONDS

In many cases, a single pair of atoms can share more than one pair of electrons, forming more than one bond. A good example is the N_2 molecule: each N atom has five electrons, and by sharing three pairs (as the N atom did in NH_3 above) each can have a filled shell, as you see in the diagram.

In this molecule, all three electron pairs are used to join only the two N atoms. Multiple bonds also occur in molecules composed of several kinds of atoms. In the compound ethylene, each of the two carbon atoms shares two pairs of electrons with its carbon neighbor and one pair with each of its two hydrogen neighbors.

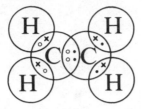

The circles around the carbon atoms and their shared electrons show how each carbon achieves a full octet of valence electrons.

EXERCISE 11.13. Draw the electron dot diagram for hydrogen peroxide, H_2O_2, in which the two oxygen atoms are connected by a double bond.

It is not always so that each of two covalently bonded atoms donates one electron to the shared electron pair. In some cases, both electrons are furnished by only one of the atoms. A shared pair that has originated from only one of the bonded atoms is called a **coordinate covalent bond.** For example, the molecule of BF_3 is deficient in electrons because each boron atom has only three electrons. Each molecule of ammonia, NH_3, however, has an unshared pair of electrons. BF_3 and NH_3 react as shown in Figure 11.22, forming a covalent B—N bond in which both electrons formerly belonged to the N atom. Coordinate covalent bonds are characteristic of several compounds, including the oxyacids, common examples of which are H_2SO_4 and HNO_3.

11.13 LEWIS STRUCTURES OF LARGER MOLECULES

Now that we have drawn Lewis diagrams and have learned more about the characteristics of covalent bonds, we can take an organized approach to making Lewis drawings for more complex molecules. Here are some general rules.

1. To draw a skeleton structure, you must know which atom is bonded to which. Experience and practice help most here, but there are also a few useful guideposts. It is helpful to know how many bonds the atoms of some common elements usually make. Knowing, for example, that each H atom makes only one bond and that each O atom usually makes two bonds leads us to draw the figure on the right, rather than that on the left.

H—H—O   H—O—H

Figure 11.22 Reaction of BF_3 and NH_3 to form a coordinate covalent bond.

2. In binary compounds that have more than one atom of one element and a single atom of another, the single atom is often the central atom, and the others are bonded to it, instead of to each other. Examples are H_2O, SO_3, and PCl_3, in which the central atoms are, respectively, O, S, and P. Such compounds can contain either simple covalent bonds, coordinate covalent bonds, or both.

3. In the oxyacids, their derivatives, and a few similar compounds, the atom appearing only once is ordinarily the central atom, the oxygens are bonded to it, and the hydrogens are bonded to the oxygens.

4. A molecule can have neither more nor fewer electrons than the sum of the electrons of the atoms in the molecules. Confining ourselves to valence electrons (because the inner electrons remain unaffected by the bonding) we take PCl_3 as an example. Each Cl has 7 electrons, making 21. The P has 5 electrons, bringing the total to 26. There are 26 valence electrons in the PCl_3 molecule. (As we will soon see, this rule must be modified for ions.)

5. The finished molecule should obey the octet rule if possible. (There are a few exceptions, such as NO_2 and BF_3, but we will not be concerned with these.) For many molecules, drawing single bonds between atoms and adding enough unshared pairs to give every atom an octet will require more electrons than the molecule has. Not all molecules, however, are constructed this way. Consider CO_2, which has 16 valence electrons altogether. If we draw a sketch with only single bonds and octets for every atom, an incorrect structure with too many electrons results. (Count them.)

Not a correct structure

Because a shared pair of electrons counts in the octet of both atoms, it is possible to gain octets with fewer electrons by making multiple bonds.

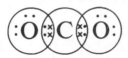

Correct Lewis diagram of CO_2

We can find out whether multiple bonds are necessary by doing electron arithmetic (Steps 1 and 2).

Now let us look at a step-by-step method for drawing Lewis diagrams of molecules.

Step 1. Decide which is the central atom and what the bonding pattern is in the rest of the molecule. Draw a preliminary structure in which the atoms will later be connected by bonds. We will use PCl_3 and SO_3 as examples.

Step 2.　Decide how many bonds must exist in the molecule, and draw them in.

    a.　Count the total number of valence electrons in all the atoms. For PCl_3 there are $5 + (3 \times 7) = 26$ valence electrons. For SO_3 there are $6 + (3 \times 6) = 24$. This gives you the number of valence electrons actually present in the molecule.

    b.　Since, according to the octet rule, each atom (except H) is to have eight electrons, multiply the number of atoms by eight (if H is present, include only two for each H atom). For PCl_3 our number is $4 \times 8 = 32$; for SO_3 it is the same.

    c.　Compare the number from Step 2b with the number of electrons *actually present* to get the number of electrons that must be shared. Because a shared pair counts as belonging to the octet of two atoms, each pair that appears in a bond can be counted as four electrons toward the molecule octet rule. For PCl_3 we actually have 26 electrons, but four octets require 32. Subtracting 26 from 32 gives 6, which is our deficit. Now, if we share 6 electrons (three pairs) by making three electron-pair bonds, 6 electrons will be doing double duty, giving us 32 electrons that can be counted in octets. We need only three bonds in PCl_3.

    d.　Draw the bonds in.

$$\begin{array}{ccc} Cl & \underset{\times\times}{} & Cl \\ & \diagdown \underset{|}{P} \diagup & \\ & Cl & \end{array}$$

To make our check easier in the first few examples, we will draw dot structures also.

We can finish our SO_3 molecule, whose total number of electrons was only 24, but whose four octets total 32. Subtracting, we get $32 - 24 = 8$ electrons, or four pairs that need to be shared: SO_3 must have four bonds. To get four bonds in the structure we have started, we will have to make one double bond. It makes no difference to which oxygen the double bond goes.

(Notice that two of the bonds in SO_3 are coordinate; that is, both electrons in each pair come from sulfur.)

Step 3.   Check and count electrons. Do this now by counting the electrons in the structures we have just finished, confirming that for PCl₃ there is a total of 26, that each atom has an octet, and that six electrons (three pairs) are doing double duty. We make the same check for SO₃, and come out with 24 electrons and four bonds.

EXAMPLE 11.2.   Draw a Lewis diagram for nitric acid, HNO_3.

Solution:

Step 1.   N is the central atom; H is attached to O.

$$
\begin{array}{ccc}
 & & O \\
H & O & N \\
 & & O
\end{array}
$$

Step 2.   a.   Count electrons.

$$
\begin{array}{lr}
1\ N & 5 \\
3\ O's\ (3 \times 6) & 18 \\
1\ H & \underline{1} \\
 & 24
\end{array}
$$

   b.   Count the number of electrons required for octets and total.

$$
\begin{array}{lr}
For\ 1\ N + 3\ O = 4 \times 8 = & 32 \\
1\ H & \underline{2} \\
 & 34
\end{array}
$$

   c.   Subtract the result of Step 2a from that of Step 2b to find the number of bonds needed.

$$34 - 24 = 10$$

Five bonds are needed.

   d.   Draw in the bonds, single bonds first, then multiple bonds as needed. Show the unshared pairs.

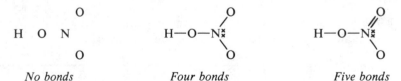

| *No bonds* | *Four bonds* | *Five bonds* |

   e.   Check the structure. If you feel it necessary at first, draw the dot structure and count. Otherwise, count pairs as shown by lines. In the structure above, 12 pairs appear, representing our 24 valence electrons. Each atom has its octet (except for the hydrogen, of course, which has only two electrons).

EXERCISE 11.14.   Draw the Lewis structure for H_3PO_4.

11.14 POLYATOMIC IONS

Many chemical compounds contain both covalent and ionic bonds. Such compounds are composed of ions, but some of the ions contain more than one atom, and the atoms within those ions are joined by covalent bonds. A simple example of such a compound is sodium hydroxide, NaOH, which is composed of Na^+ ions and OH^- ions. A hydroxide ion, OH^-, contains an oxygen and a hydrogen atom joined by a single covalent bond. In addition to the seven valence electrons that would be possessed by the neutral O and H atoms, this pair has an extra electron and thus carries a negative charge.

$$\left(\overset{\times\times}{\underset{\times\times}{\times}} \overset{}{\underset{}{\ddot{O}}} \overset{}{\underset{}{\times}} H \right)^-$$

For an example of a polyatomic ion that carries a positive charge, consider the ammonium ion, NH_4^+, which is a covalently bonded combination of an ammonia molecule, NH_3, and a hydrogen ion, H^+. In the ammonium ion, the N is the central atom; it is bonded to each of the four H atoms, and all five of the atoms share the plus charge. Count the electrons in the five neutral atoms that make ammonium ion, then count the electrons in the ion itself. Can you see why there is a positive charge?

$$\left(\begin{array}{c} H \\ {\scriptstyle \times\,o} \\ H\, {\scriptstyle\times\atop\circ}\, N\, {\scriptstyle\times\atop\circ}\, H \\ {\scriptstyle \times\times} \\ H \end{array} \right)^+$$

In drawing the structure of a polyatomic ion, we begin as though we were drawing a molecule. When we count the number of valence electrons in Step 2a, however, we must allow for the charge on the ion. For ions, Step 2a should read as follows:

Step 2a. Count the number of valence electrons in all the atoms. If the ion is negatively charged, add one electron for every charge. If the ion is positively charged, subtract one electron for every charge.

Finish the procedure just as you would the diagram of a molecule. When the structure is complete, draw brackets around it and indicate the charge.

EXAMPLE 11.3. Draw an electron dot structure for an OH^- ion.

Solution:

Step 1. O H

Step 2. a. Total number of actual valence electrons.

$$\begin{array}{lll} 1\ O & 6 \\ 1\ H & \underline{1} \\ & 7 \end{array}$$

Add 1 for one negative charge.

$$\frac{7}{1}$$
$$\frac{1}{8}$$

b. Total number needed for octet rule.

1 O   8
1 H   2
        10

c. Number of shared electrons and number of bonds.

10 − 8 = 2 shared electrons
One bond

d. Structure.

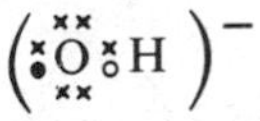

$$\left(\overset{\times\times}{\underset{\times\times}{\text{O}}}\text{H} \right)^{-}$$

Step 3. Check. O has an octet, H has two electrons, and there are eight electrons shown.

EXAMPLE 11.4. Draw a bond line diagram for SO_4^{2-} ion.

Solution:

Step 1.   O

        O   S   O

            O

Step 2. a. Valence electrons.

4 O (4 × 6) = 24
1 S   6
              30

Add two for charges.
30
2
32

b. For octets

4 O (4 × 8) = 32
S   8
              40

c. 40 − 32 = 8 electrons
Four bonds

d.

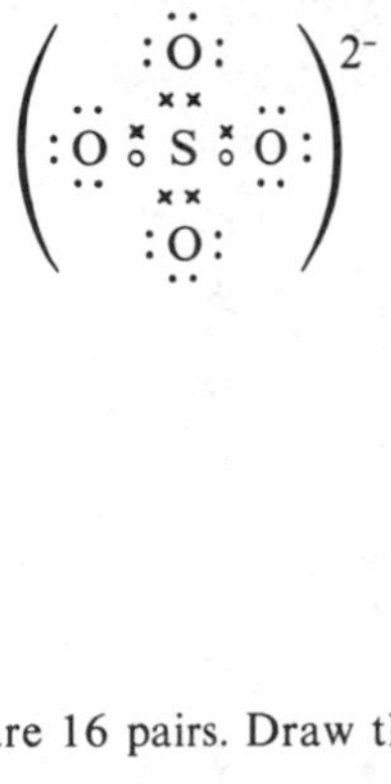

$$\left(:\overset{..}{\underset{..}{\text{O}}}\overset{\times}{\underset{\times}{\text{S}}}\overset{..}{\underset{..}{\text{O}}}: \right)^{2-}$$

Step 3. Check: each atom has an octet, and there are 16 pairs. Draw the bond line diagram.

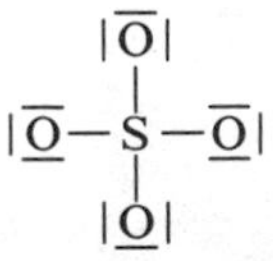

$$\begin{array}{c} |\overline{\text{O}}| \\ | \\ |\overline{\text{O}}-\text{S}-\overline{\text{O}}| \\ | \\ |\underline{\text{O}}| \end{array}$$

EXERCISE 11.5.   Draw an electron dot diagram for a NO_3 ion.

Other common polyatomic ions are carbonate, $CO_3{}^{2-}$, and phosphate, $PO_4{}^{3-}$.

$$\left(\begin{array}{c} \overset{\circ\circ}{\underset{\circ\circ}{O}} \;\; \overset{\circ\circ}{\underset{\circ}{O}} \\ \circ\circ\overset{\star}{C}\star \\ \overset{\bullet\bullet}{\underset{\circ\circ}{O}} \end{array} \right)^{2-} \qquad \left(\begin{array}{c} \overset{\circ\circ}{O} \\ \circ\circ\;\overset{\times\times}{\underset{\times}{P}}\;\circ\circ \\ O : \; : O \\ \underset{\circ\circ}{O} \end{array} \right)^{3-}$$

11.15 SHAPES OF MOLECULES

Knowing how to represent molecules and write reactions in terms of electron dot or bond line diagrams is a great step toward learning how and why reactions occur and what the results will be. Armed with our new understanding, we will now briefly examine the actual physical structures of molecules, the molecular shapes that cause many of the properties found in macroscopic samples.

The arrangement of the atoms in a molecule (which atom is bonded to which) is called the **bond structure.** As you know, a Lewis diagram shows the bond structure of a molecule. For many formulas, only one correct Lewis diagram can be drawn, and that drawing usually represents a molecule that actually exists. Whenever we can draw more than one correct Lewis diagram, there is often more than one compound, as we will see.

Although the Lewis diagram cannot tell us what *shape* a molecule has, that is, the arrangement of the atoms in space, there are other guidelines that can be used. Molecules with two atoms, for example, must form a linear shape: a straight line can be drawn through the nuclei. If there are more than two atoms, however, the nuclei may or may not lie in a straight line. It is known, for example, that the water molecule is bent. Lines drawn along the bonds from each H nucleus through the O nucleus intersect at an angle of about 105° (Figure 11.23). The angle between the bonds going from any atom to two other atoms is called the **bond angle.** The fact that water molecules are bent is enormously important; and, in fact, chemical and physical properties of molecules of all kinds depend largely on the molecular shapes that we characterize by giving bond angles. We will examine the relations between molecular shape and physical properties in Chapter 13.

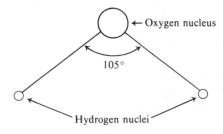

Figure 11.23   Representation of H_2O molecule, showing bond angle.

To study and understand the properties of molecules, chemists make drawings and models that display molecular architecture. What are affectionately called "tinkertoy models" show the atoms connected by long rods that represent bonds. Figure 11.24 shows tinkertoy models of H_2O and NH_3; each is shown from various directions.

Covalent bonds, however, are not long rods. Since the shared electron pair inhabits the outer shell of both the atoms, the atoms must lie close together. Although the tinkertoy models indicate the bond angles and bond structure of molecules, they cannot show what a molecule actually looks like. Other molecular models, called space-filling models, are more realistic. Notice that the atoms actually blend into one another in the space-filling models of H_2O and NH_3 (Figure 11.25).

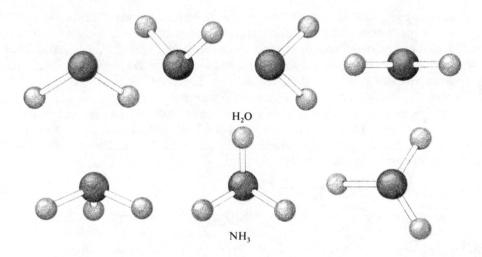

H_2O

NH_3

Figure 11.24   Different views of the tinkertoy model of H_2O and of NH_3.

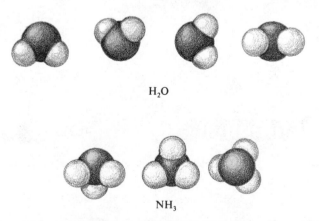

H_2O

NH_3

Figure 11.25   Space-filling models of H_2O and NH_3.

What is called the **valence shell electron pair repulsion theory (VSEPR)** allows us to predict the shapes that molecules are likely to have. Because electrons repel one another, electron pairs in orbitals stay as far from one another as possible. In the process of achieving maximum distance, valence shell electron pairs create characteristic orbital patterns. The exact angles in those patterns depend mostly on how many pairs are in the valence shells and how many of those pairs are used in bonding. We will not need to use the detailed methods of VSEPR for most of our purposes, but some familiarity with the method and with its results can lead to a better understanding of the geometry of molecules. The four simple VSEPR steps used to determine molecular structure are described in Appendix D.

11.16 MOTIONS OF ATOMS IN MOLECULES

The atoms in any sample of matter are in constant motion. Even if the sample is frozen into a solid, some motion persists. Atoms that are bonded together vibrate; that is, they constantly change their positions in relation to one another. The internuclear distances for individual bonds are well known, but those distances are only averages over time: every pair of bonded atoms vibrates constantly, lengthening and shortening its bond like two balls on a spring (Figure 11.26). Moreover, unless they are frozen in solids, larger molecules can change their shapes by rotating their various parts around bonds. Such rotation can cause significant changes in the shape of many molecules. Figure 11.27 shows butane, C_4H_{10}, in some of the forms it can assume. In gases and liquids, complex molecules are constantly changing their shapes within the limits imposed by their bonds.

11.17 WRITING SIMPLIFIED BOND STRUCTURES

Because we cannot draw a three-dimensional picture of a molecular model every time we want to show a structure, the custom of making schematic, two-dimensional drawings has arisen. One way to draw a molecule is to show only the bond structure, making little attempt to show bond angles or distances, but indicating which atoms are bonded to which atoms. In such a drawing, the NH_3 and H_2O molecules might look like the drawings shown.

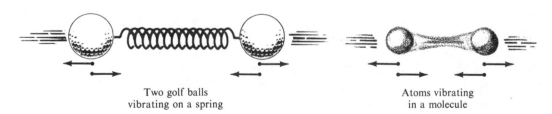

Two golf balls
vibrating on a spring

Atoms vibrating
in a molecule

Figure 11.26 Analogy between a molecule and two balls on a spring.

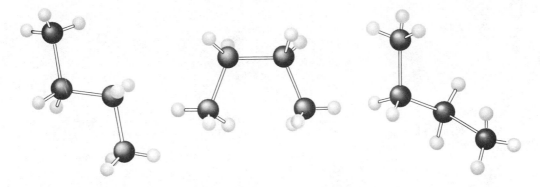

Figure 11.27 Three tinkertoy models of C_4H_{10} with different arrangements in space.

The diagram indicates the bent structure of the water molecule because it is easy to do so. In the diagram of NH_3, however, we cannot tell whether the molecule is flat or forms a pyramid. The drawing nonetheless tells us the bond structure of the molecule, and that is often all that is necessary.

The structural formula of the C_4H_{10} molecule we saw earlier in the tinkertoy pictures looks fairly simple. Compare the structural drawing with the space-filling model of the same molecule, shown in Figure 11.28.

If more information than a simple bond structure is needed, bonds can be shaped or shaded to show direction. In the drawing of CH_4, the ordinary bond lines lie in the plane of the paper, but the hydrogen with the heavy bond line lies in front of the paper and the hydrogen with the dotted bond line lies behind the paper. Compare the drawing with that of the models, which are positioned the same way (Figure 11.29).

Although structures such as that of C_4H_{10} can be indicated in more condensed ways, we will usually be content to draw bond structures that show what we need to know and leave indications of positions in space to models or more complex drawings.

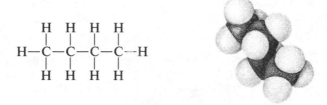

Figure 11.28 Structural drawing and space-filling model of C_4H_{10}. Compare to the tinkertoy models in Figure 11.27.

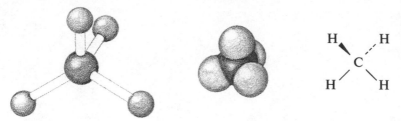

Figure 11.29   Three-dimensional models and schematic drawing of CH_4.

EXERCISE 11.16.   Draw structural formulas for C_2H_4 and $CHCl_3$. (Notice that there must be multiple bonds in one of the molecules.)

11.18 ISOMERISM

The bond structures of atoms in molecules decisively influence the characteristics of the compounds containing those molecules. The effect of molecular structure can be dramatically illustrated by a comparison of the properties of compounds whose molecules have the same numbers and kinds of atoms, but whose structures are different. The compound C_2H_6O furnishes an example. The octet rule allows us to assemble these nine atoms into two different structures:

$$
\begin{array}{cc}
\ \ \text{H} \ \ \text{H} & \ \ \text{H} \ \ \ \ \ \ \text{H} \\
\ \ | \ \ \ \ | & \ \ | \ \ \ \ \ \ \ \ | \\
\text{H--C--C--O--H} & \text{H--C--O--C--H} \\
\ \ | \ \ \ \ | & \ \ | \ \ \ \ \ \ \ \ | \\
\ \ \text{H} \ \ \text{H} & \ \ \text{H} \ \ \ \ \ \ \text{H}
\end{array}
$$

The two drawings represent **structural isomers** of C_2H_6O, *molecules that have the same numbers and kinds of atoms but different bond structures*. Samples of the two substances are entirely different. On the left is the structure of ethyl alcohol, a compound used in beverages and sometimes in attempts to cure a disease called *weltschmertz*. The structure on the right represents dimethyl ether, a rather more poisonous liquid that will boil in your hand. The physical and biological differences between the two compounds are caused entirely by the different bond patterns in their molecules. Although they share the same formula, isomers are often sufficiently unlike to be considered entirely different compounds.

EXERCISE 11.17.   Draw the structural isomers of C_4H_{10} and C_3H_8O. There are two possible structures for the first, and three for the second.

Structural isomerism, the difference in bond structures among compounds of the same formula, is not the only way in which differing arrangements of atoms affect compound properties. We will leave the study of other, more exotic kinds of isomerism to more advanced courses. We will meet structural isomerism again in Chapter 19.

11.19 SUMMARY

The understanding we have gained in this chapter will form the core around which our further study of chemistry can be organized. Let us now pull together the basic facts about chemical reaction and the formation of chemical compounds.

Although there are only about 30 or 40 common elements, there exist thousands of different substances in the world, almost all of which are compounds of the common elements. Because most of the substances we meet are compounds, we need to learn about what compounds are like and how they are created. That is why we study chemical reaction and the resulting compounds.

A compound is a stable combination of two or more elements whose atoms or ions are joined by chemical bonds. A compound is characterized by the fixed ratios of the numbers of atoms of its elements, ratios expressed in its formula. Compounds are formed by the chemical reaction of elements, of compounds, or of both.

Chemical reactions can be classed broadly as either electron transfer (oxidation-reduction) reactions or reactions that lead to electron sharing. The first kind of reaction forms ionic solids held together by the electric attractions often called ionic bonds. The second kind produces individual molecules, independent particles held together internally by shared electron pairs; that is, by covalent bonds.

The properties of compounds arise not only from the numbers and kinds of atoms in molecules, but also from the ways in which the atoms are bonded, and even from the shapes of the molecules themselves. Molecular shapes result partly from the patterns that electron pairs form in valence shell orbitals. Molecules with the same formula but different bond structures are said to be isomers and can have different properties.

QUESTIONS

Section 11.2

1. Explain the principles behind the calculation of the kernel charge for an atom.
2. Explain why the kernel charge experienced by a valence electron increases across the periodic table, even though the number of electrons is also increasing.
3. It is said that valence electrons are influenced primarily by the kernel charge, rather than by the charge of the nucleus alone. How does the average distance between the inner electrons and the nucleus justify this approximation?

Section 11.3

4. Define the terms *ionization energy* and *electron affinity*.
5. Why does it require energy to remove an electron from an atom?
6. Can an atom have both a high ionization energy and a high electron affinity? Do the atoms of greatest ionization energy also have the greatest electron affinity? Explain.

Section 11.4

7. What two factors have the greatest effect on the attraction of an atom for its valence electrons?
8. How do ionization energies vary across the periodic table and down the periodic table? Why do they vary in this way?

9. Why is the ionization energy of boron less than that of beryllium?
10. Explain why the ionization energy of oxygen is less than that of nitrogen.
11. Why is the ionization energy of sulfur less than that of phosphorus?
12. Plot the second ionization energies for the elements B through Ne. Explain why the first break in the curve (the dip) comes at C, rather than at B, as in the plot of Figure 11.4. Data: second ionization energies of Be, 417; B, 575; C, 558; N, 678; O, 803; F, 801; Ne, 1143.
13. Why is the electron affinity of iodine less than that of fluorine?
14. Predict which will have the higher *electron affinity*, oxygen or nitrogen. Why?
15. Why does the change in electron affinities generally follow the same pattern as the changes in ionization energies from one place to another in the periodic table? Why are there notable exceptions?

Section 11.5

16. Explain what happens and why during an electron transfer reaction.
17. How are ions formed?
18. Why will an atom usually gain or lose only enough electrons so that there will be eight electrons in its outer shell?
19. Explain in terms of ionization energy and electron affinity why metals form positive ions and nonmetals form negative ions.
20. Why is it possible for elements to form doubly or triply charged ions despite the fact that larger amounts of energy are required to form those ions than to form singly charged ions.
21. Why doesn't O^{3-} usually form?
22. Explain why Al^{3+} forms, but B^{3+} is unusual.
23. Why can carbon form C^{4-}, whereas lead, a member of the same family, usually forms Pb^{2+}?
24. Atoms and ions with the same electron configurations are called *isoelectronic*. Give an example of
    a. two cations that are isoelectronic
    b. two isoelectronic anions
    c. one cation and one anion that are isoelectronic
    d. an atom and an ion that are isoelectronic

Section 11.6

25. Based on the explanation of how NaCl is formed, describe the formation of MgO.
26. Define ionic solid and ionic bond.
27. Why do we not speak of ionic molecules?
28. Why is it appropriate to perform stoichiometric calculations for compounds that do not form molecules?
29. What is meant by dissociation?
30. Are negative or positive ions larger, everything else being equal? Explain.
31. What is the smallest negative ion? The smallest positive ion?
32. Why is the magnesium atom smaller than the sodium atom, even though it has more electrons and protons?

Section 11.7

33. Define: oxidizer, reducer, oxidation, and reduction.
34. Give three examples of nonmetals that can be reduced by Na. What are the oxidation states of the nonmetals after reduction?
35. Compounds containing Cu^{2+} are usually blue. Why is Cu_2O not blue in color?

Section 11.9

36. Explain how two hydrogen atoms form a molecule.
37. In a molecule of H_2, where do the electrons of the shared pair have the highest probability of being found?
38. Explain the difference between an ionic bond and a covalent bond.

Section 11.12

39. Based on the bonding of N_2, would you predict the existence of P_2? Why or why not?
40. How many bonds would there be in a molecule of C_2 if C_2 existed? Would you predict that C_2 *would* exist? Give your reasons why and why not.
41. Define the term *coordinate covalent bond*.

Section 11.15

42. Explain what is meant by the term *bond structure*.
43. Why is it important to know the structures of molecules?
44. What are the advantages and disadvantages of tinkertoy and of space-filling models?

Section 11.18

45. What is structural isomerism? Can two structural isomers have radically different properties?
46. Explain bond vibration and rotation.
47. Can rotation or vibration change the chemical and physical properties of molecules? Explain.
48. Do changes in space, as in the rotation of atoms around bonds, change the structural isomerism of a molecule?

PROBLEMS

Section 11.2

*1. What is the kernel charge experienced by a valence electron in an atom of each of the following elements:
    *a. Mg   *b. Si   c. Br   d. K
2. What is the kernel charge experienced by a valence electron in an atom of each of the following elements: Be, P, S, Ar?
3. List four elements that have a kernel charge of 6.
4. Name five elements with a kernel charge of 3.

Section 11.4

*5. For each of the following pairs of elements, predict which would have the higher first ionization energy:
    *a. Na and Be   b. Ca and K   c. P and O
6. For each of the following pairs of elements, predict which would have the higher first ionization energy: Cs and Sc, Si and As, N and S, Se and I.
*7. For each of the following pairs, predict which element will have the higher electron affinity:
    *a. Mg and Ca   c. Cl and B   e. Na and P
    b. O and S   d. K and Rb

Section 11.5

8. If an O atom were to lose two electrons, what kernel charge would the remaining valence electrons experience? If it were to gain two electrons? Gain three electrons? Draw electron dot structures for these ions.

*9. How many electrons will each of the following atoms most likely add in an electron transfer reaction: N, Cl, Br, O, S, P?

10. How many electrons will each of the following atoms most likely lose in an electron transfer reaction: K, Al, Ca, Na, Mg, Sr, Be?

11. Draw electron dot structures for the atoms and their corresponding ions shown in Problems 9 and 10.

*12. Write the symbols, including charge, for the ions of the atoms in Problem 9.

13. Write the symbols, including charge, for the ions of the atoms in Problem 10.

Section 11.6

*14. Write the electron configurations of the following ions:
    *a. Mg^{2+} c. O^{2-} e. Li^+
    b. Al^{3+} d. Cl^-

*15. List the ions and atoms in each group in order of increasing size:
    *a. Na, Na^+, P, S^{2-}, Cl   c. Li, Be^{2+}, C^{4-}, C^{4+}, O^{2-}
    b. Mg^{2+}, Al^{3+}, P^{3-}, S, Ar   d. N^{3-}, N^{3+}, O^{2-}, Ne, F^-

16. List the ions and atoms in each group in order of decreasing size:
    a. Sr^{2+}, Mg^{2+}, Ba, Mg   c. Li^+, Be^{2+}, B, C, N
    b. K, Ca, Ca^{2+}, Ag, Br^- d. O^{2-}, C^{4-}, C^{4+}, F, Li

Section 11.7

*17. For each of the following pairs of atoms predict the oxidizer, reducer, substance reduced, and substance oxidized: Mg and O, V and P, Cl and Co, Ni and O, I and Mn.

18. For each of the following pairs of atoms predict the oxidizer, reducer, substance reduced, and substance oxidized: S and Ca, Al and H, B and F, F and Sn, Br and Fe.

*19. Give the oxidation numbers for the ions formed from:
    *a. Li   c. Cs   e. I
    b. Ca   d. S   f. B

20. Write the symbols, including the oxidation states, for the ions (if any) formed from Ba, Br, Be, Bi.

Section 11.8

*21. For each of the following pairs of atoms, work out any electron transfer reactions that could occur and predict the formulas of the resulting compounds. (Assume that the elements ordinarily occurring as diatomic gases have already been split into atoms.)
    *a. Mg and O   c. Al and H
    b. Cl and Cs   d. K and S

22. For each of the following pairs of atoms, work out any electron transfer reactions that could occur and predict the formulas of the resulting compounds. (Assume that the elements ordinarily occurring as diatomic gases have already been split into atoms.) K and Se, S and Rb, F and Ba, Ba and S.

*23. What might be the formula of a molecule resulting from elements X and Y when

    *a. X has six valence electrons and Y has three valence electrons?

    b. X has seven valence electrons and Y has four valence electrons?

    c. X has five valence electrons and Y has one valence electron?

    d. X has six valence electrons and Y has two valence electrons?

*24. For each of the following compounds, find the oxidation numbers for the ions that form the compounds:

    *a. $BaBr_2$ c. MgO

    b. Al_2S_3 d. Ca_3N_2

Sections 11.12 and 11.13

*25. Draw electron dot diagrams for the following molecules: NCl_3, Cl_2O_7, SiH_4.

26. Draw electron dot diagrams for the following molecules: CF_4, P_2O_4, S_2F_2.

*27. Draw bond line structures for the molecules in Problem 25.

28. Draw bond line structures for the molecules in Problem 26.

*29. Draw electron dot diagrams for the following molecules, all of which contain multiple bonds: C_2H_4, N_2, CH_3CHO, CO_2.

30. Draw electron dot diagrams for the following molecules, all of which contain multiple bonds: HCN, C_2H_2, NOF, SO_2.

*31. Draw bond line structures for the molecules in Problem 29.

32. Draw bond line diagrams for the molecules in Problem 30.

33. Draw electron dot diagrams for the following molecules: $HClO_3$, CCl_4, SO_2Cl_2.

34. Draw electron dot diagrams for the following molecules: H_2S, NF_3, N_2O_4.

Section 11.14

*35. Draw electron dot diagrams for the following polyatomic ions: NO_2^-, CO_3^{2-}, ClO_4^-, HPO_4^{2-}.

36. Draw electron dot diagrams for the following polyatomic ions: PO_4^{3-}, BO_3^{3-}, BF_4^-, $S_2O_4^{2-}$.

Section 11.17

*37. Draw the structural formulas for the following molecules: C_2H_2, CH_3OH, CH_2O.

38. Draw the structural formulas for the following molecules: CO_2, CH_3COH, NH_3BH_3.

Section 11.18

*39. Draw all of the structural isomers of each of the following molecules: C_2H_7N, C_3H_5Cl, C_5H_8.

40. Draw all of the structural isomers of each of the following molecules: C_4H_8O, C_3H_6, $C_3H_4O_2$, C_3H_4O.

Chapter 12

Properties and Chemical Reactions of Gases

It may seem strange that such an insubstantial thing as a gas could have chemical and physical properties important enough to rate an entire chapter in a textbook, but studies of the behavior of gases have, in fact, yielded a great deal of valuable scientific knowledge. Gas is one of the three states of matter; many of the substances we see every day as solids and liquids can exist as gases while remaining the same substances. We might think that because we cannot pick up and handle a sample of gas, we cannot measure its mass as we can those of solids and liquids—but this is not the case. A sample of gas enclosed in a container has a definite set of properties that characterize it completely and that can be used to understand its behavior. Once we have examined gases and their properties, they will seem much less mysterious.

One of the most important reasons for studying the behavior of gases is that the properties of a sample of gas depend directly on the number of molecules present. Because of this, it is possible to find directly how many molecules a sample of gas has. Knowing the number of molecules present in reacting samples of substances is a great advantage to understanding the reactions themselves. For this reason, studies of gas reactions have been important historically and remain important.

We begin our study by looking at some of the properties of the gaseous state and by reviewing the theory that explains those properties. We will then learn the laws that control the behavior of gases. We will find out how to do calculations involving gas properties and the chemical reactions of gases. The chapter ends with a discussion of some of the limitations of the gas laws.

AFTER STUDYING THIS CHAPTER, YOU WILL BE ABLE TO

- list the properties that characterize samples of gases
- explain the kinetic molecular theory and list its assumptions
- explain how a barometer and a manometer measure pressure
- draw a curve describing the relative energies of molecules in a gas

- describe an experiment that verifies Boyle's law
- state verbally and mathematically the separate gas laws
- use the separate and combined gas laws to calculate the effects of changing the conditions under which samples of gas exist
- calculate the molecular weight of a gas sample from its density or from combinations of its other properties
- calculate the density of a gas from its molecular weight
- calculate the total pressure of a mixture of gases from their partial pressures
- calculate the pressure of a gas collected over water
- use the gas laws to make stoichiometric calculations from reaction equations involving gases
- list the characteristics of an ideal gas

TERMS TO KNOW

absolute temperature	ideal gas (perfect gas)	pressure
absolute zero	ideal gas law	proportionality
Avogadro's hypothesis	(combined gas law)	standard atmosphere
barometer	isolated system	standard temperature and
Boyle's law	kinetic molecular theory	pressure (STP)
Charles's law	law of Gay-Lussac	torr
controlled experiment	manometer	universal gas constant
Dalton's law	molar gas volume	vapor pressure
diffuse	partial pressure	

12.1 CHARACTERISTICS OF GASES

Because most known gases are invisible and odorless, and because gases cannot be picked up and handled like other samples, people often forget gases even exist; that is, until they get into a situation in which they cannot breathe for a moment or two. Gases are matter, characterized just as much by physical and chemical properties as other kinds of matter are. We can learn much about chemical reactions generally by studying the physical characteristics and chemistry of gases.

Volume.   A sample of gas assumes both the shape and the entire volume of its container, whereas a sample of liquid has its own volume and takes the shape of the part of the container it fills. A sample of solid has both its own shape and its own volume.

Two different gases in contact will **diffuse,** or spontaneously mix, one into the other.

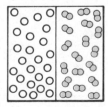
Time zero with barrier between gases

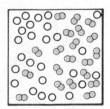
Seconds later, barrier removed, gases diffusing

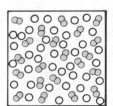
Minutes later, fully mixed

Different liquids in contact also diffuse, but the process is very much slower than the diffusion of gases.

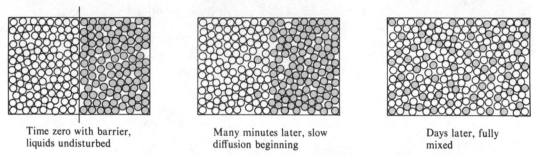

Time zero with barrier, liquids undisturbed

Many minutes later, slow diffusion beginning

Days later, fully mixed

Solids in contact do not mix with one another at a measurable rate.

Block of Au   Block of Pb

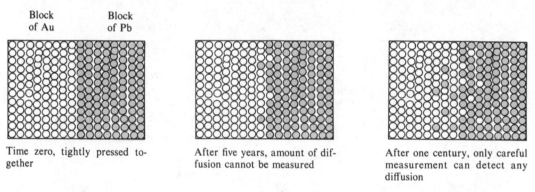

Time zero, tightly pressed together

After five years, amount of diffusion cannot be measured

After one century, only careful measurement can detect any diffusion

Gases are easily compressed; if the volume of the container holding a gas is reduced, the volume of the gas is also reduced. On the other hand, gases can also expand. If the volume of a container is doubled, the volume of the gas it contains also doubles.

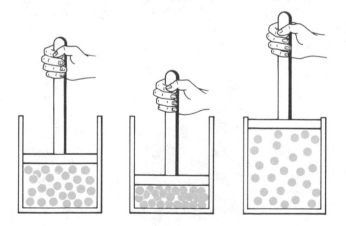

Mass.   Like samples of liquids and solids, samples of gas have mass. A glass bulb weighs less if the air is pumped out of it than if it is full of air. It weighs more filled with carbon dioxide than with the same amount of air.

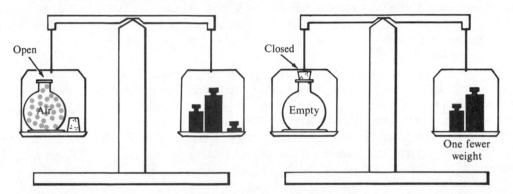

Temperature.   Like samples of other matter, samples of gas also have temperature. Because temperature change causes a greater change in the properties of a gas than in the properties of a liquid or a solid, it is particularly important to study the effects of temperature on the behavior of gases.

Pressure.   Gases have a property that solids and liquids do not have: a sample of gas exerts force on all parts of the walls of the container that holds it, and that force is the same in all parts of the container. Free-standing solids exert force only on the surface on which they rest. Liquids exert force on the parts of the containers with which they are in contact, but the amount of force depends on the depth of the liquid.

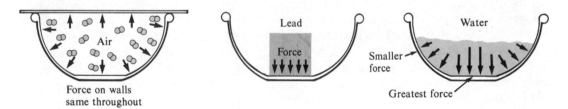

Although it is usually inconvenient to measure the total force exerted by a gas on the walls of its container, it is sufficient for most purposes to measure the force on any square centimeter or other convenient unit of area. We can do this because the force is the same on all parts of the walls. The force exerted on a unit of area is called the pressure. We are familiar with the English units of pressure, pounds per square inch, used, for example, to measure tire pressure.

The SI pressure unit of kilograms per square meter, kg/m^2, is used by some scientists. We will find, however, that grams per square centimeter, g/cm^2, is more convenient. We will also become familiar with two other pressure units, the standard atmosphere and the torr.

Before we go on, let us consider pressure for another moment. Even though a gas seems insubstantial, it can exert enormous force. Wind pushes sailboats; hurricanes drive small bits of matter through strong walls. Automobiles, steam engines, and bullets are moved by the pressure of gases acting on moveable walls of containers. Airplanes—and cars, for that matter—are held up by the pressure of gases. Although pressure is itself invisible, it can have highly visible and even violent consequences. Pressure is one of the characteristics by which gases are measured.

The physical properties by which we can characterize samples of gases are *volume, mass, temperature,* and *pressure.* Because the mass of any sample depends on the number of molecules and the mass of each, we can directly control the mass of a sample of gas by choosing to work with a certain number of moles of a gas. We can control volume simply by choosing the volume of our container. But we can control temperature only indirectly, and pressure is something that the gas itself determines according to its volume, temperature, and number of molecules. Let's find out something about how gases exert pressure.

12.2 KINETIC MOLECULAR THEORY OF MATTER

To explain the behavior of gases, scientists have devised what is called the **kinetic molecular (KM) theory.** The theory is molecular because it assumes that samples of gases are made up of individual molecules, kinetic because it proposes that the molecules are all in constant random motion. (Remember that kinetic energy is the energy of motion.) Using the same approach as for any theory, the KM theory attempts to explain gases by comparing them to an imaginary physical system that can be easily understood.

A gas exerts force on the walls of its container. To understand the forces exerted by gases, let us consider all the kinds of forces we know about. We have discussed electric forces; are gas pressures electric? It turns out that they are not: gas molecules are electrically neutral. Changing the electric charge environment of a gas does not affect its pressure. Similar tests tell us that gas pressures can be neither magnetic nor gravitational. Exposing a gas to a magnet, or turning it this way and that in the earth's gravity (or even taking it to the moon), does not change its pressure. What other kinds of forces are there? Consider momentum transfer: if you hit a tin can with a rock, or sweep up dead leaves with a garden hose, you exert force by transferring the momentum of a moving body to one that is not moving. The KM theory proposes that gases exert pressure in just this way. When gas molecules bang into the walls of their container, each collision transfers momentum, which exerts force on a wall. Let us take the individual features of the theory and justify each one by a simple physical observation.

1. *A sample of gas consists of individual molecules, all of which are in rapid, random, independent motion.* This is reasonable: gases expand rapidly to fill their containers, and easily diffuse into one another. They could not do these things if their molecules were not moving.*

---

*The kinetic theory proposes that the individual particles of solids and liquids are also in constant motion, but that the particles of a solid are quite close together, that they are held near their lattice sites, and that they move only by vibrating around those locations. Molecules in a liquid are nearly as close together as those in a solid, but can flow by crowding one another from place to place. For a given substance, molecules in the solid state have less kinetic energy than those in the liquid state; those in the gaseous state have the most energy of all.

2. *The molecules of a gas are far apart in comparison to their own sizes.* We can understand how this must be so: a sample of gas has far greater volume than the liquid or solid from which that gas came. In their change from a liquid or solid to a gas, the molecules of the sample begin by virtually touching one another but end separated by long distances. Gases are mostly empty space.

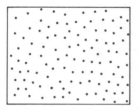

The molecules are very far apart in proportion to their size. Molecular size in relation to distance between molecules is not exaggerated here, but both are shown very large in relation to any visible container.

3. *Gases exert pressure on the walls of their containers by collisions of the molecules with the walls.* We have already learned how this can be so.

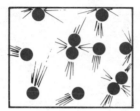

The molecules strike each other and the wall. The force on the wall is generated by molecular impacts. (In this picture, the relative size of the molecules is greatly exaggerated.)

4. *The molecules of a gas collide with one another and against the container walls without gain or loss of energy, although energy can be exchanged in collisions.* If molecules were to lose energy in their collisions, they would soon slow down and stop, and there would be no pressure. Because a sample of gas trapped in a container at constant temperature will exert pressure indefinitely, we know that the motion of molecules never ceases.

We can summarize the features of the KM theory by saying that a gas consists of large numbers of molecules that are far apart and that move rapidly in a chaotic, unorganized way. The pressure is created by the countless collisions of the molecules, the forces of the individual collisions causing a steady force on all parts of the walls.

One of the results of the kinetic theory is the understanding that the temperature of a sample, in any state, is a measure of the average kinetic energy of its particles. The greater the temperature of a sample, the more kinetic energy is possessed by its individual particles. (See Figure 12.1.) Nothing in the theory, however, says that all the particles in a sample must have the *same* kinetic energy. At any particular moment, a few of the molecules in a gas are moving at very high speeds, others are moving very slowly, and most have speeds in between.

We can best understand the different kinetic energies of the moving molecules in gases if we look at the energy distribution curve, shown in Figure 12.2. This graph shows what we would see if we could take an instantaneous picture of a typical sample of gas molecules, each molecule carrying a label saying how much kinetic energy it has. The possible energies that

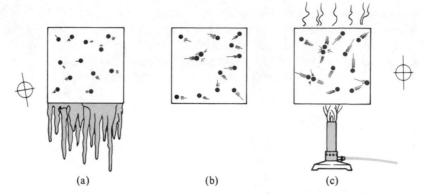

Figure 12.1 Effect of temperature change on the energies of gas molecules. In (*a*), the molecules have kinetic energy. In (*b*), their energies are greater, and in (*c*), greater still.

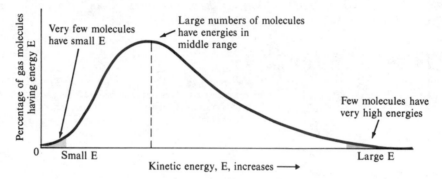

Figure 12.2 Distribution of energies among molecules in a sample of gas.

gas molecules can have are shown horizontally, with the higher energies toward the right. The percentage of molecules that actually have *any given energy* is plotted vertically. Notice that relatively few of the molecules are moving slowly (have low kinetic energy) and relatively few are moving really fast (have high energy). The energies of most molecules fall in the middle range. Because molecules constantly collide and exchange energy, any particular molecule constantly gains and loses energy. Unless the temperature of the sample is changed, however, the *number* of molecules that have any particular energy remains constant, as the curve shows.

Because an increase in temperature corresponds to an increase in the kinetic energy of the molecules, the energy distribution curve shifts to the right and flattens somewhat as the temperature is increased. As we will learn in Chapter 16, changes in the energy distribution curve have important effects on the rate of chemical reaction.

Experiments made at a single temperature but with different gases have shown that a mole of gas enclosed in a given volume exerts a particular pressure, no matter what kind of gas it is. The molecules of various gases must, therefore, have the same average energy at any given temperature, although their molecular masses may differ. The KM theory explains this fact as follows: heavy molecules move, on the average, more slowly than do light molecules at the same temperature. (Any heavy object moving slowly can have the same kinetic energy as a lighter object moving more rapidly.) In other words, *the average kinetic energy of the molecules of a gas depends only on the temperature of the sample and not on the mass of the individual molecules.*

We will soon learn that for a given number of molecules in a given container, the pressure rises if the temperature rises. A rise in temperature causes the average speed of the molecules to increase; this increase causes the molecules to hit the walls more often and harder, which in turn increases the pressure.

12.3 MEASURING GAS PRESSURE

We have learned that the important physical characteristics of a sample of gas are volume, temperature, number of molecules (as measured by mass), and pressure. Measuring the first three of these properties is a fairly straightforward matter. The volume of the sample is simply the volume of the container it occupies; the temperature can be measured by a thermometer; and the mass can be found by weighing (although it is often calculated from known values of other properties). Measurement of pressure, however, requires a little explanation.

We can measure pressure directly by determining how much force a sample of gas exerts on a flexible diaphragm. The kind of pressure gauge that you might see on a compressed air tank uses this method. (See Figure 12.3.) Laboratory work, however, relies on a different method. Understanding this method will also allow us to understand the unit of pressure used both in the laboratory and to express atmospheric pressure.

Evangelista Torricelli discovered, more than two centuries ago, that if a tube closed at one end and filled with mercury is inverted in a dish of mercury, the mercury will run out of the

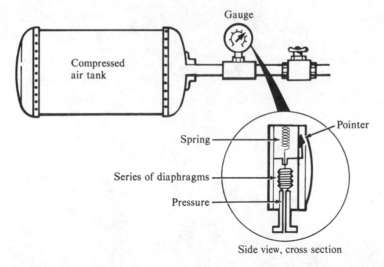

Figure 12.3 How a mechanical pressure gauge works. Because the air inside the diaphragms is at greater pressure than the atmospheric air outside, the diaphragms expand and move the pointer. A gauge like this measures only the difference between inside and outside pressures, not an absolute pressure.

tube only so far, leaving a column of mercury standing in the tube, as Figure 12.4 shows. No matter how long a tube Torricelli used, the mercury ran out until a column of a particular height was formed; the height of the column was the same for all the tubes longer than that height. If Torricelli used a tube shorter than that height, no mercury ran out at all. In a slanted or bent tube, the surface of the top of the mercury column was still the same height above the surface of the pool in the dish as it had been in a straight, upright tube (Figure 12.5).

Many theories were advanced to explain what held up the mercury in Torricelli's tubes, one of which even proposed an invisible thread attached to the surface of the liquid. The explanation is simply this: the pressure that the atmosphere exerts on the pool outside the tube holds the liquid mercury up inside the tube. This occurs because the pressure outside must be

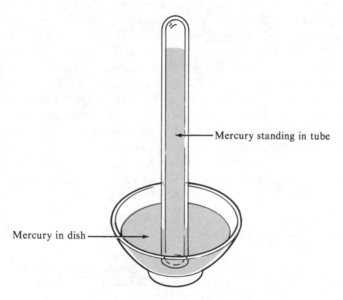

Mercury standing in tube

Mercury in dish

Figure 12.4 Simple Torricelli tube.

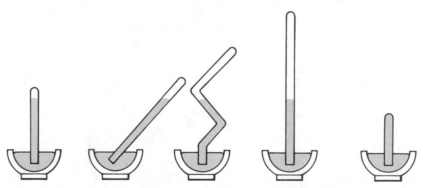

Figure 12.5 Torricelli tubes of different shapes. Notice that the mercury stands at the same height in all the tubes except the one on the far right.

balanced inside by the pressure exerted by the column of mercury, as Figure 12.6 shows. The greater the pressure of the gaseous atmosphere, the greater the height of the mercury.

The pressure of the atmosphere is reported by weather stations in inches or millimeters of mercury and is measured by a **barometer.** The barometer is a Torricelli tube with scales and adjustments attached. On an average day at sea level, the pressure of the atmosphere will cause a column of mercury to stand 760 mm high in the barometer (29.9 in.). This pressure, exactly 760 mm of mercury, has been defined as a **standard atmosphere,** or just **atmosphere (atm).**

Pressures in closed systems of gases are often measured by instruments called **manometers** and are usually stated as the height of mercury (or possibly another liquid) supported by the pressure. As you can see in Figure 12.7, a manometer works on the same principle as a barometer.

The atmosphere and the **millimeter of mercury** are the two units of pressure that we will use. In honor of Torricelli, the millimeter of mercury has been given the name **torr.** One atmosphere equals 760 torr. (Torr is both singular and plural.) Memorize this conversion factor: 1 atm = 760 torr.

EXERCISE 12.1.   Convert 720 torr to atmospheres.

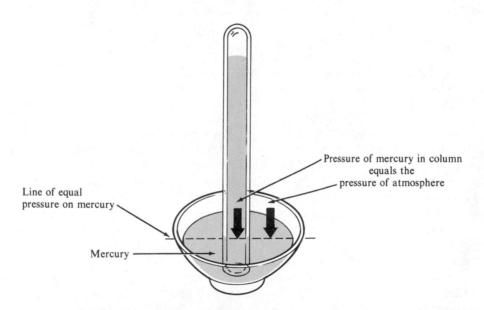

Line of equal
pressure on mercury

Pressure of mercury in column
equals the
pressure of atmosphere

Mercury

Figure 12.6 Operation of a barometer. At the level of the mercury surface shown by the dotted line, the pressure downward on the mercury in the tube must be equal to that on the surface outside the tube, otherwise the mercury will flow in or out. If the pressure of the atmosphere on the surface outside the tube increases, mercury flows into the tube, raising the level of the mercury until the pressures are again equal. If the atmospheric pressure decreases, the mercury level in the tube will fall.

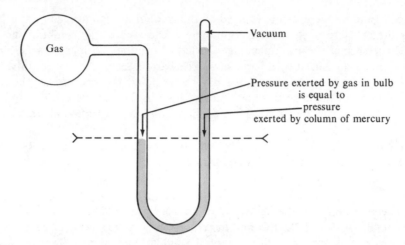

Figure 12.7 Closed tube manometer. This instrument measures pressure by taking the difference in height between the two tubes of the manometer. In the right tube, the only pressure on the mercury below the line is from the mercury above the line. That pressure matches the pressure exerted at the same level by the gas acting on the mercury surface in the left tube. Manometers in which the right tube is open to the air are in use also. Pressure readings in such instruments (like those taken with ordinary pressure gauges) must be corrected for atmospheric pressure.

12.4 PRESSURE-VOLUME RELATIONS IN GASES: BOYLE'S LAW

12.4.1 Controlled Experiments

In our earlier experiment with the tire pump, we learned that a change in the pressure of a gas caused a change in its volume. We demonstrated in at least one experiment that these properties depend on each other. All the properties of gases are, in fact, mutually interdependent: it is impossible to change the value of any property of a sample of gas without causing a change in the value of at least one other property. In this section, we will consider the relation between pressure and volume. When we study the effects of pressure and volume only on each other, we must always be sure to design our experiments so that the number of molecules in the gas sample (the mass) and the temperature of the sample are fixed, that is, held constant. Holding certain properties constant while changing others is one of the most important aspects of scientific technique, and we will consider it next.

In the most general sense, the purpose of an experiment is to find out what happens to a particular thing when we perform a particular action. The trick, then, in a scientific experiment, is to be sure that the observed effect has resulted directly and entirely from the action taken. To ensure this, we must make a **controlled experiment** with an **isolated system.** Control means making sure that the action you take produces the result you see; isolating the system from other influences is one of the best means of achieving control. Let's invent an example.

After watching a boat race in which most of the winning boats were blue, you propose the theory that blue objects experience less friction in the water than do objects of other colors. To test this theory, you drop a stone into the water at the deep end of a swimming pool and measure the time it takes to reach bottom. You then paint the stone blue, let it dry, and drop it again. It reaches the bottom sooner. Great. You have proved your theory. Or have you? You may not have noticed that the water was warmer (therefore had slightly less viscosity) the second time, that the pool pump was circulating the water and creating up-currents the first time, and that the lead in the blue paint increased the density of the stone. Your results could have depended on almost anything—except the fact that the stone was blue the second time. To do the experiment properly, you could simultaneously release two stones, identical in every way except that one was blue and one white; the white stone would act as a control, showing that any difference in the speed of the fall was due only to color. You would isolate the experiment as much as possible by dropping the stones into a still pool, where no currents or other outside effects could change the results. Even though you are using a control, you want to be sure to measure the effects of blue color on water friction, not the effects of blue color on current, temperature, or something else. A good experiment is carefully designed to answer the question being asked—and only that question.

12.4.2 An Experiment to Measure Pressure and Volume

As you learn about the classic experiments made on the properties of gases, be sure to notice which variable conditions were changed and which were kept constant in each experiment.

Although you had little control in your experiment with the tire pump, you did find that the volume of a sample of trapped gas depended on the pressure. To measure the pressure-volume relation more precisely, we will now devise another experiment with better controls. Because the quantity of a gas sample and its temperature both affect its volume and pressure, we will carefully design our experiment to keep quantity and temperature constant. Imagine a cylindrical container of gas with a moveable piston that can change the volume of the gas trapped in the cylinder. Just as we did with the tire pump (see Introduction), we will exert force on the piston by placing weights on it. We will let the area of the piston be 1 cm² so that the downward force on the piston will be the same as the upward pressure of the gas balancing it. (Remember that pressure = force/area; thus, if the area is unity, the value of the force is the same as that of the pressure.) We will also imagine that the piston itself has no weight and that it is frictionless. This will make our analysis simpler; an actual experiment would be made with a real piston, and we would have to make allowance for weight and friction. To control the temperature of the gas, we will immerse the cylinder in a bath of water held at a constant temperature by electronic controls. If we do our experiment slowly, the temperature of the gas will remain constant at that of the water. To control the amount of the gas, we will ensure that none of the sample can leak past the piston and that no gas from the outside can leak in. We will occasionally check the temperature and the amount of sample to be sure that they are unchanged. Figure 12.8 is a schematic drawing of our cylinder and piston in their water bath.

12.4.3 Boyle's Law

To perform our experiment, we simply put various amounts of weight on the piston and measure the effects on the volume of the gas. The volume can easily be measured by the position of the piston. Figure 12.9 shows our cylinder with various amounts of weight on the

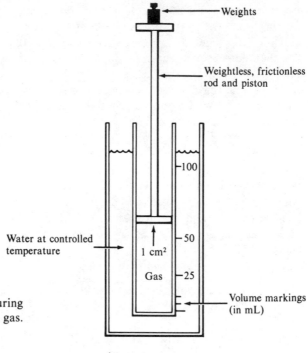

Figure 12.8 Crude apparatus for measuring pressure-volume relations in a sample of gas. (Vertical dimensions are not to scale.)

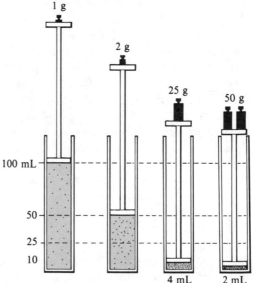

Figure 12.9 Experiments measuring the pressure-volume relation for a gas at constant temperature. The piston and rod are assumed, for simplicity, to be weightless.

piston and therefore with various pressures exerted by the gas inside. (The water bath has been left out of these drawings to make the picture easier to understand, but we remember that the measurements are being made at constant temperature.)

The data from our first series of experiments are shown in Table 12.1.

Table 12.1 Pressure-Volume Data From a Simple Experiment	
Pressure (g/cm²) (Total Mass on Piston Area of 1 cm²)	Volume of Gas (mL)
1	100
2	50
5	20
10	10
25	4
100	1

Table 12.2 Pressure-Volume Data Taken by a Student		
Pressure (torr)	Volume (mL)	Pressure × Volume (torr × mL)
1233.5	9.25	11,410
1023.7	11.18	11,445
861.5	13.22	11,389
757.8	15.07	11,420
575.9	19.75	11,374
545.5	20.92	11,412
510.2	22.41	11,434
465.4	24.55	11,426

Now: what do these data mean? One of the principal requirements of a good scientist is to be able to interpret the data once they are obtained. Methods of trying to understand the results of measurements vary from just looking at the numbers and hoping for inspiration to trying mathematical combinations. Because pictures are often easier to understand than numbers, one of the best ways to interpret data is to draw a picture, or graph. (We will construct some graphs in Section 12.4.4.) It requires little dreaming or trying-out to discover from our data, however, that the volume of the gas multiplied by its pressure always gives a constant number, in this case 100. Try it. For our experiments, P × V = 100, where P stands for the value of the pressure and V for that of the volume.

An experiment that is only slightly more complex, often used as a lecture-demonstration, yields pressure values in torr and volume in milliliters. Table 12.2 is a set of data from such an experiment run by a student. The P × V product is shown in the third column.

Notice that the P × V product in Table 12.2 is given to five significant figures and that the last two figures change from one measurement to another. To find out whether the differences in the P × V products were caused by properties of the gas itself or by errors in the experiment, the student who made the experiment removed and replaced the same set of weights on the piston over and over, reading the volume each time. The PV products for this identical, repeated measurement varied as much as those for all the measurements in the table. This convinced the student that experimental error was responsible for the differences, and that the PV values should all be rounded off to three significant figures. With this, PV = 1.14×10^4 every time, giving a precision of about 1%.

Let us look at the data again: this time, we will label the pressure and volume values for each measurement. $P_1 = 1233.5$ torr and $V_1 = 9.25$ mL (where the subscript 1 indicates that these are the P and V readings for the first measurement); $P_2 = 1023.7$ torr, $V_2 = 11.18$ mL, and so on. We have seen that the value of $P_1V_1 = 1.14 \times 10^4$, but it is also true that $P_2V_2 = 1.14 \times 10^4$, and $P_3V_3 = 1.14 \times 10^4$. For any measurement we take under these conditions, PV = 1.14×10^4. Other experiments show that within certain limits, a similar relation holds for any fixed sample of any gas at constant temperature. The pressure of the gas multiplied by its volume gives a constant number. This relation is often written as

$$PV = \text{constant} \quad \text{or} \quad PV = k$$

where k represents a constant number whose value we may or may not know. In our particular experiment, $k = 1.14 \times 10^4$ torr mL.

In general, for any gas sample of fixed mass at constant temperature, $P_1V_1 = P_2V_2 = P_3V_3$, and so on, for any number of measurements in which P and V are changed. This is a mathematical statement of **Boyle's law,** discovered by Robert Boyle more than two hundred years ago. In words, Boyle's law says that *for any fixed sample of gas at constant temperature, the value of the pressure multiplied by that of the volume gives a constant number.*

12.4.4 Boyle's Law in Graphical Form

Using the results of our measurement to find the mathematical form of Boyle's law was not too hard, but not all experiments are as easy to interpret mathematically. Graphing data—that is, drawing a picture of the numerical results of an experiment—can often help us make interpretations, especially if we can find a way to graph the data in the form of a straight line. A graph of our original PV results (Table 12.1) looks like Figure 12.10. Although this figure might suggest a mathematical relation to a trained eye, it may not be particularly useful to us. Let us try graphing different relations of the data, perhaps squaring P and graphing the relation of $(P)^2$ against V; or perhaps trying $(V)^2$ against P. Doing this gives us more curves that don't tell us much. Eventually, however, when we finally try graphing P against $1/V$, we get something more useful: a straight line (Figure 12.11).

Straight line graphs are easy to understand: a straight line is just a picture of a **proportionality.** When something is proportional (or, more precisely, directly proportional) to something else, as the one thing grows larger or smaller, the second thing becomes larger or smaller by the same relative amount. If the first doubles, so does the second; if the first is cut in half, so is the second, and so on. Mathematically, saying that A is proportional to B is the same as saying that

$$A = \text{a constant number} \times B$$

or

$$A = kB$$

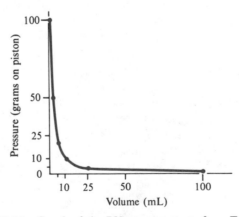

Figure 12.10 Graph of the PV measurements from Table 12.1.

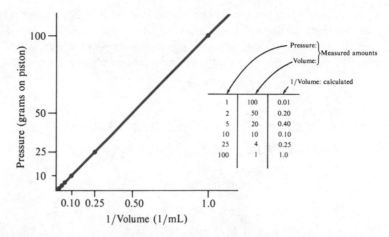

Figure 12.11   Data of Table 12.1 plotted as P versus 1/V.

where k represents the constant number. If k happens to equal 2, then the straight line graph of A = 2B shows A growing upward twice as fast as B grows to the right (Figure 12.12).

When our graph of P vs 1/V gives a straight line, we then know that P = k(1/V), or (multiplying both sides of this equation by V) that PV = k, which demonstrates Boyle's law in graphical form. We will find that graphs of data can help us interpret some of the experiments we will study later. (There is also such a thing as an *inverse proportionality*. Saying that P is proportional to 1/V is the same as saying that P and V are inversely proportional. The graphs of inverse proportions, however, are not as easily identified as such.)

12.4.5 Calculations Using Boyle's Law

Working problems that involve Boyle's law is quite easy—in fact so easy that it is not hard to become careless and go astray. Pressure-volume problems, as such, include only those involving a fixed mass of gas at a constant temperature.

If you begin with a known volume and pressure and change one of these variables, you will then want to know the effect of your action on the other variable. One way to work such a

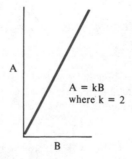

Figure 12.12   A linear graph.

problem is to use the Boyle's law equation, $P_1V_1 = P_2V_2$, in the form that will give an answer. Remembering that our calculations are good only to about 1% precision, we use our rounded values for the second and seventh measurements (Table 12.2) and see how the technique works. We will use the data and a little algebra to obtain some useful results. We begin with

$$P_1V_1 = P_2V_2$$

or

$$1.02 \times 10^3 \text{ torr} \times 11.2 \text{ mL} = 510 \text{ torr} \times 22.4 \text{ mL}$$

We then divide both sides of this equation by P_2 and find

$$\frac{P_1V_1}{P_2} = V_2$$

or

$$\frac{1.02 \times 10^3 \text{ torr} \times 11.2 \text{ mL}}{510 \text{ torr}} = 22.4 \text{ mL}$$

For convenience, we can rearrange the equation slightly.

$$V_2 = V_1 \frac{P_1}{P_2}$$

or

$$22.4 \text{ mL} = 11.2 \text{ mL} \times \frac{1.02 \times 10^3 \text{ torr}}{510 \text{ torr}}$$

This new equation, which can be used for any two measurements of the pressure and volume of a fixed quantity of gas at a constant temperature, says that if we know the volume and pressure of the gas, we can easily calculate the new volume that results from a change in pressure—provided, of course, that we know the new pressure.

To get another useful relation, we rearrange our equation once more.

$$P_2 = P_1 \frac{V_1}{V_2}$$

or

$$510 \text{ torr} = 1.02 \times 10^3 \text{ torr} \times \frac{11.2 \text{ mL}}{22.4 \text{ mL}}$$

In this form, the equation allows us to calculate a new *pressure* after *volume changes* are made.

EXAMPLE 12.1. A sample of gas weighing 2.7 g and having a volume of 3.20 L is at a pressure of 720 torr. While the temperature remains constant, the gas is brought to a new pressure of 618 torr. What is the new volume of the gas?

Solution:

Step 1. Put $P_1V_1 = P_2V_2$ into a form that will help you work this problem. If we let the initial pressure and volume be P_1 and V_1, and the new volume and pressure be P_2 and V_2, we can solve the equation for V_2, which is what we are after.

$$V_2 = V_1 \frac{P_1}{P_2}$$

Step 2. We can now substitute the numerical values.

$$V_2 = 3.20 \text{ L} \times \frac{720 \text{ torr}}{618 \text{ torr}} = 3.73 \text{ L}$$

Notice that the mass of the sample did not enter the calculation. In calculations involving pressure and volume alone, we need only be sure that the mass and temperature do not change during the experiment.

EXAMPLE 12.2. A fixed sample of gas at a constant temperature of 293 K has a volume of 447 mL and a pressure of 785 torr. If the volume is reduced to 212 mL, what is the new pressure?

Solution:

Step 1. Write Boyle's law in the proper form. What we want is P_2, the new pressure.

$$P_2 = P_1 \frac{V_1}{V_2}$$

Step 2. Insert the numbers in the equation and solve.

$$P_2 = 785 \text{ torr} \times \frac{447 \text{ mL}}{212 \text{ mL}} = 1655 \text{ torr}$$

EXERCISE 12.2. A 2.00-L sample of an ideal gas has a pressure of 1.00 atm. What would be the volume of the gas at the same temperature, but at a pressure of 0.750 atm?

As long as you are careful to start with the actual statement of Boyle's law, $P_1V_1 = P_2V_2$, you can safely apply this method. If you use a shortcut, however, and try to put the second form of the equation directly into numbers, you may get the pressure ratio upside down, and an incorrect answer.

You can best check your work by visualizing the problem as an actual physical experiment. To use Boyle's law in problems, remember that if pressure is increased, volume decreases, and vice versa. After you work a problem, look at your answer and ask yourself whether it reflects what really happened.

EXAMPLE 12.3. A cylinder with a volume of 1.00 L contains a fixed sample of gas at 22 °C and 760 torr. To what new volume must the gas be brought for the new pressure to be 1100 torr?

Solution:

Step 1. Ask yourself which variable is changing. In this case it is the volume. Begin by writing the Boyle's law equation in the proper form.

$$V_2 = V_1 \frac{P_1}{P_2}$$

Step 2.   Put in the numbers, do the arithmetic, and check the answer.

$$V_2 = 1.00 \text{ L} \times \frac{760 \text{ torr}}{1100 \text{ torr}} = 0.691 \text{ L}$$

Since 0.691 L is smaller than 1.00 L, we know that we did indeed decrease the volume to increase the pressure.

EXERCISE 12.3.   A sample of gas has a volume of 6.50 L and a pressure of 900 torr. While the temperature remains constant, the volume of the gas is increased to 10.0 L. What is the new pressure of the gas?

EXAMPLE 12.4.   A fixed sample of gas at constant temperature occupies a volume of 554 mL and has a pressure of 1.13 atm. What is its new pressure in torr if the volume is decreased to 198 mL? (Although this problem has new units and a unit change at the end, it is otherwise the same old problem.)

Solution:

Step 1.   We are looking for a new pressure. Set up the equation.

$$P_2 = P_1 \frac{V_1}{V_2}$$

Step 2.   Insert the numbers and do the arithmetic.

$$P_2 = 1.13 \text{ atm} \times \frac{554 \text{ mL}}{198 \text{ mL}} = 3.16 \text{ atm}$$

Step 3.   Convert the answer to torr, using the conversion factor 1 atm = 760 torr.

$$3.16 \text{ atm} \times \frac{760 \text{ torr}}{1 \text{ atm}} = 2.40 \times 10^3 \text{ torr}$$

EXERCISE 12.4.   The volume of a gas sample is increased from 3.1 L to 5.7 L at constant temperature. The original pressure of the gas was 695 torr. In atmospheres, what is the new pressure of the sample?

12.5 RELATION OF VOLUME TO AMOUNT OF SAMPLE: AVOGADRO'S HYPOTHESIS

If we think about it, intuition will tell us that for any particular gas, all else being equal, the larger the amount of gas, the larger the volume it will occupy. A sample containing 2 g of a gas should occupy twice as much volume as a sample weighing 1 g at the same temperature and pressure. (See Figure 12.13.) Experiments bear out our intuition. *At constant temperature and pressure, the volume of any particular gas is directly proportional to the mass of*

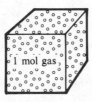
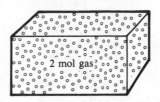

Figure 12.13   Relation between volume and amount in two samples of gas. Temperature and pressure are the same in both boxes.

the sample. This means, then, that the volume is also directly proportional to the number of moles in the sample. More gas, more volume; less gas, less volume. The equation that reflects this fact is equally simple. At constant pressure and temperature,

$$V = kn$$

The number of moles in this equation is given by n. As usual, k is a fixed number whose value we may or may not know, depending on the kind of experiment. As we did for the PV experiments, we can write the $V = kn$ equation for an experiment in which the number of moles and the volume are changed.

$$V_1 = kn_1 \quad \text{and} \quad V_2 = kn_2 \quad \text{and} \quad \frac{V_1}{n_1} = k = \frac{V_2}{n_2} \quad \text{or} \quad \frac{V_1}{n_1} = \frac{V_2}{n_2}$$

We can work problems by solving this equation for any of the variables.

EXAMPLE 12.5.   A sample of gas containing 4.66 mol has a volume of 2.25 L. Under conditions of constant pressure and temperature, the number of moles of gas is increased to 5.22. What is the new volume of the sample?

Solution:

Step 1.   The variable that is changing is V. We set up our equation to give a new V.

$$V_2 = V_1 \frac{n_2}{n_1}$$

Step 2.   With the numbers in their places and the arithmetic completed, the equation looks like this:

$$V_2 = 2.25 \text{ L} \times \frac{5.22 \text{ mol}}{4.66 \text{ mol}} = 2.52 \text{ L}$$

Since 2.52 L is larger than 2.25 L, we know that we have matched the problem to the physical change that occurred. Don't forget to do this check every time.

EXERCISE 12.5.   A 2.50-mol sample of a certain gas has a volume of 8.74 L. At constant temperature and pressure, the number of moles is decreased to 1.25. Calculate the new volume of the gas.

EXERCISE 12.6.   A sample of gas decreases in volume from 30.4 L to 18.6 L when some of the gas escapes. The final number of moles is 5.32. How many moles of gas were originally in the sample? (Temperature and pressure are constant.)

In addition to the relation between n and V for a sample of any particular gas, there is also an enormously important relation between volume and number of moles that applies to gases of *different kinds*. Let us see what that relation is.

Because we can weigh gases and can calculate the molecular weights of gases from their formulas, it is possible to prepare samples of different gases, all of which contain the same number of moles, for example, 1 mol each. We might prepare a sample of O_2 containing 32 g, one of Ar containing 40 g, one of Cl_2 containing 71 g, and so on. If we measure the volumes of these gases at the same temperature and pressure, we find that all the samples have the same volume. To a very good approximation, it is experimentally true for all gases that *gas samples containing the same number of moles at the same temperature and pressure have the same volume no matter what kinds of gases are in the samples*. This is a statement of **Avogadro's hypothesis.** Today, we can understand why the hypothesis is true, because we have access to the atomic weight scale and can measure out a mole. The formulation of the hypothesis, however, actually preceded the first measurement of atomic mass, and was an important factor in the development of the atomic theory.

Stated in another way, Avogadro's hypothesis says that *the volume of a gas sample depends on temperature, pressure, and the number of molecules in the sample, but not on the kinds of molecules.*

12.6 RELATION OF THE VOLUME OF A GAS TO ITS TEMPERATURE: CHARLES'S LAW

The law describing the effect of temperature on the volume of a sample of gas was discovered by Alexandre Charles and announced in the late 1700s. The relation is simple: *the volume of a fixed quantity of gas at constant pressure is directly proportional to its absolute temperature.* At constant pressure, Charles's law is stated mathematically as

$$V = kT \quad \text{and} \quad \frac{V_1}{T_1} = k = \frac{V_2}{T_2} \quad \text{or} \quad \frac{V_1}{T_1} = \frac{V_2}{T_2}$$

where T is the absolute temperature.

Notice that Charles's law is stated in terms of *absolute* temperature, rather than Celsius temperature. A graph of the volume of a sample of gas at different temperatures shows us why we use the absolute temperature scale for gas calculations and, in fact, where the idea of the absolute scale comes from. Figure 12.14 graphs the volume of a fixed amount of gas at constant pressure between the *Celsius* temperatures of 0° and 100°. The graph is a straight line, which indicates that a proportional relation of some kind exists. Notice, however, that when the temperature reaches 0 °C, the gas still has volume, indicating that the volume of the gas is *not* proportional to the Celsius temperature. If we look at the proportionality equation $V = kT$ and let T be the temperature on the Celsius scale, we find that the equation requires the volume of the gas to be zero when the temperature falls to zero ($V = k0 = 0$). But physically, the gas simply does not do that. The temperature in your food freezer is well below 0 °C, but the air in the freezer does not shrink to nothing. The Celsius scale clearly cannot be used in the Charles's law relation.

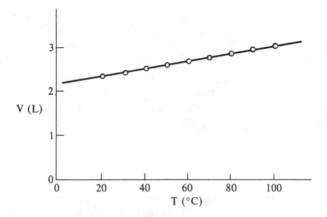

Figure 12.14 A graph of gas volume versus Celsius temperature. Notice that at 0 °C the volume of the gas is not zero.

We know that temperatures below 0 °C exist: the temperature of dry ice, for example, is −78.5 °C. Let us measure the volume of our sample at this temperature and then extend the straight line beyond this point by drawing it in (extrapolating). Figure 12.15 shows what such a graph looks like. The line goes through the point representing our last measurement and continues (dashed) until it intersects the baseline, where the value of the volume is zero. Careful extension of the temperature scale shows that the temperature at this intersection is −273.2 °C. That is the temperature at which the volume of the sample would, according to our graph, become zero. As we have already learned, the Kelvin temperature scale starts at −273 °C; we now see that the temperature-volume graph shows proportional behavior if we use that scale. The zero of absolute temperature and the Kelvin scale, sometimes called the **absolute temperature scale,** was discovered by such an extrapolation.

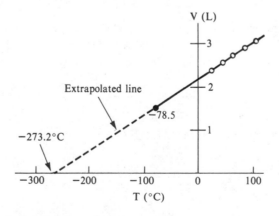

Figure 12.15 The data of Figure 12.14 replotted. A measurement taken at −78.5 °C has been added and the line has been extrapolated to zero volume of the gas.

The practical effect of the relationship we have graphed is that gas law calculations must be based on a temperature scale that is at zero when the volume is at zero. *Remember to convert all temperatures to the Kelvin scale in making gas law calculations.* If you forget to convert to the K scale and use the Celsius scale instead, you will get a wrong answer.

Technically, the zero of absolute temperature remains theoretical, because that temperature has never been achieved in the laboratory (although some experiments have come within a small fraction of a degree). Nor has any sample of gas ever been brought to zero volume. **Absolute zero** does, however, have physical meaning when interpreted by the kinetic molecular theory, and it has practical use in gas law calculations. Here is a typical Charles's law problem.

EXAMPLE 12.6.   A fixed amount of gas at a constant pressure of 750 torr has a volume of 16.0 L at a temperature of 298 K. What will be its volume if the temperature is raised to 500 K?

Solution:

Step 1.   Convert temperatures to the absolute scale if they are not already given that way. In this problem, the temperatures are already in kelvins.

Step 2.   Identify the variable whose new value is required, in this case volume. Set up a Charles's law equation in which the initial volume is multiplied by a temperature ratio that will produce the new volume.

$$V_2 = V_1 \frac{T_2}{T_1}$$

Step 3.   Do the arithmetic.

$$V_2 = 16.0 \text{ L} \times \frac{500 \text{ K}}{298 \text{ K}} = 26.8 \text{ L}$$

Check your result by visualizing a physical experiment. Decide whether the new volume should be greater or less than the original volume. (Because the temperature increased in this problem, the volume must also have increased.)

Notice that the procedure we followed for a Charles's law problem is exactly the one we used to work a Boyle's law problem, except that we added a step to change the temperature to the absolute scale. Remember, too, that Boyle's law is an inverse relation and Charles's law is a direct proportionality.

EXAMPLE 12.7.   A fixed mass of gas occupies a volume of 1.00 L at a temperature of 100 °C. At what Celsius temperature will the volume be 2.00 L?

Solution:

Step 1.   Convert the given temperature to kelvins.

100 °C + 273 = 373 K

Step 2.   Identify the variable whose new value is required, in this case temperature. (Because the variable that we are asked to find is the new *Celsius* temperature, we must remember to convert our answer back to the Celsius scale when we finish.) Set up the equation.

$$T_2 = T_1 \frac{V_2}{V_1}$$

Step 3. Do the arithmetic and convert the new absolute temperature to degrees Celsius.

$$T_2 = 373 \text{ K} \times \frac{2.00 \text{ L}}{1.00 \text{ L}} = 746 \text{ K}$$

$$746 \text{ K} - 273 = 473 \text{ °C}$$

EXERCISE 12.7.   A sample of an ideal gas at 298 K occupies a volume of 5.25 L. Calculate the volume of the sample at 500 K.

EXERCISE 12.8.   A 4.28-L sample of an ideal gas at 0 °C is cooled at constant pressure. What is the final Celsius temperature of the sample if the final volume is 2.74 L?

12.7 RELATION BETWEEN PRESSURE AND TEMPERATURE: GAY-LUSSAC'S LAW

If we think a little about Boyle's and Charles's laws, it should come as no surprise that *the pressure of a fixed sample of gas at constant volume is directly proportional to its absolute temperature.* This is the **law of Gay-Lussac,** which was discovered at about the same time as Charles's law. The algebraic statement of Gay-Lussac's law can be written in a way that looks much like the Charles's law equation.

$$P = kT \quad \text{and} \quad \frac{P_1}{T_1} = k = \frac{P_2}{T_2} \quad \text{or} \quad \frac{P_1}{T_1} = \frac{P_2}{T_2}$$

To work problems involving the effects of temperature on the pressures of gases at constant volume, we can use exactly the same method that we used to work Charles's law and Boyle's law problems.

EXAMPLE 12.8.   A sample of gas with a fixed mass of 2.4 g is contained in a flask whose volume is constant. At a temperature of 20 °C, this gas has a pressure of 745 torr. What is the new pressure if the temperature of the sample is raised to 80 °C?

Step 1. Convert temperatures to the absolute scale. The initial temperature is 293 K and the final temperature is 353 K.

Step 2. The variable whose value is to be changed is pressure. Set up the equation.

$$P_2 = P_1 \frac{T_2}{T_1}$$

Step 3. Insert the values and do the arithmetic.

$$P_2 = 745 \text{ torr} \times \frac{353 \text{ K}}{293 \text{ K}} = 898 \text{ torr}$$

Step 4.   Check the result. Because temperature and pressure are directly proportional and because the temperature is increased, the pressure must also increase. The new value of the pressure will be greater than the initial value.

EXERCISE 12.9.   At 298 K, a sample of gas occupies a container of constant volume. The measured pressure is 1.50 atm. When the gas is heated, the pressure is found to have increased to 5.00 atm. What is the new temperature?

EXERCISE 12.10.   What will be the new pressure of a sample of gas if its original pressure was 790 torr at 180 °C and the temperature is lowered to 0 °C?

12.8 THE IDEAL GAS LAW

We have seen that it is possible to change the volume of a gas by changing the temperature (Charles's law). If we then change the pressure of the gas by restoring the volume to its original value (Boyle's law), we will end with a change of temperature and pressure but none of volume (Gay-Lussac's law). In general, we can accomplish any of the physical changes in more than one step by making successive changes in other properties. All the variables that describe the state of a sample of gas are physically interdependent: if the value of any one variable is changed, the value of at least one other variable changes as well. Reflecting this fact, the individual gas laws can be combined into a single law. The four gas laws are given in algebraic form, where k_1, k_2, k_3, and k_4 all represent constant numbers, but not necessarily the same number.

$PV = k_1$	$P = k_2T$	$V = k_3T$	$V = k_4n$
Constant	*Constant*	*Constant*	*Constant*
n and T	*V and n*	*P and n*	*P and T*

The equation that combines all these laws contains all the variables, as well as the values of all the k's combined in a single constant number. The new, single constant is given the designation R and is called the **ideal gas constant** or the **universal gas constant.** The equation itself, which follows, is called the **ideal gas law** or the **combined gas equation.**

$$PV = nRT$$

Let us find the value of R. Imagine an experiment that might be done in a beginning chemistry laboratory. We want to prepare a sample of gas and measure the values of all its variables, then place those numbers in the ideal gas equation and solve for the value of R. We choose oxygen gas, O_2, because it is easily prepared and safe to work with. We will prepare 0.100 mol by heating the compound $KClO_3$.

$$2 \ KClO_3 \xrightarrow{\ Heat\ } 2 \ KCl + 3 \ O_2$$

Using the method we learned in Chapter 7, we calculate the mass of $KClO_3$ that is required to yield 0.100 mol of O_2.

$$0.100 \text{ mol O}_2 \times \frac{2 \text{ mol KClO}_3}{3 \text{ mol O}_2} \times \frac{122.5 \text{ g KClO}_3}{\text{mol KClO}_3} = 8.17 \text{ g KClO}_3$$

Now we heat the $KClO_3$ strongly in a test tube and exhaust the O_2 into a cylinder and piston arrangement. (A water bath keeps the cylinder and piston at a constant convenient temperature of 25 °C.) By moving the piston, we constantly adjust the pressure of the gas to 750 torr, the atmospheric pressure that day (Figure 12.16). Finally, when all the O_2 is driven off the $KClO_3$, we measure the volume of our 0.100 mol of gas at 750 torr and 25 °C. That volume turns out to be 2.48 L. We can now calculate the actual value of R by substituting the values of all the variables into the ideal gas equation. We must, however, first convert the temperature of 25 °C to 298 K.

$$PV = nRT \qquad\qquad R = \frac{PV}{nT}$$

$$R = \frac{750 \text{ torr} \times 2.48 \text{ L}}{0.100 \text{ mol} \times 298 \text{ K}} = 62.4 \text{ L torr/mol K}$$

Notice the units in which we have calculated R. They are L torr/mol K. If we had chosen to, we could have measured the pressure of our gas in atmospheres. In that case, the value of R would have turned out to be 0.0821 in the units of L atm/mol K. Both sets of units, L torr/mol K and L atm/mol K, are widely used in gas law calculations. You will need to have both values of R handy. Remember that the numerical value of R depends on the choice of units; always be sure to choose the proper value of R for the particular set of units you are using.

12.9 CALCULATIONS USING THE IDEAL GAS LAW

Although the ideal gas law is frequently applied in a way that uses the value of R, the equation can also be rewritten to make R disappear. If we use the equation in this form, we

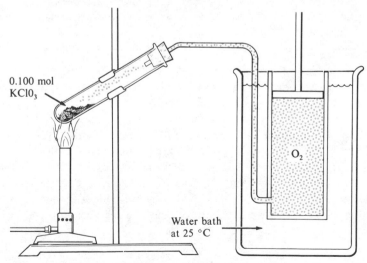

Figure 12.16 Collection of O_2 from $KClO_3$ reaction. Total pressure exerted on gas is kept constant at 750 torr by moving piston as required.

can calculate how the value of a single variable is affected by changes in the values of any or all of the other variables. Because the second approach is more often useful, we will make a few calculations using the value of R; then we will do some algebra and make R disappear.

EXAMPLE 12.9.   What is the pressure in torr exerted by 0.331 mol of the gas Ar, when the sample is contained in a flask whose volume is 400 mL and whose temperature is 20 °C?

Solution:

Step 1.   Convert temperature to the absolute scale.

$$20 \text{ °C} + 273 = 293 \text{ K}$$

Step 2.   Convert other values to the units in which the value of R appears. Because we have R in L torr/mol K, we will use those units. We must convert the 400 mL to 0.400 L.

Step 3.   Rearrange the combined gas equation to give the required variable.

$$P = \frac{nRT}{V}$$

Step 4.   Insert the values of the variables and of R and do the arithmetic.

$$P = \frac{0.331 \text{ mol} \times 62.4 \text{ L torr/mol K} \times 293 \text{ K}}{0.400 \text{ L}} = 1.51 \times 10^4 \text{ torr}$$

EXERCISE 12.11.   A flask with a volume of 250 mL contains 0.0130 mol of gas at a pressure of 1.25 atm. What is the Celsius temperature of the gas?

EXERCISE 12.12.   A 55.5-g sample of N_2 gas has a pressure of 850 torr at 273 K. What is the volume of the sample? (Hint: You will need to use the molecular weight of N_2 to find the number of moles in the sample.)

Because the temperature, pressure, and volume of a gas sample all depend in the same way on the number of gas molecules in the sample, it is particularly easy to find the number of molecules or the number of moles in a sample. One way is to measure P, V, and T, and insert those values into the combined gas equation. Of course, it is always possible to find the number of moles of a gas, just as we would that of a solid or liquid, by weighing the sample and using its mass and molecular weight to calculate the number of moles. But weighing gases is not easy, and if we do not know the formula of the gas, we cannot find the molecular weight by weighing. An advantage of working with gases, however, is that we can easily find the number of moles in a sample even without knowing the formula of the gas. This characteristic of gases was of historic importance in the development of the atomic theory.

EXAMPLE 12.10.   A sample of unknown gas has a pressure of 650 torr, a temperature of 295 K, and a volume of 3.66 L. How many moles of gas are in the sample?

Solution:

Rearrange the combined gas equation to give the number of moles, then insert the values of the properties.

$$PV = nRT \qquad \text{rearranges to} \quad n = \frac{PV}{RT}$$

$$n = \frac{650 \text{ torr} \times 3.66 \text{ L}}{62.4 \text{ L torr/mol K} \times 295 \text{ K}} = 0.129 \text{ mol}$$

EXERCISE 12.13.   A sample of gas occupies a 4.00-L flask at 50 °C and has a pressure of 1.75 atm. How many moles of gas are present?

Experimental arrangements often require that the conditions under which a gas exists be changed. In such cases, the combined gas equation is often used in a form that allows the effects of such changes to be calculated directly. We can rewrite the equation by imagining that we have a gas whose initial pressure, volume, temperature, and number of moles are P_1, V_1, T_1, and n_1, and that the values of those variables after a change in the conditions become P_2, V_2, T_2, and n_2. The combined equation applies to both situations; therefore,

$$\frac{P_1 V_1}{n_1 T_1} = R \qquad \text{and} \qquad \frac{P_2 V_2}{n_2 T_2} = R$$

Because the value of R is the same in both cases (R is a constant), we set both expressions equal to R and therefore to each other.

$$\frac{P_1 V_1}{n_1 T_1} = \frac{P_2 V_2}{n_2 T_2} = R$$

Some rearranging allows us to set up this equation to show how the value of any one variable is affected by a change in the other variables. For example, let us see how the pressure of the sample is affected by a change in volume, number of moles, and temperature. We solve our equation for P_2, the final pressure.

$$P_2 = \frac{P_1 V_1 n_2 T_2}{n_1 T_1 V_2} = P_1 \times \frac{V_1}{V_2} \times \frac{n_2}{n_1} \times \frac{T_2}{T_1}$$

The equation now gives us the final pressure in terms of the initial pressure multiplied by factors that make successive adjustments for the changes in volume, number of moles, and temperature. We can use this form of the equation to work a problem.

EXAMPLE 12.11.   A sample of He containing 0.94 mol has a temperature of 25 °C and a pressure of 696 torr. The volume of the sample is halved, its temperature is brought to 0 °C, and 0.20 mol of additional He is added. What is its new pressure?

Solution:

Step 1. Convert temperatures to the absolute scale. Our initial temperature, T_1, is 298 K and our final temperature, T_2, is 273 K.

Step 2. Rearrange the combined equation in terms of the required variable. We have already done this in the preceding explanation.

$$P_2 = P_1 \times \frac{V_1}{V_2} \times \frac{n_2}{n_1} \times \frac{T_2}{T_1}$$

Step 3. Substitute the values for the initial and final variables into the equation, remembering that $V_2 = V_1/2$. Do the arithmetic.

$$P_2 = 696 \text{ torr} \times \frac{V_1}{V_1/2} \times \frac{1.14 \text{ mol}}{0.94 \text{ mol}} \times \frac{273 \text{ K}}{298 \text{ K}} = 1547 \text{ torr}$$

EXAMPLE 12.12.   At 15 °C, a sample of gas occupies 6.00 L and has a pressure of 7.88 atm. The number of moles of gas is doubled and the volume and pressure are adjusted to 10.00 L and 800 torr. After these changes, what is the final temperature?

Solution:

Step 1. Convert the initial temperature, T_1, to the absolute scale. 15 °C becomes 288 K. This time, one of the pressures must be converted to the same units as the other. We will convert 800 torr, P_2, to 1.05 atm.

Step 2. Rearrange the combined equation in terms of the required variable.

$$T_2 = T_1 \times \frac{P_2}{P_1} \times \frac{V_2}{V_1} \times \frac{n_1}{n_2}$$

Step 3. Substitute the values for the initial and final variables into the equation, remembering that $n_2 = 2 n_1$. Do the arithmetic.

$$T_2 = 288 \text{ K} \times \frac{1.05 \text{ atm}}{7.88 \text{ atm}} \times \frac{10.00 \text{ L}}{6.00 \text{ L}} \times \frac{n_1}{2 n_1} = 32.0 \text{ K}$$

EXERCISE 12.14.   A 2.50-mol sample of gas occupies 1.30 L and has a pressure of 733 torr. The Kelvin temperature is lowered to one third its original value. After 1.35 mol of gas escapes, the final volume is 2.00 L. What is the final pressure?

EXERCISE 12.15.   A sample of gas at 0 °C and 4.00 atm occupies 1.14 L. More gas is added until the pressure has doubled. The temperature is brought to 25 °C and the volume to 2.38 L. If the final number of moles is 0.657, how many moles of gas were added?

Not every variable will necessarily be changed in a gas law problem. If one variable is unchanged, you need make no correction for it. The method is otherwise the same as the one we have been using.

EXAMPLE 12.13.   A sample of gas occupies 4.42 L at 130 °C and has a pressure of 656 torr. The temperature decreases to 40 °C. If the final pressure is 319 torr, what is the final volume?

Solution:

Step 1.   Convert the temperatures to the absolute scale. The initial temperature, T_1, is 403 K, and the final temperature, T_2, is 313 K.

Step 2.   Rearrange the combined equation to obtain the wanted variable.

$$V_2 = V_1 \times \frac{P_1}{P_2} \times \frac{n_2}{n_1} \times \frac{T_2}{T_1}$$

Since in this case the number of moles did not change ($n_1 = n_2$), the factor for the number of moles equals unity. The equation now looks like this:

$$V_2 = V_1 \times \frac{P_1}{P_2} \times \frac{T_2}{T_1}$$

Step 3.   Substitute the values for the variables and do the arithmetic.

$$V_2 = 4.42 \text{ L} \times \frac{656 \text{ torr}}{319 \text{ torr}} \times \frac{313 \text{ K}}{403 \text{ K}} = 7.06 \text{ L}$$

EXERCISE 12.16.   A sample of gas occupies 2.90 L at 962 torr. When the amount of gas is doubled under constant temperature, the final volume is found to be 7.08 L. What is the final pressure?

EXERCISE 12.17.   A sample containing 6.72 mol of gas has a pressure of 4.68 atm. The volume is reduced by half, and some of the gas is allowed to escape. The final pressure is 2.90 atm. How much gas escaped?

12.10 STANDARD TEMPERATURE AND PRESSURE

A set of conditions called the **standard temperature and pressure, STP,** has been established for convenience in reporting the results of experiments. The standard pressure is exactly 1 atm, 760 torr, and the standard temperature is 0 °C, 273 K. In scientific writing, gas volumes are usually given at STP.

We can use the combined gas law to calculate the volume of exactly 1 mol of gas at STP.

$$V = \frac{1.00 \text{ mol} \times 62.4 \text{ L torr/mol K} \times 273 \text{ K}}{760 \text{ torr}} = 22.4 \text{ L}$$

The volume of 1 mol of gas at STP, 22.4 L (Figure 12.17), is called the **molar gas volume.** Because so many data concerning gases are given at STP, the molar gas volume is a useful value to memorize.

When you use the molar gas volume in calculations, it is important to remember two things:

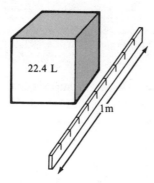

Figure 12.17   Volume of 1.00 mol of gas at STP.

1. The molar gas volume can be used only for gases, not for liquids or solids. (For example, 22.4 L of liquid water contains not just 1 mol, as for gases, but 1244 mol of water.)
2. As defined, the molar gas volume is used only at STP.

Within these restrictions, the molar gas volume is handy for several kinds of calculations, one of which is the number of moles in a sample of gas whose volume at STP is known. The molar gas volume is a conversion factor between volume at STP and the number of moles of gas.

$$\frac{22.4 \text{ L}}{1.00 \text{ mol}} \qquad \frac{1.00 \text{ mol}}{22.4 \text{ L}}$$

Once we have found the number of moles, if we also know the mass of the sample we can calculate the molecular weight.

EXAMPLE 12.14.   The mass of 1.00 L of an unknown gas at STP is 1.43 g. What is the molecular weight of the gas?

Solution:

Step 1.   number of moles = $1.00 \text{ L} \times \dfrac{1.00 \text{ mol}}{22.4 \text{ L}} = 4.46 \times 10^{-2}$ mol

Step 2.   molecular weight = $\dfrac{1.43 \text{ g}}{4.46 \times 10^{-2} \text{ mol}} = 32.0$ g/mol

You can also write this calculation as a one-line unit conversion:

$$\frac{1.43 \text{ g}}{1.00 \text{ L}} \times \frac{22.4 \text{ L}}{1 \text{ mol}} = \frac{32.0 \text{ g}}{1 \text{ mol}}$$

If the mass of a sample is reported at conditions other than STP, the molar volume can still be used to find the molecular weight. In this case, we must first correct the experimental volume to STP, using the method we learned in the last section.

EXAMPLE 12.15.   At a temperature of 25 °C and a pressure of 730 torr, a sample of gas has a volume of 3.10 L and a mass of 5.38 g. What is the molecular weight of the gas?

Solution:

Step 1.   Calculate the volume that the sample would have if it were brought to STP. (See Example 12.13.)

$$V_2 = 3.10 \text{ L} \times \frac{273 \text{ K}}{298 \text{ K}} \times \frac{730 \text{ torr}}{760 \text{ torr}} = 2.73 \text{ L}$$

Step 2.   Use the molar gas volume as a conversion factor to calculate the number of moles in the sample. You can tell which way to use the factor by watching the units.

$$\text{number of moles} = 2.73 \text{ L} \times \frac{1 \text{ mol}}{22.4 \text{ L}} = 0.122 \text{ mol}$$

Step 3.   Use the number of moles and the mass to calculate the molecular weight.

$$\text{molecular weight} = \frac{5.38 \text{ g}}{0.122 \text{ mol}} = 44.1 \text{ g/mol}$$

The gas is CO_2.

EXERCISE 12.18.   A sample of gas weighing 0.700 g has a volume of 1.23 L at 300 K and 0.500 atm. What is the molecular weight of the gas?

Remember that although the temperature, pressure, volume, and number of moles in a sample of gas all depend on each other, the molecular weight of any gas is a fixed property of that gas and remains constant regardless of the size or condition of the sample.

12.11 GAS DENSITIES AND MOLECULAR WEIGHT

Like other samples of matter, gas samples can be characterized by their densities, usually given as the number of grams of gas in 1 L at STP. Tables of gas density are used often in chemistry, and gas densities are routinely measured in the laboratory. Gases at STP are usually only about 1/1000 as dense as common liquids or solids. The density of a gas depends on temperature and pressure, but the densities of solids and liquids are little affected by changes in pressure or temperature.

If we know molecular weight, we can calculate the density of any gas at STP from its molecular weight and the molar gas volume.

EXAMPLE 12.16.   What is the density of gaseous oxygen, O_2, at STP?

Solution:

The molecular weight of O_2 is the atomic weight multiplied by 2.

$$\frac{16.0 \text{ g O}}{\text{mol O atoms}} \times \frac{2 \text{ mol O atoms}}{1 \text{ mol } O_2 \text{ molecules}} = 32.0 \frac{\text{g } O_2}{\text{mol } O_2 \text{ molecules}}$$

Multiply the molecular weight by the reciprocal of the molar gas volume to get the density. Notice how the units come out to give mass divided by volume.

$$\text{density of } O_2 = \frac{32.0 \text{ g } O_2}{\text{mol } O_2} \times \frac{1 \text{ mol } O_2}{22.4 \text{ L } O_2} = 1.43 \text{ g/L}$$

EXERCISE 12.19. What is the density of gaseous ammonia, NH_3, at STP?

To find the density of a sample of gas whose formula is known, you need to know only the conditions under which the gas exists. To make the calculation, imagine a sample of the gas that contains exactly 1 mol. Calculate the volume of that sample from the gas laws and divide the volume into the molecular weight.

EXAMPLE 12.17. What is the density of Cl_2 gas at 298 K and 750 torr?

Solution:

Step 1. To find the volume of a 1-mol sample, use the combined gas law.

$$PV = nRT \qquad \text{then} \qquad \frac{V}{n} = \frac{RT}{P}$$

$$\frac{V}{n} = \frac{62.4 \text{ L torr/mol K} \times 298 \text{ K}}{750 \text{ torr}} = 24.8 \text{ L/mol}$$

Step 2. Divide the molecular weight by the volume.

$$\text{molecular weight} = 35.5 \times 2 = 71.0 \text{ g/mol}$$

$$\text{density} = \frac{71.0 \text{ g/mol}}{24.8 \text{ L/mol}} = 2.87 \text{ g/L}$$

EXERCISE 12.20. What is the density of carbon dioxide, CO_2, at 30 °C and 800 torr?

EXERCISE 12.21. What is the density of the gas N_2O_5 at 25 °C and 0.0085 atm?

If we know the STP density of a gas, it is easy to find the molecular weight: simply use the molar gas volume as a conversion.

EXAMPLE 12.18. The density of a gas at STP is 0.714 g/L. What is the molecular weight?

Solution:

$$\frac{0.714 \text{ g}}{L} \times \frac{22.4 \text{ L}}{1 \text{ mol}} = 16.0 \text{ g/mol}$$

The gas is CH_4.

EXERCISE 12.22.   The STP density of a gas is 5.28 g/L. What is the molecular weight of the gas?

12.12 DALTON'S LAW OF PARTIAL PRESSURES

Let us consider an imaginary experiment. At 0 °C we will place 1 mol of He gas in a box that has a volume of 22.4 L. The pressure that the He exerts will be 1 atm. (Why?) Suppose instead that we place 1 mol of oxygen gas in the box. The pressure of the O_2 would also be 1 atm. The individual pressures of the two gases depend only on the number of moles in the box, the volume, the temperature, and the pressure—not on the kind of molecules (Avogadro's hypothesis again). Nothing new yet, but suppose we now put both the mole of He *and* the mole of O_2 into the box. There would be 2 mol of gas in the box, and the pressure would now be 2 atm. The O_2 molecules would exert 1 atm of pressure and the He molecules would exert 1 atm of pressure, making the total pressure the sum of the individual pressures of the two gases. This is an example of **Dalton's law of partial pressures,** which states that *when two or more gases are mixed in the same container, the total pressure is the sum of the pressures exerted by the individual gases, each behaving as if it were in the container alone.* The pressure exerted by a single kind of gas in a mixture of gases is called the **partial pressure** of that gas.

Dalton's law can be expressed mathematically.

$$P_{total} = P_1 + P_2 + P_3 \ldots$$

In this equation, the combined pressure of all the gases in the mixture is P_{total}, and the individual partial pressures of the gases (1, 2, 3) are P_1, P_2, P_3. For the case just discussed, we might write the equation this way:

$$P_{total} = P_{He} + P_{O_2}$$

The relation of the partial pressures of gases in a gas mixture is the same as the relation of the number of moles of each gas present. For example, in air, which is a mixture mainly of N_2 and O_2 in an approximate proportion of four N_2 molecules for every one O_2 molecule, the partial pressure of N_2 is approximately four times as great as that of O_2.

EXAMPLE 12.19.   What is the total pressure of a mixture of 0.972 atm of O_2, 0.012 atm of CO, and 0.133 atm of NO?

Solution:

From Dalton's law, we know that the total pressure is the sum of the partial pressures.

0.972 atm + 0.012 atm + 0.133 atm = 1.12 atm

EXERCISE 12.23.   What is the total pressure of a mixture of 15.5 g of NO_2 and 0.918 atm of SO_2, occupying a 7.00-L flask at 0 °C?

Mixtures of gases are used in many experiments and chemical reactions. In several common instructional laboratory experiments, for example, various gases are collected in bottles of water. If a gas is in contact with liquid water, there will always be some gaseous water present in the gas phase. In the experiment shown in Figure 12.18, the H_2 gas trapped in the bottle also contains some H_2O in gas form. The pressure of the gas in the bubble is caused by both the H_2O molecules and the H_2 molecules.

$$P_{total} = P_{H_2} + P_{H_2O}$$

To find the pressure of the H_2 alone, we need to measure the total pressure of the gas mixture and subtract from that value the partial pressure of the H_2O.

$$P_{H_2} = P_{total} - P_{H_2O}$$

The pressure exerted by gaseous water in the presence of liquid water is called the **vapor pressure** of the water (discussed in detail in section 13.5). Convenient tables like that in Appendix G allow you to look up the value of water vapor pressure at various temperatures.

EXAMPLE 12.20.   The total pressure of a sample of H_2 collected over water at 25 °C is 749 torr. The vapor pressure of water at 25 °C is 24 torr. What is the partial pressure of the H_2 in the sample?

Solution:

We know from Dalton's law that

$$P_{H_2O} + P_{H_2} = P_{total}$$

$$P_{H_2} = P_{total} - P_{H_2O} = 749 \text{ torr} - 24 \text{ torr} = 725 \text{ torr}$$

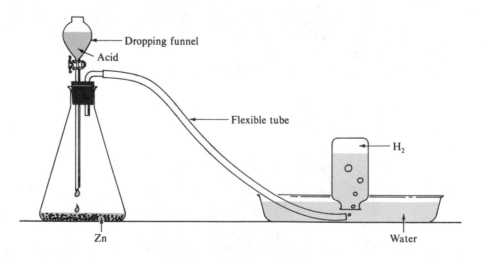

Figure 12.18   Diagram of a gas-generating and -collecting apparatus.

EXERCISE 12.24.   A sample of nitrogen is collected in a flask over water at 30 °C, at which temperature the vapor pressure of water is 32 torr. The total pressure in the flask is 0.750 atm. What is the pressure of the nitrogen?

12.13 STOICHIOMETRIC CALCULATIONS INVOLVING GASES

12.13.1 Calculations in Which Data and Results Are in Volume Units

Quite often, chemical reactions are run with gases as reactants, as products, or as both. If a reaction involves only gases, the reaction equation can be read so that calculations involving the volumes of the reacting gases become quite easy. Recall Avogadro's hypothesis, which we now state in a slightly modified way: at the same pressure and temperature, the volumes of samples of gases are directly proportional to the numbers of moles in those samples. For example, if the volume of one sample is twice as great as that of another sample, then the number of moles in the first is twice as great as that in the second. This means that we can substitute numbers of liters in a gas reaction equation for numbers of moles, then use the coefficients of the reaction equation as conversion factors to make our calculation. The process is similar to the mole ratio method. Caution: *this approach will not work for liquids or solids; it applies only to calculations in which gaseous reactants and gaseous products are under the same conditions.*

Here is the now-familiar reaction equation for the formation of NH_3 from its elements.

$$3\ H_2\ (g) + N_2\ (g) \longrightarrow 2\ NH_3\ (g)$$

The equation shows three times as many moles of H_2 as of N_2; the reaction would therefore require three times as many liters of H_2 as of N_2 at the same temperature and pressure. Corresponding relations hold for the volumes of NH_3 and the other two gases. Using liters as our unit of volume, we can write the equation in terms of volume.

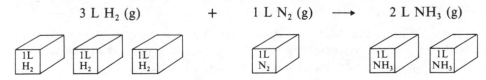

$$3\ L\ H_2\ (g) \qquad + \qquad 1\ L\ N_2\ (g) \qquad \longrightarrow \qquad 2\ L\ NH_3\ (g)$$

Volumes of gases all measured at same temperature and pressure

Expressed this way, the equation can now be used to make calculations of the reacting volumes in the NH_3 reaction.

EXAMPLE 12.21.   How many liters of NH_3 gas can be made from 15.9 L of H_2 and as much N_2 as needed? All the gases are at the same temperature and pressure.

Solution:

Use a volume-ratio conversion exactly as you used mole-ratio conversions in Section 7.2. We know from the equation that the volume ratio of H_2 and NH_3 is

$$\frac{2 \text{ L NH}_3 \text{ (g)}}{3 \text{ L H}_2 \text{ (g)}}$$

The calculation is

$$15.9 \text{ L H}_2 \times \frac{2 \text{ L NH}_3}{3 \text{ L H}_2} = 10.6 \text{ L NH}_3$$

EXERCISE 12.25.   Here is the water-gas reaction, which is run at high temperature:

$$CO \text{ (g)} + H_2O \text{ (g)} \longrightarrow H_2 \text{ (g)} + CO_2 \text{ (g)}$$

How many liters of H_2, measured at STP, can be produced from 2.96 L of CO, also measured at STP?

Even if the reaction equation includes substances that are not gases, the volume-volume method still works, provided that only gases are involved in the calculation.

EXERCISE 12.26.   In the commercial preparation of hydrogen gas, gaseous H_2O reacts with carbon (in the form of coke) at 1000 °C.

$$C \text{ (s)} + H_2O \text{ (g)} \longrightarrow CO \text{ (g)} + H_2 \text{ (g)}$$

How many liters of H_2 at 1000 °C and 900 torr can be made from 255 L of gaseous H_2O at the same temperature and pressure? (Ignore the C (solid) and work the problem in the usual way.)

12.13.2 Calculations Involving Mass as Well as Volume

In any calculation that uses a reaction equation and includes substances that are not gases, work the problem by converting the starting data into numbers of moles, so that the reaction equation can be used; then convert back into laboratory units.

Solution map: (Either the given substance or the product is a gas, whereas the other is not.)

quantity of given substance $\longrightarrow$ moles of given substance
$\longrightarrow$ moles of product $\longrightarrow$ quantity of product

---

EXAMPLE 12.22.   Metallic sodium will react vigorously with liquid water to give hydrogen gas.

$$2 \text{ Na (s)} + H_2O \text{ (l)} \longrightarrow 2 \text{ NaOH (aq)} + H_2 \text{ (g)}$$

In an experiment, 3.31 g of Na was reacted with excess H_2O to give H_2 gas. How many liters of H_2 at STP were produced?

Solution:

Step 1. Calculate the numbers of moles of the reactants. Because Na is limiting, we don't have to worry in this case about the number of moles of H_2O.

$$3.31 \text{ g Na} \times \frac{1 \text{ mol Na}}{23.0 \text{ g Na}} = 0.144 \text{ mol Na}$$

Step 2. Use the mole ratio method to convert moles of Na to moles of H_2.

$$0.144 \text{ mol Na} \times \frac{1 \text{ mol } H_2}{2 \text{ mol Na}} = 7.20 \times 10^{-2} \text{ mol } H_2$$

Step 3. Convert the number of moles of H_2 to liters at STP. Use the molar gas volume as a conversion factor.

$$(7.20 \times 10^{-2}) \text{ mol } H_2 \times \frac{22.4 \text{ L } H_2}{\text{mol } H_2} = 1.61 \text{ L } H_2$$

As a one-line unit equation, the solution is

$$3.31 \text{ g Na} \times \frac{1 \text{ mol Na}}{23.0 \text{ g Na}} \times \frac{1 \text{ mol } H_2}{2 \text{ mol Na}} \times \frac{22.4 \text{ L } H_2}{1 \text{ mol } H_2} = 1.61 \text{ L } H_2$$

EXAMPLE 12.23.   A sample of O_2 with a volume of 15.8 L at 20 °C and 755 torr is reacted with excess H_2 to form water. How many grams of water are produced?

$$2 H_2 \text{ (g)} + O_2 \text{ (g)} \longrightarrow 2 H_2O \text{ (l)}$$

Solution:

Step 1. Find the number of moles of O. This can be done in either of two ways: (1) correct the volume to STP and use the molar volume or (2) put the values directly into the combined gas equation.

$$n = \frac{755 \text{ torr} \times 15.8 \text{ L}}{62.4 \text{ L torr/mol K} \times 293 \text{ K}} = 0.652 \text{ mol } O_2$$

Step 2. Convert to moles of H_2O.

$$0.652 \text{ mol } O_2 \times \frac{2 \text{ mol } H_2O}{\text{mol } O_2} = 1.30 \text{ mol } H_2O$$

Step 3. Convert to grams of H_2O.

$$1.30 \text{ mol } H_2O \times \frac{18.0 \text{ g } H_2O}{\text{mol } H_2O} = 23.4 \text{ g } H_2O$$

EXERCISE 12.27.   Limestone, $CaCO_3$, reacts when heated to form CaO and carbon dioxide. A sample of limestone weighing 5.20 g was heated to produce reaction. How many liters of CO_2, measured at 300 K and 730 torr, were produced?

EXAMPLE 12.24. This problem is harder. A sample of $KClO_3$ mixed with nonreactive material is heated strongly until all the O_2 is driven off. The O_2 is collected over water, and the volume of the gas is found to be 0.981 L at 20 °C and a total pressure of 720 torr. What was the mass of the $KClO_3$? The vapor pressure of H_2O at 20 °C is 17.4 torr.

$$2 \ KClO_3 \longrightarrow 3 \ O_2 + 2 \ KCl$$

Solution:

Step 1. Find the number of moles of O_2 produced. Before we can use the combined gas equation or correct the volume of O_2 to STP, we must convert the temperature to K and find the partial pressure of O_2 in the collected gas sample.

The absolute temperature is 20 °C + 273 = 293 K.

The partial pressure of O_2 is the total gas pressure less the vapor pressure of the water.

(720 torr total) − (17.4 torr H_2O) = 703 torr of O_2 pressure

$$\text{mol } O_2 = \frac{PV}{RT} = \frac{703 \text{ torr} \times 0.981 \text{ L}}{62.4 \text{ L torr/mol K} \times 293 \text{ K}} = 3.77 \times 10^{-2} \text{ mol } O_2$$

Step 2. Use the mole ratio method to find the number of moles of $KClO_3$.

$$(3.77 \times 10^{-2}) \text{ mol } O_2 \times \frac{2 \text{ mol } KClO_3}{3 \text{ mol } O_2} = 2.51 \times 10^{-2} \text{ mol } KClO_3$$

Step 3. Use the molecular weight to calculate the number of grams of $KClO_3$.

$$(2.51 \times 10^{-2}) \text{ mol } KClO_3 \times \frac{122.5 \text{ g } KClO_3}{\text{mol } KClO_3} = 3.07 \text{ g } KClO_3$$

EXERCISE 12.28. Liquid pentane, C_5H_{12}, has a density of 0.626 g/mL at 25 °C. If 0.265 L of pentane is burned in an excess of oxygen, how many liters of gaseous CO_2, measured at 25 °C and 0.950 atm, .will be produced?

12.14 NONIDEALITY IN GASES

We have learned that the gas laws can be useful, helping us to understand the changes in the properties of gases and particularly allowing us to make calculations in which we need to know the number of moles of gas in a sample. Calculations based on the gas laws are ordinarily accurate to three or four significant digits. Under no conditions, however, are the gas laws perfectly exact, and under some conditions (such as the formation of a liquid from its vapor) they fail entirely. Remembering, nonetheless, that for our use the gas laws are entirely adequate and yield excellent results, let's take a brief look at why gases behave in a less-than-ideal manner.

A little careful thought about what we already know tells us that under some circumstances, the gas laws cannot operate at all. For example, we have learned that the volume of a fixed mass of gas at constant pressure is proportional to the absolute temperature, $V = kT$.

Suppose we let the temperature go to zero? True, absolute zero has never been experimentally achieved, but temperatures only a few thousandths of a kelvin away have been reached. Does the volume of a gas really disappear when the temperature reaches zero? Obviously not. Material substances cannot have zero volume: molecules of a gas take up space just as any molecules do. Because the molecules themselves have volume that will not change appreciably with pressure, conditions in which those molecules are forced to be close together (high pressures or low temperatures) cause gases to deviate from the behavior described by the ideal gas laws.

No gas behaves ideally at temperatures very close to 0 K, and many misbehave at much higher temperatures. Consider the gas H_2O, at 600 K and 1 atm. At this temperature, the gas acts in accordance with the gas laws, but it begins to misbehave as the temperature is lowered to about 400 K. At 373 K (100 °C), the volume suddenly decreases to about a thousandth of what it was, and the sample becomes a liquid. Because the molecules of the gaseous water no longer have enough kinetic energy to resist the forces that attract them to one another, they condense into liquid water.

The combined gas law is often called the ideal gas law, not because the law is ideal but because it requires the perfect behavior of gases. An imaginary gas that would obey the ideal gas law under all conditions is called an **ideal gas** or a perfect gas. No such gases exist, but many gases come close, and all gases obey the law quite closely under certain conditions. Within their generous limits, the gas laws are useful tools that we will employ again and again.

12.15 SUMMARY

Studying the chemistry of gases is rewarding, not only because gas reactions are important in themselves, but also because the reactions of gases offer us insights that further our general understanding of chemistry.

A sample of gas can be characterized by the values of four properties: pressure, volume, temperature, and mass (or number of moles). These four properties are interdependent: it is not possible to change the value of one property without causing a corresponding change in at least one of the other properties. Individual pairs of properties are related by various gas laws, such as Boyle's law (PV = a constant when T and n are fixed). The individual gas laws are often combined in a form known as the ideal gas law:

$$PV = nRT$$

The properties of gases are explained by the kinetic molecular theory, which describes gases as collections of widely separated molecules in rapid, random motion. According to the KM theory, pressure is caused by the collisions of molecules with the walls of their container, and changes in pressure by changes in the number of collisions per second and in the strength of the collisions.

Because they are commonly measured by barometers or similar devices, pressures of gases are often stated in the units of torr or millimeters of mercury. A second useful pressure unit is the standard atmosphere, which equals 760 torr.

Used properly, the gas laws enable us to calculate the value of one property of a gas if the other three values are known. We can also use the gas laws to calculate the effect on one property of changes in the values of the other properties. If we know the conditions of temperature and pressure under which a sample of gas exists, knowing its density allows us to calculate its molecular weight. This calculation is particularly easy if the STP density is known.

Application of Avogadro's hypothesis (that equal volumes of different gases contain equal numbers of molecules at the same temperature and pressure) allows us to use gas volumes directly to make calculations from the equations of gas reactions. Measuring the volumes of gases under known conditions gives us a direct way of counting molecules.

Dalton's law (that the total pressure of a sample of mixed gases is the sum of the partial pressures of the individual gases, each acting as if it were alone) allows us to make calculations for individual gases that are part of a mixture, such as gases collected along with water vapor.

Taken all together, the laws and theories about gases provide us with a powerful set of tools for the understanding of chemistry.

QUESTIONS

Section 12.1

1. List four properties common to all gases.
2. Describe how gases, liquids, and solids differ in terms of volume.
3. Gases diffuse easily. Can liquids or solids diffuse easily? Explain.
4. How does the pressure exerted by a gas differ from that exerted by a liquid or solid?

Section 12.2

5. List the four assumptions of the kinetic molecular theory.
6. State and explain the kinetic molecular theory.
7. Explain what is meant by the average kinetic energy of the molecules of a gas.

Section 12.3

8. List ways of measuring the properties of gases.
9. How does a pressure gauge differ from a barometer?
10. Explain the origin of the term *torr*.

Section 12.4

11. State Boyle's law in your own words.
12. Design and explain an experiment that can verify Boyle's law.
13. Write three different mathematical equations, all of which are statements of Boyle's law.
14. How can you determine whether two variables, such as pressure and volume, are proportional, inversely proportional, or neither?

Section 12.5

15. What law relates the volume of a gas to the amount of gas in a sample? State this law in your own words.
16. Write three different mathematical equations that describe the law in Question 15.

Section 12.6

17. What is Charles's law? How does it explain the behavior of a balloon taken out of a deep freeze on a hot day?
18. Write two mathematical equations that are statements of Charles's law.
19. Why must you use absolute temperature when using Charles's law?
20. What is the average kinetic energy of the molecules of a gas at 0 K?

Section 12.7

21. What name is given to the relationship between the pressure and temperature of a gas? Describe the relationship.
22. How does Gay-Lussac's law explain the behavior of automobile tires when they are driven for a long time on a hot day?
23. Write two mathematical statements for Gay-Lussac's law.
24. Describe an experiment that can verify Gay-Lussac's law. (You may want to include principles from Boyle's and Charles's laws.)

Section 12.8

25. Which of the gas laws are directly proportional relationships and which are inversely proportional?
26. Write three versions of the ideal gas law.
27. What is the origin of the universal gas constant? What symbol is ordinarily used for it? Give two different values of the constant, along with their corresponding units.

Section 12.10

28. What is the molar gas volume? Under what conditions is it defined?
29. What is the standard temperature and pressure, STP?
30. Give the solution map for Problem 39 using the molar gas volume to obtain the molecular weight.

Section 12.12

31. State Dalton's law of partial pressures in your own words.
32. Define the terms *partial pressure* and *vapor pressure.*
33. How do you find the pressure of a gas that has been collected over water?

Section 12.13

34. Why can we use gas volumes directly (rather than converting to numbers of moles) when we make calculations involving only gases in a chemical reaction?

Section 12.14

35. State two characteristics of an ideal gas.
36. Under what conditions do gases deviate most from ideal behavior?

PROBLEMS

Note: In the problems that follow, if a variable such as P, V, n, or T is not mentioned, its value should be assumed to be constant. In Problem 3, for example, the temperature and number of moles are constant.

Section 12.3

*1. Make the following pressure conversions:
   *a. 700 torr = mm Hg
   b. 958 mm Hg = atm
   c. 0.035 atm = torr
2. Make the following pressure conversions:
   a. 1020 torr = atm
   b. 150 mm Hg = torr
   c. 340 torr = atm

Section 12.4

*3. If a sample of gas occupies 450 mL at 850 torr, what volume will it occupy
   *a. at 524 torr   b. at 638 mm Hg   c. at 3.69 atm
*4. A 9.25-L sample of gas has a pressure of 22.86 atm. What will be the pressure if the volume is changed
   *a. to 5.42 L   b. to 2043 mL
5. In a laboratory experiment, an ideal gas had a volume of 2.50 L. When the volume was changed to 3.55 L, the pressure measured 514 torr. What was the original pressure in atmospheres?
6. A sample of gas has a volume of 7.20 L and pressure of 1.22 atm. After the volume is changed, the pressure measures 0.54 atm. What is the new volume?
7. If a sample of gas has a pressure of 134 torr and occupies 0.297 L, what will be its pressure when the volume is increased to 3.79 L?
8. A 4.958-L sample of gas has a pressure of 1.77 atm. What will be the volume at a pressure of 345 torr?
9. Graph P versus $1/V$. Use the data from Table 12.1.

Section 12.5

*10. A sample containing 3.40 mol of a certain gas occupies 148 L. What would be the volume of the gas if the sample size were increased to 5.00 mol?
11. A 3.55-L sample of gas contains 2.00 mol. What will be its volume if the number of moles is doubled?
12. Exactly 5 mol of a particular gas occupies 2.60 L. How many moles of gas are there in a 175-mL sample under the same conditions?
*13. Assuming constant temperature and pressure, how many moles of gas must be added to 1.50 L of gas to increase its volume to 3.10 L? (If you cannot answer this question as it is written, explain why and state what further data are needed, if any.)

Section 12.6

*14. Make the following temperature conversions:
   *a. 200 K = °C   c. 400 °C = K
   b. 30 °C = K   d. 145 K = °C
*15. A sample of gas occupies 6.50 L at 298 K. What will be its volume at 555 K?
*16. A 4.75-L sample of gas has a temperature of 20 °C. What would be the Celsius temperature of the same number of moles of gas if it were in a 2.55-L flask at the same pressure?

17. If a 0.500-L sample of gas increases in temperature from 150 K to 500 K, what will be its final volume?

18. By how many kelvins would you have to lower the temperature of a sample of gas to change its volume from 1.22 L to 50.0 mL if the initial temperature is 30 K? If the initial temperature is 475 K?

19. A balloon filled with gas has a volume of 2.00 L at room temperature, 22 °C. The balloon is placed in a refrigerator at a temperature of 0 °C. What will be the new volume of the balloon after it has cooled to the new temperature?

Section 12.7

*20. The pressure of the gas inside a satellite is 760 torr at sea level, where the temperature is 26 °C. The sealed satellite experiences a temperature of 534 °C during lift-off. What is the pressure inside the satellite during that time? In space, the same satellite comes to a temperature of −120 °C. What is the pressure inside at that time?

21. At 120 °C a sample of gas has a pressure equal to 2.25 atm. What will its pressure be at 25 °C?

22. A sample of gas is heated from 310 K to 420 K. If its initial pressure was 980 torr, what will its final pressure be?

*23. A sample of gas at 4.00 atm has a temperature of 200 °C. What will be its temperature at 733 torr?

24. By how many degrees kelvin would you have to raise the temperature of a gas at 175 torr to obtain a pressure of 760 torr? The initial temperature is 155 °C.

Section 12.9

25. Calculate the value of R in units of L atm/mol K, starting with 62.4 L torr/mol K.

*26. A certain sample of gas occupies 250 mL at 30 °C and 760 torr. How many moles of gas are present?

27. If a sample of gas occupies 2.20 L, has a pressure of 1.56 atm, and contains 0.0058 mol, what is its Celsius temperature?

*28. What is the pressure of 1.40 mol of gas occupying 555 mL at 15 °C?

29. What is the temperature of a 55.7-g sample of CH_4 gas occupying 10.0 L at 2.75 atm?

*30. A 1275-mL flask contains 11.3 g of oxygen at 22 °C. What pressure does the gas exert on the walls of the flask?

31. At 300 K, what volume would 13.4 g of CO_2 occupy at 1.16 atm? At 700 torr?

*32. If a 2.16-g sample of HCl has a pressure of 5.50 atm at 20 °C, what will its final mass be if the volume is halved and the temperature and pressure are changed to 50 °C and 2.00 atm?

33. A sample of gas occupies 20.6 L. Half the gas escapes. The final pressure is 468 torr, the final volume 11.3 L. What was the initial pressure if the temperature remained constant?

34. A sample containing 6.65 mol of gas occupies 7.33 L at 48 °C. What is the final temperature of the gas if the pressure is halved, 2.00 mol of gas is added, and the final volume is 25.0 L?

35. A 4.58-g sample of hydrogen occupies 6.75 L. What is the final mass of the hydrogen if the volume is doubled and the pressure is decreased to one-fourth its original value, all at constant temperature?

36. A sample of gas occupies 3.00 L at 25 °C. When the pressure is doubled, the final volume is found to be 1.00 L. What is the final temperature?

*37. At constant temperature and volume, a container of helium contains 45.0 g of He at a pressure of 2.00 atm. The container develops a leak, allowing helium to escape for several hours before it is discovered. When the leak is found, the pressure is 950 torr. How many grams of helium are left in the container?

Section 12.10

38.   How many moles of gas at STP are contained in a 500-mL container?
*39.   What is the molecular weight of a gas weighing 5.40 g and occupying 2.24 L at STP?
40.   A sample of gas occupies 5.26 L at STP. What is the molecular weight of the gas if the sample weighs 12.23 g?
41.   Chloroazide is an explosive gas. At 25 °C and a pressure of 750 torr, a sample of this gas has a volume of 3.89 L and a mass of 12.16 g. What is the molecular weight of chloroazide?
42.   What is the molecular weight of a gas weighing 0.841 g at 15.6 °C and 400 torr if its volume is 50 mL?

Section 12.11

*43.   What is the density of Ne at STP?
44.   What is the density of N_2 at STP?
*45.   What is the density of CO_2 at −10 °C and 725 torr?
46.   Calculate the density of O_2 at −20 °C and 0.444 atm.
*47.   What is the molecular weight of a gas with a density of 3.92 g/L at STP?
48.   The density of a gas is 1.3402 g/L at STP. What is its molecular weight?
49.   Calculate the molecular weight of a gas that has a density of 2.144 g/L at STP.

Section 12.12

*50.   Calculate the total pressure exerted by a gas sample containing 1.02 atm of NO, 2.50 atm of N_2O, and 0.505 atm of NO_2.
*51.   If a mixture of gases contains 833 torr CO_2, 557 torr O_2, and the rest is CO, what is the partial pressure of CO? The total pressure is 3.05 atm.
*52.   Calculate the total pressure in torr of a mixture of oxygen, hydrogen, and nitrogen. The oxygen has a partial pressure of 0.550 atm. The composition of the gas mixture is 35.4% hydrogen, 50.8% nitrogen, and the rest is oxygen.
53.   In a 5.50-L container, what is the total pressure of a mixture containing 43.7 g of O_2 and 1.00 atm of N_2 at 0 °C?
54.   Calculate the total pressure exerted by 1.57 g of CH_4 and 1.57 g of O_2 at 25 °C in a 1.00-L flask.
55.   What will be the total pressure of 16.3 g of krypton, 14.8 g of CO, and 2.37 g of helium, all of which are contained in a 20.0-L flask at 290 K?
56.   A 10.0-L flask contains NO_2F and NOCl. The total pressure is 4.00 atm at 310 K. If the flask contains 15.0 g of NO_2F, what is the partial pressure of NOCl?
*57.   What is the partial pressure of CO_2 in a mixture of CO_2 and CO with a total pressure of 2.00 atm? The 1.00-L flask contains 0.400 g of CO at 10 °C.
58.   A sample of SO_2 at 340 K in a container with a volume of 2.75 L exerts a pressure of 220 torr. Enough NO is added to increase the pressure to 367 torr. How many grams of NO were added?
*59.   The total pressure of a sample of CO_2 collected over water at 25 °C is 0.748 atm. The vapor pressure of water at 25 °C is 24 torr. What is the partial pressure of CO_2 in the sample?
60.   In an experiment, a 2.00-L sample of O_2 is collected over water at 25 °C. The vapor pressure of water at this temperature is 24 torr. How many grams of O_2 were collected if the total pressure was 0.680 atm?

Section 12.13

*61. How many liters of CO_2 and H_2O, measured together at 700 K, can be produced from the combustion of 5.75 L of CH_4 and as much O_2 as needed?

$$CH_4 \text{ (g)} + 2 O_2 \text{ (g)} \longrightarrow CO_2 \text{ (g)} + 2 H_2O \text{ (g)}$$

62. Given the reaction

$$2 H_2S \text{ (g)} + O_2 \text{ (g)} \longrightarrow 2 S \text{ (g)} + 2 H_2O \text{ (g)}$$

how many liters of H_2S are necessary to produce 6.80 L of H_2O? Both gases are measured at 350 °C and at a pressure of 1.1 atm. How many grams of sulfur will be produced?

63. Nitrogen and oxygen can produce nitric oxide according to the equation that follows.

$$N_2 + O_2 \longrightarrow 2 NO$$

How many liters of NO can be produced form 2.25 L of N_2, if reactants and products are all measured at the same T and P?

*64. Trichloromethane, $CHCl_3$, can be formed by the reaction of Cl_2 and CH_4.

$$3 Cl_2 \text{ (g)} + CH_4 \text{ (g)} \longrightarrow CHCl_3 \text{ (l)} + 3 HCl \text{ (g)}$$

If 2.40 L of Cl_2 reacts with 5.75 L of CH_4, how many liters of HCl are produced (all measured at 25 °C and 1 atm)? How many grams of $CHCl_3$ are formed? Which reactant was present in excess?

*65. How many liters of N_2 can be produced at STP from 63.2 g of KNO_3?

$$2 KNO_3 \text{ (s)} + 4 C \text{ (s)} \longrightarrow K_2CO_3 \text{ (s)} + 3 CO \text{ (g)} + N_2 \text{ (g)}$$

66. A 23.0-g sample of Ca metal reacts with 450 mL of NH_3 at STP. How many grams of Ca_3N_2 are produced?

$$6 Ca \text{ (s)} + 2 NH_3 \text{ (g)} \longrightarrow 3 CaH_2 \text{ (s)} + Ca_3N_2 \text{ (s)}$$

67. An 8.33-g sample of an unknown compound reacts with an excess of oxygen to form 4.05 L of CO_2 at STP. Give two possible molecular weights for the compound. (The equation is not balanced.)

$$X + O_2 \longrightarrow CO_2 + H_2O$$

68. At a constant temperature and pressure, 2.67 L of H_2S is mixed with 1.75 L of SO_2. According to the following equation, how many liters of H_2O can be produced, still at the same T and P?

$$H_2S \text{ (g)} + SO_2 \text{ (g)} \longrightarrow S \text{ (s)} + H_2O \text{ (g)}$$

69. A sample of phosphine, PH_3, occupying 1.50 L at 25 °C and 0.080 atm, reacts with 2.28 L of O_2 at 20 °C and 0.100 atm according to this reaction:

$$PH_3 \text{ (g)} + O_2 \text{ (g)} \longrightarrow P_4O_{10} \text{ (s)} + H_2O \text{ (g)}$$

How many grams of P_4O_{10} are produced?

70. If 4.85 L of N_2 at STP is mixed with 3.00 g of O_2, how many liters of nitrous oxide, N_2O, at 20 °C and 750 torr can be obtained?

$$2 N_2 + O_2 \longrightarrow 2 N_2O$$

*71. A sample containing 54.0 g of SO_2 reacts with 28.0 g of O_2 in a 5.00-L flask. When the reaction reaches completion at 500 °C, what is the total pressure in the flask?

$$2 SO_2 + O_2 \longrightarrow 2 SO_3$$

72. The following reaction takes place:

$$Zn \ (s) + 2 \ HCl \ (aq) \longrightarrow ZnCl_2 \ (aq) + H_2 \ (g)$$

The H_2 is collected over water in a 1.00-L flask at 22 °C and a pressure of 1.60 atm. The vapor pressure of water at 22 °C is 19.83 torr. How many grams of Zn were used?

MORE DIFFICULT PROBLEMS

*73. A balloon containing 0.26 mol of gas has a volume of 1.55 L. Some of the gas is allowed to escape. How many moles escaped if the final volume of the balloon is 0.78 L? Temperature and pressure are constant.

74. Calculate the partial pressure of hydrogen if a 1500.0-mL flask contains hydrogen and oxygen at STP. There is an equal mass of each gas.

75. A 10.0-L flask contains oxygen and hydrogen. The total pressure is 725 torr at 25 °C. If there are twice as many hydrogen molecules as oxygen molecules in the flask, what is the partial pressure of the hydrogen?

*76. Gaseous NO_2 forms N_2O_4 (also a gas) under certain conditions. If 46% of the molecules in a mixture of NO_2 and N_2O_4 are NO_2 molecules, how many grams of N_2O_4 are there in a 250-mL flask of the mixture at 25 °C and 770 torr?

77. At 105 °C and 1.00 atm in a 5.00-L flask, 45% of the molecules are NH_3 and the rest are O_2. The following (familiar?) reaction takes place:

$$4 \ NH_3 \ (g) + 5 \ O_2 \ (g) \longrightarrow 4 \ NO \ (g) + 6 \ H_2O \ (g)$$

What is the pressure in the flask at 250 °C after the reaction has finished?

Chapter 13

Condensed States: Their Properties and Energetics

Our study of the condensed phases, solids and liquids, will be rather different than our consideration of gases. The properties of gases are fairly easy to understand because there is a clear connection between the number of molecules in a gas sample and the physical properties of the sample. No such relation exists for liquids or solids. The structures of liquids and solids vary so widely and are so little understood that the theories and calculations concerning those structures are themselves complex and varied. This chapter takes a descriptive approach to the properties of liquids and solids, instead of focussing on mathematical relations between properties and structures.

In other words, Chapter 13 is rather qualitative. Having just taken a mathematical survey of substances that we cannot pick up and usually cannot even see or feel, we now take a good look at familiar substances that we see and use every day: solids and liquids. This chapter will give us some understanding of properties and processes that are familiar—so familiar that we often do not even notice them.

We will be interested in the effects of energy on the condensed phases. Science is concerned not only with what things *are,* but also with what they *do;* when anything does something, energy is involved. When they receive or give up energy, solids and liquids change either their temperature or their phase. Both kinds of change will be of interest to us. We will take particular notice of the temperatures and pressures under which phase changes occur.

We begin by reviewing some general characteristics of solids and liquids, why it is that they exist at all, what happens when heat is added to them or taken away, and how phase changes take place. We meet for the first time the important concept of dynamic equilibrium, which is involved in the control of nearly every aspect of our environment and our own biological processes. We look in detail at the forces that hold liquids and solids together. By studying the various classifications of crystalline solids, we will gain a better understanding of the physical world around us. Finally, we take a quick glance at the properties of liquids; we examine the factors that control the mixing of liquids of difference kinds, as well as those that affect the dissolving of solids in liquids. With this information, we will be thoroughly prepared for our later study of solutions.

AFTER STUDYING THIS CHAPTER, YOU WILL BE ABLE TO

- explain the reasons for the differences in the properties of gases, liquids, and solids
- show how adding energy to or taking energy from a sample can cause phase changes
- explain why adding energy to or taking energy from a sample often results *not* in a phase change, but only in a temperature change
- explain the relation of heat capacity to temperature change
- explain the difference between heat capacity and heat of vaporization or heat of melting
- calculate the amount of energy required to cause phase changes in given samples
- calculate temperature changes of samples, given heat capacities and amount of energy absorbed or lost
- explain the factors that control the rates of vaporization and condensation in a liquid that is in contact with its vapor
- show how the rates of opposing processes can become equal, resulting in a state of dynamic equilibrium
- explain the relation of vapor pressure to the temperature and nature of the sample, and make rough predictions of the vapor pressures of various substances
- show how the boiling temperature of a liquid depends on pressure, as well as on the kind of liquid
- predict, on the basis of relative electronegativity, whether a covalent bond will be nonpolar or polar, and roughly how much polarity a given bond will have
- classify crystalline substances according to the kind of particle in the crystal
- make rough predictions of the properties of substances, on the basis of the kinds of forces likely to exist among their particles

TERMS TO KNOW

boiling	equilibrium	molecular crystal
boiling point	evaporation	normal boiling point
calorie	freezing temperature	polar covalent bond
condensation	heat of fusion	specific heat
condensed state	heat of vaporization	sublimation
covalent crystal	ionic crystal	temporary dipole force
dipole force	melting temperature	vapor
electric dipole	molar heat capacity	vapor pressure
electronegativity	molar heat of fusion	

TERMS WITH WHICH TO BE FAMILIAR

body-centered cubic lattice	simple cubic lattice
London force	van der Waals forces

13.1 CONDENSED STATES

Liquids and solids are called **condensed states.** According to the dictionary, this means that they exist in a compact form. As we know, a sample of matter in liquid or solid form occupies only about a thousandth the volume that the same sample would occupy as a gas at ordinary temperature and pressure. Correspondingly, the condensed forms of a substance are typically about a thousand times more dense than the gaseous form; the volume has been reduced, but the mass remains the same. The change of phase from gas to liquid or solid is called **condensation.**

Gas samples consist mostly of empty space, but in liquids and solids, the particles are close together. Figure 13.1 is a graph of the volume of 0.1 mol of argon, plotted against the temperature of the sample. As a solid, the argon sample occupies a volume of about 2.5 cm³. At 84 K, the graph shows a little jump, corresponding to the melting of the solid argon to a liquid with a volume of about 2.8 cm³. Notice, however, that the volume change at 87.5 K is enormous; it can barely fit on the graph. The liquid has changed to a gas at 1 atm and 87.5 K, and the gas at that temperature has a volume of about 728 cm³ (or about 2260 cm³ at room temperature and pressure). A small volume change occurred, then, when the argon liquefied, and a large volume change occurred when it vaporized.

Whether they are ions, atoms, or molecules, the particles of a solid are pretty well held in place. According to kinetic theory, all particles of all substances at temperatures greater than 0 K have energy of motion, but the motion of the particles of a solid occurs only as vibration (Figure 13.2). The particles move back and forth around fixed points in the solid, but for the

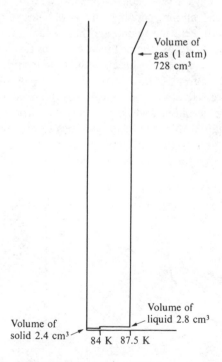

Figure 13.1   Graph of the volume of 0.1 mol of Ar as a solid, a liquid, and a gas.

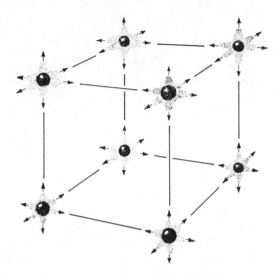

Figure 13.2 Schematic representation of molecular motion in crystalline solids.

most part they do not migrate to new places: they are literally frozen into place. Freezing, after all, is the process of becoming a solid.

In a liquid, there are always a few "holes," or empty positions. When holes are present, the particles can play musical chairs, popping into new holes and leaving holes behind. Because its particles can move in relation to one another, a liquid can flow (Figure 13.3).

When a liquid vaporizes to a gas, the distance between the individual particles increases enormously. Gases flow extremely easily because their molecules are so far apart that they hardly affect one another (Figure 13.4).

Gases are easily compressed; that is, their volumes decrease substantially with relatively small increases in pressure. Liquids and solids, however, are not especially compressible; enormous pressures are required to reduce their volumes significantly. The difference in compressibility between gases and condensed states reflects differences in the arrangement of particles: there is little empty space between the particles in condensed states but gas molecules are far apart.

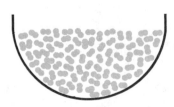

Figure 13.3 A liquid takes the shape of its container.

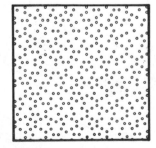

Figure 13.4 A gas fills its container.

13.2 SPECIFIC HEAT

Energy added to a sample of matter in any one phase (but not during phase changes) enters the sample as energy of motion of the individual particles. Since temperature is the measure of the average energy of the particles, the temperature rises when energy is added. When energy leaves a sample, the temperature drops. What is not as obvious, however, is that a given amount of energy produces different amounts of temperature change in samples of different substances.

Do this experiment. Put an ounce of water (about 4 tablespoonfuls) in a small plastic bag. Put it in the refrigerator. Beside it, put four quarters. The four coins have about the same mass as the water does. Let both substances get cold. Now, take out the coins and warm them in your hand. This process does not take very long, nor does it cool your hand much. Let your hand recover, then take out the water and warm that. Notice that warming the water takes much longer and leaves your hand much cooler than warming the coins. Changing the temperature of an ounce of water takes far more heat flow than making the same change in an ounce of metal.

The **calorie** is defined as the amount of heat required to raise the temperature of 1 g of water 1 °C. If a calorie of energy is applied to 1 g of aluminum, however, the temperature of the sample will rise 4.65°. One calorie of energy will raise the temperature of 1 g of uranium more than 36°.

The amount of energy gained or lost by 1 g of substance when its temperature changes by 1 °C is called the **specific heat** of the substance, and the amount of energy gained or lost when the temperature of 1 *mol* of the substance changes 1 °C is called its **molar heat capacity.** We will be concerned mostly with specific heat, and we will usually use the SI units of kilojoules for our measurements. Table 13.1 is a list of the specific heats of several elements and compounds taken at 25 °C, given in both kJ/g°C and cal/g°C.

Specific heat is defined by this equation:

$$\text{specific heat} = \frac{\text{kilojoules added or taken away}}{\text{sample mass} \times \text{degrees temperature gained or lost}}$$

Table 13.1
Specific Heats of Some Substances
(at 25 °C)

Substance	Specific Heat	
	kJ/g°C	cal/g°C
Aluminum	8.8×10^{-4}	0.21
Carbon (diamond)	5.0×10^{-4}	0.12
Magnesium	10×10^{-4}	0.24
Gold	1.3×10^{-4}	0.031
Water	48×10^{-4}	1.00
Ice*	21×10^{-4}	0.49
Benzene	18×10^{-4}	0.42
Ethyl alcohol	25×10^{-4}	0.59

*taken at 0 °C

If we let c stand for the specific heat, m for the number of grams in a sample, $\triangle T$ for the number of degrees of temperature change, and Q for the number of kilojoules of energy gained or lost, then the equation looks like this:

$$c = \frac{Q}{m \times \triangle T}$$

(For convenience, the term $\triangle T$ is sometimes written as $(T_f - T_i)$, where T_f is the final temperature at the end of the experiment and T_i is the initial temperature.)

Of the four quantities, specific heat, grams of sample, temperature change, and amount of heat, any one can be calculated if the other three are known. You need only substitute the values into the specific heat equation. In the discussion and calculations that follow, the units on $\triangle T$ are given as "deg" rather than °C. Since kelvins are the same size as °C we can use either kelvins or °C in the calculations (but not °F!).

EXAMPLE 13.1.   The specific heat of Al is 9.00×10^{-4} kJ/g deg. To a 13.1-g sample of Al is added 3.08×10^{-2} kJ of energy. How many degrees does the temperature of the sample rise?

Solution:

Step 1.   Rearrange the specific heat equation to yield degrees.

$$\triangle T = \frac{Q}{c \times m}$$

Step 2.   Insert the values and do the arithmetic.

$$\triangle T = \frac{3.08 \times 10^{-2} \text{ kJ}}{(9.00 \times 10^{-4} \text{ kJ/g deg}) \times 13.1 \text{ g}} = 2.61 \text{ deg}$$

EXERCISE 13.1.   The specific heat of liquid water is 4.8×10^{-5} kJ/g °C. To a 13.1-g sample of liquid water is added 3.08×10^{-2} kJ of energy. How many degrees does the temperature of the sample rise? (Compare your answer with that of Example 13.1. Why is there such a great difference?)

EXAMPLE 13.2.   To a 19.5-g sample of Li at 20.00 °C is added 0.173 kJ of energy, resulting in a temperature rise to 22.50 °C. What is the specific heat of Li?

Solution:

Step 1. $c = \dfrac{Q}{m \triangle T}$

Step 2. $c = \dfrac{0.173 \text{ kJ}}{19.5 \text{ g} \times (22.50° - 20.00°)} = 3.55 \times 10^{-3} \text{ kJ/g deg}$

(Notice that we have written $\triangle T$ as $T_f - T_i$.)

EXERCISE 13.2.   Exactly 4 kJ of energy is added to a 1.75-kg sample of the metal strontium, Sr, resulting in a rise in temperature from 18.51 °C to 26.10 °C. What is the specific heat of Sr?

EXAMPLE 13.3.   It is found by experiment that the product of the specific heat and the atomic weight of an unknown element is 2.47×10^{-2} kJ/mol deg. To 5.00 g of this element is added 1.21×10^{-2} kJ of energy, resulting in a temperature increase of 5.46 °C. What is the atomic weight of the element?

Solution:

To solve this problem, we must first find the specific heat of the element. Dividing the specific heat into the product of the specific heat and atomic weight then gives us the atomic weight.

Step 1.   Put the given values into the specific heat equation and do the arithmetic.

$$c = \frac{Q}{m \, \Delta T} = \frac{1.21 \times 10^{-2} \text{ kJ}}{5.00 \text{ g} \times 5.46°} = 4.43 \times 10^{-4} \text{ kJ/g deg}$$

Step 2.   Divide.

$$\frac{2.47 \times 10^{-2} \text{ kJ/mol deg}}{4.43 \times 10^{-4} \text{ kJ/g deg}} = 55.8 \text{ g/mol}$$

The element is iron.

Example 13.3 follows an experiment that was important in the early days of the atomic theory when the atomic weight scale was just being established. A theory advanced by Dulong and Petit proposed that for many elements, the specific heat multiplied by the atomic weight should be about 2.6×10^{-2} kJ/mol deg. Application of the theory to measured values of specific heats yielded valuable approximations of atomic weights.

EXERCISE 13.3.   To a 500-g sample of an unknown metal at 25 °C was added 2.15 kJ of energy. After the energy was added, the temperature of the metal was 42 °C. Using the approximation of Dulong and Petit (see Example 13.3), calculate an approximate value for the atomic weight of the metal.

13.3 CHANGES OF PHASE

At this point, it will be helpful to have a more detailed picture of the motion of particles in a condensed sample and the forces between those particles. Although what we are about to learn is entirely accurate only for pure substances whose solid phases are crystalline (discussed in Section 3.1), the processes described occur in much the same way in many other systems as well.

In a cold solid, the molecules are close together and are vibrating around their positions in the crystal. As energy is added to the solid by heating, the kinetic energy of the molecules increases and the vibration becomes more intense. If heating continues, the intensity of the vibration becomes too much for the forces holding the solid together, and the molecules begin to leave their places and migrate. Flow begins. Holes appear. The solid has melted into a liquid. As heating continues, the kinetic energy of the molecules increases still further, and the molecules bounce around more vigorously, vibrating and moving from place to place more rapidly. The energy of movement finally becomes too great for the forces holding the liquid together, and the sample vaporizes into a gas (Figure 13.5).

During the heating process, the molecules gain two kinds of energy. As the temperature of the sample rises, the molecules are gaining *kinetic* energy. While melting and vaporization are occurring, however, molecules change position in relation to one another, and the added energy does work against the forces that hold the molecules together. During the phase changes (melting and vaporization), the molecules gain only *potential* energy, not kinetic. (If you are not sure what potential energy is, turn back and reread Section 3.2 before going on.) Because the energy of motion of the molecules does not change during the phase changes, the temperature of the sample does not change.

In many substances, the possible potential energy changes are large when compared with kinetic energy changes. For example, it requires about five times as much energy to vaporize a gram of liquid water at 100 °C as it does to raise the temperature of that gram of water from 0 °C to 100 °C. We must investigate the energy of phase change a little further.

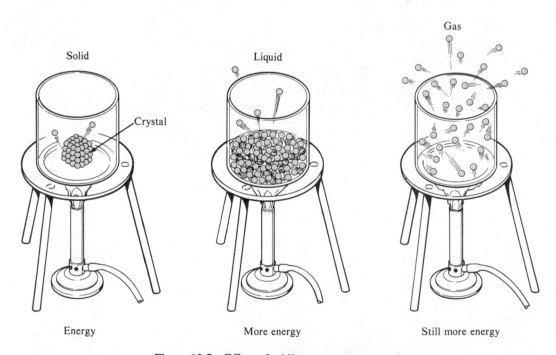

Figure 13.5 Effect of adding energy to a sample.

13.4 ENERGETICS OF MELTING-FREEZING AND VAPORIZATION-CONDENSATION

Energy is work or the capacity to do work, and work is the product of a force exerted over a distance. If we lift a brick, for example, we do work against the force of gravity. Energy, as work, is also involved in bringing together or separating molecules in the presence of attractive intermolecular forces. When the molecules of a gas come together to form a liquid, or when the molecules of a liquid come even closer together to form a solid, energy is *released*. On the other hand, as we have just seen, energy is absorbed when a solid changes to a liquid or a liquid changes to a gas.

The energy required to melt a solid into a liquid at the same temperature is called the **heat of fusion** or the heat of melting. The energy needed to turn a liquid or solid into a gas is called the **heat of vaporization.**

Values for the heat of fusion and heat of vaporization of various compounds have been measured and are given in tables of reference. The values are ordinarily given either for a gram of sample or for a mole of sample. The energy required to melt 1 g of solid H_2O into liquid is simply called the heat of fusion of H_2O. If the amount of energy is given for 1 mol of compound, it is called the **molar heat of fusion.** The **molar heat of vaporization** of Br_2, for instance, is the amount of energy required to turn 1 mol of Br_2 liquid into Br_2 gas at the boiling temperature of Br_2. To see how heats of fusion and vaporization relate to each other, look again at Figure 13.1, and read the values in Table 13.2.

EXAMPLE 13.4.   How many kilojoules of energy are required to melt 15.1 g of ice at 0 °C to liquid water at the same temperature?

Solution:

Multiply the mass of the sample by the heat of fusion (Table 13.2).

$$15.1 \times \frac{0.335 \text{ kJ}}{\text{g}} = 5.06 \text{ kJ}$$

EXERCISE 13.4.   How many kilojoules of energy are required to melt 1.35 kg of NaCl at a constant temperature?

Table 13.2
Values of Some Heats
of Vaporization and Fusion

Substance	Heat of Fusion (kJ/g)	Heat of Vaporization (kJ/g)
Water	0.335	2.26
Sodium	0.113	4.27
NaCl	0.520	2.92
Zinc	0.100	1.76

EXERCISE 13.5. How much energy is required to vaporize 21.0 g of sodium metal at a constant temperature?

The forces between the individual particles of various substances differ: some are strong, some are weak. Because the heats of fusion and vaporization represent work done against the forces between particles, substances with strong interparticle forces will have large values for the heats of fusion and vaporization, and those with weak forces will have small values. After we have learned more about the forces between particles, we can compare these forces with the values for heats of fusion and vaporization. Whether a particular substance is solid, liquid, or gas at any particular temperature and pressure depends on the forces that hold together its individual particles. If the forces are large, high temperatures will be required to disrupt the solid lattice, and the substance will be solid at room temperature. If the forces are weak, the particles will not be able to hold together at room temperature, and the substance will be gaseous. Only in a few cases will the forces be exactly right for a liquid to exist at room temperature; most substances are either solids or gases under ordinary conditions.

13.5 VAPOR PRESSURE

Under many ordinary circumstances, samples of matter can exist in more than one phase at the same time. For example, a glass of ice water contains H_2O in both the solid and liquid phase; if a cover is put on the glass, some of the water will be present as a solid, some as a liquid, and some as a gas.

Because the molecules of a sample are in constant motion, particles continually break away from the surfaces of solids and liquids to become gases. Unless those particles are contained, they will drift away, and the sample will eventually disappear. This process is called **evaporation.** (See Figure 13.6.) Both liquids and solids evaporate. An example of the evaporation of a solid is the disappearance of dry ice (solid CO_2), or the vanishing of snow at high altitudes on a sunny day. The passage of a solid directly into a gas is called **sublimation.** The heat of vaporization (or of sublimation) is the energy that the escaping molecules carry

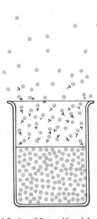

Figure 13.6 How liquids evaporate.

away. Unless more energy is supplied from outside, the condensed phase will get colder and colder: this is why evaporating perspiration cools your skin.

If a sample of liquid or solid is place in a closed container, however, the molecules that leave the surface cannot drift away. Under these circumstances, a gas phase, or what is called a vapor phase, forms in contact with the condensed phase. (**Vapor** is a word often used to describe a gas that can readily be condensed into either liquid or solid; a vapor is, nonetheless, a true gas.) When vapor and a condensed phase are in contact, molecules are constantly leaving the condensed phase to join the vapor. Simultaneously, other molecules are leaving the vapor to rejoin the condensed phase. It is instructive to study the rates at which these two processes occur.

The rate at which molecules leave the surface of a condensed phase, a liquid, for example, depends on the nature of the liquid and its temperature. At low temperatures, only a small proportion of the molecules will have enough kinetic energy to break away, and the rate of escape will be low. As the temperature rises, more molecules will be able to leave each square centimeter of surface each second. On the other hand, no particular amount of energy is required for molecules to escape from the vapor into the liquid: every molecule that strikes the liquid surface will be swallowed up. For that reason, at any given temperature and for any particular substance, the only factor that affects the rate at which vapor molecules return to the liquid is the number of vapor molecules in each cubic centimeter of the vapor.

We can now follow what happens when some liquid is put into a closed container. At first, when no vapor has yet formed, molecules will leave the liquid surface to form vapor, but none will return. As molecules are added to the vapor, the pressure of the vapor phase will continuously increase. As the number of molecules in the vapor increases, however, molecules begin to strike the liquid surface and leave the vapor. This will occur more and more frequently as more and more molecules are added to the vapor. At some point, the number of molecules rejoining the liquid every second will become as great as the number leaving the surface. From that moment, the number of molecules in the vapor will not change. Because the pressure of the vapor is directly proportional to the number of molecules in the vapor volume, the pressure of the vapor will also be constant. (See Figure 13.7.)

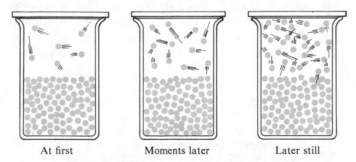

At first   Moments later   Later still

Figure 13.7 Establishment of vapor pressure. In the picture at left, liquid has just been added to the beaker. Molecules are leaving the liquid to become vapor, but little vapor has formed. In the center picture, there are more vapor molecules, and some strike the liquid and rejoin the liquid phase. In the picture at right, the vapor-liquid equilibrium has been reached: as many molecules rejoin the liquid every second as leave it. In most situations, the establishment of vapor pressure equilibrium is extremely rapid.

The constant pressure maintained by a vapor in contact with its condensed phase at any particular temperature is called the **vapor pressure** of that substance at that temperature. The vapor pressure depends on the nature of the substance and is a distinctive property of that substance. If a substance has small intermolecular forces, the molecules can easily escape, and the vapor pressure will be high. Substances that evaporate easily (have high vapor pressures) are said to be **volatile.** Carbon tetrachloride, formerly used for dry cleaning, is a volatile substance. Many substances have strong interparticle forces and correspondingly low vapor pressures. Iron is such a substance: its vapor pressure is too small to measure. Table 13.3 lists measured vapor pressures of a few substances at 20 °C. Notice the large variations in the values.

13.6 PHASE EQUILIBRIUM

A liquid or solid in contact with its vapor, after the vapor pressure has been reached at constant temperature, is said to be in **equilibrium** with its vapor. Because the equilibrium process is an important factor in chemical reaction as well as in phase changes, we will take time now to learn about it.

In the sense in which we will be using the word, equilibrium describes a situation in which two or more ongoing but opposed processes in a system balance each other in such a way that no observable changes occur in certain properties of the system. Let us observe a sample of liquid water as it comes to equilibrium with its vapor. If we put some liquid water into an empty container, the liquid begins immediately to vaporize. As the liquid disappears, vapor appears. The pressure of the vapor increases, and when the actual pressure of the vapor reaches the vapor pressure value, the rate of condensation will be equal to the rate of vaporization. No more liquid will then disappear nor will any additional vapor be formed.

$$\text{liquid} \xrightleftharpoons[condensation]{vaporization} \text{vapor}$$

At this point, liquid and vapor are in equilibrium. Although molecules continue to leave the liquid and other molecules continue to rejoin the liquid, the actual *number* of molecules in the liquid and the *number* in the vapor remain constant because as many go into the vapor every second as leave it. Although vaporization and condensation continue on the molecular level, the observable properties of the system remain constant. Because we will later apply the concept of equilibrium to opposing chemical reactions (Chapter 16), it will help us to become comfortable with it now.

The liquid-vapor equilibrium condition is a balance between the rates of the opposing processes, and is independent of the amounts of the phases as long as some of each phase is

Table 13.3
Vapor Pressures of a Few Substances

Substance	Vapor Pressure (torr at 20 °C)
Carbon dioxide (l)	4.3×10^4
Carbon tetrachloride	98.
Water	17.5
Mercury	1.2×10^{-3}

present. If a 50-g sample of liquid water is place in a closed 500-mL flask at 20 °C, it will establish a characteristic vapor pressure of 17.5 torr. If another 50-g sample of water is placed in a closed 5-L flask, it will also vaporize until the pressure of the water vapor is 17.5 torr. There will be more liquid and less vapor in the first flask than in the second, but in both, the pressure of water vapor will equal the H_2O vapor pressure of 17.5 torr (Figure 13.8).

Vapor pressure is dependent on temperature. At low temperatures, the rate of escape of molecules from a liquid or solid surface is low, and the vapor pressure is correspondingly low. At higher temperatures, more molecules leave the surface every second, and the vapor pressure is higher. Figure 13.9 is a graph of the vapor pressures of several liquids between the temperatures of −50 °C and +150 °C.

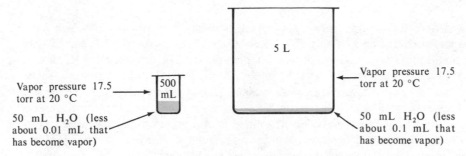

Figure 13.8 Vapor pressure is independent of container size as long as liquid remains.

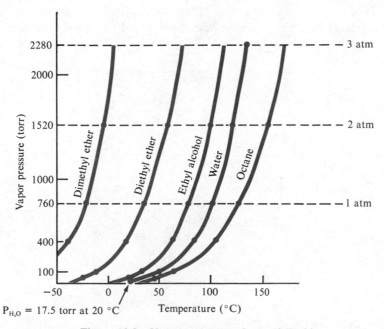

Figure 13.9 Vapor pressures of several liquids.

Equilibrium can be reached also between a solid and its vapor, and even between a solid and a liquid. If you put some ice in a glass of water at room temperature, the ice will melt, taking heat from the liquid water and cooling it. In this process, molecules leave the ice much faster than they return to it from the liquid. But when the temperature of the ice and water mixture reaches 0 °C, the rate of escape of molecules from ice to liquid is matched by the rate of return of molecules from liquid to ice. No more ice will melt unless energy is added from the surroundings. The solid water is in equilibrium with the liquid. At any particular pressure, there is only one temperature at which liquid and solid can be at equilibrium. That temperature is called either the **melting temperature** or the **freezing temperature** (or sometimes the **melting** or **freezing point**) depending on which phase is disappearing and which is growing.

13.7 BOILING POINT

In an interesting demonstration sometimes used for chemistry lectures, the gas (air and water vapor) surrounding a sample of liquid water at room temperature is suddenly removed, and the liquid literally explodes into water vapor. What happens is that the vapor pressure of the liquid is suddenly far larger than the pressure exerted on that liquid from outside, and the liquid instantly vaporizes, attempting to establish its vapor pressure as the vapor is continuously pumped away (Figure 13.10).

If liquid water is in contact with dry air at 1 atm pressure and room temperature, the pressure of the vapor that the water could form (17.5 torr) is far less than the pressure the air exerts on the surface of the liquid, and the water vapor cannot push away the air. Individual water molecules do leave the surface of the liquid, however, and find their way among the air molecules. In this way, by a slow process, the liquid establishes its vapor pressure, which is added to the pressure of the air (Figure 13.11). We have already learned that when gases are collected in the presence of water, the total pressure includes the water vapor pressure.

At higher temperatures, however, even in the presence of air, the water can **boil;** that is, it can *form vapor with enough pressure to push back the atmosphere.* To see how that happens,

Figure 13.10 Water boiling at room temperature. (Photo by author.)

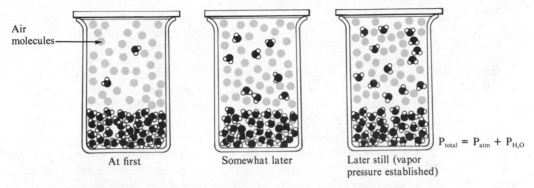

Figure 13.11 How vapor pressure is established in the presence of air.

look again at the graph of the vapor pressure of water as the temperature is raised, shown in Figure 13.9. Toward the bottom of the curve, we see that the vapor pressure of water at 20 °C is only 17.5 torr, but as we follow the graph to higher temperatures, we see that the vapor pressure rises. As a sample of liquid water is heated in a pan, its vapor pressure increases, until at some point it becomes equal to the pressure of the air on the surface of the water. When this occurs, the vapor then begins to push back the air. It is no longer necessary for vaporizing molecules to move one at a time between the air molecules as they must if the water is merely evaporating. Bubbles of vapor form, and the water is said to be boiling.

The **boiling temperature,** or **boiling point,** is that *temperature at which the vapor pressure of a liquid equals the pressure on its surface.* The boiling temperature of a liquid in the presence of air at 1 atm of pressure is called the **normal boiling point.**

Continuing to look at Figure 13.9, notice that a horizontal dotted line has been drawn at 1 atm. Each vapor pressure curve meets this line at its normal boiling point; that is, at the temperature at which its vapor can push back 1 atm of pressure. The boiling temperatures are low for the substances that have small intermolecular forces, Cl_2 for example, and higher for those that have larger intermolecular forces, H_2O for instance. Because it is directly related to the vapor pressure of a substance, the normal boiling temperature is a distinctive property of that substance. Table 13.4 gives normal boiling temperatures for several liquids.

Table 13.4
Normal Boiling Temperatures of Some Liquids

Liquid	Normal Boiling Point (°C)
Water	100
Ethyl alcohol	78.5
Octane (similar to gasoline)	125.6
Propane (LPG)	−44.5
Methane (natural gas)	−161
Ethyl ether	34.6
Oxygen	−183
Sulfur	444
Sodium	892
Chlorine	−34.6

13.8 INTERPARTICLE FORCES

13.8.1 Introduction

In Chapter 11, we studied two kinds of chemical forces: the ionic bond and the covalent bond. Chemical forces hold together many kinds of solids and a few liquids. The particles of many solids and most liquids, however, are attracted to one another by other kinds of forces, known collectively as **van der Waals forces.**

With only a few exceptions, liquids and gases consist of independent molecules. Molecules are united *internally* by chemical bonds, but the forces acting *between one molecule and another are not chemical bonds.* It is most important that you understand this distinction: forces *within* molecules are chemical bonds, but *between* molecules there are forces that are *not* chemical bonds. Chemical bonds are responsible for the existence of compounds, which are substances that can be identified by formulas. The weaker forces that act between molecules are not as strong as those that form individual molecules or solids with fixed formulas. The compound $BaCl_2$, for example, has a fixed formula. But a solid mixture of Ar and Xe held together by weak forces may have any proportion of the two kinds of atoms. In most cases, the distinction between chemical bonds and nonchemical forces is clear, but in others it is blurred. We will see examples of the second case when we discuss hydrogen bonds.

The work of Johannes van der Waals, a Dutch physicist, is an example of how the careful investigation of one phenomenon can lead to important findings about many other phenomena. Somewhat more than a hundred years ago, van der Waals became interested in finding the cause for the nonideal behavior of real gases. In 1880, he published a paper showing that nonideality was caused both by the actual volume of the individual molecules of a gas, and by the forces acting between those molecules. He described condensation as a process in which intermolecular forces caused gas molecules to stick together, and he showed that the volume of a liquid was approximately the sum of the volumes of its individual molecules. This work led van der Waals and many others to investigate the nature of intermolecular forces. Those forces, which scientists now understand, are named in honor of van der Waals. They have been classified into two categories: **dipole forces** and **temporary dipole forces.** To understand how the forces arise, we must first return to the details of the covalent bond.

13.8.2 Polar Covalent Bonds

When a covalent bond is formed by two atoms of the same element, the two electrons forming the bond are shared equally by the two atoms, both of which have the same attraction for electrons. Ionic bonds, on the other hand, are formed by the complete transfer of an electron from one atom to another entirely different atom. These two kinds of bond formation represent extremes: perfect sharing and complete transfer. Many chemical bonds, however, lie between these extremes. Electrons that form such bonds are neither transferred nor shared equally, but are shared unequally. An unequally shared pair of electrons forms a covalent bond, but the bond has distinctive properties arising from the inequality in the sharing. A covalent bond in which the electron pair is not equally shared is called a **polar covalent bond.**

Polar covalent bonds are formed because the strengths with which atoms of different elements attract valence electrons vary. (Different elements have different ionization energies and different electron affinities.) When the density of electrons is greater at one end of a pair of bonded atoms than at the other end, an **electric dipole** (or just **dipole**) results. This is where the name *polar covalent bond* comes from. A pair of atoms with an electric dipole has something of a negative charge at one end and something of a positive charge at the other, but the effect is not as great as it would be if an electron had been completely transferred.

The attraction that an atom has for a shared pair of electrons is called the **electronegativity** of that atom, and is expressed on a scale that goes from somewhat less than unity to a value of four. Electronegativity is a kind of combination of ionization energy and electron affinity. The least electronegative available element is cesium, 0.7 on the scale, and the most electronegative is fluorine, 4.0 on the scale. Electronegativity is greatest in the nonmetals that are high and to the right on the periodic table (excluding the rare gases) and least in the elements that are low and to the. left.

The amount of polarity (that is, the inequality with which the electrons are shared) in a covalent bond is indicated by the difference in the electronegativities of the two atoms. If there is no difference, the atoms have the same electron affinity and ionization energy, and sharing is equal. The more unequal the sharing, the closer the bond is to being ionic; ionic bonds are those in which there is no sharing whatever. Generally speaking, two atoms will form a covalent bond if the difference in their electronegativities is less than about 2. If the difference is greater than that, an electron will probably be transferred entirely and ions formed. (See Figure 13.12.)

Spend a few minutes studying the electronegativity values printed under the element symbols in Figure 13.13. These numbers enable you to decide what kind of bond will be formed between the atoms of any two or more elements. Along with information that we will be discussing later, the numbers will also allow you to predict many of the chemical and physical properties of compounds formed by pairs of elements.

EXERCISE 13.6.   Which element will more readily attract the electron pair in the molecule CO?
    In BrF?

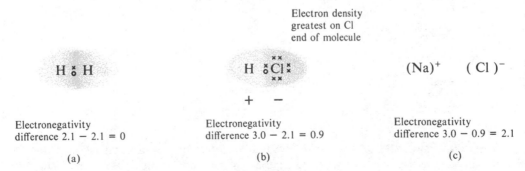

Figure 13.12 How electronegativity difference determines the kind of compound formed between two elements. (*a*) A molecule in which the electron pair is equally shared. The electronegativity difference is zero. (*b*) A molecule with an electric dipole, which arises because the two atoms have different electronegativities. In (*c*), the electron has been entirely transferred from Na to Cl. Transfer usually occurs when the difference in electronegativity is greater than about 2.

Key

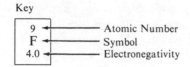

1 H 2.1							2 He —
3 Li 1.0	4 Be 1.5	5 B 2.0	6 C 2.5	7 N 3.0	8 O 3.5	9 F 4.0	10 Ne —
11 Na 0.9	12 Mg 1.2	13 Al 1.5	14 Si 1.8	15 P 2.1	16 S 2.5	17 Cl 3.0	18 Ar —
19 K 0.8	20 Ca 1.0	31 Ga 1.6	32 Ge 1.8	33 As 2.0	34 Se 2.4	35 Br 2.8	36 Kr —
37 Rb 0.8	38 Sr 1.0	49 In 1.1	50 Sn 1.8	51 Sb 1.9	52 Te 2.1	53 I 2.5	54 Xe —
55 Cs 0.7	56 Ba 0.9	81 Tl 1.8	82 Pb 1.8	83 Bi 1.9	84 Po 2.0	85 At 2.2	86 Rn —
87 Fr 0.7	88 Ra 0.9						

Figure 13.13 Relative electronegativity of some elements. The electronegativity value is given below the symbol of each element.

EXERCISE 13.7.   Fill in the table that follows.

Reacting Elements	Electronegativities of These Elements	Difference in Electronegativities	Kind of Bond Formed
Cs + F	Cs, 0.7; F, 4.0	3.3	Ionic
O + Na	_____	_____	_____
C + H	_____	_____	_____
Si + Cl	_____	_____	_____
Br + K	_____	_____	_____

A diatomic molecule in which the atoms have different electronegativity will have an electric dipole. Larger molecules, however, may or may not themselves have dipoles, even though some of the bonds in the molecule may be polar. In carbon dioxide, both the C=O bonds are polar because oxygen is more electronegative than carbon, but the molecule itself has no dipole because the two opposite dipoles cancel each other (Figure 13.14).

$$\overset{\longleftarrow}{O}=C\overset{\longrightarrow}{=}O \qquad \text{where} \quad \rightleftarrows \text{ yields no net dipole}$$

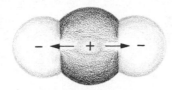

Figure 13.14   In CO_2 the dipoles cancel each other.

Because—as we have learned—water molecules are bent, the bond dipoles do not cancel (Figure 13.15). In methane, CH_4, the dipoles cancel (Figure 13.16). In CH_3Cl, they do not (Figure 13.17).

Whether a molecule has an overall dipole strongly influences its properties.

13.8.3 Dipole Forces

Consider the compound hydrogen chloride, HCl, which is a gas at room temperature. HCl liquefies at -85 °C and freezes at -115 °C, both at 1 atm. The hydrogen chloride molecule is held together internally by a single, strong covalent bond, but in liquid or solid HCl, there are no such bonds *between* molecules; there exist only dipole forces.

The HCl molecule has a moderately strong electric dipole because of the significant difference in electronegativity between H and Cl ($3.0 - 2.1 = 0.9$). (See Figure 13.18.)

Two or more electric dipoles exert attractive forces on one another because the positive end of one attracts the negative end of the other and vice versa. The attraction of two HCl molecules to each other is called a dipole-dipole attraction, or just a **dipole force** (Figure 13.19). Other HCl molecules will be attracted in the same way, and a solid or liquid will begin to form (Figure 13.20).

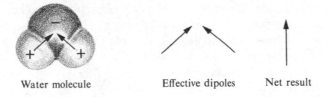

Water molecule   Effective dipoles   Net result

Figure 13.15   In H_2O the dipoles reinforce each other.

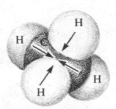

Figure 13.16   Dipoles cancel in CH_4.

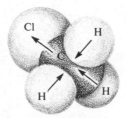

Figure 13.17   Dipoles do not cancel in CH_3Cl.

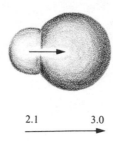

2.1   3.0

Figure 13.18 Dipole in the HCl molecule.

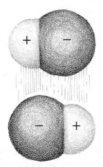

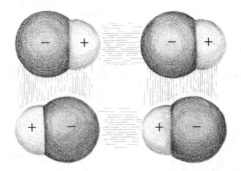

Figure 13.19 Dipole-dipole attraction between two HCl molecules.

Figure 13.20 Dipole attractions can extend beyond two molecules.

Although dipole-dipole attractive forces exist between any molecules that have overall dipoles, as we have seen, the polar bonds within some molecules cancel. For this reason, there are no dipole forces between CO_2 molecules, between CH_4 molecules, or between any molecules that do not have overall dipoles. H_2S molecules, which are bent and which have overall dipoles, do experience dipole-dipole attraction (Figure 13.21).

The forces created by dipole attraction are ordinarily much weaker than chemical bonds. Dipole attractions rarely result in stable structures for which a formula can be written, but dipole forces are nevertheless important. For example, without its strong dipole, water would not exist as a liquid under ordinary conditions.

13.8.4 Temporary Dipole Forces

Consider iodine, I_2, which is a solid at room temperature. Its molecules are clearly held together by substantial forces, but we know that they have no permanent dipoles because the two atoms in each molecule are exactly alike. Evidently, another kind of intermolecular force exists.

The intermolecular force in substances like I_2 is created by very short-lived temporary dipoles. The outer electrons in a large molecule are weakly held. Although those electrons are, on the average, uniformly distributed in the molecule, at any given instant the electron density

Dipole attractions among
H₂S molecules in small section of solid H₂S

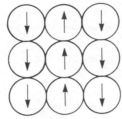

H₂S molecule showing overall dipole

Figure 13.21   Dipole attraction in hydrogen sulfide molecules.

can be greater at one end of the molecule than at the other, creating an electric dipole for a tiny fraction of a second. The electrons of any molecule that happens to be nearby will respond to the temporary dipole, and by shifting in the reverse direction, will create another short-lived dipole. The two dipoles attract each other, and this process repeated over and over results in an intermolecular force, which is usually not strong. Study the drawings in Figure 13.22. Temporary dipole forces are sometimes called **dispersion forces** or **London forces**—for Fritz London, who made the first detailed explanation of them.

Because they have more electrons and because their electrons are held more loosely, molecules made from large atoms have larger London forces than do those made from smaller atoms. The temperatures at which the halogens liquefy demonstrate the effect of molecular size. Since fluorine molecules are small and hold their electrons tightly, the forces between F_2 molecules are weaker, and fluorine must be cooled to $-199\,°C$ before it will liquefy at 1 atm pressure. Chlorine, Cl_2, on the other hand, will liquefy at $-34.6\,°C$ and Br_2 at $+59\,°C$. (Br_2 is one of the few elements that are liquid at room temperature.) I_2, of course, is solid at room temperature.

For all but the largest molecules, London forces are generally weaker than permanent dipole forces. London forces occur in all molecules, however, even in the noble gas "molecules" that contain only one atom each. Intermolecular forces are often a result of permanent dipole forces and London forces acting together.

13.8.5 Hydrogen Bonds

Hydrogen bonding is a kind of intermolecular force that is stronger than van der Waals forces but weaker than covalent bonds. Hydrogen bonds occur only among certain kinds of molecules, whereas van der Waals forces occur among molecules of all kinds.

Both molecules have even electron
distribution

In molecule at left, electrons shift
temporarily toward right

Molecule on right responds, and
force results

Figure 13.22   How a temporary dipole force develops. Density of shading indicates temporary electron density in molecules.

To see how hydrogen bonds work, let us consider water, which is an H-bonded compound. Oxygen is a strongly electronegative element. In an O—H bond, it exerts a powerful attraction on the pair of electrons shared with H. Because hydrogen atoms have no inner electrons, the shift of the electron pair in the O—H bond leaves the nucleus of the H atom somewhat exposed. This concentrates the positive end of the O—H dipole into a tiny area that can get very close to unshared electron pairs on other molecules. (Because of the presence of inner electron shells, the positive charges of other kinds of atoms at the positive ends of dipoles are not similarly exposed.)

The attraction between a partly bare hydrogen nucleus and an available electron pair at the negative end of another O—H dipole has some of the characteristics of an ionic bond, and is therefore stronger than an ordinary dipole attraction. (See Figure 13.23.)

Hydrogen bonds occur in molecules in which hydrogen atoms are bonded to oxygen, nitrogen, or fluorine, the only elements that are small enough and have enough electronegativity to have the necessary effect on hydrogen. The effects of H-bonding are extremely important in biological systems as well as in ordinary chemistry. For an example, look at Figure 13.24. The boiling points of HF, H_2O, and NH_3, are unexpectedly high because of hydrogen bonding.

13.9 CRYSTALLINE SOLIDS

In Section 3.1, we examined the nature of solid matter and the differences between the organized solids we call crystals and the unorganized glassy solids. When we discuss plastics in a later chapter, we will learn more about glassy solids. It is both useful and interesting to look now at the various kinds of crystalline solids, and see how the microscopic structure of a sample affects its measurable properties.

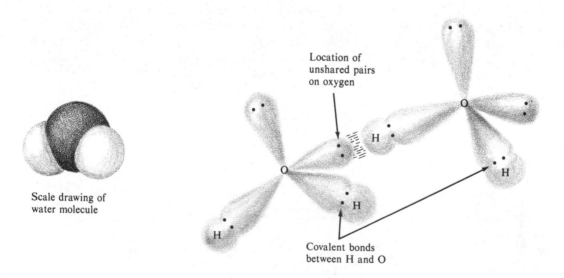

Location of
unshared pairs
on oxygen

Scale drawing of
water molecule

Covalent bonds
between H and O

Figure 13.23   Very schematic representation of hydrogen bonding.

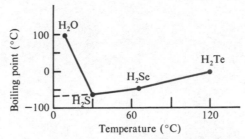

Figure 13.24 An example of the effect of hydrogen bonding on boiling points. H_2S, H_2Se, and H_2Te are not hydrogen bonded, but H_2O is. If H_2O were not, its boiling point would be on the dotted line at about −60 °C.

13.9.1 Ionic Crystals

We learned in Section 11.5 that when sodium and chlorine react, electron transfer occurs and positive and negative ions are formed. Under ordinary conditions, sodium and chloride ions combine to make crystals of sodium chloride, NaCl, which is everyday table salt. A crystal composed of ions is called an **ionic crystal.** In the sodium chloride crystal, each Na^+ is surrounded by six Cl^- ions, and each Cl^- is similarly surrounded by six Na^+ ions as its closest neighbors. Before going on, take another look at Figure 11.13 (page 216). The NaCl crystal is held together by the strong attractions between the positive electric charges on the Na^+ ions and the negative charges on the Cl^- ions. These attractive forces far outweigh the repulsive forces exerted on each ion by its next nearest neighbors, which are ions of the same charge.

The properties of ionic crystals clearly demonstrate their microscopic structures. NaCl, for example, has a high melting temperature, 1076 K, and a high heat of fusion, about 30 kJ/mol. The high melting temperature and heat of fusion are due to the strong attractions between Na^+ and Cl^-. Disrupting these strong interionic forces requires large amounts of energy. During melting, and particularly during vaporization—when the distances between the ions are being greatly increased—large amounts of work must be done.

NaCl crystals are hard, but extremely brittle: only a little force is required to break one. The hardness arises from the strong forces in the crystal, the brittleness from an interesting, different reason. The left side of Figure 13.25 shows a cross-section of a normal NaCl lattice; the section at the right shows the same lattice, after a small downward force has been applied to the right portion of the crystal. If the crystal lattice is displaced by only one ionic diameter, about 2×10^{-8} cm, the attractive forces between the two parts suddenly become repulsive, and the crystal cleaves. Because the forces uniting ionic crystals are so dependent on precise crystal structure, many ionic crystals are fragile, although hard.

Not all ionic crystals are like NaCl. Other alkali-halogen compounds, CsCl, for example, have lattice patterns in which each ion is surrounded by eight nearest neighbors. Each Cs^+ is in the center of a cube containing Cl^- ions at the corners, and vice versa, as shown in Figure 13.26. For ions with the relative dimensions of Cs^+ and Cl^-, this kind of lattice allows the ions to be most efficiently packed. The difference between the geometry of NaCl and that of CsCl is due entirely to the difference in size between Cs^+ and Na^+.

Many ionic crystals have more complicated arrangements caused by their ionic sizes, shapes, and charges. All, however, stick together because of the electric forces between the

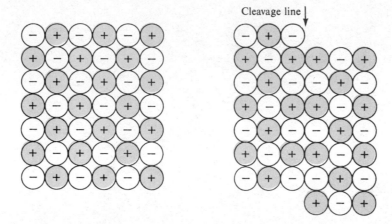

Figure 13.25 How attractive forces are replaced by repulsive forces when ionic crystals are deformed. At left: normal lattice. At right: shifted lattice.

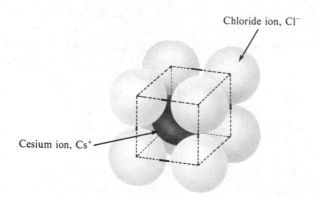

Figure 13.26 Scale drawing of the cesium chloride lattice.

ions. Many ionic compounds form crystals that are harder and tougher than those of NaCl; for example, various forms of $CaCO_3$ (limestone, marble) are used for building. These owe their strength to the strong attractive forces between the doubly charged ions of Ca^{2+} and CO_3^{2-}.

13.9.2 Covalent Crystals

Just as the atoms in a molecule are held together by covalent bonds, so can atoms be held in a much larger structure by covalent bonds. In a few substances, entire macroscopic crystals consist of atoms joined throughout by covalent bonds. In a sense, these structures could be called giant molecules, but molecules are expected to have formulas, and no two crystals have the same number of atoms. Crystals whose atoms are all joined by covalent bonds are called **covalent crystals.**

Diamond is an example of a covalent crystal. Every diamond is a crystal made of carbon atoms. Each carbon shares strong electron pair bonds with four equally close neighbors. The bonds in diamond are all extremely strong; diamond is one of the hardest substances known. Diamond also has the highest melting temperature known, about 3800 K. (The value is not precisely known, because it is difficult to find a container in which to melt the diamond crystals and because diamond sublimes.) Because all the covalent bonds of a diamond are alike and interchangeable, the possibility of creating interionic repulsive forces, as in NaCl, does not exist. The diamond crystal is thus not as brittle as NaCl, although it can be cleaved. (See Figure 13.27.)

Graphite, another form of elemental carbon, is interesting because it is soft but has a melting point close to that of diamond. High melting temperatures are usually associated with strong bonds, and softness with weak bonds; these two properties are generally inconsistent and are not ordinarily found in the same substance. Graphite, however, has both strong and weak bonds. The graphite crystal consists of sheets of carbon atoms joined internally by extremely strong covalent bonds. Because the sheets are held one to the next by rather weak bonds, force applied to the crystals easily causes the stacks of sheets to slip out of place. To change solid graphite into a liquid, however, all the bonds must be disrupted, both strong and weak; this gives graphite a high melting point. (See Figure 13.28.)

Other examples of covalent crystals are corundum, Al_2O_3, which is used for grinding, and silica, SiO_2, whose atoms are covalently connected throughout the crystal. Most kinds of glass are made from silica sand. (See Figure 13.29.)

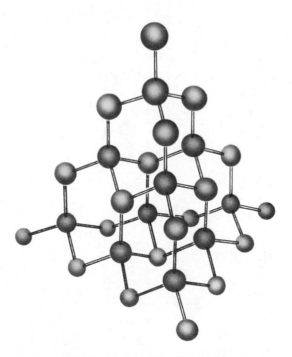

Figure 13.27   Model of the diamond lattice.

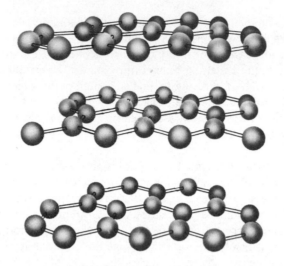

Figure 13.28 Model of the graphite lattice.

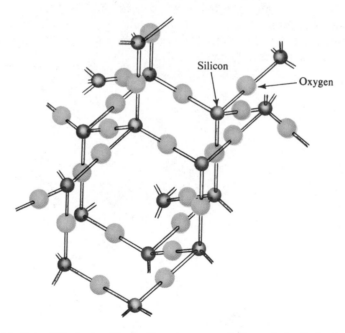

Figure 13.29 Model of the SiO_2 lattice. Each silicon is surrounded tetrahedrally by four oxygen atoms, while each oxygen is bonded to two silicon atoms.

13.9.3 Molecular Crystals

A third kind of solid, called a **molecular crystal,** is composed of individual, internally bonded molecules that are held in place in the crystal lattice only by van der Waals forces, not by chemical bonds.

Solid iodine is an example of a molecular crystal. When two iodine atoms bond to form I_2, the outer shells of the atoms are completed; thus no bonds are possible *between* iodine molecules. At room temperature, however, iodine is a solid substance because of the temporary dipole forces acting among the individual molecules. Many other common materials, such as CO_2 and H_2O, also form molecular crystals. Some, like CO_2 and I_2, are held together by temporary dipoles; other, like HCl and H_2O, are united by permanent dipoles.

The properties of a molecular crystal are not difficult to predict. Because the forces between molecules are usually very weak, we would expect to find the structure of a molecular solid much easier to destroy than that of a covalent crystal. Our prediction is confirmed by the low melting points of molecular substances, as compared with the much higher melting points of covalent and ionic crystals. Methane, CH_4, a typical molecular crystal, melts at about 90 K. Water melts at 273 K, and Br_2 at 266 K. Compare these temperatures with the melting points of typical, simple ionic substances (NaCl at 1076 K, CaO at 2850 K, K_2CO_3 at 1275 K) or with the melting point of diamond.

Not all molecular solids, however, are easily melted. Ordinary table sugar, $C_{12}H_{22}O_{11}$, can be melted only with great care. Because many dipole attractions and hydrogen bonds exist between each pair of sugar molecules, it is nearly as difficult to separate these molecules from one another as it is to decompose the molecules themselves. The melting temperature of sucrose is 460 K, a higher temperature than we might expect for molecular solids and one that causes the delicate sucrose molecules to decompose. The familiar process of caramelization, a chemical reaction in which sucrose molecules are destroyed, occurs when sucrose is heated under ordinary conditions. Caramel flavor comes from partly decomposed sugar.

As we would expect, molecular crystals are much softer than ionic or covalent crystals, a further demonstration of the weakness of the intermolecular forces.

Solid water forms a molecular crystal that has a hydrogen-bonded structure. If you look at Figure 13.30, an expanded sketch of several water molecules in an ice crystal, you can see how each H atom is closely bonded to one O atom and joined by a hydrogen bond (dotted line) to another O atom; this pattern forms little cages of water molecules, each cage having a hollow place in the center. When ice melts, most of this structure is destroyed and most of the hollow places disappear. Liquid water is therefore more dense than ice (remember that ice cubes float in liquid water), although the liquids of most substances are less dense than the solids. If ice were more dense than water, the world would be an entirely different place.

13.9.4 Metallic Crystals

Although we will not study the crystals of metals in detail, we can look briefly at their structures and properties. It is thought that when metal atoms join to form crystalline solids, each atom gives up one or more of its valence electrons, which then become the shared property of the entire metal crystal. What results is a "sea" of electrons, acting as a glue to cement the metal ions together. Because the electrons are not bound to individual atoms and are free to roam through the crystal, metals readily conduct electricity. Metals also readily conduct heat (they ordinarily feel cold to us, but when they are hot they feel very hot), because the electrons can rapidly carry kinetic energy from one place to another in the crystal.

Metal crystals bend without breaking and can be hammered into various shapes, because the metal structure does not have specific bonds, but is merely a collection of ions glued by electrons. The ions can easily change their positions within the crystal.

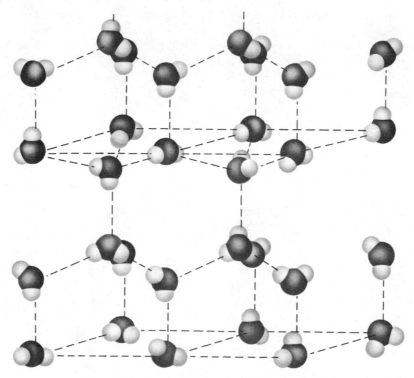

Figure 13.30 Expanded diagram of the structure of ice.

The melting and boiling temperatures of metals depend on how many electrons each atom gives to the electron sea. When more valence electrons are available, more "electron glue" is formed and greater positive charges exist on the metal ions: both factors increase the strength of the metal lattice. Al has a higher melting temperature than Mg, which, in turn, has a higher melting temperature than Na. The atoms of some metals, like those in the transition series (Fe and Cr, for example), first lose their valence electrons to their crystals, then form partial covalent bonds between the newly exposed, d-level electrons of the resulting ions. Because of these extra bonding forces, crystals of such metals are hard and have high melting points. Table 13.5 gives melting temperatures for several metals.

Table 13.5
Melting Temperatures of Some Metals

Metal	Melting Temperature (°C)
Sodium	97.8
Cesium	28.5
Magnesium	651
Barium	725
Aluminum	660
Iron	1535
Chromium	1890
Tungsten	3410

13.9.5 Properties of Different Kinds of Solids

Now that we have examined the various kinds of solid substances and their properties, we can get a good overview of what we have learned by comparing the properties of those solids. Table 13.6 lists several examples of each kind of solid and examples of typical properties. Be sure to read this table and study it carefully; it will help you tie together what you have read.

Table 13.6
Properties of Crystalline Solids

Kind of Crystal	Individual Particles	Interparticle Forces	Examples	Comments
Covalent	Atoms	Covalent bonds	Diamond Graphite SiO_2 SiC (carborundum) Asbestos	• Usually very hard and tough (Exceptions: graphite, asbestos—see text) • Very high melting points • Strongest interparticle forces • All are solids
Ionic	Positive and negative ions	Electrostatic attraction	NaCl $CaCO_3$ K_2SO_4 $Sr(NO_3)_2$ KOH	• Often quite hard, depending on charges of ions • Some very brittle • Fairly high to very high melting points • All are solids at ordinary temperatures
Molecular H-bonded	Individual molecules	H-bonds (see text)	H_2O HF NH_3 DNA (Sec. 20.6)	• Mostly liquids or gases at room temperature • Melting and boiling points much lower than those of ionic crystals • Soft, easily shattered crystals
Dipolar	Individual molecules	Dipole-dipole interaction	ICl CO_2 (dry ice) HCl	• Almost all are liquids or gases • Melting and boiling points generally lower than in H-bonded substances
London	Individual molecules	Instantaneous dipole forces	I_2 Br_2 N_2 H_2 Ar He	• Most are gases, with a few liquids and solids (Br_2, I_2, for instance) • Weakest interparticle forces of all • London forces often contribute to the attractions in other kinds of substances (as in IBr, which is mostly London but also some dipole-dipole)

13.10 SOME PROPERTIES OF LIQUIDS

As we did with crystalline solids, we can usefully classify liquids according to the forces that hold together the particles of the sample. We will exclude from our discussion the metallic liquids and melted ionic compounds, because we do not usually encounter such substances. Covalent liquids do not exist; if covalent bonds were to form between the particles, the particles would be welded into a solid. We are left, then, with liquids held together by van der Waals forces. For pure liquids, we can make a useful distinction between liquids whose molecules have dipoles, and those whose molecules do not. Although the properties of the two kinds of liquids overlap, and there are exceptions to the generalities we will make, dipolar liquids do differ somewhat from liquids whose molecules are held together only by London forces.

If we consider molecules of about the same size, those that have dipoles generally exert more attraction on one another than do those without dipoles. We can observe the effects of intermolecular attraction if we compare the normal boiling temperatures of polar and non-polar liquids. For example, HCl boils at 188.3 K at 1 atm pressure, while the normal boiling point for Ar is only 87.5 K, even though the two molecules have the same mass and the same number of electrons. The difference arises mostly from the fact that HCl has a strong dipole and Ar does not. The boiling points, freezing points, heats of vaporization and melting, as well as many other properties of nonpolar liquids, reflect the weak forces between their molecules. The two liquids with the lowest boiling points of all, He and H_2 (boiling temperatures of 4 K and 20 K), are nonpolar. Some larger and more complex molecules are also nonpolar. The hydrocarbons, compounds composed of C and H, are nonpolar because the polar C–H bonds interact so as to cancel the effects of their polarities. For this reason, compounds like C_7H_{16}, which is one of the main components in gasoline, behave in a nonpolar way. Oily substances are usually nonpolar or have few polar characteristics.

Water, which we will study in somewhat more detail in Chapter 14, is a highly polar liquid. As we know, the hydrogen bonding in liquid water gives it properties that distinguish it from nearly all other liquids. In addition to its abnormally high boiling point, water also has a high heat of vaporization, 40.5 kJ/mol, more than a hundred times that of CH_4. Such great differences in properties reflect differences in the attractive forces between molecules of liquids.

13.11 SUMMARY

Many substances can exist in any of the three states: solid, liquid, or gas. All substances can exist as solids. Whether a substance will be a solid, liquid, or gas depends on the forces between its individual particles and on its temperature. At any temperature, particles of a gaseous substance have more potential energy than particles of the same substance in the liquid form, and still more potential energy than the same particles in solid form.

The amount of energy gained when a liquid vaporizes to a gas is called the heat of vaporization. The amount gained on melting is called the heat of melting. These energies are usually quoted either for 1 g of substance or for 1 mol.

Individual particles can leave the surface of a liquid or solid to become a gas. The rate at which this process occurs depends on the forces between the particles and on the temperature.

When particles leave a liquid, drifting away and becoming lost, the process is called evaporation.

In a closed vessel, gaseous molecules form a vapor phase in contact with the liquid. The pressure of that vapor rises until the rate of return of vapor molecules to the liquid is the same as the rate of vaporization, at which time a state of equilibrium is reached. From this time, the pressure of the vapor remains constant unless the system is disturbed. The equilibrium pressure of a vapor in contact with its liquid is called the vapor pressure of that liquid, and is a property of the liquid. Solids in closed containers can also develop vapor pressures, but for any given substance the vapor pressure of the solid form is usually much smaller than that of the liquid.

The boiling point of a liquid is the temperature at which the vapor pressure of the liquid equals the pressure being exerted on the liquid surface. The normal boiling point is the temperature at which the vapor pressure equals 1 atm.

Various attractive forces can exist between particles. Solids can be held together by chemical bonds, either ionic or covalent, or by metallic bonds. Under ordinary conditions, the particles of liquids and gases are attracted to one another either by hydrogen bonds or by van der Waals forces. Solids, as well, can be united by van der Waals forces.

Although there are many electron transfer reactions, there are relatively few electron sharing reactions in which the resulting pair of electrons is equally shared by the atoms involved. Most varieties of covalent bonds exist between atoms of different elements. In such bonds, which are called polar covalent bonds, the electron pairs are not equally shared, but spend their time closer to the more electronegative atoms, thus creating electric dipoles. Polar covalent bonds can be thought of as partly ionic in character.

Van der Waals forces arise either from the mutual attraction of permanent dipoles in molecules or from the interaction of temporary dipoles. In some cases, both kinds of force operate together. Van der Waals forces are generally much weaker than chemical bonds.

Hydrogen bonds are special cases of dipole interaction. Strong forces are often created between molecules that have hydrogen atoms bonded to oxygen, nitrogen, or fluorine. Although hydrogen bonds lead to stable structural arrangements, they are not usually classed as chemical bonds because they do not lead to stoichiometric arrangements, that is, to structures with characteristic chemical formulas.

The properties of crystalline substances manifest the nature of the forces between their individual particles. The strong bonding forces, ionic and covalent, lead to hard crystals with high melting points. Solids composed of individual molecules held in place by van der Waals forces tend to be soft and to have low melting points.

QUESTIONS

Section 13.1

1. Look up the word *condensed* in the dictionary. Why are liquids and solids called condensed states? Why is the process of changing from a gas to a liquid or solid called condensation?
2. A small bedroom is 10 ft wide × 12 ft long × 8 ft high. If all the air in the room were liquefied, what would be the approximate dimensions of a container that could contain the liquid? Make a

guess, without looking up any data. Now calculate it from densities listed in tables in earlier chapters.

Section 13.2

3.   What is specific heat? How is it related to molar heat capacity?
4.   How does the definition of the calorie involve specific heat? Would the value of the calorie be different if a substance other than water were used in the definition? Why?
5.   A pan made of copper is put on a warm stove and allowed to get hot. The pan is then cooled. Water is put into the pan until the mass of the pan with the water in it is twice that of the empty pan. The pan with water is then put on the same stove and heated to the same temperature as before. Does the process take about twice as long? Explain.

Section 13.4

6.   By means of an electric heating coil, energy is added to a piece of ice that has an initial temperature of $-10\ °C$. The temperature of the ice slowly increases to $0\ °C$, at which time the ice begins to melt. Energy continues to be added, but now the temperature of the ice and water remains constant until all the ice is melted. Why does the temperature not change during melting? What becomes of the energy added during this time?
7.   A refrigerator operates by causing a liquid to boil inside coils of tubing. Explain how this causes cooling.
8.   An old-fashioned ice cream freezer causes the ice cream mixture to freeze by lowering the temperature well below $0\ °C$. To lower the temperature, rock salt is put on ice, which causes the ice to melt. Why does this produce low temperatures?
9.   Burns caused by steam are usually far more serious than those caused by boiling water. Why?
10. Why is the heat of vaporization of a substance usually larger than the heat of melting?
11. Make a rough graph showing how the temperature of a sample of water changes as energy is slowly added to it at a uniform rate. Start with a temperature of $-10\ °C$, when the water is frozen, and finish at $120\ °C$, with steam. Here is the way to set up the graph:

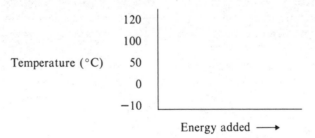

12. The heat of vaporization of water is far greater than that of liquid H_2. Why?
13. Compare the heat of vaporization and the boiling point of ammonia to those of other substances listed in Tables 13.2 and 13.4. Then explain why ammonia makes a good refrigerant.

Section 13.5

14. What is a vapor?
15. What is vapor pressure?
16. Why can't vapor pressure be established unless the substance is in a closed container?
17. Why do some substances have a higher vapor pressure than others? Name some substances with high vapor pressures and some with low. What substance is likely to have the highest vapor

pressure of all? Explain the reasons for your answer. (See also Sections 13.7 and 13.8 for help with this question.)

18. If a substance has a high vapor pressure, it is likely to have a low heat of vaporization, and vice versa. Explain.

Section 13.6

19. What is dynamic equilibrium? What processes come to equilibrium when vapor pressure is established?
20. Using the concept of dynamic equilibrium, explain why vapor pressure depends on temperature.
21. Why is the freezing point at the same temperature as the melting point? What is the difference between the two processes?

Section 13.7

22. What is boiling?
23. Why does boiling temperature depend on pressure?
24. What is the normal boiling point of a liquid?
25. In liquids, what is the relation between boiling point and the strength of the intermolecular forces?
26. How is it possible for a liquid to boil and freeze simultaneously? Explain fully.

Section 13.8

27. What are van der Waals forces? Explain at least two ways in which these forces differ from chemical bonds.
28. Explain the causes of electric dipoles.
29. With the help of diagrams, explain how dipole forces operate between molecules.
30. Explain how intermolecular forces can develop even among molecules that have no permanent dipoles.
31. Explain how a molecule can have polar bonds, yet have no dipole.
32. Name five different kinds of molecules that have permanent dipoles. Draw their structures, indicating the dipoles.
33. Draw structures for three molecules that have polar bonds but no dipoles.
34. What is a hydrogen bond, and in what circumstances does it operate?
35. How does a hydrogen bond differ from an ionic or covalent chemical bond? How does it differ from an ordinary dipole force, as found, for example, between H_2S molecules?

Section 13.9

36. Without referring to Table 13.5, construct a table that shows covalent crystals, ionic crystals, and molecular crystals. For each type, give the kinds of particles involved, the interparticle forces, the properties of the crystal, and some examples. For the molecular crystal section, break the table into three subsections: H-bonded crystals, ordinary dipole crystals, and London force crystals.

PROBLEMS

Section 13.2

*1. The specific heat of the element phosphorus is 7.94×10^{-4} kJ/g°C. How many kilojoules of energy are required to raise the temperature of a 3.75-g sample of P from 16 °C to 29 °C?

2. A sample of silver, Ag, weighing 10.0 g was heated from 25 °C to 40 °C. The energy required to accomplish this temperature change was 3.5×10^{-2} kJ. What is the specific heat of silver in this temperature range?

*3. To a sample of benzene, C_6H_6, with a mass of 257 g and a temperature of 25 °C, 7.15×10^{-2} kJ of energy is added. What is the new temperature of the sample? The specific heat of benzene is 1.8×10^{-5} kJ/g°C.

4. A sample of lithium weighing 129 g and at a temperature of 0 °C was allowed to warm to 23 °C. In this process, it absorbed 0.105 kJ of energy. What is the specific heat of lithium? What is its molar heat capacity?

5. A 100-g sample of a metallic element is heated from 20 °C to 32 °C. Accomplishing this temperature change requires 0.532 kJ. Use the approximation of Dulong and Petit to find an approximate atomic weight for this element. Which element is this metal likely to be?

Section 13.4

*6. How much energy is required to vaporize 1.35 g of water at a constant temperature of 100 °C?

7. The normal boiling point of ammonia is −33.4 °C. A sample of liquid ammonia containing 1.00 mol was vaporized at −33.4 °C, a process requiring 23.4 kJ. What is the heat of vaporization of NH_3 at its boiling point?

Section 13.8

*8. In the following molecules, which element will attract the electron pair more strongly? If the bonds are ionic instead of covalent, so state.
   * *a. CS_2 d. PBr_3 f. KH
   * *b. NI_3 e. InSb   g. SiC
   * *c. Cl_2O

9. List the following pairs of atoms in order of increasing polarity of the bonds:
   a. P and O   c. Al and I   e. F and I
   b. C and Cl   d. H and O   f. B and H

*10. For the following polyatomic molecules, which element will attract the electron pair more strongly in each of the bonds?
   * *a. HCN   c. CClHCBrF   e. CH_3OH
   * *b. CH_3NH_2 d. CH_2O f. HSO_3F

*11. What type of bond (ionic, polar covalent, or covalent) will form between the atoms of the following pairs?
   * *a. Al and N   c. Cs and I   e. O and F
   * *b. B and H   d. Mg and O   f. Si and C

*12. Based on what you know about geometry and bonding in the following molecules, predict whether an electric dipole exists in each:
   * *a. CCl_4 c. NOF   e. NF_3
   * *b. H_2S d. PCl_3 f. CH_2O

Chapter 14

Water and its Elements

It is silly to try to name the compound or element that is most important to the human race, because there are dozens of substances without which we could not exist. It does make sense, however, to ask which substance is essential in the greatest variety of ways: the substance is water.

The human body is composed mostly of water. Body chemistry takes place in water solution and usually directly involves water molecules. The energetics of the body, sensory action, fuel and waste transport, and numberless other processes all require the presence of water.

An entire kind of chemistry, the reactions of ionic substances, takes place almost exclusively in water solutions. Such reactions are especially important to biological systems.

Our environment is fashioned by water. Eighty percent of the earth's surface is covered with water, and the little that is left has been shaped by water. The oceans and icecaps, which contain 99% of the water on the surface of the earth, control our weather. We cannot escape or overestimate the importance of water.

The elements that make up water are important in their own right. Oxygen powers the energy-producing chemistry of the body, combines with nearly every other element, and is, by weight, the most abundant element in the earth's crust.

By number of atoms, however, hydrogen is the most abundant element. The unusual chemical properties of hydrogen are responsible for equally unusual properties in some of its compounds, of which water is a prime example. In the important chemistry of acids and bases, the ions of hydrogen are the acid species in water solutions.

In this short chapter, we first consider some of the chemistry of oxygen, of hydrogen, and of water itself. We investigate natural waters and the difficulty of maintaining adequate supplies of acceptable water. Finally, we glance at the energetics of water molecules and at the possible use of hydrogen as a fuel.

AFTER STUDYING THIS CHAPTER, YOU WILL BE ABLE TO

- discuss the history, properties, preparation, uses, and chemistry of oxygen and hydrogen
- explain how ozone is essential to life on earth although it is harmful to living tissues
- tell how the ozone layer of the atmosphere could be destroyed, and what the consequences would be
- explain what makes much of the chemistry of hydrogen unique
- explain the energetics of the formation and decomposition of water and apply the explanation to the concept of hydrogen as a fuel
- name some of the impurities found in water supplies
- write reactions for the formation of hard water and explain some of the problems it causes

TERMS TO KNOW

allotrope	hydrogenation	water gas
hard water	ozone	
hydride ion	ozone layer	

14.1 PROPERTIES AND HISTORY OF OXYGEN

It was once widely believe that when substances burned, they released something called phlogiston, a substance that, curiously enough, must have had negative mass, because the objects always weighed more after burning. We know today that substances gain mass when they are burned because they combine with oxygen to form oxides, but we came to this knowledge only after many false starts and much wasted effort.

The discovery of oxygen in its elemental form is credited to Joseph Priestley, who reported his experiments in 1774. Priestley prepared oxygen gas in fairly pure form by decomposing mercuric oxide. He thought the oxygen gas was a new kind of air.

$$2\ HgO \longrightarrow 2\ Hg + O_2$$

Working shortly after Priestley, Lavoisier was able to show that oxygen was a component of the atmosphere, accounting for about 20% by volume.

Elemental oxygen is a colorless and odorless gas composed of diatomic molecules, O_2. At STP it has a density of 1.4 g/L. Liquid oxygen, sometimes called LOX, boils at 90.2 K and is used in several processes in which concentrated sources of O_2 are needed.

Most oxygen exists not as O_2 but in a combined state as oxides and other compounds. The waters of the earth, for instance, are about 88% oxygen by mass. About half the mass of the lithosphere, the earth's crust, comes from oxygen atoms, primarily in minerals like the carbonates ($CaCO_3$, $MgCO_3$) and in those minerals containing silicon, the simplest of which is SiO_2, ordinary sand.

Oxides have been prepared for all the elements except the lighter noble gases. The atomic weight scale was, in fact, originally based on oxygen, because it was easily available and could combine with so many other elements.

14.2 PREPARATION AND COMMERCIAL USE OF OXYGEN

Oxygen gas can easily be prepared in the laboratory by decomposing any of several oxygen-containing compounds. In addition to the historic HgO, other compounds, such as $KClO_3$, will serve. We have seen the reaction before.

$$2 \; KClO_3 \xrightarrow{\text{Heat}} 2 \; KCl + 3 \; O_2$$

The electrolytic decomposition of water is often used in the laboratory to obtain O_2.

$$2 \; H_2O \xrightarrow[\text{energy}]{\text{Electric}} 2 \; H_2 + O_2$$

Such methods are too slow and expensive, however, for commercial use. Electrolytic decomposition requires an enormous amount of power, and compounds such as KNO_3 and $KClO_3$ are expensive. Large-scale production of O_2 is usually accomplished by the fractional distillation of liquid air. The process requires heavy machinery, but is fairly inexpensive once the original investment is made. In areas where electric power is available at low cost from hydroelectric plants, electric energy is still used to prepare oxygen, primarily because H_2 is also obtained in the process.

The largest single industrial use of O_2 is in the making of steel. More and more steelmakers are abandoning open hearth and Bessemer converters (see Section 18.14), and using oxygen to burn excess carbon out of freshly smelted iron. Steelmaking accounts for about a third of all the pure O_2 prepared.

Oxyacetylene welding has been practiced for decades throughout the world. Oxygen prepared for this purpose is bottled under high pressure in the familiar, 5-ft steel cylinders, and is shipped everywhere. Gaseous O_2 is used also to produce hot flames in other kinds of torches.

The petroleum industry uses O_2 in the preparation of chemical products for large-scale uses, including the manufacture of plastics.

Highly purified O_2 is used for breathing. At high altitudes and in space, LOX is vaporized and piped into breathing systems. Sometimes O_2 is used with He in self-contained underwater breathing apparatus. In medicine, oxygen is used in anaesthesia apparatus and to enrich the atmosphere in oxygen tents.

14.3 OZONE

Although nearly all the elemental oxygen on earth occurs as diatomic O_2, there is another form, or **allotrope,** of elemental oxygen, called **ozone.** The formula of ozone is O_3. (Elements that appear in more than one form are said to occur in different allotropes.)

Ozone occurs naturally in small concentrations in the atmosphere. It is created by the action of sunlight on oxygen in certain circumstances, and by electric discharges, such as lightning. In the upper atmosphere, at altitudes of about 25 to 75 km, there is an unusually high concentration of O_3, called the **ozone layer.** Without the ozone layer, life on earth would probably not be possible. O_3 absorbs the sun's most energetic short-wavelength ultraviolet light before it reaches the surface of the earth. These highly energetic photons are deadly because they can destroy the cells of living tissues and decompose the carbon compounds that are the building blocks of living creatures. Probably only deep sea life would survive the destruction of the ozone layer. There is presently serious discussion regarding the use of certain fluorine-containing compounds as propellants in spray cans. Such compounds readily decompose ozone into O_2, without being destroyed themselves. Continued escape of the compounds into the atmosphere may weaken and finally destroy the ozone layer, leading first to a high incidence of cancer in humans and finally to the disappearance of life.

Although O_3 at high altitudes is necessary to life because of its ability to filter sunlight, it is harmful at the surface of the earth. Ozone is a strong oxidizing agent that attacks plastics, rubber, and living tissues. One of the components of photochemical smog (the kind found in Los Angeles, and other places with many autos and much sunlight) is ozone, which is harmful to eyes and lungs. Ozone also plays a role in the formation of other air pollutants.

14.4 PROPERTIES OF HYDROGEN

Hydrogen is a unique element in several ways. The neutral hydrogen atom is composed of a proton and a single $1s$ electron. Hydrogen is the only element that in ordinary chemical reaction can lose its entire electron shell (one electron) to become a bare nucleus. Because of this capacity, the element has several interesting chemical properties.

Hydrogen is the lightest of all atoms, and its properties reflect this fact. The colorless and odorless gas has a density at STP of 2 g/22.4 L, or about a tenth of a gram for every liter. At its boiling point, about 20 K, liquid hydrogen is the second lightest liquid at 0.07 g/cm³ (water is nearly 15 times as dense). Solid H_2, only slightly more dense than the liquid, has a density about the same as that of the lightweight plastic foam of which Christmas-tree balls are sometimes made. It is the lightest crystalline substance.

Only weak dispersion forces act between H_2 molecules. Because each molecule has only two electrons, both tightly held, the forces caused by temporary distortions of the electron clouds are quite small. The result is that H_2 has a boiling point of 20 K, the lowest of any liquid except helium.

14.5 PREPARATION AND USE OF H_2

In the laboratory, hydrogen gas is most commonly prepared by the reaction between an active metal like Zn and acid in water solution, for example,

$$Zn \text{ (s)} + H_3O^+ \text{ (aq)} \longrightarrow Zn^{2+} \text{ (aq)} + H_2 \text{ (g)} + H_2O \text{ (l)}$$

Because H_2 is nearly insoluble in water, the H_2 gas bubbles off and can be collected, as shown in Figure 12.18.

Commercially, H_2 is prepared at a high temperature by what is called the **water gas** reaction. To run the reaction, water is passed over incandescent coke or coal. The mixture of CO and H_2 is used industrially in large quantities.

$$C \text{ (s)} + H_2O \text{ (g)} \xrightarrow{1000 \ K} CO \text{ (g)} + H_2 \text{ (g)}$$

A similar reaction is run with natural gas, CH_4. In this process, the reactants are passed over a catalyst to speed up the reaction.

$$CH_4 \text{ (g)} + H_2O \text{ (g)} \xrightarrow{Catalyst} CO \text{ (g)} + 3 \ H_2 \text{ (g)}$$

To separate the H_2 from the CO, the gases are mixed with steam and passed over another catalyst at 775 K. Under these carefully controlled conditions, the CO is converted to CO_2, in what is called the water gas shift reaction.

$$CO\ (g) + H_2O\ (g) \xrightarrow{\ Catalyst\ } CO_2\ (g) + H_2\ (g)$$

The CO_2 and H_2 mixture is then passed into liquid water where the soluble CO_2 is trapped by the water and the H_2 bubbles off and is collected.

Several decades ago, H_2 was chiefly used to fill balloons and dirigibles. As other uses developed, the aeronautical use of H_2 simultaneously declined. It ceased almost entirely after the Hindenburg disaster in 1937, when the enormous amount of H_2 in the famous dirigible caught fire and a number of persons perished. Lighter-than-air craft now rely almost entirely on helium.

H_2 is sometimes used for torches; in controlled circumstances, H_2 and O_2 react to produce an extremely hot flame. In the preparation of certain highly purified metals, H_2 is used as an atmosphere, both to prevent oxidation and to react with oxides already present.

Hydrogen is employed most extensively in industry. It is often used at high pressures for **hydrogenation** reactions in which it combines directly with other molecules. An example is the large-scale preparation of methyl alcohol.

$$2\ H_2\ (g) + CO\ (g) \xrightarrow[300\ atm,\ 575\ K]{\ Catalyst\ } \underset{Methyl\ alcohol}{H_3COH\ (g)}$$

One of the most important uses of H_2 is in the Haber process, which produces ammonia.

$$N_2\ (g) + 3\ H_2\ (g) \xrightarrow[High\ temperature]{\ Catalyst\ } 2\ NH_3\ (g)$$

Gaseous H_2 is also widely used to hydrogenate fats, as in the preparation of margarine. This process, which involves adding H_2 to C=C double bonds (Section 19.8.1), converts edible oils to more saturated forms that have higher melting points and are solids at room temperature.

14.6 UNIQUENESS OF HYDROGEN

Like the atoms of the alkali metals, each H atom has only one electron in its outer shell; but the resemblance does not extend to many chemical reactions. The alkali atoms invariably react by losing one electron per atom to become singly charged positive ions, whereas hydrogen atoms usually form covalent bonds, and rarely take the form of positive ions in pure samples of their compounds. Because of the proximity of the H electron to the nucleus, the H atom has a high ionization energy and moderately high electron affinity. These characteristics account for the differences between H atoms and atoms of the alkali metals. Compare, for example, the kernel of H with that of Na. The H kernel has a radius of only 10^{-3} Å, whereas the diameter of the Na kernel is about 1 Å: yet both have the same kernel charge, +1. The positive charge of H is concentrated in an extremely small volume; that of an alkali kernel exercises its influence from within a larger volume. Although this concentration of charge takes hydrogen, in effect, out of the alkali family, it still does not make the H kernel as attractive to electrons as are the kernels of oxygen, chlorine, or other elements of high kernel charge. When H forms bonds with atoms of high electron affinity, in HCl, for example, the electron pair bond is unequally shared, and the molecule has a dipole. In many HCl reactions, such as that with H_2O, the H comes away without its electron.

$$HCl\ (g) + H_2O\ (l) \longrightarrow H_3O^+\ (aq) + Cl^-\ (aq)$$

In some circumstances, H atoms can complete their outer shells by adding electrons and becoming H$^-$ ions, called **hydride ions.**

$$H_2\ (g) + 2\ Na\ (s) \longrightarrow 2\ NaH\ (s)$$

NaH is an ionic compound whose crystal lattice is made up of Na^+ and H$^-$ ions. Compounds containing hydride ions are quite reactive and are not found in nature.

14.7 HYDROGEN PEROXIDE

Although much of the hydrogen on earth is locked up with oxygen in the form of water, these elements also form another compound, hydrogen peroxide, H_2O_2. Peroxides are distinguished by their bonding pattern: two oxygen atoms are joined by a single covalent bond, and each oxygen is then covalently bonded to yet another atom. In hydrogen peroxide, the other atoms are hydrogen, resulting in the following structure:

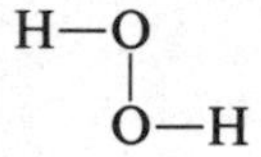

$$
\begin{array}{l}
H - O \\
\quad\ \ | \\
\quad\ O - H
\end{array}
$$

Hydrogen peroxide is a syrupy liquid that freezes at $-1.7\ °C$ and has a density of 1.44 g/mL. It is not often prepared in the pure state because it is unstable and spontaneously decomposes to water and oxygen.

$$2\ H_2O_2\ (l) \longrightarrow 2\ H_2O\ (l) + O_2\ (g)$$

In World War II, the first German V2 rockets used pure H_2O_2 as a source of oxygen to burn their kerosene fuel.

Hydrogen peroxide is commonly sold mixed in small concentrations with water. In such solutions, it is a useful antiseptic and mild bleaching agent.

14.8 WATER

Pure water is an odorless, colorless, tasteless compound that is liquid at room temperature and pressure. It is a common compound, present on earth in generous amounts. Water displays unusual properties when compared with other compounds that have molecules of about the same size. As we have learned, hydrogen bonding gives water its surprisingly high melting and boiling temperatures. We will find also that the polarity of water molecules strongly affects the relations of water to other substances.

Nearly every chemistry student has seen the lecture-demonstration in which H_2 and O_2 react chemically with a violent explosion.

$$2\ H_2 + O_2 \longrightarrow 2\ H_2O + energy$$

If so much energy is released when water molecules are formed, a like amount of energy must be required to decompose water back into its elements. In other words, water is a stable substance not easily decomposed. Because reactions that form water do release large amounts of energy, and because the release of energy is one of the things that makes reactions go well, chemists have taken their cue from nature and incorporated the production of water into the design of chemical reactions. The burning of natural gas (which produces water) is an example that can be taken to represent any of the familiar combustion reactions that propel our technological society.

$$CH_4 \text{ (g)} + 2 O_2 \text{ (g)} \longrightarrow CO_2 \text{ (g)} + 2 H_2O \text{ (g)} \qquad \triangle H = -439 \text{ kJ}$$

The oxidation of glucose, which furnishes the energy that runs your body, is a similar reaction, also resulting in the formation of water.

Our earth has a diameter of about 13,000 km. On most of its surface is a thin film of water, no more than about 12 km thick at its deepest, making up the oceans and the ice caps. In response to energy provided by the sun, some of the water in the oceans evaporates, later condensing over the land to form rain. From the rain comes our small supply of fresh water. By runoff and evaporation, the fresh water again finds its way to the sea. This circulation is what keeps the water supply usable for the biological inhabitants of the earth. Not only human life, but all life on earth depends on the unique properties of water and on the continual refreshment of the waters of the earth by the seas.

Water reacts chemically with many other substances and will dissolve many other compounds. In later chapters, we will often discuss the reactions of water and its use in solutions.

14.9 OUR ENDANGERED WATER SUPPLIES

Most of us take a dependable supply of fresh water for granted. Although we use water for countless purposes—drinking, cooking, hygiene, flushing away waste, manufacturing, agriculture, power, and so on—we rarely stop to consider exactly what it is we are using, where it comes from, or how precious it really is.

What we call fresh water is never pure water, and in many places is so tainted as to be almost undrinkable. Even in the absence of pollutants from industrial and urban sources, natural fresh waters contain numerous contaminants. From the atmosphere, lakes and streams pick up dust and dissolve various gases and solids. From the earth, water supplies draw along small pieces of clay, sand, and other soil particles, and dissolve numerous minerals as well. Living organisms and the products of their decomposition add yet more foreign substances to natural waters. It is not unusual to find significant amounts of Na^+, Ca^{2+}, Mg^{2+}, Fe^{2+}, and NH_4^+, along with Cl^-, HCO_3^-, SO_4^{2-}, NO_3^-, and CO_3^{2-} in city water supplies. Usually, small amounts of algae, viruses, bacteria, and dissolved organic compounds are also present. Even the water in the clearest mountain stream is far from pure. Not all contaminants in water supplies are necessarily bad, but some substances—such as microscopic asbestos fibers—that were not considered undesirable in the past are now known to be harmful. Other contaminants now accepted without concern may someday be found to be dangerous.

Although what is called **hard water** is probably not harmful to health, it is a serious problem in water supply systems. When exposed to air, water supplies can dissolve small quantities of carbon dioxide, CO_2. If the water later runs over limestone ($CaCO_3$ and $MgCO_3$ mostly), the dissolved CO_2 makes it possible for some of the metal ions in the limestone to dissolve in the form of bicarbonates such as $Ca(HCO_3)_2$. When the resulting **hard water** is later used for washing, the dissolved calcium and magnesium ions react with soap to make an insoluble, scummy compound that settles on sinks, cooking utensils, and laundry. When hard water is heated before use, it can undergo a different reaction:

$$Ca(HCO_3)_2 \text{ (aq)} \longrightarrow CaCO_3 \text{ (s)} + CO_2 \text{ (g)} + H_2O \text{ (l)}$$

$CaCO_3$ is insoluble and forms deposits on the insides of pipes and water heaters, finally filling them entirely (Figure 14.1). The presence of Ca^{2+} or Mg^{2+} ions is the cause of most hard water. To make soft water, the Ca and Mg ions must be removed or replaced with Na^+ ions, which do not cause the same problems.

Much of our fresh water is recycled. Waste water from industrial processes and water from sewage find their way into groundwater and eventually into metropolitan water supplies. The amount of water used in the United States for sewage disposal alone is roughly equivalent to the flow of the Colorado River. Because of this constant cycle of use and reuse, it is important that the water be adequately purified during each part of the cycle.

Natural processes provide some purification. Impure water that flows to streams and lakes is filtered during its passage through sand and rocks. Microorganisms in the water consume organic matter from sewage, and some contaminants are oxidized by contact with air. There are limits, however, to the amount of purification that natural processes can provide. Bacteria use the oxygen dissolved in the water as they consume organic material. This

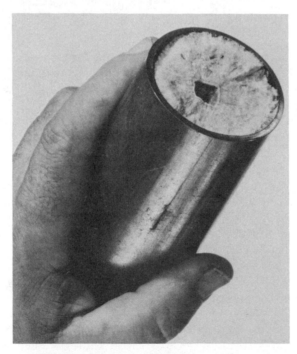

Figure 14.1 Boiler scale deposited in a section of hot-water pipe. The pipe was in use for only two years. (Photo courtesy of The Permutit Company.)

oxygen must be replenished from the air. If the oxygen is used more rapidly than it can be absorbed from air, the bacteria die, and the purification process stops.

Insufficient oxygen has other serious consequences. When the oxygen level falls to less than four parts per million, water becomes unsuitable for many kinds of aquatic life. As the oxygen content falls still lower, the character of the water changes visibly: it becomes turbid and dark, and aquatic life is further disrupted. Finally, when dissolved oxygen is no longer available, the decomposition of organic matter can take place only by anaerobic (no air) processes that release carbon dioxide, methane, hydrogen sulfide (the gas that gives rotten eggs their smell), and foul-smelling organic sulfur compounds. The hydrogen sulfide reacts with metal ions to form a layer of black scum on the surface, thus the descriptive term, "black water." To prevent oxygen depletion, the influx of organic matter, such as sewage, must be kept to a minimum. Modern sewage disposal techniques can accomplish this, but many large American cities are hopelessly behind in the race to treat the growing quantities of sewage adequately.

14.10 HYDROGEN AS A FUEL

Because of concern about the depletion of fossil fuel supplies, there have been several proposals, often mentioned in the popular press, for using H_2 as a fuel in powerplants, and especially in automobiles. Several hydrogen cars have been built and successfully operated.

The electrolysis of water is ordinarily used to prepare H_2. Apart from the cost of the electric power, the process is cheap and efficient. Considering the inevitable energy losses that occur in even the most efficient industrial processes, however, it must be clear that an overall energy loss occurs when water is electrolyzed to give H_2, and the H_2 is then burned with O_2 to give water again. A popular argument for the use of H_2 as a fuel is that so much water is available, we need only use it to obtain H_2 and we will have plenty of power. It is often forgotten that more energy must be put into such a cycle than is gotten out. If H_2 could be prepared without the sacrifice of other valuable fuel, by the direct use of sunlight for example, it would immediately become practical as a fuel.

Although it is light, H_2 is not especially portable, because it is a low-boiling gas. As a gas, it can be stored in high-pressure cylinders of the kind used by welders, but one cylinder weighs about 40 kg and carries only enough H_2 for a few miles of driving. Present technology allows H_2 to be liquefied only at temperatures too low for practical use in automobiles.

Some experiments have been made using metal hydrides as storage media for H_2. Lithium hydride, for example, can be decomposed by any of several methods to give H_2; so far, however, the processes of preparing and using the metal hydrides are slow, unwieldy, and expensive. To use H_2 economically for motor fuel will require technical and engineering breakthroughs that are not now in sight.

QUESTIONS

Section 14.1

1. Explain how and why it was once believed that there existed a substance that had negative mass.
2. Why was the atomic mass scale once based on oxygen, rather than on ^{12}C as it now is?

Section 14.2

3. Name three methods for producing O_2. Give equations for any reactions involved.
4. Give four large-scale uses of oxygen in its pure or nearly pure form.

Section 14.3

5. What is ozone?
6. Explain why ozone is essential to life on earth.
7. Explain how the use of fluorinated hydrocarbons, as in spray-can propellants, could destroy the ozone layer.
8. Explain the harmful effects of ozone.

Section 14.4

9. Why are many of the reactions and properties of hydrogen different from those of the other elements?
10. Why does hydrogen have such a low melting and boiling temperature?

Section 14.5

11. What is water gas? Write the water gas reaction.
12. Name two sources of H_2 other than the water gas reaction.
13. Name three large-scale uses of H_2.

Section 14.6

14. What is a hydride ion? Under what circumstances can hydride ions be formed?

Section 14.8

15. Why are many reactions designed to form water as a product? Name two important reactions of which H_2O is a product.

Section 14.9

16. Name some contaminants often found in water supplies.
17. What is hard water? Write reaction equations showing how hard water is formed.
18. Write equations for and explain the effects of two reactions in which hard water is harmful.
19. Why is it important that water supplies contain dissolved oxygen? What happens when they do not?

Section 14.10

20. What are some of the advantages and disadvantages of H_2 as a fuel?
21. Even though water is cheap and plentiful, why is it not a good idea at present to use the hydrogen in water as a fuel?

Chapter 15

Chemical Reactions
in Solution

A bumper sticker that occasionally appears on automobiles at college and university campuses says "Chemists Have Solutions"—which is true. Chemistry could hardly have begun without solutions.

Numerous chemical reactions, especially those performed in experimental laboratories, are carried out in solution. Many kinds of chemical measurements are made in solution. Compounds are prepared and sold in solution. Many phenomena important to chemistry occur only in solution. Chemists *do* have solutions.

In this chapter, we first learn something of the language of solutions, then investigate why and how substances mix and dissolve in one another. We learn which substances will make solutions and which will not, and that there are limits to solubility. We see how ionic compounds go into solution by dissociating into their individual ions and how those ions can enter chemical reaction. We will see that many solutions are created by chemical reaction rather than by simple mixing. We define and begin to use two important terms, *acid* and *base*. Finally, we learn how to make quantitative measurements by reacting two solutions together.

Once we have studied and understood solutions and reactions in solution, we will have obtained some of the most important mental equipment we need to practice the art of chemistry.

AFTER STUDYING THIS CHAPTER, YOU WILL BE ABLE TO

- classify a mixture as a solution or a suspension
- determine which solutes will dissolve in polar or nonpolar solvents
- determine whether a compound will dissolve in water
- label a compound as an acid or a base and as weak or strong
- write neutralization equations for a given acid-base pair
- calculate the weight percent of a solution, given mass data
- calculate the molarity of a solution

- use molarity and weight percent in reaction equation calculations
- make concentration calculations involving the dilution of a solution and the mixing of solutions
- make calculations based on titration data
- calculate the concentration of H_3O^+ from the concentration of OH^- and vice versa
- calculate pH from concentrations of H_3O^+ or OH^- and vice versa

TERMS TO KNOW

acid	hydrogen ion	solute
acidic solution	hydrolysis	solvated ion
aqueous solution	hydronium ion	solvent
base	hydroxide ion	strong acid
basic solution	ionizing solvent	strong base
colloid	liquid aerosol	strong electrolyte
concentrated solution	molarity	supersaturation
concentration	net ionic equation	suspension
diluent	neutralization	titration
dilute solution	neutral solution	weak acid
dissociation	pH	weak base
electrolyte	proton transfer reaction	weak electrolyte
emulsion	salt	weight fraction
foam	saturated solution	weight percent
hydrate	solid aerosol	
hydrated ion	solubility	

15.1 LANGUAGE OF SOLUTIONS

We learned in Chapter 1 that a solution is a homogeneous mixture of two or more substances. Solutions can exist in any of the three phases: air is a gaseous solution of N_2 and O_2; a brass doorknob is a solid solution of copper and zinc; and gasoline, beer, seawater, tea with sugar, tap water—and most of the other liquids with which we come into contact—are solutions of one kind or another.

Solutions and samples of pure compounds are often confused with each other because they are both homogeneous substances. You can distinguish between the two if you remember that every compound contains its elements in definite proportions according to a formula. The components of a solution, however, can be present in various proportions.

Two other kinds of systems have some of the properties of solutions and are often confused with them, but are not solutions at all; these are either **suspensions** or **colloids.** In a suspension, particles of one phase are mixed heterogeneously with a second phase, usually a liquid or gas. Suspensions exist only temporarily; given time, the suspended particles will settle out, and the mixture will separate itself. Flour shaken with water forms a suspension, but most of the flour particles will eventually settle to the bottom of the container. (See Figure 15.1.) There are, however, similar systems in which the particles are so small that they remain suspended indefinitely: such systems are called colloids. In a colloid, the dispersed particles can be solid, liquid, or gas. The resulting colloids often have individual names. For

Figure 15.1   Dirt suspended in water will eventually settle out. (Photo by author).

example, a system of insoluble liquid particles dispersed in another liquid (like milk) is called an **emulsion.** A system of gas dispersed in a liquid is a **foam.** A cloud (liquid particles dispersed in gas) is a **liquid aerosol.** Smoke is a **solid aerosol.** (See Figure 15.2.)

The distinctions between solutions, collloids, and suspensions depend on particle size. Particles larger than about 10^{-4} cm in diameter will usually not remain mixed throughout a liquid or gas indefinitely; the suspensions formed by such particles require some stirring to persist. Particles whose diameters fall roughly between 10^{-4} and 10^{-6} cm can form colloids that are definitely of two phases and that remain in stable suspension. Particles of atomic or molecular size form the homogeneous, single-phase systems we call solutions.

Before we go on to learn how solutions work, we need a few definitions. Solutions are composed of **solvents** and **solutes.** In most cases it is easy to say which is solute and which is solvent. If substances of different phases are involved, the one that has the same phase as the resulting solution is called the solvent. For example, if solid sugar is dissolved in liquid water to form a liquid solution, the water is said to be the solvent and the sugar the solute. If all the substances in a solution are of the same phase, then the substance present in the greatest proportion is said to be the solvent. In beer, the solvent is water because more than nine tenths of the solution is water. In a few cases, the distinction between solvent and solute is difficult to make (and of little importance), but we will be concerned only with straightforward situations, usually with solids dissolved in liquids.

A solution is said to be **concentrated** if the proportion of solute in relation to solvent is high. If there is only a small proportion of solute, the solution is **dilute.** There is no particular proportion of solute beyond which every solution is concentrated; the use of the two terms depends on the kind of solution and the circumstances of its use. Any solution can be made less concentrated if it is diluted with more solvent. A solvent used to make a solution more dilute is often said to be a **diluent.** The quantitative relation between the amounts of solute and solvent in a solution is called the **concentration** of the solution.

(a)   (b)

Figure 15.2   The scattering of light by colloids (the Tyndall effect) can be used to detect nearly invisible particles in a colloidal suspension. (a) A beam of light shining through a flask of pure water; most of the light goes right through. (b) The particles of a colloid scatter the light in the beam and cause it to be strongly outlined. (Photos by author.)

We usually characterize solutions by saying what the solute is, what the solvent is, and what the value of the concentration is. Many sets of units can be used to state concentration. Solutions in which the solvent is water are called **aqueous solutions.** Because water is the solvent in most solutions, the solvent in a solution is usually understood to be water unless otherwise stated.

15.2 KINDS OF SUBSTANCES THAT DISSOLVE IN ONE ANOTHER

Water is able to dissolve more substances than any other liquid. It is not easy to predict exactly which substances will dissolve which, but a rule of thumb helps in most cases. Chemists are fond of saying that *like dissolves like.* The word "like" refers to the relative polarities of the substances involved. Substances whose particles have large dipoles, for example, tend to mix easily with similar substances but not with substances in which no dipoles are present. Let's see why this is so.

Molecules with dipoles are attracted to other polar molecules by relatively strong dipole-dipole forces. Molecules with no dipoles attract one another only with much weaker temporary dipole forces. Only weak forces exist between molecules that have dipoles and molecules that have none. If two polar substances are mixed, all the molecules attract one another with strong forces, and the substances mix well. If two nonpolar substances are mixed, little intermolecular attraction occurs anywhere, and the two substances can mix well. But if a polar substance is mixed with a nonpolar substance, the polar molecules attract one another and exclude the nonpolar molecules, for which they have little attraction. Water is polar and oil is not. Try to mix these two. The water molecules attract one another strongly and form a closed society, refusing to separate and allow the oil molecules to mix with them. Try to dissolve a drop of water in oil: the nonpolar oil molecules have no objection, but the water molecules stay in their drop, reluctant to leave each other. Sugar, on the other hand, dissolves easily in water, because sugar molecules have numerous dipoles with which to attract water molecules. Paraffin will dissolve in gasoline, because neither of these nonpolar substances will exclude the other. In terms of polarity, water is somewhat like sugar, and paraffin is somewhat like gasoline. Water is not like gasoline.

Knowing the structures of two substances will allow you to make some good guesses about whether they will dissolve in each other.

15.3 SOLUBILITY AND SATURATED SOLUTIONS

The amount of a substance that can dissolve in a given amount of another substance is often limited. A cup of water will dissolve a teaspoonful of sugar, or even 2 teaspoonfuls if you stir for awhile, but if you put in 10 teaspoonfuls of sugar, most of the sugar will remain as a solid in the bottom of the cup, no matter how long you stir. A solution that has dissolved as much as it can of a given solute is said to be **saturated** with that solute. Saturated solutions are usually prepared by adding to the solvent more solute than can be dissolved, mixing until no more solute disappears, then leaving some excess solute in contact with the solution. Ordinarily, the higher the temperature of a solution, the greater the amount of solute it can contain at the saturation point. More sugar will dissolve in a cup of hot water than in a cup of cold.

The maximum amount of solute that can be dissolved in a given amount of solvent to give a saturated solution at a particular temperature is called the **solubility** of that solute in that solvent. For example, at 20 °C, 65.2 g of KBr will dissolve in 100 g of water, so the solubility of KBr in water at 20 °C is said to be 65.2 g/100 g H_2O. At 80 °C, the solubility of KBr in water is 95.0 g/100 g H_2O.

Suppose that you prepare a warm, saturated solution, using a solute that has greater solubility in warm water than in cold and leaving some solid solute in contact with the solution. As the solution cools, some of the solute will come out of the solution, forming additional solid. The solution will then be saturated at the new temperature.

In the proper circumstances, however, **supersaturated** solutions can be created. Such solutions contain more dissolved solute than they would in the ordinary saturated form. Supersaturation occurs when the extra amount of solute that would ordinarily come out of solution as crystalline solid has no pattern on which to initiate a crystal lattice.

Supersaturated solutions are unstable and can exist only if none of the solid form of the solute is present. A popular lecture-demonstration begins with a cool, supersaturated solution of sodium acetate into which a microscopic crystal, a "seed," of solid sodium acetate is dropped. Crystals of the solid compound grow rapidly on the seed, and within seconds the container is full of solid sodium acetate (Figure 15.3).

15.4 IONIC SOLUTIONS

Water readily dissolves NaCl, but NaCl is insoluble in nonpolar liquids. We might have predicted this behavior because we know that an NaCl pair is extremely polar; the Cl is present as a negative ion and the Na as a positive one. Actually, there is more to it than that: ionic substances dissolve in water by separating completely into their individual ions, a process called **dissociation.**

Figure 15.3 Demonstration of crystal growth in a supersaturated solution. A seed was added to a supersaturated solution of sodium acetate. The entire sequence lasts less than a minute. (Photos by author.)

Think of the surface of the NaCl crystal exposed to a liquid, then look closely at one individual ion, say a Cl^-. The Cl^- is surrounded by attractive Na^+ ions that hold it in place in the crystal. If the surrounding liquid is nonpolar, the Cl^- will be constantly jostled by the molecules of the liquid, but it will have little reason to leave the crystal. If the liquid molecules are polar, as in H_2O, the situation is different. The negative chloride ion will exert attractive force on the positive ends of the water molecules, while the nearby Na^+ ions attract the negative ends of the same molecules. The continuous vibrations of the water molecules and of the ions themselves will now have an effect. Because the dipolar water molecules can, in a manner of speaking, form a bridge for the forces between the ions of the crystals, they can easily penetrate the crystal surface, surround individual ions, and remove them. Study Figure 15.4, which shows how ions are removed from the surface of a crystal.

Notice that when the ions are finally taken into solution, they remain in the form of individual ions. Liquids that cause dissociation of ionic crystals into individual ions are called **ionizing solvents.** Water is the most important ionizing solvent, and much of the chemistry we will study will be that of ions in water solution.

The chemistry of ionic solutions occurs as reactions between individual ions, regardless of the source of the ions themselves. For example, a solution of calcium ions, Ca^{2+}, will form solid calcium sulfate when mixed with a solution containing SO_4^{2-}, whether the SO_4^{2-} came from H_2SO_4, from Na_2SO_4, or from some other soluble sulfate. In Section 15.10, we will learn a set of rules that will, in many cases, allow us to predict when ionic reactions are likely to occur and when they are not. Right now, however, we will find it helpful to begin writing reaction equations in **net ionic form;** that is, including only the ions that actually react.

For the reaction of $Ca(NO_3)_2$ and Na_2SO_4, we might write

$$Na_2SO_4 \text{ (aq)} + Ca(NO_3)_2 \text{ (aq)} \longrightarrow CaSO_4 \text{ (s)} + 2 \, NaNO_3 \text{ (aq)}$$

We would understand from this formula that $Na(NO_3)_2$ (aq), for instance, refers to a solution made from sodium nitrate and containing Na^+ ions and NO_3^- ions rather than dissolved $NaNO_3$ in molecular form ($NaNO_3$ is completely dissociated in water solution).

If our purpose, however, is to indicate only what kind of reaction takes place, we could also write

$$Ca^{2+} + SO_4^{2-} \longrightarrow CaSO_4 \text{ (s)}$$

This shorter form is accurate overall, because the Na^+ and the NO_3^- do not react. They were in the solution before reaction occurred and remain in solution afterward. Because the Na^+ and NO_3^- do not participate, we need not show them. It would be proper to indicate that the ions existed in water solutions by writing (*aq*) after each ion, but since nearly all reactions of ions occur in water solutions, the (*aq*) is usually left out for ions.

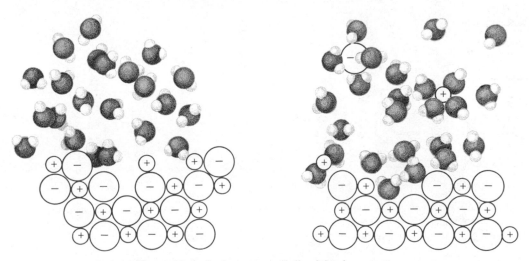

Figure 15.4   Ionic compound dissolving in water.

In the rest of the book, you will see some reaction equations written out fully and others in ionic form. You will find occasions to use both forms.

An ionic substance that will dissolve and dissociate in solution is called an **electrolyte,** because solutions of ions will conduct electric current. Electric current is the movement of charged particles. If a substance is to conduct electric current, it must contain charged particles, and those particles must be able to travel through the substance. An aqueous solution of sucrose (table sugar) will not conduct because it contains no charged particles; sucrose is not ionized. An aqueous solution of NaCl does conduct, even though solid NaCl will not. That NaCl solutions conduct tells us that charged particles, ions, exist in the solution; we would conclude that those ions came from the dissociation of NaCl. (See Figure 15.5.)

Ionic substances that dissociate completely are called **strong electrolytes;** sodium chloride is a good example. Some compounds form solutions in which only a small proportion of the compound is in the form of ions; such substances are called **weak electrolytes.** To the extent to

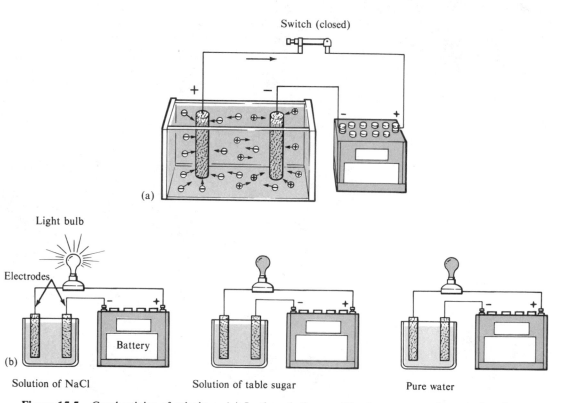

Figure 15.5   Conductivity of solutions. (*a*) In the solution, positive ions move to the negative electrode and negative ions to the positive electrode. (*b*) NaCl is an electrolyte. The lighted bulb shows that NaCl conducts electric current. Neither sugar nor water is an electrolyte.

which they are soluble, nearly all ionic substances are strong electrolytes; however, not all ionic substances are necessarily very soluble. Calcium sulfate, for example, is a strong electrolyte that is only very slightly soluble in pure water. The forces between the Ca^{2+} ions and the SO_4^{2-} ions are so strong that water cannot disrupt the crystal lattice. The ions of $CaSO_4$ are completely dissociated in solution, but solutions of $CaSO_4$ do not conduct well because they contain relatively few ions.

Many compounds form ionic solutions but are not themselves ionic. These compounds dissolve by means of a chemical reaction that creates ions where there were none before. An example is hydrogen chloride, HCl, which is covalent, yet is a strong electrolyte.

$$HCl\ (g) + H_2O\ (l) \longrightarrow H_3O^+ + Cl^-$$

We will learn more about reactions of this kind and about H_3O^+ in a later section.

When a dissolving ion leaves the surface of its crystal, it carries with it a layer of solvent molecules (look back at Figure 15.4) and is said to be **solvated.** If the solvent happens to be H_2O, the solution is said to be **hydrated.** Sometimes the water molecules are loosely attached to the ion, as in solutions containing Na^+, but in many cases they are strongly attracted to the ion and form stable structures called **hydrates.** For example, in many of its compounds, Cu^{2+} is present in a structure in which four H_2O molecules surround each ion. The Cu^{2+} ions keep their water molecules even when they join with negative ions to form ionic crystals.

In crystalline ionic compounds, either or both of the ions often carry water of hydration into the crystal. In solid $CuSO_4$ as it is ordinarily found, both the Cu^{2+} and the SO_4^{2-} are hydrated, the Cu^{2+} having four H_2O molecules, and the SO_4^{2-} having one H_2O. To represent crystalline compounds that contain water of hydration in formulas, we write the formula of the compound, followed by a dot and the number of water molecules associated with one unit of the formula. The hydrate of cupric sulfate, for example, would appear as $CuSO_4 \cdot 5\ H_2O$.

15.5 SOLUBILITY RULES

Now that we know something about solutions of ionic compounds in water, it will be helpful to look at some of the experimentally determined solubilities of ionic compounds in water. Table 15.1 gives us some general rules.

The solubility rules of Table 15.1 are rough approximations. **Soluble** can be taken to mean that several tens of grams of the compound will dissolve in 100 g of water at room temperature. **Moderately soluble** might mean from 5 to about 15 g of compound in 100 g of water. Only a gram or so of a **slightly soluble** compound and much less still of a **very slightly soluble** compound will dissolve in 100 g of water at room temperature.

It is not necessary to memorize all these rules at once, but you should study them now, then go through them again when we discuss the chemical reactions of ionic solutions. Mastery of chemistry lies in an understanding of fundamental concepts and relations, but that learning must be reinforced by factual knowledge derived from laboratory results.

EXERCISE 15.1.   Which of the following compounds are soluble in water: Hg_2S, K_2SO_4, $BiCl_4$, Ag_3PO_4?

Table 15.1
Solubilities of Ionic Compounds in Water

Compounds	Solubility	Exceptions
Compounds of the alkali metals and of ammonium ions	Soluble	A few Li compounds
Hydroxides (OH^-) and sulfides (S^{2-})	Most are insoluble	Hydroxides and sulfides of the alkali metals are soluble. Ca, Sr, and Ba have moderately soluble hydroxides and sulfides
Nitrates (NO_3^-), acetates ($H_3CCO_2^-$), and chlorates (ClO_3^-)	Soluble	Few exceptions
Chlorides, bromides, and iodides	Soluble	Exceptions are those of Ag, and a few others of Pb, Hg, Bi, and Sb
Sulfates (SO_4^{2-})	Soluble	Exceptions are those of Ca, Sr, Ba, Ag, Hg, and Pb, which are slightly to moderately soluble
Sulfites (SO_3^-), carbonates (CO_3^{2-}), and phosphates (PO_4^{3-})	Insoluble	Exceptions are those of NH_4^+ and the alkali metals

15.6 ACIDS AND BASES

In the vocabulary of chemistry are two common and important words, *acid* and *base,* that apply to two categories of reactive substances. Although widely used, the two words are not widely understood. How many people listening to a television commercial and hearing that their stomach may be "too acidic," or that their car engine makes "this much" acid, know what the words mean? The meanings, however, are easily grasped, and we will use the terms often.

In the most general sense, an **acid** is any species (atom, ion, or molecule) that will be strongly attracted to an unshared electron pair. A **base** is any species that has an easily available electron pair. If we confine our use of the terms to water solutions, we say that an acid is any substance that will release H_3O^+ ions, or **hydronium ions,** to a water solution. Individual **hydrogen ions,** H^+, are acids because they are attracted to unshared electron pairs; however, H^+ ions cannot exist individually in the presence of water, because water molecules have unshared pairs to which the H^+ ions stick.

$$H^+ + \underset{\underset{H}{|}}{O}-H \longrightarrow \left(H-\underset{\underset{H}{|}}{O}-H \right)^+$$

H^+ is always present in water solutions as a hydronium ion, H_3O^+, which makes additional loose associations such as $H_5O_2^+$ and $H_7O_3^+$. Because it is the most common practice, we will refer to H^+ in water solution as H_3O^+, but some chemists use H^+ simply for convenience.

For our purposes, a base is any substance that will release **hydroxide ions,** OH^-, in aqueous solutions. OH^- ions are basic because their negative charge enhances the attraction of their unshared electron pairs to H^+ ions.

Aqueous solutions of acids and bases react vigorously to yield water.

$$H_3O^+ + OH^- \longrightarrow 2\ H_2O$$

In the preceding reaction, a proton, H^+, was transferred from the H_3O^+ to the OH^-. Similar proton transfers are typical of acid-base reactions in aqueous solutions. Because of this, acid-base reactions are often called **proton transfer reactions.**

15.7 AQUEOUS SOLUTIONS OF ACIDS AND BASES

For our present purposes, we can say that a solution is **acidic** when there are more H_3O^+ ions in it than OH^- ions and that it is **basic** when there are more OH^- ions than H_3O^+ ions. The reaction of gaseous hydrogen chloride, HCl, with water is an example of an acid-base reaction that produces H_3O^+. Reactions in which water breaks up another species are called **hydrolysis** reactions (a term formed by the greek suffix *lysis,* meaning loosening or decomposing, and the prefix *hydro,* meaning water). In the following reaction, HCl is **hydrolyzed.**

$$HCl + H_2O \longrightarrow H_3O^+ + Cl^-$$

The reaction is complete; the water has acted as an ionizing solvent. No HCl molecules remain in the water, and the pressure of HCl gas is reduced essentially to zero. If a little hydrogen sulfate is mixed with water, a similar reaction occurs.

$$H_2SO_4 + H_2O \longrightarrow H_3O^+ + HSO_4^-$$

This reaction is followed by a second step.

$$HSO_4^- + H_2O \longrightarrow H_3O^+ + SO_4^{2-}$$

The first step of the decomposition goes essentially to completion, but at the end of the second step, a fraction of the HSO_4^- remains undissociated.

An acid, such as HCl or H_2SO_4, that is essentially completely dissociated in water solution is called a **strong acid.** There are some acids, however, of which only a fraction is dissociated in water: these are called **weak acids.** Acetic acid, mentioned earlier, is a weak acid. If 1 mol of acetic acid is added to 1 L of water, only about 4×10^{-3} mol will react.

$$\underset{\textit{Acetic acid}}{H_3CCO_2H} + H_2O \longrightarrow \underset{\textit{Acetate ion}}{H_3CCO_2^-} + H_3O^+$$

Nearly all of the acetic acid will remain in the solution as undissociated acetic acid molecules. (Why such reactions do not go to completion is the subject of the next chapter.)

Water reacts with certain metals to form hydroxides and liberate H_2. The reaction with Na is an example.

$$2\ H_2O\ (l) + 2\ Na\ (s) \longrightarrow 2\ OH^- + 2\ Na^+ + H_2\ (g)$$

The hydroxides of the alkali metals are highly soluble and totally dissociated. Other metals react to give only partially soluble hydroxides. Magnesium, for example, reacts with steam.

$$Mg\ (s) + 2\ H_2O\ (g) \longrightarrow Mg(OH)_2\ (s) + H_2\ (g)$$

If we were to mix 1 mol of NaOH with 1 L of water, we would find that all of the NaOH would dissolve and dissociate. If we were to make a similar mixture of $Mg(OH)_2$ and water, however, the result would be that only a small portion of the $Mg(OH)_2$ dissolves, but the dissolved portion would nonetheless be totally dissociated into its ions. Because they are totally dissociated, both NaOH and $Mg(OH)_2$ are said to be **strong bases,** regardless of solubility.

Another class of compounds is known as **weak bases.** Although these may be quite soluble in water, they exist only partially in ionic form, leaving most of their molecules unreacted. An example is ammonia, NH_3.

$$NH_3 \text{ (g)} + H_2O \longrightarrow NH_4^+ + OH^-$$

Because at any given time only a small proportion of the NH_3 is in the form of ions (most remaining as NH_3), NH_3 is classified as a weak base.

Strong acids and bases are strong electrolytes, and weak acids and bases are weak electrolytes. Table 15.2 lists some strong acids and bases, all in aqueous solution. Examples of weak acids and bases, with indications of their strengths, appear in Table 16.1 on page 393.

15.8 ACIDIC AND BASIC OXIDES

The oxides of metals and of nonmetals react with water in characteristically different ways. Metals usually have basic oxides.

$$Na_2O \text{ (s)} + H_2O \longrightarrow 2\,Na^+ + 2\,OH^-$$
Sodium oxide

$$BaO \text{ (s)} + H_2O \longrightarrow Ba(OH)_2 \text{ (s)}$$
Barium oxide

Because NaOH is soluble and ionized, we write it in ionic form. Because $Ba(OH)_2$ comes out of the reaction as a solid (it is only very slightly soluble), we write it as $Ba(OH)_2$, rather than in ionic form. The tiny amount of $Ba(OH)_2$ in solution, however, is entirely ionized. $Ba(OH)_2$ is a strong base.

Although basicity is a typical property of metal oxides, there are some metals that form acidic oxides in certain circumstances. Other metals, such as aluminum, have oxides that can be either acidic or basic, depending on the conditions in the solution. We will learn more about this kind of behavior in Chapter 18.

The oxides of most nonmetals react with water to form acidic solutions. Sulfur trioxide is an example.

$$SO_3 \text{ (g)} + 3\,H_2O \longrightarrow (H_2SO_4) \longrightarrow H_3O^+ + HSO_4^-$$

Table 15.2
Common Strong Acids and Strong Bases in Aqueous Solutions

Strong Acids	Strong Bases
HCl	NaOH
H_2SO_4*	KOH
HNO_3	$Ba(OH)_2$
$HClO_4$	$Ca(OH)_2$
HBr	$Sr(OH)_2$
HI	

*In dilute solutions, HSO_4^- also behaves like a strong acid.

Nonmetal oxides do not necessarily form *strong* acids, however. The oxide of phosphorus, P_4O_{10}, forms H_3PO_4, a weak acid.

15.9 NEUTRALIZATION AND THE FORMATION OF SALTS

When aqueous solutions of acids and bases are mixed, the H^+ and OH^- react instantly to yield water, and the other ions often remain unchanged. This process, which is typical of acids and bases, is called **neutralization,** because the active acid and base have reacted to give an unreactive substance, water. A **neutral solution** is one which is neither acidic nor basic, that is, one in which the numbers of H^+ and OH^- ions are equal.* An example of a neutralization is the reaction of hydrochloric acid, HCl, with sodium hydroxide, NaOH.

$$H_3O^+ + Cl^- + Na^+ + OH^- \longrightarrow 2\ H_2O + Cl^- + Na^+$$

In net ionic form, the neutralization looks particularly simple.

$$H_3O^+ + OH^- \longrightarrow 2\ H_2O$$

Total neutralization occurs only if the number of H_3O^+ ions is the same as the number of OH^- ions. If one of the two species is added in excess, the solution will become either acidic or basic; that is, not neutral at all.

If the water is evaporated after a neutralization reaction, the remaining ions will be left as a crystalline solid. In the reaction of HCl and NaOH, for example, the Cl^- and Na^+ ions will be left behind in the form of NaCl. The ionic substances that result from the reaction of acids and bases are called **salts;** thus, for example, NaCl is a salt, as are KCl, Na_2SO_4, LiBr, BaI_2, and NH_4F.

15.10 REACTIONS OF IONS IN SOLUTION

This is a qualitative section that is easy to read—but you should not think that it is unimportant. Now that we know something about solutions of ions, we can learn a principle that will let us predict the results of many reactions.

In some cases, mixtures of ions in solution are unreactive. For example, $BaCl_2$ and $NaNO_3$ can be dissolved in the same solution, or solutions of the two compounds can be mixed. The result in either case will be a solution that contains all four kinds of ions: no chemical reaction occurs. If the water in the solution is evaporated, a mixture of crystals of NaCl, $NaNO_3$, $BaCl_2$, and $Ba(NO_3)_2$ remains. Since two new compounds are present, a chemical reaction can be said to have taken place when the water was evaporated, but as long as the compounds were in the solution, no reaction occurred.

There are, however, four general ways in which ions will react while still in solution; that is, there are four ways in which ions can be removed from solution or changed chemically.

1. Ions can combine to form a solid, insoluble compound, which separates from the solution.
2. Ions can combine to form a gas, which bubbles out of the solution.

---

*At this point you may be wondering why the first paragraph in this section opens by saying that OH^- and H_3O^+ react instantly, yet later says that in a neutral solution the numbers of OH^- and H_3O^+ ions are equal. This question is examined in Section 15.15.

3. Ions can combine to form an un-ionized or only slightly ionized compound. That compound can stay in solution, but some or all of the ions are no longer present as ions.

4. Ions can unite with other ions or molecules to form species with coordinate bonds. Such species are called complex ions. (We will study complex ions in a later chapter.)

Here are some examples:

1. Formation of an insoluble solid: mixing solutions of $AgNO_3$ and $NaCl$.

$$Ag^+ + NO_3^- + Na^+ + Cl^- \longrightarrow Na^+ + NO_3^- + AgCl \text{ (s)}$$

The Cl^- reacts with the Ag^+, which leaves the solution as solid $AgCl$. Notice that the Na^+ and the NO_3^- do not react, but stay in solution.

2. Formation of a gas: mixing solutions of Na_2CO_3 and HCl.

$$2 Na^+ + CO_3^{2-} + 2 H_3O^+ + 2 Cl^- \longrightarrow 2 Na^+ + 2 Cl^- + 2 H_2O + CO_2 \text{ (g)}$$

The CO_2 gas bubbles off and leaves the solution. (This reaction proceeds for two reasons: because the CO_2 leaves the solution and because an un-ionized product, H_2O, is formed; see Example 3.)

3. Formation of an un-ionized product: acid-base reaction.

$$H_3O^+ + Cl^- + Na^+ + OH^- \longrightarrow Na^+ + Cl^- + 2 H_2O \text{ (un-ionized)}$$

The ions of H^+ and OH^- form water and cease to be ions.

4. Formation of a complex ion: mixing solutions of $Fe(NO_3)_3$ and $KCNS$.

$$Fe_3^+ + 3 NO_3^- + 6 K^+ + 6 CNS^- \longrightarrow Fe(CNS)_6^{3-} + 6 K^+ + 3 NO_3^-$$

Although the Fe_3^+ is still in the solution as part of another ion, $Fe(CNS)_6^{3-}$ has different properties than those of Fe_3^+. Complex ions are discussed in detail in Section 18.13.

These examples, along with the solubility rules of Section 15.5, will allow you to predict the outcome of many reactions.

EXAMPLE 15.1. Will mixing solutions of the following pairs of compounds result in reaction? Write the formulas of any products.

Solution:

Compounds	Results
$NaOH + MgCl_2$	Reacts, forming $Mg(OH)_2$ (insoluble)
$NH_4SO_4 + Ba(ClO_3)_2$	Reacts, forming $BaSO_4$ (insoluble)
$K_2S + CuCl_2$	Reacts, forming CuS (insoluble)
$NH_4I + KOH$	No reaction
$NaOH + HCl$	Reacts, forming $NaCl$ solution (H_3O^+ and OH^- disappear, forming H_2O)

EXERCISE 15.2.   Will mixing solutions of the following pairs of compounds result in reaction? Write the formulas of any products.

$K_3PO_4 + SrCl_2$
$K_2SO_4 + MgCl_2$
$Ba(OH)_2 + HClO_3$
$Cu(NO_3)_2 + Na_2S$
$AgNO_3 + Na_2SO_4$

The classic reaction that removes ions from solution by forming an un-ionized product is the acid-base reaction that forms water. We will meet a few other reactions that proceed for the same reason. The formation of a gas from ionic solutions is not common, but several industrial processes are designed to take advantage of the formation of gases from ions.

15.11 CONCENTRATION

Many experiments in chemistry are done in solutions. In most of these experiments, it is necessary to know how much solute the solution contains. The amount of solute dissolved in a given amount of solvent or in a given amount of finished solution is called the **concentration** of that solute in the solution (or often just the concentration). Depending on the units used for the amount of solute and solvent or solution, concentration can be expressed in various ways.

15.11.1 Weight Percent of Solute

The concentration of solute in a solution is often given by stating *the relative masses of the solute and the entire solution*. Because mass is usually measured by weighing, such a concentration designation is ordinarily called the **weight fraction** or the **weight percent** of solute. Weight fraction and weight percent are primarily used where the chemistry of the solution is not under study and it is not important to know how many moles of solute are present in a sample of solution.

As an example of how weight percent works, we might consider a solution in which we would dissolve 1 g of sugar in 3 g of water, to make 4 g of solution. Then one fourth of the total mass of that solution would be sugar. The weight fraction of sugar would be ¼, and the weight percent of sugar 25%. To calculate weight fraction, divide the weight of the solute by the weight of the solution. To calculate weight percent, multiply the weight fraction by 100%.

EXAMPLE 15.2.   A solution is made by dissolving 33 g of glucose in 66 g of water. What is the weight fraction of glucose in the solution? The weight percent?

Solution:

Step 1.   Find the total weight of the solution by adding the weights of the solute and the solvent.

   33 g glucose + 66 g water = 99 g solution

Step 2.   Divide the weight of the solute by the total weight of the solution.

$$\frac{33 \text{ g glucose}}{99 \text{ g solution}} = \frac{1}{3} \text{ glucose}$$

Step 3. If weight percent is required, multiply the weight fraction by 100%.

$$\frac{1}{3} \text{ glucose} \times 100\% = 33.3\% \text{ glucose}$$

EXAMPLE 15.3.   A solution is prepared by dissolving 12.4 g of ethylene glycol in 78.0 g of water. What is the weight percent of ethylene glycol in this solution?

Solution:

Step 1.   Find the total weight.

12.4 g ethylene glycol + 78.0 g water = 90.4 g solution

Step 2.   Find the weight fraction of solute.

$$\frac{12.4 \text{ g ethylene glycol}}{90.4 \text{ g solution}} = 0.137 \text{ ethylene glycol}$$

Step 3.   Find the weight percent.

0.137 ethylene glycol × 100% = 13.7% ethylene glycol

EXERCISE 15.3.   What is the weight percent of $NaHPO_4$ in a solution prepared by dissolving 21.3 g of $NaHPO_4$ in 450 g of H_2O?

EXAMPLE 15.4.   Sodium carbonate is sometimes obtained as the decahydrate $Na_2CO_3 \cdot 10 \ H_2O$. How many grams of $Na_2CO_3 \cdot 10 \ H_2O$ must be used to prepared 250 g of a 10% solution of Na_2CO_3?

Solution:

Step 1.   Find the weight of Na_2CO_3 needed.

250 g solution × 10% Na_2CO_3 = 25.0 g Na_2CO_3

Step 2.   Find the weight of $Na_2CO_3 \cdot 10 \ H_2O$ needed.

$$25.0 \text{ g } Na_2CO_3 \times \frac{286 \text{ g } Na_2CO_3 \cdot 10 \ H_2O}{106 \text{ g } Na_2CO_3} = 67.5 \text{ g } Na_2CO_3 \cdot 10 \ H_2O$$

EXERCISE 15.4.   How many grams of $SnCl_2 \cdot 2 \ H_2O$ are needed for 75 g of an 8.0% solution of $SnCl_2$?

15.11.2 Molarity

To describe concentration, most chemists and others who run chemical reactions in solution use the relation between the number of moles of solute and the number of liters of solution: the number of moles per liter. When the concentration of a solution is expressed in moles per liter, it is called **molarity.** Molarity allows us to convert numbers of moles of reacting substance to laboratory volume units. Molarity is so widely used that it has its own symbol and abbreviation: the molarity of a substance in a solution is designated by a numerical value following a pair of brackets surrounding the symbol or name for an element or a compound. The capital letter M means molar.

[A] designates the molarity of A

The adjective that describes molarity is *molar;* for example, if a solution has a molarity of 1 mol/L, it is said to be 1 molar. Chemists often say that a solution has a molarity *in* a species, meaning that the molarity of that species is the stated value. The statement, "The solution is 3.0 M in OH^-" means that $[OH^-] = 3.0$.

EXAMPLE 15.5.   Interpret the two equations that follow:

$$[C_2H_6O] = \frac{2.3 \text{ mol } C_2H_6O}{L \text{ solution}} = 2.3 \text{ M}$$

$$[Na^+] = \frac{0.065 \text{ mol } Na^+}{L \text{ solution}} = 0.065 \text{ M}$$

Solution:

The concentration of C_2H_6O in the solution is 2.3 mol/L of solution, or 2.3 M.

The concentration of sodium ion is 0.065 M.

Although molarity is used for solutions with all kinds of solvents, it ordinarily refers to aqueous solutions. If no solvent is designated in a concentration term, it is usually safe to assume that the solvent is water. Because the symbol [ ] has its own designated set of units, the units are sometimes not explicitly stated when the symbol is used, as in the concentration of Cl^-: $[Cl^-] = 0.10$.

Be careful about two things when you use molar concentrations: first, be sure to remember that the volume refers to the number of liters of *solution,* not solvent. If you dissolve 2.3 mol of C_2H_6O in 1 L of water, you will finish with more than 1 L of solution, and the molarity will not be 2.3. To make a solution that is 2.3 molar in C_2H_6O, you would place 2.3 mol of C_2H_6O in a container and add water until the volume of the solution reached 1 L. (See Figure 15.6.) Second, notice that the molar designation is used to refer to substances that are actually present in the solution, as well as substances with which the solution is made. Carelessness about this point can lead to considerable confusion. For example, if you were to place 1 mol of HCl in a container and add enough water to make 1 L of solution, the solution would not actually be 1 M in HCl, because all the HCl would have reacted with the water to make H_3O^+ and Cl^-. The solution would be 1 M in each of these ions, but not in HCl. In common usage, it

Empty volumetric flask | Same flask with 2.3 mol C_2H_6O | To complete preparation of solution, enough water is added to make a total of exactly 1 L

Figure 15.6 How to use a volumetric flask.

is acceptable to say that a solution is 1 M in HCl, meaning that 1 mol of HCl was used to prepare 1 L of solution. For the same solution, it would also be correct to say that it was 1 M in H_3O^+ or in Cl^-. This distinction may seem fussy, but when we discuss chemical equilibrium in the next chapter, we will use molarity to refer only to substances actually present in solution.

To find the molarity of any given solution, simply divide the number of moles of solute by the total number of liters of solution. If the number of moles of solute is not known, it must be calculated.

EXAMPLE 15.6.   A solution is made by dissolving 0.88 mol of C_2H_6O in enough water to make a total of 5.1 L. What is the molarity of C_2H_6O in the solution?

Solution:

Divide the number of moles of solute by the number of liters of solution.

$$\frac{0.88 \text{ mol } C_2H_6O}{5.1 \text{ L}} = 0.17 \text{ M}$$

EXAMPLE 15.7.   In preparing a KI solution, 9.1 g of KI was dissolved in water, making 225 mL of solution. What was $[K^+]$? (The question assumes that you remember that KI (s) $\longrightarrow K^+ + I^-$. KI is completely dissociated; 1 mol of KI yields 1 mol of K^+.)

Solution:

The unit conversion method will help here.

Solution map: grams of KI $\longrightarrow$ moles of KI $\longrightarrow$ molarity of K^+

Step 1.   Calculate the number of moles of KI.

$$9.1 \text{ g KI} \times \frac{1 \text{ mol KI}}{166 \text{ g KI}} = 5.5 \times 10^{-2} \text{ mol } K^+$$

Step 2. Divide the number of moles of K^+ by the solution volume. Be sure to put in a conversion to make the volume come out in liters. It was given in milliliters.

$$\frac{5.5 \times 10^{-2} \text{ mol } K^+}{225 \text{ mL}} \times \frac{1000 \text{ mL}}{L} = 0.24 \text{ mol } K^+/L$$

Here is the entire chain of unit conversions in one equation.

$$9.1 \text{ g KI} \times \frac{1 \text{ mol } K^+}{166 \text{ g KI}} \times \frac{1}{225 \text{ mL}} \times \frac{1000 \text{ mL}}{L} = 0.24 \text{ mol } K^+/L$$

EXERCISE 15.5.   Calculate the molarity of a solution prepared by dissolving 40.3 g of $Fe(ClO_4)_2$ in enough water to make 750 mL of solution.

When you make calculations involving solutions, use the conversions between grams and moles in the ordinary way.

EXAMPLE 15.8.   A solution is 0.155 M in sucrose, $C_{12}H_{22}O_{11}$. (Sucrose is not ionized.) How many grams of sucrose are there in 45 mL of this solution?

Solution:

Solution map: milliliters of solution $\longrightarrow$ liters of solution
$\longrightarrow$ moles of sucrose $\longrightarrow$ grams of sucrose

Here is the problem worked out in a single unit conversion. The abbreviation *su* is used for sucrose.

$$45 \text{ mL solution} \times \frac{1 \text{ L}}{1000 \text{ mL}} \times \frac{0.155 \text{ mol su}}{L \text{ solution}} \times \frac{331 \text{ g su}}{\text{mol su}} = 2.3 \text{ g su}$$

EXAMPLE 15.9.   A solution of Na_2SO_4 is 0.85 M. How many grams of Na^+ are contained in a 350-mL sample of this solution? (Assume that the Na_2SO_4 dissociates entirely.)

Solution:

Solution map: milliliters of solution $\longrightarrow$ liters of solution
$\longrightarrow$ moles of Na_2SO_4 $\longrightarrow$ moles of Na^+ $\longrightarrow$ grams of Na^+

$$350 \text{ mL solution} \times \frac{1 \text{ L solution}}{1000 \text{ mL}} \times \frac{0.85 \text{ mol } Na_2SO_4}{L \text{ solution}}$$

$$\times \frac{2 \text{ mol } Na^+}{\text{mol } Na_2SO_4} \times \frac{23 \text{ g } Na^+}{\text{mol } Na^+} = 13.7 \text{ g } Na^+$$

EXERCISE 15.6.   Find the number of grams of $CoCl_3$ in 137 mL of solution that is 0.25 M in Cl^-.

Notice that the value of the molarity of species actually in solution may differ from the molarity based on the way the solution was prepared.

EXAMPLE 15.10.   What is the molarity of H_3O^+ in a solution that was prepared by mixing 1.0 mol of H_2SO_4 with enough water to make 1.0 L of solution?

Solution:

Step 1.   Write the equation for any reaction or dissociation that occurs when the solution is made.

We make the approximation that in this solution, the H_2SO_4 will dissociate fully in both steps.

$$H_2SO_4 + 2\ H_2O \longrightarrow 2\ H_3O^+ + SO_4^{2-}$$

Step 2.   Calculate the required molarity.

$$\frac{1.0\ \text{mol}\ H_2SO_4}{1.0\ \text{L solution}} \times \frac{2\ \text{mol}\ H_3O^+}{\text{mol}\ H_2SO_4} = \frac{2.0\ \text{mol}\ H_3O^+}{\text{L solution}}$$

or $[H_3O^+] = 2.0$

EXERCISE 15.7.   What is the molarity of Na^+ in 2.0 M Na_3PO_4, in which the Na^+ is totally dissociated?

EXERCISE 15.8.   A solution was made by dissolving 119.3 g of beryllium phosphate, $Be_3(PO_4)_2$, in enough water to make 1.73 L of solution. In this solution, what were the values of $[Be^{2+}]$ and $[PO_4^{3-}]$? (Assume that complete dissociation occurred and that the dissolved species were Be^{2+} and PO_4^{3-}.)

15.12 DILUTION

Adding solvent to a solution dilutes that solution; that is, it reduces the concentration of the solute in the resulting solution. Some of the operations of chemistry involve adding diluents to solutions, or mixing solutions containing different solutes. (Mixing two such solutions dilutes both.) The concentration of diluted solutions is not hard to calculate.

To make our calculations more convenient, let us label our solutions so that a subscript 1 designates the original solution and a subscript 2 designates the diluted solution. For example, the volume of the original solution would be V_1, and the molarity of the diluted solution would be M_2. The equations relating the concentrations of the two solutions can be written as follows:

$$M_1V_1 = \text{number of moles in solution 1}$$

$$M_2V_2 = \text{number of moles in solution 2}$$

Remember that *the number of moles in a solution is not changed by dilution;* that is, the new solution will contain the same amount of solute as did the original solution:

number of moles in solution 1 = number of moles in solution 2

Therefore,

$$M_1 V_1 = M_2 V_2$$

We can rearrange this equation to give us the molarity of the new solution after dilution.

$$M_2 = M_1 \frac{V_1}{V_2}$$

This little equation, which looks something like the equations we used for changes in the conditions of gases, says that we can find the molarity of a diluted solution by multiplying the original molarity by a ratio of the volumes. We know which volume goes in the numerator because we know that the new solution will not be as concentrated as the original; we put the smaller volume on top. (See Figure 15.7.)

EXAMPLE 15.11.   A sample containing 88 mL of solution that is 0.228 M in $BaCl_2$ is diluted to 134 mL. What is the molarity of the new solution?

Solution:

$$\text{new molarity} = M_2 = 0.228 \text{ M} \times \frac{88 \text{ mL}}{134 \text{ mL}} = 0.150 \text{ M}$$

EXAMPLE 15.12.   To 500 mL of a solution that is 1.33 M in C_2H_6O is added 200 mL of water. What is the molarity of the new solution?

Solution:

Because we were not told otherwise, we assume that the solvent of the original solution was water and we assume that adding 200 mL of water to the original sample made its new volume 700 mL.

$$M_2 = 1.33 \text{ M} \times \frac{500 \text{ mL}}{700 \text{ mL}} = 0.95 \text{ M}$$

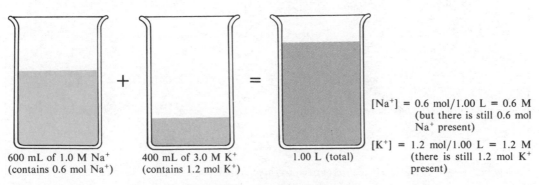

600 mL of 1.0 M Na^+
(contains 0.6 mol Na^+)

400 mL of 3.0 M K^+
(contains 1.2 mol K^+)

1.00 L (total)

$[Na^+]$ = 0.6 mol/1.00 L = 0.6 M
(but there is still 0.6 mol Na^+ present)

$[K^+]$ = 1.2 mol/1.00 L = 1.2 M
(there is still 1.2 mol K^+ present)

Figure 15.7   Example of dilution.

EXERCISE 15.9.   What is the final molarity of a solution created by adding enough water to 43 mL of 2.00 M $Mn(SO_3)_2$ to make 155 mL of solution?

To calculate the effect of diluting a solution whose molarity is known, we assumed that the number of moles of solute is not changed by dilution. In one sense, this is correct. Adding water to a solution cannot change the amount of solute with which the solution was originally made. In some quite important cases, however, adding more water to a solution *can change the way in which the solute dissociates.* In such cases, a simple dilution calculation will not be satisfactory. If the solute is a strong electrolyte, then the dilution calculation will work for the ions, as well as for the total amount of solute. If the solute is a weak electrolyte, however, you must consider the degree of its dissociation. We will learn about solutions of weak electrolytes in Chapter 16.

15.13 CALCULATIONS FROM REACTIONS IN SOLUTION

Many important reactions begin with ions or other species in solution. To make stoichiometric calculations for those reactions, it is often necessary to start with solution concentrations. Using the language of molarity usually simplifies the problem because the given volumes and concentrations can so easily be translated into numbers of moles.

EXAMPLE 15.13.   Solid NaCl will react with Ag^+ to produce insoluble AgCl. How many grams of NaCl are required to react exactly with 250 mL of 0.0950 M $AgNO_3$ solution?

$$Cl^- + Ag^+ \longrightarrow AgCl \ (s)$$

Solution:

Solution map: milliliters of solution $\longrightarrow$ liters of solution
$\longrightarrow$ moles of Ag^+ $\longrightarrow$ moles of NaCl $\longrightarrow$ grams of NaCl

Step 1.   Find the number of moles of Ag^+.

$$250 \text{ mL solution} \times \frac{1 \text{ L solution}}{1000 \text{ mL solution}} \times \frac{0.095 \text{ mol } Ag^+}{\text{L solution}} = 2.375 \times 10^{-2} \text{ mol } Ag^+$$

Step 2.   Find the number of moles of NaCl. From the reaction equation and the formula for NaCl, we know that 1 mol of NaCl reacts with 1 mol of Ag^+.

$$(2.375 \times 10^{-2}) \text{ mol } Ag^+ \times \frac{1 \text{ mol NaCl}}{1 \text{ mol } Ag^+} = 2.375 \times 10^{-2} \text{ mol NaCl}$$

Step 3.   Use the molecular weight of NaCl to find the number of grams of NaCl.

$$(2.375 \times 10^{-2}) \text{ mol NaCl} \times \frac{58.45 \text{ g NaCl}}{\text{mol NaCl}}$$

$$= 1.39 \text{ g NaCl} \qquad \text{(rounded to three significant figures)}$$

EXAMPLE 15.14.   Metallic zinc reacts readily with strong acids.

$$Zn + 2 H_3O^- \longrightarrow Zn^{2+} + H_2 + 2 H_2O$$

How many liters of 0.33 M H_3O^+ solution are required to react exactly with a sample containing 4.4 g of Zn metal?

Solution:

Solution map: grams of Zn $\longrightarrow$ moles of Zn
$\longrightarrow$ moles of H_3O^+ $\longrightarrow$ liters of solution

Step 1.   Find the number of moles of Zn in the sample.

$$4.4 \text{ g Zn} \times \frac{1 \text{ mol Zn}}{65.4 \text{ g Zn}} = 6.73 \times 10^{-2} \text{ mol Zn}$$

Step 2.   Find the number of moles of H_3O^+ required.

$$(6.72 \times 10^{-2}) \text{ mol Zn} \times \frac{2 \text{ mol } H_3O^+}{\text{mol Zn}} = 1.34 \times 10^{-1} \text{ mol } H_3O^+$$

Step 3.   Find the number of liters of solution needed. Notice that we must invert the molarity to do this step.

$$(1.34 \times 10^{-1}) \text{ mol } H_3O^+ \times \frac{1 \text{ L solution}}{0.33 \text{ mol } H_3O^+} = 0.41 \text{ L}$$

Here is the entire problem done in one operation:

$$4.4 \text{ g Zn} \times \frac{1 \text{ mol Zn}}{65.4 \text{ g}} \times \frac{2 \text{ mol } H_3O^+}{\text{mol Zn}} \times \frac{1 \text{ L solution}}{0.33 \text{ mol } H_3O^+} = 0.41 \text{ L}$$

EXAMPLE 15.15.   How many milliliters of 1.04 M H_2SO_4 will be required to react with 12 g of metallic aluminum? Remember that H_2SO_4 is a strong acid.

$$2 Al + 6 H_3O^+ \longrightarrow 2 Al^{3+} + 3 H_2 + 6 H_2O$$

Before you look at the solution map, write down the steps required to change grams of Al to milliliters of solution.

Solution:

Solution map: grams of Al $\longrightarrow$ moles of Al $\longrightarrow$ moles of H_3O^+
$\longrightarrow$ moles of H_2SO_4 $\longrightarrow$ liters $\longrightarrow$ milliliters

Here is the calculation accomplished in one step:

$$12 \text{ g Al} \times \frac{1 \text{ mol Al}}{27 \text{ g Al}} \times \frac{6 \text{ mol } H_3O^+}{2 \text{ mol Al}} \times \frac{1 \text{ mol } H_2SO_4}{2 \text{ mol } H_3O^+}$$

$$\times \frac{1 \text{ L solution}}{1.04 \text{ mol } H_2SO_4} \times \frac{1000 \text{ mL solution}}{1 \text{ L solution}} = 6.4 \times 10^2 \text{ mL solution}$$

EXERCISE 15.10.   Using the following reaction equation, calculate how many milliliters of 3.50 M HNO_3 solution are required to react with 2.83 g of Cu.

$$3 \, Cu + 8 \, HNO_3 \longrightarrow 3 \, Cu(NO_3)_2 + 2 \, NO + 4 \, H_2O$$

EXERCISE 15.11.   A 550-mL sample of 1.55 M $Ba(OH)_2$ solution is neutralized exactly by the addition of pure H_2SO_4. How many grams of the acid are required?

First, write the reaction equation. Depending on whether you write it in ionic form, you can make your calculation in either of two ways. Here are both solution maps:

milliliters of solution $\longrightarrow$ liters of solution $\longrightarrow$ moles of $Ba(OH)_2$
$\longrightarrow$ moles of OH^- $\longrightarrow$ moles of H_3O^+ $\longrightarrow$ moles of H_2SO_4 $\longrightarrow$ grams of H_2SO_4

or

milliliters of solution $\longrightarrow$ liters of solution $\longrightarrow$ moles of $Ba(OH)_2$
$\longrightarrow$ moles of H_2SO_4 $\longrightarrow$ grams of H_2SO_4

Now set up and do the calculation.

15.14 TITRATION

Titration, one of the standard methods of analysis in laboratory chemistry, is used to find the concentration of a given solution (the **unknown**) by adding to it precisely the right amount of another solution (the **known**) to react with the unknown exactly. The method is best explained by an example.

EXAMPLE 15.16.   To a sample of 125 mL of NaOH solution of unknown concentration was added 88.1 mL of 0.117 M HCl solution, which reacted exactly with all the OH^- in the NaOH solution. What was the concentration of the NaOH solution?

$$H_3O^+ + OH^- \longrightarrow H_2O$$

Solution:

We have the reaction equation, and we know that there is 1 mol of H_3O^+ in every mole of HCl and 1 mol of OH^- in every mole of NaOH. We can therefore say that 1 mol of HCl reacts with 1 mol of NaOH. Knowing the concentration and volume of the HCl solution, we can calculate the number of moles of HCl, which must be the same as the number of moles of NaOH. From the number of moles of NaOH and the volume of the NaOH solution, we can calculate the concentration of the NaOH solution.

Solution map: milliliters of HCl $\longrightarrow$ liters of HCl $\longrightarrow$ moles of HCl
$\longrightarrow$ moles of NaOH $\longrightarrow$ molarity of NaOH

For clarity, we will do the calculation in two steps.

Step 1. $88.1 \text{ mL HCl solution} \times \dfrac{1 \text{ L}}{1000 \text{ mL}} \times \dfrac{0.117 \text{ mol HCl}}{\text{L HCl solution}}$

$\times \dfrac{1 \text{ mol NaOH}}{1 \text{ mol HCl}} = 1.03 \times 10^{-2} \text{ mol NaOH}$

Step 2. $\dfrac{1.03 \times 10^{-2} \text{ mol NaOH}}{125 \text{ mL NaOH solution}} \times \dfrac{1000 \text{ mL solution}}{\text{L solution}} = 0.0824 \text{ mol NaOH/L}$

Notice that the unit conversion from milliliters to liters in Example 15.16 cancels out. If you are careful to work in milliliters both at the beginning and the end of the problem, you can leave out the unit conversions from milliliters to liters. In such a case, instead of having moles of HCl and NaOH in the middle of your solution map, you would have millimoles (each millimole being 1/1000 of a mole). As long as you are careful to match up the units, you need not make this distinction between milli-units and units.

EXAMPLE 15.17. How many milliliters of 0.227 M H_2SO_4 solution are required to titrate 25.0 mL of 0.550 M KOH?

For this problem, remember that 1 mol of H_2SO_4 yields 2 mol of H_3O^+.

Solution map: milliliters of KOH $\longrightarrow$ millimoles of KOH $\longrightarrow$ millimoles of H_2SO_4 $\longrightarrow$ milliliters of H_2SO_4

Solution:

$25.0 \text{ mL KOH solution} \times \dfrac{0.550 \text{ millimol KOH}}{\text{mL KOH solution}} \times \dfrac{\text{millimol } H_2SO_4}{2 \text{ millimol KOH}}$

$\times \dfrac{\text{mL } H_2SO_4 \text{ solution}}{0.227 \text{ millimol } H_2SO_4} = 30.3 \text{ mL } H_2SO_4 \text{ solution}$

The main difference between this problem and example 15.16 is in the proportions of the reactant: 1 mol of H_2SO_4 is equivalent to 2 mol of KOH. Notice that we made the calculation directly in millimoles.

Figure 15.8 shows a titration operation.

EXERCISE 15.12. How many milliliters of 0.500 M $HClO_4$ are required to titrate 500 mL of a solution of $Ba(OH)_2$ that is 1.00 M?

To titrate a solution, we first arrange a way to determine exactly when the unknown has been completely used up. A common method is to add an indicator dye to the solution; the dye changes color as soon as any excess of the known solution is present. In a measured way, the known solution is carefully added to the unknown until the reaction is finished. The volume of known solution used is then recorded and any necessary calculations made.

15.15 IONIZATION OF WATER

There is a limit to the strength of chemical bonds, even that of the strongest covalent ones. At any given moment, in any collection of many molecules, a mole, for example, at least a few molecules will have been knocked apart by the motion of the molecules around them. Covalent bonds can be broken in such a way that the fragments are ions (that is, that one member of the pair keeps both electrons from the shared pair), or the bond can be broken evenly, so that the shared electron pair is split. If the molecules that have been broken are

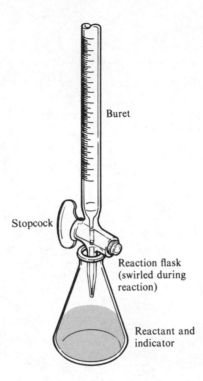

Buret

Stopcock

Reaction flask
(swirled during
reaction)

Reactant and
indicator

Figure 15.8 Set up for titration. The solution in the buret is allowed to drip slowly into the flask until the indicator changes color. The buret readings before and after give the volume required for the completed reaction..

simple, the bonds may rejoin and the molecule be restored. If complicated molecules are broken, they may stay broken because the fragments are unable to reorganize. The slow deterioration of substances over time often represents the breaking of covalent bonds that remain unrepaired.

In any sample of water, some of the molecules are fragmented, or dissociated into ions.

$$H_2O \longrightarrow H^+ + OH^-$$

Since the bare proton at once joins another water molecule temporarily, we could, if we wished, write the dissociation of water as a reaction between two water molecules.

$$2\ H_2O \longrightarrow H_3O^+ + OH^-$$

A hydroxide ion, however, has a very strong attraction for the positively charged proton, so the reaction is reversible, and the water molecules are quickly restored. Nevertheless, at any particular instant, in any sample of liquid water, there are some dissociated molecules. When water dissociates, one OH^- ion is made for every H_3O^+ ion; the concentrations of the two ions in a sample of pure water are therefore equal: $[H_3O^+] = [OH^-]$. (The same cannot usually be said of solutions, however.) The amount of dissociation is small: only 10^{-7} mol/L of water is dissociated at any time in a sample of pure water at room temperature. For pure water at 25 °C

$$[H_3O^+] = [OH^-] = 1 \times 10^{-7} \text{ M}$$

One ten millionth of a mole per liter seems like a small amount to be concerned with, but the dissociation of water plays an important part in many chemical and physical processes. We will encounter it again later.

15.16 THE pH SCALE

In laboratory reactions, industrial processes, living creatures, and countless other systems in which solution chemistry is important, the level of solution acidity is often crucial. Acidity in water solutions, which is determined by the concentration of H_3O^+, is so often measured and reported that a special language, called the **pH scale,** has been developed to describe it. The concentration of H_3O^+ can vary widely in solution, from several moles per liter to almost nothing.

The value of the pH of a solution is defined as the negative logarithm (base 10) of the H_3O^+ concentration. (Remember that the logarithm of a number is the power to which 10 must be raised to give that number. Thus, the logarithm of 10^4, for example, is 4, because that is the power to which we must raise 10 to get 10^4.) If the H_3O^+ value is an integer power of 10, it is easy to find the pH. For example, if $[H_3O^+]$ is 1×10^{-3}, or simply 10^{-3}, we take the log of 10^{-3}, which is -3, and change its sign to get the negative log. This gives us a pH of 3.

EXAMPLE 15.18. The concentration of H_3O^+ in a solution is 10^{-8} M. What is the pH of that solution?

Solution:

Step 1. Take the logarithm of the H_3O^+ concentration.

The log of 10^{-8} is -8.

Step 2. Change the sign of the log, which gives the pH.

The pH of the solution is $-(-8)$, or 8.

The pH scale is most often used for $[H_3O^+]$ values between 1×10^{-1} M and 1×10^{-14} M; that is, between pH values of 1 and 14. Because pure water has a $[H_3O^+]$ of 1×10^{-7} M and a $[OH^-]$ of 10^{-7} M, it is therefore neutral (neither acidic nor basic), and has a pH of 7. A pH value of less than 7 means that $[H_3O^+]$ is greater than 10^{-7} and also greater than $[OH^-]$, the solution is therefore acidic. A pH value greater than 7 means that $[H_3O^+]$ is less than 10^{-7} and less than $[OH^-]$. Such a solution is basic. As the value of the pH becomes smaller than 7, it indicates an increase in acidity. A pH value of 1 or less indicates that the solution is very acidic. Go over this until you are sure that you understand why smaller pH values mean larger acidities. (It's caused by the negative sign in the definition of pH.) Table 15.3 is a tabular version of the pH scale.

Table 15.3
Tabular Version of the pH Scale

$[H_3O^+]$	pH	$[OH^-]$	$[H_3O^+]$	pH	$[OH^-]$
10^0	0	10^{-14}	10^{-4}	4	10^{-10}
10^{-1}	1	10^{-13}	7.9×10^{-5}	4.1	1.3×10^{-10}
10^{-2}	2	10^{-12}	6.3×10^{-5}	4.2	1.6×10^{-10}
10^{-3}	3	10^{-11}	5.0×10^{-5}	4.3	2.0×10^{-10}
10^{-4}	4	10^{-10}	4.0×10^{-5}	4.4	2.5×10^{-10}
10^{-5}	5	10^{-9}	3.2×10^{-5}	4.5	3.2×10^{-10}
10^{-6}	6	10^{-8}	2.5×10^{-5}	4.6	4.0×10^{-10}
10^{-7}	7	10^{-7}	2.0×10^{-5}	4.7	5.0×10^{-10}
10^{-8}	8	10^{-6}	1.6×10^{-5}	4.8	6.3×10^{-10}
10^{-9}	9	10^{-5}	1.3×10^{-5}	4.9	7.7×10^{-10}
10^{-10}	10	10^{-4}	10^{-5}	5	10^{-9}
10^{-11}	11	10^{-3}			
10^{-12}	12	10^{-2}			
10^{-13}	13	10^{-1}			
10^{-14}	14	10^{-0}			

Note: as $[H_3O^+]$ increases (reading *up* on the left side), pH decreases (middle scale), and $[OH^-]$ similarly decreases. Low pH means large acidity. High pH means high basicity. The right-hand portion of the table is a more detailed version of the part of the scale between pH 4 and 5. It shows concentration values for fractional pH's.

Notice that it is possible to have negative pH values, although these are seldom used. If, for example, $[H_3O^+] = 10$ M ($= 10^1$ M), then the pH of the solution is the negative log of 10^1, or -1.

As the pH value changes from 7 to larger values, it indicates a growing value of the basicity of the solution. A pH value of 14 or greater indicates a very strongly basic solution.

If the value of the $[H_3O^+]$ concentration of a solution is not an integer power of 10, the value of the pH will not be an integer number. For example, if $[H_3O^+] = 4.5 \times 10^{-3}$, the value of the pH will be the log of that number. Because $4.5 \times 10^{-3} = 10^{-2.35}$, the pH of the solution would be 2.35.

A three-step sequence can be used to find noninteger pH values.

EXAMPLE 15.19.   A solution has $[H_3O^+]$ of 7.0×10^{-4}. What is its pH?

Solution:

Step 1.   Find $[H_3O^+]$ as a power of 10. To do this, take the log of the number by which the 10 is multiplied.

$[H_3O^+] = 7.0 \times 10^{-4}$ mol/L   The log of 7.0 is 0.85.

Step 2.   Add that log to the exponent of the 10. Watch the sign. (The exponent is an integer with as many significant figures as you need.)

$-4.00 + 0.85 = -3.15$

$[H_3O^+] = 10^{-3.15}$

Now take the log of $[H_3O^+]$: $\log [H_3O^+] = \log 10^{-3.15} = -3.15$

Step 3. Change the sign of the log, which gives the pH.

$$-(-3.15) = 3.15$$

The pH of the solution is 3.15.

EXERCISE 15.13.   A solution has $[H_3O^+] = 4.0 \times 10^{-2}$. What is its pH?

If $[OH^-]$ is given, it is first necessary to convert that to $[H_3O^+]$. To do this, you need to know that $[H_3O^+] [OH^-] = 10^{-14}$ M^2; we will learn the details of this relation in the next chapter.

EXAMPLE 15.20.   A solution of NaOH has $[OH^-]$ of 10^{-3} mol/L. What is its pH?

Solution:

Step 1.   Find $[H_3O^+]$.

$$[H_3O^+] = \frac{10^{-14}}{[OH^-]} = \frac{10^{-14}}{10^{-3}} = 10^{-11} \text{ mol/L}$$

Step 2.   Find pH in the usual way.

$$pH = -\log [H_3O^+] = -\log 10^{-11} = 11$$

EXAMPLE 15.21.   In a solution of $Ba(OH)_2$, $[OH^-] = 4.1 \times 10^{-4}$ mol/L. What is the pH of the solution?

Solution:

Step 1. $[H_3O^+] = \dfrac{10^{-14}}{4.1 \times 10^{-4}} = 2.4 \times 10^{-9}$ mol/L

Step 2. $\log 2.4 \times 10^{-9} = -9.0 + 0.38 = -8.62$

Step 3.   pH = 8.62

EXERCISE 15.14.   Find the pH of a 3.00 M KOH solution.

EXERCISE 15.15.   What is the pH of a solution with $[OH^-]$ equal to 3.3×10^{-9} M?

It is frequently necessary to know the value of $[H_3O^+]$ or $[OH^-]$ when only the pH is given. The process of finding concentration from pH is the reverse of what we have just done.

EXAMPLE 15.22.   The pH of a solution is 5.2. Is the solution acidic or basic? What is the concentration of H_3O^+ in the solution?

Solution:

Since the value of the pH is less than 7, the solution is acidic. The calculation of $[H_3O^+]$ can be done in two steps.

Step 1.   Convert the pH to concentration as a power of 10.

Because pH is the negative log of $[H_3O^+]$ and because a log is a power to which we raise 10, we first change the sign of the pH value, then use it as an exponent of 10.

Change of sign: -5.20
Use as exponent: $10^{-5.20}$

$[H_3O^+]$, then, is equal to $10^{-5.20}$.

Step 2.   Rewrite the $[H_3O^+]$ value so that it does not have a fractional exponent.

$$10^{-5.20} = (10^{0.80})\,(10^{-6})$$

Notice that we write the term with the fractional exponent as a product of 10 raised to a positive, fractional exponent and 10 raised to a negative integer. We can then rewrite the first term as an ordinary number.

$$10^{0.80} = 6.3$$

$$[H_3O^+] = 6.3 \times 10^{-6} \text{ mol/L}$$

EXAMPLE 15.23.   What is the hydroxide ion concentration of a solution that has a pH of 11.71?

Solution:

Remembering that $[H_3O^+]\,[OH^-] = 10^{-14}$, proceed as you did earlier to find $[H_3O^+]$, then find $[OH^-]$.

Step 1. $10^{-11.71} = 10^{0.29} \times 10^{-12}$

Step 2. $10^{0.29} = 1.9$

$$[H_3O^+] = 1.9 \times 10^{-12} \text{ mol/L}$$

Step 3. $[OH^-] = \dfrac{10^{-14}}{[H_3O^+]} = \dfrac{10^{-14}}{1.9 \times 10^{-12}} = 5.1 \times 10^{-3} \text{ mol/L}$

EXERCISE 15.16.   What is $[H_3O^+]$ in a solution that has a pH of 9.5?

EXERCISE 15.17.   What is $[OH^-]$ in a solution that has a pH of 12.84?

It is often important to know the acidity level of aqueous solutions. There are several good ways to determine directly the pH of a solution. Electronic instruments rely on electrodes dipped into a solution to give immediate, accurate readings of pH levels. (See Figure 15.9.) Paper strips impregnated with mixtures of dye will give instantaneous pH approximations when they are wet with a solution; these are improved versions of the old and honored litmus test, in which the litmus paper turned pink to indicate acidity and blue to indicate basicity.

The level of acidity in solution is crucial in many kinds of chemical reactions, especially those of biological systems. The pH of human blood, for example, is 7.4, corresponding to a $[H_3O^+]$ of 4×10^{-8} M. Deviation from 7.4 by even one tenth of a pH unit can cause serious consequences, and deviations of more than two tenths of a pH unit can cause death. Plants can grow only in soils in which the pH of the moisture lies close to an optimum value. Farmers and horticulturists must therefore regulate soil pH. We will encounter pH and the regulation of acidity again and again as we continue.

15.17 INDUSTRIAL PREPARATION OF H_2SO_4 and NaOH

Our study so far has been focussed on the principles of chemistry, although we have had some exposure to practical chemistry along the way. There is a third aspect of chemistry, however, which is most important, and that is industrial chemistry.

The present human population (about 4.5 billion) can survive only if the supply of necessary food and other materials continues to be produced at a high rate. We no longer have the choice of going back to the agrarian society of the nineteenth century. Like it or not, most of the world is industrialized and must remain so unless the population is drastically reduced.

Figure 15.9   Instrument for measuring pH. (Photo courtesy of Pope Scientific, Inc., Menomonee Falls, Wisc.)

Industrial societies depend on a continuous supply of enormous amounts of fertilizers, fuels, medicines, fabrics, and innumerable other materials. The production of these depends, in turn, on the supply of what are called heavy chemicals (heavy in terms of the total amount produced, not in density). To learn a little about how chemistry helps maintain our society, we will study the ways in which two heavy chemicals, sulfuric acid and sodium hydroxide, are produced.

15.17.1 Sulfuric Acid

Sulfuric acid is a colorless, viscous, extremely corrosive liquid with a density of 1.8 g/cm^3. The total mass of H_2SO_4 produced in the United States every year is about twice the mass of all the men, women, and children in the country, but almost none of the acid reaches the consumer market because it is used in manufacturing processes.

Sulfuric acid is used primarily as a cheap, concentrated source of hydrógen ions. Somewhat more than a third of the available H_2SO_4 is used to make fertilizers. An example is the preparation of the moderately soluble calcium dihydrogen phosphate, $Ca(H_2PO_4)_2$, from the natural mineral $Ca_3(PO_4)_2$. Because it is nearly insoluble, $Ca_3(PO_4)_2$ is virtually useless to plants.

$$Ca_3(PO_4)_2 + 2\ H_2SO_4 \longrightarrow Ca(H_2PO_4)_2 + 2\ CaSO_4$$

H_2SO_4 is used also in the manufacture of other chemicals, as well as in the production of steel, petroleum, nonferrous metals, paint, and plastics.

Except for one stage, the manufacture of H_2SO_4 is straightforward. Sulfur is burned (or metal sulfides are roasted in air) to form sulfur dioxide, a reaction that occurs readily.

$$S + O_2 \longrightarrow SO_2\ (g) + heat$$

The next step is more difficult, because the reaction does not go quickly. Pure oxygen, high pressure, and a catalyst make it practical. (See Sections 16.4 and 16.5.)

$$2\ SO_2 + O_2 \longrightarrow 2\ SO_3$$

The SO_3 can then be made to react with water to give H_2SO_4. Figure 15.10 is a diagram of the process.

$$H_2O + SO_3 \longrightarrow H_2SO_4 + heat$$

15.17.2 Sodium Hydroxide

Sodium hydroxide, NaOH, is the most commonly used soluble strong base; its common name is lye. It is prepared in large quantities by the electrolysis (reaction brought about by electric current) of brine, NaCl solution (Figure 15.11).

$$NaCl + 2\ H_2O \xrightarrow{\textit{Electric power}} 2\ Na^+ + 2\ OH^- + Cl_2 + H_2$$

The manufacture is complicated by the fact that Cl_2 reacts in basic solutions to form ClO^- and Cl^-.

$$Cl_2 + 2\ OH^- \longrightarrow ClO^- + Cl^- + H_2O$$

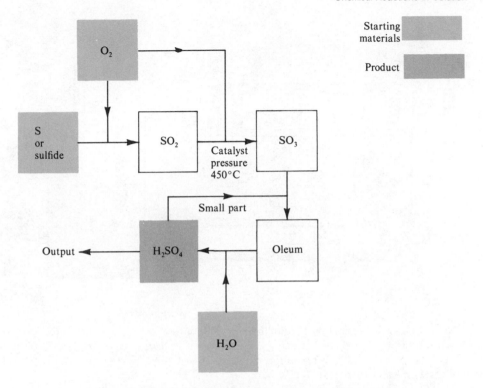

Figure 15.10   Flowchart for the synthesis of sulfuric acid.

Ingenious methods are used, however, to keep the OH⁻ and the Cl₂ separate. The electric energy is the most significant cost of production. Sodium hydroxide is used in large quantities to produce petroleum, soap, and paper. A small fraction of the mixture produced by the electrolysis is allowed to continue to react, and the NaClO is sold as bleach.

15.18 SUMMARY

Solutions provide a convenient means of running chemical reactions in a controlled and measurable way, and many reactions can be run only in solution. We can characterize any solution by naming its solvent, solute, and concentration. Among the several ways of expressing concentration, the one most convenient to chemists is the molar concentration. The molarity of a solution is determined by the number of moles of solute in a liter of solution.

The solubility of a solute at a given temperature is the maximum amount of the solute that will dissolve in a given amount of solvent at that temperature. Solubilities vary widely, and depend on the structures of both solute and solvent. Because water is the best all-around solvent, much chemistry is carried out in aqueous solution. Although ionic compounds usually dissociate when they dissolve, the solubilities of ionic compounds vary greatly.

Acids and bases are important to chemistry, to industries, and to biological systems. A rough and convenient (but not entirely general) definition of an aqueous acid solution is one

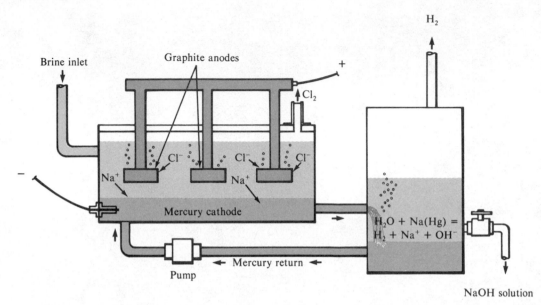

Figure 15.11 Industrial preparation of sodium hydroxide. In the electrolytical cell at the left, metallic Na forms at the surface of the cathode and goes into solution in the mercury. Cl_2 bubbles off from the anode. The mercury is circulated between the electrolytical cell and a reaction cell, at the right, where the Na reacts with water to give NaOH.

that contains more H_3O^+ than OH^-. A solution that contains more OH^- than H_3O^+ is said to be basic. If, however, the concentrations of H_3O^+ and OH^- are equal, as they are in pure water, the solution is neither acidic nor basic, but neutral. The pH scale is often used to assign values to the acidity or basicity of a solution. The pH is the negative logarithm of the H_3O^+ concentration. Acids and bases readily react, creating water and salts.

To calculate the amounts of solutes reacting in solutions, it is most convenient to use molar concentration values. Solution concentration can be measured by titration, a technique in which measured amounts of a known solution are added to another solution whose concentration is unknown.

QUESTIONS

Section 15.1

1. Explain the differences between solutions and suspensions.
2. List all the possible kinds of solutions made up of combinations of the three states of matter.

Section 15.3

3. What effect does raising the temperature usually have on solubility? What might be the reason for this effect?

4. Think of an example of a supersaturated solution that can be made in the kitchen and that results in something edible when seeded.

Section 15.4

5. Explain the process of hydration. Draw a picture of a hydrated Br^- ion.
6. How can you experimentally determine whether a compound is a strong or weak electrolyte?

Section 15.6

7. The hydrogen ion, H^+, does not exist in aqueous solution. Explain why.

Section 15.7

8. Nonchemists frequently confuse the terms *concentrated* and *strong*. Define each term.
9. Explain how a concentrated solution can be weakly acidic.
10. Explain how a solution of a strong base can be only slightly basic.

Section 15.10

11. What are the four general ways by which ions can be removed from solution?

Section 15.11

12. What are the differences between molarity and weight percent? (There are more than one.)

Section 15.14

13. Explain the technique of titration.

Section 15.15

14. Explain the process of ionization of water. In pure water, what fraction of the molecules are ionized?

Section 15.16

15. Explain the relationship between pH and the concentrations of H_3O^+ and OH^-.
16. What is the pH of pure water?

PROBLEMS

Section 15.1

1. Give two examples each of a suspension and a colloid. Give three examples of solutions that can be found in an ordinary kitchen.
2. Name the solute and the solvent in the following systems. Explain the reasons for your choices.
    a. seawater
    b. radiator water and coolant
    c. 6 g of mercury and 3 g of sodium (an amalgam)
    d. 14 karat gold (pure gold is 24 karat)
    e. bronze (copper and tin)

Section 15.2

3. Name a common polar solvent. Which of the following substances will dissolve in polar solvents: sugar, gasoline, vegetable oil, Na_2CO_3, H_2SO_4, H_2O, NH_3, CH_4?

Section 15.4

4. Which of the following compounds are strong electrolytes: sugar, acetic acid, alcohol, KI, H_2O, HCl, NH_3, $Ca(OH)_2$, AgCl, $BaSO_4$?
5. Arrange the following compounds in order of their increasing strength as electrolytes: sucrose, acetic acid, HCl, H_2O, CaS, $PbSO_4$. Identify any that are equally strong or weak.

Section 15.5

6. Which of the following ionic compounds will dissolve readily in water: NH_4NO_3, $Y(OH)_3$, $Th(CO_3)_2$, $PbSO_4$, KBr, BaS, NiS, Hg_2CO_3, $MnPO_4$?

Section 15.7

7. Label the following compounds as weak or strong acids, weak or strong bases, or none of these: acetic acid, ethyl alcohol, $Mg(OH)_2$, KOH, HF, H_2SO_4, HSO_4^-, HI, H_2O, HPO_4^{2-}.

Section 15.9

*8. Write out the neutralization equations for the following combinations of acids and bases. In which cases, if any, are there no acid-base reactions?
*a. $KOH + H_3PO_4$ e. $Ba(OH)_2 + H_3CCOOH$
*b. $HNO_3 + NaOH$ f. $HNO_3 + Mg(OH)_2$
*c. $Ca(OH)_2 + HCl$ g. $RbI + Sr(OH)_2$
d. $H_2SO_4 + KOH$ h. $CO_2 + H_2O + LiOH$

Section 15.10

*9. State whether mixing solutions of the following pairs of compounds results in a reaction. Give the products, if any.
*a. $KOH + FePO_4$ d. $NH_4NO_3 + PbCO_3$
*b. $CsCl + AgNO_3$ e. $CaS + (NH_4)_2 CO_3$
c. $BaSO_4 + CuBr_2$

Section 15.11.1

*10. What is the weight percent of a solution prepared by mixing 65 g of C_3H_6O into 110 g of ethyl alcohol?
11. What is the weight percent of a solution prepared by dissolving 21.10 g of NaOH in 55 mL of H_2O? (Remember that the density of water is 1.00 g/mL.)
12. The weight percent of a $Ca(OH)_2$ solution is 15%. How many grams of $Ca(OH)_2$ was it necessary to mix with 105 g of H_2O to make the solution?
13. A solution was made by dissolving 1.00 mol of K_2SO_4 in 95.0 g of water. What was the weight percent of the finished solution?
14. How many grams of solvent would be necessary to prepare a 30% solution containing 23 g of KNCS?

*15. It is necessary to make an aqueous solution that is 25.0% $LiNO_3$ and contains 0.71 mol of $LiNO_3$. How much water must be used to make the solution?

*16. A solution that weighed 19.1 g was evaporated to dryness; care was taken that no solute was lost. After the evaporation was complete, 3.25 g of NaCl remained. What was the weight percent of NaCl in the solution?

17. An aqueous solution has a density of 1.08 g/mL and contains 14.0% by weight of HNO_3. How many milliliters of the solution must be taken to obtain 22 g of HNO_3? To obtain 1.0 mol of HNO_3?

*18. How many grams of $NiCl_2 \cdot 6 H_2O$ must be used to prepare 150 g of a 5.0% solution of $NiCl_2$?

Section 15.11.2

*19. What is the molarity of a solution prepared with 15 g of C_5H_{12} and enough solvent to make 250 mL of solution?

*20. How many grams of $CoCl_2$ are necessary to prepare 550 mL of a 1.50 M solution of Co^{2+}?

21. How many grams of Hg_2Cl_2 are necessary to prepare 840 mL of a 1.65 M solution of Cl^-?

22. Cupric sulfate is usually sold as the pentahydrate $CuSO_4 \cdot 5 H_2O$. How many grams of this substance must be used to prepare 75 mL of 0.55 M Cu^{2+}?

23. Sulfuric acid is sold in concentrated form at 18 M. What volume of the concentrated acid must be used to prepare 325 mL of a solution that is 1.10 M in H_2SO_4?

24. What volume of concentrated sulfuric acid must be used to prepare 500 mL of a solution in which $[H_3O^+] = 2.25 \times 10^{-2}$? Assume that

$$H_2SO_4 + 2 H_2O \longrightarrow 2 H_3O^+ + SO_4^{2-}$$

25. How many moles of Ba^{2+} are contained in 175 mL of a solution that was prepared to be 0.33 M in $BaCl_2$? How many moles of Cl^- are in this solution?

*26. Gaseous HBr is needed to prepare 300 mL of 4.00 M HBr. What volume of gas will we use at 20 °C and 750 torr?

*27. What is the $[H_3O^+]$ of 2.50 L of solution containing 1.43 g of H_3PO_4? Assume that the following reaction occurs when the solution is prepared:

$$H_3PO_4 + H_2O \longrightarrow HPO_4^{2-} + 2 H_3O^+$$

28. A solution contains 19.6 g of $MnSO_4$ and 150 g of water. The density of the solution is 1.12 g/mL. What is the weight percent of $MnSO_4$ in the solution? What is the molarity of $MnSO_4$ in the solution?

29. To prepare a photographic fixing solution, 45 g of solid sodium thiosulfate ($Na_2S_2O_3$) is put into a 250-mL volumetric flask. Then 200 mL of hot water is added, and the solid is brought into solution. After the solution cools, enough water is added to fill the flask to exactly 250 mL. What is the molarity of $Na_2S_2O_3$ in the final solution?

Section 15.12

*30. A 2.5-mL sample of 1.50 M NaOH is diluted to 1.00 L. What is the molarity of the final solution?

31. A solution is 0.625 M in $MgBr_2$. A 125-mL sample of this solution is diluted to 500 mL. What is $[Br^-]$ in the final solution?

32. A 2.50 M solution of $CrBr_2$ was allowed to mix with enough solvent to make 3.60 L of solution. If we started with 750 mL of solution, what was the final molarity? What was the molarity of Cr^{2+} in that solution? Of Br^-?

33. Concentrated HCl from suppliers is usually an aqueous solution that has a density of 1.198 g/mL and contains 40% HCl by weight. What volume of this HCl must be used to prepare 1.00 L of a 0.200 M solution?

Section 15.13

*34. How many grams of Cr are necessary to react with 250 mL of 0.44 M HCl?

$$Cr + 2\ HCl \longrightarrow CrCl_2 + H_2$$

How many liters of H_2 measured at 760 torr and 20 °C will be produced by the reaction?

35. An excess of Na_2CO_3 solution is added to 230 mL of a $BaCl_2$ solution that is 0.55 M in Ba^{2+}. What precipitates and how many grams of it result?

*36. To a 50-mL solution of 0.00211 M $Ca(NO_3)_2$ is added 25 mL of 1.10 M H_2SO_4. How many grams of $CaSO_4$ are precipitated?

37. How many milliliters of 2.70 M $AgNO_3$ are necessary to form 16 g of Ag_2CrO_4?

$$AgNO_3\ (aq) + K_2CrO_4\ (aq) \longrightarrow Ag_2CrO_4\ (s) + 2\ KNO_3\ (aq)$$

How many milliliters of 3.50 M K_2CrO_4 will be used in the process?

38. How many liters of hydrogen at 50 °C and 1.05 atm can be produced from 15.6 g of Mg and 150 mL of 5.40 M $HClO_4$, according to the following equation?

$$Mg\ (s) + 2\ HClO_4\ (aq) \longrightarrow Mg(ClO_4)_2\ (aq) + H_2\ (g)$$

Section 15.14

*39. What is the molarity of a HNO_3 solution if 53.2 mL was required to titrate 100.0 mL of 2.05 M KOH?

40. A 10-mL sample of a $Ca(OH)_2$ solution that was 1.00 M in OH^- was used to prepare 500 mL of a solution. The solution was then titrated with 23.6 mL of a HCl solution of unknown concentration. What was the concentration of the HCl solution?

41. In a titration, 15.4 mL of a solution with a pH of 3.50 was used to titrate 10 mL of a KOH solution. Calculate the concentration of the KOH.

42. Because it is easily purified and handled and reacts readily, potassium acid phthalate, $KHC_8H_4O_4$, is used to find the concentration of solutions to be used in titrations. A sample of $KHC_8H_4O_4$ weighing 3.176 g reacts exactly with 43.25 mL of a solution of Na_2CO_3. What is the molarity of CO_3^{2-} in the solution? The molecular weight of $KHC_8H_4O_4$ is 204.26 g/mol.

$$2\ KHC_8H_4O_4\ (aq) + Na_2CO_3\ (aq) \longrightarrow 2\ NaKC_8H_4O_4\ (aq) + H_2O\ (l) + CO_2\ (g)$$

43. A 25.00-mL sample of a solution that was 0.442 M in H_3PO_4 was titrated with 0.113 M $Ba(OH)_2$. What reaction occurred? What volume of the $Ba(OH)_2$ solution was needed?

Section 15.16

*44. What is $[H_3O^+]$ when $[OH^-]$ is 5.5×10^{-3}? When it is 7.5×10^{-9}?

*45. What is the pH of each of the following?
   *a. $[H_3O^+] = 8.4 \times 10^{-2}$ M   d. $[H_3O^+] = 6.7 \times 10^{-2}$ M
   *b. $[OH^-] = 1 \times 10^{-4}$ M   e. $[OH^-] = 2.4 \times 10^{-11}$ M
   c. $[OH^-] = 9.5 \times 10^{-6}$ M

46. Which solutions in Problem 45 are acidic?

*47. Find the $[H_3O^+]$ of solutions with the following pH values:
   *a. 3.82   c. 10.5   e. 7.2
   b. 12.9   d. 4.7

*48. Find the $[OH^-]$ of solutions with the following pH values:
   *a. 12.7   c. 5.2   e. 9.8
   b. 3.4   d. 13.5

49. A 15-g sample of Ba metal was reacted with enough water to form 2.00 L of solution.

$$Ba \ (s) + 2 \ H_2O \longrightarrow Ba^{2+} + 2 \ OH^- + H_2 \ (g)$$

Calculate the final pH of the solution.

*50. How many grams of Li_2O are necessary to make 250 mL of a solution with pH = 9.40?

$$Li_2O \ (s) + H_2O \longrightarrow 2 \ Li^+ + 2 \ OH^-$$

Chapter 16

Reaction Rate and Chemical Equilibrium

In previous chapters, we learned how to write equations that describe the results of chemical reactions. Given the atoms of two different elements, we can now predict what kind of chemical bond those atoms will form. Knowing those things, we can make some reasonable guesses about the properties of a reaction product. If a reaction goes to completion, we know how to calculate the amount of product formed. We are beginning to get a certain feel for chemistry.

There is an important *if* in the first paragraph. "*If* a reaction goes to completion," but many reactions do not go to completion. Some substances will not react at all. Other systems react so slowly that lifetimes would pass before any product could be detected. We have learned about the reactions of individual atoms of the elements, but most atoms on earth are no longer individual, but are already combined. To understand a chemical reaction as it usually happens, to be able to predict the yield or rate of a reaction—indeed, to predict whether it will even occur at all—we must study the factors that determine the rate of reaction. We need to apply to chemical reaction the concept of dynamic equilibrium, which we first met in the context of vapor pressure. Our familiarity with the general concept of dynamic equilibrium will enable us to understand the characteristics of chemical equilibrium.

Chemical equilibrium is an important controlling factor in a large proportion of all chemical reactions. Equilibrium conditions control not only the chemical and physical processes of all living organisms, but also the conditions that make life possible on earth. Indeed, they regulate much of the operation of the entire universe.

We begin the chapter by examining chemical equilibrium as the balance between the reaction rates of a set of opposing chemical reactions. We explore the factors that control reaction rates, and how those factors affect the equilibrium. We learn how to make calculations about the relative amounts of reactants and products in an equilibrium system, and finally apply these techniques to real reaction systems.

AFTER STUDYING THIS CHAPTER, YOU WILL BE ABLE TO

- define dynamic equilibrium and chemical equilibrium
- predict relative rates of chemical reactions from relative activation energies
- explain the effects of concentration changes on reaction rates
- explain the effects of temperature changes on reaction rates
- write an equilibrium equation, given a reaction equation
- calculate an equilibrium constant from experimental data
- calculate the concentrations of products and reactants after equilibrium has been reached, given the initial concentrations and the equilibrium constants in the reaction system
- use Le Chatelier's principle to explain the effects of a change in the conditions under which an equilibrium system exists
- predict the effects of temperature changes on systems at chemical equilibrium
- calculate ionization constants from experimental data
- calculate H_3O^+ concentrations in solutions of weak acids and bases, given molar concentrations and ionization constants
- calculate solubility product constants, given experimental data
- calculate the concentrations of ions present in saturated solutions of slightly soluble electrolytes, given the solubility product constants

TERMS TO KNOW

activation energy	equilibrium constant	rate constant
buffer system	equilibrium equation	solubility product
catalyst	ionization constant	
chemical equilibrium	Le Chatelier's principle	

16.1 COMPLETION IN CHEMICAL REACTION

In our work with stoichiometry and in our calculations from reaction equations, we have assumed that when samples of reacting substances are mixed, reaction occurs and continues until one or all of the reactants are used up. For many kinds of chemical reactions, this assumption is not even approximately true. In many systems, reaction occurs rapidly at first, but soon seems to cease—even though only a small proportion of the reactants have actually been converted to products. Here are two examples.

If 1 mol of acetic acid, CH_3CO_2H, is mixed with enough water to make 1 L of solution, the acetic acid will react with the water to form hydronium ion and acetate ion. (Only one hydrogen leaves the acetic acid molecule in its reaction as an acid. To reflect this fact, the formula for acetic acid is often written as CH_3CO_2H.)

$$CH_3CO_2H \text{ (aq)} + H_2O \text{ (aq)} \longrightarrow H_3O^+ \text{ (aq)} + CH_3CO_2^- \text{ (aq)}$$
  Acetic acid   *Acetate ion*

After this reaction has taken place and no further changes can be observed, only about 1/500 of the acetic acid has been converted to products. Additional time produces no further results, nor does adding more CH_3CO_2H have a significant effect. The reaction has gone as far as it will.

When gaseous H_2 and I_2 are mixed at high temperatures, the following reaction occurs:

$$H_2 (g) + I_2 (g) \longrightarrow 2 HI (g)$$

Although the reaction is not fast, it eventually reaches a point at which product no longer forms. If the reaction is run at 425 °C, for instance, starting with 1 mol each of H_2 and I_2 in a 1-L vessel, the reaction proceeds until about one sixth of the H_2 and of the I_2 still remain. The reaction has gone as far as it will.

Why do these reactions not go to completion? Why do the reactants not react entirely?

The answer is that the reaction system has in each case reached a state of chemical equilibrium, a situation in which two opposing chemical reactions balance each other, just as the two opposing processes of vaporization and condensation balance each other when vapor pressure equilibrium is reached.

Most chemical reactions are reversible in practice, and all reactions are in theory. For each of the two reactions we just saw, there is an opposing, reverse reaction.

$$H_3O^+ (aq) + CH_3CO_2^- (aq) \longrightarrow CH_3CO_2H (aq) + H_2O (aq)$$

$$2 HI (g) \longrightarrow H_2 (g) + I_2 (g)$$

Reaction equations in which both the forward and the reverse reactions are significant are often written with a double arrow.

$$H_2 (g) + I_2 (g) \longleftrightarrow 2 HI (g)$$

The terms *reactant* and *product* are merely convenient names for the substances written on the left and right sides of the reaction equation. They are not always descriptive of how the reaction actually occurs. For example, if we were to start with 2 mol of HI in a 1-L vessel and heat it to 425 °C, the HI would react to form gaseous H_2 and I_2. After awhile, we would have a mixture of H_2, I_2, and HI in the same proportions as we did when we started with H_2 and I_2. In the first experiment, we wrote H_2 and I_2 on the left side of the equation and called them the reactants. In the second case, we would write HI on the left and call it the reactant.

$$2 HI (g) \longleftrightarrow H_2 (g) + I_2 (g)$$

The substances taking part in the reactions do not care how we write our equations or what we call the reactants and products. The reactions proceed as usual, regardless of how they are written.

We can follow each of our reactions graphically. We begin with 1 mol each of H_2 and I_2 in our liter vessel, and plot their disappearance as the reaction runs. As the H_2 and I_2 disappear, HI is simultaneously created. The number of moles of HI in our liter is graphed in Figure 16.1 as a dotted line.

If, instead, we begin with 2 mol of HI, the curves look like the one in Figure 16.2, where the decrease of HI is shown as a solid line and the increase of H_2 and of I_2 as a dotted one.

When a reverse reaction occurs along with the forward reaction, the result is a dynamic balance that limits the extent of the reaction. Let us look at this pattern more closely.

16.2 DYNAMIC EQUILIBRIUM REVISITED

In our study of vapor pressure, we found that the measured vapor pressure of a liquid resulted from the dynamic equilibrium established between the two competing processes of

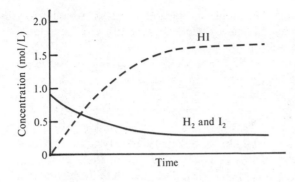

Figure 16.1 Reaction of H_2 and I_2. One mol each of H_2 and I_2 are placed in a 1-L vessel at 425 °C. The concentration of HI grows as the concentrations of H_2 and I_2 decrease. After a while, no more change occurs.

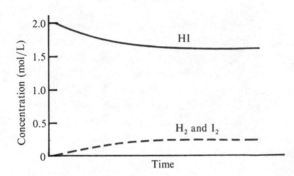

Figure 16.2 Decomposition of HI. The concentration of HI decreases as the concentrations of H_2 and I_2 increase.

vaporization and condensation. At first, when no vapor was present, there were no molecules to return to the liquid, the rate of vaporization was more rapid than the rate of condensation, and the population (pressure) of vapor molecules increased rapidly. But as the number of molecules in the vapor continued to increase, vapor molecules began to strike and rejoin the liquid, and the rate of condensation increased. When the rates of the two processes finally became the same, no further change in the pressure of the vapor occurred, and equilibrium was established. With small differences in detail, the establishment of equilibrium between the rates of forward and reverse chemical reactions follows the same pattern. Equilibrium processes control many events in the physical and chemical universe. The levels of lakes and seas, the concentration of calcium in human blood, the temperature of the earth—all are controlled by equilibrium processes. In a reaction system, the state of equilibrium between the forward and reverse reactions is called **chemical equilibrium.** We can tell that a reaction system has reached chemical equilibrium because the relative amounts of reactants and products have ceased to change. Remember, however, that although equilibrium may have been reached and the system does not appear to be active, chemical reaction, forward and reverse, continues.

16.3 COLLISION THEORY AND ACTIVATION ENERGY

To understand chemical equilibrium, we must first understand the rates of the reactions that control the equilibrium. To understand the rates, we need to know something of the way in which individual particles react. Although the details of the processes by which many reactions occur are thought to be quite complicated and to include large numbers of individual steps, we need not worry about such complexities.

Before molecules can react, they must come into contact. Consider a system of two gases that have been mixed. We know that in a gas under ordinary conditions, each molecule collides with another about 10^{10} times every second. If only 1 collision in every 1000 were to produce reaction, reactions between gases would still be complete within millionths of a second. Some gas reactions, however, take years. This is because only a small proportion of molecular collisions actually result in reaction. Liquids are often similarly slow to react. Generally, several factors determine whether a molecular collision will result in reaction: two of these are the energy with which the collision occurs, and the requirements that must be met during the collisions. Even after decades of research, the details of the way in which molecules must come together to react are not well understood.

The requirements for reaction vary widely. When molecules react with other molecules, covalent bonds must be broken as well as made; this is sometimes a difficult process. Some reactions of ions, however, have low requirements. A hydrogen ion, for example, needs only to encounter a hydroxide ion for a bond to be formed and a water molecule created.

The energy needed for successful reaction depends largely on the kind of reaction and its requirements. In many cases, as in our example of H_2 and O_2, collisions between molecules must be highly energetic for reaction to occur. During the collision, the electron clouds of the molecules must be so mingled that there is as much likelihood that new bonds will form as that the original bonds will survive. The collision must squeeze the atoms in the molecules so close together that it matters little which pairs of atoms were originally bonded. Because the negative outer shells of atoms repel each other, considerable energy is required to make them intermingle. The energy that colliding particles must have to be able to react is called the **activation energy.** The chief reason that most collisions do not result in reaction is that the colliding particles do not possess the required activation energy.

Schematically, we can think of the activation energy as a barrier across which reactants must pass in order to become products. Figure 16.3 will help you to visualize activation energy in this way.

16.4 EXTERNAL FACTORS THAT AFFECT REACTION RATES

Of the many conditions that can affect the rate of a reaction, and therefore the equilibrium between competing reactions, we will be most interested in concentration and temperature.

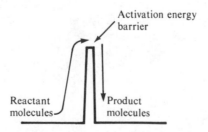

Figure 16.3   Representation of the activation energy barrier.

16.4.1 Temperature

We learned in Chapter 12 that the temperature of a system is a measure of the *average kinetic energy* of its molecules, and we can see graphically that not all the molecules in a system have the same kinetic energy, as shown in Figure 16.4 (which appeared earlier as Figure 12.2). Most molecules have energies close to the average, but at any given instant a few have especially high energies. In reactions with large activation energies, only the molecules with the highest energies are able to react. For an understanding of why this is so, study Figure 16.5 and its explanation.

What happens to a reaction rate when the temperature of the reactants is changed? A small increase in the temperature of a system leads to a small increase in the *average* kinetic energy of the particles in that system. Because of the shape of the energy distribution curve, however, one of the effects of a small increase in average kinetic energy is that the proportion of molecules with especially large energies changes far more than does the average kinetic energy itself. Let's say that in a team of 100 track runners, only 1 runner can achieve a 4-min mile, and 9 others can run a 4.5-min mile. If everyone's performance were to improve by only half a minute, then ten runners could do the mile in 4 min. The number of exceptional runners, judged on the basis of a 4-min mile, would increase tenfold, although the average performance would have increased by only about 10%. Changes in particle energy follow a similar pattern. A small rise in the average kinetic energy of a system (temperature) often leads to a large increase in the proportional number of particles with sufficient kinetic energy to overcome a large activation energy barrier. Moreover, an increase in kinetic energy means that there is an increase in the average speed of the particles, which in turn means that there will be more collisions per particle per second. When there are more frequent collisions and also a higher proportion of very energetic collisions, the rate of reaction between particles must then be greater.

What all this means is that reaction rates are sometimes highly sensitive to temperature. A rule of thumb used by chemists is that under ordinary conditions, room temperature, for example, the rate of many reactions approximately doubles every time the temperature of the system is raised by 10 °C. Although there are many exceptions, the very existence of the rule shows how much influence temperature can have on reaction rate. On the other hand, if the activation energy is low, and most molecules will already be able to pass the barrier, a temperature change will have little effect. Look at Figure 16.5 again to see why this is so.

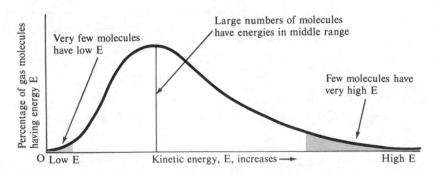

Figure 16.4 Molecular energy distribution curve.

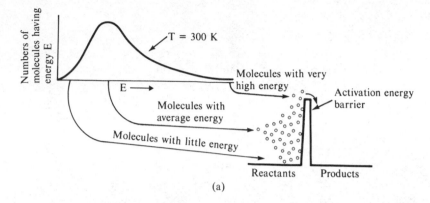

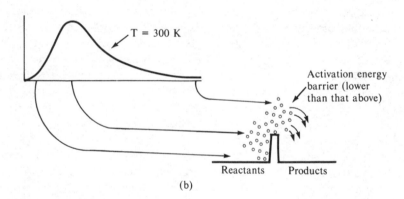

Figure 16.5 Comparisons of energy distribution curves and activation energy barrier. Activation energies of different sizes are compared to a molecular energy distribution curve like that of Figure 16.4. In part (*a*), the activation energy is high in comparison with the energies of most of the molecules. Because at any one time only relatively few molecules have enough energy to cross the barrier and become products, the reaction is slow. Part (*b*) compares the same distribution curve with a reaction with a lower activation energy: more molecules can cross the barrier every second. All other things being equal, reactions that require lower activation energies are faster than those that require higher activation energies. (In these sketches, we are not yet concerned with the reverse reaction of products returning to reactants.)

In a reversible system, the reaction in each direction has an activation energy requirement, and these requirements often differ. For example, in Figure 16.6, we see the system

$$2\ SO_2 + O_2 \longrightarrow 2\ SO_3$$

To react, the SO_3 molecules must climb an energy wall that is higher than that climbed by SO_2 and O_2 molecules. Given equal concentrations of reacting molecules at equal tempera-

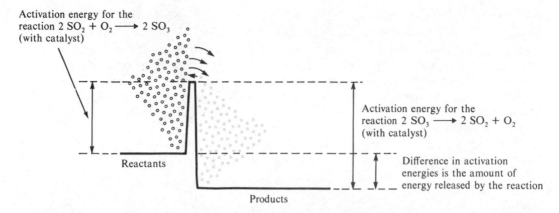

Activation energy for the
reaction 2 SO$_2$ + O$_2$ ⟶ 2 SO$_3$
(with catalyst)

Activation energy for the
reaction 2 SO$_3$ ⟶ 2 SO$_2$ + O$_2$
(with catalyst)

Reactants

Difference in activation
energies is the amount of
energy released by the reaction

Products

Figure 16.6 Activation energies for the formation and decomposition of SO$_3$. As we learned in Section 15.17, one of the problems in the manufacture of sulfuric acid is that the exothermic reaction of SO$_2$ and O$_2$ to form SO$_3$ is ordinarily too slow to be commercially useful. Raising the pressures, that is, the concentrations, of SO$_2$ and O$_2$ is some help, but the reaction becomes commercially feasible only when a catalyst is used. The catalyst provides a way for the reaction to proceed at a lower activation energy, which makes it possible for a much larger proportion of the O$_2$ and SO$_2$ molecules to react at any given temperature. Catalysts for other reactions operate in a similar way. (See Section 16.5)

tures, the reaction to the right in this system is therefore faster than that to the left. We will soon see the effect of this difference on the equilibrium between reactants and products.

Temperature change often affects the forward and reverse reactions in a system differently. An increase in temperature may increase the rate of one reaction more than that of the other. Cooling the system may slow down one reaction more than the other. Because changes in the relative rates of forward and reverse reactions also change the equilibrium between reactants and products, temperature changes often produce equilibrium changes.

We can learn something else of value from Figure 16.6. Notice that after the reacting SO$_2$ and O$_2$ molecules have climbed the activation energy barrier and reacted, the resulting SO$_3$ molecules drop into a deeper energy well. The reverse reaction has a higher activation energy than does the forward reaction. This happens because the molecules of product actually *have* less energy than those of the reactants: *that is, the reaction is exothermic.* In an exothermic reaction, the activation energy of the products is greater than that of the reactants. In an endothermic reaction, the activation energy of the reactants is greater. Consider these relations, and study Figure 16.6 until you understand them.

16.4.2 Concentration

The concentrations of reacting substances are important factors in determining the rates of reactions. To understand the effects of concentration on rate, let us take a simple imaginary reaction

$$A + B \longrightarrow C + D$$

For A and B to react, they must at some point collide. Suppose you had a 1-L vessel that contained 10 molecules of A and 500 molecules of B. The reaction to form AB would not be fast, because the B molecules would have difficulty finding A molecules with which to react. If there were 100 A molecules, however, the B molecules would have to search, on the average, only a tenth as long. The reaction rate would be increased tenfold. The rate of the reaction is proportional to the number of A molecules in the liter (to the concentration of A). If we raise the number of B molecules from 500 to 5000, the effect is the same: the rate of reaction is also proportional to the concentration of B. To sum up: when concentration increases, the number of collisions per second increases; when the collision rate increases, the rate of the reaction increases.

If the rate of our reaction is proportional to the concentration of A and also to that of B, it must be proportional to the product of the two. This can be expressed as an equation:

reaction rate = k (concentration of A) (concentration of B)

We have described the concentration of A and B as the number of molecules in 1 L, but the unit of moles is more convenient. We can rewrite our rate equation in molar language.

reaction rate = k [A] [B]

In this equation, k represents what is called a **rate constant.** For any given reaction system at any constant temperature, k is a constant number which depends in part on the activation energy for the A + B reaction. It increases as temperature increases and decreases as temperature decreases.

To understand the effect of concentration on reaction rate, we have taken a simple, one-collision system. Many, perhaps most, reactions, however, apparently occur in complex sequences of several steps. Nonetheless, the rates of reactions do depend on the concentrations of the reactants in such a way that valid rate equations, similar to the one we have seen, can be written.

16.5 CATALYSTS

Many systems are affected by the presence of **catalysts,** substances that change the rates of reactions but do not themselves react in the sense that they are chemically changed. Catalysts operate by providing mechanisms that are easier than those by which reactions ordinarily occur. This action has the effect of lowering the activation energy, as represented in Figure 16.7.

The most remarkable instances of catalyzed reactions are in biological systems, in which complex reactions are caused to proceed at rapid rates in specific ways. The entire chemistry of living beings operates with the assistance of catalysts called enzymes (Section 20.7).

Although catalysts can affect reaction rate, they affect both forward and reverse reactions in the same way and do not change the position of chemical equilibria.

16.6 THE EQUILIBRIUM CONSTANT AND EQUILIBRIUM EQUATION

In a reaction system, a state of chemical equilibrium is reached when the forward and reverse reaction rates are equal. Because the reaction rates depend, in turn, on the concentrations of the reacting substances, the equilibrium requires that there also be a relation between the concentrations themselves, regardless of the complexity of the individual reactions. In the H_2-I_2 system at equilibrium at 425 °C, for example, the following is found experimentally to be true.

$$\frac{[HI]\ [HI]}{[H_2]\ [I_2]} = 54$$

Study this remarkable statement for a few minutes. It says that if you measure the actual concentration of HI in moles per liter, then square that value and divide it by the product of the concentrations of H_2 and I_2, you will get the number 54. It does not say how much H_2 or I_2 you started with or whether you might have added some HI or taken some out. It simply says that if the system has reached equilibrium, the relations of the concentrations of HI, H_2, and I_2 will be as stated—no matter what amounts were put into or removed from the system.

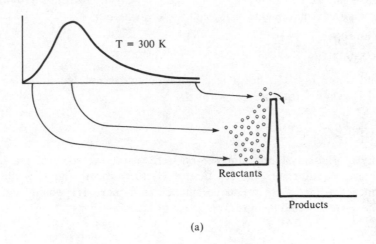

(a)

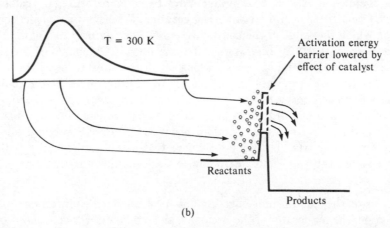

(b)

Figure 16.7   Rate of a reaction without (*a*) and with (*b*) a catalyst. (*a*) Without catalytic assistance, the reaction runs slowly because few reactant molecules have enough energy to cross the activation energy barrier. (*b*) Because a catalyst can lower the activation energy barrier, it can speed up a reaction. Notice, however, that the catalyst lowers the barrier for reaction in both directions, and therefore does not affect the position of the equilibrium.

Suppose you begin with 1 mol each of H_2 and I_2, place them in a 1-L vessel, heat to 425 °C, and wait until equilibrium is established. If you then measure the concentrations of all the species, you will find that $[I_2]$ = 0.214, $[H_2]$ = 0.214, and $[HI]$ = 1.572. Let us put this set of values into the equilibrium equation for the reaction.

$$\frac{[HI]^2}{[H_2][I_2]} = \frac{(1.572)^2}{0.214 \times 0.214} = 54.0$$

Within the limits of our significant figures, that experiment behaved according to the equilibrium equation. Let's try again. This time we will begin our reaction with exactly 10.0000 mol of H_2 and 0.50000 mol of I_2 in a 1-L vessel. When we measure our concentrations after equilibrium is reached, we find that $[H_2]$ = 9.5019, $[I_2]$ = 0.00194, and $[HI]$ = 0.9961. The system again meets the requirements of the equilibrium equation.

$$\frac{(0.9961)^2}{9.5019 \times 0.00194} = 53.8$$

If you could find some *pure* HI, put it in a vessel of any volume whatever, and heat it to 425 °C, H_2 and I_2 would form until

$$\frac{[HI]^2}{[H_2][I_2]} = 54$$

Now, in any of these systems, suppose you were to remove some of the HI after the system had come to equilibrium. The reaction of HI to form H_2 and I_2 would slow down because the concentration of HI would be smaller, and more HI would be formed than disappeared. The system would soon restore its equilibrium and

$$\frac{[HI]^2}{[H_2][I_2]} = 54$$

For any chemical system at equilibrium at any fixed temperature, an **equilibrium constant** exists and can be written with its **equilibrium equation.** We have written the equation for the HI system in which the value of the constant was 54. *The equilibrium constant of a reaction system is a fixed property of that system.* No matter what amounts of the substances are initially present, added, or taken away, as long as all the substances of the equilibrium are present, the reaction will run until the concentration relations given by the equilibrium equation are met. *For each temperature, the equilibrium constant for a reaction will have a particular value,* which can ordinarily be determined experimentally.

There is a rule for writing equilibrium equations. Without worrying too much about exactly why the rule is true, but with confidence that every equilibrium equation so written will be a true statement, you should learn how to write equilibrium equations according to the rule.

> To write the equilibrium equation for a reaction at equilibrium, form a fraction: the numerator of the fraction is the product of the concentrations of the substances appearing on the right side of the reaction equation (each concentration raised to the power corresponding to the coefficient of that substance in the reaction equation) and the denominator of the fraction is the product of the concentrations of the substances on the left side of the reaction equation, again raised to the appropriate powers. (In certain cir-

cumstances, not all the concentration factors are used. We will consider these instances in Sections 16.11 and 16.13.)

Although we have written an explicit value, 54, for the equilibrium constant of the HI system, the equilibrium equation often appears with the letter K standing for the constant, especially if the numerical value is not known. We again write the equilibrium equation for the HI reaction, following the recipe and using a capital K. Since the value of the constant is known, we add it also.

$$K = \frac{[HI]^2}{[H_2][I_2]} = 54$$

EXAMPLE 16.1. Write the equilibrium equation for the NH_3 reaction.

$$3\ H_2\ (g) + N_2\ (g) \longleftrightarrow 2\ NH_3\ (g)$$

Solution:

Step 1. Form the denominator.
  a. Write the concentration of H_2 in moles per liter, using the [ ] symbol. Raise that concentration to a power that is the same as the number of times that H_2 appears in the reaction equation, namely 3.

$$[H_2]^3$$

  b. Write the concentration of N_2 similarly. In this case, the exponent for the concentration is 1 and is omitted.

$$[N_2]$$

  c. The denominator is the product of these two.

$$[H_2]^3\ [N_2]$$

Step 2. Form the numerator. In this case, it is the NH_3 concentration raised to the second power.

$$[NH_3]^2$$

Step 3. Form the fraction, using the denominator and numerator as written. Set that fraction equal to K.

$$\frac{[NH_3]^2}{[H_2]^3\ [N_2]} = K$$

Although we have written the equilibrium constant for the NH_3 reaction, we do not necessarily know its value at any temperature. That information must be experimentally obtained.

Notice that the equilibrium constant for this reaction has units.

$$\frac{\left(\dfrac{mol\ NH_3}{L}\right)^2}{\left(\dfrac{mol\ H_2}{L}\right)^3 \left(\dfrac{mol\ N_2}{L}\right)} = \frac{\left(\dfrac{mol}{L}\right)^2}{\left(\dfrac{mol}{L}\right)^4} = \left(\dfrac{L}{mol}\right)^2$$

Whether the equilibrium constant has units depends on the form of the reaction. We will demonstrate this to ourselves in the next exercise or two, then, as most chemists do, we will dispense with the writing of units on equilibrium constants.

Many of our examples of equilibrium will involve systems of reacting gases; such systems are often simpler than reactions in the liquid phase. We will use the units of mol/L in our discussions, even when we speak of gases, but we should keep in mind that gas concentrations (unlike those of liquids) can be and often are measured in terms of pressure. Remember that $PV = nRT$, and thus $n/V = P/RT$. If the temperature is known, gas pressures can be converted directly to concentrations in mol/L. For many purposes, pressures, rather than concentrations, can be used with reactions of gases. In liquid systems, of course, pressures cannot substitute for concentrations. (Notice that the *value* of the equilibrium constant for a gas reaction may not be the same in pressure units as in concentration units.)

EXAMPLE 16.2.   The reaction to prepare methyl alcohol from hydrogen and carbon monoxide is

$$2\ H_2\ (g) + CO\ (g) \longleftrightarrow H_3COH\ (g)$$

Write the equilibrium equation for this reaction.

Solution:

The numerator is $[H_3COH]$.
The denominator is $[H_2]^2$ multiplied by $[CO]$, or $[H_2]^2\ [CO]$.
The equilibrium equation is

$$\frac{[H_3COH]}{[H_2]^2\ [CO]} = K\ \frac{L^2}{mol^2}$$

EXERCISE 16.1.   Write the equilibrium equation for the following reaction. What are the units on the constant?

$$2\ NO\ (g) + Cl_2\ (g) \longleftrightarrow 2\ NOCl\ (g)$$

EXERCISE 16.2.   As we have learned, ozone, O_3, is an extremely important component of our atmosphere. Write the equilibrium equation for the formation of ozone from oxygen molecules. What are the units on this constant?

$$3\ O_2\ (g) \longleftrightarrow 2\ O_3\ (g)$$

Because the value of K depends not only on the nature of the reactions, but also on temperature, we must always specify the temperature for which the equilibrium constant is given. For example, the value of K for the HI reaction is 54 at 425 °C, 45 at 490 °C, and 59 at 394 °C. (Notice that the equilibrium constant for the HI reaction has no units.)

16.7 FINDING THE VALUES OF EQUILIBRIUM CONSTANTS

As we have seen, many chemical reactions stop running long before they go to completion; they reach characteristic states of equilibrium in which significant amounts of both products and reactants are present. An equilibrium constant is simply a way of writing the quantitative relation that exists between the concentrations of reactant and product in a system that has reached equilibrium. The constant is the same for a given system at any specific temperature, no matter what amounts of reactant and product were originally present. A reaction will run until it reaches its characteristic equilibrium.

We need to understand what equilibrium constants tell us about the reaction systems they represent. We will also want to learn how to use the constants quantitatively. To accomplish this, we will make a few calculations, then compare the results with the constants themselves.

The values of equilibrium constants can be determined experimentally by letting a reaction system come to equilibrium and then measuring the concentrations of all the species in the system.

EXAMPLE 16.3.   In the gaseous state, phosphorus pentachloride decomposes reversibly according to the following reaction:

$$PCl_5 \ (g) \longleftrightarrow PCl_3 \ (g) + Cl_2 \ (g)$$

A sample of PCl_5 was placed in a vessel and heated to 250 °C. After equilibrium was reached, the concentrations of the three substances in the vessel were

$$[PCl_5] = 8.55 \times 10^{-3} \ mol/L$$
$$[PCl_3] = 1.90 \times 10^{-2} \ mol/L$$
$$[Cl_2] = 1.90 \times 10^{-2} \ mol/L$$

What is the value of the equilibrium constant for this reaction?

Solution:

Step 1.   Write the equilibrium equation.

$$\frac{[PCl_3] \ [Cl_2]}{[PCl_5]} = K$$

Step 2.   Insert the measured concentration values and do the arithmetic.

$$K = \frac{[PCl_3] \ [Cl_2]}{[PCl_5]} = \frac{(1.90 \times 10^{-2}) \ (1.90 \times 10^{-2})}{(8.55 \times 10^{-3})} = 4.22 \times 10^{-2}$$

EXERCISE 16.3.   A group of students made a series of measurements to find the equilibrium constant, K, for the following reaction at 25 °C:

$$2 \ NO \ (g) + Br_2 \ (g) \longleftrightarrow 2 \ NOBr \ (g)$$

The final results of the experiments, averaged together, were the concentrations of the three gases at equilibrium.

$$[NOBr] = 0.936 \ mol/L$$
$$[NO] = 0.053 \ mol/L$$
$$[Br_2] = 0.110 \ mol/L$$

Calculate the equilibrium constant for this reaction at 25 °C.

Where can we find the values of equilibrium constants? There are tables giving data for hundreds of constants. In most cases, the values come from laboratory data, although some have been calculated from theory.

16.8 INTERPRETING THE VALUES OF EQUILIBRIUM CONSTANTS

What do the values of equilibrium constants mean? In our imaginary experiment in Section 16.6, we started with 1 mol each of H_2 and I_2; approximately three quarters of each reactant was converted to HI. This result matches the rough guess we might have made just by looking at the value of the equilibrium constant. Since that number is larger than unity, the numerator of the equilibrium constant must be larger than the denominator. The reaction therefore runs until [HI] is much larger than either [H_2] or [I_2], and that is the result we calculated. Without making calculations, we can sometimes look at the values of equilibrium constants and guess how reactions will go, especially if the constants are particularly large or small.

Equilibrium constants have a wide range of values. The equilibrium constant for the reaction of H_2 and Cl_2 to form HCl, for example, is 2×10^{33} at room temperature. If we were to start with 1 mol each of H_2 and Cl_2 in a 1-L vessel and bring them to equilibrium at 25 °C, there would be left only 4×10^{-16} mol of H_2 and Cl_2, or 0.0000000000000004 mol. Because the equilibrium constant is so large, we know that the reaction runs nearly to completion. In fact, if we had started with exactly equal amounts of each substance, it would be impossible to detect any unreacted Cl_2 or H_2 at equilibrium. A large value for an equilibrium constant means that the numerator is much larger than the denominator and that the concentrations of the products will be much greater than that of at least one of the reactants. The reaction will go nearly to completion to the right as written because it is the forward reaction that is favored.

If the equilibrium constant is small, we can assume that hardly any products will be present at equilibrium, and that nearly all the reactants will remain. The reverse reaction is the favored one. For example, the equilibrium constant for the reaction of O_2 and N_2 to give NO_2 is very small (which is fortunate; otherwise there would be no oxygen in the air).

$$2 \, N_2 \, (g) + O_2 \, (g) \longleftrightarrow 2 \, NO_2 \, (g)$$

$$K = \frac{[NO_2]^2}{[N_2]^2 \, [O_2]} = 8.3 \times 10^{-10} \text{ at } 25 \text{ °C}$$

When the constant is this small, the concentration of NO_2 in equilibrium with O_2 and N_2 must be small as well. Fairly significant concentrations of NO_2 are sometimes present in smoggy air, however, because NO_2 decomposes quite slowly, and equilibrium is slow to arrive.

When equilibrium constants are very small, we can easily and quite accurately estimate the amount of product formed when the reactants come to equilibrium. Let's say we start with 2 mol of O_2 and 2 mol of N_2 in a 1-L flask and wait (a long time) until they reach equilibrium. Because the amount of NO_2 formed is tiny, we can approximate, and say that almost none of the reactants was used up, that is, that there are still 2 mol each of the reactants in the vessel. We place these values in the equilibrium equation and solve for NO_2.

$$\frac{[NO_2]^2}{2 \times 2} = 8.3 \times 10^{-10}$$

therefore,

$$[NO_2]^2 = 8.3 \times 10^{-10} \times 4 = 3.3 \times 10^{-9}$$

$$[NO_2] = 5.7 \times 10^{-5} \text{ mol/L}$$

A concentration of 5.7×10^{-5} mol/L is very small. Our assumption that very little O_2 and N_2 was used is valid.

EXERCISE 16.4.   The equilibrium constant for the formation of ozone (see Exercise 16.2) has the value of 1×10^{-48} at 25 °C. If 1.00 mol of O_2 is placed in a 2.00-L container at 25 °C, what will the concentrations of O_2 and O_3 be at equilibrium?

16.9 DRIVING A CHEMICAL REACTION

It is possible to make a reaction run in a desired direction by shifting the position of its equilibrium. There are two common methods of accomplishing this effect.

16.9.1 Concentration Changes

Concentration adjustments are often used to cause a reaction to reach a desired result. To find out the effects of concentration changes on equilibria, we need first to find out how those changes affect the forward and reverse rates of reaction.

Remember that the position of an equilibrium (that is, the relative concentrations of reactants and products) is determined by the competing rates of the forward and reverse reactions. For example, if the forward reaction is quite fast (has a low activation energy) and the reverse reaction is slow (has a high activation energy), there must be a much larger concentration of products than of reactants before as many molecules per second cross the barrier to the left as to the right. Study Figure 16.8. It is also true that a slow forward reaction and fast reverse reaction result in a larger concentration of reactants than products. Both these effects must be considered to understand how concentration changes affect equilibrium.

Suppose, now, we add some additional HI to a HI system at equilibrium.

$$H_2 \text{ (g)} + I_2 \text{ (g)} \longleftrightarrow 2 \text{ HI (g)}$$

The reverse reaction, the one that uses up HI, will be speeded up. Because that reaction will temporarily be faster than the forward reaction, some HI will disappear and more H_2 and I_2 will be formed. But as the concentration of HI becomes smaller and the concentrations of H_2 and I_2 grow, the decomposition of HI will progress at a slower rate and the forward reaction at a faster rate until the two are again the same and the equilibrium equation is satisfied. By adding the HI, we cause some (but not all) of the additional HI to be used up and some more H_2 and I_2 to be created. The equilibrium shifts so as to reduce the concentration of the added ingredient; that is, to reduce the effect of the attempted change.

More practical is the question of what would happen in the HI equilibrium system if we were to *remove* some HI. The rate of the forward reaction would not be immediately affected because the concentrations of H_2 and I_2 would not initially be changed. The rate of the reverse

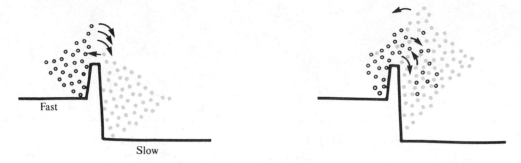

Fast

Slow

Figure 16.8   Forward and reverse activation energy barriers determine the position of equilibrium. At left, the system is not at equilibrium. The rate of forward reaction (to right) is greater than the rate of reverse reaction. At right, the rates are the same, and the system is at equilibrium. Most of the reactants have become products. The forward reaction is slower than it was at first, because there are now fewer reactant molecules. The reverse reaction is faster because there are now more product molecules. Equilibrium occurs when the two rates are the same.

reaction would drop, however, because the withdrawal would have made the concentration of HI smaller. In the final result, more HI would be formed, more H_2 and I_2 would be used up, and the equilibrium would again be restored. The system would respond so as to reduce the effect of the withdrawal of HI. If we were to continue to remove HI, the H_2 and I_2 would continue reacting to replace it until nothing of those two substances remained. Isolating and removing a product is one way to cause equilibrium reactions to proceed to completion in a desired direction.

If we look at an equilibrium equation and imagine adding or taking away one of the members of the system, we can understand arithmetically what the effects will be. We will again use the concentration values from the first HI problem. In the flask at equilibrium the concentrations in moles per liter were

$$[H_2] = 0.214$$

$$[I_2] = 0.214$$

$$[HI] = 1.57$$

$$K = \frac{(1.57)^2}{0.214 \times 0.214} = 53.8$$

Suppose we were suddenly to take away nine tenths of the HI. The value of the numerator in the equation would become $(0.157)^2$, and the requirement that $K = 54$ would no longer be satisfied. The system would not be at equilibrium.

$$\frac{(0.157)^2}{0.214 \times 0.214} = 0.538$$

The new figure is not even close to 54. For the equilibrium requirement again to be satisfied, the numerator would have to become larger and the denominator smaller: HI would have to be formed at the expense of H_2 and I_2. Reaction would occur until $K = 54$. If we were to measure the values of the concentrations at the new equilibrium, we would find that they had adjusted themselves to $[H_2] = 0.0630$, $[I_2] = 0.0630$, and $[HI] = 0.462$, and that

$$\frac{(0.462)^2}{0.0630 \times 0.0630} = 53.8$$

Compare the new values of the concentrations with those present just after the HI was removed. At first, we were left with 0.157 mol/L of HI; reaction then occurred until 0.462 mol/L was present (more than 0.157, but less than the original 1.57). Meanwhile, $[H_2]$ and $[I_2]$ were reduced from 0.214 mol/L each to 0.0630 mol/L. *Removing HI caused some of the H_2 and I_2 to be used up and more HI to be formed.*

EXAMPLE 16.4. Ammonia is at equilibrium with its elements in a vessel at 650 °C.

$$N_2 \text{ (g)} + 3 H_2 \longleftrightarrow 2 NH_3 \text{ (g)}$$

To this equilibrium system is added as much H_2 as was already present. Qualitatively, what will be the effect of this addition on the concentrations of N_2, H_2, and NH_3?

Solution:

From what we have just learned, we know that adding H_2 to this system should cause reaction to occur to the right (the direction that uses up some of the added H_2). Not enough reaction will occur, however, to use up all the added H_2. Therefore, the new concentration of H_2 will still be greater than it was originally. The reaction will create NH_3, so the new concentration of NH_3 will be greater than before. The new concentration of N_2, however, will be less than before, because some of the N_2 originally present will be used up.

EXERCISE 16.5. The gaseous equilibrium system described by the following equation is present in a container at 1000 K.

$$H_2O \text{ (g)} + CO \text{ (g)} \longleftrightarrow CO_2 \text{ (g)} + H_2 \text{ (g)}$$

To this system is added additional H_2O. What is the effect of this change on each of the concentrations?

16.9.2 Temperature Change

The details of the relationship between temperature change and equilibrium are not difficult to understand, and are, in fact, the subject of a problem at the chapter end. But for the purpose of making quick decisions about the effect of temperature change on equilibrium systems, we can take some shortcuts.

Almost all chemical reactions are exothermic or endothermic; in only a few reactions is energy neither absorbed nor released. Remember (Section 5.6) that if $\triangle H$ is positive, the reaction is endothermic (energy is absorbed) and that if $\triangle H$ is negative, the reaction is exothermic (energy is produced). A quick way to decide how temperature will affect an equilibrium is simply to consider the enthalpy as either a reactant or product. If energy is absorbed by the reaction ($\triangle H$ positive), then adding or taking away energy has the same effect as adding or taking away a reactant. If energy is released ($\triangle H$ negative) then adding or taking away energy has the same effect as adding or taking away a product. Raising the temperature is the same as adding energy. Lowering the temperature is the same as taking away energy.

Consider an endothermic reaction

$$A + B \longleftrightarrow C + D \qquad \triangle H = +100 \text{ kJ}$$

Adding energy to this endothermic system (that is, raising the temperature) will have the same effect as adding either A or B: it will shift the equilibrium to the right. If we run this reaction at higher temperatures, we will obtain a higher yield of C and D than we would have if we had run it at lower temperatures.

The reaction of phosphorus trichloride with chlorine yields phosphorus pentachloride and is exothermic.

$$PCl_3 + Cl_2 \longleftrightarrow PCl_5 + 92.8 \text{ kJ} \qquad (\triangle H = -92.8 \text{ kJ})$$

We can see that adding PCl_5 to this equilibrium system would drive the reaction to the left. Adding energy (raising the temperature) would have the same effect.

EXAMPLE 16.5.   Nitrogen dioxide, NO_2, reacts with itself to produce N_2O_4, nitrogen tetroxide.

$$2 \text{ NO}_2 \longleftrightarrow N_2O_4 + 58 \text{ kJ} \qquad (\triangle H = -58 \text{ kJ})$$

What will happen if we raise the temperature of the system of NO_2-N_2O_4 at equilibrium?

Solution:

Raising the temperature of this equilibrium system reduces the equilibrium concentration of N_2O_4, which increases that of NO_2.

An interesting lecture-demonstration is based on this reaction. NO_2 is a brown gas; N_2O_4 is colorless. A flask containing the equilibrium mixture at room temperature is pale brown in color. If the flask is heated, the brown color intensifies because the equilibrium shift causes an increase in the concentration of NO_2.

EXERCISE 16.6.   The decomposition of HBr into its elements is endothermic.

$$2 \text{ HBr} \longleftrightarrow H_2 + Br_2$$

What effect will raising the temperature have on the concentration of HBr in an equilibrium system?

16.10 LE CHATELIER'S PRINCIPLE

A system at equilibrium tends to respond to an externally caused change by reacting in a direction that produces results opposite to those of the change itself. This tendency is formalized in a statement called **Le Chatelier's principle,** which was first proposed by Henri Le Chatelier (France, 1850-1936).

If conditions are changed in a reaction system at dynamic equilibrium, reaction will take place in the direction that tends to counteract the change and bring the system again to equilibrium.

Notice that the response to change in a system at equilibrium never proceeds far enough to counteract *all* the change. The response to a given change can be determined precisely only by calculation or experiment.

Both in the laboratory and in industry, Le Chatelier's principle is applied to drive reactions in a desired direction. Concentrations of reactants are kept as high as practical, and products are removed continuously whenever possible. With exothermic reactions, energy is removed whenever possible; endothermic reactions are run at high temperatures.

So we see that Le Chatelier's principle applies to temperature and pressure changes as well as to changes in concentration. Before we leave the subject, we should remember an important distinction between temperature change and other kinds of changes in conditions. *The value of the equilibrium constant of a reaction system is a property of that system at a particular temperature. Changing the conditions of concentration (or pressure in a gaseous system) will not change the value of the equilibrium constant.* The system reacts to such changes so as to satisfy the equilibrium equation for the original value of the constant (as we saw it do in the HI system). *If the temperature is changed,* however, *the value of the constant can change.* A change in the constant, of course, also shifts the equilibrium.

16.11 THE IONIZATION CONSTANT

We first encountered weak electrolytes in Chapter 15. Most weak electrolytes are either weak acids or weak bases. We now know enough to understand why weak electrolytes behave as they do. As our example we take ammonia, a weak base. Ammonia reacts with water to form NH_4^+ and OH^- ions.

$$NH_3 + H_2O \longleftrightarrow NH_4^+ + OH^-$$

This is an equilibrium reaction in which most of the ammonia remains in the un-ionized state. In other words, the equilibrium constant for the reaction of ammonia with water is small.

$$K = \frac{[NH_4^+]\,[OH^-]}{[NH_3]\,[H_2O]} = 3.1 \times 10^{-7} \text{ at } 25\ ^\circ C$$

Because it yields relatively few OH^- ions to a solution, ammonia is classed as a weak base.

Weak acids and bases are usually used in dilute aqueous solutions, that is, in solutions that are mostly water. A liter of water contains 55.5 mol of water.

$$1\ L\ H_2O \times \frac{1000\ g\ H_2O}{L\ H_2O} \times \frac{mol\ H_2O}{18.02\ g\ H_2O} = 55.5\ mol\ H_2O/L$$

or

in pure water, $[H_2O] = 55.5\ M$

As a good approximation, we can say that in a dilute aqueous solution of a weak electrolyte (in which the concentration of water is therefore large), the amount of water used up in reaction with the weak electrolyte will not significantly change the concentration of the water.

That is, the concentration of H_2O in such reactions can be assumed constant at about 55.5 M. Let us rewrite the equilibrium equation for the ionization of ammonia.

$$3.1 \times 10^{-7} = \frac{[NH_3^+][OH^-]}{[NH_3][H_2O]} = \frac{[NH_4^+][OH^-]}{[NH_3](55.5\ M)}$$

So that we will not have to rewrite 55.5 M every time we use an equilibrium equation for a weak electrolyte, we multiply both sides of the equation by 55.5 M, which gives us all the constant numbers on the left and all the variable concentrations on the right.

$$(55.5\ M)(3.1 \times 10^{-7}) = \frac{[NH_4^+][OH^-]}{[NH_3]} = 1.7 \times 10^{-5}\ mol/L\ (25\ °C)$$

The equilibrium equation for the ionization of a weak electrolyte is usually written in this way, with the concentration of the water included as part of the constant. Written this way, equilibrium constants for the ionization of weak electrolytes are called **ionization constants,** generally given the symbol K_i. Ionization constants for weak acids are given the special symbol K_a, and those for weak bases K_b. For ammonia, K_b is written as follows.

$$K_b = \frac{[NH_4^+][OH^-]}{[NH_3]} = 1.7 \times 10^{-5}$$

Ionization constants are usually determined by making experimental measurements and calculations based on the values found. Table 16.1 lists some weak acids and their ionization constants.

EXAMPLE 16.6.   Formic acid, HCO_2H, reacts with water to form H_3O^+ and formate ion, HCO_2^-.

$$HCO_2H + H_2O \longleftrightarrow HCO_2^- + H_3O^+$$

If a solution is made with 1.000 mol of formic acid and enough water to make 1.000 L of solution, $[H_3O^+]$ is found to be 1.32×10^{-2} and $[HCO_2H]$ 0.988. What is the ionization constant for formic acid?

Solution:

Step 1.   Write the equilibrium equation.

$$K_a = \frac{[HCO_2^-][H_3O^+]}{[HCO_2H]}$$

Step 2.   Substitute the values for the concentrations. Notice that because one hydrogen ion is formed when each formate ion is formed, $[HCO_2^-] = [H_3O^+]$. Now, do the arithmetic.

$$K_a = \frac{(1.32 \times 10^{-2})(1.32 \times 10^{-2})}{0.988} = 1.76 \times 10^{-4}$$

EXERCISE 16.7.   Calculate the ionization constant for the third ionization of H_3PO_4 when a 0.001 M solution in Na_2HPO_4 has a pH of 7.68.

$$HPO_4^{2-} + H_2O \longleftrightarrow PO_4^{3-} + H_3O^+$$

Table 16.1
Some Weak Acids and Their Ionization Constants

Acid	Formula	Ionization Reaction	K_a
Chlorous	$HClO_2$	$HClO_2 + H_2O \longleftrightarrow ClO_2^- + H_3O^+$	1.0×10^{-2} M
Phosphoric	H_3PO_4	$H_3PO_4 + H_2O \longleftrightarrow H_2PO_4^- + H_3O^+$	7.1×10^{-3} M
Nitrous	HNO_2	$HNO_2 + H_2O \longleftrightarrow NO_2^- + H_3O^+$	4.5×10^{-4} M
Acetic	HAc*	$HAc + H_2O \longleftrightarrow Ac^- + H_3O^+$	1.8×10^{-5} M
Hydrogen sulfide	H_2S	$H_2S + H_2O \longleftrightarrow HS^- + H_3O^+$	9.1×10^{-8} M
Hypochlorous	HClO	$HClO + H_2O \longleftrightarrow ClO^- + H_3O^+$	1.1×10^{-8} M
Hydrogen cyanide	HCN	$HCN + H_2O \longleftrightarrow CN^- + H_3O^+$	2.0×10^{-9} M

*Because of space limitations, acetic acid, H_3CCO_2H is often abbreviated as HAc.
Notes: 1. Some of these acids can ionize a second time or more, as in

$$H_2PO_4^- + H_2O \longleftrightarrow HPO_4^{2-} + H_3O^+ \qquad K_{a2} = 6.2 \times 10^{-8}$$

followed by

$$HPO_4^{2-} + H_2O \longleftrightarrow PO_4^{3-} + H_3O^+ \qquad K_{a3} = 4.4 \times 10^{-13}$$

2. There is only one widely used weak base. It is ammonia.

$$NH_3 + H_2O \longleftrightarrow NH_4^+ + OH^- \qquad K_b = 1.7 \times 10^{-5}$$

For this reason, we will not require a table of ionization constants of weak bases.

We can solve problems that involve weak acids and bases in the same way that we solve other equilibrium problems.

EXAMPLE 16.7.   What is $[OH^-]$ in a solution of ammonia that is prepared by adding enough water to 1.0 mol of NH_3 to make 1 L of solution? K_b for NH_3 is 1.7×10^{-5}.

Solution:

Step 1. Write the reaction equation and the equilibrium equation.

$$NH_3 + H_2O \longleftrightarrow NH_4^+ + OH^-$$

$$K_b = \frac{[NH_4^+] \, [OH^-]}{[NH_3]} = 1.7 \times 10^{-5}$$

Step 2. If the problem is not already stated in terms of concentration, calculate the initial concentrations.

In this problem, we have been given the number of moles of NH_3 placed in 1 L. We can therefore use the numbers of moles directly as concentration values.

Step 3. Substitute terms into the equilibrium equation. Notice that for every mole of NH_3 that reacts, 1 mol of each of the ions is formed. Because of this, it must be true that

$$[NH_4^+] = [OH^-]$$

We will use the letter x for the number of moles of NH_3 that react.

number of moles of NH_3 remaining in 1 L = $[NH_3]$ = $(1.0 - x)$

number of moles of ions created = $[NH_4^+] = [OH^-] = x$

Then

$$\frac{(x)\,(x)}{(1-x)} = K_b = 1.7 \times 10^{-5}$$

Step 4. Solve the equilibrium equation for the value of the unknown.

To find an exact solution to this equation we would have to use the quadratic formula because the unknown, x, appears both in the first and second power. As we found in Section 16.8, however, we can often use approximations to solve such problems more easily.

If we look at the equilibrium equation, we find that the value of K_b is very small. This means that the amounts of the ions formed will ordinarily be very small when compared to the amount of un-ionized ammonia. The amount of ammonia used up in the reaction will therefore be very small compared with the amount originally present. If that is the case, we can approximate and say that the amount of ammonia left is the same as the amount we started with and that $(1 - x) \approx 1$. When we substitute this into the equilibrium equation, the equation becomes easy to solve.

$$K_b = 1.7 \times 10^{-5} = \frac{(x)\,(x)}{(1-x)} \approx \frac{(x)\,(x)}{1} = x^2$$

$$[OH^-] = x = \sqrt{1.7 \times 10^{-5}} = 4.1 \times 10^{-3} \text{ mol/L}$$

We can now go back and see if our approximation was valid. We said that $(1.0 - x) \approx 1.0$; we now find that the value of x is 4.1×10^{-3}, or 0.0041. If we subtract 0.0041 from 1.0, we get 0.9959, which is indeed close to 1.0. That tells us that our approximation was legitimate.

Although the approximation we just used is appropriate for most of the problems we will encounter, we need a way to decide whether to try it. If the values of the concentration terms in a problem are at least 10,000 times larger than the equilibrium constant to be used, it is ordinarily safe to assume that those concentrations will not be significantly changed by any substances that are created or used by the reaction. In Example 16.7, the initial concentration of NH_3 was 1.0 M, and the value of the equilibrium constant was 1.7×10^{-5}, or 0.000017; that is, the concentration was 50,000 times larger than the constant.

EXAMPLE 16.8. What is $[H_3O^+]$ in a solution in which 0.03 mol of hypochlorous acid, HOCl, is dissolved in enough water to make 3 L of solution? The ionization constant, K_a, of HOCl is 3.5×10^{-8}.

$$HOCl + H_2O \longleftrightarrow H_3O^+ + OCl^-$$

Solution:

Step 1. Write the equilibrium equation.

$$K_a = \frac{[H_3O^+]\,[OCl^-]}{[HOCl]}$$

Step 2. Convert given values to initial concentrations.

$$\text{initial concentration of HOCl} = \frac{0.03 \text{ mol HOCl}}{3 \text{ L}} = 0.01 \text{ mol/L}$$

Steps 3 and 4.

Let x stand for the concentrations of H_3O^+ ion and of OCl^- ion. Notice that even though the initial concentration of HOCl is only 0.01 mol/L, the ionization constant is still about 300,000 times smaller. We will make the approximation that the concentration of HOCl is unchanged by the reaction and remains 0.01 mol/L. We substitute this value into the equilibrium equation and do the arithmetic.

$$K_a = 3.5 \times 10^{-8} \text{ M} = \frac{x^2}{0.01}$$

$$x^2 = (3.5 \times 10^{-8})(0.01) = 3.5 \times 10^{-10}$$

$$[H_3O^+] = x = 1.9 \times 10^{-5} \text{ M}$$

As a check, we compare $[H_3O^+]$ with the initial [HOCl]. Less than 1% of the HOCl reacted to make H_3O^+, therefore we can say that the equilibrium concentration of HOCl $\approx$ 0.01.

EXERCISE 16.8.   Calculate $[H_3O^+]$ in an aqueous solution of 0.55 M H_3CCO_2H.

Suppose that we were asked to calculate the hydrogen ion concentration in a solution prepared by adding 0.01 mol of nitrous acid, HNO_2, to enough water to make 1 L of solution. The K_a of HNO_2 is 4.5×10^{-4}. In this solution, the concentration is only about 200 times larger than the ionization constant, so we know that the problem must be solved by means other than approximation, possibly by use of the quadratic formula. We cannot assume in this instance that the concentration of HNO_2 was unchanged by its dissociation.

16.12 BUFFER SYSTEMS

In many reaction systems, including industrial applications and crucial biological systems, chemical equilibria are used to exert automatic, precise control of the concentrations of chemical species. Such control is made possible by what are called **buffer systems.**

An example of an important buffer system is the one that controls the pH of human blood. For the blood to function properly, the bloodstream must maintain a pH between about 7.3 and 7.5. If the pH level ranges even slightly beyond these limits, severe malfunction occurs, usually followed by death. In other words, a change in $[H_3O^+]$ of only one part in a million is usually fatal. How does blood maintain its precise pH, even though acid is continually added or removed? By means of buffer systems.

To see how a buffer system operates, we will use the ammonia-ammonium ion system in water.

$$NH_3 \text{ (aq)} + H_2O \longleftrightarrow NH_4^+ \text{ (aq)} + OH^- \text{ (aq)}$$

$$K_b = \frac{[NH_4^+][OH^-]}{[NH_3]} = 2 \times 10^{-5}$$

Using NH_3 and NH_4Cl, we set up a system that at equilibrium contains 1 mol/L of NH_3 and 1 mol/L of NH_4^+. (We remember that NH_4Cl is soluble and is a strong electrolyte. We ignore the Cl^-.) Now we do some simple arithmetic.

$$[OH^-] = K_b \times \frac{[NH_3]}{[NH_4^+]} = K_b \frac{1}{1} = 2 \times 10^{-5}$$

The pH is 9.3.

As long as the ratio

$$\frac{[NH_3]}{[NH_4^+]}$$

is close to unity, $[OH^-]$ *must* be close to 2×10^{-5}. Suppose now that we were to try to bring the pH to 9 by adding about 10^{-3} mol/L of NaOH; that is, by increasing $[OH^-]$ fiftyfold. What would happen? The added OH^- would not be enough to affect the concentrations of NH_3 and NH_4^+ appreciably, and

$$\frac{[NH_3]}{[NH_4^+]}$$

would still be nearly 1. Therefore, $[OH^-]$ would still be required to be close to 2×10^{-5}. To meet this requirement, the system would use up OH^- by running the equilibrium reaction to the left, increasing $[NH_3]$ to about $(1 + 0.001)$ mol/L and similarly decreasing $[NH_4^+]$ to $(1 - 0.001)$ mol/L. Notice that

$$\frac{1 + 0.001}{1 - 0.001} \approx \frac{1}{1}$$

The action taken by the system to reduce the effect of additional OH^- is an excellent example of Le Chatelier's principle. A buffer is simply a system designed to apply Le Chatelier's principle to the precise control of a concentration.

A buffer can be prepared to control the concentration of any component of an equilibrium system, provided that the concentrations of all the other species on both sides of the reaction system are large in comparison to the concentration of the controlled component. In the system A + B $\longleftrightarrow$ C + D, we could buffer a small concentration of any component by arranging to maintain large concentrations of the other three. In the system E + F $\longleftrightarrow$ G, however, we could buffer only E or F. Buffers are most often employed in acid-base systems in which the pH values are within a few units of 7.

Buffers are most effective if the concentration of the buffered substance is numerically close to the value of the equilibrium constant for the buffering system. We could, for instance, set up a hydroxide ion buffer at $[OH^-] = 6 \times 10^{-5}$ using the ammonia system, simply by setting $[NH_3] = 3$ mol/L and $[NH_4^+] = 1$ mol/L.

$$[OH^-] = \frac{[NH_3]}{[NH_4^+]} \times K_b = \frac{3}{1}(2 \times 10^{-5}) = 6 \times 10^{-5}$$

But if we were to attempt to set OH^- at 10^{-7} mol/L, then

$$[OH^-] = 10^{-7} = K_b \times \frac{[NH_3]}{[NH_4^+]}$$

or $\quad \dfrac{[NH_3]}{[NH_4^+]} = \dfrac{10^{-7}}{K_b} = \dfrac{10^{-7}}{2 \times 10^{-5}} = \dfrac{1}{200}$

We would not be able to put in enough NH_3 to do much buffering. In that case, we would not use the NH_3 system, but would find one with a K value close to 10^{-7}.

EXERCISE 16.9.   What would be $[H_3O^+]$ in a buffer system that contained 0.10 mol/L of HNO_2 and 0.10 mol/L of NO_2^-? What would the pH be? $K_a = 4.5 \times 10^{-4}$

$$HNO_2 \text{ (aq)} + H_2O \longleftrightarrow NO_2^- \text{ (aq)} + H_3O^+ \text{ (aq)}$$

EXERCISE 16.10.   What would be the pH of a buffer system containing 3.2 mol/L of HNO_2 and 0.5 mol/L of NO_2^-? (For data, see Table 16.1.)

EXERCISE 16.11.   What would be the pH of a system containing 1.5 mol/L of HCN and 0.5 mol/L of KCN? $K_b = 5.0 \times 10^{-6}$

$$CN^- + H_2O \longleftrightarrow HCN + OH^-$$

16.13 SOLUBILITY PRODUCT

In Section 15.5, we learned that some solid substances dissolve easily in water, whereas others are insoluble or only slightly soluble. Although we did not then find out what it is that limits the solubility of substances, we now know enough to answer that question. When a compound is at its limit of solubility, an equilibrium exists between the ions on the surface of the solid and those in solution. Under these conditions, as many ions per second leave the surface of the solid as leave the solution to join the solid. We can take as an example the very slightly soluble compound silver chloride, AgCl, for which we can write what looks like a reaction equation to indicate the equilibrium between the solid and the ions in solution.

$$AgCl \text{ (s)} \longleftrightarrow Ag^+ + Cl^-$$

As we can for any equilibrium, we write an equilibrium equation for the process.

$$K = \frac{[Ag^+] [Cl^-]}{[AgCl]}$$

In this equilibrium equation, however, there is a problem. Since the solution contains no undissociated AgCl, what does the concentration term [AgCl] really mean? In an ordinary equilibrium equation, the denominator would give the concentration of *undissociated* compound in solution. There is no such thing for solid ionic compounds, because everything that is in solution is dissociated. As long as solid is present, the term [AgCl], whatever it represents, will remain unchanged, therefore we can simply place it on the left side of our equilibrium equation. (In a similar way, we placed the concentration of water on the left side when we wrote ionization constants.) We define a new constant that is the product of the original constant and the unchanging term [AgCl].

$$K_{sp} = K [AgCl] = [Ag^+] [Cl^-]$$

K_{sp} is the **solubility product** (sometimes called the *ion product*) for a system in which solid compound is in equilibrium with a solution of the ions of that compound. Solubility products are used in exactly the same way that other equilibrium constants are used, except that in using solubility products, the problems are somewhat simplified because the only variable terms in the equation are the concentrations of the ions in solution.

It is not uncommon for reactions to occur between solids and substances in solution, and such reactions all have equilibrium constants. Reactions also occur between gases and solids. In every case in which a solid is involved in a reaction, the solid is not included in the equilibrium equation. We will see another example of this practice in Section 17.8.

When we use solubility products (or, for that matter, any equilibrium constants), we need to remember that their values apply only to equilibrium situations and that all the members of the equilibrium system must be present and at equilibrium for the constants to be valid. Solubility product constants, then, apply only to saturated solutions in contact with undissolved solid. Like most other equilibrium constants, solubility products are temperature-dependent; they usually increase as temperature increases.

Table 16.2 lists some slightly soluble compounds and their solubility products.

Table 16.2
Some Slightly Soluble Compounds
and Their Solubility Products

Solid	Solubility Product
Magnesium carbonate, $MgCO_3$	2.6×10^{-5}
Calcium carbonate, $CaCO_3$	8.7×10^{-9}
Barium carbonate, $BaCO_3$	8.1×10^{-9}
Lithium carbonate, Li_2CO_3	1.7×10^{-3}
Calcium sulfate, $CaSO_4$	3.2×10^{-4}
Barium sulfate, $BaSO_4$	1.1×10^{-10}
Silver chloride, $AgCl$	1.6×10^{-10}
Silver bromate, $AgBrO_3$	5.7×10^{-5}
Copper sulfide, CuS	8.7×10^{-36}

EXAMPLE 16.9.   Solid AgCl is shaken with enough water to make exactly 1 L of finished solution. When the molar concentrations of the dissolved ions are measured, it is found that

$$[Ag^+] = 1 \times 10^{-5} \text{ mol/L}$$

$$[Cl^-] = 1 \times 10^{-5} \text{ mol/L}$$

Calculate the value of the solubility product for AgCl.

Solution:

Step 1.   Write the appropriate equations.

$$AgCl \longleftrightarrow Ag^+ + Cl^-$$

$$K_{sp} = [Ag^+][Cl^-]$$

Step 2. Substitute the values and do the arithmetic.

$$K_{sp} = (1 \times 10^{-5})(1 \times 10^{-5}) = 1 \times 10^{-10}$$

The units on this particular K_{sp} are mol^2/L^2.

EXERCISE 16.12.   What is the value of the solubility product of $PbCrO_4$ if the $CrO_4{}^{2-}$ concentration is 2.12×10^{-3} M in a saturated solution of the compound? Hint: If $[CrO_4{}^{2-}]$ is 2.12×10^{-3}, what must $[Pb^{2+}]$ be?

EXERCISE 16.13.   A saturated solution of CaF_2 is found to have $[Ca^{2+}] = 2.15 \times 10^{-4}$ M. What is the value of the solubility product of CaF_2? (Remember that when one Ca^{2+} ion is formed in solution, two F^- ions also form.) What are the units on K_{sp}?

EXAMPLE 16.10.   The value of the solubility product of silver acetate, CH_3CO_2Ag, is 2×10^{-3} M^2. At 25 °C, what is the molar concentration of Ag^+ in a saturated solution of CH_3CO_2Ag? (This is the same as asking the solubility in moles per liter of CH_3CO_2Ag at 25 °C.)

Solution:

Step 1. Write the equations.

$$CH_3CO_2Ag \longleftrightarrow Ag^+ + CH_3CO_2{}^-$$

$$K_{sp} = [Ag^+][CH_3CO_2{}^-] = 2 \times 10^{-3}$$

Step 2. Let $x = [Ag^+] = [CH_3CO_2{}^-]$.

(Why are the concentrations of the two ions equal?)

Then $K_{sp} = (x)(x) = x^2$

$x^2 = 2 \times 10^{-3}$

$x = \sqrt{2 \times 10^{-3}} = 4.5 \times 10^{-2} = [Ag^+]$

$[Ag^+] = 4.5 \times 10^{-2}$ M

EXERCISE 16.14.   Calculate the molar concentration of Cu^+ in a saturated solution of CuI if the value of the solubility product of CuI is 4×10^{-5}.

16.14 SUMMARY

If a set of substances can react chemically to form products, then the products can also react, in the reverse reaction, to form the original substances. All chemical reactions are reversible. In some, the degree of reversibility is so small as to be unmeasurable, but in others, the reverse reactions occur to such an extent as to exert substantial influence.

The state of balance eventually achieved by competing forward and reverse reactions is called chemical equilibrium. When a reaction system reaches equilibrium, the relative concentrations of products and reactants cease to change. Equilibria limit the extent to which many reactions proceed.

At equilibrium, the relative concentrations of products and reactants can be expressed by the equilibrium equation, the constant of which is called the equilibrium constant. The equilibrium constant is a property of the system at any particular temperature. We can use the equilibrium equation to calculate the concentrations of reactants and products under various conditions. Equilibrium constants for particular kinds of systems are written in special ways and given individual names, such as *ionization constant* and *solubility product*.

Systems at equilibrium respond to changes in conditions, for example, the addition or removal of reacting substances, or changes in temperature or pressure. Such responses are the subject of Le Chatelier's principle, which states that the response of an equilibrium system to change is to reverse, to some extent, the effects of the change. Le Chatelier's principle is often applied to obtain wanted results from equilibrium reactions.

An automatic and practical use of Le Chatelier's principle can be seen in buffer systems, which control the (small) concentration of one component in an equilibrium system by means of relatively large concentrations of other components. Buffer systems are essential to biological systems and industrial chemistry.

QUESTIONS

Section 16.1

1. Explain why many reactions do not go to completion. Does any reaction truly go to completion? Explain.
2. What distinguishes reactants from products in a reaction equation? Is this a fundamental distinction? Explain.

Section 16.3

3. Why, in some reaction systems, do only a few of the collisions between reactants result in reaction?
4. What is activation energy?
5. Other things being equal, why are reactions requiring high activation energy slower than those that require low activation energy?

Section 16.4

6. Explain how and why temperature affects the rates of reactions.
7. How does an increase in temperature affect the rate constant of a reaction? A decrease in temperature?
8. Why do small changes in temperature often produce large changes in reaction rates?
9. Explain how a rise in temperature would affect an endothermic reaction. An exothermic reaction.
10. How are the concentrations of reactants and products involved in the equilibrium process?

Section 16.5

11. Using activation energy diagrams, explain how a catalyst can increase the rate of a chemical reaction.

12. From reading you have done in other chapters, give an example of the use of a catalyst in an industrially important reaction (other than the preparation of sulfuric acid).

Sections 16.6

13. From memory, give the procedure for writing an equilibrium equation.

Sections 16.7 and 16.8

14. How are the values of equilibrium constants usually determined?
15. If the value of an equilibrium constant is small, what can be said about how far the reaction will proceed? If it is large, what can be said?

Section 16.9

16. Explain how adding or removing one reactant or product can affect the concentrations of all the species in an equilibrium system.
17. How do changes in temperature affect an endothermic reaction system at equilibrium? An exothermic reaction system?

Section 16.10

18. Various activation energy barriers and relative energies of the reactants (R) and products (P) in equilibrium systems are shown in the figures that follow. All the systems are at the same temperature.

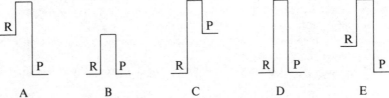

A   B   C   D   E

a. In which system or systems will the equilibrium favor the reactants? Why?
b. In which system or systems will the equilibrium favor the products? Why?
c. In which system or systems will the concentrations of reactants and products be about the same at equilibrium? Why?
d. Which reactions are likely to reach equilibrium soonest?
e. Which reactions are exothermic? If there is more than one of these, which is more exothermic?
f. Which reactions are endothermic? If there is more than one of these, which is more endothermic?
g. In which reaction, if any, is there essentially no enthalpy change?

19. The following reactions are at equilibrium: (Reaction 1 is endothermic.)
   1. $2 HF (g) \longleftrightarrow H_2 (g) + F_2 (g)$
   2. $H_2O (g) + CO (g) \longleftrightarrow CO_2 (g) + H_2 (g)$
   3. $NH_4Cl (s) \longleftrightarrow NH_3 (g) + HCl (g)$
      a. How will the equilibrium of reaction 1 be affected if H_2 is removed from the system? If HF is added to the system? If the temperature of the reaction mixture is lowered?
      b. In reaction 2 the concentration of CO rises as the temperature is raised. Is the reaction exothermic or endothermic? How would the equilibrium be affected if both CO_2 and H_2 were removed?

c. Reaction 3 must be brought to a high temperature before enough NH_3 is present to be measured. What does this tell you about the reaction?

20. What is Le Chatelier's principle?
21. Using the concentrations of A, B, C, and D in the equilibrium reaction system shown, describe in detail why Le Chatelier's principle must in fact be valid if an equilibrium constant is to remain constant. As an example, suppose you were to remove some D from the system at equilibrium. *Why* would the other concentrations change?

$$A + B \longleftrightarrow C + D$$

$$K = \frac{[C] \, [D]}{[A] \, [B]}$$

Section 16.11

22. Why does the concentration of water not appear in equilibrium equations for ionization reactions?
23. Give an example of an equilibrium system in which the concentration of water must appear. Why must it appear here, and not in the reactions mentioned in Question 22?
24. What are K_a and K_b?
25. In calculating concentrations from equilibrium constants, when and where is it possible to ignore the amount of a substance that has reacted? Why is this approximation possible?

Section 16.12

26. What is a buffer system?
27. Explain how a buffer system operates.
28. Why must a buffer system include large concentrations of all the members of the equilibrium except the controlled species?
29. In the design of a buffer system, what two factors determine the concentration of the controlled species? Explain and give examples.
30. From other reading, try to list several systems in which buffers play an essential part.

Section 16.13

31. Explain the operation of the equilibrium in a saturated solution of a slightly soluble salt in contact with its solid.
32. When we write equilibrium constants for systems like that of Question 31, why do we not show the term for the concentration of the solid?

MORE DIFFICULT QUESTIONS

33. Other things being equal, is a small temperature change likely to have a greater effect on a fast reaction or on a slow reaction? Explain.
34. Suppose you assemble a reaction system in which the equilibrium will be very much in favor of the products. You measure the rate of reaction according to how much product appears every second. How will the rate of reaction change, if at all, between the time the reactants are mixed and the time equilibrium is reached? Explain the reasons for your answer.
35. Does a catalyst affect the position of an equilibrium? In other words, will adding a catalyst to a system at equilibrium change the relative amounts of reactants and products, even if it does affect the reaction rates? Explain, using appropriate diagrams.

PROBLEMS

Section 16.4

1. The table gives the characteristics of five reactions, 1, 2, 3, 4, 5, the rates of which are to be compared when the two reactants are first mixed. Assume that no products have yet appeared; thus there are not yet any reverse reactions.

Reaction	Activation Energy	Temperature	Concentrations of Reactants
1	low	low	1 mol/L each
2	high	low	1 mol/L each
3	low	high	1 mol/L each
4	low	low	1 mol/L of one reactant 3 mol/L of second reactant
5	low	low	3 mol/L of each

Which reaction is faster? Explain each answer.

a. 1 or 2   c. 1 or 4   e. 4 or 5
b. 2 or 3   d. 1 or 5   f. 1 or 3

Section 16.6

*2. Write equilibrium equations for the following reactions:
*a. $2 NO (g) + Br_2 (g) \longleftrightarrow 2 NOBr (g)$
b. $SO_2 (g) + Cl_2 (g) \longleftrightarrow SO_2Cl_2 (g)$
c. $CH_4 (g) + O_2 (g) \longleftrightarrow CO_2 (g) + 2 H_2O (g)$
d. $4 NH_3 (g) + 7 O_2 (g) \longleftrightarrow 4 NO_2 (g) + 6 H_2O (g)$
3. Write equilibrium equations for the following reactions:
a. $2 CO (g) + O_2 (g) \longleftrightarrow 2 CO_2 (g)$
b. $2 NO (g) + O_2 (g) \longleftrightarrow 2 NO_2 (g)$
c. $4 HCl (g) + O_2 (g) \longleftrightarrow 2 Cl_2 (g) + 2 H_2O (g)$
d. $4 NH_3 (g) + 5 O_2 (g) \longleftrightarrow 4 NO (g) + 6 H_2O (g)$
*4. At 300 K, the pressure of a sample of O_2 gas is 1.15 atm. What is its concentration in mol/L?
5. The concentration of gaseous HNO_3 in a tank at 750 K is 0.172 mol/L. What is its pressure?

Section 16.7

6. In the preparation of sulfuric acid, O_2 and SO_2 react to give SO_3.

$$O_2 (g) + 2 SO_2 (g) \longleftrightarrow 2 SO_3 (g)$$

In one experiment, O_2 and SO_2 were put into a vessel and allowed to come to equilibrium at 1350 K. When the concentrations of the three substances were measured, $[SO_3] = 1.20$ M, $[O_2] = 0.45$ M, and $[SO_2] = 1.80$ M. Calculate the equilibrium constant for this reaction at this temperature.
*7. Given $[CO] = 4.53$ mol/L, $[O_2] = 5.17$ mol/L, and $[CO_2] = 1.56 \times 10^{-3}$ mol/L at a certain temperature, calculate the equilibrium constant for this reaction in which all the reacting species are gases.

$$2 CO + O_2 \longleftrightarrow 2 CO_2$$

8. The dissociation of H_2S gas is in equilibrium at high temperature.

$$2 H_2S (g) \longleftrightarrow 2 H_2 (g) + S_2 (g)$$

At equilibrium, the concentrations are: $[H_2S] = 0.50$ mol/L, $[H_2] = 0.10$ mol/L, $[S_2] = 0.40$ mol/L.

What is the value of the equilibrium constant for this reaction at this temperature?

9. The ammonia reaction is run in a 2.50-L vessel at a certain temperature.

$$N_2 \text{ (g)} + 3 H_2 \text{ (g)} \longleftrightarrow 2 NH_3 \text{ (g)}$$

The reaction results in the following numbers of moles of each substance: NH_3, 1.00 mol; N_2, 0.50 mol; H_2, 0.75 mol. What is the value of K for this reaction at this temperature?

Section 16.8

10. A 1-L vessel contains 0.015 mol of Br_2, 0.037 mol of NO, and 0.027 mol of NOBr. What is the value of the equilibrium constant? (See Problem 2a.)

*11. Calculate the equilibrium constant for the reaction

$$SO_2 \text{ (g)} + Cl_2 \text{ (g)} \longleftrightarrow SO_2Cl_2 \text{ (g)}$$

At equilibrium at 300 K, a vessel containing this system has the following partial pressures: $P_{SO_2 Cl_2} = 0.0830$ atm, $P_{Cl_2} = 1.91$ atm, $P_{SO_2} = .0664$ atm.

12. For the reaction in Problem 7, given that $[CO] = 1.60$ mol/L and $[O_2] = 4.50$ mol/L, calculate the concentration of CO_2 at equilibrium. $K = 4.27 \times 10^{-2}$.

13. The equilibrium constant for the following reaction is 6.28×10^{-2}.

$$4 HCl \text{ (g)} + O_2 \text{ (g)} \longleftrightarrow 2 Cl_2 \text{ (g)} + 2 H_2O \text{ (g)}$$

Given the following partial pressures, calculate the partial pressure in atmospheres of Cl_2 in the equilibrium system at 400 K: $P_{HCl} = 1.52$, $P_{O_2} = 1.24 \times 10^{-2}$, $P_{H_2O} = 2.75 \times 10^{-4}$.

14. If the equilibrium constant for the following reaction is 9.81, what is the concentration of CH_4 when $[CO_2] = 4.66$ mol/L and $[H_2O] = 2.75$ mol/L?

$$CH_4 \text{ (g)} + O_2 \text{ (g)} \longleftrightarrow CO_2 \text{ (g)} + 2 H_2O \text{ (g)}$$

Section 16.11

*15. Write equilibrium equations for the following reactions:
*a. $H_2CO_3 \text{ (aq)} + H_2O \text{ (l)} \longleftrightarrow H_3O^+ \text{ (aq)} + HCO_3^- \text{ (aq)}$
 b. $HNO_2 \text{ (aq)} + H_2O \text{ (l)} \longleftrightarrow H_3O^+ \text{ (aq)} + NO_2^- \text{ (aq)}$

*16. Calculate $[H_3O^+]$ in an aqueous solution of the following weak acids, each of which was prepared as a 1.00 M solution.
*a. H_3BO_3 $K_a = 4.4 \times 10^{-7}$ (consider only the first stage of ionization of H_3BO_3)
 b. HCN $K_a = 2 \times 10^{-9}$
 c. HF $K_a = 6.7 \times 10^{-4}$
 d. HIO $K_a = 3 \times 10^{-11}$

*17. Calculate $[H_3O^+]$ in an aqueous solution of the following weak acids, each of which was prepared as a 2.50 M solution.
*a. H_2S $K_a = 9.1 \times 10^{-8}$ (consider only the first stage of ionization)
 b. H_2O_2 $K_a = 2.4 \times 10^{-12}$
 c. HN_3 $K_a = 1.2 \times 10^{-5}$
 d. HCO_2H $K_a = 1.7 \times 10^{-4}$

18. Calculate the NH_4^+ concentration in a 1.25 M solution of NH_3. What is the pH of the solution? $K_b = 1.7 \times 10^{-5}$.

*19.   What is the pH of a 0.0250 M solution of HCN?

*20.   Calculate K_a for the sweetener saccharin if a 0.875 M solution of the compound has a pH of 5.87. (For a weak acid, the formula of which you do not know, write the ionization equation as HA + $H_2O \longleftrightarrow H_3O^+ + A^-$.)

21.   Lysine is one of the amino acids. Find the K_a of lysine if a 0.225 M solution has a pH of 5.59.

22.   For glycine, another amino acid, $K_a = 1.67 \times 10^{-10}$. What is the pH of a 1.00 M solution?

23.   Hydroxylamine, NH_2OH, is a weak base. Calculate its K_b if a 1.50 M solution has a pH of 10.1.

24.   Find the pH of the weak base hydrazine in a 1.00 M solution. $K_b = 1.7 \times 10^{-6}$.

*25.   Calculate K_a for the second ionization of H_2CrO_4 if a 0.010 M solution of $NaHCrO_4$ has a pH of 4.2.

26.   Calculate K_a for the second ionization of H_3AsO_4 if a 0.005 M solution of NaH_2AsO_4 has a pH of 4.54.

*27.   Calculate the pH of each of the solutions for which buffer concentrations are given in the table.

Concentrations	K_a
*a. $COOH^-$, 1.0 M; HCOOH, 1.0 M	1.7×10^{-4}
b. $COOH^-$, 0.3 M; HCOOH, 1.5 M	1.7×10^{-4}
c. $COOH^-$, 2.0 M; HCOOH, 0.5 M	1.7×10^{-4}
d. N_3^-, 1.0 M; HN_3, 1.0 M	1.2×10^{-5}
e. N_3^-, 1.0 M; HN_3, 0.10 M	1.2×10^{-5}

28.   Calculate the pH of each of the solutions for which buffer concentrations are given in the table.

Concentrations	K_a
a. F^-, 1.0 M; HF, 1.0 M	6.7×10^{-4}
b. F^-, 0.5 M; HF, 2.5 M	6.7×10^{-4}
c. F^-, 1.8 M; HF, 0.3 M	6.7×10^{-4}
d. HSO_3^-, 6.0 M; H_2SO_3, 0.6 M	1.0×10^{-2}

Section 16.12

29.   What would be $[H_3O^+]$ in a buffer system containing 0.01 mol/L of NaH_2PO_4 and 0.01 mol/L of Na_2HPO_4? What would be the pH?

$$H_2PO_4^- \text{ (aq)} + H_2O \text{ (l)} \longleftrightarrow HPO_4^{2-} \text{ (aq)} + H_3O^+ \text{ (aq)}$$

$$K_a = 6.3 \times 10^{-8}$$

*30.   Calculate the pH of a buffer system containing 0.100 M $K_2H_2P_2O_7$ and 0.300 M $Na_3HP_2O_7$. $K_a = 1.7 \times 10^{-6}$.

31.   Assume that XOH is a weak base with $K_b = 1.1 \times 10^{-4}$. What would be $[OH^-]$ in a buffer system containing 1.50 M XOH and 0.550 M X^+?

*32.   A buffer system consisting of 0.95 M hydroxylamine, NH_2OH, and 0.085 M NH_3OH^+ has a pH of 7.50. Calculate K_b.

*33.   You wish to make a buffer containing 1.25 M hypochlorous acid, HClO, with a pH of 6.5. What concentration of KClO must you use? $K_a = 2.95 \times 10^{-8}$.

34.   Select reasonable concentrations of HNO_2 and $NaNO_2$ needed to make a buffer with a pH of 5.5. $K_a = 4.5 \times 10^{-4}$.

*35.   Given access to the weak acids and their ions listed in the table, tell how to set up buffer systems with the following pH values:
   *a.   pH = 3.2   b.   pH = 2.3   c.   pH = 5.3

Weak Acid/Ion Combinations	K_a
H_3BO_3 and $H_2BO_3^-$	4.4×10^{-7}
H_3CCOOH and H_3CCOO^-	2×10^{-5}
H_3COOH and H_3COO^-	1.7×10^{-4}
H_2SO_3 and HSO_3	1.0×10^{-2}

36. Using any of the weak acid systems of Problem 35, set up a second buffer to give each pH value. Consider the two buffer systems for each pH. Which of the two is the better system (if there is a clear choice)? Give the reasons for your answer.

Section 16.13

Problems 37–41 pertain to saturated solutions at equilibrium.

*37. Calculate the solubility product for $MgCO_3$ if $[Mg^{2+}] = 2.37 \times 10^{-3}$, and for FeS if $[S^{2-}] = 2.24 \times 10^{-9}$.

38. Calculate the solubility product for Ag_2SO_4 if $[SO_4^{2-}] = 2.52 \times 10^{-2}$.

39. Calculate the solubility product of $Mn(OH)_2$ if $[Mn^{2+}] = 1.09 \times 10^{-4}$.

*40. Calculate the molar concentration of the metal ions for the following compounds:
  *a. $CaSO_4$ $K_{sp} = 2 \times 10^{-5}$ c. $MgCO_3$ $K_{sp} = 5.6 \times 10^{-6}$
  b. $CuBr$ $K_{sp} = 4 \times 10^{-8}$ d. $SrCrO_4$ $K_{sp} = 3 \times 10^{-5}$

41. Calculate the molar concentration of the metal ions for the following compounds:
  a. $SrSO_4$ $K_{sp} = 2.8 \times 10^{-7}$ c. $AgBrO_3$ $K_{sp} = 5.2 \times 10^{-5}$
  b. $PbCO_3$ $K_{sp} = 3 \times 10^{-14}$ d. $CuCl$ $K_{sp} = 1 \times 10^{-6}$

MORE DIFFICULT PROBLEMS

42. Use the procedures and suggestions provided to explain how raising the temperature of an exothermic reaction system reduces the equilibrium constant. Explain why the reverse is true for endothermic systems.
  a. After studying Figures 16.3 and 16.4, prepare two separate sketches comparing the molecular energy distribution curves with the activation energies of the forward and reverse reactions of the equilibrium system containing SO_2, O_2, and SO_3. Because the three gases are together in the system, the reactants and products will have the same temperature and therefore the same distribution curve. Assume a temperature and distribution curve that allows about one third of the reactant molecules (and therefore only a small fraction of the product molecules) to pass the barrier.
  b. Keeping in mind the discussion of Section 16.4.1 and referring to your sketches, estimate the relative effects of raising the temperature of the system on rates of the forward and reverse reactions.
  c. Translate the changes of rate into a shift of the entire equilibrium.

43. Gaseous phosphorus pentachloride dissociates as discussed in Example 16.3.

$$PCl_5 \ (g) \longleftrightarrow PCl_3 \ (g) + Cl_2 \ (g)$$

The value of the equilibrium constant for this reaction, as expressed in concentration terms, is

$$K = \frac{[PCl_3] \ [Cl_2]}{[PCl_5]} = 4.22 \times 10^{-2} \text{ at } 250 \ °C$$

  a. Write an equilibrium constant for this reaction as expressed in the partial pressures of the gases. Use the standard symbol for the partial pressures of the gases.
  b. What is the value and what are the units of this equilibrium constant at 250 °C?

Chapter 17

Oxidation-Reduction Reactions

Chemical reaction always involves the outer electrons of atoms. Often, electrons are actually transferred from one particle to another. Reactions in which electrons are transferred are called oxidation-reduction (redox) reactions.

Electron transfer reactions are in some ways similar to proton transfer (acid-base) reactions. We use special language to describe both kinds of reactions, and both kinds are worth studying. One difference between proton transfer and electron transfer is that electrons can travel through wires, whereas protons cannot; electron transfer reactions can therefore generate electric current and be controlled by electric current.

In this chapter, we review and strengthen our knowledge of oxidation-reduction reactions. We learn how to assign oxidation numbers and how to use those numbers to balance difficult reaction equations. We also study the equations of some significant oxidation-reduction reactions. We then learn a few rules for naming oxidation states. At the close of the chapter, we find out how electric current can bring about electron transfer reactions, and how electric batteries work.

AFTER STUDYING THIS CHAPTER, YOU WILL BE ABLE TO

- explain the process of oxidation-reduction
- determine the oxidation number of an atom in a molecule or ion
- find out whether a reaction is an oxidation-reduction
- determine which species have been oxidized and which have been reduced in an oxidation-reduction reaction
- write oxidation and reduction half reactions, given a redox reaction
- balance redox reactions both for the number of atoms and the charges
- write balanced reaction equations for some common and important redox reactions
- name compounds, using either the Stock system or the "old" system

- describe how a battery functions
- use the activity series to determine whether a reaction will occur between two given species

TERMS TO KNOW

activity series	Daniell cell	electrode
anode	disproportionation	electrolysis
battery	electric circuit	oxidation number system
cathode	electricity	redox

17.1 REVIEW OF OXIDATION-REDUCTION REACTIONS

We learned in Section 11.7 that atoms or other particles can be oxidized or reduced. When an atom gains an electron in an electron transfer reaction, that atom is said to be reduced; when an atom loses an electron, the atom is said to be oxidized. An electron transfer requires both an electron donor and an electron receiver. Therefore, in an oxidation-reduction reaction, there must be reduction every time there is oxidation. Oxidation-reduction reactions are often called **redox** reactions.

Anything that takes electrons from a substance oxidizes that substance and is called an oxidizing agent. Anything that donates electrons to a substance reduces that substance and is said to be a reducing agent. As it oxidizes something, an oxidizing agent gets reduced; as it reduces something, a reducing agent gets oxidized. If you go back to Section 11.8 and review these definitions and principles in detail, the rest of this chapter will be much easier.

In some redox reactions, electron transfer occurs in such a way that it is easy to see where the electron came from and where it went; that is, to find out which substance was the oxidizing agent and which the reducing agent. In other reactions, however, it is not as obvious where electrons came from or where they went—or even how many electrons were involved in the reaction. In yet other redox reactions, electrons are not transferred entirely; there is merely a change in the pattern of electron sharing. An electron assignment system, developed for redox reactions and called the **oxidation number system,** makes such reactions easier to understand. Knowing how to use oxidation numbers also makes it possible to balance reaction equations systematically and to understand reactions in which electric current plays a part.

17.2 OXIDATION NUMBER SYSTEM

The reaction of Na and Cl atoms is a simple electron transfer that is easy to understand.

$$Na + Cl \longrightarrow Na^+ + Cl^-$$

Some reactions, however, are less easy to follow, although oxidation and reduction have clearly occurred. Dichromate ion, $Cr_2O_7^{2-}$, and ethyl alcohol, C_2H_6O, give us an example.

$$Cr_2O_7^{2-} + 3\ C_2H_6O + 8\ H_3O^+ \longrightarrow 3\ C_2H_4O + 2\ Cr^{3+} + 15\ H_2O$$

We know that this is an electron transfer reaction, but where did the electrons come from and where did they go? Oxidation numbers give us an organized method of answering such questions.

The oxidation number system is a counting mechanism that keeps track of electrons in elements and compounds. Even in covalent compounds, in which the electrons are shared, it assigns the electrons to the bonded atoms. Oxidation numbers allow you to follow the paths of electrons during reaction and to say which substances were oxidized and which reduced. In many cases, the assignment of oxidation numbers is somewhat arbitrary, but because the system is only a kind of bookkeeping, the arbitrary assignments cause no problem—as long as the system is consistently applied.

The oxidation number system assigns oxidation numbers to individual atoms in elements or compounds, bonded or not. Oxidation numbers can be positive or negative and are sometimes fractions. *Losing electrons* (being oxidized) *causes the oxidation number* of a species to *increase algebraically*.* For example, when sodium atoms are oxidized, they lose their electrons to become Na^+ ions, and the oxidation number of sodium increases from 0 to +1. *Gaining electrons* (being reduced) *makes the oxidation number* of a species *decrease algebraically*. The reduction of Cl atoms to Cl^-, for example, decreases the oxidation number of the chlorine from 0 to −1. Oxidation numbers are assigned according to a set of rules. The rules that follow include most of the generally accepted set, although other sets could be devised. The third rule is the most essential.

Rule 1.   The oxidation number of hydrogen in its compounds is +1 (except in the metal hydrides, such as NaH; hydrides are uncommon and occur only with active metals).

Rule 2.   The oxidation number of oxygen in its compounds is −2 (except in those compounds containing O—O bonds, peroxides, such as H_2O_2; peroxides have the word *peroxide* in their names).

Rule 3.   The sum of the oxidation numbers of all the atoms in a molecule or ion must equal the charge on that molecule or ion. If the particle is a neutral molecule, the sum of oxidation numbers is 0. In N_2, for example, each N atom has the oxidation number 0; the sum of the oxidation numbers in N_2 is therefore 0. In H_2O, the oxidation number of each H is +1, that of O is −2, and the sum is 0. In sulfate ion, SO_4^{2-}, the oxidation numbers add up to −2 (+6 for S and −2 for each O).

Rule 4.   With only a few exceptions, such as I_3^-, the oxidation number of the atoms in any element not combined with another element is 0.

Rule 5.   To assign oxidation numbers to individual atoms in covalent compounds, assign the electrons in each bond to the more electronegative atom sharing those electrons. (Remember that the sum of the oxidation numbers of neutral molecules must be 0.)

Rule 6.   The algebraic sum of the oxidation numbers of a group of atoms (such as a polyatomic ion) usually remains constant if those atoms remain together during the reaction in question. For example, in reactions of H_2SO_4 in which the sulfate ion is not altered, the entire ion can be assigned the oxidation number −2, which is the sum of −8 for the oxygens and +6 for the sulfur. See Example 17.3.

---

*The words *algebraic* and *algebraically* refer here to operations along the number line, as explained in Appendix A3. To increase algebraically means to move toward the right on the number line, regardless of whether the actual numbers themselves become larger or smaller. A move from −6 to −3, for instance, is an algebraic increase, as is a move from 4 to 7, or from −3 to 1.

Although we need to be wary of exceptional cases, we can often assign oxidation numbers by relying on our general knowledge of chemistry and considering the position of the elements in the periodic table. Here are some guidelines.

Atoms of the metallic elements, because they lose electrons in reactions, generally form positive ions and therefore have positive oxidation numbers. Atoms or groups of atoms of the nonmetals generally have negative oxidation numbers in their ionic compounds, but in covalent compounds can have positive or negative oxidation numbers. The number of electrons gained or lost in the formation of ionic compounds is often a characteristic of an elemental family, so the positions of elements in the table can sometimes help us determine oxidation numbers. Table 17.1 lists the possible oxidation numbers for the Group A elements. Many elements and ions have invariant oxidation numbers. For instance, Na always has an oxidation number of +1 in its compounds. The number is positive because Na is a metal, and it has a value of 1 because, in chemical reaction, each atom of Na can lose only one electron. Similarly, Mg always has an oxidation number of +2 in its compounds. A list of some useful, invariable oxidation numbers follows.

1. The elements of the first column of the periodic table (H through Cs) have +1 oxidation number in their compounds. An exception is H in hydride compounds, with a number of −1. Ag^+ and NH_4^+ are always assigned oxidation number +1.
2. In their compounds, the elements Mg, Ca, Sr, Ba, Cd, and Zn have the oxidation number of +2. Al has oxidation number +3 in its compounds.
3. Hydroxide and cyanide ions, OH^- and CN^-, have an oxidation number of −1.
4. In binary compounds with metals, the halogens (F through I) have an oxidation number of −1. Fluorine has an oxidation number of −1 in *all* its compounds.

Although electrons are not entirely transferred when covalent bonds are formed, it is often convenient to assign oxidation numbers to covalently bonded atoms. For example, consider the reaction of bromide ion with chlorate ion in acid solution.

$$ClO_3^- + 6\ Br^- + 6\ H_3O^+ \longrightarrow Cl^- + 3\ Br_2 + H_2O$$

In this reaction, bromide ions obviously had to lose electrons to become Br_2. (If this is not clear to you, draw electron dot pictures for Br^- and Br_2 and count the electrons. Do not go on until you have the pictures firmly in mind.) But we want to know where those electrons went. We can find out by assigning oxidation numbers and watching the numbers change during the reaction.

To assign oxidation numbers to covalently bonded atoms, we first make assignments to those atoms that have invariant oxidation numbers. Then we can often complete the process

Table 17.1
Oxidation Numbers Possible to the
Elements of Groups IA Through VIIA

Group	IA	IIA	IIIA	IVA	VA	VIA	VIIA
Possible oxidation numbers	+1	+2	+3	+4 to −4	+5 to −3	+6 to −2	+7 to −1

by using Rule 3 and summing to zero. In SiH_4, for example, we would assign +1 to each of the H atoms, which would require Si to be assigned −4; that is, $(4 \times +1) -4 = 0$. In cases where there are no invariant oxidation numbers to guide us, we begin by assigning both electrons in each shared electron pair to the more electronegative of the bonded pair of atoms. The oxidation number of each atom is then the same as the charge on that atom would be if the electrons were really on the atoms to which we assigned them. (To review electronegativity, see Section 13.8.2.)

EXAMPLE 17.1.   What are the oxidation numbers of H and S in H_2S?

Solution:

We can most easily approach this problem by remembering that H always has an oxidation number of +1 in its compounds (except hydrides). This gives a total of +2 for the two H atoms, and requires S to have a number of −2 for the total to come out to 0.

　　To familiarize ourselves with the electronegativity approach, let us make the assignment also on that basis. In the electron dot drawing, we see that S shares a pair of electrons with each of the two H atoms. Consulting Figure 13.13, we find that the electronegativity of H is 2.0 and that of S 2.5. We therefore assign both electrons in each pair to S, which gives the S atom a total of eight electrons in its outer shell. Since a neutral S atom has only six electrons in its outer shell, S in H_2S has two extra electrons, giving it an assigned charge of −2, as well as an oxidation number of −2. Each of the H atoms must then have an oxidation number of +1.

EXAMPLE 17.2.   What are the oxidation numbers of carbon and chlorine in carbon tetrachloride, CCl_4?

Solution:

In this case, we have no invariant numbers to use as a guide, so we rely on electronegativities instead. From Figure 13.13 we find that Cl is significantly more electronegative than C. The electron dot drawing shows that the carbon shares one pair of electrons with each chlorine. If we then assign both electrons in each pair to chlorine, the carbon has no electrons in its outer shell and a resulting oxidation number of +4. Each chlorine receives one extra electron and is assigned an oxidation number of −1.

EXERCISE 17.1.   What are the oxidation numbers of phosphorus and of bromine in PBr_3? In PBr_5?

EXERCISE 17.2.   In chloroform, $CHCl_3$, what are the oxidation numbers of carbon, chlorine, and hydrogen?

In larger covalent molecules, the assignment of oxidation numbers becomes quite arbitrary and rather complex. Sometimes different atoms of the same element have different oxidation numbers in the same molecule, depending on the structures into which they are bonded. Oxidation numbers occasionally turn out to be fractions. We will come across some unusual cases in Chapter 19, when we learn about the oxidation of compounds that contain more than one carbon atom.

Although we will sometimes need to count electrons and assign our own oxidation numbers to individual atoms, as we did with sulfur atoms, we will usually be able to determine oxidation numbers by relying on the invariant oxidation numbers of a few elements and doing a little arithmetic.

EXAMPLE 17.3. What are the oxidation numbers of the atoms in sulfate ion, SO_4^{2-}?

Solution:

Sulfate ion has no O—O peroxide bonds, so each oxygen atom must have an oxidation number of −2. Remember that the sulfate ion has a double negative charge and that the sum of the oxidation numbers of all the atoms must therefore be −2 (Rule 3). We can write this relation:

sum of oxidation numbers of four oxygen atoms + oxidation number of S = −2

therefore

$$4 (-2) + \text{oxidation number of S} = -2$$

$$\text{oxidation number of S} = -2 + 8 = +6$$

EXAMPLE 17.4. What are the oxidation numbers of the atoms in sulfurous acid, H_2SO_3?

Solution:

There are neither hydride ions nor peroxides in H_2SO_3, therefore we assign oxidation numbers of +1 to hydrogen and −2 to oxygen. Because H_2SO_3 is a neutral molecule, the sum of its oxidation numbers must be 0.

$$2(+1) + 3(-2) + \text{oxidation number of S} = O$$

$$\text{oxidation number of S} = 6 - 2 = +4$$

EXAMPLE 17.5. Sulfurous acid is placed in water solution and ionizes to SO_3^{2-}. What is the oxidation number of sulfur after this occurs?

Solution:

$$3(-2) + \text{oxidation number of S} = -2$$

$$\text{oxidation number of S} = -2 + 6 = +4$$

Notice that the ionization of the acid did not change the oxidation number of S in H_2SO_3. Since ionization and dissociation of a compound is not a redox reaction, no oxidation numbers change.

In three of the last four examples, the oxidation number for sulfur has assumed three different values: −2, +4, and +6. Sulfur is one of the many elements that can have any of several oxidation numbers, depending on the nature of the compound it is in. Table 17.2 gives the oxidation number values most commonly found for several elements.

EXERCISE 17.3.   What are the oxidation numbers of nitrogen and oxygen in NO_3^-?

EXERCISE 17.4.   What is the oxidation number of nitrogen in NO_2^-?

17.3 BALANCING REDOX EQUATIONS

We have learned in previous chapters how to balance simple reaction equations. Some equations are so easy that you can write the answer immediately; others require a little thought and perhaps some trial and error. Many equations, especially those involving redox reactions, are so difficult to balance that the trial and error method can take hours. To save ourselves time, we need an organized approach.

Complex redox equations can be balanced by any of several methods, all of which depend on one fact: in an electron transfer reaction, no electrons are created or destroyed. The number of electrons gained by the atoms of one group must equal the number of electrons lost by the atoms of another group. One of the easiest methods is based on the following steps. (Along with the explanation we will balance the equation for the oxidation of aluminum by chlorine.)

Step 1.   Below the symbol for each kind of atom in both the reactants and products, write the oxidation number for that species of atom. Make a written note of which oxidation numbers change and how much they change.

$$Al + Cl_2 \longrightarrow AlCl_3$$
$$0 \quad0 \quad+3-1$$

The oxidation number of Al changes from 0 to +3.
The oxidation number of Cl changes from 0 to −1.

Table 17.2
Principal Oxidation Numbers of Some Common Elements

Element	Oxidation Numbers	Element	Oxidation Numbers
As	+3, +5	Cl	−1, +1, +3, +5, +7
Au	+1, +3	Br	−1, +1, +3, +5, +7
Cr	+2, +3, +6	I	−1, +1, +3, +5, +7
Cu	+1, +2	S	−2, +4, +6
Co	+2, +3	N	−3, +1, +2, +3, +4, +5
Hg	+1, +2	P	−3, +3, +5
Pb	+2, +4	C	*Several possibilities
Sn	+2, +4		

*See Chapter 19.

Step 2. Divide the reaction equation into two half reactions, one for the substance that is being oxidized (the one that has atoms in which the oxidation number increases) and one for the substance that is being reduced. Include in the half reactions only the atoms being oxidized and reduced. Write the oxidation number of each atom as you would write the charge on an ion. (The oxidation number of the ion itself is the same as its charge.)

$$Al \longrightarrow Al^{3+} \text{ (oxidation)}$$
$$Cl_2 \longrightarrow 2\ Cl^- \text{ (reduction)}$$

Step 3. Remembering that oxidation is the loss of electrons and reduction the gain of electrons, add to each half reaction the number of electrons required to bring about the oxidation number change that occurs in that half reaction. (It is convenient to use the symbol e^- for an electron.) In the half reaction in which reduction occurs, the electrons are added to the left side. In the oxidation equation, they are added to the right. If, for example, in a half reaction, Mg is oxidized to Mg^{2+}, then $2\ e^-$ must be added to the right side of the equation.

$$Al \longrightarrow Al^{3+} + 3\ e^- \text{ (oxidation)}$$
$$Cl_2 + 2\ e^- \longrightarrow 2\ Cl^- \text{ (reduction)}$$

Notice that the charges and oxidation numbers balance. (In this example, the total of the charges and oxidation numbers on each side of the equation happens to be zero.)

Step 4. Multiply all the terms in each half reaction by a number that will make the number of electrons used in each equation the same. (In many cases, that number will be unity for one of the equations.)

$$2\ Al \longrightarrow 2\ Al^{3+} + 6\ e^- \text{ (oxidation)}$$
$$3\ Cl_2 + 6\ e^- \longrightarrow 6\ Cl^- \text{ (reduction)}$$

Step 5. Add together the two new half reactions and cancel out the terms that appear on both sides. If the process has been done correctly, the numbers of electrons will cancel each other and will not appear in the final equation. Now replace and balance any substances left out when the half reactions were first written.

$$2\ Al + 3\ Cl_2 + \cancel{6\ e^-} \longrightarrow 2\ AlCl_3 + \cancel{6\ e^-}$$

Step 6. Check the numbers of atoms of each kind on both sides of the equation, to be sure no mistake has been made. In a reaction equation, it is necessary to balance not only the numbers of atoms of each kind, but also the total number of electrons. This means that if there are ions involved in the reaction, the algebraic totals of all the charges on each side of the reaction must be equal. When checking an equation for balance, be sure to count charges, too.

$$2\ Al \longrightarrow 2\ Al$$
$$6\ Cl \longrightarrow 6\ Cl$$

Charge adds up to zero on both sides.

EXAMPLE 17.6. Balance the following reaction equation:

$$KClO_4 + H_3AsO_3 \longrightarrow KCl + H_3AsO_4$$

The **carbonates,** solid salts of carbonic acid, are of tremendous importance. Much of the earth's crust is composed of limestone, a crystalline form of $CaCO_3$ and $MgCO_3$. Acids readily attack limestone in a reaction that is the reverse of the one in which CO_2 dissolves in H_2O.

$$CaCO_3 + 2\ H_3O^+ \longrightarrow Ca^{2+} + CO_2 + 3\ H_2O$$

Pearls and seashells are primarily carbonates. Sodium carbonate, Na_2CO_3, and sodium hydrogen carbonate, $NaHCO_3$ (better known as sodium bicarbonate, or baking soda), are important industrial chemicals.

18.6.3 Silicon

Silicon has the same outer electron population as carbon, but its properties are different, primarily because the outer shell of carbon can contain only 8 electrons, whereas the outer orbitals of silicon can accommodate 18 ($3s^2$, $3p^6$, $3d^{10}$). Samples of elemental silicon have a dull, metallic luster reminiscent of lead; however, silicon is much less dense than lead and behaves chemically like the nonmetal it is.

We are interested in silicon primarily because it is the second most abundant element in the earth's crust. A large part of the earth is made up of siliceous rocks—feldspar ($KAlSi_3O_8$), for example. Beach sand is primarily silicon dioxide, SiO_2.

Silicon finds an important but limited use in the production of synthetic materials, such as silicone rubber. These have attractive properties (resistance to heat and solvents, for instance) not possessed by similar materials not containing silicon. Glass is made mostly from SiO_2. Concrete is a mixture of carbonates and silicates.

18.7 SOME GENERALIZATIONS ABOUT NONMETALS

Our discussion of nonmetals has focussed on the characteristics of elemental families and on their variations within a group. As a summary, let's take a brief overview of the entire class of nonmetals and look at a few examples of the effects of kernel charge.

Of all the elements in the periodic table, the elements that are plainly nonmetallic are in the minority, numbering about 18 in a list of more than 100. Of these, six elements are essentially inert noble gases, which leaves about a dozen active nonmetals.

Of that dozen, five are halogens and seven are not; we have examined the halogens thoroughly. Since selenium is uncommon, that leaves only six: H, C, O, N, P, and S. This group is small, but of enormous importance. If you understand the chemistry of these six elements along with that of the halogens and the general chemistry of the metals, you will have a good, fundamental background in descriptive chemistry.

As a final example of the effects of atomic structure on the properties of the nonmetals, let us review the properties of three nonmetal hydrides. The hydrides of N, O, and F show a range of behavior that clearly displays the effect of kernel charge. Examine the structures of the compounds.

$$\overset{\displaystyle H}{\underset{\displaystyle H}{H:\overset{\displaystyle ..}{N}:}} \qquad \overset{\displaystyle H}{H:\overset{..}{\underset{..}{O}}:} \qquad H:\overset{..}{\underset{..}{F}}:$$

All have at least one pair of unshared electrons. In aqueous solution, ammonia acts as a base and readily accepts a proton to become NH_4^+. Under the same circumstances, hydrogen fluoride acts as an acid and gives up its proton. The kernel charge accounts for this difference in properties. Fluorine attracts all four electron pairs so strongly that the H—F bond is very polar. By contrast, the N—H bonds are stronger; in ammonia, the unshared pair is weakly held and is available for bonding. Water lies between these extremes.

The hydrides of the other members of the families of N, O, and F show similar trends, but NH_3, H_2O, and HF form a distinct set because of their unique tendency to form hydrogen bonds. If we are careful, it is possible to make similar comparisons in other classes of compounds, but there are many exceptions and special cases.

18.8 CHARACTERISTICS OF THE CHEMISTRY OF METALS

We now leave the nonmetals to consider the metals, the elements on the left and lower portions of the periodic table. The behavior of the metals exhibits less variety than that of the nonmetals. The chemistry of the metals does not, for the most part, depend on the subtle influences that determine the character of covalent bonds.

If metals have any one characteristic that is so typical as to be definitive, it is their behavior in chemical reaction. In reaction, each atom of a metallic element usually gives up one or more electrons to become a positive ion.

Atoms of metals cannot *accept* electrons to become negative ions. If atoms of a typically metallic element are mixed with atoms of another metal—even one that is not as strongly metallic—both elements will still give up electrons and create metallic bonding in the solid sample. If both elements are truly metallic, neither will accept electrons.

Some metals do, however, participate in covalent bonding. A very few metallic compounds ($HgCl_2$ for example) are covalent. Covalency can also occur in solid metallic crystals between inner shell electrons of ions that have donated their outer shell electrons to the "electron sea." Such tendencies are found in the transition elements and are responsible for the strength and hardness of elements like iron and chromium.

The reason that atoms of metallic elements give up electrons is, of course, that they have low ionization energies and low electron affinities. The lack of strong attraction between the kernels of metallic atoms and electrons is one of the fundamental causes of electron transfer. It also profoundly determines the properties of the compounds these reactions form.

18.9 ALKALI METAL FAMILY

Metallic characteristics occur in their perfect form only in an imaginary "ideal metal," because every kernel of a real atom has at least some attraction for electrons. A few metals, however, do approach ideality, and the heavier members of the alkali* metal family come closest of all.

Technically, the alkali metal family (Table 18.5) includes all the elements that lie in the leftmost column of the periodic table, all of which have one electron in the outer shell of each

---

*The word *alkali* comes from an Arabic word meaning *bush,* because the ashes of bushes were found to be rich in compounds of what we now know as Na and K. In ordinary usage, alkali came very early to be associated with basic compounds such as the carbonates of Na and K found in deserts.

Table 18.5
The Alkali Metals

Name	Symbol	Atomic Number	Atomic Weight	Melting Point (°C)	Boiling Point (°C)	Atomic Radius (Å)	Ionization Energy (kJ/mol)
Lithium	Li	3	6.939	186	1336	1.55	522
Sodium	Na	11	22.990	97.5	880	1.90	493
Potassium	K	19	39.102	62.3	760	2.35	415
Rubidium	Rb	37	85.47	38.5	700	2.48	406
Cesium	Cs	55	132.91	28.5	670	2.67	377

neutral atom. As we already know, however, hydrogen is different from the elements that lie directly below it, and is not considered an alkali metal.

The alkali metals, then, include the elements Li, Na, K, Rb, Cs, and Fr. The first five are easily available in the laboratory and are thoroughly characterized. Francium, Fr, is radioactive and does not occur in nature. Its chemistry is less well known, but is similar to that of the rest of the family.

Except for some minor deviations, as in lithium, which has a small kernel and a fairly concentrated kernel charge, the chemical properties of the alkali metals are almost identical and are typified by sodium, each of whose atoms readily gives up one electron in chemical reaction. Chemically, the ions of the alkali metals are essentially inert, because they have closed shell structures and little attraction for electron pairs on other atoms. Because of the tendency of their ions to be inert and because they are so much alike, the alkali metals below Li are difficult to separate chemically from one another.

Although alkali metal ions are inert, the metals themselves are among the most chemically active elements known. Their great reactivity arises from the fact that neutral atoms of the alkali metals cannot defend their outer electrons from attack and seizure by other atoms and molecules. Because they are so willing to give up electrons and because the world is full of electrophilic substances, all naturally occurring alkali metals have long since reacted and become ions; they are not found as metals in nature.

The nonmetals show a large gradation of properties within families. Although the properties of metals depend on their atomic radii, family gradations are not large, and, once ions are formed, often become undetectable. In the alkali metal family, the reactivity of the metals with water increases somewhat with atomic weight. Sodium reacts rapidly with water, K a little more rapidly, and so on to Cs, which reacts most rapidly and vigorously of all. The difference in reactivity illustrates the decrease in ionization energy as the kernel charge becomes increasingly diffuse in the larger atoms. When we consider the ions, however, the story is different. The strong resemblances of Na^+, K^+, Rb^+, and Cs^+ make them quite difficult to separate chemically.

18.10 ALKALINE EARTH METALS

Next to the alkalis, in the second column of the periodic table, is the family called the alkaline earth metals (Table 18.6). All the members from Be through Ra fall to the left of the boundary between metals and nonmetals, but the lighter members of the family approach the

Table 18.6
The Alkaline Earth Metals

Name	Symbol	Atomic Number	Atomic Weight	Melting Point (°C)	Boiling Point (°C)	Atomic Radius (Å)	Ionization Energy (kJ/mol)
Beryllium	Be	4	9.012	1278	2970	1.21	898
Magnesium	Mg	12	24.31	650	1100	1.60	732
Calcium	Ca	20	40.08	810	1439	1.97	589
Strontium	Sr	38	87.62	752	1377	2.15	541
Barium	Ba	56	137.3	830	1737	2.31	502

boundary so closely that they no longer act as ideal metals ($BeCl_2$, for example, is almost covalent). Early chemists called powdery, nonmetallic substances "earths." Since the oxides of the family were discovered before the metals, the "earth" name stuck. The present name was coined later, after resemblances to the alkali metals had been observed.

In chemical reaction, the alkaline earth metals typically lose two electrons per atom, forming ionic compounds with formulas like $MgCl_2$, BaO, and $Ca(OH)_2$. It is not as easy, however, to remove electrons from atoms of the alkaline earth elements as it is from those of the alkali metals. For example, it requires about 493 kJ to ionize 1 mol of Na, but 732 kJ to produce 1 mol of Mg^+. To remove a second electron from a magnesium atom requires about 1440 kJ/mol. This reflects the fact that the removal of the first electron is facilitated by the repulsion between it and its neighbor in the $3s$ shell. If these values seem large, compare them with the ionization energy of nearly 8 million kJ for the *third* electron of Mg (representing penetration of a closed shell), or 1500 kJ for the first electron of argon.

If the second electron is so much more difficult to remove than the first, why do the alkaline earth elements always appear in their compounds as doubly charged ions? Some of the energy required to remove the second electron is furnished by the energy advantage of creating a doubly charged ion, smaller than the neutral atom, close to which electron pairs of other species can come. In assessing the probability of any particular chemical process, we must consider the total energy change of an entire process, not just the gain or loss that occurs in one step.

The fact that the ionization energies of the alkaline earth elements are higher than those of the alkali metals causes the alkaline earth metals to be less reactive. The dependence of reactivity on the strength of electron binding, however, is exhibited more clearly by the alkaline earth elements than by the alkali metals. Magnesium will not react with cold water, but produces H_2 slowly from hot water and more rapidly from steam. Calcium will react slowly with cold water, rapidly with warm water. Barium reacts rapidly with cold water, in about the same way that sodium does. Look at the ionization energies listed in Tables 18.5 and 18.6, and correlate these reactivities with the values you see there.

The alkaline earth elements are not used extensively as metals, except for Mg, which makes light, strong alloys used in aircraft construction, racing cars, stepladders, and elsewhere. Calcium, an important component of the earth's crust, is found in limestone, calcite, and marble, which are primarily $CaCo_3$. Calcium compounds are responsible for the rigidity and strength of bones. Radium, all isotopes of which are radioactive, has medical application where small sources of intense radioactivity are needed.

18.11 ALUMINUM

Of the trivalent (or three-electron) metals, only aluminum is of special interest to us. Aluminum is one of about a dozen elements that have brought about large changes in our way of life, but whose contribution is nevertheless taken for granted.

Aluminum metal was first prepared less than 100 years ago. It is one of the most widely used metals, both in industry and by consumers, yet it would be virtually useless except for one interesting, almost accidental, characteristic. Aluminum is a highly active metal; it is a good reducing agent. Even though Al reacts readily with oxygen in the air and with water, structures made of aluminum last for years, and we cook in aluminum pans. The cause of this apparent contradiction is that aluminum forms an oxide layer that adheres tightly to the metal, effectively sealing it off from any reactant except those that attack the oxide itself. Our aluminum pans are covered by a thin, almost invisible film of the oxide. If the film is scratched away, it forms again within seconds. Without its oxide layer, aluminum would be a mere laboratory reagent.

18.11.1 Chemistry of Aluminum

Like other metals we have studied, aluminum loses electrons in chemical reaction. Because it is triply charged, however, the positive Al ion is not easily separated from the negative products of the reaction. The ion Al^{3+} does not remain in solution as such. For instance, if the hydroxide is prepared in the presence of water, the product of the reaction is insoluble $Al(OH)_3$. The attraction of Al^{3+} for electrons is so great that in the presence of sufficient water, the hydroxide is actually a stable hydrate with the formula $Al(OH)_3(H_2O)_n$, where n is typically 3 or more. If the water of hydration is driven off by heating, the hydrogens rearrange, and the hydroxide decomposes to the oxide.

$$2\ Al(OH)_3(H_2O)_3 \longrightarrow Al_2O_3 + 9\ H_2O$$

This reaction is reversible. To prepare the hydroxide from the oxide, simply add water.

Aluminum hydroxide has properties that give evidence of the proximity of Al to the nonmetals in the periodic table. The hydroxide is capable of either acidic or basic behavior. In the presence of acids, the solid $Al(OH)_3(H_2O)_3$ dissolves as a base.

$$3\ H_3O^+ + Al(OH)_3(H_2O)_3 \longrightarrow Al(H_2O)_3^{3+} + 6\ H_2O$$

In the presence of bases, however, it acts as an acid and donates protons to the base.

$$Al(OH)_3(H_2O)_3 + OH^- \longrightarrow Al(OH)_4(H_2O)_2^- + H_2O$$

The acidic reaction occurs because the aluminum ion attracts the electron pairs of the bound water so greatly that the O—H bonds in the water molecules are weakened, and protons can be pulled from them by strong bases. Aluminum metal itself will react with NaOH, liberating hydrogen.

$$2\ Al + 2\ OH^- + 2\ H_2O \longrightarrow 2\ AlO_2^- + 3\ H_2$$

$$AlO_2^- + 2\ H_2O \longrightarrow Al(OH)_4$$

Substances that can act either as acids or bases are said to be **amphoteric.** Some other amphoteric hydroxides are $Zn(OH)_2$, $Pb(OH)_2$, and $Cr(OH)_3$. Because of this behavior, one should not put lye (NaOH or KOH) in galvanized (zinc-coated) or aluminum vessels.

18.11.2 Metallurgy of Aluminum

Although aluminum is the third most abundant element in the earth's crust, metallic Al was nothing more than an expensive laboratory curiosity until the end of the nineteenth century. The metal is so reactive that ordinary reducing agents (C, CO, for example) cannot be used to extract it from ores. Moreover, since a fresh surface of the metal reacts with water, pure Al cannot be obtained by the electrolysis of aqueous solutions containing the metal ion. Commercial production was made possible only by the electrolytic reduction of a molten salt, as in the case of the production of Na from molten NaCl.

Unfortunately, the commonly used aluminum ore bauxite, Al_2O_3, not only has too high a melting point for electrolytic application, but is also a nonconductor. Only after competent chemists and metallurgists the world over had spent years in a fruitless search for an efficient way to produce aluminum did a 22-year-old student, Charles Hall, discover that bauxite is soluble in molten cryolite, Na_3AlF_6, and that metallic Al can be produced by the electrolysis of the bauxite-cryolite solution.

In present practice, Al_2O_3 is carefully purified by wet chemical methods, then dried and added continuously to electrolytic cells containing molten cryolite (Figure 18.6). The cryolite is thought to ionize into Al^{3+} and F^- ions, which then migrate to the electrodes, where the processes are then thought to be

$$Al^{3+} + 3\ e^- \longrightarrow Al$$

$$2\ F^- \longrightarrow F_2 + 2\ e^-$$

followed by

$$2\ Al_2O_3 + 6\ F_2 \longrightarrow 4\ Al^{3+} + 12\ F^- + 3\ O_2$$

which restores the cryolite.

No economical way has been found to obtain aluminum metal from its more plentiful silicate ores. Bauxite occurs only in certain places in the world; what we use in the United States comes from Arkansas, South America, and Jamaica. The scarce cryolite is imported from Greenland or manufactured synthetically as needed.

In the United States alone, about 2 million tons of aluminum are manufactured yearly; it is far from being a scarce metal. As the richer iron ores are used up, the light metals industries will continue to grow and take a larger share of the metals market. Pound for pound, aluminum is more expensive than steel, but for many uses, the lower density of Al offsets the price difference.

18.12 TRANSITION METALS

We have so far discussed elemental families that occupy vertical columns in the periodic table. Vertical families exist because the elements in any given column are characterized by a fixed number of electrons in the outer shells of their atoms. There are, however, families that occupy *horizontal* rows in the table. The existence of such groups is a consequence of the energy overlap between main shells.

Let us review the sequence of energy level, beginning with the $1s$ level. After $1s$ is filled with its two electrons, a new main shell is begun. The 2 level accommodates eight electrons in the $2s$ and $2p$ subshells. Level 3 also takes eight electrons in the $3s$ and $3p$ subshells. Although

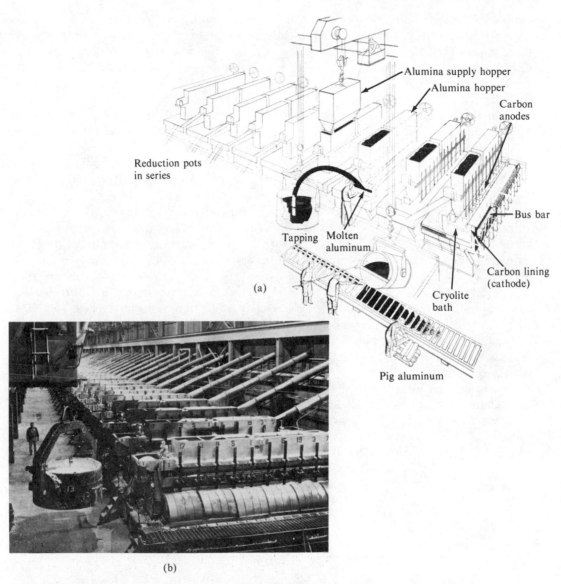

(a)

(b)

Figure 18.6 (*a*) A schematic drawing of the electrolytic smelting of aluminum. (Photo courtesy of Aluminum Company of America.) (*b*) Electrolytic cells in an aluminum smelter plant. (Photo courtesy of the Aluminum Association.)

ten more electrons can be placed in the five orbitals of the 3*d* subshell, these levels have a slightly higher energy than that of the 4*s* level.

Because of the overlap between 4*s* and 3*d*, the 4*s* subshell fills before the 3*d* subshell; and, in the periodic table, the elements potassium and calcium come before scandium, Sc. As a result, the ten elements following calcium have their outer electrons in the fourth shell. Correspondingly, the ten elements (Y through Cd) following strontium, Sr, have a five-level outer shell. The elements of a third set of ten, beginning with element 71, lutecium, and going through mercury, Hg, have six-level outer shells but incomplete inner levels. Each of these three series of ten elements represents a sequence in which a d subshell is filling beneath an already occupied s subshell of next higher principal quantum number. (The filling of f subshells creates two more, similar sequences of 14 elements each. We will not be interested in these last 28 elements as a family, but we will discuss their radioactivity in Chapter 21.)

The three groups of elements, Sc through Zn, Y through Cd, and Lu through Hg, make up the transition series. Because the elements in the groups have $4s$, $5s$, and $6s$ outer shells, they are much alike in their properties and form three families (or, for that matter, could even be considered one family). The elements in these groups bear certain resemblances to the alkali metals and alkaline earth metals, but they also have striking characteristics of their own.

Consider the structures of the transition elements. For the d electrons, the atoms have kernel charges of at least $+11$ (Sc, Y, Lu), and the elements late in the series have colossal kernel charges of up to $+20$ (Zn, Cd, Hg). For the outer s electrons, however, all of the atoms have kernel charges of $+2$. This means that in chemical reaction, the transition elements can easily lose two outer electrons to become divalent ions. Once the valence electrons are gone, one, two, or more of the inner d-shell electrons can also be lost, resulting in ions with as many as five or six positive charges. For this reason, many of the transition elements are **polyvalent,** which means that they can form compounds in which they occur in different atomic ratios. For example, when iron forms $FeCl_2$ and $FeCl_3$, its oxidation states are $+2$ and $+3$. Table 18.7 shows oxidation states of the transition metals. Polyvalency greatly influences the chemistry of the transition elements.

Although the transition elements are clearly metals when they are in their elemental state, they are more difficult to classify once chemical reaction has occurred and the $4s$ electrons have been lost. The loss of the outer electrons exposes an atomic kernel with a substantial charge. Loss of the s electrons also exposes one or more inner-shell electrons, some of which may be unpaired. This makes it possible for transition elements to introduce cova-

Table 18.7
Oxidation States of the Elements Scandium Through Copper in Compounds*

Oxidation States	Elements								
	Sc	Ti	V	Cr	Mn	Fe	Co	Ni	Cu
+7					□				
+6				□	○	○			
+5			□	○	○	○			
+4		□	□	○	□	○	○	○	
+3	□	○	□	□	□	□	□	○	○
+2		○	□	○	□	□	□	□	□
+1			○	○	○	○	○	○	○
0		○	○	○	○	○	○	○	○
−1		○	○	○	○		○	○	
−2				○		○			
−3					○				

*□ designates the most common states. The ○ designates a state that is known to exist, but that may not be stable.

lency into their bonding. As an example, consider the permanganate ion, MnO_4^-, in which the manganese exists in a +7 oxidation state. If the bonds between manganese and oxygen were purely ionic, we would expect the MnO_4^- structure to ionize in aqueous solutions, giving Mn^{7+} and oxygen ions, but it does not do so. A tendency to form covalent bonds also occurs in some of the metals as elements. After the s electrons depart to join the electron sea in the metallic crystal, the d electrons tend to bond covalently. This makes the metals hard and tough. (There are exceptions, of course; mercury at room temperature is one.)

Although some of them are fairly rare, the transition metals are extremely important commercially. Cr, Mn, V, and Ni are widely used to harden and toughen steel. Titanium is beginning to replace steel in certain applications, such as in aircraft metal. Tantalum has medical applications. Tungsten, W, is used in the manufacture of steel and the filaments of electric lights. Iron is so crucial to our way of life that we devote a section to it, as we do to the noble metals, Ag, Au, and Pt.

18.13 COMPLEX IONS AND THE TRANSITION ELEMENTS

Our discussion of chemical bonding has so far been fairly restricted to situations in which atoms gain, lose, or share electrons in such a way as to cause their outer shells each to contain eight electrons (octet rule behavior). This pattern is followed strictly by the elements in the second row of the periodic table and is displayed by many other elements throughout the table. There are, however, other important ways in which chemical bonding occurs. Although we have not discussed them, we have already seen a few examples (PCl_5, for instance, in Chapter 16). Elements beyond the second row of the table can, and frequently do, form structures in which an atom makes five or six covalent bonds. Such structures often arise from coordinate covalency (Section 11.12), in which one atom furnishes both members of a pair of electrons in a bond.

Coordinate covalency occurs in many systems, particularly where one species has an available unshared pair and the second species has strong attraction for electrons. Notable among the species that attract electron pairs to form coordinate covalent bonds are many of the transition metals. Such widely used compounds as $KMnO_4$ and K_2CrO_4 are examples of compounds that contain coordinately bonded transition metals.

A large class of coordinate covalent compounds includes **complex ions,** which are species that have most of the characteristics of compounds but that are not compounds in the strictest sense. A complex ion is a combination of a positive ion that is highly attractive to electron pairs and one, or usually several, ordinarily independent molecules or ions that have available, unshared electron pairs. The positive ion is called the **central ion,** and the electron pair donors are called **ligands.** A good example is given by the ferrocyanide ion (iron (II) cyanide ion), which has the formula $Fe(CN)_6^{4-}$.

$$
\begin{array}{ccc}
 & \overset{\displaystyle N}{\underset{\displaystyle}{C}} & \\
CN & \cdot \; \cdots \; \cdot & CN \\
 & \cdot \; Fe^{2+} \; \cdot & \\
CN & \cdot \; \cdots \; \cdot & CN \\
 & \underset{\displaystyle N}{\overset{\displaystyle C}{}} &
\end{array}
$$

The iron (II) ion, Fe^{2+}, not only has a positive charge but also has a substantial, partly exposed kernel charge. Let's check that. A neutral iron atom has two electrons in its $4s$ subshell and six in its $3d$ subshell. The Fe^{2+} ion has no $4s$ electrons. The atomic number of iron is 26. If we subtract from 26 the charges in the $1s^2$, $2s^2$, $2p^6$, $3s^2$ and $3p^6$ electrons, 18 in all, we have a net positive charge of +8. The $3d$ electrons are exposed to this charge, and to some extent, so are any electron pairs that come close to the iron ion. The effect of the actual +2 charge on the ion, added to that of the partly exposed kernel, results in a powerful attraction for available electron pairs.

The cyanide ion, on the other hand, is electron rich. It has two unshared pairs, both of which are made somewhat more available by the repulsion of the crowd of six electrons that overflow the triple bond.

$$:C:::N:$$

Because carbon is less electronegative than nitrogen, its unshared pair is less tightly held. It is this pair that is attracted to the iron ion. Cyanide makes a good ligand.

The most successful theory of complex ion chemistry assumes that neither the orbitals of the central ion nor the electrons in those orbitals are directly involved in coordinate bonding. (After all, a coordinate bond is one in which the electron pairs are furnished by only one participant in the bond, in this case the ligand.) But the electrons and the outer orbitals of the central ion are certainly *affected* by the presence of the electron pairs of the ligands. The energies of some of the d orbitals of the central ion are affected more than the energies of others, so that the d orbitals are split into two groups that have different energies.

In many complexes, electrons of the central ion can make transitions from the d orbitals of one energy to those of another. The energies of those transitions create visible spectral effects. Many of the brilliant colors observed in the chemistry laboratory and in natural minerals are due to electron transitions between the split d levels of coordinate complexes. The complex formed by Fe (III) and thiocyanate, $Fe(CNS)_6^{3-}$, for example, is deep red. In fact, one of the tests for the presence of Fe (III) in a solution is to add a small amount of KCNS. The resulting complex is so deeply colored that mere traces of Fe (III) can be detected.

The colors of the complexes of metal ions are directly related to the structures of both the central ion and the ligand. As an example of the various effects of ligands, Table 18.8 shows the colors of complexes of Co^{3+} with Cl^- and NH_3.

Even in crystalline solids, it is unusual for ions of the heavier metals to be present in a noncomplexed form. For example, the compounds of many metals form crystals that include water as a ligand, what is sometimes called water of hydration. Copper gives an example. In solutions, Cu (II) is blue; the color arises from the effect of four water molecules complexing the ion $Cu(H_2O)_4^{2+}$. In many of its solid compounds, Cu (II) crystallizes as the $Cu(H_2O)_4^{2+}$ complex, giving a deep blue color to the crystals; copper sulfate pentahydrate, $CuSO_4 \cdot 5\,H_2O$,

Table 18.8
Colors of Some Cobalt Complexes

Formula	Color
$Co(NH_3)_6^{3+}$	Yellow
$Co(NH_3)_5Cl^{2+}$	Purple
$Co(NH_3)_4Cl_2^+$	Green

is an example. (The presence of water of hydration in a crystal is sometimes indicated by the formula of the compound followed by a dot and the appropriate number of water molecules.) In $CuSO_4 \cdot 5\ H_2O$, each Cu (II) is surrounded by four coordinated H_2O molecules. The fifth H_2O is attached to the sulfate ion.

Equilibria exist between ions and their ligands. Depending on the value of the equilibrium constants, some complexes are far more stable than others. For example, when $Cu(H_2O)_4{}^{2+}$ solution is added to ammonia, the light blue $Cu(H_2O)_4{}^{2+}$ complex ions exchange their water molecules for NH_3 molecules, which bind to the Cu (II) more strongly. The solution then takes on the deep, rich blue color of $Cu(NH_3)_4{}^{2+}$. This complex ion can be crystallized with any of several anions, forming intensely colored blue crystals.

If $CuSO_4 \cdot 5\ H_2O$ crystals are heated strongly, the water of hydration is driven off, and most of the color is lost. **Anhydrous** (without water) copper sulfate, $CuSO_4$, is pale greenish-blue.

18.14 METALLURGY OF IRON

The time is nearly past when it is possible to scoop up fairly pure iron oxide from the surface of open pits. Once there were many large deposits of hematite, Fe_2O_3, in grades of purity up to 90%. Since most of the hematite is now gone, the industry has developed ways to use lower-grade ores, such as taconite. Large amounts of taconite are still available.

The reduction of iron oxide is economically accomplished by reaction with carbon monoxide. The CO is produced from the burning of coke, which is essentially carbon; some of the iron oxide is directly reduced by carbon itself. The overall reaction is fairly simple, but the actual smelting process is more complicated than the equation indicates.

$$Fe_2O_3 + 3\ CO \longrightarrow 2\ Fe + 3\ CO_2$$

Ore is mixed with coke and limestone and fed in batches into the top of a blast furnace, as shown in Figure 18.7. Hot air is forced into the bottom of the furnace, where the coke burns, producing temperatures up to about 1600 °C.

$$2\ C + O_2 \longrightarrow 2\ CO + heat$$

Bathed in CO, the iron oxide is reduced in steps, beginning in the cooler regions near the top of the furnace.

$$3\ Fe_2O_3 + CO \longrightarrow 2\ Fe_3O_4 + CO_2$$

$$Fe_3O_4 + CO \longrightarrow 3\ FeO + CO_2$$

$$FeO + CO \longrightarrow Fe + CO_2$$

Each reaction occurs in a successively lower part of the furnace, until the iron is finally melted to form a pool at the bottom. Before the metal can be used, however, it must be freed of its silicon-containing impurities. This is accomplished in a blast furnace by the limestone, $CaCO_3$, which is quickly converted by heat to CaO. The CaO reacts with any SiO_2 that is present either as sand or as other substances.

$$CaO + SiO_2 \longrightarrow CaSiO_3$$

Slag

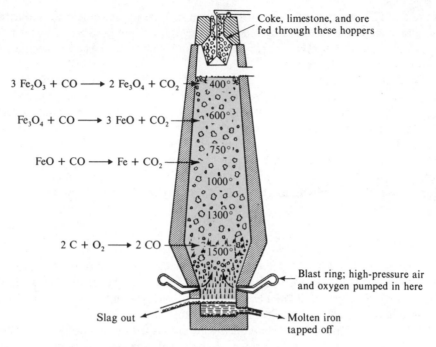

$$3 \text{ Fe}_2\text{O}_3 + \text{CO} \longrightarrow 2 \text{ Fe}_3\text{O}_4 + \text{CO}_2$$

$$\text{Fe}_3\text{O}_4 + \text{CO} \longrightarrow 3 \text{ FeO} + \text{CO}_2$$

$$\text{FeO} + \text{CO} \longrightarrow \text{Fe} + \text{CO}_2$$

$$2 \text{ C} + \text{O}_2 \longrightarrow 2 \text{ CO}$$

Coke, limestone, and ore fed through these hoppers

400° 600° 750° 1000° 1300° 1500°

Blast ring; high-pressure air and oxygen pumped in here

Slag out

Molten iron tapped off

Figure 18.7   Diagram of a blast furnace.

The CaSiO$_3$ melts and floats on the molten iron, protecting it from oxidation by the incoming air blast.

Until recently, blast furnaces were tapped about every six hours. Modern steel mills, however, often use pure O$_2$, which burns the coke more efficiently and eliminates the waste involved in heating the unreactive nitrogen that comes with air. This modification in the process allows the furnaces to be tapped at much shorter intervals. From the furnace, the molten metal is run into bottle-shaped tank cars and taken to be made into steel (Figure 18.8).

When it comes from the blast furnace, the iron still contains a few impurities that make it unsuitable for steel. The impurities were formerly removed by blasting with air from below in a Bessemer converter. They are now usually reacted away in open-hearth or oxygen furnaces. In the open hearth, the molten metal forms a shallow pool, heated by flames directed on the surface. Pure iron oxide is put on the lining of the hearth and floated on the metal surface. It oxidizes the carbon, phosphorus, sulfur, and silicon impurities, which are then removed either as gases or slag.

In the oxygen furnace method, molten metal and limestone to make slag are put in a large pot, and pure oxygen is blown into the metal at high velocity (Figure 18.9). The oxygen blasts aside the slag and converts some of the iron to the oxide, which reacts with impurities as it does in the open hearth. The violent mixing produced by the oxygen stream makes it possible to complete the refining process in about half an hour, compared with the 6 to 8 hr required

Figure 18.8   A blast furnace being tapped. The molten steel runs through a channel by the feet of the workman and flows into a huge ladle. (Photo courtesy of the American Iron and Steel Institute.)

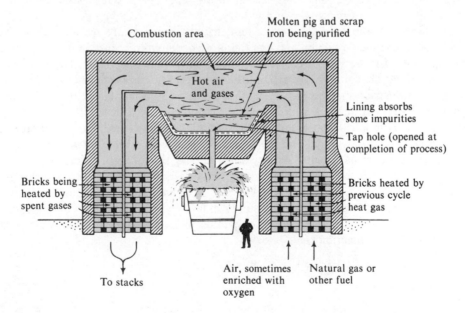

Figure 18.9   An open hearth.

for open-hearth refining. Because of the time it requires, the hearth method is slightly more expensive. However, it is more easily controlled because samples can be removed and analyzed during the process.

Because pure iron is not particularly useful as a metal, most of the iron produced today is converted to steel, which contains up to 2% carbon. The trick in refining is to remove all the P, S, and Si, but to leave just the right amount of carbon to make the desired kind of steel. The oxidation is often allowed to proceed nearly to completion, then the proper amount of carbon is added later. In addition to carbon, small amounts of manganese, vanadium, or other metals are used to alloy steel, depending on its intended purpose. The trace components are usually added during either the spectacular tapping of the open hearth or the pouring from the oxygen furnace.

Because it is so large and so fundamental to the way we live, the steel industry is an economic weathervane. As steel goes, so goes the economy. A few production figures will give you an idea of the size of this important industry. In recent years, steel production in the United States has been around 100 million tons yearly, about 1000 lb per capita. At an approximate wholesale price of $40 per ton, this amounts to a total of about $4 billion yearly, a significant percentage of the entire gross national product.

18.15 COINAGE METALS

A group of three transition elements (possibly four) deserves a brief glance because of the place these elements hold in history and because of their unusual position in our society today. These are the so-called **noble,** or **coinage, metals:** silver, gold, and platinum. Copper might also be included, because it is somewhat similar to the other three, but it is not rare and has not been as highly prized for coinage and jewelry.

The term *noble metal* refers to the fact that the elements are resistant to oxidation and other kinds of chemical reaction and can therefore be found in nature in metallic form. Before the beginnings of modern metallurgy, these were the only metals known, and they consequently acquired great desirability. We are familiar with the use of platinum in jewelry, but Pt also finds many industrial applications in which its high melting point, low reactivity, and ability to serve as a catalyst make it useful. In addition to being used in jewelry, gold and silver are used industrially in a wide variety of ways, especially in electronics. The photography industry depends on the element silver. The more plentiful Cu is used for electric wires and for engineering applications that require its resistance to oxidation and corrosion.

Much of the production of noble metals, especially Au, is still accomplished by mining sources of the "native" metal. Copper, however, is ordinarily mined as one of its compounds and smelted into the metal. This is usually a fairly simple process. For instance, the ore chalcocite is roasted in air, and a series of reactions takes place, the overall result of which can be expressed by the reaction that follows.

$$Cu_2S \quad + O_2 \longrightarrow 2\ Cu + SO_2$$

Chalcocite

(What is reduced and what is oxidized in this reaction?)

18.16 SUMMARY

In this chapter, we have reviewed the properties of the classes of elements and of several individual elements. We have discussed the chemistry of several compounds, and of a few elements. We have learned also about the industrial preparation and use of several compounds and elements.

We can especially profit by looking back at what we have learned about the chemistry of covalent bonds. In reactions of covalent substances, some bonds are broken and other bonds are made. The relative strengths of bonds determine, to a large extent, which reactions will occur and which will not.

Bond strength depends on the attraction exerted by the two bonded atoms on the shared electron pair. Attraction for a shared pair of electrons depends on the nature of the atoms, their kernel charges, sizes, and internal electron shell structure. The attraction of an atom for electrons depends also on the influences of nearby atoms. Bonded neighbors can supply or withdraw electrons, thereby strengthening or weakening bonds. All such factors are significant in determining chemical characteristics such as acid and base strength or strength as an oxidizing or reducing agent. Chapter 18 has shown us a few trends in chemical behavior, and we will come to understand more as we continue.

QUESTIONS

Section 18.1

1. Discuss the characteristics of an atom that affect its ability to attract a shared pair of electrons.
2. Why can the halogens form both ionic and covalent compounds whereas the alkali metals can form only ionic compounds?
3. Explain why F_3CCOOH is a stronger acid than $ClCH_2COOH$.
4. Write reactions showing the behavior of a metal oxide as a base and a nonmetal oxide as an acid. Explain, in terms of electron pair attraction, the difference in behavior. (Reviewing Section 15.8 might help you answer this question.)
5. Nitric acid, HNO_3, is strong, but nitrous acid, HNO_2, is weak. Explain why this is so.

Section 18.2

6. Explain why the noble gases are very nearly chemically inert.
7. Only a few noble gas compounds have been prepared, none at all from neon. Why are noble gas compounds limited to the heavier members of the family?
8. The boiling temperature of krypton is $-152\ ^\circ C$ and that of xenon is $-107\ ^\circ C$. Estimate the boiling temperatures of radon and neon.

Section 18.3

9. It is said that dipole-dipole attractions are generally stronger than London forces, yet HCl is a gas at room temperature and Br_2 is a liquid. Discuss in detail.
10. Why is there no way to prepare F_2 by chemical reaction?

11. Sodium hypochlorite is used as a germicide in swimming pools, but ClO^- is not especially harmful to bacteria. Moreover, $NaClO$ is not especially effective unless the pH of the water is in the correct range. Discuss, giving reaction equations when appropriate.

12. List the hydrogen halides in increasing order of acid strength. Why do the acid strengths follow this pattern?

13. Explain why the acid strengths of the following compounds increase in the order given: HOI, HOBr, HOCl.

14. Why do the oxyacids of the halides increase in acid strength as the oxidation state of the halogen increases?

15. Not much is known about astatine. What would you predict its chemical and physical properties to be?

Section 18.4

16. Explain the Frasch process for mining sulfur.

17. Name a few of the uses of sulfur.

18. Use principles you already know to explain why sulfur can be found in both positive and negative oxidation states, but oxygen is found only in oxidation states O^- and O^{2-}.

19. If you could make H_2TeO_3, would it be a stronger or weaker acid than H_2SeO_3? Give your reasoning.

Section 18.5

20. Draw an electron dot diagram of the N_2 molecule. Why is the strength of this bond so important in the chemistry of nitrogen?

21. Write reaction equations for the manufacture of nitric acid, starting with N_2. Explain any special conditions under which each reaction is run.

22. What are a few of the uses of compounds containing nitrogen?

Section 18.6

23. What are the elemental forms of carbon? What are some of the uses of each form?

24. How is elemental carbon obtained industrially?

25. Give some properties and reactions of CO_2.

26. What are carbonates? Name several important carbonates.

Section 18.9

27. List some of the properties of the alkali metals. How do the properties change within the alkali family?

28. Why are the alkali elements difficult to separate and purify?

Section 18.10

29. The first ionization energy of magnesium is 732 kJ and the second is about 1440 kJ, yet Mg appears as a doubly charged ion in its compounds. Explain why there are no compounds containing Mg^+.

30. List some of the properties of the alkaline earth metals. How do these properties differ from those of the alkalis?

Section 18.11

31. Although aluminum is an active metal, aluminum articles last for years despite contact with air and water. Why?
32. Aluminum is *amphoteric*. Explain what that term means and use reaction equations to show how aluminum displays amphoteric behavior.
33. Explain the Hall process used to prepare aluminum.

Section 18.12

34. In the periodic table, elemental families occupy vertical columns, yet the horizontal series of transition metals are called families. Explain.
35. The transition metals commonly form ions with charges of only +1, +2, or +3, but often appear in oxidation states of up to +7. How can this be?
36. Why are transition metal ions particularly attractive to unshared electron pairs?
37. The ions of many transition metals behave like nonmetals. Give an example and explain why the ion you selected behaves that way.

Section 18.13

38. What is a complex ion? How does a complex ion differ from an ordinary compound?
39. What is a ligand? What are the characteristics of a good ligand? Name some molecules and ions that are good ligands.
40. Why are complex ions of the transition metals often intensely colored?
41. What is water of hydration?

Section 18.14

42. How is iron ore reduced to the element? Give equations.
43. How is iron purified? What impurities are often found in freshly smelted iron?
44. What is steel? How is steel made from iron?

Section 18.15

45. What are some of the so-called noble, or coinage, metals? Why are they sometimes used as a medium of exchange?
46. Copper can be smelted merely by the heating of Cu_2S, whereas iron requires a reducing agent and aluminum must be electrolyzed. Explain the reasons for the differences.

Chapter 19

Carbon and the Compounds of Carbon

Although we took time in the preceding chapter to review some of the interesting and useful chemistry of several elements, most of the elements escaped even a mention because they have little direct importance to us. We are now about to devote an entire chapter to the study of a single element and its chemistry. What is so important about that one element? Why does it deserve so much attention?

Carbon and carbon compounds are essential to life and to the functioning of modern society. The number of different compounds for which it furnishes the structure makes carbon the most versatile element. All life depends on carbon compounds for crucial structural, chemical, and energetic functions. The hand with which you are holding this book is made mostly of carbon compounds. The book itself is made of carbon compounds. The energy with which you turn the pages and move your eyes and, for that matter, see and think, comes from the oxidation of carbon compounds. The energy with which the book was printed and delivered to you was most likely derived from the oxidation of carbon compounds. Most of the new technology of synthetic materials (plastics, fabrics, foods, building materials, fuels, and so on) is founded on carbon compounds.

Carbon compounds are so numerous, varied, interesting, and important that an entire branch of chemistry is devoted entirely to their study. We will nonetheless be able, in just one chapter, to get a good idea of what organic chemistry is all about and why carbon atoms are so versatile.

The study of carbon compounds is called organic chemistry. The chemistry of all the other elements, almost by default, is called inorganic chemistry.

AFTER STUDYING THIS CHAPTER, YOU WILL BE ABLE TO

- list the properties of carbon that make carbon chemistry unique
- list the properties of multiple bonds in carbon compounds
- name the saturated hydrocarbons, methane through decane

- explain the differences between saturated and unsaturated hydrocarbons
- explain the process of hydrogenation
- give the names and describe the structures of important functional groups
- name many kinds of organic compounds, given their structures
- draw the structures of many kinds of organic compounds, given their names
- give the common names and describe the structures of several important organic compounds
- list the properties of alkanes, alkenes, alkynes, alcohols, carboxylic acids, amines, and benzene
- explain the differences between most organic reactions and the reactions of ionic compounds
- explain and give examples of substitution and elimination reactions
- predict the products of several kinds of organic reactions
- explain the difference between ordinary polymers and network polymers
- write equations for reactions that produce various kinds of common polymers
- make simple predictions of polymer properties, based on the structures of the polymers
- describe the nature of petroleum and the refining of petroleum products

TERMS TO KNOW

addition reaction	double bond	paraffin
aldehyde	electron delocalization	polyamide
alkane	elimination	polymer
alkene	ester	polymerization reaction
alkyl group	free radical	R group
alkyne	functional group	saturated hydrocarbon
amide	hydrocarbon	substitution
amine	IUPAC system	tetrahedral angle
aromatic compound	ketone	tetrahedral carbon
benzene	monomer	triple bond
carbonyl	network polymer	unsaturated hydrocarbon
carboxylic acid	organic chemistry	vinyl polymer

TERMS WITH WHICH TO BE FAMILIAR

bifunctional molecule	linear molecule	secondary alcohol
condensed ring	polyfunctional molecule	trigonal planar
diene	primary alcohol	

19.1 CARBON

What makes carbon so unusual that it should have a chemistry all its own? The answer has to do with principles we already know about. Nothing in the chemistry of carbon conflicts with what we have learned about other kinds of atoms.

19.1.1 Properties of the Carbon Atom

The uniqueness of carbon arises from several of its characteristics. Of these, the most important are the values of its electron affinity and ionization energy and the nature and population of its outer electron shell. There are four electrons in the valence shell of a carbon atom; likewise, there are four vacancies. By forming four covalent bonds, a carbon atom can fill its outer shell with eight electrons. So far, nothing surprising; silicon can do the same.

The electron affinity and ionization energy of carbon are about halfway between the highest and lowest known values. Carbon atoms do not attract electrons strongly enough to take them away from other atoms, but neither will carbon willingly give away any of its electrons. Carbon atoms have just the right amount of electron attraction to form strong covalent bonds with a wide assortment of other kinds of atoms and with other carbon atoms as well. The ability of the carbon atom to make strong covalent bonds with one or more other carbon atoms is especially important; it is responsible for much of the versatility of carbon in forming compounds.

19.1.2 Tetrahedral Carbon

In most of their compounds, carbon atoms each form four covalent bonds. Because electrons repel one another, the most efficient way for four pairs of electrons to surround an atom is to be at the greatest possible distance from one another. That means that the bonds formed by four electron pairs should be at the maximum possible angle. If you begin with a point in space and draw four lines coming out of it, all at the maximum angle from one another (and therefore at equal angles), you will draw the lines 109.5° apart. This is called the **tetrahedral angle,** because the lines point to the four corners of a regular tetrahedron, a four-sided solid. Carbon atoms frequently form their four bonds at the tetrahedral angle; carbon bonded at this angle is called **tetrahedral carbon** (Figure 19.1).

To understand why carbon atoms are almost the only atoms that can form strong, stable skeletons for large complex molecules, we need to look at the ways in which other kinds of molecules can be broken by chemical attack. To build a strong molecule, it is first necessary to begin with atoms that make strong covalent bonds. A molecule that is weakly bonded will be knocked apart simply by the kinetic energy of its own atoms and by collisions with nearby molecules. But bond strength alone is not enough to bring stability to molecules, because even the strongest bonds can be broken by attack from outside. For example, the strong bonds between hydrogen and chlorine atoms in HCl molecules are broken instantly when HCl is mixed with water. The hydrogens in the HCl molecules are attracted by the unshared electron pairs of oxygen atoms on one or more of the water molecules jostling the HCl, while the

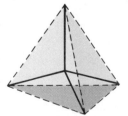 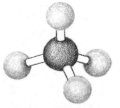

Figure 19.1   Tetrahedral carbon atom.

hydrogen atoms of other water molecules crowd in and try to get close to the unshared pairs of the chlorine atom. Because the electron pair of the H—Cl bond is already more than half owned by the Cl atom, and because there is plenty of room around both the Cl and the H atoms for water molecules to intrude, the HCl molecule is easily destroyed by water. The resulting H^+ ion is content to share the electrons of oxygen atoms in the water molecules, and the highly electronegative chlorine keeps both the electrons from the broken HCl bond.

For comparison, consider methane, CH_4. The hydrogen atoms in methane are not looking for unshared pairs of electrons on other molecules because they already have a considerable ownership in the pairs that bond them to carbon. After all, carbon is not as electronegative as chlorine, and it shares its electrons generously with its hydrogen atoms. Moreover, in tetrahedral carbon, there is no easy way for attacking molecules to approach from the back side of the carbon atom and attract the electron pair away from a C—H bond. Because the carbon atom is surrounded by other atoms strongly bonded to it, all the electron pair bonds are protected. We see why methane and similar molecules are fairly immune to attack from almost all other kinds of substances. At all but very high temperatures, methane is inert.

19.1.3 Multiple Bonds in Carbon Molecules

Not all bonds formed by carbon atoms are tetrahedral. Carbon forms many kinds of molecules in which some of the pairs of atoms share more than one bond. For example, in the compound ethylene, C_2H_4, the carbon atoms share two pairs of electrons, forming a **double bond.**

$$\begin{array}{ccc} H & & H \\ & \ddot{} & \\ & C::C & \\ & \ddot{} & \\ H & & H \end{array}$$

Ethylene

Similarly, in formaldehyde, CH_2O, the carbon and oxygen atoms form a double bond.

$$\begin{array}{cc} H & \\ \ddot{} & \ddot{} \\ C & :: O \\ \ddot{} & \ddot{} \\ H & \end{array}$$

Formaldehyde

And in acetylene, C_2H_2, the carbon atoms share three electron pairs, making a **triple bond.**

$$H:C:::C:H$$

Acetylene

Although multiple bonds are stronger than single bonds, they are also more vulnerable to attack by other molecules. Two pairs of electrons cannot occupy the space between two atoms. In a multiple bond, only one electron pair lies along the line between the nuclei; the other pair is forced to lie somewhat outside. In a multiple bond, only the central electron pair forms a strong bond. The other is weaker, and can be rather easily disrupted by molecules or other particles that are attracted to electron pairs. (See Figure 19.2.) Later, we will see examples of some of the reactions of multiple bonds.

The geometry of carbon atoms that have multiple bonds differs from that of tetrahedrally bonded carbon atoms. A double-bonded carbon atom can be thought of as an atom with three single bonds, to one of which is added a second bond. The three single bonds lie in a plane, with approximately equal bond angles of 120°: the result is a flat molecule with a geometry known as **trigonal planar** (Figure 19.3).

A triple-bonded carbon atom can be thought of as having two single bonds, one of which is accompanied by two extra shared pairs of electrons. The single bonds lie at an angle of 180°, forming a *linear* molecule. In acetylene, for example, all four of the atoms lie in a straight line (Figure 19.4).

19.1.4 Number and Variety of Carbon Compounds

Carbon atoms can bond with each other to make long, stable chain structures. One common arrangement is called a straight chain.

Straight chain

In this use, *straight* means *not branched,* rather then *linear.* A straight carbon chain has 109.5° tetrahedral bends in it, because of the 109.5° carbon bond angle. Carbon atoms, however, can also form branched chains, or even rings.

Branched chain   *Ring structure*

Although there is almost no limit to the possible length of a carbon chain, carbon atoms rarely form rings that have more than about eight members.

Molecules containing carbon can be very large; in some fields of study, molecules formed by a hundred or more atoms are considered small. Molecules that do not contain carbon, on

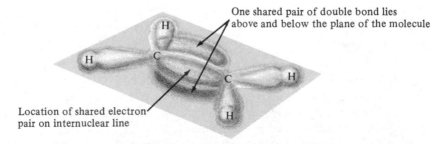

Figure 19.2 Representation of electron densities in the double bond of an ethylene molecule.

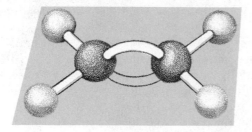

Figure 19.3 Tinkertoy and scale models of a planar molecule.

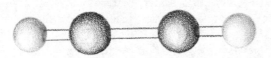

Figure 19.4 Tinkertoy and scale models of a linear molecule.

the other hand, tend to not be large, and rarely contain more than a few atoms. Besides bonding to each other, carbon atoms can also form strong covalent bonds to hydrogen, oxygen, nitrogen, sulfur, or the halogens. Atoms that bond to carbon can form different arrangements, that is, different isomers. (Recall Section 11.18, in which we saw two different compounds with the formula C_2H_6O.)

All these properties combine to enable carbon atoms to form an almost endless number of different compounds. There are more known compounds that contain carbon than there are compounds of all the other elements put together. Yet anyone who knows a little chemistry can easily write the structures of any number of carbon compounds, some of which may not even have been discovered.

Because there are so many known carbon compounds and so many possibilities that remain to be studied, a system has been developed for naming and classifying carbon compounds. The naming system is designed so that if the name of the compound is known, the structure can immediately be written. Conversely, the name of a compound is directly determined by the structure. Although we need not go into the details of the naming system, we will learn how it works and how to name several hundred moderately complex compounds.

Carbon chemistry is so vast, so special, and so important that it has its own name. Early chemists thought that carbon compounds could be made only by living creatures, and called carbon compounds *organic;* that is, related to life. We now know that almost any carbon compound can be prepared in the laboratory, but **organic chemistry** is still used to refer to the chemistry of carbon compounds.

We begin our study of organic compounds with the simplest family of carbon compounds, the hydrocarbons.

19.2 HYDROCARBONS

Most carbon compounds contain hydrogen. The simplest, those that contain only carbon and hydrogen, are called **hydrocarbons.** A hydrocarbon in which each of the carbon atoms forms bonds to four other atoms is called a **saturated hydrocarbon.** We will see the reason for this name in a moment.

The simplest saturated hydrocarbon is methane, which we have already met several times. The methane molecule is three-dimensional; it has tetrahedral bonding. Because it is not convenient to draw three-dimensional pictures every time we want to show the bond structure of a molecule, we will continue to use flat diagrams, as we did in Section 11.17. We must not forget, however, that the atoms in real molecules are not usually arranged in a flat plane and that there are almost never 90° angles between the bonds of carbon atoms. To learn how to interpret flat diagrams, we will compare them to scale models and to tinkertoys. (See Figure 19.5.)

Methane is a colorless, odorless gas. When cooled to −161 °C at 1 atm pressure, it condenses to a clear liquid with about half the density of water. Natural gas is mostly methane. Methane occurs also as a by-product of several processes, including the digestive processes of mammals. Like other saturated hydrocarbons, methane is nearly inert. At high temperatures, it will react vigorously with oxygen in combustion. At room temperature, however, it will not react even with concentrated sulfuric acid, which rapidly decomposes nearly every other kind of organic molecule (including human skin).

We have seen that carbon atoms can bond to other carbon atoms, and we have seen several examples of compounds containing C—C bonds. Now let's take a closer look.

The simplest saturated hydrocarbon after methane is ethane, C_2H_6, which contains one C—C bond (Figure 19.6). The properties of ethane are much like those of methane, except that ethane has a somewhat higher boiling temperature.

The compound that has one more carbon and two more hydrogen atoms is propane, C_3H_8.

$$H-\overset{\overset{\displaystyle H}{|}}{C}-\overset{\overset{\displaystyle H}{|}}{\underset{\underset{\displaystyle H}{|}}{C}}-\overset{\overset{\displaystyle H}{|}}{\underset{\underset{\displaystyle H}{|}}{C}}-H$$

Propane

Because it is useful to know their structures, organic molecules are not always described by empirical formulas but are often sketched. Most chemists simplify the sketches by leaving out the hydrogen atoms attached to carbon atoms and showing only the carbon skeletons and

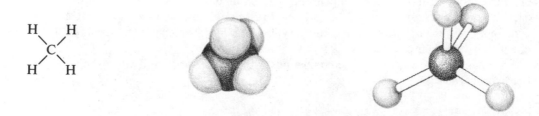

Figure 19.5 Three representations of methane: flat diagram, scale model, and tinkertoy.

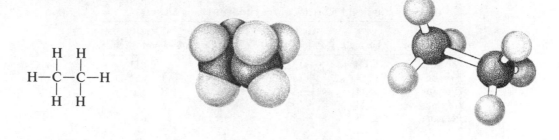

Figure 19.6 Three representations of ethane.

other significant features. As you view and draw such sketches, remember that each carbon atom forms four covalent bonds and that you can complete each drawing by adding bonded H atoms until every C has four bonds. Using this shorthand, butane (C_4H_{10}) would be drawn as you see in either the left or the center sketch. The right sketch is an even more abbreviated version of butane. We will use all these methods to indicate structure.

Three representations of butane

The next largest compound in the alkane series is pentane, C_5H_{12}, followed by hexane, C_6H_{14}, heptane, C_7H_{16}, octane, C_8H_{18}, nonane, C_9H_{20}, and decane, $C_{10}H_{22}$. Notice that in the formulas for these compounds the number of hydrogen atoms is always twice the number of carbon atoms plus two. If the number of carbon atoms in the formula is n, the number of hydrogens is 2n + 2, C_nH_{2n+2}. The eleventh compound in the series, for example, would have 11 C atoms and (2 × 11) + 2 = 24 hydrogen atoms, or $C_{11}H_{24}$. We could go on to write formulas for hydrocarbon molecules with any number of carbon atoms. For any formula we write, the molecule exists or can theoretically be prepared. For that matter, any molecule that we might propose could also be named, but the names of very large molecules tend to be equally large. We will discuss the preparation and properties of very large molecules in a later section.

The family of saturated hydrocarbons is called the **alkane,** or **paraffin,** family. (What is commonly called *paraffin* is a mixture of long-chain, mostly saturated, hydrocarbon compounds.) Table 19.1 lists the properties of some of the alkanes.

A hydrocarbon structure attached to some other group of atoms is called an **alkyl group.** The alkyl groups are named after their hydrocarbons. For example, H_3C— is a *methyl* group, H_5C_2— is an *ethyl* group, and so on.

The alkanes are part of everyday life. Methane is natural gas. Propane, sold as liquefied petroleum gas (LPG), is used for heating and motor fuel. Gasoline is mostly a mixture of hydrocarbons whose molecules contain six to eight carbon atoms; the molecules of kerosene are in the C_{10} to C_{17} range; those of fuel oil are in the C_{20} to C_{30} range. Mineral oil is much

Table 19.1
Some Physical Properties of the Straight-chain Alkanes

Substance	Formula	Melting Point (°C)	Boiling Point (°C)	Density of Liquid (g/cm³)
Methane	CH_4	−182	−161	0.45*
Ethane	C_2H_6	−183	−89	.65*
Propane	C_3H_8	−188	−42	.50
Butane	C_4H_{10}	−138	−1	.58
Pentane	C_5H_{12}	−130	36	.63
Hexane	C_6H_{14}	−95	69	.66
Heptane	C_7H_{16}	−91	98	.68
Octane	C_8H_{18}	−57	126	.70
Nonane	C_9H_{20}	−54	151	.72
Decane	$C_{10}H_{22}$	−30	174	.73
Pentadecane	$C_{15}H_{32}$	10	271	.77
Eicosane	$C_{20}H_{42}$	37	343	.79
Triacontane	$C_{30}H_{62}$	66	447	.81

*At melting point. Other densities at 20 °C.

like fuel oil, but is more highly purified. Heavy oils and greases have still larger molecules, and the solid paraffin used in canning fruits contains alkanes in the 60 to 70 carbon range.

Because of isomerization, the number of compounds in any family of organic compounds far exceeds the number of different possible empirical formulas. Methane, ethane, and pro-pane each have only one isomer. But butane, which has only one empirical formula, C_4H_{10}, has two isomers. Pentane has three.

$$C-C-C-C-C \qquad\qquad \begin{matrix} C-C-C-C \\ | \\ C \end{matrix} \qquad\qquad \begin{matrix} C \\ | \\ C-C-C \\ | \\ C \end{matrix}$$

The three isomers of pentane

Nonane, C_9H_{20}, has 35 isomers, and it has been calculated that there are 366,319 isomers for $C_{20}H_{42}$.

Be sure you learn how to distinguish between different structures and identical structures that have been drawn differently. The sketches that follow all represent the same structure.

$$\begin{matrix} C-C-C \\ | \quad | \\ C \quad C \\ | \\ C \end{matrix} \qquad\qquad \begin{matrix} C \\ | \\ C-C-C-C \\ | \\ C \end{matrix} \qquad\qquad \begin{matrix} C \quad\ C \\ | \quad\ | \\ C-C-C \\ | \\ C \end{matrix}$$

Three sketches of 2-methyl pentane

EXERCISE 19.1.   Draw all five isomers of hexane.

19.3 UNSATURATED HYDROCARBONS

A hydrocarbon that contains a double or triple carbon-carbon bond somewhere in its chain is said to be **unsaturated.** In other words, the chain does not contain as many H or other groups as it could if there were no multiple bonds: the chain is not *saturated* with H atoms. The simplest unsaturated hydrocarbon is **ethene,** commonly called **ethylene.**

$$\begin{array}{ccc} H & & H \\ \diagdown & & \diagup \\ & C=C & \\ \diagup & & \diagdown \\ H & & H \end{array}$$

Ethene

Ethene is the smallest member of the **alkene** family. The six atoms of ethene lie in a plane; the ethene molecule is flat.

The alkenes have names like those of the alkane family except that the endings differ: the alkane names end in *-ane,* the alkene names in *-ene.* Thus we have propene, C_3H_6, butene, C_4H_8, and so on. The alkenes have the formula C_nH_{2n}. Like those of the alkanes, the first few members of the alkene family are colorless gases. The alkenes, however, are not odorless. The alkenes are of little direct use to consumers, although they are widely used in laboratories and industrial processes. The higher, more complex unsaturates are common in biological organisms.

Molecules can have more than one double bond; those with two double bonds are called **dienes.** Butadiene is an example.

$$\begin{array}{ccccc} H & & H & & \\ \diagdown & & \diagup & & \\ & C=C & & & H \\ \diagup & & \diagdown & & \diagup \\ H & & & C=C & \\ & & \diagup & & \diagdown \\ & & H & & H \end{array}$$

Butadiene

A *triene* contains three double bonds and a *tetraene* four. There are names for molecules with even larger numbers of double bonds.

If a triple carbon-carbon bond exists in a molecule, the name **alkyne** is used; the *-yne* ending shows the presence of a triple bond. The smallest alkyne is ethyne, commonly called **acetylene.**

$$H-C\equiv C-H$$

Acetylene

Alkynes are less common than either alkanes or alkenes.

19.4 RING HYDROCARBONS

Both saturated and unsaturated hydrocarbons exist as ring structures. Because there are no end carbons in a ring, a ring hydrocarbon has two fewer hydrogens than the corresponding open chain hydrocarbon. For example, saturated rings have the formula C_nH_{2n}.

Hydrocarbons form rings of several varieties. The most stable and common saturated ring hydrocarbon is **cyclohexane** (*cyclo* meaning "ring" and *hexane* meaning "six saturated carbon atoms").

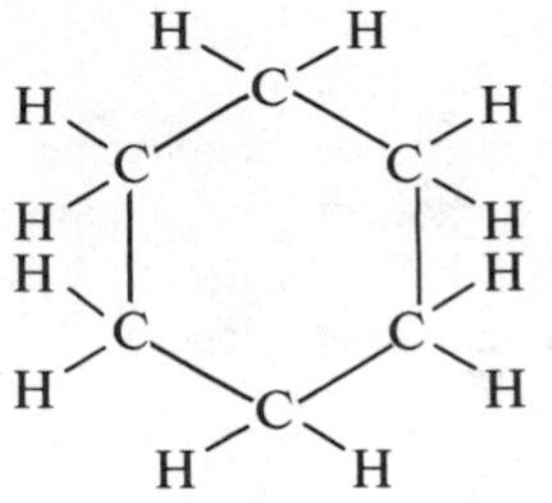

Cyclohexane

The most stable ring structures are those that contain from four to seven carbon atoms. Small rings, like cyclopropane, do exist, but they are unstable.

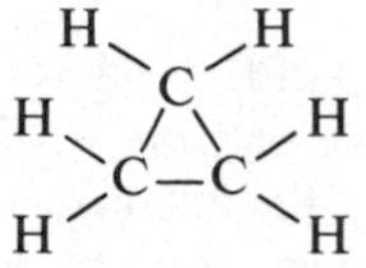

Cyclopropane

The angle between carbon atoms in cyclopropane, for example, is only 60°; to ask a tetrahedral carbon atom to make a 60° bond angle is to ask for an unstable molecule. Unless a multiple bond is involved, the more that a bond angle deviates from the ideal 109.5°, the less stable the molecule will be.

Cyclobutane has a bond angle of 90°.

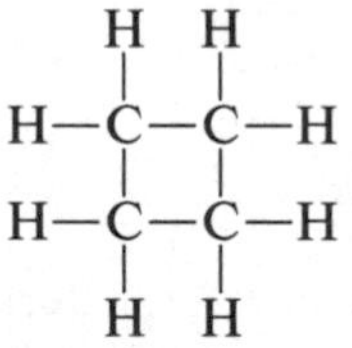

Cyclobutane

This is closer to the 109.5° tetrahedral angle, and cyclobutane is correspondingly more stable than cyclopropane. Even so, four-member rings are not common. A pentagon has interior angles of 108°, which is quite close to 109.5°. We would expect cyclopentane to be fairly stable, and it is.

The more atoms there are in a flat ring, the larger the carbon-carbon bond angle will be. For a large ring, it will approach 180°. This is rather far from 109.5°, and we might expect that large ring structures would be rather unstable. However, rings with as many as 30 carbons have been prepared. Such rings attain reasonable bond angles by puckering and are thereby stabilized. For example, the cyclohexane molecule has actual bond angles close to 109°, although if drawn flat, the ring has 120° angles. Cyclohexane molecules take either of two puckered forms, the "boat" form or the "chair" form (Figure 19.7).

Large numbers of stable *unsaturated* ring compounds exist. Even cyclopropene can be prepared, but because so many electrons are now crowded into a small space, it is unstable and tends to be explosive even at room temperature.

Cyclopropene

There is one especially stable compound in which everything fits so well that the molecule exists without even a pucker. It is C_6H_6, universally called **benzene.**

Benzene

The angle between a double bond and a single bond is ordinarily 120°, and the interior angles of a hexagon are also 120°, so benzene molecules experience no strain. Benzene is even more stable than might be expected because of an interesting phenomenon called *delocalization of electrons*. For benzene, it is possible to draw two structural formulas that differ only in the locations of the double bonds in the ring.

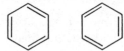

Two representations of a benzene ring

For the benzene molecule to change from one structure to the other, no atoms need move. The electrons in the double bonds can arrange themselves into either structure without changing anything else in the molecule. The electrons are not restricted to only three double bonds but have the entire ring in which to roam. Because electrons repel one another and tend to move apart as far as possible, any structure that provides extra room for electrons is particularly favored, and the benzene molecule has such a structure. The electrons in the three double bonds of benzene are so delocalized that it is not correct to draw a benzene ring with double bonds; the molecule should be shown with the six electrons distributed around the entire ring. Many chemists prefer to use a hexagon with a circle inside to indicate benzene, but such a drawing cannot show the number of electrons.

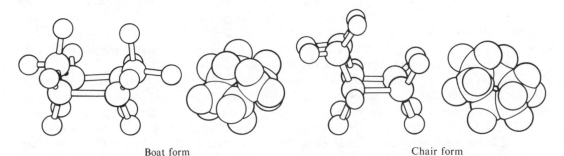

Boat form   Chair form

Figure 19.7   Tinkertoy and scale models of the two forms of cyclohexane.

Benzene symbol indicating delocalized electrons

So that we can keep track of the electron count, we will include the double bonds in our drawings of benzene and other delocalized electron molecules, *but we must remember that such drawings do not accurately represent the actual structure of the molecules.*

Two or more ring structures can have a common side, forming a *condensed ring* structure. Both saturated and unsaturated ring structures form condensed rings, but the unsaturated compounds are more common and of more interest. A common condensed ring system is naphthalene, $C_{10}H_{18}$.

Naphthalene

Condensed ring systems made up of large numbers of rings have been synthesized.

Ring compounds that contain the alternating double and single bond structure seen in benzene and naphthalene are classed as **aromatic compounds.** The name was originally based on the characteristic odors of certain members of the family. Aromatic hydrocarbons are extremely important in industry, where they are used as starting compounds for the synthesis of a variety of chemical products.

19.5 FUNCTIONAL GROUPS

A single bond between a pair of carbon atoms is strong and stable. Since carbon-carbon bonds are nonpolar, and carbon-hydrogen bonds are only slightly polar, the bonds in a saturated hydrocarbon are all much alike. The structure of a saturated hydrocarbon molecule does not offer any particularly weak place to attack.

The saturated hydrocarbons, however, are only a small proportion of the entire family of organic compounds. In most organic compounds, one or more of the hydrogen atoms have been replaced by another kind of atom or group of atoms. Such a substitution usually creates a bond or group of bonds that, for one reason or another, is more reactive than a carbon-carbon bond or a carbon-hydrogen bond. The result is that *the chemistry of most organic compounds depends on the structures that are attached where H atoms would be in hydrocarbons.* Such structures (including single atoms) are called **functional groups;** double or triple carbon-carbon bonds are also considered to be functional groups.

The chemistry of organic compounds, then, is largely the chemistry of functional groups. In the absence of high temperatures, which can knock any bonds apart, changes in the carbon skeletons of organic molecules are almost always caused by the behavior of functional groups.

The different kinds of structures that form functional groups have been given names, and organic compounds are named after their functional groups. Table 19.2 lists the common functional groups and their structures. All the structures are shown with terminal bonds that

Table 19.2
Functional Groups and Their Names

Structure	Name	Structure	Name
—C=C—	Double bond or alkene	—C≡C—	Triple bond or alkyne
—C—OH	Alcohol	—C—O—C—	Ether
—C=O, H	Aldehyde	—C, C=O, —C	Ketone
—C—NH$_2$	Amine	—C(=O)—OH	Carboxylic acid
—C(=O)—O—C—	Ester	—C(=O)—N—C—	Amide

indicate where the structures are attached to hydrogen or to a carbon chain of some kind; in some cases, one carbon of the chain is shown with the functional group. Many organic compounds have two or more functional groups, either alike or different. Halogens should be added to the list of functional groups because a Br or a Cl atom bonded to carbon has characteristic properties and reactions, just as other functional groups do. Halogens attached to organic molecules, however, do not have special functional group names.

19.6 NAMING ORGANIC COMPOUNDS

Hundreds of thousands of organic compounds are known, and the number of other possible structures seems limitless. If the name of a compound is to be of much use, it should describe the structure. Think of the confusion that would result if half a million compounds were named at random. In the early days of organic chemistry, however, organic compounds were not systematically named. The first person to isolate or prepare a compound would simply give it a name, usually one that described its origin (*formic acid* from ants, Latin *formica*), or its properties (*morphine* from Morpheus, the Greek god of sleep). Occasionally, a compound would become known by the name of its discoverer (Micheler's ketone). Names of this sort gave little clue to the structure of a compound.

After many decades of confusion, a naming system has been accepted worldwide; it is a kind of international language. In 1921, the International Union of Pure and Applied Chemistry commissioned delegates from many countries to meet in Geneva to work out what is now known as the **IUPAC system,** or the **Geneva system.** In this language, a completely descriptive, unique name exists for every structure, whether discovered or not, and only one structure can be described by each name. Given the name, anyone familiar with the system can write the structure, and vice versa. The only disadvantage of the system is that names for complex compounds are rather long, so that for many compounds, the short, informal names are still used. It is simpler to speak of citric acid than of 3-carboxy-3-hydroxy-pentanedioic acid. To be fluent in organic chemistry, it is necessary not only to know the Geneva system but also to memorize several familiar names. Although it is not necessary that we learn all the details of the Geneva system, we will learn the basis of the system and how to use it for simple structures.

To name a compound in the Geneva system, you first identify and name the main chain, the longest C—C chain in the structure. The chain length designations are taken from the names of the hydrocarbons. For example, the syllable *pent* means that the longest unbranched carbon chain in the structure contains five carbon atoms. Table 19.3 lists designations for chain lengths.

After the main chain has been designated, branches in the chain are identified by the addition of prefixes based on the names of alkyl groups. To name an alkyl group, use the length designation for a group of that size and add the ending *-yl*. To add a group containing two carbons to hexane, you would say ethyl hexane.

If one or more of the hydrogen atoms on the main chain or the branches has been replaced with a functional group, the main chain designation is given an ending that identifies one of the groups; otherwise, the ending *-ane* is used, which means that the hydrocarbon is saturated. Multiple bond structures are considered to be functional groups: the *-ene* indicates double bonds, and the *-yne* triple bonds. Table 19.4 lists name endings for functional groups. The names of esters are derived from the names of the corresponding acids.

Each of the two isomers of the compound shown has three carbons in its chain, so its name must contain the syllable *pro-*. Because it is an alcohol, its name must end in *-ol*. The two structures are propanols.

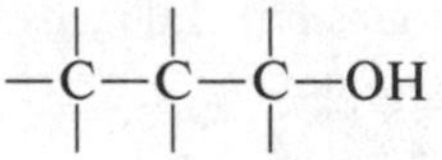

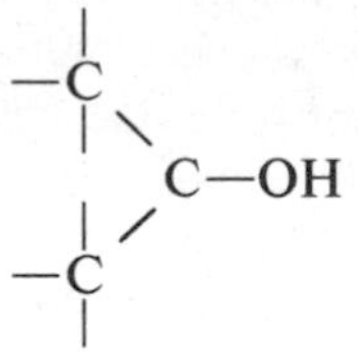

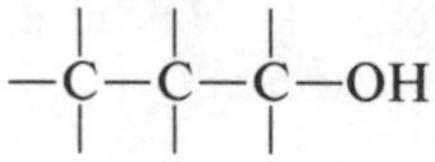

If more than one functional group appears in a structure, the groups not identified by the ending must be identified by prefixes. Of the structures that follow, the structure on the left

Table 19.3 Chain Length Designations
meth- = 1
eth- = 2
pro- = 3
but- = 4
pent- = 5
hex- = 6
hept- = 7
oct- = 8
non- = 9
dec- = 10

Table 19.4
Name Endings for Functional Groups

Group	Ending
Double bond	-ene
Triple bond	-yne
Aldehyde	-al
Ketone	-one
Alcohol	-ol
Carboxylic acid	-oic acid
Amide	-oamide
Ether	ether
Amine	amine

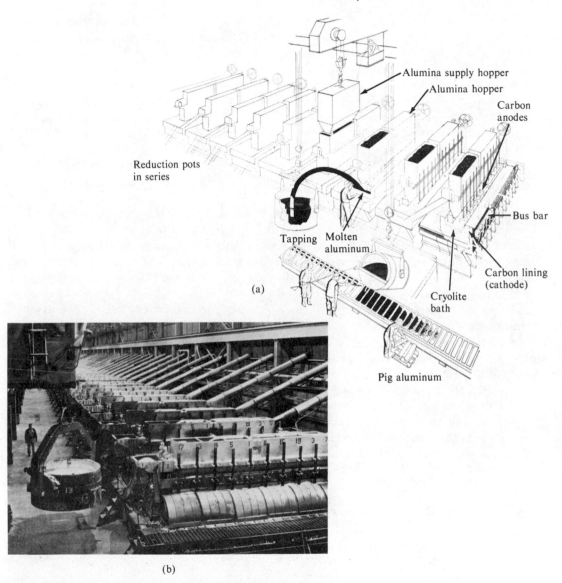

Alumina supply hopper

Alumina hopper

Carbon anodes

Reduction pots in series

Bus bar

Tapping Molten aluminum

Carbon lining (cathode)

(a)

Cryolite bath

Pig aluminum

(b)

Figure 18.6 (*a*) A schematic drawing of the electrolytic smelting of aluminum. (Photo courtesy of Aluminum Company of America.) (*b*) Electrolytic cells in an aluminum smelter plant. (Photo courtesy of the Aluminum Association.)

ten more electrons can be placed in the five orbitals of the $3d$ subshell, these levels have a slightly higher energy than that of the $4s$ level.

Because of the overlap between $4s$ and $3d$, the $4s$ subshell fills before the $3d$ subshell; and, in the periodic table, the elements potassium and calcium come before scandium, Sc. As a result, the ten elements following calcium have their outer electrons in the fourth shell. Correspondingly, the ten elements (Y through Cd) following strontium, Sr, have a five-level

particles that are attracted to electron pairs. (See Figure 19.2.) Later, we will see examples of some of the reactions of multiple bonds.

The geometry of carbon atoms that have multiple bonds differs from that of tetrahedrally bonded carbon atoms. A double-bonded carbon atom can be thought of as an atom with three single bonds, to one of which is added a second bond. The three single bonds lie in a plane, with approximately equal bond angles of 120°: the result is a flat molecule with a geometry known as **trigonal planar** (Figure 19.3).

A triple-bonded carbon atom can be thought of as having two single bonds, one of which is accompanied by two extra shared pairs of electrons. The single bonds lie at an angle of 180°, forming a *linear* molecule. In acetylene, for example, all four of the atoms lie in a straight line (Figure 19.4).

19.1.4 Number and Variety of Carbon Compounds

Carbon atoms can bond with each other to make long, stable chain structures. One common arrangement is called a straight chain.

Straight chain

In this use, *straight* means *not branched,* rather then *linear.* A straight carbon chain has 109.5° tetrahedral bends in it, because of the 109.5° carbon bond angle. Carbon atoms, however, can also form branched chains, or even rings.

Branched chain   *Ring structure*

Although there is almost no limit to the possible length of a carbon chain, carbon atoms rarely form rings that have more than about eight members.

Molecules containing carbon can be very large; in some fields of study, molecules formed by a hundred or more atoms are considered small. Molecules that do not contain carbon, on

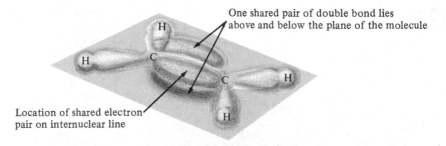

One shared pair of double bond lies above and below the plane of the molecule

Location of shared electron pair on internuclear line

Figure 19.2 Representation of electron densities in the double bond of an ethylene molecule.

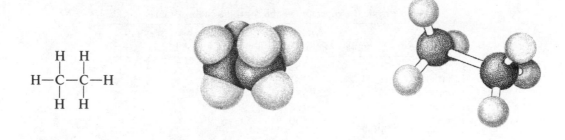

Figure 19.6   Three representations of ethane.

other significant features. As you view and draw such sketches, remember that each carbon atom forms four covalent bonds and that you can complete each drawing by adding bonded H atoms until every C has four bonds. Using this shorthand, butane (C_4H_{10}) would be drawn as you see in either the left or the center sketch. The right sketch is an even more abbreviated version of butane. We will use all these methods to indicate structure.

$$-\overset{|}{\underset{|}{C}}-\overset{|}{\underset{|}{C}}-\overset{|}{\underset{|}{C}}-\overset{|}{\underset{|}{C}}- \qquad\qquad C-C-C-C \qquad\qquad \diagup\!\!\diagdown\!\!\diagup$$

Three representations of butane

The next largest compound in the alkane series is pentane, C_5H_{12}, followed by hexane, C_6H_{14}, heptane, C_7H_{16}, octane, C_8H_{18}, nonane, C_9H_{20}, and decane, $C_{10}H_{22}$. Notice that in the formulas for these compounds the number of hydrogen atoms is always twice the number of carbon atoms plus two. If the number of carbon atoms in the formula is n, the number of hydrogens is 2n + 2, C_nH_{2n+2}. The eleventh compound in the series, for example, would have 11 C atoms and (2 × 11) + 2 = 24 hydrogen atoms, or $C_{11}H_{24}$. We could go on to write formulas for hydrocarbon molecules with any number of carbon atoms. For any formula we write, the molecule exists or can theoretically be prepared. For that matter, any molecule that we might propose could also be named, but the names of very large molecules tend to be equally large. We will discuss the preparation and properties of very large molecules in a later section.

The family of saturated hydrocarbons is called the **alkane,** or **paraffin,** family. (What is commonly called *paraffin* is a mixture of long-chain, mostly saturated, hydrocarbon compounds.) Table 19.1 lists the properties of some of the alkanes.

A hydrocarbon structure attached to some other group of atoms is called an **alkyl group.** The alkyl groups are named after their hydrocarbons. For example, H_3C- is a *methyl* group, H_5C_2- is an *ethyl* group, and so on.

The alkanes are part of everyday life. Methane is natural gas. Propane, sold as liquefied petroleum gas (LPG), is used for heating and motor fuel. Gasoline is mostly a mixture of hydrocarbons whose molecules contain six to eight carbon atoms; the molecules of kerosene are in the C_{10} to C_{17} range; those of fuel oil are in the C_{20} to C_{30} range. Mineral oil is much

Table 19.2
Functional Groups and Their Names

Structure	Name	Structure	Name
$-C=C-$	Double bond or alkene	$-C\equiv C-$	Triple bond or alkyne
$-C-OH$	Alcohol	$-\overset{\mid}{C}-O-\overset{\mid}{C}-$	Ether
$-C=O$ with H below	Aldehyde	Ketone structure with $C=O$	Ketone
$-\overset{\mid}{C}-NH_2$	Amine	$-C\overset{O}{\diagup}\underset{OH}{\diagdown}$	Carboxylic acid
Ester structure $-C\overset{O}{\diagup}\underset{O-C-}{\diagdown}$	Ester	Amide structure $-C\overset{O}{\diagup}\underset{N}{\diagdown}C-$	Amide

indicate where the structures are attached to hydrogen or to a carbon chain of some kind; in some cases, one carbon of the chain is shown with the functional group. Many organic compounds have two or more functional groups, either alike or different. Halogens should be added to the list of functional groups because a Br or a Cl atom bonded to carbon has characteristic properties and reactions, just as other functional groups do. Halogens attached to organic molecules, however, do not have special functional group names.

19.6 NAMING ORGANIC COMPOUNDS

Hundreds of thousands of organic compounds are known, and the number of other possible structures seems limitless. If the name of a compound is to be of much use, it should describe the structure. Think of the confusion that would result if half a million compounds were named at random. In the early days of organic chemistry, however, organic compounds were not systematically named. The first person to isolate or prepare a compound would simply give it a name, usually one that described its origin (*formic acid* from ants, Latin *formica*), or its properties (*morphine* from Morpheus, the Greek god of sleep). Occasionally, a compound would become known by the name of its discoverer (Micheler's ketone). Names of this sort gave little clue to the structure of a compound.

After many decades of confusion, a naming system has been accepted worldwide; it is a kind of international language. In 1921, the International Union of Pure and Applied Chemistry commissioned delegates from many countries to meet in Geneva to work out what is now known as the **IUPAC system,** or the **Geneva system.** In this language, a completely descriptive,

contains a chlorine atom in place of one of its hydrogens, so it is given the prefix *chloro-*. It is one of the isomers of chloropropanol. The structure on the right has both chlorine and bromine substituents; it is one of the bromochlorobutanols.

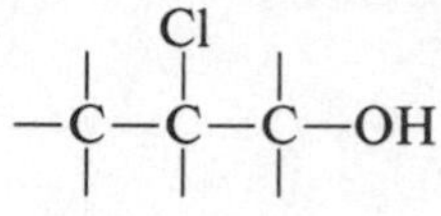

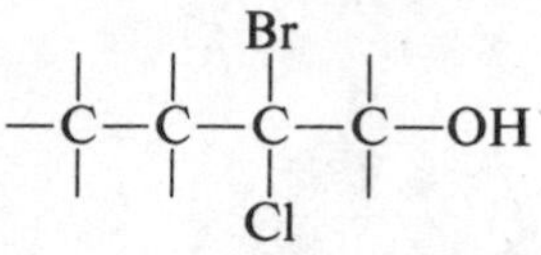

$$
\begin{array}{c}
\quad\;\; \overset{\displaystyle Cl}{|} \\
-C-C-C-OH \\
\end{array}
\qquad\qquad
\begin{array}{c}
\qquad\quad \overset{\displaystyle Br}{|} \\
-C-C-C-C-OH \\
\qquad\quad \underset{\displaystyle Cl}{|}
\end{array}
$$

Most organic compounds can have several different isomers. To differentiate among these, it is necessary to identify the positions of all the groups that have been substituted for hydrogen on the main chain (including alkyl groups). Numbers are used to designate positions, starting with a carbon at one end as number one. In the structure that follows, we call the end carbon that has the —OH group number one, which gives us 3-bromo-2-chloro-1-propanol. (Dashes are used to separate numbers from letters, commas to separate numbers. Carboxylic acid and aldehyde groups must be given position one.)

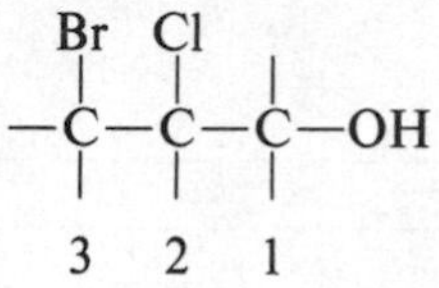

$$
\begin{array}{c}
\overset{\displaystyle Br}{|}\;\; \overset{\displaystyle Cl}{|} \\
-C-C-C-OH \\
\;3\quad 2\quad 1
\end{array}
$$

EXAMPLE 19.1.   Name the following structure:

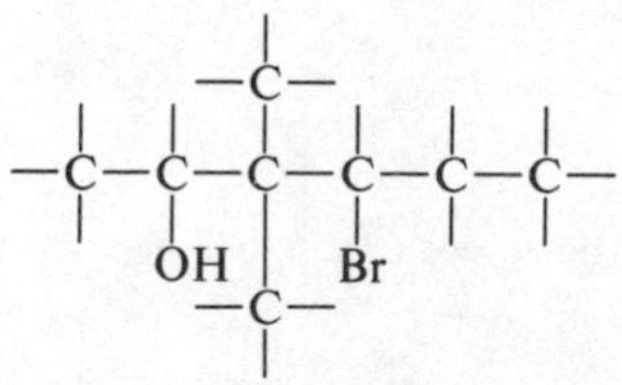

$$
\begin{array}{c}
\quad\;\; |\;\; \\
\;\;\; -C- \\
\quad\;\; |\;\; \\
-C-C-C-C-C-C- \\
\;\;\;\; | \quad\;\; | \\
\;\;\; OH \;\;\; Br \\
\quad\;\;\;\;\;\; | \\
\quad\;\;\; -C- \\
\quad\;\;\;\;\;\; |
\end{array}
$$

Solution:

Step 1.   Find the longest unbranched carbon chain, circle it for convenience, and name it. (In most structures to follow, bonds from C to H are not shown unless necessary.) To help us keep track of the main chain, it is shown in color.

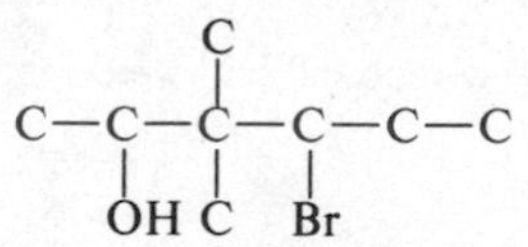

$$
\begin{array}{c}
\quad\quad\;\; C \\
\quad\quad\;\; | \\
C-C-C-C-C-C \\
\quad\;\; | \;\; | \\
\quad OH\; C \;\; Br
\end{array}
$$

Our compound will have the syllable *-hex* in its name. Because the compound is saturated, the letters *-an* will also appear. (If there had been a double bond, the letters *-en* would have been used, and if a triple bond, *-yn*.)

    -hexan-

Step 2.   Find the functional group for which the compound will be named. We will name this compound for the —OH group, using the ending *-ol*. We have a hexanol. (Like alkyl groups, halogens are described by prefixes rather than endings. At this stage, we ignore the bromine.)

    -hexanol

Step 3. Number the carbons on the chain, starting at the end nearest the functional group (to keep the numbers as small as possible). For this molecule, we start at the left.

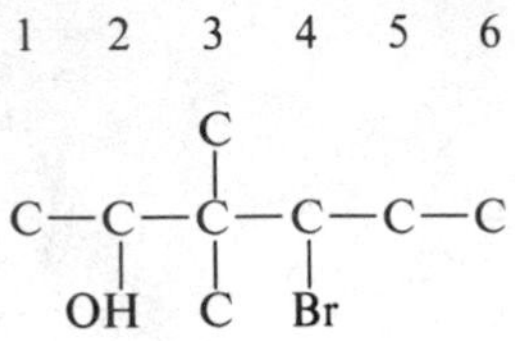

Step 4. Using the numbers you just assigned, begin to write the full name by designating the numbers and positions of the substituents on the main chain. In this structure there is a bromine on the fourth carbon and two methyl groups on the third carbon; each methyl group requires a number.

4-bromo-3,3-dimethyl-hexanol

Step 5. Finish the name by designating the position of the substituent for which the ending is used. The —OH group is on the second carbon.

4-bromo-3,3-dimethyl-2-hexanol

EXAMPLE 19.2. Name the following structure:

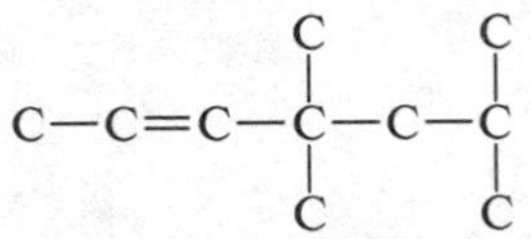

Solution:

Step 1. Here the longest continuous chain is seven carbons, even though it is not written as a straight line. We use the letters -*hept*.

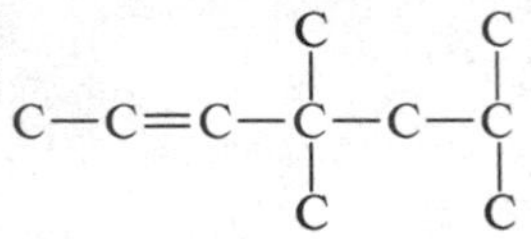

-hept-

Step 2. The only functional group is the double bond. The compound will have the hydrocarbon ending of the double bond, -*ene*.

-heptene

Step 3. Starting at the end nearest the double bond, number the carbons.

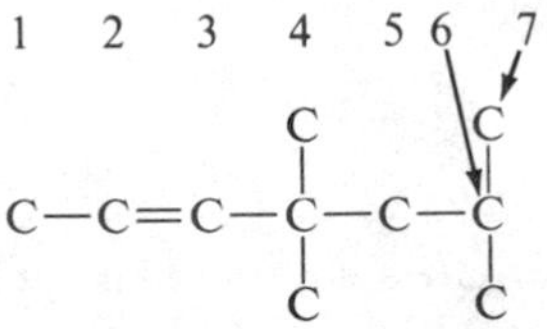

Step 4. Assign the substituents their places. In this structure, all are methyls. Because there are three methyl groups, the prefix *tri-* will appear in front of the name *methyl*. Each methyl receives a position designation.

   4,4,6-trimethyl- -heptene

Step 5. Designate the position of the substituent for which the ending is given. The double bond lies between the second and third carbons, so we have a -2-heptene.

   4,4,6-trimethyl-2-heptene

EXAMPLE 19.3.   Name the following structure:

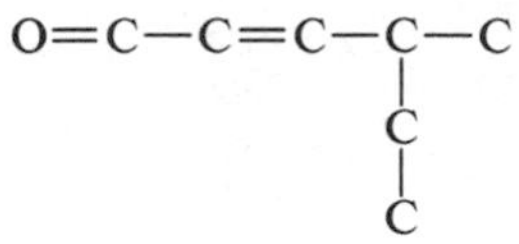

$$O{=}C{-}C{=}C{-}C{-}C$$
$$\begin{array}{c} | \\ C \\ | \\ C \end{array}$$

Solution:

Step 1. Beginning at the aldehyde group, the longest chain is six carbons. Use the letters *-hex-*.

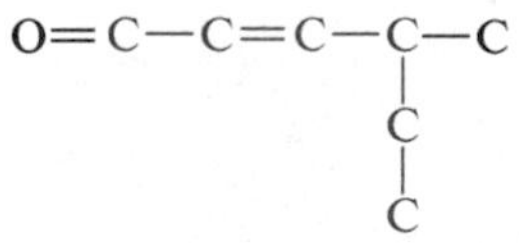

$$O{=}C{-}C{=}C{-}C{-}C$$
$$\begin{array}{c} | \\ C \\ | \\ C \end{array}$$

          -hex-

Step 2. There are three functional groups, two of which can be shown in the ending. The double bond gives us the letters *-en-*, and the aldehyde gives the ending *-al*.

   -hexenal

Step 3. In numbering the chain, we automatically start at the aldehyde end because aldehydes are always considered to be on the first carbon.

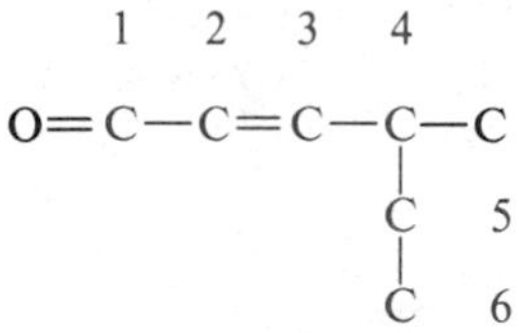

$$\overset{1}{O}{=}\overset{2}{C}{-}\overset{3}{C}{=}\overset{4}{C}{-}C{-}C$$

        1   2   3   4
$$O{=}C{-}C{=}C{-}C{-}C$$
$$\begin{array}{c} | \\ C \quad 5 \\ | \\ C \quad 6 \end{array}$$

Step 4. The methyl group is attached to the fourth carbon. The double bond lies between the second and third carbons. (No number needs to be given for the aldehyde.) Finish the name.

   4-methyl-2-hexenal

Look at the entire structure again, to see how a relatively short name gives a great deal of precise information.

When you name a compound, be sure to check the following:

1. Have you chosen the longest chain, even if it zigzags?
2. Have you started numbering from the end closest to the functional group, regardless of whether the numbers go from right to left or top to bottom? Both the structures in the following diagram are 1-butene.

$$C=C-C-C \qquad C-C-C=C$$

1-butene

3. Does the name show all the groups and positions in the structure, even the hydrocarbon groups?

EXERCISE 19.2.   Name the following compound:

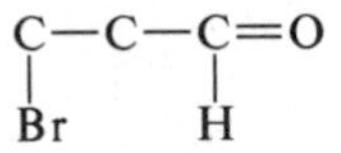

EXERCISE 19.3.   Name the following compound:

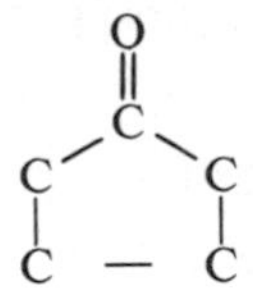

EXERCISE 19.4.   Name the following structure:

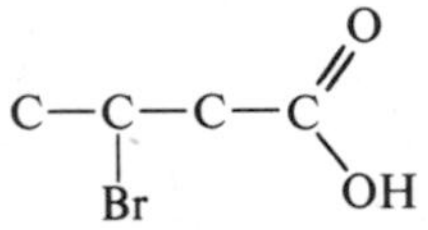

If you are given a name and you want to draw the structure, start at the end of the name to find the length of the main chain. Write a chain of carbon atoms of this length, then number it, starting at whichever end you choose. Add the substituent groups in the positions indicated, being careful to include those indicated in the ending as well as those that are spelled out. Finally, add enough hydrogen atoms to each carbon so that it has four bonds. Until you have had some practice in reading and writing names, it is best to not leave out H atoms.

EXAMPLE 19.4.   Write the structure of 3-hydroxy-3-isopropylpentanoic acid.

Solution:

Step 1.   Write the main chain, including the acid group as the first carbon.

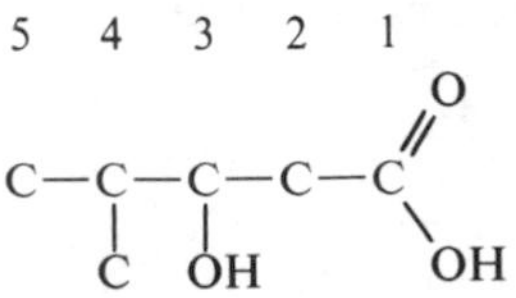

Step 2. Write in the substituents at the designated positions. (It does not matter whether substituents are placed above or below the chain, as long as they are bonded to the proper carbon atom. The prefix *iso-* means "*branched.*")

$$
\begin{array}{c}
\text{C}\\
|\\
\text{C}\text{O}\\
|\parallel\\
\text{C}-\text{C}-\text{C}-\text{C}-\text{C}\\
||\backslash\\
\text{C}\text{OH}\text{OH}
\end{array}
$$

Step 3. Add hydrogen atoms to complete the structure. Carbon number five has only one bond, and needs three hydrogens. Carbon number four has three bonds and needs one hydrogen. Neither carbon number one nor carbon number three needs more bonds, but carbon number two requires two hydrogens. Be sure to put hydrogens on all the alkyl groups.

EXERCISE 19.5.   Write the structure of 5-ethyl-4-methyloctanoic acid.

EXERCISE 19.6.   Write the structure of 2,4,6-trimethylcycloheptanone.

19.7 PROPERTIES AND REACTIONS OF ORGANIC COMPOUNDS

As we have seen, a saturated hydrocarbon is not easily dismantled. The carbon-carbon bonds and carbon-hydrogen bonds are strong and not especially polar. The bonds are shielded from outside influences because the carbon atoms are surrounded on all sides by strongly bonded atoms. Most reactions of organic molecules take place either within a functional group or because a functional group is being replaced or eliminated. Although carbon-carbon bonds *can* be broken in organic reactions, such breaks are usually helped by a nearby functional group.

19.7.1 The R Shorthand

As a convenient shorthand, organic chemists frequently write the structures of molecules to emphasize some particular functional group. The practice is much like writing only the reacting ions in an ionic reaction. The shorthand method substitutes one or more letters *R* for inactive portions of molecules. For example, an organic chemist interested only in alcohols and not caring *which* alcohols were involved, would write the alcohols as R—OH; R could be methyl, ethyl, or any more complex group. R could even contain other functional groups, as long as they were not involved in the reaction in question. Any of the following structures, for example, might be shown as R—OH (R shown in color).

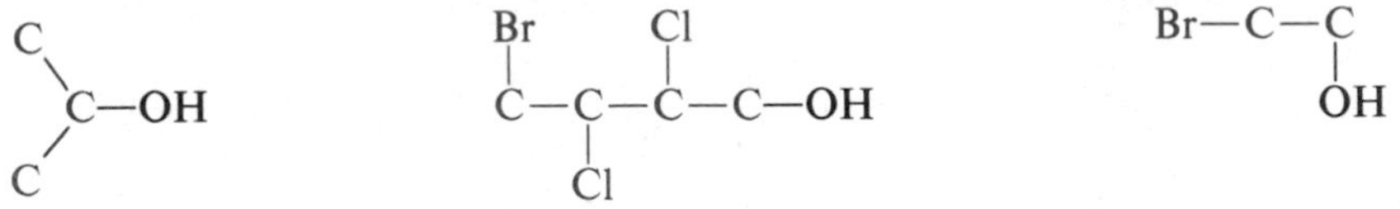

$$
\begin{array}{ccc}
\ce{C} & & \\
\ \ \backslash & & \\
\quad\ \ \ce{C-OH} & \quad\quad \ce{\overset{\displaystyle Br}{|}C-\overset{\displaystyle Cl}{|}C-\underset{\displaystyle Cl}{|}C-C-OH} & \quad\quad \ce{Br-C-\overset{|}{\underset{OH}{C}}} \\
\ / & & \\
\ce{C} & &
\end{array}
$$

On the other hand, a chemist interested only in the reactions of bromine substituents would write the structure shown on the right as R—Br. Of course, the R's in these two examples would not represent the same structures. We will find it convenient to use the R language.

19.7.2 How Reactions Occur in Organic Molecules

The reactions of most organic compounds involve molecules. Although some reactions occur between organic ions, and others involve what are called free radicals, we can gain the most insight to organic reactions by first discussing what must occur in the reactions of molecules as such.

Molecules are held together by covalent bonds, bonds that must be at least moderately strong to keep their structures intact. If a molecule is to be altered—if it is to react—covalent bonds must be broken. The breaking of covalent bonds usually occurs simultaneously with the formation of other bonds. Only occasionally do organic reactions react by fragmenting and recombining. (Certain kinds of organic molecules do ionize, however. We will study this in a later section.)

Organic chemists classify most reactions of organic molecules as either **substitutions** (when a functional group in a molecule is replaced by a different group), **additions** (when two or more groups are added to a molecule that has a double or triple bond), or **eliminations** (when two or more groups leave a molecule, usually causing the creation of a double bond).

A good example of a reaction in which bonds are simultaneously broken and created is the substitution of a cyanide group for a bromine atom, to form what is called a nitrile molecule (Figure 19.8). Because the Br atom is more electronegative than the C atom, the C—Br bond is not very strong and is quite polar. The polarity of the bond leaves the C somewhat positive and attractive to the CN⁻, which attacks from the back side. The approach of CN⁻ further releases the electrons in the C—Br bond, allowing the Br to take both electrons from the C—Br bond and leave as Br⁻. The result is a bond between a CN group and a C.

19.8 REACTIONS OF UNSATURATED MOLECULES

Many unsaturated compounds (most with double rather than triple bonds) occur in nature, and more are made industrially. Reactions of double bonds are important biologically and in manufacturing. We have learned that in a double carbon-carbon bond, one pair of electrons spends most of its time somewhat outside the space directly between the bonded atoms. The second shared pair forms a bond that is weaker and more easily broken than the bond formed by the centrally located pair. In one of the common reactions of double bonds, the second pair is caused to split. One electron then goes to each carbon atom and is used as half of a covalent bond to some other atom.

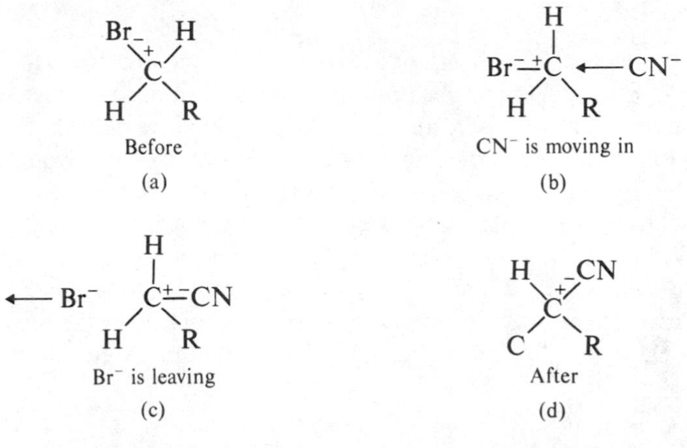

(a) Before

(b) CN^- is moving in

(c) Br^- is leaving

(d) After

Figure 19.8 Attack on a C—Br bond by a CN⁻ ion.

19.8.1 Additions to Double Bonds

The most characteristic reaction of double bonds is the addition of two atoms, or groups of atoms, to form a compound with single bonds. H_2 furnishes an important example.

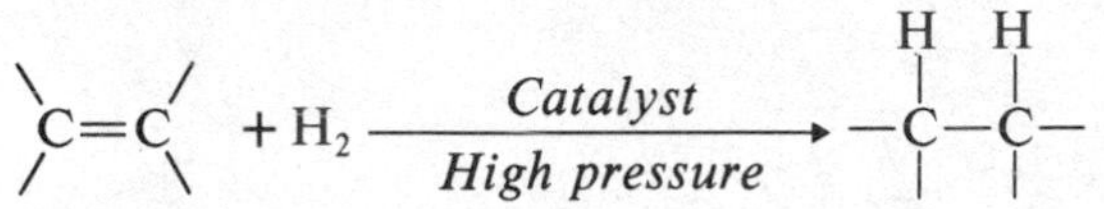

Crude oil contains molecules with double bonds. These compounds are undesirable for a variety of reasons, but they can be made into useful saturated products by hydrogenation. Hydrogenation is useful also in agriculture. The manufacturers of many foodstuffs, including most margarines, depend on the conversion of unsaturated vegetable oils, such as cottonseed or corn oil, into saturated compounds that have higher melting points and can be sold as solids.

Other substances add different groups to the two atoms joined by a double bond. An example is the synthetic preparation of ethanol from ethylene (ethene), a byproduct of petroleum refining. The compound used to make the conversion is H_2SO_4. The intermediate molecule is reacted with water to give the alcohol.

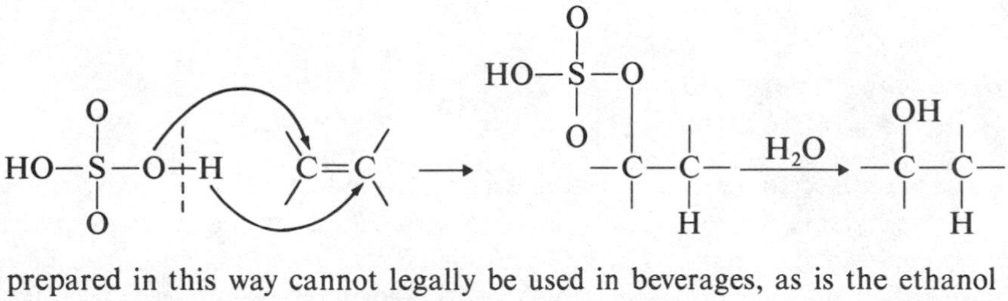

Ethanol prepared in this way cannot legally be used in beverages, as is the ethanol made by fermentation, but it is widely used as a solvent and as a starting product for the synthesis of other compounds.

19.8.2 Free Radical Reactions

A **free radical** is *an atom or a group of atoms that has an unshared electron,* that is, an odd number of electrons. A nonorganic example of a free radical is any molecule that contains an odd number of nitrogen atoms, NO_2, for example. Organic free radicals form when a bond breaks in such a way that each fragment carries away one electron from the pair, not a common reaction. Free radicals are quite active chemically. Because they react with one another to form electron pair bonds, free radicals can be prepared only in dilute solutions.

A free radical will attack a carbon-carbon double bond and form a bond to one of the carbons. We can see how this happens by drawing electron dot diagrams. We use R to represent the free radical because we don't care what the structure of the radical is, only that it has an unshared electron. (In the drawings that follow, the unshared electron is shown as a large dot.)

Notice that the result of this reaction is itself a free radical; there is still an unshared electron. This new free radical will react readily with another unsaturated molecule, to form still another, longer, free radical.

Because reactions of this kind cannot eliminate the unshared electron, the reaction continues for thousands of steps until the supply of unsaturated molecules is gone or until two free radicals happen to meet and share their electrons. Reactions that are repeated over and over to produce very large molecules are called **polymerizations,** *poly* meaning "many" and *mer* meaning "piece" or "fragment." A polymerization makes one large molecule, a **polymer,** out of many individual **monomer** molecules (*mono* means single).

Free radical polymerizations are responsible for many of the thousands of synthetic substances popularly called plastics. The most common plastics of this kind are the vinyl polymers. **Vinyl** is another name for ethylene in which one of the hydrogens has been replaced by another atom or group. We can represent a segment of a vinyl polymer by using X for the substituted H. Although we are looking at only a few links in the vinyl chain, the chains of useful vinyl polymers are from fifty thousand to several hundred thousand units long.

Vinyl polymers are used to make an incredible array of products. If X is merely hydrogen, the polymer is polyethylene, which is a soft, inert material used for plastic bags, plastic sheets, raincoats, and soft molded items. Polyvinylchloride, PVC, is the polymer in which X is Cl. PVC is harder than polyethylene and softens at a higher temperature; it is used for molding and extrusion, as in PVC water pipe. If X is a benzene ring, the polymer is called polystyrene and can be used to make toys, taillight lenses, pens, and numberless other products. Polystyrene molds easily and is cheap to manufacture. If X is another vinyl group, the monomer is called butadiene and is used to make synthetic rubber for automobile tires.

$$-C-C=C-C-C-C=C-C-C-C=C-$$

The butadiene polymer chain structure

During our discussion of the halogens, we learned that fluorine is highly electronegative and able to make covalent bonds. Like the other halogens, fluorine can take the place of hydrogen in hydrocarbons. The long-chain vinyl polymer in which fluorine replaces all the atoms except the carbons of the main chain is called Teflon. Teflon is highly resistant to chemical attack and has little attraction for other molecules of any kind. There are two reasons for the unique properties of this polymer: first, the unshared pairs on the fluorine atoms are tightly held by the large, concentrated fluorine kernel charges and are consequently not available for sharing or ionization. Second, because fluorine atoms are much larger than hydrogen atoms, they effectively occupy all the space around the carbon main chain, protecting the carbon-carbon and carbon-fluorine bonds from attack (Figure 19.9). As a result of its structure, Teflon is extremely inert and stable, properties that are put to good use in applications from cookware to lubricants to containers for corrosive fluids. Teflon is a good example of a useful, man-made material that is quite unlike anything seen on earth before the advent of synthetic organic chemistry.

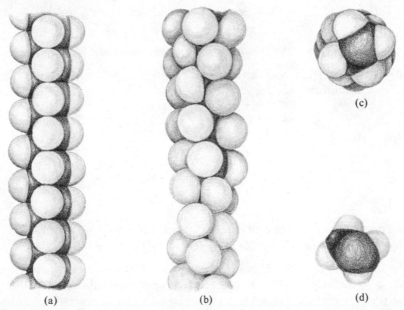

Figure 19.9 Scale models of Teflon and polyethylene chains: (*a*) Polyethylene. (*b*) Teflon. (*c*) Teflon, view end-on. (*d*) Polyethylene, view end-on.

19.9 REACTIONS OF AROMATIC COMPOUNDS

The aromatic hydrocarbons include benzene and the multiple ring systems built on the benzene model. Although these compounds are not saturated, they do not react as do compounds that have ordinary double bonds. Ordinary unsaturated compounds will add bromine readily; the bromine addition reaction is a standard test for unsaturation.

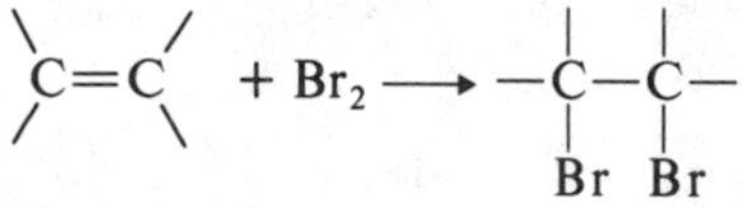

$$\overset{\diagdown}{\underset{\diagup}{C}}=\overset{\diagup}{\underset{\diagdown}{C}} \ + \ Br_2 \longrightarrow -\overset{|}{\underset{|}{C}}-\overset{|}{\underset{|}{C}}- \\ \qquad\qquad\qquad\quad Br \ \ Br$$

Aromatic compounds, however, add bromine only extremely slowly, and under certain conditions, will instead substitute bromine for hydrogen. Electron delocalization causes the properties of aromatic hydrocarbons to differ from those of ordinary unsaturates.

There are hundreds of useful compounds in which one or more of the H atoms on benzene have been substituted by functional groups. Compounds in which all six of the hydrogens have been substituted are not uncommon. Of the mono-substituted benzenes (those in which a single hydrogen has been replaced), a few of the more important are shown in the diagram that follows.

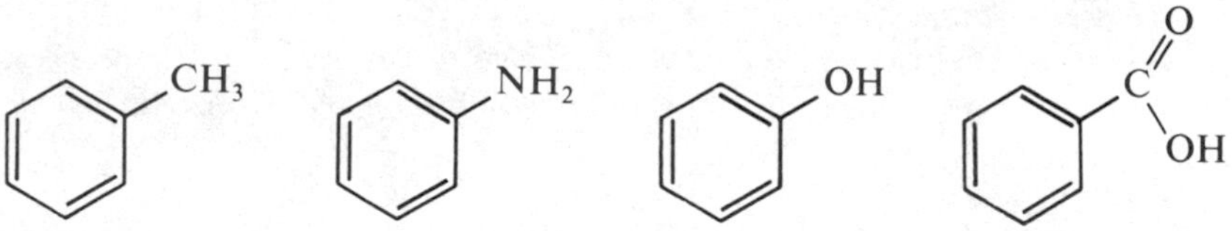

| *Toluene* | *Aniline* | *Phenol*
(Carbolic acid) | *Benzoic acid* |

Some mono-substituted benzene compounds

If two hydrogens are replaced, the resulting molecule can be arranged in three possible ways. The three dichlorobenzenes are examples.

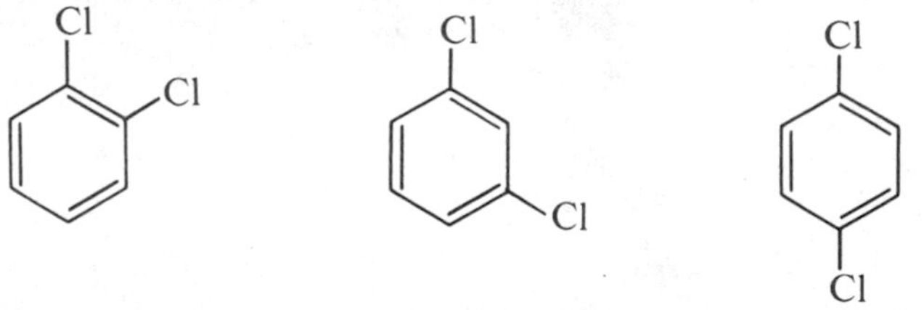

o-dichlorobenzene   *m-dichlorobenzene*   *p-dichlorobenzene*
1,2-dichlorobenzene   *1,3-dichlorobenzene*   *1,4-dichlorobenzene*

These are the three possible isomers of dichlorobenzene. Notice that each structure has been given two names, reflecting the fact that there are two systems for naming ring compound isomers. The prefixes *ortho-, meta-, para-* (indicated by the abbreviations *o-, m-,* and *p-*) indicate the positions of substitution on the benzene ring. Starting with any substituent in position X as shown in the following diagrams, the letters *o, m,* and *p* indicate the positions other substituents can take.

The system that uses numbers is more formal, but more versatile. Starting with the position of a substituent X at carbon number one, the carbons on the ring are numbered in order around the ring, and positions of other substituents are given by number. Two examples are shown.

2-bromo-5-nitrophenol   *5-amino-2-hydroxy-3-nitrotoluene*

Because different isomers of some of the substituted benzenes have different properties, it is important to know the positions of the substituents. For example, *para*-dichlorobenzene (sometimes used to kill moths) is a solid that melts at 53 °C, whereas *meta*-dichlorobenzene is a liquid that freezes at −25 °C.

We have seen that rings can have common sides, that is, can be fused together. The most common fused ring system is the two-ring pair called naphthalene, $C_{10}H_8$. Three rings can fuse to give $C_{14}H_{10}$, which has two isomers.

Naphthalene   *Anthracene*   *Phenanthrene*
$C_{10}H_8$ $C_{14}H_{10}$ $C_{14}H_{10}$

Naphthalene is used in mothballs and as a starting ingredient for the synthesis of many dyes and perfumes. The larger fused ring systems appear primarily in drugs and complex biological molecules.

19.10 ALCOHOLS AND THEIR OXIDATION REACTIONS

As we have learned, the compounds in which a hydrogen atom on a hydrocarbon (except in aromatics) has been replaced with a OH group are called alcohols. The character of an alcohol depends on the length of the hydrocarbon chain and the amount of branching in it. Methyl alcohol, CH_3OH, the smallest molecule in the series, mixes with water in all proportions and is quite polar. It might be thought of as a water molecule in which one of the hydrogens has been replaced with methyl. As the hydrocarbon chain gets larger, the effect of the OH group becomes less important. The higher alcohols behave more like the hydrocarbons from which they came; 1-hexanol is only slightly soluble in water, and 1-octanol is insoluble. Because the OH group is bonded to carbon with a strong covalent bond, the alcohols do not ionize in water.

Methanol, ethanol, and 2-propanol (often called isopropyl alcohol) are the most commonly used alcohols. Methanol and ethanol can be mixed with gasoline and used as automobile fuel; under proper conditions, either can be used alone. Both alcohols are used widely as solvents and as starting materials for the synthesis of other compounds. Ethanol, of course, is used in beverages, but the tonnage used industrially far exceeds that sold to consumers. Methanol is poisonous.

The familiar rubbing alcohol is 2-propanol. The compound is also used as a solvent and as a starting material for synthesis.

Alcohols can be oxidized by any of several methods. A **primary alcohol,** in which the OH group is on an end carbon, yields an aldehyde from the first step of the oxidation. Further oxidation leads to a carboxylic acid, and total oxidation results in CO_2.

$$RC\text{—}OH \longrightarrow RC{=}O \longrightarrow RC\overset{\displaystyle O}{\underset{\displaystyle OH}{\big/\!\!\big/}} \longrightarrow CO_2 + H_2O$$

Primary alcohol   *Aldehyde*   *Carboxylic acid*

Oxidation of a **secondary alcohol,** one in which the OH group is on a carbon bonded to two other carbons, leads to a ketone and then to CO_2. Because carbon can make only four bonds, a carboxylic acid carbon cannot have two other carbons bonded to it. Secondary alcohols cannot oxidize to carboxylic acids.

$$\underset{C}{\overset{C}{\diagdown}}\!\!C\!\!\underset{OH}{\overset{H}{\diagup}} \qquad \underset{C}{\overset{C}{\diagdown}}C{=}O$$

2-propanol   *Dimethylketone*
(Acetone)

Alcohols can be oxidized by oxygen and a catalyst, or by an ionic oxidizing agent such as permanganate, MnO_4^-, or dichromate, $Cr_2O_7^{2-}$.

19.11 ALDEHYDES AND KETONES

Aldehydes and ketones belong to the class of compounds known as **carbonyls.** In both kinds of compound, carbon atoms are double bonded to oxygen atoms. In **aldehydes,** the oxygen is always bonded to an end carbon, and there is also a hydrogen on that carbon. In **ketones,** the oxygen is bonded to a carbon that is not first in the chain, that is, to a carbon that is bonded to two other carbon atoms.

$$-\underset{\underset{\text{H}}{|}}{\text{C}}=\text{O} \qquad \underset{\text{C}}{\overset{\text{C}}{\diagdown}}\text{C}=\text{O}$$

Aldehyde   *Ketone*

Methanal, the smallest aldehyde, is often called formaldehyde because it oxidizes to formic acid. Methanal is a gas at room temperature, but is usually sold as formalin, a 40% solution of methanal in water. The compound is used as a germicide and fumigant, but its largest use is as one of the compounds out of which the durable plastics Bakelite and Melmac are made. (See Section 19.14.) For industrial purposes, methanal is prepared by the cataylzed oxidation of methanol.

Ethanal, more commonly called acetaldehyde, and the higher aldehydes are used mostly in the laboratory and in industry.

Of the ketones, the most well-known are dimethylketone (acetone) and methylethylketone (MEK). Both compounds are excellent solvents and are used in paints, varnishes, cements, paint removers and drugs and as industrial solvents.

Aldehydes and ketones can be reduced to their corresponding alcohols by the catalytic addition of hydrogen at high pressure or by means of strong inorganic reducing reagents.

19.12 AMINES

Amines are organic bases. We can think of amine molecules as ammonia molecules in which one or more of the hydrogen atoms have been replaced by organic groups, often alkyl groups. Like ammonia molecules, amine molecules offer unshared electron pairs and are attractive to acids.

$$\text{H}-\underset{\underset{\text{H}}{\diagdown}}{\overset{\overset{\text{H}}{\diagup}}{\text{N}}} \qquad\qquad \text{H}-\underset{\underset{\text{Me}}{\diagdown}}{\overset{\overset{\text{Me}}{\diagup}}{\text{N}}}$$

Ammonia   *Dimethyl amine*

$$\text{H}-\underset{\underset{\text{CH}_3}{\diagdown}}{\overset{\overset{\text{H}}{\diagup}}{\text{N}}} \qquad\qquad \text{Me}-\underset{\underset{\text{Me}}{\diagdown}}{\overset{\overset{\text{Me}}{\diagup}}{\text{N}}}$$

Methylamine   *Trimethyl amine*

Amines are classed as primary, secondary, and tertiary, depending on whether there are one, two, or three carbon atoms bonded to the nitrogen. Methylamine, for example, is a primary amine, dimethyl amine is a secondary amine, and trimethyl amine is tertiary.

The nitrogen atom in an amine, like that in ammonia, can have four atoms bonded to it, but it must then exist as an ion. On accepting protons, amine molecules become ions; that is, they form salts. Not many amines are soluble in water but most have soluble salts. Many drugs contain amine groups. Because such drugs often have large, complex molecules, they are sold in the form of amine salts so that they can be dissolved easily.

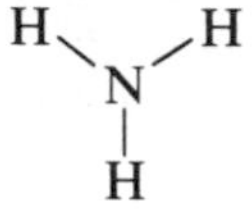

Ammonia 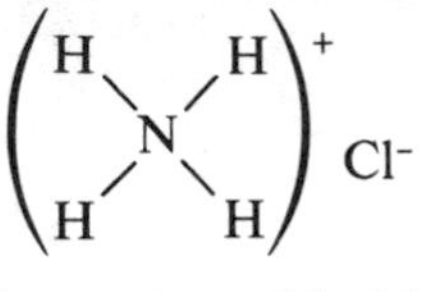   *Ammonium chloride*

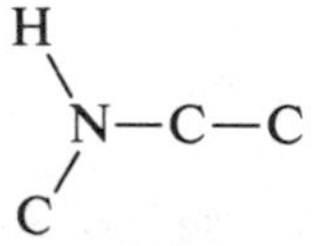

Methylethylamine 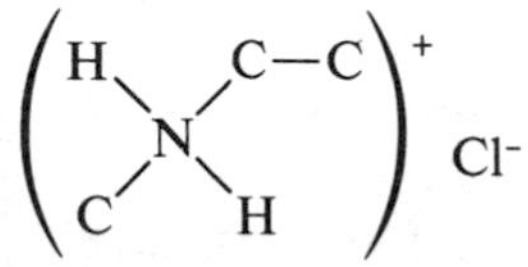   *Methylethyl ammonium chloride*

19.13 CARBOXYLIC ACIDS, ESTERS, AND AMIDES

By examining the **carboxylic acid** group, we can find out why it is acidic. The doubly bonded oxygen is highly electronegative, as are all oxygen atoms. Because it is held close by the short double bond, the oxygen makes its eletronegativity felt beyond the carbon atom. It exerts its pull on electrons even as far away as the O—H bond and weakens this bond by making the electron pair less available to the hydrogen atom. The hydrogen can break away from the molecule, but it leaves the electron pair behind and goes away as H^+. Anything that supplies H^+, of course, is an acid. (See Figure 19.10.)

We have seen acetic (ethanoic) acid several times. It is a typical and common carboxylic acid.

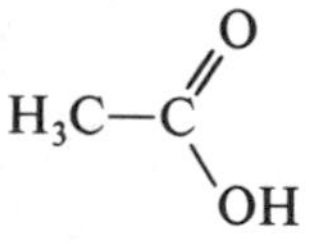

Acetic acid

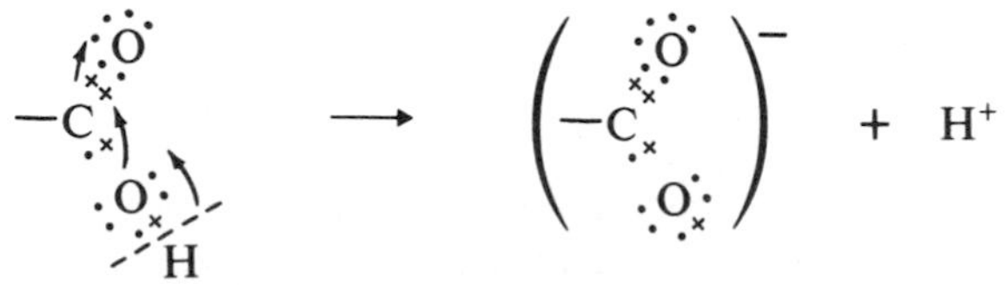

Figure 19.10   How electron attractions cause the carboxyl group to be acidic. The drawing on the left shows how the doubly bonded oxygen exerts attraction on electrons as far away as those in the O—H bond. The electron shift weakens the bond and allows the hydrogen to ionize as H^+. The drawing on the right shows the resulting ion pair.

Acetic acid is weak, but there are stronger organic acids. Trichloroacetic acid, for example, is nearly as strong as nitric acid because of the power of the electronegative chlorine atoms to withdraw electrons. By attracting electrons, the Cl atoms further weaken the O—H bond.

Many carboxylic acids are known by common names because they were among the first organic compounds studied. Citric acid, for example, is found in citrus fruits. And, in turn, the names of some hydrocarbons are derived from the names of acids. The name *butane* comes from butyric acid, which is related to compounds found in butter. Benzene is named after benzoic acid.

Carboxylic acids of low molecular weight have pungent and often unpleasant odors. The smell of rancid butter comes from butyric acid; that of dirty socks usually from valeric acid (pentanoic acid). Acetic acid has a sharp odor that in concentrated form will take your breath away. Like all acids, carboxylic acids taste sour. Acetic acid is what makes vinegar sour.

Although ionization is the most characteristic reaction of carboxylic acids, these acids undergo other important reactions as well. A carboxylic acid can combine with an alcohol molecule, eliminating a molecule of water in the process and forming an **ester.**

$$H_3C-C\overset{\displaystyle O}{\underset{\displaystyle OH}{\big\|}} \quad HO-CH_3 \longrightarrow H_3C-C\overset{\displaystyle O}{\underset{\displaystyle O-CH_3}{\big\|}} \quad + H_2O$$

Acetic acid *Methanol* *Methyl acetate*
(methyl ester of acetic acid)

Esters are pleasant-smelling compounds that are not usually very water soluble. The fragrances of fruits and flowers are often esters of moderate molecular weight. The somewhat sweet smell of fingernail polish remover and of lacquer comes from ethyl acetate or pentyl acetate (banana oil). Esters are used for perfumes, flavorings and solvents and in industry.

Animal fats are the esters of long, straight-chain carboxylic acids, such as stearic acid (16 carbons). The salts of such acids are soaps. Soaps act as they do because their long hydrocarbon tails dissolve in oily particles, while the acidic ends are attracted to water molecules. As a result, the oily particles become surrounded by water molecules and form emulsions (Figure 19.11). (For a review of emulsions, see Section 15.1.) Most modern soaps are sodium salts, but until the early part of this century, soaps were made by using lard and lye (KOH) made from wood ashes. These materials produced potassium salts, which formed "soft soap."

The sodium and potassium salts of long-chain carboxylic acids are soluble, but magnesium and calcium salts are not (because the double charges on the Ca^{2+} and Mg^{2+} ions attract the acid ions strongly and will not allow them to dissociate). Hard water contains calcium and magnesium ions. When soap is dissolved in hard water, some of the soap reacts with the Ca^{2+} and Mg^{2+}, forming an undesirable, insoluble scum. Modern synthetic detergents, which are not carboxylic acid ions, have long chains terminated by ionic groups. Because the ionic groups are designed not to attract Ca^{2+} and Mg^{2+}, detergents form little or no scum.

When a carboxylic acid reacts with ammonia or an amine, the result is called an **amide.**

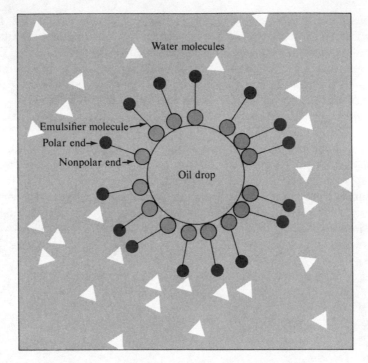

Figure 19.11 How an emulsifier works. Schematic drawing of emulsifier molecules oriented around a globule of nonpolar substance. The emulsifier molecules surround the globule and form a coat of polar groups compatible with the other liquid.

The only amides we will have occasion to study are the amide polymers, an important class that includes the proteins and the synthetics, such as nylon.

19.14 CONDENSATION POLYMERS

It is possible for carbon chains to carry more than one functional group, and to undergo more than one reaction. Molecules that have two reactive groups are said to be **bifunctional,** and **polyfunctional** if they have more. Examples of bifunctional molecules are adipic acid (1,6-hexanedioic acid), a six-carbon chain with carboxyl groups on each end, and 1,6-diaminohexane.

1,6-hexanedioic acid   *1,6-diaminohexane*

Bifunctional molecules are of special interest: a reaction on one end leaves the other end open for further reaction. When two bifunctional molecules join, they form a molecule that is still bifunctional.

$$\underset{HO}{\overset{O}{\parallel}}C-C-C-C-C-\underset{\underset{H}{\overset{|}{N}}}{\overset{O}{\overset{\parallel}{C}}}C-C-C-C-C-C-NH_2$$

The condensation reaction between adipic acid and 1,6-diaminohexane can continue for another step, two more steps, or an indefinite number of steps, forming a long-chain polymeric amide, a **polyamide.** The polymer formed by these two compounds is called nylon. In a popular lecture-demonstration called the "nylon rope trick," the plastic forms at the intersection of solutions containing the two reactive substances (Figure 19.12). Nylon fibers are used to make all kinds of fabrics from strong sailcloth canvas to sheer hosiery. Because it is inexpensive, easy to mold, strong, durable, self-lubricating, and noncorrosive, nylon is excellent for molding into gears, bearings, and various kinds of mechanical parts. Nylon is one of the most versatile plastics.

Terephthalic acid and ethylene glycol are both bifunctional.

$$\underset{HO}{\overset{O}{\parallel}}C-\!\!\!\!\!\!\bigcirc\!\!\!\!\!\!-\underset{OH}{\overset{O}{\parallel}}C \qquad\qquad HO-C-C-OH$$

Terephthalic acid   *Ethylene glycol*

Figure 19.12   Nylon rope trick. Nylon is formed at the junction of the solutions of ethylenediamine and adipyl chloride.

These two compounds react to form a polyester called mylar.

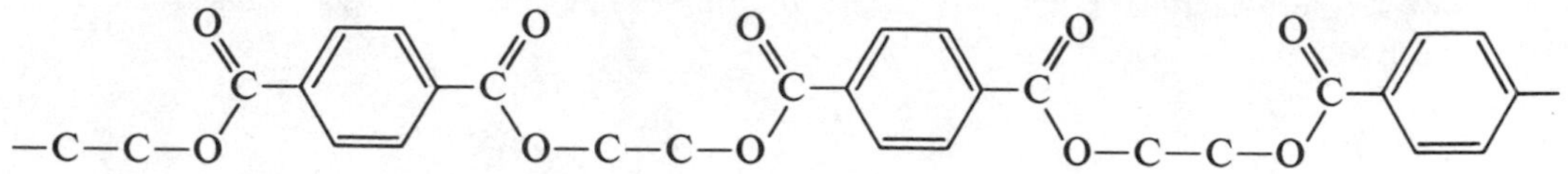

A portion of the mylar polymer

Mylar is a tough, transparent polymer often sold as sheets to be used for plastic windows and packaging. Potato chip packages are often made of mylar, which is sometimes vacuum coated with a thin film of aluminum to keep out light.

Depending on the structures of their monomers and on the temperature, the long molecules in a solid sample of a straight-chain polymer can shift position somewhat in relation to each other. If such shifts occur, the solid will be flexible and can soften when heated. Polyethylene is soft and melts easily; mylar can bend; Lucite softens when heated. All are straight-chain polymers. Instead of making straight chains, however, polyfunctional monomers can form **network polymers,** structures that are bonded in three-dimensional patterns. The individual units of network polymers are held rigidly in place by bonds to surrounding atoms in all directions and are unable to shift position. When the units of a polymer are locked in place, the solid is hard and stiff and will not soften when its temperature is raised.

The reaction of methanal and phenol furnishes an example of how network polymers are formed. Each molecule of phenol (hydroxybenzene) has five hydrogen atoms attached to the carbons on its ring. Because of the presence of the OH group, which withdraws electrons from the ring and weakens the C—H bonds, the hydrogen atoms of phenol are somewhat more reactive than those of ordinary benzene. The hydrogen atoms of adjacent phenol rings tend to react with methanal (Figure 19.13). The reaction continues, forming a network polymer. A typical example of a phenol-methanal polymer is Bakelite, a product first manufactured in the early part of this century (Figure 19.14). Bakelite is a hard, dark-brown solid used for pan handles and electrical parts such as switches, plugs, and sockets. In Bakelite, each phenol ring is usually linked to three or more rings.

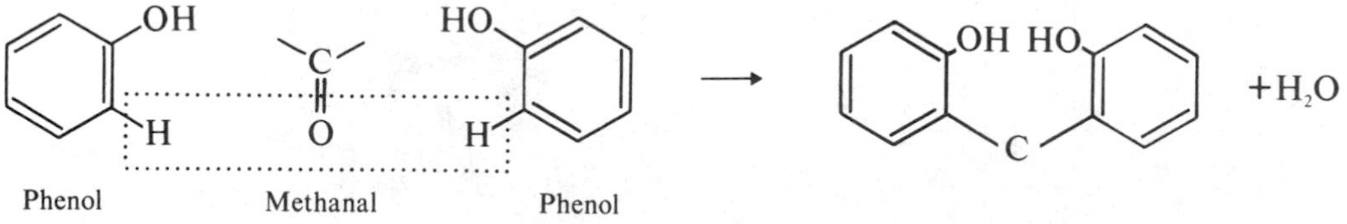

Figure 19.13 The reaction of two phenol molecules and a methanal molecule. The doubly bonded oxygen of methanal reacts with hydrogen atoms on the two phenol rings, eliminating H_2O (enclosed by dotted lines). In the resulting molecule, the two rings are linked by the carbon atom formerly in the methanal molecule.

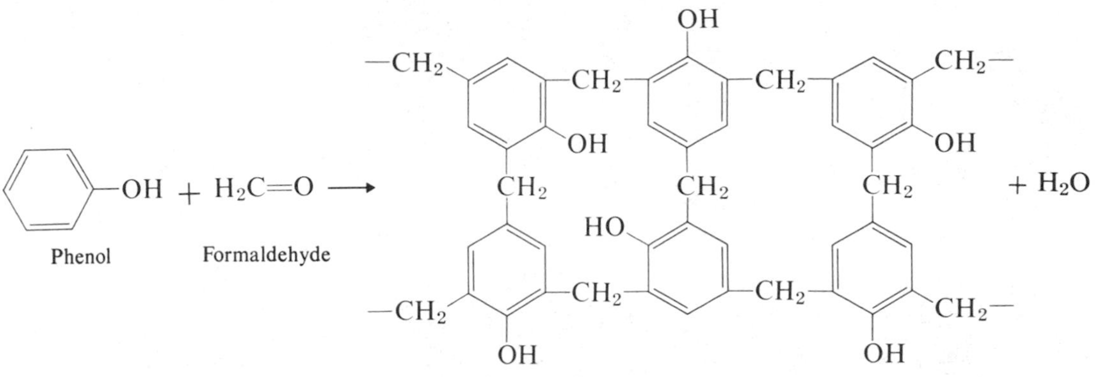

$$\text{Phenol} + \text{Formaldehyde} \longrightarrow \text{(Bakelite structure)} + H_2O$$

Figure 19.14   Portion of the Bakelite structure.

Methanal reacts with the hydrogens of the trifunctional ring amine called melamine.

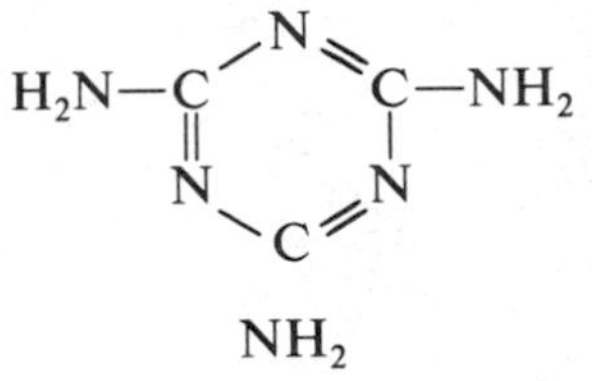

Melamine

The result of this reaction is the hard, durable solid called Melmac. Melmac is used for lightweight dinnerware and similar household applications.

Protein, the polymer from which animal tissues are made (your skin, for instance) is a polyamide with a structure similar to that of nylon. We will discuss proteins in Chapter 20.

19.15 UNUSUAL STRUCTURES

During the past few years, organic chemists have been challenged by the possibility of making new structures. Although in some cases this has become almost a game, every new structure has yielded important facts about bond angles, molecular strain, stability, and the relation of structure to physical properties. (See Figures 19.15 and 19.16.)

19.16 PETROLEUM

Petroleum is the name given to mineral deposits of hydrocarbons produced millions of years ago by the transformation of vast amounts of marine sediment. The sediment was composed primarily of the remains of living plants and animals.

19.16.1 The Nature of Petroleum

Petroleum occurs in the form of liquid crude oil, oil sands and oil shales and in deposits of asphalt, a tarry mixture of organic compounds of high molecular weight. Frequently associat-

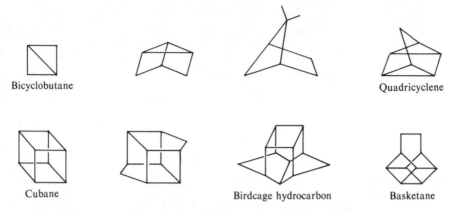

Bicyclobutane

Quadricyclene

Cubane

Birdcage hydrocarbon

Basketane

Figure 19.15 Some unusual organic molecules that have been synthesized.

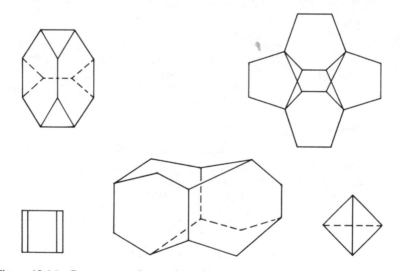

Figure 19.16 Some unusual organic molecules that have not yet been synthesized.

ed with petroleum are deposits of subsurface trapped gas composed mostly of methane (natural gas) and smaller amounts of ethane, propane, and butane. Liquid petroleum is made up primarily of somewhat larger hydrocarbons ranging from pentane through octane and decane, up to high-boiling compounds with skeletons containing 15 to 18 carbon atoms. Along with a few unsaturates, all these compounds are present in most or all of their isomers. Petroleum is a complex mixture.

19.16.2 Refining

The refining process separates the various petroleum compounds into groups classed according to chain length or structure. During refining, less valuable compounds are chemically converted to useful substances.

Separations are accomplished mostly by distillation, in which various "fractions" or "cuts" contain molecules of approximately the same molecular weight. In order of increasing molecular weight, the fractions are usually classed as naphthas, gasolines, kerosenes, fuel oils, lubricating oils, and greases. Because there is greater demand for the gasolines than for compounds of either greater or smaller chain length, some of the smaller molecules are polymerized or "alkylated" (the addition of alkyl groups) into larger molecules that can be used for gasoline. Large molecules are catalytically "cracked" into smaller ones for the same purpose.

19.16.3 Other Petroleum Products

For decades, hydrocarbons with chain lengths of from two to four carbons, particularly the unsaturates, were considered worthless and were discarded into the atmosphere or burned. Now, however, research has produced so many ways to use these compounds that a vigorous demand exists for them. Besides being used as the starting materials for plastics, and directly as fuel (propane, LPG), these compounds with low molecular weight are the basis for the synthesis of many organic heavy chemicals. Ethylene, for example, is used to prepare ethanol.

19.17 SUMMARY

Carbon atoms have a unique set of properties that give them remarkable versatility in forming bonds with atoms of other elements, and especially with other carbon atoms. More than half the different compounds on earth are compounds of carbon. It was once thought that carbon compounds could be synthesized only by living organisms, which is why the chemistry of carbon was given the name "organic chemistry."

A system of naming and classification has been developed to cope with the colossal number and variety of organic compounds. According to the system, which has been adopted worldwide, the name of a compound uniquely represents the structure of the molecules of the compound. A molecular structure can be written from the information in a name, and vice versa. A set of rules makes it possible for anyone to use the system.

The chemistry of organic molecules is largely confined to the functional groups attached to the carbon skeletons. There are several kinds of functional groups. An organic molecule can have no functional groups or several. The entire set of physical and chemical characteristics of any organic compound depends on the nature of its functional groups and on their positions in the molecules.

In the past few decades, synthetic, long-chain, organic polymers (plastics) have changed the way of life throughout the world. Studies of structure and of the effects of structure on properties have made it possible to synthesize materials to conform to predetermined sets of requirements. There are now in wide use many materials totally unlike any that existed 40 years ago.

We depend on petroleum for energy and as a source material for organic compounds. The refining of petroleum and the use of many kinds of petroleum products depend on the applications of organic chemistry. There are well-informed people who believe that we should not use petroleum as fuel because we are wasting materials that will be needed to make organic chemicals in future years. It can only be hoped that before we run out of petroleum, organic chemists will have found ways to substitute other materials from which to make the myriads of compounds necessary to our society.

Table 19.5 summarizes the kinds of compounds we have discussed.

Table 19.5
A Few of the More Important Kinds of
Organic Compounds and Some of Their Reactions

Kind of Structure	Generic Formula	Naming	Reactions Discussed
Saturated hydrocarbons	C_nH_{2n+2} (except for saturated rings)	• Forms basis for naming system • Names end in -*ane*	• Combustion • Hydrocarbons are not very reactive
Unsaturated hydrocarbons	$\diagdown\diagup$ C=C $\diagup\diagdown$ $\diagdown\diagup$ C≡C $\diagup\diagdown$	• Like saturated hydrocarbons except for endings • Use -*ene* for C=C and -*yne* for C≡C	• Additions
Aromatics	One or more rings with alternating C=C bonds	• Various names • C_6H_6 is benzene • $C_{10}H_{10}$ is naphthalene	• Additions to double bonds • Substitutions for hydrogens
Alcohols	$\underset{\mid}{\overset{\mid}{-C-}}$OH	• Hydrocarbon beginning with -*ol* ending as in *butanol*	• Oxidation to aldehydes or ketones
Aldehydes	$-\underset{H}{\overset{\mid}{C}}$=O	• Hydrocarbon beginning with -*al* ending as in *propanal*	• Reduction to alcohols • Polymerization with other compounds
Ketones	C $\diagdown$ O=C $\diagup$ C	• Two alkyl group names with -*one* ending as in *methylethylketone*	• Reduction to alcohols
Amines	$\underset{\mid}{\overset{\mid}{-C-}}NH_2$	• Alkyl group name plus *amine* as in *methylamine*	• Hydrogens can be substituted by alkyl groups • Amines are organic bases • Used in condensation polymerizations
Carboxylic acids	$-C\overset{\displaystyle O}{\underset{\diagdown OH}{\diagup\!\!\!\parallel}}$	• Hydrocarbon beginning plus -*oic acid* as in *butanoic*	• Organic acids • Used extensively in condensation polymerizations • Can form esters and amides • Form water-soluble salts

QUESTIONS

Section 19.1

1. What properties make carbon unique?
2. Why does carbon form strong covalent bonds to many other kinds of atoms?
3. Draw a sketch of a tetrahedral carbon atom.
4. Why can C atoms form long-chain structures while Cl atoms cannot?
5. Explain why multiple bonds are vulnerable to attack by other molecules, even though multiple bonds are stronger than single bonds.
6. Draw a sketch of the bonds of a trigonal planar carbon atom.
7. List three properties of carbon that make it possible for carbon atoms to form an endless number of different compounds.

Section 19.2

8. What are hydrocarbons? Saturated hydrocarbons?
9. Draw three different versions of the structure of one isomer of pentane.

Section 19.3

10. What is the general formula for an alkane? An alkene? An alkyne?
11. Explain the difference between saturated and unsaturated hydrocarbons.
12. What are the differences between alkanes, alkenes, and alkynes?
13. Distinguish between the terms *diene*, *triene*, and *tetraene*.

Section 19.4

14. Why are ring compounds of fewer than five carbons unstable? How do ring compounds formed by larger numbers of carbon atoms maintain bond angles of approximately 109°?
15. What makes benzene so stable?

Section 19.5

16. Why are functional groups of interest to organic chemists?
17. Identify the functional groups outlined by dotted lines.

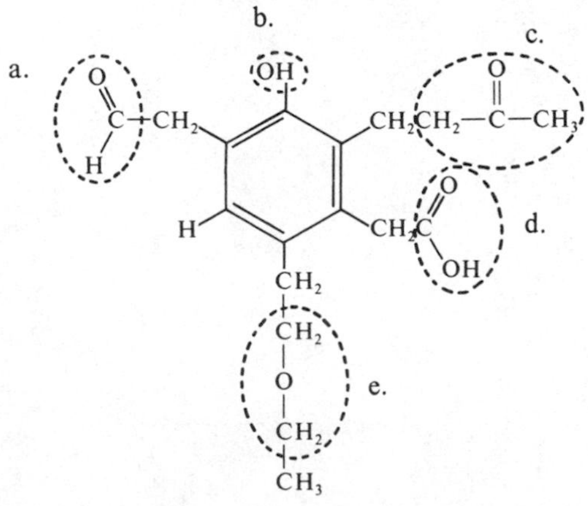

Section 19.6

18. What advantage does the IUPAC system have over the use of common names for organic compounds?
19. List the steps used in the IUPAC system to name organic compounds.

Section 19.7

20. What is the R shorthand? Why do organic chemists use it?
21. Describe in detail a substitution reaction.

Section 19.8

22. Give examples of an addition reaction and a free radical reaction. Illustrate with electron dot diagrams.
23. How are polymers created?
24. What factors determine the characteristics of a polymer such as polystyrene or Teflon?

Section 19.9

25. Why does benzene not add halogens, although many other unsaturated molecules readily do so?
26. There are three compounds that could be described as dimethyl benzenes, yet they have different properties. What structural differences cause the properties to differ?

Section 19.10

27. What two structural properties of an alcohol determine its characteristics? Explain in detail.
28. Why can a secondary alcohol not be oxidized to a carboxylic acid?

Section 19.13

29. Explain why a carboxyl group can produce H^+ ions, whereas alcohols rarely do so.
30. Explain why the ionization constant of the first carboxylic acid is about 100 times less than that of the second and why that of the second is about 100 times less than that of the third.

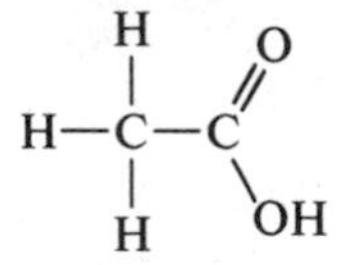

$$K_i = 1.8 \times 10^{-5}$$

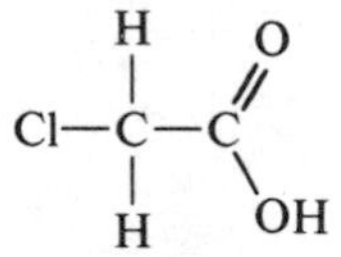

$$K_i = 1.5 \times 10^{-3}$$

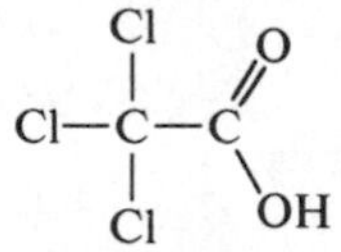

$$K_i = 2.0 \times 10^{-1}$$

31. Show how esters are formed from carboxylic acids and alcohols.
32. How do soaps work? What causes soap scum?

Section 19.14

33. Write out the reaction for the formation of nylon (one step).
34. What is the difference between linear and network polymers?

PROBLEMS

Section 19.2

1. The following are drawings of some of the structural isomers of C_7H_{16}. Label the drawings of *different isomers* with letters A, B, C, and so on. Label drawings of the identical isomers with the same letter.

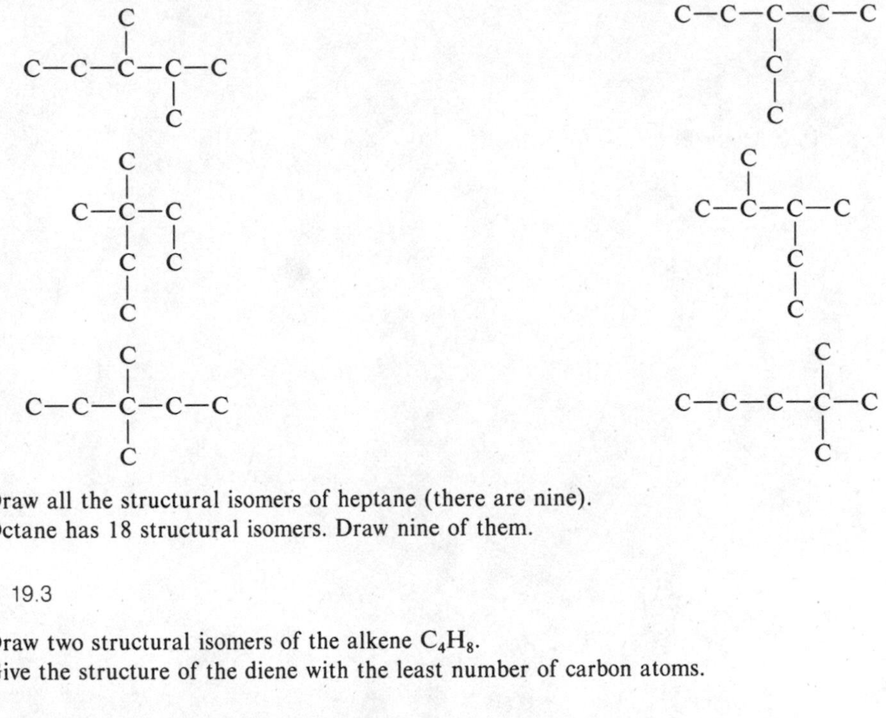

```
          C                              C—C—C—C—C
          |                                    |
    C—C—C—C—C                                  C
          |                                    |
          C                                    C


          C                                    C
          |                                    |
      C—C—C                              C—C—C—C
          |  |                                 |
          C  C                                 C
          |                                    |
          C                                    C


          C                                    C
          |                                    |
    C—C—C—C—C                          C—C—C—C—C
          |                                    |
          C                                    C
```

*2. Draw all the structural isomers of heptane (there are nine).
3. Octane has 18 structural isomers. Draw nine of them.

Section 19.3

4. Draw two structural isomers of the alkene C_4H_8.
5. Give the structure of the diene with the least number of carbon atoms.

Section 19.4

*6. Draw all the ring compounds that have the formula C_5H_{10}.
7. Draw as many ring structures for C_5H_8 as you can.
*8. Write a structural formula for one open-chain and one ring isomer of each of the following:
    *a. C_6H_8 b. C_7H_8 c. C_7H_{12}
    For at least one of these, write an open-chain structure that has only one multiple bond.
*9. Write electron dot structures for the following compounds:
    *a. CH_3CH_3 e. $(CH_3CH_2)_2O$
    b. $(CH_3)_3COH$ f. CH_2O
    c. CH_3NH_2 g. CH_3COOH
    d. $(CH_3)_2CO$

*10. For each compound in Problem 9, give the name of any functional group (except alkyl groups).

*11. Give all of the structural isomers for each of the following:

    *a. C_4H_9Br c. C_3H_7Cl

    b. C_3H_6O d. C_3H_7OBr

Section 19.6

12. Give the IUPAC names of the structural isomers found in Problem 2.

*13. Name the structural isomers you drew in Problem 6.

*14. Give the IUPAC names for the following:

15. Give the IUPAC names for the following:

*16. Name the following compounds:

*17. Write structural formulas for each of the following compounds:
   *a. 2-bromo-3-chloro-3-methylpentanol   c.   methyl ester of ethanoic acid
    b. 1,3-dichloropropyne   d.   2-methylpropanamide
 18. Write structural formulas for each of the following compounds:
    a. 1,2-dichloroethene   c.   orthodibromobenzene
    b. 2-methyl-2,4-cyclopentadienone   d.   2-chloro-3-methylbutanal

Section 19.7

*19. Complete the following reactions:
   *a. $CH_3CH_2Br + CN^- \longrightarrow$
    b. $(CH_3)_2CHCH_2CH_2OH + Br^- \longrightarrow$
    c. $CH_3CH_2CHCH_3 + HCl \longrightarrow$
             |
             OH

*20. Fill in the missing reactant.
   *a. $CH_3CH_2CH_2Cl +$ _____ $\longrightarrow CH_3CH_2CH_2F$
    b. _____ $+ PBr_3 \longrightarrow 3\ CH_3CHCH_3$
                                            |
                                            Br

Section 19.8

*21. Complete the following reactions:
   *a. $-C\equiv C- + H_2 \longrightarrow$

    b. $CH_3CH=CHCH_3 + H_2O \xrightarrow{H_2SO_4}$

    c. $3\ CH_2=CH_2 + \cdot CH_3 \longrightarrow$

    d. $CH_2CHCH_3 + Cl_2 \longrightarrow$

*22. Give the missing reactant.
   *a.
                                                              Br
                                                              |
            _____ $+ H_2 \longrightarrow$?;   ? $+ Br_2 \longrightarrow CH_3CHCHCH_3$ (two steps)
                                                              |
                                                              Br

    b. $2\ CH_3CH=CH_2 +$ _____ $\longrightarrow CH_3CHCH_2CHCH_2CH_3$
                                                      | |
                                                      CH_3   CH_3

    c. $CH_2=CH_2 +$ _____ $\xrightarrow{H_2SO_4} CH_2-CH_3$
                                                       |
                                                       OH

    d. _____ $+2\ H_2 \longrightarrow$

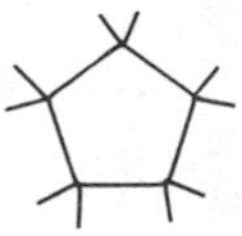

 23. Give the missing reactant.

    a.  $+$ _____ $\longrightarrow$

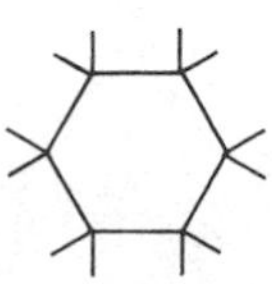

b. $+ H_2 \longrightarrow CH_3CH{=}CH_2$

c. $2\ CH_2{=}CH_2 +$ $\longrightarrow CH_3CH_2CH_2CH_2CH_2 \cdot$

Section 19.10

*24. Give the product of the first oxidation and of the second oxidation for the following alcohols:

    *a. CH_3CH_2OH b. $(CH_3)_2\underset{\underset{\displaystyle H}{|}}{C}OH$

25. Give the product of the first oxidation and of the second oxidation for the following alcohols:

    a. $CH_3\underset{\underset{\displaystyle OH}{|}}{CH}\underset{\underset{\displaystyle OH}{|}}{CH}CH_3$ b. $CH_3\underset{\underset{\displaystyle OH}{|}}{CH}CH_3$ c. $(CH_3)_3COH$

*26. Complete the following reactions:

    *a. $H_3C{-}CH_2{-}CH_2{-}CH_2OH \xrightarrow{\ MnO_4^-\ }$
                     $\underset{\underset{\displaystyle CH_3}{|}}{}$

    b. $CH_3\underset{\underset{\displaystyle OH}{|}}{CH}CH_3 \xrightarrow{\ Cr_2O_7^{2-}\ }$

    c. $CH_3CH_2\overset{\overset{\displaystyle O}{\|}}{C}{-}H \xrightarrow{\ MnO_4^-\ }$

*27. Give the missing reactant.

    *a $\xrightarrow{\ Cr_2O_7^{2-}\ } CH_3\overset{\overset{\displaystyle O}{\|}}{C}CH_2CH_3$

    b. $\xrightarrow{\ MnO_4^-\ }$

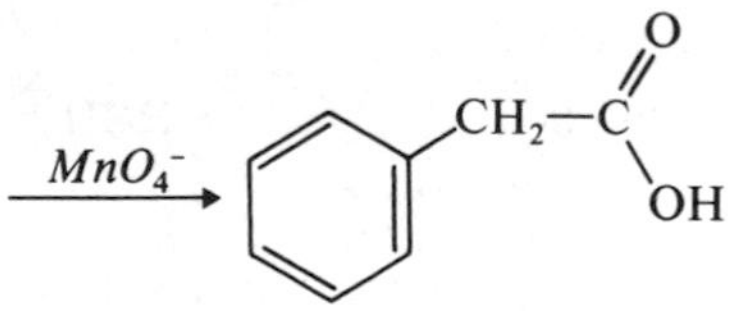

    c. $\xrightarrow{\ MnO_4^-\ } CH_3\underset{\underset{\displaystyle O}{\|}}{C}CH_2\underset{\underset{\displaystyle O}{\|}}{C}CH_3$

Section 19.13

*28. Complete the following reactions:

    *a. $CH_3{-}CH_2C\overset{\displaystyle O}{\underset{\displaystyle OH}{\Big<}} + CH_3CH_2OH \longrightarrow$

    b.

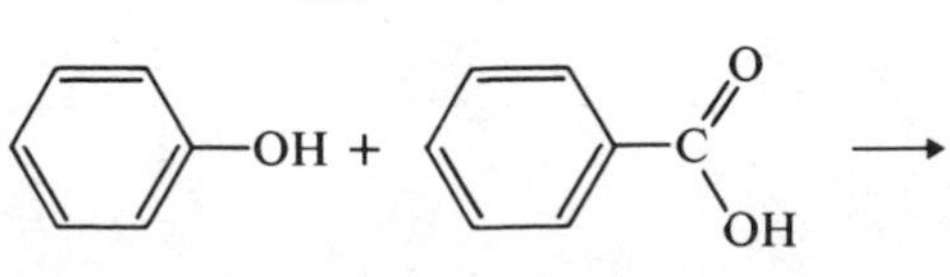

c.

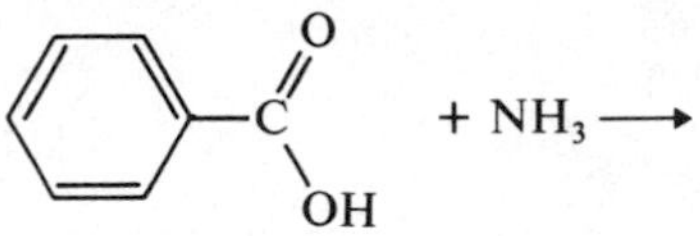

$+ NH_3 \longrightarrow$

d.

$CH_3CH_2NH_2 + CH_3\overset{O}{\underset{OH}{C}} \longrightarrow$

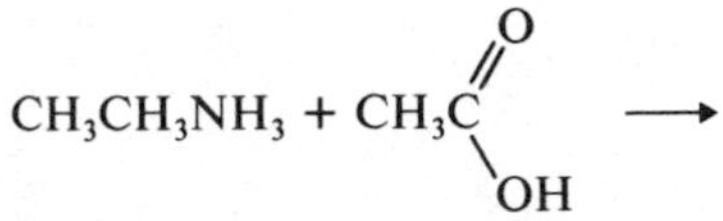

*29. Give the missing reactant.

*a.

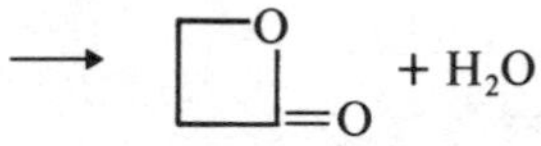

$\longrightarrow$ $+ H_2O$

b.

$CH_3CH_2OH +$ $\longrightarrow CH_3CH_2O\overset{O}{\overset{\|}{C}}CH_2CH_3 + H_2O$

c.

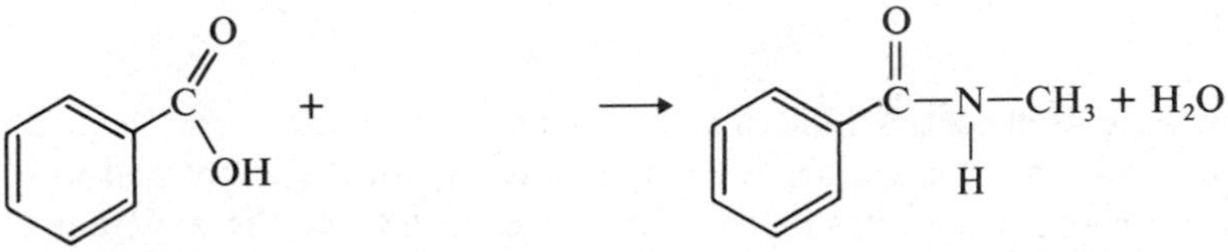

$+$ $\longrightarrow$ $+ H_2O$

Section 19.14

*30. Draw structures containing at least four monomer units of polymers of the following pairs of
compounds (hint: either leave out the H atoms attached to carbon or show saturated chains as 
$\bigwedge\!\!\bigvee\!\!\bigwedge$, as in $H_2N-C\bigwedge\!\!\bigvee\!\!\bigwedge C{=}O$). In words, explain the differences between the two
structures.

*a.

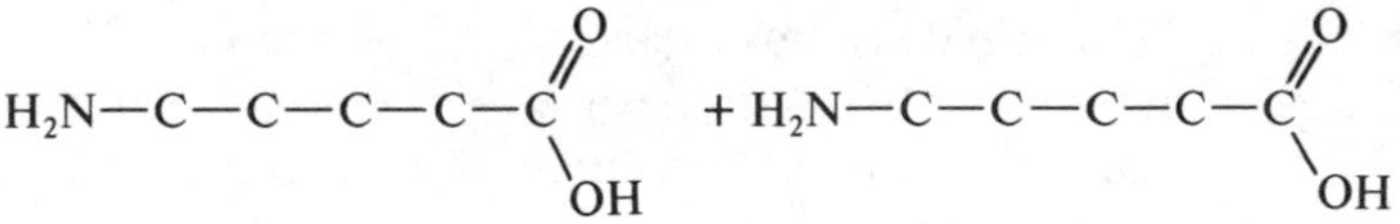

$H_2N-C-C-C-C-\overset{O}{\underset{OH}{C}} + H_2N-C-C-C-C-\overset{O}{\underset{OH}{C}}$

b.

$H_2N-C-C-C-C-NH_2 + \underset{HO}{\overset{O}{C}}-C-C-C-C-\overset{O}{\underset{OH}{C}}$

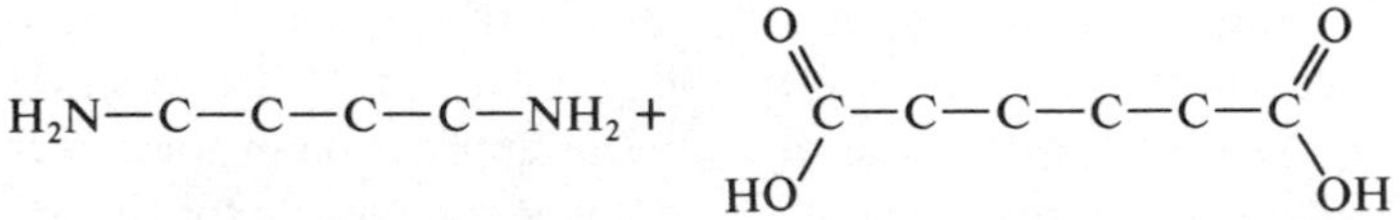

31. Give the structure of the missing reactant.

$+ HOCH_2CH_2OH \longrightarrow$

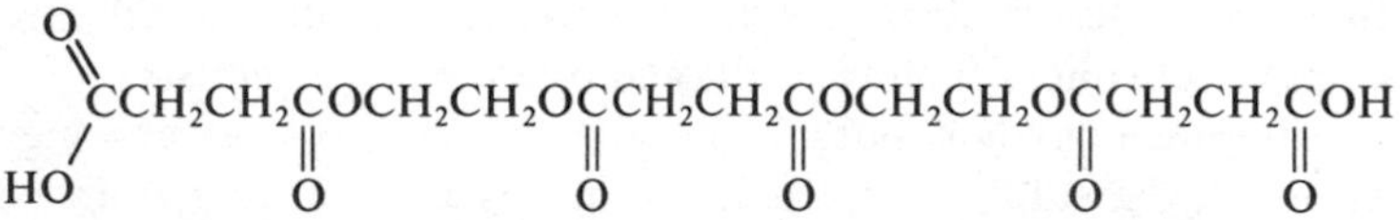

$\underset{HO}{\overset{O}{\overset{\|}{C}}}CH_2CH_2\overset{\|}{\underset{O}{C}}OCH_2CH_2O\overset{\|}{\underset{O}{C}}CH_2CH_2\overset{\|}{\underset{O}{C}}OCH_2CH_2O\overset{\|}{\underset{O}{C}}CH_2CH_2COH$

Chapter 20

Chemistry of
Living Organisms

Thus far, we have studied the fundamentals of chemical reaction, of bonding, and of molecular structure. We have applied our learning to a few natural systems and to some man-made materials. Now we are going to extend our studies to include the most amazing of all natural phenomena, the processes of life and the structures of living organisms.

Our own bodies and, for that matter, the substances of all living creatures, are made of atoms, mostly atoms of elements that we have already studied in some detail. However, biological tissues are so wonderfully constructed and perform their functions in such an elegant and economical way, that for centuries, they were thought to be made of a kind of matter fundamentally different from that of nonliving substances.

Biological structures are indeed different from the kinds of matter that we have discussed, but the difference lies only in the tremendous molecular complexity and chemical efficiency that have evolved in living creatures since life first began. The complex structures and efficient chemical agents are nevertheless created and maintained by chemical reactions. The nature of such reactions is now partly understood.

The creation of biological structures is only one of the thousands of different kinds of chemical reactions carried on by even the simplest living creatures. Other processes include countless kinds of energy transformations, from the efficient application of solar energy to the use of muscles to the blinking of a firefly. Chemical reaction is responsible for the delicate and intricate action of nerves and brain tissue. An involved series of chemical reactions, rigidly controlled, supplies the body heat of warm-blooded animals. Finally, chemistry is necessary to the management of all these biological processes. The control mechanisms that tell body functions when to start or stop, or how much to do, are all chemical. A living organism is a complex chemical factory, built and operated by chemistry.

Because biological organisms are the most complex and highly organized form of matter on earth, the study of biological chemistry is a correspondingly complex science that involves ideas and terminology not used elsewhere. We do not need to become fluent in biochemical language to appreciate how living tissues are planned, constructed, and operated. The concepts and language we have learned so far are quite applicable to our purposes.

AFTER STUDYING THIS CHAPTER, YOU WILL BE ABLE TO

- list the parts of a cell and describe the function of each part
- describe the process of photosynthesis
- describe the process of respiration
- explain how the ADP-ATP system stores and releases biological energy
- show how amino acids polymerize to form proteins
- list the levels of protein structure and describe each kind of structure
- list several kinds of proteins and give their functions
- outline the way in which DNA stores and transmits information
- sketch the structure of DNA
- explain how enzymes function
- list the kinds of molecules used as sources of food energy and draw the general structural features of those molecules

TERMS TO KNOW

adenosine diphosphate, ADP	deoxyribonucleic acid, DNA	nucleic acid
		photosynthesis
adenosine triphosphate, ATP	double helix	protein
	enzyme	protein primary structure
alpha amino acid	essential amino acid	protein secondary
alpha helix	essential fatty acid	structure
carbohydrate	fat	respiration
cell	fatty acid	saturated fat
cell nucleus	gene	starch
cellulose	globular protein	unsaturated fat
chlorophyll	hemoglobin	
chromosome	hormone	

TERMS WITH WHICH TO BE FAMILIAR

cell membrane	Krebs cycle	protein quaternary structure
cytoplasm	lipid	
disaccharide	mitochondrion	protein tertiary structure
glycerol	monosaccharide	ribosome
glycogen	polysaccharide	sickle-cell anemia
hydrogenation		

20.1 CHEMISTRY IN THE LABORATORY AND IN LIVING ORGANISMS

Scientists once believed that the processes of living organisms were not ruled by the physical laws that applied to the chemistry and physics of the laboratory. Most thought that there existed a vital force, a supernatural power that directed the functioning of living tissue.

For example, it was believed that the organic compounds of carbon (all carbon compounds except the oxides, the metal carbides, and a few others) could be made only by living organisms, that is, could not be prepared in the laboratory.

In the past few decades, even complex organic compounds have been artificially synthesized. Until recently, however, it had nevertheless been assumed that no living organism could be prepared in the laboratory entirely from nonliving matter. This belief is now under attack; a few researchers are beginning to predict that life will someday be synthesized. To prepare even a single living cell, however, would be such a long and difficult process that the task is beyond any technology existing today. What nature does effortlessly is, for any practical purpose, still beyond the reach of science. (We should remember, however, that it took several hundred million years for nature to create life. Human beings have been working in laboratories for only about 200 years.)

Although we cannot yet duplicate many of the products of nature, we are now beginning to understand the methods of nature. Living organisms accomplish their remarkable chemical operations by building for each reaction a specific, superefficient chemical tool, a catalyst called an **enzyme.** Enzymes bring about difficult and complex reactions rapidly and without need for the harsh reagents that are often necessary in the laboratory. An enzyme can accomplish in one concerted step a reaction that would require many reagents and several steps in the laboratory. Enzyme chemistry, the chemistry of living organisms, is rapid, efficient, and economical of energy and material, but it obeys the same laws of chemistry that apply in the laboratory. The chemistry of living creatures is not fundamentally different from laboratory chemistry; it is just far more elegant and efficient.

20.2 THE LIVING CELL

All living organisms (with the exception of viruses, which are considered by some scientists not to be living) are constructed primarily of fundamental building units called **cells.** The cell is the smallest unit that can independently carry on the processes of life. In a way, the cell is to living tissue what the atom is to ordinary matter. Some kinds of cells are large, but the average cell is quite small. Many thousand ordinary cells could be stacked on the head of a pin.

Cells vary enormously in structure and function, but most cells have several structures in common. We can understand a little of how a cell functions by looking at a few of its major parts (Figure 20.1).

The cell **membrane** serves as the outer cover of the cell and also encloses the **nucleus** of the cell. Cell biologists sometimes refer to the membrane as a "butter sandwich" because it is believed to be constructed of two layers of protein enclosing a layer of lipids (fatty substances). In addition to acting as containers, the cell membrane has several important functions in the chemistry of the cell. It responds to various chemical commands by permitting passage of desired molecules while denying passage to other kinds of molecules; the membrane acts somewhat the same way that valves and pumps do in the laboratory. Researchers are beginning to understand the nature of the commands that control the membrane. Chemical messengers, hormones and enzymes, released in other places tell the membrane what to do.

The **cytoplasm** is the cellular fluid in which the internal organs of the cell are suspended. Cytoplasm contains a supply of chemical building blocks that are used to construct needed molecules of various kinds.

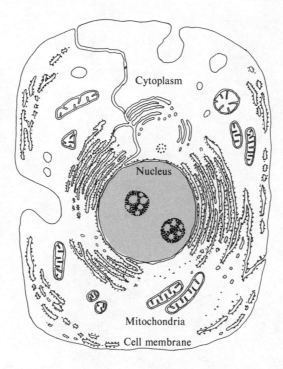

Figure 20.1 Diagram of a typical cell. The ribosomes are shown as dots in the cytoplasm.

The nucleus of the cell is primarily a collection of **chromosomes,** threadlike bodies that contain the information the cell needs to be able to function and reproduce itself. Each chromosome contains several **genes.** A gene is composed of DNA, a kind of a molecular library that we will later examine in detail.

Spread around the cytoplasm are numerous bodies called **ribosomes.** These serve as tiny factories, synthesizing protein according to instructions given them by chemical messengers from the nucleus. Also distributed in the cytoplasm are large numbers of particles called **mitochondria,** which are the chemical power stations from which the cell derives the energy to perform its functions.

Cells perform the biological chemistry that makes the parent organism operate. Cells synthesize tissue, both for growth and for repair and replacement of existing tissue. Most of the thousands of different kinds of complex molecules necessary to body chemistry are supplied by cells. The secretory organs, such as the liver, stomach lining, pancreas, and salivary glands; the hormone-producers, such as the thyroid and adrenals; and the blood-making machinery of the bone marrow all operate by means of cells. Figure 20.2 is a schematic drawing of the way in which it is thought that the cells of the pancreas manufacture the hormone insulin.

20.3 BIOLOGICAL ENERGETICS

A living cell is a complex, highly organized unit. To carry on its multitude of tasks, it must have a source of energy. How cells obtain and use energy is such an important subject

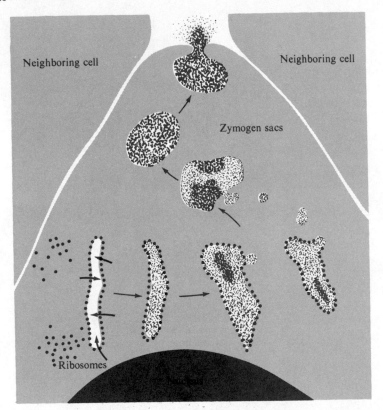

Figure 20.2 Representation of the process of globular protein secretion by the pancreas. The process of synthesis takes place in zymogen sacs that are formed by membrane tissue. The ribosomes seem to play a large part in the process, the sequence of which is indicated by arrows.

that all living organisms can be usefully described on the basis of energy functions. Certain cells of plants receive, use, and store energy from the sun. Cells of animals also use energy but must ultimately depend on the fuels stored by plants. Let us examine how plants convert the energy of sunlight into chemical fuel, then we will consider how animals use that fuel.

20.3.1 Photosynthesis

Photosynthesis is the process that enables a plant to receive, use, and store energy from sunlight. More important than knowing how photosynthesis occurs is to know what it does.

The overall photosynthetic process accomplishes two functions: (1) creates a biological fuel, glucose, from water and carbon dioxide, and (2) simultaneously converts a compound called ADP to one called ATP, a process that is something like charging a chemical battery. Although photosynthesis is an extremely complex series of reactions, the overall result of glucose synthesis can be simply shown.

$$6 \; H_2O + 6 \; CO_2 + \text{light energy} \longrightarrow 6 \; O_2 + C_6H_{12}O_6$$

Glucose

In most cases, the glucose produced by photosynthesis is polymerized into either starch or cellulose.

Taken altogether, the bonds in CO_2 and H_2O are stronger than those in glucose. In the photosynthesis process, more energy must go into breaking bonds than is recovered when the new bonds in glucose and O_2 are made; that is, energy is required to synthesize glucose from water and carbon dioxide. That energy is absorbed from sunlight by plant cells containing chlorophyll. **Chlorophyll** is a compound with large, complex molecules that can absorb light energy and use it to run chemical reactions. Efforts of laboratory workers to duplicate the action of chlorophyll have generally been unsuccessful. Unlocking and duplicating the performance of chlorophyll and plant cells would rapidly solve the energy problems of the world, and would end our dependence on fossil fuels.

20.3.2 Respiration

All living organisms use energy. When your eye moves to read these words, you use energy. A cat dreams about a mouse and its paws twitch. The nerve impulses that are the dreams use energy, and the commands that make the paws twitch use energy. A firefly blinks in the night; it uses energy to make the flash. You digest a piece of cheese and use its amino acids to build cells; that process uses energy. When a tree pumps sap from its roots to the top of its crown, energy is required. Awake or asleep, our bodies constantly use energy. All living organisms, plants and animals, oxidize glucose to obtain needed energy. That statement deserves repetition. *All living organisms, plants and animals, oxidize glucose to obtain needed energy.*

In the laboratory, 1 mol of glucose (180 g, or about two thirds of a cupful) can be burned in oxygen, and the process will yield about 2900 kJ.

$$C_6H_{12}O_6 + 6\ O_2 \longrightarrow 6\ CO_2 + 6\ H_2O \qquad \triangle H = -2884\ kJ$$

In living cells, however, burning cannot occur; oxidation takes place gradually through a series of steps. But the total amount of energy released is the same whether the sugar is burned or is oxidized in the body. Even when oxidized in living cells, glucose must be thought of as fuel.

The process by which cells oxidize glucose is called **respiration.** In the great energy cycle of living things, respiration is the reverse of photosynthesis. Although animal cells also use energy from reactions of fats and proteins, the energy of any of these reactions was originally obtained from the sun and stored in glucose molecules. Let us see how the oxidation of glucose proceeds in living cells.

Glucose and O_2 are carried to cells by various means, one of which is the bloodstream of animals. Once in the cell, the glucose is stored and awaits use. When energy is needed, oxidation occurs, primarily in the mitochondria. A mitochondrion is a miniature chemical factory, shaped like a grain of rice and perhaps 30,000 Å long. A single cell can have thousands of mitochondria.

Respiration is not a single violent process like burning but takes place in a number of steps. Taken together, the steps are called the **Krebs cycle,** in honor of Sir Hans Krebs who won the 1953 Nobel prize for working out its details. The first step of the cycle is the transformation of glucose to lactic acid.

Glucose → Lactic acid

Just this first step requires 11 separate chemical reactions, each of which requires its own highly specific and complex biological catalyst. The first step releases about 85 kJ of useful energy to the cell.

The rest of the oxidation process occurs in about nine additional steps, each of which is as complicated as the first. In the total process, about 1500 kJ of the available 2800 kJ/mol is released as energy that the cell can use; the rest is released as heat. Respiration, then, is about 50% efficient, compared with an efficiency of about 20% for the burning of hydrocarbons by an automobile engine. And respiration produces neither noise nor air pollution.

20.3.3 ADP-ATP System

The oxidation of glucose is not a rapid process, but the energy demands of a biological system vary with time and are often sudden. Because of this, nature has devised a way of storing the energy so that it can be released quickly. All living organisms, even plants, use the system.

As far as is known, every cell of every living organism contains the ion of a compound called **adenosine triphosphate,** $C_9H_{16}O_{13}N_4P_2$, which is universally abbreviated as **ATP.** When given the proper chemical signal, the ATP ion reacts with water to lose a phosphate ion, PO_4^{3-}, and become the ion of **adenosine diphosphate, ADP.** In this process, about 40 kJ of energy is released that can be directly and efficiently used by the cell. Later, in the process of respiration, the ADP molecules are changed back into ATP molecules and energy is again stored. The oxidation of one molecule of glucose in the Krebs cycle will convert about 38 molecules of ADP to ATP.

When any cell needs to use energy for any purpose, it calls on the ATP-ADP system. No thinker thinks, no bird flies, no tree grows without energy derived from the conversion of ATP to ADP. Do this experiment to make the concept more understandable. Hold your hand where you can see it clearly. Move your index finger one centimeter. Now the experiment is finished. You have expended about a thousandth of a kilojoule of energy, or about a hundred millionth of a piece of pie. To accomplish the movement, you converted about 10^{19} ATP ions to ADP ions.

20.4 PROTEINS

Ordinarily when we hear the word *protein,* we think of one of the essential foods. Proteins do fulfill this function, but just now, let us focus on the fact that we ourselves are made mostly of protein. Look at the skin on your hand: protein. Lift your arm: the muscle that moved it is protein. You have just read this line. The photographic material in your eye is protein, not silver, as it would be in camera film. Thoughts are going through your brain and commands are being transmitted to your muscles. Brain and nerve tissue are protein. Without the protein in your bloodstream, you could not use the oxygen you breathe. We require protein in our diet because we ourselves are protein, and we need supplies to repair our protein structures. What is protein, anyway?

Protein is one of the biological polymers. In our study of man-made polymers, we learned about nylon, a long-chain polyamide formed by the reaction of carboxylic acid groups with amine groups on monomer molecules. Proteins are polyamides much like nylon, except that each monomer unit carries, in addition to the carboxylic acid and the amine group, a third functional group that becomes a side group attached to the main polymer chain. The monomer from which a protein polymer is made is called an **alpha amino acid.**

An alpha amino acid is a carboxylic acid that has an amino group, $-NH_2$, bonded to the "alpha" carbon, the carbon atom lying next to the carboxyl group. The simplest alpha amino acid is glycine.

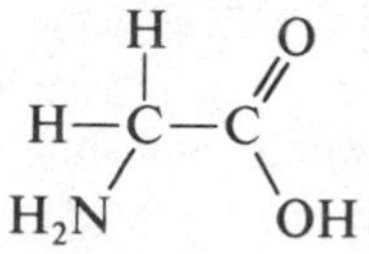

Glycine

If we draw the glycine structure with the letter *R* in the place of one of the hydrogen atoms on the alpha carbon, we have the fundamental structure of the amino acids from which all proteins are constructed. In this representation, R is the third functional group.

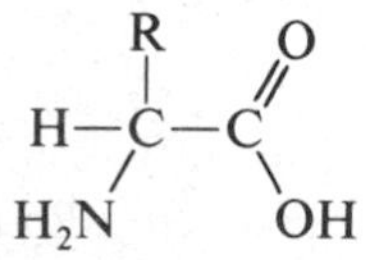

Fundamental amino acid structure

Two amino acid molecules can condense to form what is called a dipeptide.

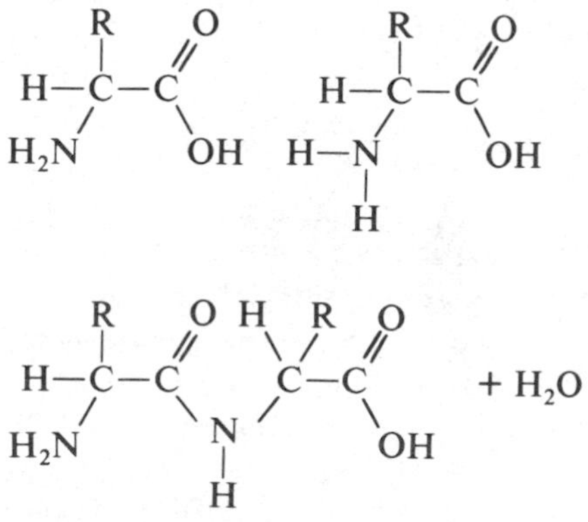

Dipeptide

Three amino acids will give a tripeptide, and several make a polypeptide. A **protein** is a polypeptide that contains more than about 50 amino acid units. Although there are a few smaller polypeptides that are biologically important, most polypeptides can be classified as proteins.

The protein structure is arranged so that an R group is attached to every third atom in the main chain, but there are no R groups within the chain itself. Somewhat more than 25 amino acids, differing only in the structures of their R groups, have been isolated from proteins, and about 20 of these are common. Table 20.1 shows the names and structures of the 20 common amino acids. Notice that the basic structures, shown on the right of each drawing, are identical in all the compounds. The chemistry of the human body is able to synthesize all but 8 of these 20 amino acids. Those eight, marked with an asterisk in the table, must be obtained from outside sources and are called the **essential amino acids.**

20.5 STRUCTURE OF PROTEINS

Just as the position of the oxygen atom causes the great differences between ethyl alcohol and dimethyl ether, so do the structures of different proteins cause those proteins to have different individual properties. As we continue, we should keep in mind that such different substances as fingernails, muscle tissue, nerves, and red blood cells are all made of exactly the same kind of long-chain amino acid polymers, and that their enormous differences in properties are caused fundamentally by the different arrangements of their R groups.

Protein molecules are large and complex, and their structures are correspondingly intricate. Although protein structure is usually classified into four levels, primary, secondary, tertiary, and quaternary, we will be concerned with the details of only the first two.

20.5.1 Primary Structure of Proteins

If we were to begin with the 20 amino acid monomers listed in Table 20.1 and attach only 2 together, making a dipeptide, we would be able to choose any of the 20 for the first position and any of the 20 for the second. That would give us the possibility of 20×20, or 400, different combinations. All 400 dipeptides would have the same chain structure, but would differ in the R groups attached to the chain. There are $20 \times 20 \times 20$, or 8,000, ways in which we could make a tripeptide, 160,000 ways to make a polymer with four units, and so on. It should be plain that there must be an enormously large number of possible R group arrangements on a long protein chain. With all those different possibilities, and each different polymer having its own characteristic properties, you could be tempted to think that all of the protein molecules in, say, the human body might be different. That is far from the actual case.

Stop and think about this astounding fact: in any normal biological organism, all the protein molecules that have exactly the same function also have exactly the same number of amino acid units and exactly the same sequence of R groups along their chains.

Let's look at an example. In mammals and many other species, oxygen is carried from the lungs to other parts of the body by a protein called **hemoglobin,** which is found in red blood corpuscles. In all human beings, each hemoglobin molecule is composed of four amino acid chains, two of one kind and two of another. What is called the *beta* hemoglobin chain has 146 amino acid units. This number of units presents the possibility of $(20)^{146}$ different arrangements, but in all people with normal hemoglobin, the beta chain sequence is exactly the same. This sequence is shown schematically in Figure 20.3. Is this sequence important? Look at the

Table 20.1
Twenty Amino Acids Commonly Found in Proteins

Chemical Structure	Amino Acid	Chemical Structure	Amino Acid
	Glycine		**Aspartic acid**
	Alanine		**Glutamic acid**
	Valine*		**Lysine***
	Leucine*		**Arginine**
	Isoleucine*		**Histidine**
	Serine		**Cysteine**
	Threonine*		**Methionine***
	Tyrosine		**Asparagine (AspNH$_2$)**
	Phenylalanine*		**Glutamine (GluNH$_2$)**
	Tryptophan*		**Proline**

Note: Structures within shaded blocks are R groups.

*Indicates essential amino acids.

figure and find the sixth amino acid unit from the —NH$_2$ end. It is glutamic acid (glu). A small proportion of humans have hemoglobin beta chains in which that sixth unit is not glutamic acid but is valine (val). Such people have a condition called **sickle-cell anemia.** They are never really strong and often die young. The hemoglobin beta chain in gorillas is the same

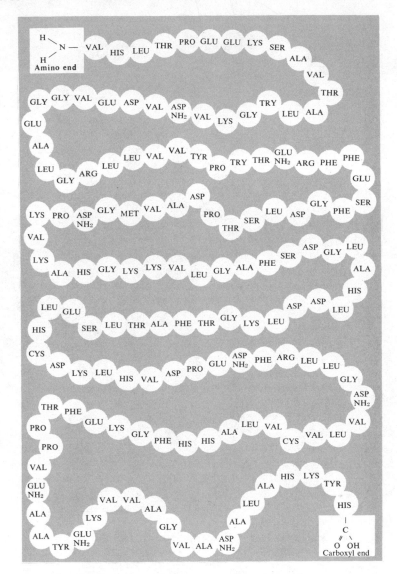

Figure 20.3 Sequence of amino acid residues in the beta chain of human hemoglobin.

as that in humans except for one different unit; that of horses differs from ours by about 18 units. Humans can use neither gorilla nor horse hemoglobin.

The sequence of R groups along a protein chain is called the **primary structure** of the protein. All the other structural features and all functioning of the protein depend on the primary structure. Until about 30 or 40 years ago, scientists knew little about the structures of proteins, and it was believed that little would ever be known. Finding the primary structure of a protein is like working a giant jigsaw puzzle; much time and labor have gone into what we now know, and for the most part, take for granted.

20.5.2 Secondary and Higher Structure in Proteins

In nylon, a polyamide like protein, there is little organized arrangement of the various parts of the polymer chains. In most cases, the nylon solid could be compared to a collection of countless pieces of string, randomly tangled together. Protein molecules, by contrast, are highly organized. The parts of each chain relate to each other in specific ways, and different chains are joined together in specific patterns.

The form that a single protein chain takes in space, its shape, is called its **secondary structure.** Long a subject of debate and research, the secondary structure of a protein was first discovered by Linus Pauling, who is said to have been lying in bed with a cold at the time the inspiration came to him. (It must be noted, however, that Pauling had been laboring over protein structure for several years before the answer finally came.)

The structure that Pauling discovered is now called the **alpha helix** and is found in many kinds of proteins. In the alpha helix, the individual chains are arranged in a helical form; that is, they are twisted into a spiral. The helix is precisely formed and held together by hydrogen bonds between amino groups and carboxyl groups in successive coils. Study Figure 20.4, which is a schematic drawing of a few coils of an alpha helix. Notice how specific the relations are between the double-bonded oxygen atoms and the hydrogen atoms on the amino groups. The proteins of hair, connective tissue, and hemoglobin, as well as other kinds of protein, have the alpha helix structure. Since Pauling's discovery, other kinds of secondary structure have been found.

The overall shape of a single protein chain, that is, the relation of different parts of the chain to one another, is called the **tertiary structure** of the protein. The beta chain of hemoglobin, for example, assumes a particular shape held together by
along the chain, as shown in Figure 20.5.

There is a last level, called the **quaternary structure** of proteins. Quaternary structure is the highly specific relation of different chains of protein that combine to form a functional molecule. The quaternary structure of hemoglobin includes two alpha chains, two beta chains, and another structure called a prosthetic group. The entire assembly is one hemoglobin molecule, shown in Figure 20.6. In keratin, a protein found in connective tissue, the alpha helices make successively larger sets of coils in much the same way that we make cables (Figure 20.7).

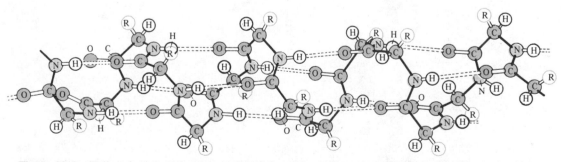

Figure 20.4 Ball-and-stick molecular model of the alpha helix structure in proteins. The heavy black lines represent the backbone bonds. Hydrogen bonds between NH groups and—C=O groups are shown as dotted lines.

Figure 20.5   Tertiary structure of the beta chain of hemoglobin.

Figure 20.6   Complete hemoglobin structure.

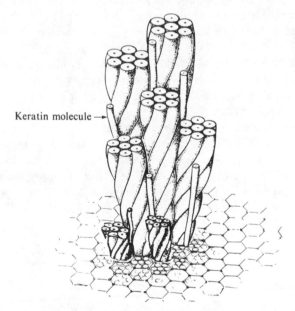

Keratin molecule →

Figure 20.7   Representation of the molecular structure of a tissue composed of keratin. The individual molecules have alpha helix configurations, each of which is shown here as a single twisted rod. Frequently, fibrous proteins contain seven-stranded cables, composed of a central molecule with six others wound around it. A composite of such cables makes up the tissue fiber.

If the higher levels of structure in proteins are destroyed, the proteins are said to be denatured. A good example of denaturation is given by egg white, albumin. If albumin is heated, the highly organized protein chains unravel and tangle. The result is cooked egg white, a polymeric solid with no more organization than solid nylon. To tear apart the tangle and reassemble the parts into normal albumin is a task far beyond the ability of any existing laboratory. Only a chicken can do it.

20.5.3 Kinds of Proteins and Their Functions

In living organisms, particularly in animals, proteins have a wide range of functions. Proteins not only provide much of the physical structure of animals, but are the primary agents for controlling the complicated and elegant chemistry that is the basis of life.

Most structural proteins can be classified as either fibrous or muscular material. The former includes the keratins, of which hair, skin, feathers, claws, and wool are examples, and the collagens, which make up tendons and connective tissues. Muscular proteins include the substances that compose cell walls in ordinary tissues and the contractile proteins that give muscle power.

The nonstructural **globular proteins** exist in living organisms as essentially unconnected, individual entities. These serve a wide variety of functions, ranging from molecular synthesis to direct protection of the organism from invasion by foreign substances. Globular proteins include the following, some of which are discussed in later sections.

Enzymes	Highly specific catalysts, essential to the chemistry of living organisms.
Respiratory proteins	Used for O_2 and CO_2 transport and storage. An example is hemoglobin.
Nucleoproteins	Function together with DNA and RNA to store and transmit the information of heredity.
Antibodies	Protect the organism from harmful substances such as viruses and the byproducts of bacterial metabolism.
Hormones	Important regulatory proteins that act as controls for all the chemical processes of the body.

20.6 NUCLEIC ACIDS

In addition to proteins, biological organisms create several other important kinds of polymers, among which are the **nucleic acids.** The "nucleic" in nucleic acids has little direct connection to atomic nuclei; the compounds were so named because they were first identified in the nuclei of biological cells. Moreover, although the nucleic acids are somewhat acidic, their biological function has little to do with the donation of hydrogen ions.

20.6.1 Function of Nucleic Acids

The function of nucleic acids is to store and communicate all the information needed to create and operate biological organisms. The nucleic acids of a cat, for instance, contain all the information necessary to create and operate a cat; the nucleic acids of humans perform the same function for humans. Imagine what the English-language version of the owner's manual for your body would look like if it contained all the information about how to build every part of your body: the skin, eyes, brain, circulatory organs, digestive system, and all the rest. Now, suppose that it also had all the instructions about how to make nerves work, how to see, how to think, how to operate the blood chemistry, and so on. The owner's manual would be an entire library. Add to that information the special design characteristics that make any one person different from all others: facial features, height, length of fingers, hair and eye color, and so on. The library gets bigger. Now, get ready for this. Read it, and stop to think about it a moment: The entire owner's manual, all the information about how to build and operate a biological organism, yourself for example, is stored in the nucleus of *every cell* in that organism. Your body has billions of individual cells, and each cell contains all that information. Your life began when two cells merged into one and combined their genetic information; now every cell of your body contains the set of genes with that information. A gene is a molecule of **DNA, deoxyribonucleic acid.** (Another kind of nucleic acid, RNA, has similar functions, but we will confine our discussion primarily to DNA.)

If all this seems impossible to believe (and if it does not, you are not thinking about it), you should know that a fairly simple calculation yields the estimate that all the words in about 100 million large college libraries could be stored in a thimbleful of DNA. Next to this, a hand calculator that knows foreign languages, plays music, and remembers telephone numbers is simple stuff. Given this capability for information storage density, DNA can easily store your owner's manual in a tiny speck.

20.6.2 Structure of DNA

DNA molecules are long-chain polymers made from individual units called nucleotides. Each nucleotide is composed of a phosphate group, a sugar molecule, and a base R group (Figure 20.8).

The structure of the nucleic acid backbone is shown schematically in Figure 20.9. Notice that the phosphate and sugar groups form the chain itself and that the R groups lie outside the chain, as do the R groups in protein. Just as the arrangement of the side groups is what gives protein molecules different functions, so the set of structures attached to the main chain is the operative part of a DNA molecule. To each sugar ring in the chain is attached an R group consisting of an organic base, usually one of the four structures shown in Figure 20.10.

Just as a word in the English language is composed of a sequence of letters drawn from an alphabet of 26 letters, so is a DNA "word" composed of a sequence of bases drawn from an alphabet of the four base structures shown in the figure. Our alphabet has 26 letters, some have more, some fewer. Many computers use an alphabet of only two letters. DNA uses an alphabet of four letters. The sequence of bases along the DNA chain is the primary structure.

The secondary structure of DNA is similar to that of the alpha helix of protein, except that a DNA molecule contains two chains coiled together, a **double helix,** Figure 20.11.

Study the figure. Notice that the two helices are held together in their double spiral by bridges formed from the base groups, represented in this drawing by their abbreviations. The bases are hydrogen-bonded together in a highly specific way, a lock-and-key system that allows only particular pairs to occur: adenine pairs only with thymine and cytosine pairs only with guanine. If cytosine occurs at a particular place on one chain, for example, guanine must appear at the corresponding link on the opposite chain. The dotted lines in Figure 20.11 show how the hydrogen bonds link the pairs of bases.

20.6.3 Operation of DNA

In its inactive form, DNA remains in the shape of a double helix, but when it transmits information, the DNA molecule unzips to expose the base sequence code that is ordinarily

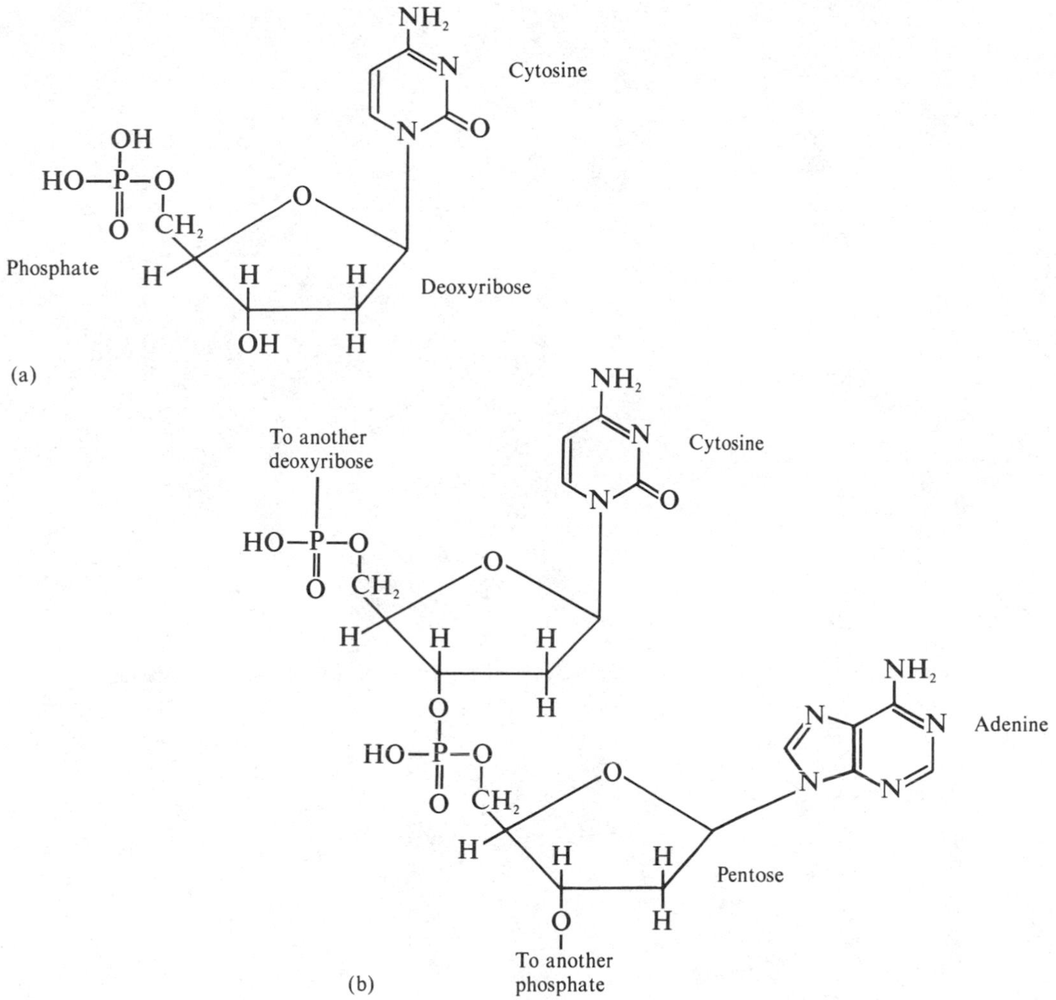

Figure 20.8   Nucleotide structures. (*a*) Nucleotide unit, cytosine deoxyrobonucleotide. (*b*) Two units in a DNA chain.

closed up in the center of the helix. DNA helices unzip also when they reproduce themselves, or *replicate*. DNA replication, a process that occurs when cells divide, enables each new cell to take away a full set of DNA molecules. During replication, each half of a double helix assembles its own new matching half by taking prefabricated individual nucleotides from the cytoplasm and mating them to its R group bases. When the replication is finished, a complete DNA molecule, identical to the original, has been created by each of the two individual chains, as shown in Figure 20.12.

The principal function of nucleic acids is to create protein molecules, which the organism then uses to construct cells, enzymes, globular proteins, and so on. DNA and RNA work jointly to create protein molecules, but each has a different function. DNA stores the information that is used to make the proteins, and RNA reads the information and uses it to build particular proteins. One kind of RNA takes information from DNA, and another kind of

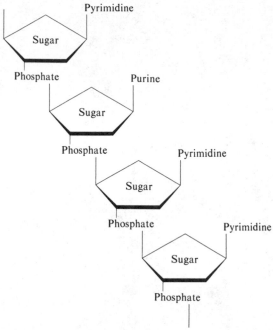

Figure 20.9   Simplified schematic drawing of a portion of a nucleic acid molecule.

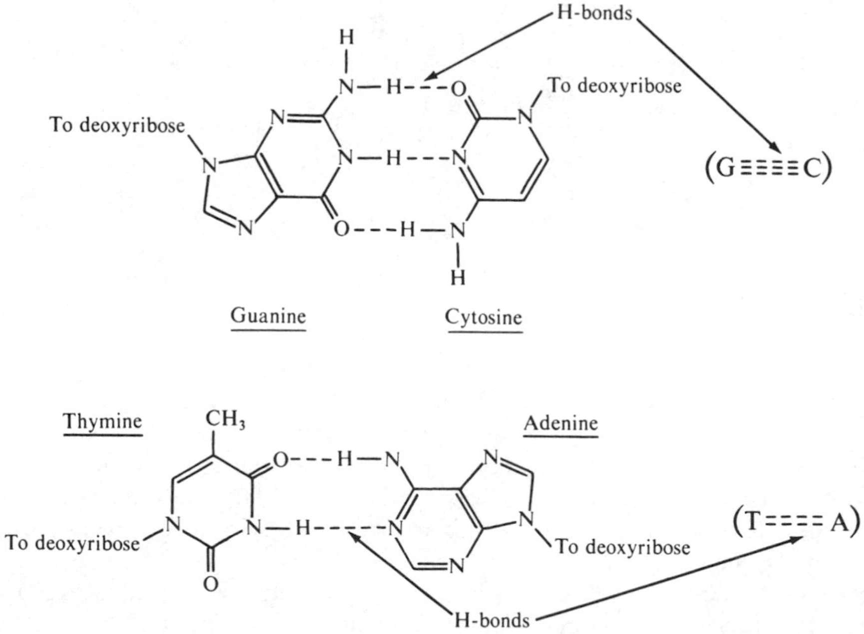

Figure 20.10   The four bases found in DNA. Instead of thymine, the RNA molecule contains uracil, a base with a structure similar to that of thymine.

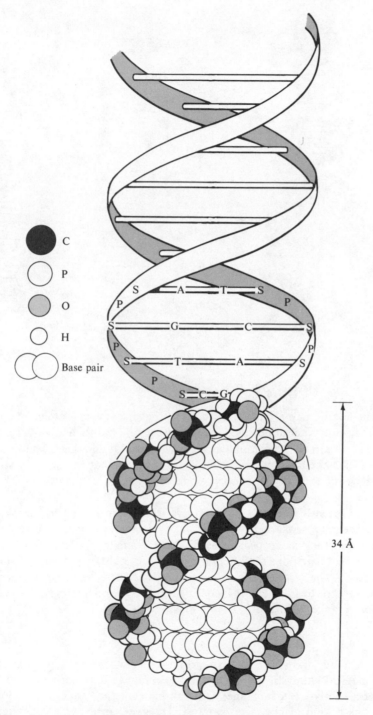

Figure 20.11 Secondary structure of DNA.

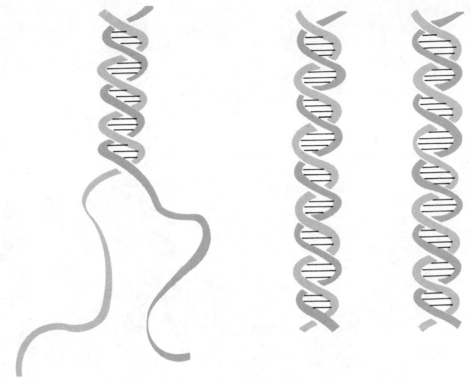

Figure 20.12   Replication of nucleic acids.

RNA assembles the appropriate amino acids to make that particular protein. The process of making a protein molecule is not understood in complete detail, but the individual DNA codes that correspond to certain individual amino acids are known. Quite a lot is known also about the overall way in which the proteins are assembled. It is clear that the primary structure of each protein, the R group sequence, is precisely related to the base sequence of some portion of a DNA molecule.

Biochemists have recently learned how to remove DNA from cell nuclei, break the DNA chains, recombine them in new arrangements, and replace them in the nuclei. When the treated cells reproduce, they make organisms different from themselves. This method has allowed scientists to create new kinds of life in the laboratory. Although it has been successful so far only on bacteria, the technique is so promising that there has been a court decision allowing research workers to patent new kinds of life. Imagine applying for a patent on a living organism. Dr. Frankenstein should see us now.

20.7 ENZYMES

Earlier in the chapter, we saw that respiration and photosynthesis were accomplished with the aid of biological catalysts, and from time to time, we have seen the word *enzyme*. Remembering that a catalyst is a substance that can increase the rate of a reaction, we can regard an enzyme as a kind of biological catalyst. However, an enzyme compares with an ordinary catalyst such as MnO_2 about as well as an automobile compares with a roller skate.

Without reacting themselves, enzymes efficiently and rapidly bring about complex reactions that can take months to accomplish in the laboratory—if indeed, they can be done in the laboratory at all.

Enzymes are large, complex molecules that are designed to perform specific kinds of chemical operations in programmed ways. In highly automated factories, such as those that make automobile engines, certain large machines can pick up pieces of metal, and, while holding them and turning them this way and that, perform a series of operations that result in the formation of complex parts. Enzymes work the same way in biological systems, as Figure 20.13 shows. An enzyme molecule approaches and enfolds the molecule on which it will operate (called the substrate molecule). While holding the substrate, the enzyme stretches and eventually breaks particular substrate bonds. At the same time, different parts of the enzyme molecule pick up other ions or molecules, bring them to the reaction site, and attach them to the substrate. The reaction process is so precise and sequential, it almost seems as if the enzyme were alive or directed by a computer.

Enzymes have been known to exist since the early nineteenth century, when Berzelius found that starch could be decomposed by enzymes from potatoes far faster than by sulfuric acid. Today, the list of well-characterized enzymes and their reactions would fill several pages, and new examples are constantly being added.

The names of enzymes usually come from the reaction they assist plus the ending *-ase*. Thus, a transaminase is an enzyme that transfers an amine group from one molecule to another, and an esterase is an enzyme that transfers an amine group from one molecule to another, and an esterase is an enzyme that operates on ester groups. Exceptions are the enzymes that were discovered earliest, such as pepsin and trypsin, which are digestive enzymes.

Enzymes are characterized by their speed and often by their specificity. The enzyme lipase, which breaks ester linkages in lipids, can catalyze the decomposition of the ester, ethyl butyrate, ten thousand times faster than if the decomposition were catalyzed by hydroxide ions and a billion times faster than if catalyzed by hydrogen ions.

$$H_2O + -\overset{|}{\underset{|}{C}}-\overset{|}{\underset{|}{C}}-O-\overset{O}{\overset{\|}{C}}-\overset{|}{\underset{|}{C}}-\overset{|}{\underset{|}{C}}-\overset{|}{\underset{|}{C}}- \longrightarrow -\overset{|}{\underset{|}{C}}-\overset{|}{\underset{|}{C}}-OH + \underset{HO}{\overset{O}{\overset{\|}{C}}}-\overset{|}{\underset{|}{C}}-\overset{|}{\underset{|}{C}}-\overset{|}{\underset{|}{C}}-$$

Ethyl butyrate   *Ethanol*   *Butyric acid*

Because enzymes are so rapid and efficient, only infinitesimal amounts are needed to cause reaction. The turnaround rate for enzyme catalysis, that is, the number of molecules of a substrate that can be handled in one second by one molecule of enzyme, is incredibly great. A typical enzyme molecule can process, one at a time, several hundred substrate molecules every second. Because of their speed and efficiency, some of the most important enzymes are present in the body in only trace amounts.

Enzymes are often highly specialized. Many enzymes can operate only on certain substrates and can perform only one kind of reaction. Urease, for example, performs the single function of decomposing urea. Such specificity is essential in places like cell nuclei, where it is important that detailed reactions of a particular kind be carried out without disturbing quite similar chemical systems existing in the same solution. Where broad-spectrum enzymes are needed, however, as in digestion, the enzyme molecules are less discriminating. The digestive

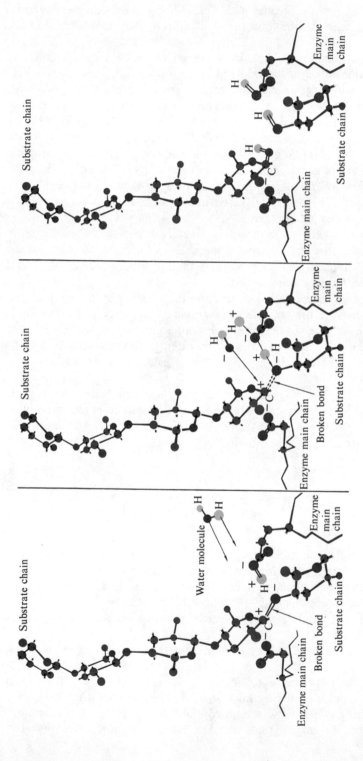

Figure 20.13 Detailed case of the splitting of a molecule by an enzyme. A model of a portion of a long-chain molecule to be split is shown vertically. It is embedded in the much larger enzyme molecule, only two branches (the active groups) of which are shown (stubby bonds are to H atoms not shown). In the reaction sequence, a water molecule moves in from the right and is ionized. The proton goes to the acid oxygen of an enzyme carboxyl group, which simultaneously loses a different proton. This transfers to an oxygen on the O—C bond in the substrate. The oxygen had already been made negative and attractive to the proton because of the nearness of the negatively charged ionic oxygen of the enzyme (left active site). This O⁻ had partly bonded to the C (marked C⁺) on the main chain, thus weakening the C—O bond in the substrate. The reaction finishes when the OH⁻ from the original water molecule reaches and bonds to the dangling carbon on the substrate chain.

enzyme pepsin, for example, can decompose most kinds of proteins. Figure 20.14 shows schematically how an enzyme can react with some molecules while rejecting others.

20.8 HORMONES

The complex machine that is the human body grows from a single fertilized germ cell. One cell. Although it is known that all the information required to build and operate a human body is stored in the DNA of every cell nucleus, it is still not understood in detail just how that information is used. While a baby is being formed in the womb, some cells are making skin, others are making brain tissue, and still others are making lenses for eyes. All those cells have the same set of DNA, but how do cells know which part of the DNA information to use in each case? For that matter, how does the skin of an adolescent male know when to start growing a beard and where to put the beard? Why do the skin cells of women not grow beards, although women do grow hair on their heads, just as men do? How does a broken bone know how to knit? Questions like these are important to biology and physiology, and the answers are, in every case, ultimately chemical.

Biological processes, the construction, repair, and operation of living organisms, are controlled by chemical messengers called **hormones.** Hormones are the primary mechanism of control in the growth and functioning of animals. The injection of a small amount of a fairly simple compound, for example, can cause a baby rooster to start growing spurs long before he ordinarily would. Hormones somehow activate enzymes or trigger specific DNA molecules,

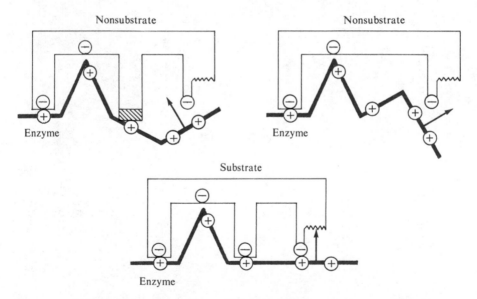

Figure 20.14 Specificity in enzyme reactions. The drawings suggest how a substrate molecule must fit an enzyme molecule before reaction can occur. The small circles represent areas of positive and negative charge that bind the enzyme to the substrate. The first two pictures show why the enzyme cannot react with a nonsubstrate molecule (one for which it is not designed). The third shows a situation in which the substrate and enzyme fit each other. When the molecules match, the active site of the enzyme (arrow) is held next to the bond to be broken (wavy line on substrate).

but we do not really know exactly how hormones work, or why particular hormones work only on certain cells. One thing is clear, however: the action of hormones is chemical. The long series of events that causes a single fertilized cell to grow into a lion rather than a sparrow is chemical.

Although scientists do not know precisely how hormones operate, they do know the details of the structures and functions of several hormones. Hormones and hormonelike molecules have even been synthesized. Many hormones, insulin, for instance, are polypeptides; others are part of the family of compounds called steroids, some of which are shown in Figure 20.15.

Hormones are manufactured in several endocrine glands that are scattered throughout the body and probably in other places as well. In a number of studies, many of which make fascinating reading, hormones have been supplied externally or hormone supplies have been cut off. Although these experiments can tell us *what* hormones do, they have not yet given much information about *how* they work.

20.9 FOOD AND DIGESTION

To grow, operate, and repair their bodies, animals must have a constant supply of many different substances that we classify as foods. Proteins, carbohydrates, and fats are needed in large quantities, but only small amounts of other, equally essential foods, such as vitamins, minerals, and enzymes, are required. Foods also contain nonessential but nevertheless

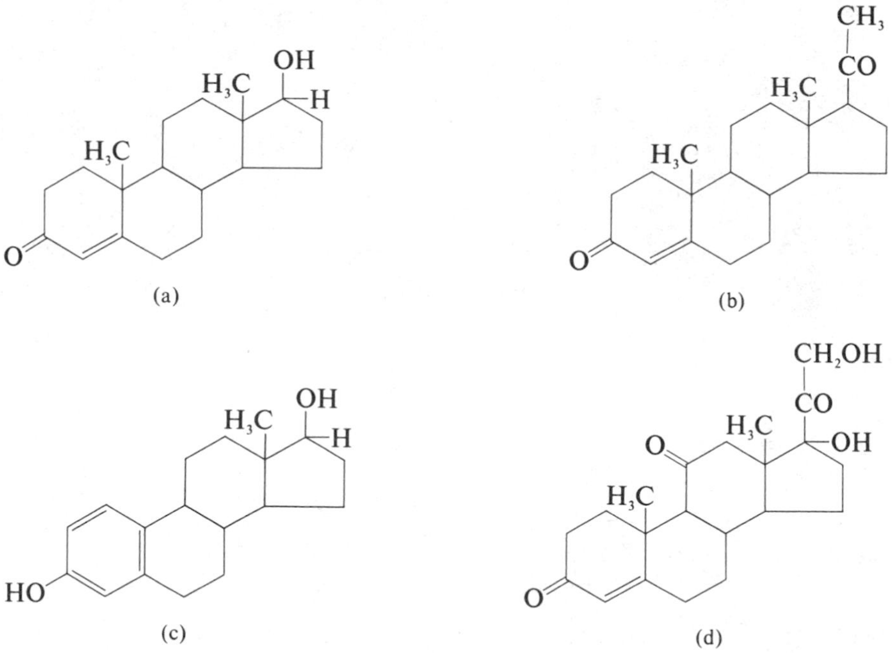

Figure 20.15 Structures of four hormones: (*a*) Testosterone is the male sex hormone. (*b*) Progesterone is a female sex hormone secreted during the latter half of the menstrual cycle; it exerts an antiovulary effect and, in its derivatives, is used as an oral contraceptive. (*c*) Estradiol is also a female sex hormone. (*d*) Cortisone is an adrenal hormone that influences many metabolic processes.

important substances that provide flavor, texture, and color. Although we have discussed proteins in some detail, we have not yet given much attention to other kinds of food.

20.9.1 Carbohydrates

About half the diet of the average American is made up of **carbohydrates,** substances that have the formula $C_xH_{2y}O_y$. The name comes from an early theory that the compounds were made of carbon and water. (The formula can be rewritten as $C_x(H_2O)_y$.) Today we know that carbohydrates are primarily aldehydes or ketones and that they have large numbers of alcohol groups.

Carbohydrates are the cheapest foods. In some parts of the world, they make up as much as 80% of the average diet, and come mostly from cereal grains, potatoes, sugarcane, and beets. Although carbohydrates are an important food, they make up less than 1% of body tissue. After digestion, most carbohydrates are used immediately to create energy, or are converted to fats and stored.

The simplest carbohydrates are the **monosaccharides,** or simple sugars. Most of these compounds fall into one of two classifications, the pentoses ($C_5H_{10}O_5$) or the hexoses ($C_6H_{12}O_6$). The ending -*ose* on the name of a compound means that it is a sugar; the pentoses have five carbons in each molecule and the hexoses six. There are many isomers of each of these kinds of sugar, and each isomer has somewhat different biological properties.

Sugar molecules arrange themselves either into straight chains or rings, most often rings. The structural formulas for the two forms of glucose, a sugar we have already discussed, are shown in the diagram that follows. Notice that the ring contains five carbon atoms and one oxygen atom and that one of the carbon atoms appears as a methyl group attached to the ring.

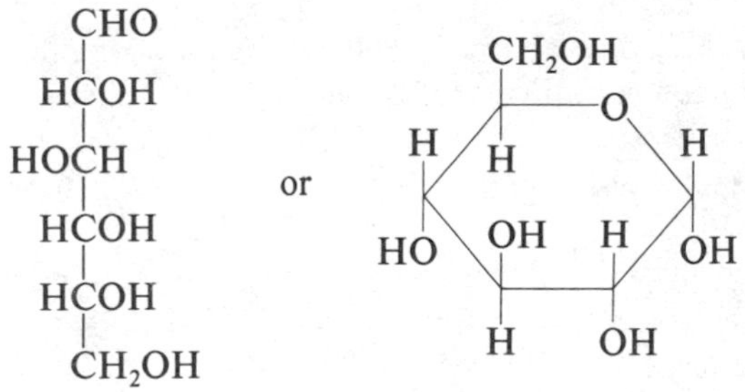

Glucose

Fructose is another hexose that is important in the diet. In its ring form, fructose has only four carbons and one oxygen in the ring, with two carbon atoms outside.

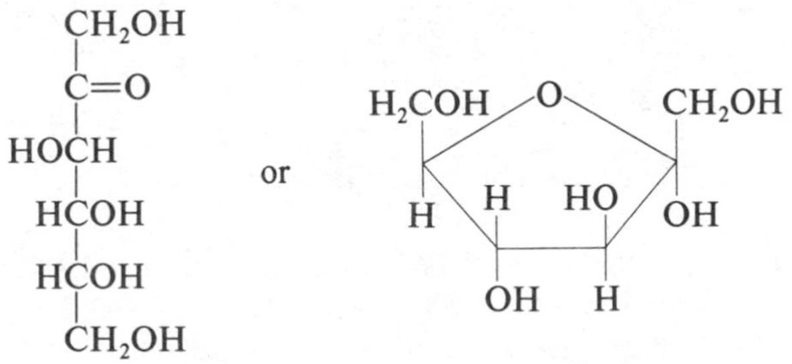

Fructose

Many sugars are **disaccharides,** whose molecules are dimers made from smaller sugars. Sucrose (the common table sugar made from sugarcane or sugar beets) is the dimer of glucose and fructose.

Sucrose

The disaccharides are easily decomposed into their monomers by the catalytic action of an acid or base, or (much more efficiently) by the action of the appropriate enzyme. It has been estimated that nearly one fourth of the carbohydrate in the American diet comes from sucrose. Other disaccharides are lactose, the dimer of glucose and galactose found in mammalian milk, and maltose, which contains only glucose.

The sugars have varying degrees of sweetness. On a sweetness taste scale in which sucrose is ranked 100, glucose ranks 74 (only three fourths as sweet) and fructose ranks 173. In the making of jelly and jam, the acid in the fruit causes part of the added sucrose to decompose into its monosaccharide components. Each molecule of sucrose that reacts supplies a molecule of glucose and one of fructose, making the product more than twice as sweet as did the original sucrose. Sucrose that has been partly decomposed in this way is called invert sugar. Many products, especially candies, use invert sugar because it produces more sweetness at less cost.

Monosaccharides not only form disaccharides, but also **polysaccharides,** which are large polymers of the sugars and the most common carbohydrates found in nature. Three especially important polysaccharides are **starch, cellulose,** and **glycogen.** All are polymers of glucose. Starch and cellulose are interesting examples of the effect of small differences in structure on the properties of molecules.

The diagram that follows shows a portion of a glucose polymer. Both starch and cellulose can be represented by this drawing, in which the part in brackets represents a structure repeated over and over in the polymeric chain.

Glucose polymers can have different arrangements in space. In starch molecules, the carbon groups attached to the bridging oxygen are mirror images of those found in the cellulose molecules. This difference is too subtle to show in a flat drawing, and is not immediately apparent even in a molecular model. This tiny difference in structure nevertheless makes an enormous difference in biological properties.

With the aid of the enzymes in their digestive juices, animals can decompose starch into its component glucose molecules and can then oxidize the glucose for energy. The enzymes of animals cannot, however, break down the cellulose polymer. Cellulose as such is therefore useless as food, even though it is made of glucose molecules in much the same way that starch is made. Fortunately, the grazing animals such as horses, cattle, and sheep maintain in their stomachs colonies of bacteria that can digest cellulose. That is why cows can digest grass but we cannot; grass is mostly cellulose.

Glycogen is a carbohydrate stored in animal tissues to provide glucose when it is needed for respiration. Glycogen is a glucose polymer similar to starch but with somewhat shorter chain lengths. Enzymes present in the cells can quickly convert glycogen to glucose without waiting for a digestive process to occur.

20.9.2 Fats and Oils

Edible fats and edible oils are the same kind of compound; the difference is that fats are solid at room temperature, whereas oils are liquid. We will refer to both simply as fats.

Like carbohydrates, fats are widely distributed in nature and are an important source of food energy. Fats belong to the general class of organic substances called lipids, plant and animal products that are soluble in ether rather than in water.

Most **fats** are the esters of **fatty acids.** A fatty acid is a long-chain carboxylic acid in which there are one or more double bonds in the chain. Table 20.2 shows the formulas for some common fatty acids, both saturated and unsaturated. The alcohol of which fats are the esters is **glycerol** (glycerine). Each molecule of glycerine contains three hydroxy groups.

$$
\begin{array}{ll}
\begin{array}{c}
\text{H} \\
| \\
\text{HO}-\text{C}-\text{H} \\
| \\
\text{HO}-\text{C}-\text{H} \\
| \\
\text{HO}-\text{C}-\text{H} \\
| \\
\text{H}
\end{array}
&
\begin{array}{c}
\overset{\displaystyle O}{\overset{\|}{CH_3(CH_2)_{16}-C}}-O-CH_2 \\
\overset{\displaystyle O}{\overset{\|}{CH_3(CH_2)_{16}-C}}-O-CH \\
\overset{\displaystyle O}{\overset{\|}{CH_3(CH_2)_{16}-C}}-O-CH_2
\end{array}
\\
\textit{Glycerol} & \textit{A typical fat}
\end{array}
$$

Animal fats are usually far more saturated than vegetable fats, as a look at Table 20.3 will show. Notice that the unsaturated fatty acids (the last three listed) make up more than 75% of the composition of the vegetable oils such as corn oil.

Because the saturated fats melt at higher temperatures than do the unsaturated fats, the first are usually solid and the second usually liquid at room temperature. We can learn the reason for this difference in melting point by considering the structural differences between the two kinds of fat. The long chains of the saturated fats are regular in shape and can fit together well, but the kinks and bends of the unsaturated chains do not fit together as efficiently (Figure 20.16). When the arrangement is efficient, the dipoles of neighboring molecules lie close together and exert strong attraction on one another. When the stacks of fatty acid cannot assemble efficiently, as in the unsaturated fats, intermolecular forces are smaller.

Table 20.2
Fatty Acids

Saturated	
Butyric	$CH_3(CH_2)_2COOH$*
Caproic	$CH_3(CH_2)_4COOH$
Caprylic	$CH_3(CH_2)_6COOH$
Capric	$CH_3(CH_2)_8COOH$
Lauric	$CH_3(CH_2)_{10}COOH$
Myristic	$CH_3(CH_2)_{12}COOH$
Palmitic	$CH_3(CH_2)_{14}COOH$
Stearic	$CH_3(CH_2)_{16}COOH$
Arachidic	$CH_3(CH_2)_{18}COOH$

Unsaturated	
Oleic	$CH_3(CH_2)_7CH{=}CH(CH_2)_7COOH$
Linoleic	$CH_3(CH_2)_4CH{=}CHCH_2CH{=}CH(CH_2)_7COOH$
Linolenic	$CH_3CH_2CH{=}CHCH_2CH{=}CHCH_2CH{=}CH(CH_2)_7COOH$

*In the condensed language used in this table, adjacent

$$-\overset{\displaystyle H}{\underset{\displaystyle H}{C}}-$$

groups in a chain are shown by $(CH_2)_x$, where x is the number of adjacent groups. For example, $CH_3(CH_2)_7{=}CH(CH_2)_7COOH$ is

Table 20.3
Approximate Percentage Fatty Acid Composition in Some Common Fats and Oils

	Butterfat	Lard	Olive Oil	Cottonseed Oil	Peanut Oil	Coconut Oil	Soybean Oil	Corn Oil	Safflower Oil
Butyric	3								
Caproic	2								
Caprylic	1					7			
Capric	2					7			
Lauric	3					46			
Myristic	10	1				19			
Palmitic	30	28	8	21	8	10	9	6	5
Stearic	11	8	2	2	4	2	2	2	1
Arachidic				1	3		1	1	1
Oleic	30	56	82	29	55	8	32	37	20
Linoleic	3	12	8	47	25	1	53	54	70
Linolenic							3		3

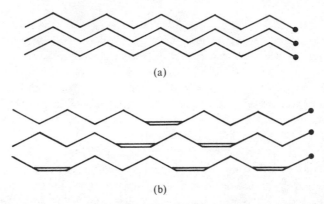

(a)

(b)

Figure 20.16 Schematic sketches: (*a*) Saturated chains can stack efficiently. (*b*) Unsaturated chains stack less efficiently. Forces between molecules diminish when close stacking is not possible.

As we learned in Chapter 13, small intermolecular forces lead to low melting points and large forces lead to high melting points. Because they have somewhat different properties, we have become accustomed to thinking of animal fats and vegetable oils as entirely different kinds of substances.

Since many people prefer solid fats to liquid ones, there has developed a large industry for the conversion of unsaturated fats to saturates. The process is accomplished by **hydrogenation,** in which H_2 is catalytically added to double bonds. In this way, for example, the oleic acid ester of glycerol can be converted to the stearic acid ester.

$$CH_3(CH_2)_7CH{=}CH(CH_2)_7{-}\overset{\overset{\displaystyle O}{\|}}{C}{-}O{-}CH_2$$

$$CH_3(CH_2)_7CH{=}CH(CH_2)_7{-}\overset{\overset{\displaystyle O}{\|}}{C}{-}O{-}CH \ + \ 3\ H_2$$

$$CH_3(CH_2)_7CH{=}CH(CH_2)_7{-}\overset{\overset{\displaystyle O}{\|}}{C}{-}O{-}CH_2$$

Hydrogenation of a fat

Hydrogenation is accomplished on a large scale by dissolving H_2 in hot vegetable oil, say corn oil, under high pressure, then adding a powdered catalyst, often nickel. After the reaction finishes, the resulting fat is filtered, cooled, flavored, packaged, and sold, usually as margarine.

It was once thought that the function of fats in biological organisms was merely to store energy. It is true that any excess intake of protein, or especially carbohydrate, in the diet is likely to be converted to body fat and that body fat can be burned for energy. Fairly recent research has shown, however, that the growth of baby rats ceases completely if all fats are removed from the diet. With small additions of linoleic, linolenic, and arachidonic acids, however, growth resumes. These fatty acids, now called **essential fatty acids,** evidently have a role in metabolism beyond that of simply storing energy.

20.10 SUMMARY

Living creatures such as ourselves grow from microscopic cells into mature, functioning organisms by a series of chemical reactions. We move, see, think, digest, fight disease, and function in hundreds of other ways through chemical reaction. Every living thing, plant or animal, is, in a real sense, an operating chemical factory. But biochemical processes are not the same as the chemistry we learned in the first 19 chapters of this book. The fundamentals are the same, but in complexity, versatility, and efficiency, the chemistry carried on by living organisms is far beyond ordinary laboratory chemistry.

Living organisms are largely constructed of polymeric organic molecules. Of these, we have paid particular attention to two kinds, proteins and nucleic acids.

Although protein polymers are constructed, for the most part, from about 20 simple building blocks—the biological amino acids—there exists an almost infinite number of kinds of proteins that perform separate and widely varied functions. Proteins differ fundamentally in the sequence in which amino acids appear in the protein chain and in chain length, but these two factors lead to a series of more complex kinds of structures and thence to versatility in function. We have studied the functions of several kinds of proteins.

Nucleic acids are nature's way of storing and transmitting genetic information. The nucleic acid DNA stores information in an unbelievably compact way. All the information, physical as well as chemical, that is necessary to build an entire human being, for example, is contained in a speck of matter too small to see with the naked eye.

Life on earth depends on an exchange of energy between plants and animals. Through photosynthesis, plants store energy from sunlight in organic polymers (mostly starches and sugars), using CO_2 and releasing O_2. Animals eat and metabolize the products of photosynthesis, gaining energy to operate their bodies, releasing CO_2, and using O_2. In an ideal ecological situation, the two sets of processes balance each other.

An enzyme is a kind of molecule that is specifically designed to catalyze biological chemical reactions. Enzymes enable biological chemistry to achieve its miracles. We have learned how enzymes operate and about the functions of several kinds of enzymes.

QUESTIONS

Section 20.2

1. List the major parts of a cell and give the functions of each part.
2. Explain how genes, chromosomes, and DNA are related.

Section 20.3

3. Give the overall reactions of photosynthesis and respiration.
4. What is the role of chlorophyll in photosynthesis?
5. What is the Krebs cycle? Why is it essential to life?
6. Describe how the ADP-ATP system stores and releases energy.

Section 20.4

7. Using the R shorthand, draw the general structure of an alpha amino acid.

8. Using structural formulas, write a reaction equation for the condensation of two amino acids to form a dipeptide.
9. List the essential amino acids. Draw the structures of their R groups.

Section 20.5

10. Explain how the primary, secondary, tertiary, and quaternary structures of proteins differ from one another.
11. Explain how the primary structure of a protein is involved in the ultimate function of the protein.
12. Sketch the alpha helix of a protein.
13. What does it mean to denature proteins?
14. List five globular proteins and give their functions.

Section 20.6

15. Sketch a nucleotide.
16. What are the operative parts of a DNA molecule?
17. How does the double helix of DNA stay together?
18. The letters that follow represent the code sequence in a strand of DNA. Write out the sequence for the complementary strand.

       A T A C G T A A C T T G C T A G G A C T G A

19. Draw a diagram of the replication process of DNA.
20. How do the functions of DNA and RNA differ?

Section 20.7

21. What is the function of an enzyme?
22. Describe how an enzyme functions.
23. What is meant by turnaround rate?

Section 20.8

24. What is the principal function of a hormone?
25. List several examples of hormones.

Section 20.9

26. What is the general formula of a carbohydrate?
27. Explain the difference between monosaccharides, disaccharides, and polysaccharides.
28. Draw the structure of sucrose and label the glucose and fructose.

Section 20.10

29. What is a fatty acid? List the essential fatty acids.
30. What is the difference between a saturated and an unsaturated fat?
31. What is the general reaction for hydrogenation?
32. Name some carbohydrates, fats, monosaccharides, polysaccharides, saturated fats, and unsaturated fats.

Chapter 21

Chemistry and the Atomic Nucleus

Because chemistry is directly concerned with the properties and reactions of substances and because substances contain entire atoms rather than only nuclei, the study of the nucleus is really the proper concern of physics rather than chemistry. However, nuclear events usually have profound effects on substances and therefore eventually involve chemistry. Moreover, the social consequences of nuclear developments are so great that as many people as possible should be informed about the workings of nuclear technology. For these reasons, we finish our study with a consideration of nuclear science and technology.

Nuclear science is a relatively new field. Less than a century has elapsed since the discovery of radioactivity, and until the Second World War, little information about nuclear studies had reached the general public. Now, because of the enormous consequences of the use of nuclear energy for both peaceful and warlike purposes, decisions regarding the future of nuclear technology must be made for social as well as scientific reasons. Depending on how nuclear energy is used, our future may be brighter than it has ever been, or we might have no future at all.

We begin our study from a historical perspective, then learn some new terms and concepts. Our purpose is to gain enough understanding of nuclear science to be able to make intelligent choices among the alternatives presented by the possible applications of nuclear energy.

AFTER STUDYING THIS CHAPTER, YOU WILL BE ABLE TO

- list the names and discoveries of several of the pioneers of nuclear science
- name the three most common kinds of radioactive radiation and the characteristics of each
- explain how a radioactive isotope differs from one that is not radioactive
- explain the term *half-life* and the methods of measuring half-life

- explain how radioactive isotopes can be used to date archeological specimens or rocks
- write balanced nuclear reaction equations
- write equations for the artificial transmutation of some elements
- list several uses of natural radioactive isotopes
- name several artificial isotopes and discuss their uses
- explain quantitatively, by means of calculations, why nuclear events release such large amounts of energy
- discuss in everyday terms and calculate mathematically the amounts of energy that are equivalent to various small samples of matter
- distinguish among radioactive decay, fission, and fusion reactions
- explain the functioning of a nuclear reactor
- list some of the advantages and disadvantages of nuclear power plants
- discuss the problems involved with nuclear waste

TERMS TO KNOW

alpha rays	Geiger counter	nuclear waste
beta rays	half-life	radioactive decay
binding energy	mass defect	radioactivity
chain reaction	nuclear fission	radiocarbon dating
critical mass	nuclear fusion	transmutation
gamma rays	nuclear reactor	

TERMS WITH WHICH TO BE FAMILIAR

betatron	fuel element	scintillation
cosmic rays	linear accelerator	scintillation counter
Curie	parent isotope	synchrotron
cyclotron	rem, millirem	Wilson cloud chamber
daughter isotope	Roentgen	

21.1 EARLY DEVELOPMENTS IN NUCLEAR SCIENCE

Radioactive emission was first discovered by a French scientist in 1896. Henry Becquerel, who was working on an entirely different problem at the time, found that photographic plates became blackened when they were left standing near an impure sample of a compound of the element uranium. Unknown rays from the sample had passed through the black paper that protected the plates from exposure to visible light. Somewhat earlier, by a similar accident, Roentgen had discovered X rays, which are not nuclear but which have the same effect.

Soon after Becquerel's discovery, a student of his named Marija Sklodowska began what was to be a lifetime study of the new phenomenon. Following her marriage to one of Becquerel's assistants, she and her husband devoted many years of patient effort to studying radioactivity. They finally isolated two elements from which radioactivity came, and named these elements polonium and radium. For her monumental efforts, this tireless worker, later known to the world as Madame Curie, became one of the most celebrated scientists of all time.

Shortly afterward, Ernest Rutherford, whom we have already met, discovered that if the invisible rays from radium were allowed to strike a screen coated with zinc sulfide, a glow of visible light would result. Upon microscopic examination, the glow resolved itself into numbers of small, individual flashes of light, or **scintillations,** which suggested that the rays were not a continuous flow of energy but might instead be streams of individual particles.

Remembering Thomson's experiments with the effect of magnetic fields on flying electrons (Section 8.2), Rutherford directed a beam of radium rays through a magnetic field and separated the beam into three parts (Figure 21.1). Rutherford named the three components of the beam alpha (α), beta (β), and gamma (γ) rays. The beta rays were most affected by the magnet and the gamma rays were not affected at all.

The nature of alpha, beta, and gamma rays eventually became known. In the famous experiment that revealed the atomic nucleus, Rutherford used alpha rays.

> **Alpha rays** are streams of high-velocity particles of matter, helium nuclei, with a mass number of 4 and a charge of +2. In some later experiments, Rutherford trapped alpha particles in an evacuated space and showed that after stopping, each of the particles acquired two electrons and became a helium atom. Alpha particles are easily stopped by matter because the +2 charge of each particle interacts strongly with the charged electrons and protons of atoms with which the particle comes into contact. Alpha particles can be stopped by a piece of paper.

> **Beta rays** are streams of high-speed electrons. Because electrons have far less mass than helium nuclei, a magnetic field has greater effect on beta rays than on alpha rays. Beta particles are less easily stopped than alpha particles, partly because the flying electrons have less charge than alpha particles do. Beta particles can usually penetrate many sheets of paper, but can be stopped by a sheet of metal such as tin cans are made of.

> **Gamma rays** are not particles of matter, but are streams of extremely high-energy photons (visible light is a stream of far less energetic photons). Gamma rays have great penetrating power because photons have no charges and do not interact easily with the electrons and nuclei of atoms they meet. Thick layers of lead are used to screen out gamma rays. (See Figure 21.2.)

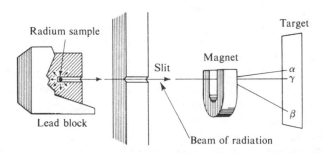

Figure 21.1 Illustration of the effect of a magnetic field on radioactive emanation from radium.

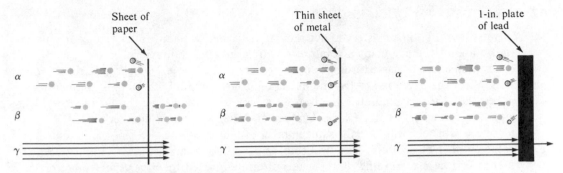

Figure 21.2   Penetrating power of alpha, beta, and gamma rays.

The streams of penetrating rays from the atoms of certain elements became known as radioactivity, or radioactive radiation. (See Table 21.1.) It is now known that many elements have radioactive isotopes. Other kinds of rays have also been discovered, but are less common than α, β, and γ rays.

Table 21.1
Kinds of Nuclear Radiation

Kind of Radiation	Symbol	Electric Charge	Mass (amu)	Penetration (See Figure 21.2)	Some Sources[1]
Alpha particles	α, $^4_2$He	+2	4	• Only a few centimeters of air • Will not penetrate skin • Stopped by paper	$^{210}_{84}$Po $^{238}_{92}$U $^{131}_{53}$I $^{226}_{88}$Ra
Beta particles	β, $^0_{-1}$e	−1	$\dfrac{1}{1837}$	• Paper or thin foil • Up to about 1 cm of tissue • Depends on energy	$^{90}_{38}$Sr $^{137}_{55}$Cs $^{228}_{88}$Ra $^{32}_{15}$P $^3_1$H $^{131}_{53}$I $^{14}_6$C
Gamma rays	γ	0	0	• Up to 1 or more cm of lead • Can pass entirely through body • Depends on energy	$^{60}_{27}$Co $^{226}_{92}$Ra $^{131}_{53}$I $^{137}_{55}$Cs
Neutrons	$^1_0$n	0	1	• Large penetrations of most kinds of matter	Usual sources are fission or nuclear reactions like the Ra/Be source[2]
Positrons	$^0_{+1}$e	+1	$\dfrac{1}{1837}$	• Same as β particles	$^{18}_9$F $^{138}_{57}$La[3]

(1) Although there are hundreds of radioactive isotopes, a few are noteworthy because of their wide use, their place in history, or the dangers they present. A few of these are listed in the table. (2) Naturally occurring neutron emission is not common. Nearly all sources of significant intensity are man-made. (3) Positron emission is not common, and naturally occurring positron emitters are extremely rare.

21.2 HOW RADIOACTIVITY OCCURS

There still exists some mystery about radioactivity, but, even so, an enormous amount is now known about it, much being a result of research done for military purposes.

In Chapter 8, we discussed isotopes in some detail (Section 8.7). Let's just refresh our memories. With the exception of $_1^1H$, the nuclei of atoms contain both protons and neutrons. Remember that the mass number is given as a superscript and the atomic number as a subscript. For example, $_3^7Li$ has a mass number of 7 and an atomic number of 3. Different isotopes of the same element have the same number of protons in the nuclei of their atoms, but a different number of neutrons. The two isotopes of He, for example, are $_2^4He$ and $_2^3He$. For convenience and ease of reading, isotopes are often designated by either the symbol or the name of the element followed by a dash and the mass number. For example, $_{92}^{238}U$ can be written as U-238 or uranium-238. In this chapter, you will see all three kinds of notation used, the subscript-superscript language being reserved mostly for equations of nuclear reactions.

In the light elements, the most common ratio of protons to neutrons is about one to one, as in He-4, which has two neutrons and two protons in its nucleus. In the heavier elements, however, the proportion of neutrons increases until it is about one and one half times the atomic number, as in Au-197. (See Figure 21.3.)

Although most elements have more than one isotope, not all isotopes are stable. Some combinations of protons and neutrons last indefinitely, but other combinations tend to disintegrate, either losing one or more of their particles or flying apart entirely. Of the hundred or so known elements, only 82 have any stable isotope at all; these 82 elements have a total of about 275 stable isotopes. The 20 or so elements of highest atomic number have only unstable isotopes, and most of the lighter elements have one or more unstable isotopes in addition to their stable isotopes. All told, about 800 unstable isotopes of all the elements are known.

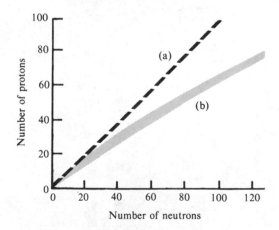

Figure 21.3 Nuclear stability curve. (*a*) Linear curve that would result if the number of neutrons in nuclei always equals the number of protons. (*b*) The proton/neutron ratios of actual elements all fall within the colored area of the graph.

Samples of unstable isotopes are said to undergo **radioactive decay** and to be **radioactive;** that is, the atoms of unstable isotopes disintegrate in one way or another, usually to form other isotopes of the same element or atoms of another element. At the same time, radioactivity of one kind or another is usually emitted, ordinarily in the form of α, β, or γ rays.

21.3 HALF-LIFE

If a nuclear disintegration produces radioactivity, as it usually does, the intensity of the radiation (the number of particles that will be emitted from decaying atoms in a sample of given size in a given time) depends on how much of the radioactive isotope is present in the sample and how unstable the isotope is. (Grapes are more unstable than apples; given a sample of 100 grapes and 100 apples, you'll see the decay of more grapes than apples every week. But given 10,000 apples and only 100 grapes, you would see more apples decay every week than grapes.)

Highly unstable isotopes produce intense radioactivity but do not last long, whereas more stable isotopes are long-lived but produce less intense radioactivity. The stability of isotopes can therefore be measured by the rate at which they decay. Since the amount of an isotope in a sample is often not known, the rate of decay is usually given in terms of the **half-life,** which is the amount of time that it takes for half the isotope to decay, no matter what the amount was to start with. Because the radioactive intensity of a sample is proportional to the amount of active isotope present, when the intensity of radiation from an isotope sample has dropped to half of what it was at some previous time, half of that isotope must then have decayed. (A half-life is not difficult to measure when a sample contains only one radioactive isotope, but when more than one is present, the measurement and calculation are more difficult.)

For example, if we measure the number of particles emitted every second by a sample containing a quantity of a single radioactive isotope and find twenty minutes later that only half as many particles are emitted every second, we would say that the half-life of that isotope is twenty minutes. The half-life is a characteristic property of an isotope, and half-lives vary enormously. For instance, the half-life of Ra-226 is 1602 years, that of Co-60 5.3 years, that of tritium, H-3, 12.26 years, and that of No-254 about three seconds. Iodine has about twenty-five known radioactive isotopes whose half-lives range from about ten minutes to about ten million years (I-129 decays by emitting a β particle and has a half-life of 1.72×10^7 years).

21.4 RADIOCARBON DATING

The atmosphere (and the earth's inhabitants) are bombarded by various kinds of high-energy particles from outer space (cosmic rays). Approximately one particle per minute strikes every square centimeter of the earth's atmosphere. Among these particles are fast neutrons, which can convert nitrogen nuclei to those of C-14 (see Section 21.5). Because of this, the atmosphere contains a small but constant amount of radioactive C-14, which decays by emitting β particles. The half-life of C-14 is 5730 years.

The carbon contained in the compounds of living plants and animals is mostly C-12 but includes a small proportion of C-14 derived from the atmosphere. That proportion has presumably been the same in living organisms for many thousands of years.

When an organism dies, however, the C-14 in its compounds decays without being replaced by that from the atmosphere. If the intensity of the radioactivity of a sample of carbon from a contemporary source is compared with that of a sample from a formerly living organism, an estimate of when that organism lived and died can be obtained.

One gram of pure carbon prepared from contemporary biological sources contains enough C-14 to yield about 16 β particles per minute. Suppose that we take a sample of pure carbon from a fossil tree trunk and find that 1 g of that carbon yields eight β particles per minute. We could conclude that half of the C-14 originally contained in the wood has decayed, that the tree died exactly one half-life, or 5730 years, ago.

EXAMPLE 21.1. One tenth of a gram of carbon from a fossil tree yields 12 β particles in 1 hr. How old is the fossil?

Solution:

Step 1. Calculate how many β particles would be emitted from 1 g of fossil carbon sample in 1 min.

$$\frac{12 \ \beta \ \text{particles}}{0.1 \ \text{g} \times \text{hour}} \times \frac{1 \ \text{hr}}{60 \ \text{min}} = \frac{2 \ \beta \ \text{particles}}{\text{g} \times \text{min}}$$

Step 2. Compare the result of Step 1 with the emission from a fresh sample, 16 particle/g min. Calculate the fraction of the C-14 that remains.

$$\frac{\left(\dfrac{2 \ \beta \ \text{particles}}{\text{g} \times \text{min}}\right)}{\left(\dfrac{16 \ \text{particles}}{\text{g} \times \text{min}}\right)} = \frac{1}{4}$$

Step 3. Calculate the number of half-lives and the age.

If only one fourth of the C-14 is left, then two half-lives must have passed since the tree died, or 5730 × 2 = 11,460 years.

It is not difficult to calculate the age of a sample, even if the amount of elapsed time is not an integer number of half-lives; we will not need to make such calculations, however, to understand the method of radiocarbon dating.

21.5 MEASUREMENT OF RADIOACTIVITY

When fast particles from radioactive isotopes strike ordinary matter, most of the interactions take place between the particle and the electrons of the target (recall the Rutherford nuclear atom experiment). (See Figure 21.4.) Usually, electrons are knocked off the atoms of the target material, leaving short-lived positive ions. The **Geiger counter** takes advantage of this fact to measure radioactive emissions.

The operative part of a Geiger counter is a thin-walled metal tube filled with a low-pressure gas, mostly argon. A fast particle, an α particle perhaps, will pass through the tube, as shown in Figure 21.5, leaving behind a trail of ionized gas molecules. These ions are then accelerated by an electric field and strike other gas molecules, causing even more ionization. All the ions are then collected by electrodes; this creates a tiny, instantaneous pulse of current that produces a click in a speaker. A rapid succession of pulses can convince a meter that the

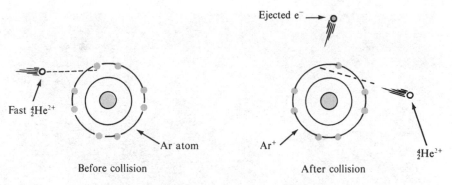

Figure 21.4   How radiation causes ionization.

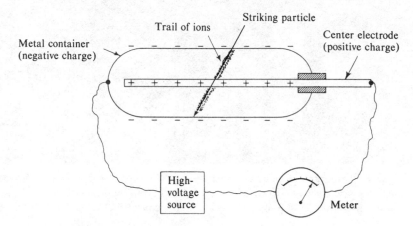

Figure 21.5   How a Geiger counter works. The trail of ions, formed during the passage of the striking particle, permits a pulse of current to flow, indicated by the meter.

flow of current is constant. The intensity of the current will be proportional to the intensity of the radioactivity striking the Geiger tube.

Another useful instrument employs a photomultiplier, a sensitive device that can see and measure the tiny scintillations produced when radioactivity strikes fluorescent material (again we are reminded of Rutherford's experiment). The photomultiplier converts each scintillation into an electric pulse that can be observed in any of several ways and recorded. Scintillation counters are used when there is little radioactivity to be measured or when it is important to discriminate among different, but simultaneous, radioactive events.

The passage of fast particles can be made visible by what is called a Wilson cloud chamber. In this instrument, small liquid droplets form in the track of each particle as it passes through a supercooled vapor. Figure 21.6 is a photograph of typical cloud chamber results. Other, more sophisticated methods are also available for detecting the passage of radioactive radiation. Detectors designed for special purposes are sometimes as large as a big room.

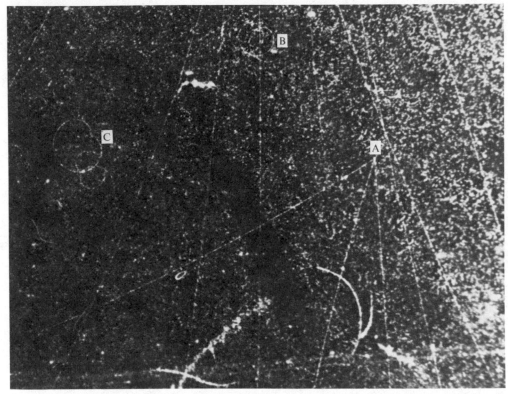

Figure 21.6   Typical photograph of cloud chamber results. The three-track apex at A represents three particles resulting from a collision between a fast neutron and a proton. The circular paths B and C are caused by fast electrons spiralling in a magnetic field. (Photo courtesy of Brookhaven National Laboratory.)

Although it is convenient in many kinds of research to measure radioactivity in units of events per second, other units are used for special purposes. The **Curie,** for example, is the amount of radioactive substance that will undergo 3.7×10^{10} disintegrations per second. In biological work, radioactive radiation is often measured by its effects. The **Roentgen** was defined many decades ago as the amount of X ray radiation that would produce a particular number of ions per second in a given volume of matter under certain conditions. Today, radioactivity is discussed in terms of **rems** (Roentgen equivalent in man) or **millirems,** units similar to the Roentgen.

21.6 WRITING NUCLEAR EQUATIONS

In the days of alchemy during the Middle Ages, great efforts were made to change various elements into others; gold or silver was usually the objective. Little was then known about elements and compounds, and because it was common to see various substances changed into other substances by chemical reaction, it was thought that elements might be changed into other elements in a process called **transmutation.** (The fundamental difference between elements and other substances was not then understood.) However, all efforts to transmute elements were totally unsuccessful. Somewhat later, Dalton's atomic theory, with the view that atoms of all kinds were entirely indestructible and unchangeable, put a virtual end to alchemy and the search for synthetic gold. We now know, however, that in many cases, atoms of one kind can be made to change into atoms of another kind, either through natural radioactive processes or by artificial means. In many cases, atoms of a radioactive element can experience successive changes into atoms of several other elements before achieving stability. Bearing in mind that a change in atomic number is a change of element, let us follow some elemental transmutations.

The unstable nuclei of U-238 decay spontaneously by emitting α particles. We can write an equation for that process, balancing for mass and charge just as we balance chemical equations. From the mass number of the uranium isotope we must subtract that of the α particle, namely 4. The mass number of the resulting atom is then $238 - 4 = 234$. From the atomic number of uranium, which is 92, we must subtract that of the α particle, namely 2, which leaves an atomic number of 90. In the periodic chart, we find that the element with atomic number 90 is thorium, Th. Now we write our equation.

$$^{238}_{92}\text{U} \longrightarrow {}^{4}_{2}\text{He} + {}^{234}_{90}\text{Th}$$

EXAMPLE 21.2. Write the equation for the process in which the 210 isotope of lead emits a β particle.

Solution:

Step 1. Write the symbol for the isotope of the element that is undergoing the decay.

$$^{210}_{82}\text{Pb}$$

Step 2. Begin the equation by placing an arrow after the symbol of the decaying isotope. Follow the arrow with the symbol of the particle that is being emitted. (For a β particle, use the symbol for an electron; the mass number of the electron can be taken to be 0 because the electron has a negligibly small mass.

$$^{210}_{82}\text{Pb} \longrightarrow {}^{0}_{-1}\text{e}$$

Step 3. Now write the mass number and atomic number (nuclear charge) of the product isotope. In this case, the mass number is unchanged, $210 - 0 = 210$. The atomic number becomes one unit larger than that of the decaying isotope, because the emitted particle has a negative charge. $82 - (-1) = 82 + 1 = 83$.

$$^{210}_{82}\text{Pb} \longrightarrow {}^{0}_{-1}\text{e} + {}^{210}_{83}$$

Step 4. Finally, look up the name and symbol of the element that has been created. Write the symbol in the equation. In this decay, the product has the atomic number 83, which belongs to bismuth.

$$^{210}_{82}\text{Pb} \longrightarrow {}^{0}_{-1}\text{e} + {}^{210}_{83}\text{Bi}$$

Notice that nuclear equations must be balanced for both the sum of the mass numbers and that of the atomic numbers. In the example, the mass number on the left is 210, and the sum of the mass numbers on the right is $0 + 210 = 210$. The atomic number on the left is 82 and that on the right is $83 + (-1) = 82$. When you finish writing a nuclear equation, always check to be sure it is balanced.

EXERCISE 21.1.   Write the equation for the process in which Bi-199 loses an α particle.

EXERCISE 21.2.   Write the equation for the process by which neutrons create C-14 from N-14. What else is produced?

In radioactive transformations, the isotope that decays is called the **parent isotope** and the product isotope is called the **daughter.** In natural radioactive processes, the daughter itself is often unstable and decays to give yet another daughter. In some cases, the process of decay continues through many different steps until a stable isotope finally results. The decay of U-238 is one such series; the isotope passes through 15 different disintegrations, finally producing a stable isotope of lead, Pb-206. This process is shown in graphic form in Figure 21.7. The various isotopes of the elements involved are positioned according to atomic number in a vertical scale and mass number in a horizontal scale. Study the chart and write the decay equations for the processes that occur. In addition to the series beginning with U-238, there are two other naturally occurring series, one in which Th-232 decays to Pb-208 and one in which U-235 decays to Pb-206.

The isotope listed last in Figure 21.7 is Pb-206, which is stable. All of the Pb-206 in nature comes from the decay of U-238. If it can be assumed that when a rock was formed, there was no Pb-206 in it, it is possible to estimate the age of a rock that contains U-238 by finding the ratio of the amounts of U-238 and Pb-206.

EXAMPLE 21.3.   Analysis of a rock sample shows the following ratio of the number of atoms of U-238 to the number of atoms of Pb-206:

$$\frac{\text{atoms } ^{238}_{12}\text{U}}{\text{atoms } ^{106}_{88}\text{Pb}} = \frac{1}{3}$$

What is the age of the rock?

Solution:

Assuming that all the Pb-206 in the sample came from U-238 originally in the rock, we reason that three fourths of the uranium has decayed into lead. In one half-life, half of the uranium would change to lead, giving a ratio of 1:1. After a second half-life, half the *remaining* uranium would have decayed to lead, leaving only one fourth of the original uranium, the other three fourths now being lead.

The age of the rock is two half-lives of U-238, or 2×4.5 billion years $= 9$ billion years. (No rocks this old have been found on earth. The age of the earth is estimated to be somewhat more than 3 billion years.)

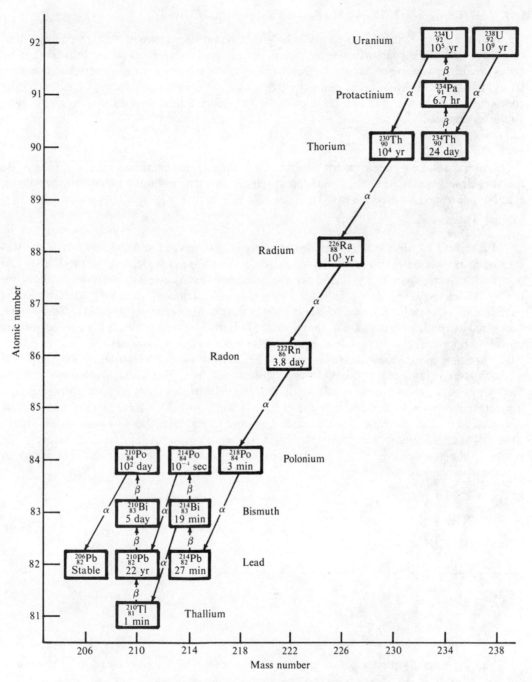

Figure 21.7 Decay series U-238 to Pb-206.

21.7 INDUCED TRANSMUTATION AND ARTIFICIAL RADIOACTIVITY

Nuclear events can be caused artificially by bombarding atoms with high-energy particles. In 1919, eight years after he proposed the nuclear atom, Rutherford showed that he could create an isotope of oxygen by bombarding nitrogen atoms with α particles. It is now thought that the reaction takes place through an unstable isotope of fluorine created by the combination of the nitrogen and the α particle. The fluorine then disintegrates to give oxygen and a fast proton.

$$^{14}_{7}N + {}^{4}_{2}He \longrightarrow {}^{18}_{9}F \longrightarrow {}^{17}_{8}O + {}^{1}_{1}H$$

A useful source of neutrons for experiments is the radium-beryllium source, a tube of Be metal enclosing a sample of Ra-226. Alpha particles from the radium bombard the beryllium, ejecting neutrons and producing C-12 in the process.

$$^{9}_{4}Be + {}^{4}_{2}He \longrightarrow {}^{12}_{6}C + {}^{1}_{0}n$$

Rutherford's artificial transmutation of nitrogen into oxygen and the reaction of the Ra-Be source are examples of reactions accomplished by natural radioactivity, in these cases fast α particles. Now, there are large, expensive machines that accelerate ordinary particles for use in nuclear studies. The celebrated **cyclotron,** an instrument that magnetically directs particles into a spiral track and accelerates them to very high energies, was invented by E. O. Lawrence at the University of California in 1932. Early cyclotrons were able to produce fast protons whose energies were a few times greater than those of most natural nuclear events. There are now more sophisticated machines, the **synchrotron** and **betatron,** for example, that are a thousand times more powerful than the first cyclotron. The **linear accelerator** at Stanford University, a mile-long assembly of complex electronics and vacuum plumbing, can accelerate electrons to nearly the speed of light. (See Figure 21.8.) High-speed accelerators are used not only to bring about nuclear transmutations, but also to knock nuclei apart. Studying the resulting fragments helps give researchers clues to the fundamental nature of the nucleus. The following are some nuclear transmutations that have been brought about by artificially accelerated particles:

$$^{23}_{11}Na + {}^{1}_{1}H \longrightarrow {}^{23}_{12}Mg + {}^{1}_{0}n$$

$$^{209}_{83}Bi + {}^{2}_{1}H \longrightarrow {}^{210}_{84}Po + {}^{1}_{0}n$$

$$^{2}_{1}H + {}^{2}_{1}H \longrightarrow {}^{3}_{1}H + {}^{1}_{1}H$$

EXERCISE 21.3.   Fill in the blank.

$$^{96}_{42}Mo + {}^{2}_{1}H \longrightarrow {}^{1}_{0}n + \,?$$

Many artificially produced isotopes are themselves unstable and radioactive. The production of radioactive isotopes has become a vigorous industry. Such isotopes are used widely in medicine. Co-60, for example, is used to produce γ rays for the bombardment of cancer cells; a radioisotope of iodine, I-131, is used to map the thyroid gland. Radioactive isotopes are used in numerous ways in industry: to make the numerals on digital watches visible in the dark, to

(a)   (b)

Figure 21.8   E. O. Lawrence holds the first cyclotron, which he designed, in his hand (*a*). Contrast this with the mile-long Stanford Accelerator shown in (*b*). (Photo (*a*) courtesy of University of California, Lawrence Berkeley Laboratory; photo (*b*) courtesy of the Stanford Linear Accelerator Center.)

monitor the flow of oil through pipelines, to scan metal castings for flaws, to provide quality control of paper thickness, and so on.

As we have seen, none of the elements beyond bismuth, element 83, has any stable isotope, although some of the long-lived unstable elements still exist in nature (uranium, for instance). However, none of the series Am, Cm, Bk, Cf, Es, Fm, Md, No, Lw, or Element 105 is found in nature. Although the existence of these elements was predicted by the periodic table, none of their atoms then existed on earth. By the late 1960s, however, all had been prepared and identified, even though some have extremely short lifetimes.

The first of these actinides to be synthesized was a neptunium isotope prepared at the University of California by McMillan and Abelson in 1940. The isotope was prepared in a cyclotron by bombarding uranium with high-speed deuterons (H-2).

$$^{238}_{92}\text{U} + ^{2}_{1}\text{H} \longrightarrow ^{239}_{92}\text{U} + ^{1}_{1}\text{H}$$

$$^{239}_{92}\text{U} \xrightarrow{\textit{Natural decay}} ^{239}_{93}\text{Np} + ^{0}_{-1}\text{e}$$

Soon after, plutonium was identified from a similar reaction.

$$^{238}_{92}\text{U} + ^{2}_{1}\text{H} \longrightarrow ^{238}_{93}\text{Np} + 2\,^{1}_{0}\text{n}$$

$$^{238}_{93}\text{Np} \longrightarrow ^{238}_{94}\text{Pu} + ^{0}_{-1}\text{e}$$

The three elements most recently discovered are elements 104, 105, and 106, all prepared by artificial transmutation. Because there is some question about who first discovered them and has the right to name them, the names for these elements are still tentative.

21.8 ENERGY FROM NUCLEAR REACTIONS

The greatest impact of nuclear research has undoubtedly come from the discovery that enormous quantities of energy can be released by nuclear reaction. The availability of such energy, for both useful and destructive purposes, is new to civilization. In only a few decades, nuclear energy has already changed the world and will change it far more in the future.

Although the nucleus is tiny compared with the entire atom, it contains all the positive charge of the atom. Positive charges repel one another, and the closer they are, the greater the force of repulsion. Since protons nevertheless remain close together in nuclei, the forces between particles in a nucleus must be extraordinarily strong. If nuclear particles subject to those forces are to be moved closer together or farther apart, the energy involved (force × distance = work) must be correspondingly large.

Suppose that we dismantle an H-2 nucleus into a proton and a neutron. We first measure the mass of the H-2 nucleus. After we have separated the proton from the neutron, we measure the mass of each. The results of the measurements are:

mass of H-2 nucleus	2.01355 amu
mass of proton	1.00727 amu
mass of neutron	1.00866 amu
total mass: proton + neutron as separate particles	2.01593 amu

The mass of the H-2 nucleus is less than the mass of its parts. What has happened? Where did the extra mass come from?

We can answer that question more easily by asking some other questions. When we separated the proton and neutron, we did work. To make such a separation for 1 mol of H-2 atoms (only 2 g), we would have to do about 2×10^8 kJ of work, roughly the amount of energy required to boil away a large swimming pool of water or melt all the metal in 100 large automobiles. Where did all that energy go? It is stored as potential energy in the now separate protons and neutrons. But where *is* it, actually? What *is* potential energy, anyway?

The answers to these questions involve the conversion of mass to energy and the reverse, processes that obey the celebrated Einstein equation, $E = mc^2$. In this equation, E is the amount of energy in joules equivalent to an amount of mass, m, in kilograms, and c is the speed of light. The constant, c, has the very large value of 3×10^8 m/sec. A small amount of mass, m, is therefore equivalent to a very large energy.

Recall that our proton and neutron had greater mass than did the H-2 nucleus from which they came. *The work we did to separate them was converted to additional mass during the separation process.* In general, it is believed that potential energy of all kinds is stored as mass. If we lift a brick against the attraction of the earth's gravity, the earth and the brick have greater potential energy when separate than when together. The increase in potential energy is stored as increased mass in the brick-earth system. In this case, however, the amount of energy is so small and the mass of the brick (not to mention that of the earth) is so great that the change in mass could never be measured. Because nuclear particles exist under such strong forces, energy changes involving those forces are large enough that the equivalent masses become measurable.

Any nucleus of any element (except that of the single proton of H-1) has a mass less than the combined masses of its protons and neutrons taken as separate particles. This difference in mass is called the **mass defect,** or, when converted to energy units, the **binding energy** of the nucleus. The conversion factor between mass and energy (c^2) can be written in a way that is useful to us. It is simply 1.5×10^{-13} kJ/amu or $9 \times 10^{+13}$ kJ/kg.

The energies released by nuclear reactions are changes in binding energy, corresponding to differences in mass defect between reactants and products.

EXAMPLE 21.4.   One of the nuclear reactions that powers the sun and the stars is

$$\text{}_1^2\text{H} + \text{}_1^2\text{H} \longrightarrow \text{}_2^4\text{He}$$

The mass of an H-2 nucleus is 2.01355 amu and that of a nucleus of He-4 is 4.00260 amu. How much energy is produced when two H-2 nuclei react to form one He-4 nucleus? How much energy is produced when 1 mol of H-2 reacts to form He-4?

Solution:

Step 1.   Find out how much matter is converted to energy; that is, what the change in mass defect is.

$$
\begin{array}{lr}
\text{mass of 2 }_1^2\text{H nuclei}\ 2 \times 2.01355 = & 4.02710 \text{ amu} \\
\text{mass of }_2^4\text{He nucleus} & \underline{4.00260 \text{ amu}} \\
\text{mass defect change} & 0.0245\ \ \text{ amu}
\end{array}
$$

Step 2.   Convert the change in mass defect to energy.

$$0.0245 \text{ amu} \times (1.5 \times 10^{-13} \text{ kJ/amu}) = 3.7 \times 10^{-15} \text{ kJ}$$

To find the energy released when 1 mol of H-2 nuclei reacts to form He-4, we observe that the result we just obtained was for the reaction of two individual nuclei. To find the energy release for 1 mol of H-2 nuclei, we divide by 2 and multiply by N_A.

$$\frac{3.7 \times 10^{-15} \text{ kJ}}{2 \text{ H-2 nuclei}} \times \frac{6.02 \times 10^{23} \text{ H-2 nuclei}}{\text{mol H-2}} = 2.2 \times 10^9 \text{ kJ/mol}$$

Our answer is about ten times greater than the value we saw earlier for the binding energy of a single H-2 nucleus.

EXERCISE 21.4.   The nuclear fusion weapon, the so-called H-bomb, is said to use the following as one of its reactions:

$$\text{}_3^7\text{Li} + \text{}_1^1\text{H} \longrightarrow 2\ \text{}_2^4\text{He}$$

Calculate the energy released when 2 mol of He-4 nuclei is produced by this reaction. Use the value of 7.01546 amu for the Li-7 nucleus and 1.00727 for H-1.. Values for the other mass is given in Example 21.4.

The mass defect taken as an average for all the protons and neutrons in a nucleus (the binding energy per particle) is greatest in the elements of low atomic number, decreases to a minimum with iron, then rises again in elements of higher atomic number. The pattern is shown in Figure 21.9.

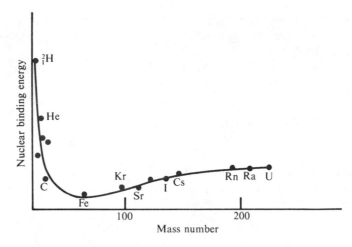

Figure 21.9 Mass defect curve for some of the elements.

The curve tells us that if the particles of nuclei of less mass than that of iron can re-arrange themselves to form larger nuclei, energy can be released. The particles of large nuclei, on the other hand, will release energy upon reacting to form smaller nuclei. These two effects are the bases for fission and fusion reactions.

21.9 NUCLEAR FISSION

The fission reactions of uranium and plutonium show how particles with large nuclei can lose energy by forming smaller nuclei with a smaller average mass defect. Two isotopes of uranium, U-235 and U-233, and one of plutonium, Pu-239, are fissionable; that is, they can be bombarded with slow neutrons and made to disintegrate into smaller nuclei. The reaction that follows is one of the several ways in which uranium can undergo fission.

$$\ce{^{235}_{92}U} + \ce{^1_0n} \longrightarrow \gamma + 2\ce{^1_0n} + \ce{^{141}_{55}Cs} + \ce{^{93}_{37}Rb}$$

Counting the decay energies of radioactive products as well as the kinetic energy imparted to the neutrons and the new nuclei, this fission reaction yields about 15 billion kJ/mol, or about 75 times as much energy as that required to separate 1 mol of H-2 nuclei into protons and neutrons.

Just as dropping several identical teacups on a concrete floor will result in several break-age patterns and give fragments of many sizes, so do uranium nuclei break in different ways. The fission of 1 mol of U-235 results in a mixture of various isotopes of several elements, many of which are radioactive.

Of the three fissionable isotopes U-235, U-233, and Pu-234, only U-235 occurs naturally. It has a geonormal abundance of about 1% in naturally occurring uranium, but can be separated only at great expense and by means of sophisticated methods. Both U-233 and Pu-239 are produced artificially, plutonium in particular being formed in a two-step reaction from U-238, which is easily available in large quantities.

$$^{238}_{92}\text{U} + ^{1}_{0}\text{n} \longrightarrow ^{239}_{92}\text{U}$$

$$^{239}_{92}\text{U} \longrightarrow ^{239}_{93}\text{Np} + ^{0}_{-1}\text{e}$$

$$^{239}_{93}\text{Np} \longrightarrow ^{239}_{94}\text{Pu} + ^{0}_{-1}\text{e}$$

Notice that the fission of each U-235 nucleus produces two neutrons, although only one neutron was required to cause the reaction. Under proper circumstances, each of the released neutrons can cause the fission of another uranium nucleus, which in turn will release two more neutrons, which will cause more fission, and so on. A fission reaction that follows this pattern is self-propagating, and is called a **chain reaction** (Figure 21.10). Chain reactions of U-235 and Pu-239 were used in the first atomic bombs.

EXERCISE 21.5.   Fill in the blank.

$$^{240}_{94}\text{Pu} \longrightarrow ^{135}_{51}\text{Sb} + ? + 2\,^{1}_{0}\text{n}$$

The size and shape of a sample of fissionable metal strongly affect the probability of a nuclear chain reaction. In a small sample of metal, most of the neutrons escape through the

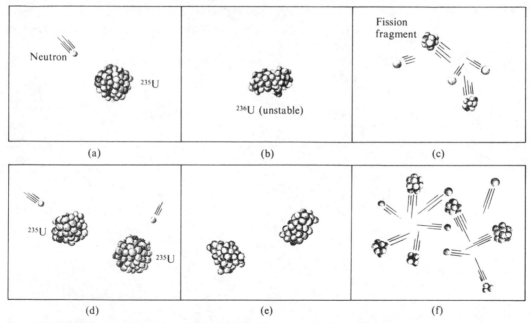

Figure 21.10   Fission and chain reaction of $^{235}_{92}\text{U}$. (*a*) A neutron strikes a U-235 atom, creating (*b*) an unstable nucleus that soon (*c*) disintegrates (fissions), giving three neutrons and smaller nuclei. (*d*) Neutrons emitted (in *c*) strike other U-235 atoms (*e*) creating unstable atoms that then (*f*) themselves fission, yielding more neutrons to continue the reaction. Each fission produces two or more fission fragments and three neutrons. Most of the emitted neutrons are then usually captured by other $^{235}_{92}\text{U}$ nuclei, continuing the chain reaction.

surfaces of the sample before they have a chance to cause further fission; only samples of a size larger than what is called the **critical mass** can sustain a chain reaction and explode. One of the first atomic bombs was made to explode when two subcritical pieces of U-235 were brought together; one was fired from a small cannon into the other. A small fission bomb begins with only a few kilograms of fissionable metal.

In addition to the frightful destruction they cause instantly at the point of impact, fission bombs also create serious hazards that last for centuries and are felt at great distances. Many fission fragments are, or produce, long-lived radioactive isotopes of biologically important elements. The dangerous isotopes that result from nuclear bombs are called fallout, and the list of dangerous fallout isotopes is a long one. Some of these, which are either primary fission products or the daughters of such products, are listed in Table 21.2.

21.10 NUCLEAR FUSION

If they are brought together in a strong collision, certain light nuclei can be made to combine, forming heavier nuclei and releasing energy in the process. As we have seen, 1 mol of H-2 reacting to give He-4 releases about 2 billion kJ.

The combination of two light nuclei to produce a larger nucleus is called nuclear fusion. Fusion reactions occur naturally in stars, in our sun for example, and can be generated artificially, as in the hydrogen bomb. The reactions can take place, however, only at extremely high temperatures because the colliding nuclei are mutually repelled by their positive charges and must be travelling at exceptionally high speeds in order to collide hard enough to fuse. In other words, the activation energy of the reaction is extremely high. Successful, small-scale nuclear fusion experiments have nevertheless been run in laboratories, and huge sums are

Table 21.2
Dangerous Isotopes Released by Fission Bombs

Isotope	Emission	Half-life	Characteristics
Strontium-90	β	27.7 years	• Collects in bones as a substitute for calcium • Found in plants and milk • Causes bone cancer
Carbon-14	β	5,730 years	• Becomes a part of any kind of organic molecule in the body, primarily proteins and fats
Iodine-131	β, γ	8 days	• Collects in the thyroid gland • Because of its short life, I-131 has intense radioactivity • Is a major source of dangerous fallout
Plutonium-239	α, γ	24,390 years	• One of the deadliest substances known • Collects in bones where its emission does enormous damage

presently being spent in the effort to learn how to run larger-scale reactions in a safe, controlled way.

EXERCISE 21.6.   Fill in the blank.

$$2 \, {}_2^3\text{He} \longrightarrow {}_2^4\text{He} + ?$$

21.11 NUCLEAR POWER

Not all nuclear chain reactions result in explosions. In an apparatus called a **nuclear reactor,** fission reactions can be controlled and made to run slowly. The first nuclear reactor (Figure 21.11) was secretly constructed in an unused stadium at the University of Chicago, during the research that led to the first nuclear weapon.

The reactor is an assembly of **fuel elements,** rods or cylinders of metal containing small proportions of fissionable isotopes, often uranium. Because ordinary U-238 is not fissionable, the uranium used for the fuel elements must be enriched; that is, the proportion of fissionable U-235 must be increased from the natural 0.7% to about 4% before the reactor can sustain a chain reaction. The enriched fuel elements are separated from one another by the moderator, a material that slows the fast neutrons emitted by fissioning atoms to a speed that will most efficiently cause fission in other atoms. The reactor is controlled by rods made from a neutron-absorbing material, such as cadmium. To achieve critical mass under controlled circumstances and with highly dilute fissionable fuel requires careful attention to the design of the reactor.

Figure 21.11   A drawing of the first nuclear reactor. (Courtesy of the U.S. Department of Energy.)

Reactors are operated for many different purposes. Although they are most often used to produce energy, they also produce artificial isotopes for medicine and industry as well as supplies of fissionable elements, such as Pu-239.

Nuclear power plants are reactors whose fission energy is captured and converted to electric power. In such a reactor, the energetic particles produced by fission lose their energy through collisions with atoms in the reactor core, thereby producing heat. The heat is carried away from the reactor by the coolant, usually water, which is turned into steam as it passes through the reactor. Turbines and generators convert the energy of the steam to electric power.

Nuclear power plants have several advantages. They use no fossil fuel, and in ordinary operation produce no gases, ashes, or smoke to pollute the atmosphere. With the increased price of oil, the fissionable isotopes that run the reactors cost less than an equivalent amount of oil. But there are also serious disadvantages. Although the design of nuclear power plants is such that there is only a remote chance of serious accident, such an accident could be disastrous. While a nuclear power plant cannot physically explode like an atomic bomb, it is possible for the reactor to get out of control and damage or even melt down its fuel elements. If the contents of a large reactor were dumped into the environment, many deaths would result, and large sections of the surrounding country would be unlivable for decades. The contamination from a catastrophic accident would release large amounts of isotopes similar to those in fission bomb fallout. Moreover, if the reactor were fueled with plutonium, large amounts of this extremely dangerous element would also be released. Although the safety record of nuclear reactors is generally excellent (even considering Three Mile Island, an accident in which no one was seriously injured), there is always the possibility that sabotage or acts of war could bring about a catastrophe.

Less immediate, but nonetheless important, is the fact that no successful solution has yet been proposed to the problem of what to do with **nuclear waste.** Because fission products are highly radioactive and long-lived, they must be removed from the environment and stored safely for hundreds of years in such a way that allows no possibility of their ever being released. Although nuclear power plants have been in wide use for little more than two decades, there is already enough nuclear waste to pose a serious threat to life on earth. Much of this radioactive material is stored in liquid form in tanks, some of which have leaked, depositing their contents in the earth where they then come in contact with groundwater. The only sure way to rid our immediate environment permanently of this dangerous material is to launch it into outer space, and even then there are serious risks. It is frightening to think about what would happen if a large rocket containing nuclear waste were to malfunction and fall back to earth.

Despite its disadvantages, the production of nuclear power is growing rapidly in many parts of the world, even though it has been slowed in the United States. In an energy-hungry world, there seems to be no other way to produce power while we wait for something better to be developed. Nuclear fusion power plants, for example, would have few risks and would produce no long-lived, dangerous wastes. (The principal radioactive byproduct, H-3, has a half-life of only about 12 years, versus thousands of years for some fission wastes.)

More controversial than ordinary nuclear power plants are *breeder* reactors. A properly designed reactor will, by nuclear transmutation, make more nuclear fuel than it burns, mostly by producing Pu-239. Because of the toxicity of plutonium, and especially because it can easily be used to make crude atomic weapons, the proposed use of breeder reactors in power plants has created a storm of argument. Whatever the decision of the United States, many of whose citizens are concerned about the environment and about the proliferation of nuclear weapons, breeder power reactors will surely be built and operated by other countries.

For the first time, nuclear technology has brought to the world the undesirable necessity of making life-or-death decisions about a legacy of danger that may affect people yet unborn for hundreds of generations to come.

21.12 SUMMARY

The study of radioactive isotopes is not even a hundred years old, yet the use of unstable isotopes has attracted more attention and concern than perhaps any other aspect of science.

While some combinations of protons and neutrons form nuclei that last indefinitely, others form unstable nuclei. The last few elements in the periodic table have no stable isotopes whatever. When the nuclei of unstable isotopes disintegrate, they usually emit some form of radioactivity. The common forms are α, β, and γ emission, the first two of which are high-energy particles. Gamma rays are highly energetic photons.

The relative stability of an isotope is measured by its half-life, the period during which half the unstable nuclei in a given sample disintegrate. Half-times range from extremely small fractions of a second to thousands of years. Simple calculations can be used to relate the intensity of radioactive emission to half-life. The half-lives of certain isotopes can be used to determine the ages of rocks and of organic matter.

The energies produced by nuclear events are uncommonly large, and the availability and harnessing of such energies has presented society with grave problems. What to do about the proliferation of nuclear weapons or, for that matter, about their very existence, is perhaps the most difficult question that we have ever had to face. We also have to choose among several undesirable alternatives for supplying the energy our society will need in the coming decades. Although nuclear power can furnish a substantial part of that energy (eventually, perhaps most of it), nuclear power also brings enormous disadvantages, chief among which are the chances of disastrous accidents and the difficulty of disposing of dangerous, long-lived fission wastes.

QUESTIONS

Section 21.1

1. What are the differences between α, β, and γ rays?
2. Why are α and β rays affected by magnets, whereas γ rays are not?
3. Why do some kinds of radioactivity penetrate matter to a greater depth than others? Name two factors that influence penetration.

Section 21.2

4. What is the difference between a stable isotope and an unstable isotope?
5. Why might the heavy elements (those with large atomic number) usually have more isotopes than the light elements?

Section 21.3

6. What is the definition of half-life? Explain why half-life is a particularly convenient language to use in characterizing radioactive decay rates.

Section 21.4

7. Explain how radiocarbon dating works. What is the source of the C-14 used in the method?

Section 21.5

8. What does a Geiger counter do? How does it work?
9. Name and briefly explain two other ways to detect radioactive radiation.

Section 21.6

10. Why is it best to use the full description of an isotope, such as $^{238}_{92}U$, when writing equations for nuclear reactions?
11. What is a radioactive decay series?
12. Explain how rocks can be dated using the U-238/Pb-206 ratio. On what assumption is this method based?
13. The earth is more than 3 billion years old. Radium-226 has a half-life of only 1600 years, yet it can be found in nature. How can that be?

Section 21.7

14. Explain what the process of artificial transmutation involves. Why can we change elements to other elements, although early alchemists could not?
15. Is it possible and practical to produce gold by artificial transmutation? Explain the pros and cons of your answer.
16. Write balanced nuclear reactions for the preparation of three artificial isotopes.

Section 21.8

17. Why do the particles of both large nuclei and smaller nuclei gain stability by forming nuclei of moderate sizes?
18. What is the relation between binding energy and mass defect?
19. Where did the name *mass defect* come from?

Sections 21.9 and 21.10

20. Explain the differences between fission and fusion reactions.
21. Why is it necessary to use a fission reaction to set off a fusion reaction on earth?
22. It is said that H-bombs are "cleaner" than fission weapons. Explain.

MORE DIFFICULT QUESTION

23. Find a second answer to Question 13.

PROBLEMS

Section 21.3

*1. A sample of an element is known to contain 2.5×10^7 atoms of one of its isotopes that is a γ emitter with a half-life of 7.2 hr. How many atoms of the isotope will remain in the sample 36 hr later?

Section 21.4

*2. A sample of pure carbon weighing 0.311 g is prepared from a fragment of mummified tissue. When the β activity is measured, it is found that the sample emits 2.5 counts per minute. What is the age of the tissue?

Section 21.6

*3. Write equations for the following processes:
   *a. $^{23}_{10}Ne$ loses a β particle   c. $^{227}_{93}Np$ emits an α particle
    b. ^3_1H emits a β particle   d. $^{30}_{15}P$ loses a positron

4. Write equations for the following processes:
    a. $^{239}_{94}Pu$ loses an α particle   c. $^{32}_{15}P$ emits a β particle
    b. $^{35}_{16}S$ loses a β particle   d. $^{22}_{11}Na$ loses a positron

*5. Complete the equations for the following nuclear reactions:
   *a. $^{13}_6C + ? \longrightarrow {}^{14}_6C$ e. $^{137}_{53}I \longrightarrow {}^1_0n + ?$
    b. $^{59}_{26}Fe \longrightarrow {}^{0}_{-1}e + ?$ f. $^{26}_{13}Al \longrightarrow {}^{25}_{12}Mg + ?$
    c. $^{40}_{19}K \longrightarrow {}^{0}_{-1}e + ?$ g. $? \longrightarrow {}^{24}_{12}Mg + {}^{0}_{-1}e$
    d. $^{246}_{96}Cm + {}^{12}_6C \longrightarrow 4\,{}^1_0n + ?$

6. Complete the equations for the following nuclear reactions:
    a. $? \longrightarrow {}^{0}_{-1}e + {}^{87}_{37}Kr$ e. $^{27}_{13}Al + ? \longrightarrow {}^{25}_{12}Mg + {}^4_2He$
    b. $^{27}_{13}Al + ? \longrightarrow {}^{30}_{15}P + {}^1_0n$ f. $^{40}_{20}Ca + {}^1_0n \longrightarrow ? + {}^4_2He$
    c. $? + {}^2_1H \longrightarrow {}^1_0n + {}^{97}_{43}Tc$ g. $^{104}_{47}Ag \longrightarrow ? + {}^{0}_{-1}e$
    d. $^{32}_{16}S + {}^1_0n \longrightarrow {}^1_1H + ?$

7. Complete the equations for the following nuclear reactions:
    a. $? \longrightarrow {}^{227}_{89}Ac + {}^4_2He$ c. $^{250}_{98}Cf + ? \longrightarrow 4\,{}^1_0n + {}^{257}_{103}Lr$
    b. $^{53}_{24}Cr + {}^4_2He \longrightarrow ? + {}^{56}_{26}Fe$ d. $^{231}_{91}Pa \longrightarrow {}^{227}_{89}Ac + ?$

*8. Pu-241 decays to Bi-209. How many α and β particles are emitted during this process?

9. U-235 decays to Pb-207 via the following successive emissions:

   $\alpha\ \beta\ \alpha\ \beta\ \alpha\ \alpha\ \alpha\ \alpha\ \beta\ \beta\ \alpha$

   Write symbols for each isotope formed during this process.

10. You are starting with an unknown element $^{300}_{118}G$ and constructing part of a table like Table 21.2. Using the emission or capture of any of the following particles, $^{0}_{-1}e$, ^4_2He, ^1_1H, ^1_0n, and $^{0}_{+1}e$, write nuclear reaction equations to fill as many squares as possible nearby your unknown isotope. Make a diagram of your equations similar to the diagrams used in Figure 21.7.

*11. Strontium-90 and iodine-131 are two dangerous waste products in the fallout of fission bombs. Strontium-90 is formed from the primary fission product bromine-90, while I-131 comes from another primary fission product via three successive reactions in which three β particles are emitted. Write the equation for the formation of Sr-90. What is the original fission product from which I-131 comes?

12. If fluorine-19 could be caused to fission into boron-10 and beryllium-9, would this process release energy? Calculate any energy change that would occur. Masses in *amu* are: F-19, 18.99840; B-10, 10.0129; Be-9, 9.01218.

Appendix A

Mathematics Review

This appendix reviews the mathematics you will need to understand the concepts and to work the problems in the text. You will have to perform quickly and confidently the operations of simple arithmetic, manipulate fractions, and set up and solve simple algebraic relations.

Not everyone who is beginning chemistry has had practice in the quantitative treatment of data. For such individuals, I have included sections on positive and negative numbers, percentages, significant figures, logarithms, exponents, and scientific notation. Familiarity with these topics is a prerequisite to understanding of some of the concepts presented in the text.

AFTER STUDYING THIS APPENDIX, YOU WILL BE ABLE TO

- perform the basic operations of arithmetic (If your arithmetic is rusty, brush it up. You will want to use it without having to stop and think.)
- use exponential numbers, including fractional and negative exponents
- express numerical values in the decimal system, in percentage language or exponentially, and convert easily one of these forms to another
- perform calculations with numbers in any of these forms
- use negative numbers confidently
- distinguish between relative and absolute measurements
- express ideas in the form of algebraic equations, using symbols to represent known and unknown quantities
- solve simple algebraic equations and systems of equations
- use logarithms in calculations and as a part of scientific language

TERMS TO KNOW

common denominator	linear equation	percent
constant	log	positive
denominator	logarithm	power
division	multiplication	root
exponent	negative	significant figures
improper fraction	numerator	variable

A.1 ARITHMETIC

By this time, you have already learned arithmetic. It is important, however, to look again at a few arithmetical operations and see how these fit into science.

A.1.1 Basic Operations

Multiplication is the addition of a number to itself for a specified number of times. **Division,** the reverse process, is used to find out how many times a particular number is contained in another number. Multiplying 5 by 6 means that 5 is added to itself six times to give 30. In dividing 30 by 6, we find out that 30 is made up of five 6's. The signs $\times$ for multiplication and $\div$ for division are familiar to you, but in science, other methods are also often employed to indicate these operations. A dot $\cdot$ is often used for multiplication or sometimes no sign at all appears, in which case, the things to be multiplied are simply enclosed in parentheses ( ) and placed together. Here are examples of multiplication:

$$7 \times 3 = 7 + 7 + 7 = 21$$

Since 7×3 is the same as 3×7,

$$3 + 3 + 3 + 3 + 3 + 3 + 3 = 21$$

The first multiplication could also be shown as

$$7 \cdot 3 = 21$$

or

$$(7)(3) = 21$$

Expressions using the equal sign are called equations. An equation says that whatever appears to the left of the equal sign has the same value as whatever appears to the right. Similar kinds of statements can indicate that two sides of a relation have approximately the same value. Three signs indicate approximation: $\sim$, $\approx$, or $\simeq$. You should become accustomed to reading an equation just as you read a sentence written in words. In an equation, the equal sign is the verb.

In division, the numbers to be divided are often shown as a fraction, where the number above the line is divided by the number below the line.

$$7\overline{)42} \quad \text{is the same as } 42 \div 7 = 6$$

which is the same as $\dfrac{42}{7} = 6$

Each of these forms says that 42 can be divided into seven equal parts of 6 each. (Multiplying 7 by 6, or adding 7 to itself six times, gives 42.)

Sometimes, in word problems, the word *per* is used to indicate division. For example, if there are six students and 42 apples, how many apples are there per student?

$$\frac{42 \text{ apples}}{6 \text{ students}} = 7 \text{ apples per student}$$

Division is often shown by slanted lines instead of horizontal ones.

$$\frac{42 \text{ apples}}{6 \text{ students}} \qquad \text{is shown as} \qquad 42 \text{ apples}/6 \text{ students}$$

Notice that simply because they are in the equation, the words *apple* and *student* are also treated like numbers. Words or abbreviations used this way are called units. The use of units is discussed in Chapter 1.

You will occasionally be instructed to invert a number or to take its reciprocal. These mean the same thing, namely to divide that number into 1, to construct a fraction in which the number to be inverted appears below the line.

Instruction:	Invert 2.	Answer:	1/2.
Instruction:	Invert 372.	Answer:	1/372.
Instruction:	Take the reciprocal of 372.	Answer:	1/372.

If what is to be inverted is already a fraction, simply turn it upside down.

Instruction: Invert 3/4.   Answer: 4/3.

Multiplication can be performed after inversion.

Instruction: Invert 7 and multiply by 3.   Answer: $(1/7) \times 3$ or 3/7.

In this example, we divided 3 by 7. Division itself is just the process of inverting and multiplying.

Instruction: Divide 39 by 13.
Solution: 1.   Invert 13 to give 1/13.
　　　　　　　2.   Multiply 39 by the result. $39 \times (1/13) = 39/13 = 3$

A.1.2 Working With Fractions

As we have seen, a fraction is a symbol showing that one thing is to be divided by another. For instance, the fraction $\frac{2}{3}$, often shown as 2/3, says the same thing as $2 \div 3$, and indicates that 2 is to be divided by 3. What does this mean? As an example, cut an apple into three equal pieces and pick up two of them. You have in your hand 2/3 of the apple. You divided the apple by three, then "multiplied" the result (which was 1/3 apple) by two. In a fraction, the number above the line is called the **numerator** and the number below the line the **denominator.** Numerators indicate multiplication and denominators indicate division. Another example may help clarify that. Suppose you want to "take" 2/3 of 6, that is, to multiply 6 by the fraction 2/3. To do this, you can multiply 6×2, giving 12, then divide 12 by 3, giving 4. Numerators multiply; denominators divide.

You can multiply fractions together by multiplying the numerators together and the denominators together.

$$\frac{3}{2} \times \frac{4}{5} = \frac{3 \times 4}{2 \times 5} = \frac{12}{10}$$

As we have seen, fractions are often shown on a single line. The problem we just looked at would often be written as

$$(3/2)(4/5) = (3)(4)/(2)(5) = 12/10$$

This practice need not confuse you. You will see both forms used, sometimes even in combination.

If one fraction is to be divided by another, the operation itself may be shown as a fraction.

$$(3/2) \div (4/5) = \frac{3/2}{4/5} = \frac{15}{8}$$

(When dividing fractions by fractions, it is handy to use both the / sign and the — sign.) To solve a problem like this one, simply invert (turn upside down) the denominator and multiply it by the numerator:

$$\frac{3/2}{4/5} = (3/2)(5/4) = 15/8$$

Many of the fractions we have seen have numerators larger than their denominators. Such a fraction has a value (or can be worked out to be) greater than unity (1) and is sometimes called an **improper fraction.**

$$\frac{15}{8} = 15 \div 8 = 1 + \frac{7}{8}$$

Improper fractions often come up in scientific work. No matter what you have learned, there is nothing wrong with "improper" fractions, and they are sometimes conveniently used as is. Since it is not usually necessary to divide them out, don't bother unless you have to.

Adding and subtracting fractions is usually harder than multiplying them, because fractions must have the same denominator, or a **common denominator,** before they can be added or subtracted. An example will make this easy to understand. Suppose you want to add 5/2 to 7/2. Because it makes no difference in what order you perform the operations (add then divide, or divide then add) you could add the 5 to the 7, then divide the result by 2. But this will not work unless the two numerators are divided by the same numbers. (If the denominators were different, which would you divide by?)

EXAMPLE A.1.   Add 1/4 + 1/2 + 5/6.

Solution:

Step 1.   Convert all the fractions to a common denominator. By examining them, we find that all the denominators can be divided evenly into 12; that is, each can be converted to 12 if it is multiplied by some whole number.

Step 2.   Multiply the numerator and denominator of each fraction by the number that will convert the denominator of that particular fraction to the selected common denominator, in this case 12. (Multiplying anything by a fraction that has the same numerator and denominator leaves it unchanged, because a fraction with the same figure on the top and bottom is simply equal to one.)

$$\frac{1}{4} \times \frac{3}{3} = \frac{3}{12} \qquad \frac{1}{2} \times \frac{6}{6} = \frac{6}{12} \qquad \frac{5}{6} \times \frac{2}{2} = \frac{10}{12}$$

Step 3.   Now that the fractions all have 12 as a denominator, add their numerators.

$$\frac{1}{4} + \frac{1}{2} + \frac{5}{6} = \frac{3}{12} + \frac{6}{12} + \frac{10}{12} = \frac{(3 + 6 + 10)}{12} = \frac{26}{12} \text{ or } \frac{13}{6}$$

EXERCISE A.1.   Add 1/2 + 7/12 + 5/8.

EXAMPLE A.2.   Subtract 1/7 from 1/2.

Solution:

Step 1.   Find the common denominator.

$$\frac{7}{14} - \frac{2}{14}$$

Step 2.   Subtract.

$$\frac{7}{14} - \frac{2}{14} = \frac{5}{14}$$

EXERCISE A.2.   Subtract 1/9 from 1/4.

A.1.3 Decimal System

In science, the decimal system is often used to express fractions. In the decimal system, the denominator is some power of ten. (A *power of ten* is ten multiplied by itself some number of times. See Section A.5 for a discussion of powers of numbers.) In decimal fractions, the denominator is not written out as a number, but is indicated by the position of the decimal point, a period shown with the numerator. Here are some ordinary fractions converted to decimals:

$$1/10 = 0.1 \qquad 1/100 = 0.01 \qquad 1/1000 = 0.001$$
$$3/100 = 0.03 \qquad 22/1000 = 0.022$$

If the denominator is already a power of ten, changing an ordinary fraction into a decimal fraction requires only one simple step. Simply write down the numerator, then place the decimal point by starting at the right side of the numerator and counting left as many places as there are powers of ten in the denominator; then put the decimal point there.

EXAMPLE A.3.   Convert 33/100 to a decimal fraction.

Solution:

$$\frac{33}{100} = \frac{33}{10^2} = 0.33$$ Start to the right of the final 3,
and count two places to the left.

EXAMPLE A.4.   Convert 4267/1000 to a decimal fraction.

Solution:

$$\frac{4267}{1000} = \frac{4267}{10^3} = 4.267$$ Move the decimal three places.

EXERCISE A.3.   Convert 225/100 to a decimal fraction.

Sometimes leading zeroes must be added to give the proper number of places.

EXAMPLE A.5.   Convert 23/10,000 to a decimal fraction.

Solution:

$$\frac{23}{10,000} = \frac{23}{10^4} = 0.0023$$

EXERCISE A.4.   Convert 37/1000 to a decimal fraction.

To convert fractions to decimals when the denominators are not integer powers of ten, first multiply both numerator and denominator by a number that will make the denominator a power of ten. Then place the decimal in the usual way.

EXAMPLE A.6.   Convert 3/4 to a decimal fraction.

Solution:

Step 1.   Multiply numerator and denominator by a number that will make the denominator a power of ten.

$$\frac{3}{4} = \frac{3 \times 25}{4 \times 25} = \frac{75}{100}$$

Step 2.   Make the conversion in the usual way.

$$\frac{75}{100} = 0.75$$

EXERCISE A.5.   Convert 3/5 to a decimal fraction.

If the denominator cannot be multiplied by a whole number to get a power of ten, a decimal number can be used.

EXAMPLE A.7.   Convert 200/300 to a decimal fraction.

Solution:

$$\frac{200}{300} = \frac{200 \times 3.33}{300 \times 3.33} = \frac{666}{1000} = 0.666$$

Notice that 300×3.33 is actually 999, not 1000. The question of how precise to be in such cases is discussed in Section A.9.

EXERCISE A.6.   Convert 313/500 to a decimal fraction.

Decimal notation can also be used to show numbers greater than unity.

$$\frac{7}{2} = \frac{7 \times 5}{2 \times 5} = \frac{35}{10} = 3.5$$

$$\left(\frac{143}{5}\right) - \left(\frac{143 \times 20}{5 \times 20}\right) = \frac{2860}{100} = 28.60$$

To add and subtract decimal numbers, line up the decimal points one above another, then add in the ordinary way.

EXAMPLE A.8.   Add $3.2515 + 87.11 + 2001.772 + 5.31$.

Solution:

$$
\begin{array}{r}
3.2515 \\
87.11 \\
2001.772 \\
5.31 \\
\hline
2097.4435
\end{array}
$$

(This number should now be rounded off to 2097.44. The reason for this is discussed in Section A.9.)

EXAMPLE A.9.   Subtract 51.120 from 772.53.

Solution:

$$
\begin{array}{r}
772.53 \\
-\ 51.120 \\
\hline
721.41
\end{array}
$$

EXERCISE A.7.   Add 44.1212 + 3.721 + 0.011 + 325.4445.

To multiply decimal numbers, (1) first multiply as though the numbers were not decimals, and then (2) count the total number of digits to the right of the decimal point in all the numbers that have been multiplied together. The answer should have that number of digits to the right of its decimal. (Do not forget zeroes when multiplying.)

EXAMPLE A.10.   Multiply 2.513 × 4.30.

Solution:

$$2.513 \ \times \ 4.30 \ = 10.80590$$

3 digits + 2 digits = 5 digits

EXERCISE A.8.   Multiply 22.414 × 6.02.

To divide decimal numbers, (1) first write the operation as a fraction, (2) then multiply the top and bottom by a number that will make the denominator a whole number (not a decimal fraction), and (3) divide in the usual way, placing the decimal point directly above the place in which it appears in the number being divided.

EXAMPLE A.11.   Divide 135.921 by 3.71.

Solution:

Step 1.   Write the operation as a fraction and make the denominator a whole number.

$$\frac{135.921}{3.71} = \frac{135.921 \times 100}{3.71 \times 100} = \frac{13592.1}{371}$$

Step 2.   Divide.

```
          36.6
371) 13592.1
     1113
     ‾‾‾‾
     2462
     2226
     ‾‾‾‾
      2361
      2226
      ‾‾‾‾
      1450
```

The answer is 36.6 (again rounded for the reasons discussed in Section A.9).

EXERCISE A.9.   Divide 62.4 by 1.3.

Although your calculator will do these operations for you, you still need to understand what the calculator is doing. In the section on scientific notation, we will see another way to multiply and divide decimal numbers.

A.2 PERCENT

Closely related to the decimal system is the practice of writing fractions in terms of percent, fractions whose denominators are always 100. In fractions written this way, the denominator is not stated, but is understood always to be 100.

In each of the following examples, the steps are the same. First, form a fraction that has 100 as a denominator. Then discard the denominator and place a percent sign, %, after the numerator.

EXAMPLE A.12.   Change 5/10 to percent.

Solution:

Step 1.

$$\frac{5}{10} = \frac{5}{10} \times \frac{10}{10} = \frac{50}{100}$$

Step 2.   Discard the denominator and place a % after the numerator, which gives 50%.

EXAMPLE A.13.   Change 0.00072 to percent.

Solution:

Step 1.

$$0.00072 = \frac{0.00072 \times 100}{100} = \frac{0.072}{100}$$

Step 2.   Now change this to 0.072%.

EXAMPLE A.14.   Change 3.00/7.00 to percent.

Solution:

Step 1.

$$\frac{3.00}{7.00} = \frac{3.00}{7.00} \times \frac{14.3}{14.3} = \frac{3.00 \times 14.3}{100} = \frac{42.9}{100}$$

Step 2.

$$\frac{42.9}{100} = 42.9\%$$

In Example A.14, we first had to find out what number both the numerator and denominator were to be multiplied by. This was done by dividing 100 by 7.

EXERCISE A.10. Change each of the following to percent: 3/10, 0.22, 0.009, 2.0/9.

We have learned the fundamental operations needed to calculate percent using arithmetic. Now that we understand these, let's see how to take shortcuts. If you are using a calculator and you want to change an ordinary fraction to a percent, (1) divide the numerator by the denominator, producing a decimal fraction, (2) now convert this decimal fraction to a percent by moving the decimal two places to the right and adding a % sign. (Moving the decimal gives a fraction with a denominator of 100. This is discarded when the % is added.)

EXAMPLE A.15. Using a calculator, convert 3.00/7.00 to percent.

Solution:

Step 1.

$$\frac{3.00}{7.00} = 0.429$$

Move decimal to the right, giving

$$\frac{42.9}{100}$$

Do the last part of Step 1 in your head.

Step 2. Convert the decimal fraction to percent by discarding denominator and adding %.

The answer is 42.9%.

EXAMPLE A.16. Change 12.8/90 to percent.

Solution:

$$\frac{12.8}{90} = 0.142 = 14.2\%$$

EXERCISE A.11. Convert each of the following to percent: 42.0/59.0, 33.1/52.7.

We will express fractions in terms of percent, especially when we are discussing experimental error.

A.3 WORKING WITH POSITIVE AND NEGATIVE NUMBERS

We are all accustomed to using positive numbers to make calculations of various kinds, but negative numbers are less familiar. Because negative numbers are essential to scientific work, we will explore them carefully.

To appreciate the importance of negative numbers (as well as positive ones) in science, we need to understand that science often depends on experimental measurements involving numerical values taken in relation to a fixed point. Let's look into this a little.

Suppose we are asked to measure the height of a stack of books on a table. Our first question should be, the height in relation to what? The tabletop? The ground? Sea level? The center of the earth? The center of the solar system? When we say how high, we either state or imply a reference point; there is no such thing as absolute height. Experimental determination of height is an example of a measurement that can give only a relation between two or more things. (On the other hand, there is also such a thing as an *absolute* experimental result. Count the fingers on one hand. You've just made an absolute determination. You didn't measure the number of fingers in relation to some other hand, you simply counted them.)

Let's go on with our height example. We choose the tabletop as our reference point. Next, by measuring, we find that the top of our stack is 1.03 ft above the tabletop, and we say that its height is **positive** 1.03 ft. (Or just 1.03 ft. If a number is given no sign, it is assumed to be positive.) In saying this, we have defined the word *positive* as meaning "in the up direction."

Attached to the side of the table is a shelf. We may want to put the books on the shelf, so we decide to measure its height, *using our same reference point, the tabletop*. With our measuring tape, we find that the shelf is 2.65 ft *below* the tabletop. Because the shelf is at a height *less* than that of the tabletop, we need some way of indicating this fact. We could, of course, simply say that the shelf is 2.65 ft below the tabletop, or we could say that the height of the shelf is **negative** 2.65 ft. If we do the latter, we will have used a negative number to indicate, from our reference point, a direction opposite to the direction we have already defined as positive. Although they may not be used often for shelves, negative numbers are essential to science.

To help you visualize arithmetic using negative numbers, let's look at the **number line,** on which the positive and negative numbers are arranged in opposite directions, starting from zero.

$$-10 \quad -9 \quad -8 \quad -7 \quad -6 \quad -5 \quad -4 \quad -3 \quad -2 \quad -1 \quad 0 \quad 1 \quad 2 \quad 3 \quad 4 \quad 5 \quad 6 \quad 7 \quad 8 \quad 9 \quad 10$$

In our illustration, distances below tabletop would be described by the numbers to the left of zero and distances above by those on the right.

Now, for numbers of the same sign, *adding* one number to another is a movement *away* from zero along our number line. *Subtracting* is movement *toward* zero. To add 3 to 2, we start at 2 and move three numbers to the right (away from zero), ending at 5. To subtract 4 from 7, we start at 7 and move four spaces to the left (toward zero), which takes us to 3.

On the negative side of the number line, the process is similar. To add -3 to -5, we begin at -5 and move three spaces left, to -8. (Notice that movement to the *left* now is *away* from zero.) To subtract -3 from -5, we start at -5 and move three numbers to the right, which gives us -2.

Let us now add and subtract numbers of different signs. Remember that the direction of addition for positive numbers (movement to the right) is the same as that for subtraction of negative numbers, and vice versa. This means that *adding* a *negative number* to a positive number is the same as *subtracting* that amount from the positive number. Similarly, adding a *positive* number to a *negative* one is the same as subtracting that amount (i.e., it is a movement *toward* zero from a position on the left).

EXAMPLE A.17.   Add −3 to 6.

Solution:

Simply subtract +3 from 6.

$$6 - 3 = 3$$

EXAMPLE A.18.   Add −17 to −33.

Solution:

Subtract +17 from −33, remembering that the second number is already negative and that you must move away from zero, starting from −33.

$$-33 - 17 = -50$$

Look at these examples for a moment until you are sure you understand them.

EXERCISE A.12.   Add the following pairs of numbers:

      a.   3 + (−4)   b.   (−9) + (−24)   c.   −41 + 90

Now we will work on subtracting negative numbers. As you read the rules that follow, work it out on the number line: to *subtract* a *negative number* from another number, (1) *change the sign* of the negative number and (2) *add* it to the other number. First, let's do it once, then we'll see why it works.

EXAMPLE A.18.   Subtract −4 from 8.

Solution:

Step 1.   Change the sign of −4 to 4.

Step 2.   Add the result to 8.

$$8 + 4 = 12$$

Let us see exactly what happened in that example. Subtracting −4 means moving to the right along the number line for four numbers, which is exactly what you would do if you were adding +4.

Follow the operations in Example A.19 on the number line to see if you agree.

EXAMPLE A.19.

Subtract −2 from 10.   Answer: 12.
Subtract −3 from −5.   Answer: −2.
Subtract −5 from −3. Careful.   Answer: 2. (Why?)

EXERCISE A.13.   Do the following subtractions:

a. $7 - (-5)$ c. $(-25) - (4)$
b. $(-31) - (-15)$ d. $(-25) - (-80)$

A.4 ALGEBRA

In science it is often necessary to work with mathematical tools that are more powerful than simple arithmetic. Although the list of such methods is long, embracing fields such as calculus and vector analysis, we will be able to manage quite well using a few techniques from algebra.

In algebra, letters or other symbols take the place of the quantities being calculated, and the relations between those quantities are expressed in the form of equations written with the symbols. The symbols are either **variables,** standing for quantities that can change, or **constants,** which stand for fixed quantities. The last few letters of the alphabet, x, y, and z, are often used for variables, while the earlier letters, along with the letter k or K, are often used for constants. There are also many exceptions to these practices.

As we have learned, an equation says that one mathematical expression has the same value as another, or that the two expressions are the same in amount and kind. Both sides of an equation must refer to the same kinds of things; that is, they must have the same units. Although the quantities may be equal, it is incorrect to say that three cats equal three dogs.

A simple equation is $y = 1 + 3$, which says that the quantity on the left of the equation, y, is the same as the quantity on the right, $1 + 3$. In this case, the value of y is 4, and whatever y refers to, there must be four of them.

The operations of algebra consist primarily of setting up equations with symbols representing both known and unknown quantities, then solving the equations to find the values of the unknowns. We will be doing many algebraic operations in our study of chemistry.

Section 12.4.4 introduces the concept of proportionality. In that section, we learn that an equation that is a proportionality graphs as a straight line. Such an equation is called a **linear equation.** If an equation has more than two variables, and if all the possible pairs of the variables are linear, then the equation itself is said to be linear. Another way to describe a linear equation is to say that it is one in which none of the variables has an exponent greater than unity. Since many equations are linear, we will take a brief look at the methods for solving linear equations; that is, for finding the value of one of the variables—either absolutely or in relation to other variables.

To reach a solution, we must change the equation to isolate the desired variable on one side of the relation without making the equality statement untrue. There are several ways to make such changes:

1. Both sides of the equation can be multiplied by the same entity (we take *entity* to mean a number, constant, or variable).
2. Both sides of the equation can be divided by the same entity.
3. The same entity can be added to both sides of the equation.
4. The same entity can be subtracted from both sides of the equation.
5. Both sides of the equation can be raised to the same power.

The following steps describe what is generally the best procedure for solving a linear equation. Although the steps will guide you, what is most essential is plenty of practice.

Step 1.   To clear the equation of fractions, multiply both sides by an entity that will accomplish this.

Step 2.   Group the parts of the equation so that all the parts containing the desired unknown appear on only one side.

Step 3.   Finish by adding, subtracting, or multiplying as needed.

EXAMPLE A.20.   Solve for x.

$$\frac{2x - 5}{x + 5} = \frac{1}{3}$$

Solution:

Step 1.   To clear the fractions, multiply both sides by $x + 5$ and by 3.

$$\frac{(x + 5)3(2x - 5)}{(x + 5)} = \frac{(x + 5)3}{3}$$

Notice that on the left side, $(x + 5)$ now appears both as numerator and denominator. Because any entity divided by itself is equal to unity and because any entity multiplied by unity is unchanged, we can simply "cancel out" the $(x + 5)/(x + 5)$ on the left, and the $3/3$ on the right. This leaves us with

$$3(2x - 5) = x + 5$$

Perform the multiplication indicated on the left side.

$$6x - 15 = x + 5$$

Step 2.   Rearrange the equation, placing all the terms containing x on the left and all those not containing x on the right. Do this by subtracting x from both sides and adding 15 to both sides.

$$6x - x - 15 + 15 = x - x + 5 + 15$$

$$5x = 20$$

Step 3.   Divide both sides by 5.

$$x = 4$$

EXERCISE A.14. Solve for y.

$$\frac{2(y - 5)(3y + 7)}{2y} = 3y - 3$$

Equations often contain symbols for constants and variables other than the one being solved for. The solution to such an equation will contain many or all of those symbols.

EXAMPLE A.21. Solve for x.

$$\frac{x}{2} + \frac{y}{3} = 5a$$

Solution:

Step 1. Give the fractions on the left a common denominator. The lowest common denominator of 3 and 2 is 6, so we use 6.

$$\frac{3x}{(3)(2)} + \frac{2y}{(2)(3)} = 5a$$

$$\frac{3x}{6} + \frac{2y}{6} = 5a$$

Multiply by the denominator and cancel the 6's on the left.

$$3x + 2y = 30a$$

Steps 2 and 3. Isolate x and solve.

$$3x = 30a - 2y = 2(15a - y)$$

$$x = \frac{2}{3}(15a - y)$$

EXERCISE A.15. Solve for z.

$$b(4z + 16x) = 12a + 20d$$

Although some of the systems discussed in Chapter 16 lead to equations that are not linear, you will be able to solve all the examples in the text by using the methods you have just learned and by making certain approximations. For this reason, we will not be concerned with nonlinear relations, such as quadratic equations.

A.5 EXPONENTS

An exponent is a simple way to show that a number has been repeatedly multiplied by itself. The number of times a number has been multiplied by itself, the **power** to which it has been raised, is indicated at the top right of the number as a superscript. Here are examples.

$$2 \times 2 \times 2 = 2^3$$

$$2 \times 2 \times 2 \times 2 = 2^4$$

$$2 \times 2 \times 2 \times 2 \times 2 = 2^5$$

In the first example, three 2's are multiplied together, thus the superscript, or **exponent,** is 3. The exponents for the other two examples are 4 and 5. Any number raised to a power of unity (1) is simply the number itself, and any number raised to a power of zero is equal to one.

$$8^1 = 8$$

$$8^0 = 1$$

The examples we have looked at so far all have positive exponents, but an exponent can also be negative. Repeatedly multiplying a *fraction* by itself leads to a negative exponent. For example,

$$\frac{1}{9} = 9^{-1}$$

$$\frac{1}{2} \times \frac{1}{2} \times \frac{1}{2} = \frac{1}{2 \times 2 \times 2} = \frac{1}{2^3} = 2^{-3}$$

The reverse of raising a number or fraction to a power is called taking the **root.** Asking, "What is the fourth root of 16?" is equivalent to asking, "What is the number that, multiplied by itself four times, equals 16?" The fourth root of 16 is 2 because $2^4 = 16$. The symbol for a root is a radical sign, $\sqrt{}$. A fractional exponent is an easier way to express the root. An example looks like this:

$$\sqrt[4]{16} = 2 \qquad \text{or} \qquad 16^{\frac{1}{4}} = 2$$

An exponent can indicate both a power and a root at the same time. The exponent 3/2 directs us to take the second root (often called the square root) and then raise that answer to the third power. Here is an example. It's easy.

$$9^{\frac{3}{2}} = (9^{\frac{1}{2}})^3 = (\sqrt[2]{9}\;)^3 = (3)^3 = 27$$

A.6 SCIENTIFIC NOTATION

In chemistry we meet some exceptionally large and some exceptionally small numbers. Two such numbers are:

602,000,000,000,000,000,000,000 and 0.000000000000000000000000000901.

It would clearly be inconvenient to write such numbers out in full. A system called scientific notation was developed to make long numbers easier to write. The system makes use of the number 10 raised to appropriate powers. To convert decimal numbers to scientific notation:

1. Move the decimal point in the original number to the right or left to convert it to a number lying between 1 and 10.
2. If you moved the decimal to the left, multiply the new number by 10 raised to a power equal to the number of places you moved the decimal. If the decimal was moved to the right, multiply the new number by 10 raised to a power equal to the number of spaces you moved the decimal, then place a negative sign in front of the exponent. Here are some examples. Follow each one.

$$1000 = 1 \times 10^3$$
$$1/100 = 0.01 = 1 \times 10^{-2}$$
$$35.68 = 3.568 \times 10^1$$
$$0.000075 = 7.5 \times 10^{-5}$$
$$48{,}391 = 4.8391 \times 10^4$$

Go back to the numbers at the beginning of this section and express them in scientific notation. You will be seeing them again.

A.7 CALCULATIONS USING EXPONENTIAL NOTATION

Because we will often use scientific notation, we need to learn how to do arithmetic with exponential numbers. Multiplying the same number raised to different exponents is easy: just add the exponents. To divide the same number raised to different exponents, subtract the exponent of the denominator from that of the numerator. For example,

$$(2^2)(2^3) = 2^{(2+3)} = 2^5 = 32$$

$$\frac{10^2}{10^3} = 10^{(2-3)} = 10^{-1}$$

We can show that this process works by solving for the numbers first, then multiplying.

$$(2^2)(2^3) = (4)(8) = 32 = 2^5$$

A number raised to a power can then be raised to another power. In these circumstances, multiply the exponents.

$$(2^2)^3 = 2^{(2\times3)} = 2^6$$

$$(9^{\frac{1}{2}})^3 = 9^{\frac{3}{2}}$$

The second example should look familiar. Work out the first example to see if it is correct.

To multiply and divide numbers in scientific notation, follow the same basic rules of exponents, multiplication, and division.

1. Multiply or divide the coefficient (the number lying between 1 and 10).
2. Multiply the result by 10 raised to the sum (if multiplying) or the difference (if dividing) of the exponents.
3. If the coefficient of the answer does not fall between 1 and 10, move the decimal to make it do so and change the exponent of 10 accordingly.

EXAMPLE A.22.   Divide:

$$\frac{3.56 \times 10^5}{6.78 \times 10^8}$$

Solution:

Step 1. $\dfrac{3.56}{6.78} = 0.525$

Step 2. $\dfrac{10^5}{10^8} = 10^{(5-8)} = 10^{-3}$

Step 3. $0.525 \times 10^{-3} = 5.25 \times 10^{-4}$

EXERCISE A.16.   Divide 2.241×10^1 by 6.24×10^5.

Adding and subtracting exponential numbers is just as straightforward. All the numbers, however, must be the same before the addition or subtraction can be performed. To make all the exponents the same, follow the rules for scientific notation.

EXAMPLE A.23.   Add $(3.75120 \times 10^3) + (1.2150 \times 10^2) + (3.2 \times 10^{-1})$.

Solution:

Step 1. Change all the numbers so that the exponents of all the 10's are the same. We will change them all to 10^2.

$$(37.5120 \times 10^2) + (1.215 \times 10^2) + (0.0032 \times 10^2)$$

Step 2. Add the coefficients. (Be sure to align the decimals.)

$$\begin{array}{r} 37.5120 \\ 1.2150 \\ \underline{0.0032} \\ 38.7302 \end{array}$$

Step 3. Multiply by 10^2.

$$38.7302 \times 10^2$$

Step 4. Move the decimal to make the coefficient fall between 1 and 10.

$$38.7302 \times 10^2 = 3.87302 \times 10^3$$

(According to the rules in Section A.9, we should round this answer to 3.9×10^3.)

EXERCISE A.17.   Add $(5.998 \times 10^7) + (3.900 \times 10^5) + (1.1 \times 10^3)$.

A.8 LOGARITHMS

To serve as a brief, handy way to express certain kinds of numerical values, a system has been developed in which numbers can be given entirely in terms of exponents of a chosen number, commonly ten. A number expressed by this system is called a **logarithm.** In chemistry several important kinds of information, such as pH, are expressed almost exclusively with logarithms.

When the system uses the exponents of the number 10, the exponent of 10 is called the **common logarithm,** or **log.** To learn how to find a logarithm, study the following example.

EXAMPLE A.24.   What is the log of 1000?

Solution:

Step 1.   Put the number in scientific notation.

   $1000 = 10^3$

Step 2.   The log of 10^3 is 3. This is expressed as

   $\log 1000 = 3$

EXAMPLE A.25.   What is the log of 2000?

Solution:

Step 1.   Put 2000 into scientific notation.

   $2000 = 2 \times 10^3$

Step 2.   Find the power to which ten must be raised to give the coefficient.

   $2 = 10^{0.3}$

Step 3.   Replace the coefficient in Step 1 with the answer from Step 2 and add the exponents.

   $10^{0.3} \times 10^3 = 10^{3.3}$

Step 4.   The resulting exponent is the log.

   $\log 10^{3.3} = 3.3$ therefore $\log 2000 = 3.3$

EXERCISE A.18.   What is the log of 90,000? Of 500?

In the last example, we found the power to which 10 had to be raised to give the number 2; $2 = 10^{0.3}$. Fractional powers of 10 can be looked up in tables or obtained from calculators that have log functions. Notice that 0.3 is simply the log of 2, so if we ask our calculator for log 2, it will give us the power to which 10 must be raised to give 2.

EXAMPLE A.26.   What is the logarithm of 325? (What is log 325?)

Solution:

Step 1. $325 = 3.25 \times 10^2$

Step 2. $3.25 = 10^{0.512}$

Step 3. $10^{0.512} \times 10^2 = 10^{2.512}$

Step 4. $\log 10^{2.512} = 2.512$

therefore $\log 325 = 2.512$

If you have the log of a number, you can find the number itself by taking what is called the **antilog;** simply use the log as an exponent of 10.

EXAMPLE A.27.   What is the number of which 3 is the log? (What is the antilog of 3?)

Solution:

Raise 10 to the power of 3.

$10^3 = 1000$

This is expressed as

antilog $3 = 1000$

What we did was the reverse of taking the log.

EXAMPLE A.28.   What is the antilog of 2.512 (from our earlier example)?

Solution:

$10^{2.512} = 325.08$

(Notice that the answer is not exactly 325, because we earlier rounded the log of 325 from 2.511883361 . . . to 2.512.)

antilog $2.512 = 325$

EXERCISE A.19.   What is the antilog of 1.773? Of 5.331? Of 2.666?

Logarithms can be used to multiply, divide, and make exponential calculations. Certain kinds of chemical problems require the use of logarithms, therefore you will need to know the rules that follow. Study the rules, then prove to yourself that they are correct.

Rules:

1. $\log(a \times b) = \log a + \log b$
2. $\log (a/b) = \log a - \log b$
3. $\log a^b = b (\log a)$
4. $\text{antilog} (a + b) = (\text{antilog } a) \times (\text{antilog } b)$

EXAMPLE A.29.   Find $(0.00000047)^{\frac{1}{4}}$.

Solution:

Step 1.   Put the number into scientific notation.

$$(4.7 \times 10^{-7})^{\frac{1}{4}}$$

Step 2.   Find the log of 4.7×10^{-7}, which is -6.33.

Step 3.   Multiply $1/4$ times -6.33, to give -1.58.

Step 4.   Find the antilog of -1.58. The answer is 2.62×10^{-2}.

EXERCISE A.20.   Find the values of the following. Give the answers in scientific notation.

a. $(4.332)^{\frac{1}{6}}$ b. $\sqrt[3]{2738}$ c. $(7.13)^9$

We could ask our calculator to give us the log of any number we wish. For instance, it could tell us directly the log of 325, which we just worked out. That, of course, is easier than working it out, and you may ask why you should bother learning how logs work when your calculator will figure them for you. The answer is that certain kinds of scientific language we will meet must be expressed in logarithms, and the data from which the logarithms come cannot be put on a hand calculator simply by asking it for a log. If you do not understand the working of logarithms, you will not understand certain scientific language or be able to work certain problems.

In some systems, the logarithm is the exponent of a number other than 10. One such system uses 2.71828, ordinarily called **e.** Logarithms using e are called natural logarithms and are given the symbol **ln.** In this textbook there are no problems that require natural logarithms.

A.9 SIGNIFICANT FIGURES

Appendix B contains a detailed discussion of *precision* applied to measurements or combinations of measurements as, for example, in the results of experiments. It is extremely important that scientists report their results no more precisely than their instruments allow. A language called the system of **significant figures** has been developed to indicate the precision of a reported value. Experimental precision is shown by giving an appropriate number of digits in a numerical value, that is to say, by giving the correct number of significant figures.

In many kinds of measurements, and especially in calculations made with calculators and computers, excess digits can be generated. We need to know which of these digits are significant and which are spurious, that is, which are relevant and which are merely the result of manipulation.

All digits in a numerical value are to be considered significant, subject to the rules that follow. Reports of numerical values must be rounded to include only the correct number of significant figures.

1. Zeroes are significant if they are to the left of the decimal point, as in the number 2000. for example.

2. If there are *only* zeroes to the left of a digit, as in 0.002 for instance, those zeroes are only placeholders and therefore not significant.

3. Similarly, if there are *only* zeroes to the right of a digit, they are also placeholders, unless preceded by a decimal. For instance, 2000 has only one significant figure, the 2, and would be expressed in scientific notation as 2×10^3. However, in the number 2.000 there is no need for placeholders, but the zeroes have been shown nonetheless. The zeroes in 2.000 are significant.

4. In many cases, a final decimal can be used to distinguish between placeholders and significant zeroes. For example, 2000. has four significant figures, while as we have seen, 2000 has only one.

5. An integer, an exact number that might perhaps result from a counting of individual objects, has an infinite number of significant digits. To express such a number, we use periods to indicate that the digits continue indefinitely. Exactly two, for example, would be written as 2.00. . . . Occasionally an exact number is written as a word, such as *three* or *twenty*. The subscripts in chemical formulas, such as the 2 in H_2O, are taken to be exact because they represent ratios between counted numbers of individual atoms.

When you multiply and divide numbers, the number with the smallest number of significant figures determines the number of significant figures in the answer.

EXAMPLE A.30.   Perform the operation and round the answer to the correct number of significant figures: 2.0×0.003

Solution:

$$2.0 \times 0.003 = 6 \times 10^{-3}$$

The first number has two significant figures, but 0.003 has only one, therefore the answer should have only one. Here are more examples. Figure out why the answers are as they are.

$$\frac{9.00}{3.0} = 3.0$$

$$1.21785 \times 2.0 = 2.4$$

$$1.21785 \times 2000 = 2400 = 2 \times 10^3$$

EXERCISE A.21.   Perform the following multiplications:

$4.42 \times 36257 \times 0.8843 \times 19$

$0.81 \times 788.3 \times 0.000344$

In addition and subtraction, the number with the smallest number of decimal places determines the number of significant figures in the answer.

EXAMPLE A.31.   Round the answer to the correct number of significant figures.

$$\begin{array}{r} 3.4560 \\ 221.75 \\ 9.421 \\ \underline{9.00} \\ 243.6270 \end{array}$$

Solution:

The number 221.75 determines the number of significant figures in the answer. The answer 243.6270, therefore, can have only two decimal places. Round it to 243.63.

EXERCISE A.22.   Perform the following operations:

a. $33.482 + 0.88 + 2.9993 + 410.33$ c. $51.334 - 7.8832$
b. $553.1 + 0.22 + 25.334$ d. $344.28544 - 91$

Rounding numbers before or during a calculation can lead to rounding errors. Round after the calculation is finished. The rules for rounding follow:

1.   If the number after the last significant figure is greater than five, round up.
2.   If the number after the last significant figure is less than five, leave the digit to its left unchanged.
3.   If the number after the last significant figure *is* five, increase the value of the digit to the left by one if the last significant figure is odd and leave it unchanged if the last significant figure is even. Study these examples in which each number has only three significant figures.

$43.82 \longrightarrow 43.8$ $43.85 \longrightarrow 43.8$
$43.78 \longrightarrow 43.8$ $43.75 \longrightarrow 43.8$

PROBLEMS

Section A.1.2

*1.   Calculate the numerical values of the following:
  a. 3^4 c. $4^{\frac{1}{2}}$ e. $27^{\frac{2}{3}}$ g. 5^{-2}
  b. 5^3 d. $8^{\frac{1}{5}}$ f. $(64^{\frac{1}{2}})^{\frac{1}{3}}$ h. 2^{-3}

2. Calculate the numerical values of the following:
   a. 4^3 e. $243^{\frac{2}{5}}$
   b. 7^2 f. 4^{-1}
   c. $125^{\frac{1}{3}}$ g. 3^{-3}
   d. $81^{\frac{1}{4}}$

*3. Calculate the numerical values of the following:
   a. $2^2 \times 2^3$ c. $8^2 \times 8^{\frac{1}{3}}$ e. $5^3 \times 5^{-2}$
   b. $3^3 \times 3^2$ d. $3^3 \cdot 3^2$ f. $2^3/2^{-2}$

4. Calculate the numerical values of the following:
   a. $3^3 \times 3^3$ d. $5^8 \times (5^{-4} \times 5^{-1})$
   b. $7^5 \times 7^{-3}$ e. $9^{\frac{1}{2}}/27^{\frac{1}{3}}$
   c. $2^5 \times 4^{\frac{3}{2}}$

*5. Perform the following multiplications:
   *a. $\dfrac{1}{2} \times \dfrac{1}{3}$ d. $\dfrac{2}{7} \times \dfrac{1}{2}$

   *b. $\dfrac{1}{3} \times \dfrac{1}{4} \times \dfrac{1}{5}$ e. $3 \times \dfrac{4}{5} \times \dfrac{1}{2}$

   *c. $\dfrac{2}{3} \times \dfrac{3}{2}$

6. Perform the following multiplications:
   a. $\dfrac{1}{4} \times \dfrac{1}{3}$ d. $\dfrac{2}{9} \times \dfrac{1}{2}$

   b. $\dfrac{1}{5} \times \dfrac{1}{7} \times \dfrac{1}{5}$ e. $5 \times \dfrac{4}{5} \times \dfrac{1}{7}$

   c. $\dfrac{3}{7} \times \dfrac{3}{5}$

*7. Perform the following divisions:
   *a. $\dfrac{\frac{1}{4}}{\frac{1}{2}}$ *b. $\dfrac{\frac{1}{4}}{\frac{1}{7}}$ c. $\dfrac{\frac{1}{3}}{\frac{1}{9}}$ d. $\dfrac{\frac{1}{9}}{\frac{1}{3}}$

8. Perform the following divisions:
   a. $\dfrac{\frac{1}{6}}{\frac{1}{3}}$ b. $\dfrac{\frac{1}{6}}{\frac{1}{9}}$ c. $\dfrac{\frac{1}{4}}{\frac{1}{8}}$ d. $\dfrac{\frac{1}{8}}{\frac{1}{4}}$

*9. Find common denominators for the following:
   *a. $\dfrac{1}{2}, \dfrac{1}{8}$ d. $\dfrac{1}{3}, \dfrac{5}{6}, \dfrac{6}{24}$
   *b. $\dfrac{1}{4}, \dfrac{1}{12}, \dfrac{1}{6}$ e. $\dfrac{1}{6}, \dfrac{1}{5}, \dfrac{1}{3}$
   c. $\dfrac{1}{4}, \dfrac{3}{8}, \dfrac{1}{12}$

10. Find common denominators for the following:
    a. $\dfrac{1}{3}, \dfrac{1}{12}$ c. $\dfrac{1}{5}, \dfrac{1}{3}, \dfrac{2}{7}, \dfrac{4}{28}, \dfrac{1}{3}$
    b. $\dfrac{1}{3}, \dfrac{2}{9}$ d. $\dfrac{1}{4}, \dfrac{1}{2}, \dfrac{2}{5}$

*11. Find numerical values for the following:
    *a. $\dfrac{1}{3} + \dfrac{1}{5}$ *b. $\dfrac{2}{5} + \dfrac{4}{3} + \dfrac{1}{2}$ *c. $\dfrac{4}{5} - \dfrac{1}{10} - \dfrac{1}{6}$

12. Find numerical values for the following:
    a. $\dfrac{1}{4} + \dfrac{1}{5}$
    b. $\dfrac{3}{5} + \dfrac{1}{3} + \dfrac{1}{15}$
    c. $\dfrac{2}{3} - \dfrac{1}{6} - \dfrac{1}{12}$

Section A.1.3

*13. Convert the following fractions to decimal fractions:
    *a. $\dfrac{1}{4}$
    *b. $\dfrac{2}{6}$
    *c. $\dfrac{3}{10}$
    d. $\dfrac{7}{110}$
    e. $\dfrac{33}{7}$

14. Convert the following fractions to decimal fractions:
    a. $\dfrac{1}{2}$
    b. $\dfrac{2}{8}$
    c. $\dfrac{2}{10}$
    d. $\dfrac{9}{215}$
    e. $\dfrac{41}{9}$

The answers to many of the following problems require significant figure analysis and, in many cases, rounding. For the present purposes, however, simply make the calculation and leave the answer as it comes out. We will return to these problems when we do the calculations requested under Section A.9.

*15. Perform the following operations:
    *a. 0.311×1.82
    *c. $0.08206/22.414$
    *b. 0.0033×0.14
    *d. $(760 \times 22.4)/(1.97 \times 274)$

16. Perform the following operations:
    a. 0.221×11.45
    c. $22.414/0.08206$
    b. 0.000673×77.8
    d. $(0.0821 \times 298)/22.4$

*17. Perform the following operations:
    *a. $3.175 + 221.01 + 15.33 + 0.002$
    c. $11.355 + 9.18 + 0.007 - 71.001$
    *b. $431.25 - 0.07$
    d. $0.152 - 3.41$

18. Perform the following operations:
    a. $22.414 + 0.009 + 1.233 + 441.10$
    c. $11.355 + 9.18 + 2.077 - 0.08$
    b. $78.6001 - 0.00377$
    d. $0.1102 - 22.03$

Section A.2

*19. Change each of the following fractions to percent:
    *a. $\dfrac{1}{3}$
    *b. $\dfrac{1}{6}$
    *c. $\dfrac{1}{9}$
    d. $\dfrac{2}{5}$
    e. $\dfrac{5}{9}$

20. Change each of the following fractions to percent:
    a. $\dfrac{1}{4}$
    b. $\dfrac{1}{7}$
    c. $\dfrac{1}{5}$
    d. $\dfrac{3}{7}$
    e. $\dfrac{9}{11}$

*21. Change each of the following decimal fractions to percent: 0.067, 0.121, 0.000335, 9.3/11.5.

22. Change each of the following decimal fractions to percent: 0.197, 0.000022, 0.055, 2.2/5.3, 113.3/442.11.

Section A.3

*23. Perform the following operations:
    *a. $12 - 7$
    c. $(355 + 22) - (14 - 95)$
    e. $-117 + 205$
    *b. $(-12) - 7$
    d. $-28 + 11$

24. Perform the following operations:
    a. $41 - 33$
    c. $(12 + 350) - (115 - 277)$
    e. $-94 + 155$
    b. $(-5) - 3$
    d. $-42 + 33$

25. Perform the following operations:
    a. 22×-4
    d. $(33.5 \times -0.008211)/(3 \times 441)$
    b. $-12/3$
    e. $22/(35 \times -0.18)$
    c. -22.1×-330

Section A.4

*26. Solve the following for the unknown variable, x, y, or z:
    *a. $2(4x - 5) = 5x - 1$
    c. $z(z - 1/2) = 2z$
    *b. $4y = y + (10y - 7)/3$
    d. $x + a = 4x - 51/7$
27. Solve the following for the unknown variable, x, y, or z:
    a. $3(y - 1) = 4y - 6$
    b. $7x = 2(x + 1) + 1$
    c. $5z + b = 3(2z + b/5)$
*28. Solve the following for the indicated variable:
    *a. Solve for x in terms of a, b, and y
       $ax/b = 3a^2b + y/2b$
    b. Solve for p in terms of n, K, and z
       $ap = K(2na^2/31 + za)$
    c. Solve for v in terms of n, M, j, and x
       $(v + 2)8M = (4M + j)/16 + nx + M^{-1}$
*29. Put the following numbers into scientific notation: 0.225; 253; 44,163; 2,000,019; 0.00000321.
30. Put the following numbers into scientific notation: 0.44; 415.32; 2115; 502,331; 0.00000000991.
*31. Put the following fractions into scientific notation:
    *a. $\dfrac{12}{335}$
    *b. $\dfrac{441}{2.3}$
    c. $\dfrac{11}{(33 \times 215)}$
    d. $\dfrac{515 \times 1032}{0.0011}$
32. Put the following fractions into scientific notation:
    a. $\dfrac{9}{33}$
    b. $\dfrac{29}{0.11}$
    c. $\dfrac{52}{(188 \times 117)}$
    d. $\dfrac{3122 \times 655}{0.00055}$
*33. Perform the following calculations. Put the answers into scientific notation.
    *a. $(1.15 \times 10^{-3})(2.4 \times 10^5)$
    d. $\dfrac{1.11 \times 10^{-4}}{5.32 \times 10^7}$
    *b. $(3.5 \times 10^{-7})(2.55 \times 10^{-3})$
    *c. $\dfrac{2.15 \times 10^3}{9.93 \times 10^5}$
    e. $\dfrac{(2.24)(763)}{(5.2)(8.21 \times 10^{-1})(273)}$
34. Perform the following calculations. Put the answers into scientific notation.
    a. $(3.32 \times 10^2)(1.1 \times 10^{-1})$
    d. $\dfrac{4.14 \times 10^{-3}}{5.2 \times 10^5}$
    b. $(9.0 \times 10^{-28})(6.0 \times 10^{23})$
    c. $\dfrac{22.797}{6.0 \times 10^{23}}$
    e. $\dfrac{2.95 \times 10^{-7}}{1.64 \times 10^{-23}}$

Section A.8

*35. Find the logarithm to the base 10 of the following: 10, 10^3, 10^{A3}, $10^{0.221}$, 1000, 0.0001, 3.1623, 316.23.
36. Find the logarithm to the base 10 of the following: 100, 1/10, 10^4, 10^{-5Q}, 0.00001, 5, 0.005.
*37. Find the logarithm to the base 10 of the following: 3.16×10^4, 7.7×10^{-6}, 0.00012. Write each answer as the sum of two logarithms.
38. Find the logarithm to the base 10 of the following: 5.0×10^6, 1.99×10^{-7}, 310,000. Write each answer as the sum of two logarithms.
*39. Find the antilogs of the following logarithms to the base 10: 3.5, −6.7, 2.02, −1.2. Hint: Write the negative logs as the sum of a negative whole number and a positive decimal fraction.

40. Find the antilogs of the following logarithms to the base 10: 2.4, −5.5, −3.09, 0.6.

Section A.9

*41. How many significant figures are there in each of the following numbers: 225.0; 1000; 0.0003210; 0.000312; 1,000,000.; 0.121000; 2.00001?

42. How many significant figures are there in each of the following numbers: 41.7; 22.150; 0.0030; 3000; 1,000,000.1; 300.0?

*43. Round your answers to Problem 13 to the appropriate number of significant figures.

44. Round your answers to Problem 14 to the appropriate number of significant figures.

45. Round your answers to Problem 15 to the appropriate number of significant figures.

46. Round your answers to Problem 16 to the appropriate number of significant figures.

*47. Round your answers to Problem 17 to the appropriate number of significant figures.

*48. Perform the indicated arithmetic operations, and round the results to the appropriate number of significant figures.

*a. 77.981×2.33

*b. 4×0.0665

*c. $88.7/32$

d. $441.332/0.224$

e. $11.31 + 453.2 + 0.22 + 0.0003 + 7.7$

f. $22.414/(0.082 \times 273)$

g. $(4.1 - 0.0093)(0.21 + 0.19)$

h. $(319.61/35.67) + 123$

i. $(4001 + 0.043 + 3.3)(0.0001)$

j. $(24/6.002)(0.000056 + 35.742)$

The following two problems give practice in drawing graphs. For discussion of graphs, see Section 12.4.

49. Construct a graph with the vertical axis numbered from 0 to 100 in steps of 10 and the horizontal axis numbered from 0 to 20 in steps of 5. Plot the following pairs of numbers and draw the curve that best fits them.

Vertical	Horizontal	Vertical	Horizontal
5	0.9	55	9.9
16	2.7	60	10.1
27	5.0	72	13.0
31	5.9	81	14.6
40	7.2	97	17.5

50. Construct a graph from the following pairs of numbers. Choose numbers for axes so that the curve is properly displayed.

Vertical	Horizontal	Vertical	Horizontal
75	0.043	2850	1.63
650	0.38	4090	2.34
1420	0.80	4900	2.80

Appendix B

Relative Precision in Scientific Measurement

We know from Sections 1.1 and 1.2 that measurements cannot be exact, and that our confidence in the result of a measurement depends on the precision of the measurement itself.

To find out how precise a method of measurement is, it is necessary to make the same experiment several times and compare the results of each measurement. There are many kinds of sophisticated statistical ways to compare experimental results, but we can learn the basic principles by studying a few examples of simple methods.

Suppose that we are weighing an automobile and that we make several weighings to get the best value possible. Here are our results.

Weighing	Result (lb)
1	3717
2	3725
3	3701
4	3698
5	3731
Average	3714

We calculate the average weight by adding the five results and dividing their sum by 5. If we take the average to be our best guess at the real weight, we find that our *deviations from the average* (the differences between the individual weights and the average) ranged from 3 lb in the first measurement (3717 − 3714 = 3) to 17 lb in the last (3731 − 3714 = 17). It is useful to study the deviations in a set of experimental data because the sizes of the individual deviations, as well as their average size, give us an indication of how precise our experiment was.

When we found the sizes of the deviations, we needed to know only how far each result differed from the average weight, not whether it was larger or smaller than the average. We can therefore change all the negative deviations to positive to obtain the **absolute value** of the deviations. (An example is 3701 − 3714 = −13, which is changed to 13.) We then average the deviations.

Weighing	Result (lb)	Absolute Deviation (lb)
1	3717	3
2	3725	11
3	3701	13
4	3698	16
5	3731	17
Average absolute deviation		12

The average deviation is 60/5 = 12. The result of the weighing can now be reported as 3714 lb plus or minus 12 lb, or as 3.71×10^3 lb. The second report would give the reader less information, however.

To decide whether a deviation is important, we should compare it with the measured value, and ask what fraction or what percent of the value the deviation is. The fraction formed by dividing the average deviation by the average value of the result is called the **relative average deviation,** sometimes called **relative precision,** or simply precision. The relative precision is usually expressed in percent form. For our weighing,

$$\frac{12 \text{ lb}}{3714 \text{ lb}} = 3.2 \times 10^{-3}$$

and

$$3.2 \times 10^{-3} \times 100\% = 0.32\%$$

Here, the relative precision was 3.2×10^{-3} or 0.32%. Suppose that we weighed a 50-lb sack of flour and found that the average deviation was 12 lb, the same as in the car weighing. The relative average deviation for the flour weighing would have been almost 25%. A 12-lb error in a weighing of nearly 4000 lb is acceptable, but a 12-lb error in a 50-lb weighing is not; that is sloppy work. To decide whether a given level of experimental precision is acceptable, the deviation must be considered in a *relative* way; that is, it must be compared with the size of the measured value. In much scientific work, a relative precision of 1% or 2% is considered acceptable. A relative precision of 25% is hardly acceptable in any kind of work. (See Figure B.1.)

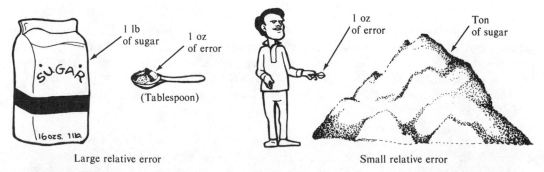

Figure B.1   Relative and absolute error. In both cases, the absolute (or actual) error is the same.

EXAMPLE B.1. Several workers measured the number of gallons of gasoline in an underground tank at a service station. They made the measurements by dipping a marked pole through a port in the top of the tank. The following table lists the results. Calculate the average number of gallons, the absolute average deviation, and the relative precision of the measurements.

Measurement	Gallons
1	2210
2	2101
3	1998
4	2044
5	2151
6	2035
7	2158

Solution:

Step 1. To calculate the average number of gallons of gasoline, add the individual values and divide the result by seven.

The sum of the individual values is 14,693 gal. The average is 14,693/7 gal or 2099 gal.

Step 2. Calculate the deviation for each measurement by subtracting the average from each reading. Make a deviations column.

Measurement	Gallons	Deviation
1	2210	111
2	2101	2
3	1998	−101
4	2045	−54
5	2151	52
6	2035	−65
7	2158	59

Sum 14,693

Average 2099

Step 3. Add the absolute values of the deviations; that is, add the *amounts* of the deviations, taking them all to be positive. Calculate the average deviation.

The sum of the absolute deviations is 443 gal. The average deviation is 443/7 gal or 63 gal.

Step 4. Divide the average deviation by the average number of gallons. This will give you the relative average deviation (relative precision).

$$\frac{63 \text{ gal}}{2099 \text{ gal}} = 0.030$$

Step 5. Now, multiply this number by 100% to put it into percent form.

$$0.030 \times 100\% = 3.0\%$$

Three percent of about 2000 gal is about 60 gal. If you happen to own the service station, that is a large amount. You may or may not be willing to accept a result with that level of precision.

EXERCISE B.1. A highway construction crew is paid by the foot of finished roadway. Several engineers have measured the length of roadway needed between two overpasses. Five values were obtained: 10,243 ft, 9981 ft, 10,173 ft, 9996 ft, and 10,048 ft.

Calculate the absolute average deviation and the relative precision of the measurements.

EXERCISE B.2. A class of chemistry students is measuring the volume of a container, using the units of milliliters, mL. The results of nine measurements, all in milliliters, are 213.4, 210.0, 212.0, 214.2, 103.2, 211.7, 208.9, 212.1, and 215.0.

Calculate the percent relative precision for these values. (Although we should always be extremely careful about discarding any data, we can see that the value of 103.2 mL must be an error. Discard it, and base your calculation on the other values.)

Every carefully written numerical report of a measurement should include an assessment of the precision of that measurement, whether it be an actual statement of the precision or an indication within the reported value itself. As we have seen, the number of significant figures in a reported value implies something about the precision. For example, suppose you were told that 50,000 people went to a football game. Would you interpret this to mean that exactly 50,000 people went, not 50,001 or 49,999? Of course not. On the other hand, if you were told that 49,342 people went, you might properly believe the count to be accurate within a person or two.

In reporting the weight of the car at the beginning of this appendix, we might have used the average weight. If so, we probably would have given a numerical evaluation of the precision of the weighing.

weight of car: 3716 ± 12 lb

If the value is reported this way, anyone reading the weight will have a good idea of how trustworthy the figure is.

PROBLEMS

1. In evaluating the precision of an experiment, why is it necessary to consider the deviations in relative form?
*2. Calculate the average of each of the following sets of numbers:
   *a. 372, 221, 87, 1333, 600, 12
   *b. 0.102, 1.300, 1.551, 2.044, 0.903
   *c. 73.1, 77.8, 79.9, 77.2
   d. 9.458, 10.493, 8.769, 11.823, 10.049
   e. 1048, 953, 1165, 984
   f. 0.0349, 0.0347, 0.0343, 0.0350, 0.0346, 0.0353
*3. For each of the sets in Problem 2, calculate the deviation of each number from the average for that set. Calculate the average deviation for each set.

*4. For each of the sets in Problem 2, calculate the relative average deviation in percent form.

*5. Here is a set of results for successive measurements of the highway distance between Portland and Seattle, all in miles: 184, 180, 197, 178, 177, 185, 184, 179.

   *a. Calculate the average distance.

   *b. Calculate the average deviation.

   *c. Calculate the percent relative average deviation.

6. To evaluate the scale in a grocery store, a brass cylinder whose weight is known to be 7.00012 ± 0.00009 lb is weighed six times on the scale. The results, all in pounds, are 6.98, 6.99, 6.98, 7.00, 7.01, 6.98.

   a. Calculate the average deviation and the relative average deviation.

   b. Examine the data carefully, then briefly discuss the probable sizes of random and systematic errors in the measurements.

*7. The concentration of a solution of NaOH is being determined by titration. In five successive titrations, each of 25.00 mL of NaOH solution, the following volumes of HCl solution are used:

Titration Number	Volume
1	35.12
2	34.90
3	34.75
4	35.41
5	35.00

   *a. What is the average volume used?

   *b. Calculate and list the deviations from the average.

   *c. What is the average deviation?

   *d. What is the percent relative deviation (the precision in percent)?

8. To determine the heat capacity of a sample, a measured amount of energy is added and the temperature increase, $\triangle T$ is measured. The values of $\triangle T$ from six successive experiments are

Experiment	$\triangle T$ (degrees)
1	12.66
2	12.75
3	12.50
4	12.59
5	12.81
6	12.70

   a. What is the average $\triangle T$?

   b. Calculate and list the deviations from the average.

   c. What is the average deviation?

   d. What is the percent relative deviation?

9. Samples of rock were weighed in two independent sets of experiments. The same balance was used in both sets. The consecutive values for six weighings of each sample were:

Sample A	Sample B
0.51 g	28.72 g
0.57 g	28.69 g
0.49 g	28.78 g
0.52 g	28.67 g
0.48 g	28.72 g
0.51 g	28.81 g

a. Calculate the percent relative deviation for each of the sets.
b. By referring to the results of the two sets of weighings, discuss the effect of sample size on the precision of a measurement.

10. The density of a liquid is to be determined by weighing a measured volume. Five different samples of the liquid are measured out and weighed. Here are the results:

Sample Number	Volume (mL)	Mass (g)
1	25.4	44.25
2	25.0	43.44
3	24.5	42.87
4	25.1	44.13
5	25.2	43.84

a. What are the experimental values of the density?
b. What is the percent relative deviation in the density results?

11. In an effort to improve the experiment of Problem 10, five successive weighings were made of the first sample. The results, all in grams, were 44.25, 44.28, 44.19, 44.27, and 44.25.
a. What is the percent relative deviation in these weighings?
b. To improve the results of Problem 10, would it be more productive to improve the quality of the weighings or of the volume measurement? Explain your answer.

Appendix C

Temperature Scales and Conversions

In Section 2.6, we learned the relations among the Celsius, Kelvin, and Fahrenheit scales. For easy reference, sketches of the three scales are shown in Figure C.1.

In our study of chemistry, we rarely have occasion to convert Celsius temperatures to Fahrenheit temperatures, or vice versa. It is nevertheless useful to know how such conversions are made. Practice in making temperature conversions can improve our general understanding of the arbitrary scales of values that are essential to science.

We will first see how conversions look graphically, then we will do some arithmetically.

EXAMPLE C.1.   Choose a temperature on the Celsius scale, say 20 °C. What is it on the Fahrenheit scale? On the Kelvin scale?

Solution:

On the Fahrenheit scale, look opposite the Celsius temperature of 20°. It is just under 70 °F; 20 °C is nearly 70 °F. Now look opposite 20 °C on the Kelvin scale. It is just above 290 K.

EXERCISE C.1.   What is the normal Celsius temperature of the human body? It is 98.6° on the Fahrenheit scale.

There are two disadvantages to using a graph to convert temperatures. It is not likely that a picture of the two scales will always be handy, and even at best, you get only an approximate answer. An arithmetic conversion is more portable and more accurate.

Arithmetic temperature conversions can be done conveniently in three steps:

Step 1.   Start at a known physical point (the boiling and freezing temperatures of water are good points) and record how many degrees the sample temperature is above or below that point on the given scale.

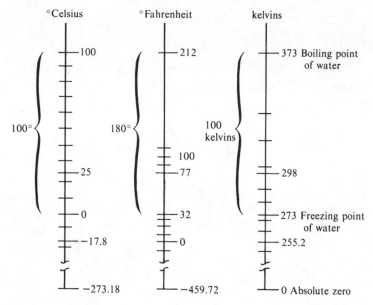

Figure C.1 Celsius, Fahrenheit, and Kelvin temperature scales.

Step 2. Convert that number of degrees to the corresponding number of degrees in the desired scale.

Step 3. Add the number of degrees in the new scale to the temperature of the known point on the new scale. (Subtract if the sample temperature was below the known point.) Now round the new value to the same number of significant figures as there were in the old temperature. Check your answer to see if it is reasonable.

EXAMPLE C.2.   Convert 20 °C to degrees Fahrenheit. (This is the arithmetic version of Example C.1)

Solution:

Step 1.   20 °C is 20 Celsius degrees above the freezing point of water.

Step 2.   An interval of 20 Celsius degrees represents about twice that number of Fahrenheit degrees. To be exact,

$$20 \text{ C degrees} \times \frac{9 \text{ °F}}{5 \text{ °C}} = 36 \text{ F degrees}$$

Remember that because the Fahrenheit degrees are smaller than the Celsius degrees, there will be more of them in the interval. For every 100 Celsius degrees, there will be 180 Fahrenheit degrees. To convert to Fahrenheit degrees, we multiply the number of Celsius degrees by the ratio 180/100, or 9/5, a number larger than unity that will have the desired effect of increasing the number of degrees. We would invert that ratio to go from Fahrenheit degrees to Celsius degrees.

Notice that this part of the calculation gives only the number of Fahrenheit-sized degrees above the freezing point of water. That is not the same as the Fahrenheit temperature, because water freezes at 32 °F, and the temperature is 36 degrees above that. We must add 32 degrees to our 36 degrees to find the Fahrenheit temperature.

Step 3. In the Fahrenheit scale, now add the 36 degrees to the Fahrenheit freezing point of water, 32 °F.

$$32 \text{ °F} + 36 \text{ more F degrees} = 68 \text{ °F}$$

Room temperature is usually not far from 70 °F. The answer seems reasonable, and the number of significant figures is acceptable.

EXAMPLE C.3.   Convert 98.6 °F to degrees Celsius.

Solution:

Step 1. 98.6 °F − 32.0 °F = 66.6 Fahrenheit degrees. (We say "66.6 Fahrenheit degrees" rather than "66.6 °F" to emphasize that this particular value does not represent an actual temperature, but a number of degrees.)

Step 2. $66.6 \text{ F degrees} \times \dfrac{5 \text{ C degrees}}{9 \text{ F degrees}} = 37.0 \text{ C degrees}$

Step 3.   0.0 °C + 37.0 C degrees = 37.0 °C

Compare this answer with your answer for Exercise C.1.

The arithmetic method, which requires that you understand the scales and think your way through each step, will work for any two temperature scales whatever and is superior to memorizing equations. You can successfully use equations without understanding what they are about, but it is risky: if you make a mistake, you may never know it. Don't substitute equations for understanding. Now that you have been cautioned, here are two equations: once you understand them, they can be used for rapid conversion from degrees Celsius to degrees Fahrenheit and back. They simply follow the steps that we have just followed. Although they will be useful to you, they cannot solve problems involving temperature scales other than Celsius and Fahrenheit. For the others, you will need to think out the problem.

From Celsius to Fahrenheit:

$$\text{°F} = \left(\frac{9 \text{ F degrees}}{5 \text{ C degrees}}\right) \text{°C} + 32 \text{ °F}$$

From Fahrenheit to Celsius:

$$\text{°C} = (\text{°F} - 32 \text{ °F}) \left(\frac{5 \text{ C degrees}}{9 \text{ F degrees}}\right)$$

EXERCISE C.2.   The normal boiling temperature of ethyl alcohol is 78.5 °C. What is it in Fahrenheit?

PROBLEMS

1. To define a temperature scale, why is it necessary to designate two actual fixed temperatures?
*2. Make the following temperature conversions:
   * 331 °C = K   −221 °C = K   11.4 °C = K
   *   285 K = °C   33 K = °C   520 K = °C
   *   23 °C = °F   −12 °C = °F   122 °C = °F
   *   95 °F = °C   25 °F = °C   −42 °F = °C
   *−444 °F = K   273 °F = K   75 °F = K
*3. In each of the following groups, which is the highest temperature?
   *a.   44 °F, 6.0 °C, 277 K   *b.   12 °C, 20 °F, 200 K   c.   −15 °C, 7 °F, 260 K
4. In each of the following groups, which is the lowest temperature?
   a.   −11 °F, −22 °C, 250 K   b.   148 °C, 250 °F, 325 K   c.   1000 °F, 800 °C, 1100 K
5. On a certain mountaintop, water boils at 96.2 °C. What is the Fahrenheit equivalent of this temperature?
6. The liquid form of the element helium boils at 4 K. What is the Fahrenheit equivalent of this temperature?
*7. At what temperature is the numerical reading on the Fahrenheit scale the same as that on the Celsius scale?
8. A ruling prince of ancient Egypt ordered his court scientist to invent a temperature scale, which he decreed would be called the Fish scale, after his coat of arms. Since no one had seen ice, the freezing temperature of water was not known. Instead, it was decided that the zero of the Fish scale would be at the boiling point of water (what we know as 100 °C) and that 100° Fish would be the melting point of glucose (refined from honey). The melting point of glucose on the Celsius scale is 146°.
   a.   Are Fish scale degrees larger or smaller than Celsius degrees?
   b.   How many degrees in the Fish scale are equal to 100 Celsius degrees?
   c.   On the Fish scale, what is the freezing temperature of water?
   d.   On the Celsius scale, what is the equivalent of 25° Fish?

Appendix D

Valence Shell Electron Pair Repulsion Theory

In an atom that is covalently bonded as a part of a molecule, the valence electrons, both shared and unshared, occupy orbitals that cause the electrons to spend most of their time in particular regions of the atom. The shapes of those orbitals dictate the relative locations of the shared pairs in the covalent bonds and, ultimately, the shape of the molecule. Because the total number of valence electron pairs in an atom determines what kind of a set of orbitals will be used, we can predict the shape that a molecule is likely to have if we know how many valence electron pairs are on its atoms. To make such predictions, we use the valence shell electron pair repulsion (VSEPR) theory.

Although all the electron pairs in the valence shell have a role in determining the particular geometry of the molecule, only the positions of the atoms themselves can be detected experimentally. Since unshared electron pairs are not a part of the molecular shape, the geometry of a molecule itself is sometimes different than the geometry of all the electron pairs. Figure D.1 shows a space-filling model of water along with a schematic sketch of the arrangement of the orbitals in the molecule.

VSEPR theory predicts the angles of bonds (and therefore the positions of the atoms in space) on the basis of the total number of shared and unshared electron pairs in the valence shell of the central atom of a molecule. With the help of Table D.1, geometry can be predicted in four easy steps.

Step 1. Draw an electron dot diagram of the molecule.

Step 2. Find the central atom. (In molecules that contain two or more atoms, each of which is bonded to two or more other atoms, it will be possible to determine the geometry around more than one atom.)

Step 3. Find the electron group number of the atom.
   a. Count the number of atoms bonded to the central atom. If one of the bonded atoms happens to have a double or triple bond, it still counts as only one bonded atom.

b. Count the number of unshared electron pairs on the central atom. In this count, do not include any pairs that are parts of bonds.

c. Add together the results of Steps a and b. The sum is called the *electron group number* of the atom.

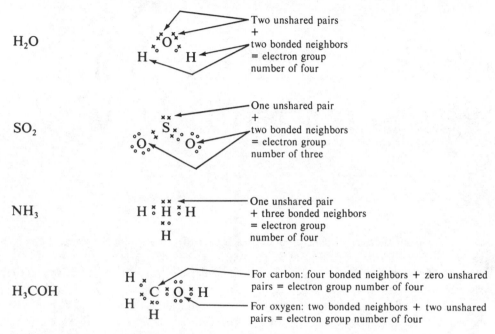

Step 4. From Table D.1, find the bonding geometry of the central atom. Notice that the geometry of the electron pairs will be the same as the bonding geometry only if all the pairs are used in bonding; that is, only if the count in Step 3b is zero.

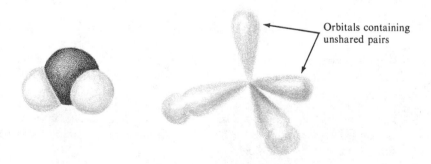

Figure D.1   Schematic drawing of water molecule, showing relative positions sp^3 orbitals that contain unshared pairs of electrons.

Table D.1
Predicted Geometry and Bond Angles of Molecules

Electron Group Number*	Electron Pair Geometry	Number of Atoms Bonded to Central Atom	Molecular Geometry (Shape)	Bond Angle
Two	Linear	1	Linear	
Two	Linear	2	Linear	180°
Three	Triangular	2	Bent	Less than 120°
Three	Triangular	3	Triangular	120°
Four	Tetrahedral	2	Bent	Less than 109°
Four	Tetrahedral	3	Pyramidal	Less than 109°
Four	Tetrahedral	4	Tetrahedral	109°

*Similar tables exist for electron group numbers five and six, but we will not be concerned with the geometry of such molecules.

EXAMPLE D.1. Find the geometry of the molecules shown in Step 3.

Solution:

For H_2O, we found that the electron group number is four. Since there are two atoms bonded to the central atom, the molecule is bent and should have a bond angle of less than 109°. (Experiments show that the water molecule is indeed bent and that it has a bond angle of 105°.)

H $\overset{O}{\leftrightarrow}$ H

Less than 109°

It is thought that the bond angle in water molecules is less than 109° because the mutual repulsion of the electron pairs in the O—H bonds is reduced by the bonded protons.

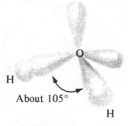

About 105°

For SO_2 the electron group number is three, even though there are four pairs of electrons around the central sulfur atom. The number of bonded atoms is two. The molecule is bent, with a bond angle of less than 120°.

O $\overset{S}{\leftrightarrow}$ O

Less than 120°

For NH_3 the electron group number is four, and there are three bonded atoms. The electron pair geometry is tetrahedral, but the molecular geometry is pyramidal. The predicted bond angle is less than 109°, and the measured angle turns out to be 107°.

All the N-H bond angles are equal and are about 107°

H_3COH has two atoms for which we can predict bond geometry, C and O. The electron group number and the number of bonded atoms around the C atom are both four. The geometry is tetrahedral around C, with bond angles close to 109°. The electron group number for O is four, but there are only two other atoms bonded to O. The bond angle of C—O—H is bent, with an angle of somewhat less than 109°.

The bond angles around carbon (seen in this flat reproduction as 90°) are close to 109°. The C—O—H bond angle is somewhat less than 109° because of unshared pair repulsion.

EXERCISE D.1. What is the geometry of H_2CO?

PROBLEMS

*1. Make a table giving the number of atoms and unshared electron pairs attached to the central atom, the electron group number, the electron pair geometry, the molecular geometry, and the expected bond angles for the following molecules and ions: CO_2, BH_3, NO_3^-, SO_3, H_2SO_4, PO_4^{3-}.

2. Make a table giving the number of atoms and unshared electron pairs attached to the central atom, the electron group number, the electron pair geometry, the molecular geometry, and the expected bond angles for the following molecules and ions: CO, NO_2^-, NH_4^+, NH_2^-, CCl_4, SO_2, H_2SO_3.

Appendix E

Naming Ionic Compounds

1. To name ionic compounds, we use combinations of the names of the individual ions. The name of the cation is given first. Examples: sodium chloride, NaCl; ferric hydroxide, $Fe(OH)_3$; potassium sulfate, K_2SO_4.

2. If the anion contains only one atom, the ending *-ide* is used. (There are, however, a few *-ide* compounds with polyatomic anions such as the hydroxides and cyanides.) Examples: sodium bromide, NaBr; aluminum oxide, Al_2O_3; potassium nitride, K_3N; magnesium cyanide, $Mg(CN)_2$.

3. If the cation can appear in more than a single oxidation state, the name of the compound must indicate which oxidation state is present. (See Table E.1.) There are two ways to indicate oxidation states in cations:

Table E.1
Some Metals That Have More Than One
Common Oxidation State

Element	Oxidation State	Old Name for Ion	Examples of Compounds
Cu	I	Cuprous	Cu_2O, CuBr
	II	Cupric	$CuSO_4$, CuO, $CuCl_2$
Fe	II	Ferrous	$FeCl_2$, $Fe(OH)_2$
	III	Ferric	$Fe(OH)_3$, $Fe(NO_3)_3$, FeI_3
Cr	II	Chromous	$CrCl_2$
	III	Chromic	$CrCl_3$, Cr_2O_3
	VI	Chromate	K_2CrO_4, H_2CrO_4*
Sn	II	Stannous	$SnCl_2$, SnF_2**
	IV	Stannic	$SnCl_4$

*Chromium (VI) exists only in the oxyanion CrO_4^{2-} and related ions.
**SnF_2 is the stannous fluoride found in some toothpastes.

a. The endings *-ous* and *-ic* are used to mean *lower* and *higher*. Examples: ferrous chloride, $FeCl_2$, iron in $+2$ state; ferric chloride, $FeCl_3$, iron in $+3$ state; stannous chloride, $SnCl_2$, tin in $+2$ state; stannic chloride, $SnCl_4$, tin in $+4$ state.

b. The Stock system is an unambiguous way to designate oxidation states. To give the oxidation state, place a Roman numeral after the name of the element. Examples: iron (II) chloride for $FeCl_2$, iron (III) sulfate for $Fe_2(SO_4)_3$, tin (IV) iodide for SnI_4.

4. Oxyacids: the relative oxidation state of the central atom in the anion of the oxyacids is indicated by the beginning and ending combinations given in Table E.2 in order of decreasing oxidation number (decreasing number of attached oxygen atoms).

5. To name salts of the oxyacids, give the name of the cation followed by the name of the anion. (See Tables E.2 and E.3.) Examples: barium nitrate, $Ba(NO_3)_2$; sodium hydrogen carbonate (sodium bicarbonate), $NaHCO_3$; calcium dihydrogen phosphate, $Ca(H_2PO_4)_2$; potassium sulfite, K_2SO_3.

Oxidation states of the central atoms in ions and of the salts of oxyacids are designated in a way that is similar to that used with the acids themselves.

Table E.2
Prefixes and Suffixes for Oxyacids

Oxidation States	Name Combinations	Examples of Acid
Highest	per——ic	Perchloric acid, $HClO_4$
↑	-ic	Chloric acid, $HClO_3$
↓	-ous	Chlorous acid, $HClO_2$
Lowest	hypo——ous	Hypochlorous acid, $HClO$

Table E.3
Prefixes and Suffixes for Salts of the Oxyacids

Oxidation States	Name Combinations	Examples of Salt
Highest	per——ate	Sodium perchlorate, $NaClO_4$
↑	-ate	Sodium chlorate, $NaClO_3$
↓	-ite	Sodium chlorite, $NaClO_2$
Lowest	hypo——ite	Sodium hypochlorite, $NaClO$

Table E.4
Some Common Oxyacids and Their Ions

Acid	Name as an Acid	Ion	Name of Ion
HNO_3	Nitric acid	NO_3^-	Nitrate ion
HNO_2	Nitrous acid	NO_2^-	Nitrite ion
H_2SO_4	Sulfuric acid	SO_4^{2-}	Sulfate ion
H_2SO_3	Sulfurous acid	SO_3^{2-}	Sulfite ion
H_3PO_4	Phosphoric acid	PO_4^{3-}	Phosphate ion
H_2CO_3	Carbonic acid	CO_3^{2-}	Carbonate ion
$HClO_4$	Perchloric acid	ClO_4^-	Perchlorate ion
$HClO_3$	Chloric acid	ClO_3^-	Chlorate ion
$HClO_2$	Chlorous acid	ClO_2^-	Chlorite ion
$HClO$	Hypochlorous acid	ClO^-	Hypochlorite ion
H_3CCO_2H	Acetic acid	$H_3CCO_2^-$	Acetate ion

Table E.5
Names of Ions From the Partial Dissociation
of Some Oxyacids

Acid	Ions From Successive Steps of Dissociation	Name	Common Name
H_2SO_4	HSO_4^-	Hydrogen sulfate ion	Bisulfate ion
	SO_4^{2-}	Sulfate ion	Sulfate ion
H_3PO_4	$H_2PO_4^-$	Dihydrogen phosphate ion	Diphosphate ion
	HPO_4^{2-}	Hydrogen phosphate ion	Biphosphate ion
	PO_4^{3-}	Phosphate ion	Phosphate ion
H_2CO_3	HCO_3^-	Hydrogen carbonate ion	Bicarbonate ion
	CO_3^{2-}	Carbonate ion	Carbonate ion

Table E.6
Common Names of Some Compounds

Common Name	Formula	Chemical Name
Baking soda, sodium bicarbonate	$NaHCO_3$	Sodium hydrogen carbonate
Blue vitriol	$CuSO_4 \cdot 5\ H_2O$	Copper (II) sulfate pentahydrate
Borax	$Na_2B_2O_7 \cdot 10\ H_2O$	Sodium tetraborate decahydrate
Cream of tartar	$KHC_4H_4O_6$	Potassium hydrogen tartrate
Epsom salts	$MgSO_4 \cdot 7\ H_2O$	Magnesium sulfate heptahydrate
Freon	CCl_2F_2	Dichlorodifluoromethane*
Grain alcohol	C_2H_5OH	Ethanol
Lye	$NaOH$	Sodium hydroxide
Muriatic acid	HCl solution	Hydrochloric acid
Rubbing alcohol	$(CH_3)_2CHOH$	Isopropyl alcohol
Sugar	$C_{12}H_{22}O_{11}$	Sucrose
TSP	Na_3PO_4	Trisodium phosphate
Wood alcohol	CH_3OH	Methanol

*There are several other freons.

Appendix F

Vapor Pressure of Water

Vapor Pressure of Water at Various Temperatures

Temperature (°C)	Pressure (torr)	Temperature (°C)	Pressure (torr)
−10.0	2.15	40.0	55.3
0.0	4.58	50.0	92.5
10.0	9.21	75.0	289.1
20.0	17.5	100.0	760.0
25.0	23.8	110.0	1075
30.0	31.8		

Appendix G

Glossary

Absolute temperature Temperature expressed in Kelvins and on a scale that begins at absolute zero.

Absolute zero On a graph of P vs V of an ideal gas at constant volume, the extrapolated temperature at which V becomes zero. Absolute zero is -273.2 °C.

Acceleration Any change in velocity. Quantitatively, the rate of change of velocity.

Accuracy Closeness with which a measured result approaches the true value.

Acid Substance that yields hydrogen ions in aqueous solution; substance that donates protons in a chemical reaction.

Acidic solution Aqueous solution in which the concentration of hydrogen ions (primarily as H_3O^+) is greater than that of hydroxide ions.

Activation energy In a chemical reaction, the energy barrier that reactants must overcome in order to become products.

Activity series Table of the relative strengths of chemical species as oxidizers and reducers; chemical species are usually given in order of increasing strength as reducers (weakest reducing agents, strongest oxidizers, at the top).

ADP/ATP Adenosine diphosphate-triphosphate system used by biological systems as a quick-response energy storage system.

Alcohol Compound containing an —OH group covalently bonded to an alkyl group: ROH.

Aldehyde Compound, each molecule of which contains an oxygen atom double-bonded to a carbon atom, which in turn is bonded to a hydrogen atom and either an alkyl group or another hydrogen.

Alkali metals Column of elements forming the leftmost family in the periodic table, Group IA. Each atom of an alkali metal has one valence electron.

Alkaline earth metals Family of elements immediately to the right of the alkali metals (which see), Group IIA. Each atom of an alkaline earth metal has two valence electrons.

Alkane Family of hydrocarbons, the molecules of which have only single bonds between carbon atoms.

Alkene Family of hydrocarbons in which each molecule has at least one carbon-carbon double bond.

Alkyl group Group of atoms consisting of an alkane molecule with one hydrogen atom removed.

Alkyne Family of hydrocarbons in which each molecule has at least one carbon-carbon triple bond.

Allotrope One of two or more different forms of an element in the same state.

Alpha amino acid Carboxylic acid containing an amine group attached to the alpha carbon (the carbon next to the carboxyl group).

Alpha helix   Secondary protein structure in which the chain is coiled into a helix that has slightly more than three amino acid groups in each turn.

Alpha particle   High-speed helium nucleus, usually emitted by a radioactive element.

Alpha rays   Stream of alpha particles.

Amine   Organic compound, each molecule of which contains an —NH group attached to a carbon atom. An organic base.

Amphoteric compound   Substance that can act either as an acid or a base, depending on conditions.

Anaerobic   Able to live in the absence of oxygen.

Anhydrous   Without water.

Anion   Negative ion.

Anode   In an electrolytic cell the electrode to which anions migrate. In an electronic device, the electrode to which electrons flow.

Aqueous solution   Solution in which the solvent is water.

Aromatic ring   Unusually stable organic ring compound, the molecules of which are formally characterized by alternating single and double bonds.

Artificial isotope   Radioactive isotope created artificially, usually by bombardment with neutrons or charged particles from a cyclotron or other accelerator.

Atmosphere   Unit of pressure based on the pressure of the atmosphere at sea level. One atmosphere equals 760 torr.

Atom   Smallest particle of matter that has the characteristics of an element.

Atomic kernel   The part of an atom that includes the nucleus and all the electrons except those in the valence shell.

Atomic mass scale   Relative scale of atomic masses based on a C-12 mass of exactly 12.

Atomic mass unit   Unit of mass that is exactly 1/12 of the mass of an atom of C-12.

Atomic number   Number of protons in the nucleus of an atom of an element.

Avogadro's hypothesis   Theory stating that equal volumes of gases at the same temperature and pressure contain equal numbers of molecules.

Avogadro's number   Number of atoms in exactly 12 g of C-12. The number of individual items in 1 mol of those items. The value of Avogadro's number is 6.0225×10^{23}.

Balanced equation   Reaction equation in which the same number of electric charges and of atoms of each kind appears on both sides of the equation.

Barometer   Device for measuring atmospheric pressure.

Base   A substance able to accept protons in a chemical reaction. Substance that yields hydroxyl ions in aqueous solutions.

Basic solution   Aqueous solution containing more OH^- ions than H^+ ions.

Beta particle   High-speed electron.

Beta ray   Stream of beta particles.

Binary compound   Compound consisting of two elements.

Binding energy   Energy difference between a nucleus and the separated individual particles that constitute that nucleus. The mass defect taken in energy units.

Bohr atom   Theoretical model of the atom, shown to be incorrect, in which the electrons were thought to orbit the nucleus like tiny satellites.

Boiling point (boiling temperature)   Temperature at which the vapor pressure of a liquid equals the pressure exerted on it by the atmosphere or other gas and the liquid boils.

Bond angle   Angle between the internuclear lines of two covalent bonds extending from an atom.

Bond structure   Designates which atoms are bonded to which in a molecule. A diagram showing the bond structure of a molecule.

Boyle's law   States that for a sample of an ideal gas at constant temperature the volume is inversely proportional to the pressure.

calorie   Amount of energy required to raise the temperature of 1 g of pure water from 14.5 °C to 15.5 °C.

Carbohydrate   Organic compound containing carbon, hydrogen, and oxygen. Carbohydrates are polyhydroxy aldehydes or ketones that contain H and O in the ratio of two atoms of H to one of O.

Carbonyl   Organic compound, each molecule of which contains an oxygen atom double-bonded to a carbon atom; the other two bonds are either to carbon or to hydrogen atoms.

Carboxyl group   Organic acid structure in which a carbon atom is bonded to an OH group, and double-bonded to another O atom.

Catalyst Substance that increases the rate of a reaction by lowering the activation energy. A catalyst remains unchanged by the reaction.

Cathode In an electrolytical cell, the electrode to which the cations migrate. In an electronic device, the electrode that furnishes electrons.

Cation Ion carrying one or more positive charges.

Celsius scale Temperature scale that defines the freezing temperature of water at 0° and the boiling temperature at 100° (both at 1 atm).

Centimeter One hundredth of a meter.

Chain reaction Chemical or nuclear reaction in which the products cause further reaction to occur.

Change of state Process in which a solid, liquid, or gas is changed to either of the other two states.

Charles's law States that the volume of a sample of gas at constant pressure is proportional to the absolute temperature. $P = kT$.

Chemical analysis Measurement to determine what elements are in a compound or what the components of a mixture are (qualitative analysis) or to determine what the relative amounts of the components are (quantitative analysis).

Chemical bond Force that holds the atoms together in a chemical compound.

Chemical equilibrium Dynamic state in which the forward and reverse reactions of a reversible reaction occur at the same rate, such that the concentrations of the reactants and products do not change.

Chemical family Group of elements appearing in a vertical column in the periodic table and having similar sets of properties. Identical valence electron configurations cause familial resemblances.

Chemical formula Combination of element symbols and subscripts representing the ratio of elements in a compound or molecule.

Chemical reaction Process in which one or more substances (reactants) are transformed into other substances (products) by the making and breaking of chemical bonds.

Chromatography Separation technique that depends on the strength by which surfaces adsorb different substances.

Coefficient In a reaction equation, the number preceding the symbol of a chemical species.

Coke Nearly pure carbon prepared by heating coal with insufficient oxygen.

Colloid Liquid in which a second phase is dispersed in aggregates too small to settle out but too large for the system to be a solution.

Compound Substance formed by the chemical combination of two or more elements and characterized by a chemical formula.

Concentration Designation of the relative amounts of solvent and solute in a solution.

Condensed state Either of the two states of matter in which the particles are close together; either solid or liquid.

Conversion factor Quantitative mathematical relation that changes a quantity given in one set of units to the equivalent in a different set.

Coordinate covalent bond Covalent bond in which both electrons in the shared pair are furnished by only one of the bonded atoms.

Covalent bond Chemical bond in which a pair of electrons is shared by both the bonded atoms.

Covalent crystal Crystal in which all the atoms are connected by covalent bonds.

Critical mass In a fissionable material, the minimum mass in which more neutrons create fission than escape through the surface of the sample.

Crystal Substance in which the component particles (atoms, ions, or molecules) are arranged in a repeating pattern and held in place by chemical bonds or other forces.

Cubic centimeter A measure of volume equivalent to a cube with sides of 1 cm. The same as a milliliter.

Cubic meter The SI standard of volume. A cube with sides of 1 m.

Cytoplasm Fluid in which the interior parts of biological cells are immersed.

Dalton's law States that in a mixture of gases, the pressure exerted by each gas equals the pressure that the gas would exert if alone in the container at the same temperature. $P_t = P_1 + P_2 + P_3 \cdots$

Density Ratio of the mass of a substance to its volume.

Deoxyribonucleic acid, DNA Biological polymer that stores genetic information in living organisms.

Diatomic molecule   Molecule that contains two atoms.

Diluent   Additional solvent added to a solution.

Dilute solution   Solution that is not concentrated. Solutions containing less than a few grams per liter of solute are ordinarily considered dilute.

Dipeptide   Condensation product of two amino acid molecules.

Dipole-dipole interaction   Intermolecular force caused by the dipoles of two or more molecules.

Disaccharide   Molecule containing two simple sugar molecules (such as sucrose, which contains one glucose molecule and one fructose molecule).

Disproportionation   Reaction in which the same substance is both oxidized and reduced.

Dissociation   Usually refers to the transformation of ionic solids into individual ions upon going into aqueous solution.

Distillation   Purification process that involves a change of state from liquid to vapor and back to liquid.

Double bond   Two covalent bonds (two shared electron pairs) between one set of two atoms.

Double helix   The secondary, screw-shaped structure of nucleic acids.

Dynamic equilibrium   State of rest that results when the rates of two opposing, ongoing processes are equal.

Electric charge   Fundamental property of matter such that an electrically charged particle will exert a force on another electrically charged particle.

Electric circuit   Continuous loop of matter that conducts electricity.

Electric current   Flow of electric charge through a conductor.

Electric dipole   Separation of electric charge in a particle, such that one portion has a positive charge and another portion a negative charge.

Electrode   In a battery, electrolysis cell, or electronic device, a conductor from which or into which electrons flow.

Electrolysis   Chemical reaction caused by electric energy.

Electrolyte   Substance that conducts electric current when dissolved in water.

Electron   Fundamental particle with about 1/1838 the mass of a proton and a single negative electric charge of 1.6×10^{-19} coulombs.

Electron affinity   Energy lost by an electron that comes from a great distance and joins a neutral, gas-state atom to create a negative ion.

Electron configuration   List of the numbers of electrons in the successive subshells of an atom.

Electron dot diagram   Schematic sketch of a molecule or atom showing all the valence electrons, especially those in covalent bonds.

Electron shell   Everyday term for the energy level of an electron in an atom. Corresponds to any particular value of the principal quantum number N.

Electron transfer   Chemical reaction in which an electron is transferred from one species (atom, molecule, or ion) to another.

Electronegativity   Somewhat arbitrary designation of the relative attractiveness of an atom for the pair of electrons by which it is covalently bonded to another atom.

Element   Substance that cannot be decomposed into simpler substances by chemical means. A collection of atoms all with the same atomic number.

Elimination   Chemical reaction in which two molecules join together and produce another simple molecule (such as water or HCl) in the process.

Empirical formula   Represents the smallest integer ratio of atoms in a compound.

Endothermic process   Reaction or other process that absorbs energy.

Energy   Work or the ability to do work.

Energy level   One of the quantized, allowable amounts of energy that an electron in an atom can possess. All energy, whether in atoms or otherwise, is quantized.

Enthalpy   Energy content of a sample of matter.

Enzyme   Biological catalyst, usually a protein.

Equilibrium   See *dynamic equilibrium.*

Equilibrium constant   Numerical value that indicates the extent to which a reversible reaction proceeds. A temperature-dependent function containing the concentrations of the reactants and products of the reaction.

Error   Difference between a measured value and the true value of what is measured.

Essential amino acid   One of the alpha amino acids from which proteins are formed and which the human body cannot synthesize.

Ester Organic compound formed by the elimination reaction between a carboxylic acid and an alcohol.

Evaporation Slow vaporization of a liquid into a surrounding gas.

Exothermic process Reaction or other process that releases energy.

Extensive property A property, the value of which depends on the amount of the sample taken.

Fahrenheit scale Temperature scale defined as having the freezing point of water at 32° and the boiling point at 212° (both at 1 atm).

Faraday Amount of electric charge possessed by 1 mol of electrons. A Faraday will, for example, cause the deposit of 1 mol of silver metal from a solution of silver ions.

Fatty acid One of the long-chain acids that can form a fat through esterification with glycerol.

Fluoridation Addition of fluoride compounds to water supplies with the objective of reducing the incidence of tooth decay.

Frasch process Process by which sulfur is mined. Steam is forced into wells, bringing melted sulfur to the surface.

Free radical Molecule or molecular fragment characterized by having an unpaired electron.

Functional group Any of several groups of atoms that can be attached to a hydrocarbon chain (replacing a hydrogen atom) and can determine some of the characteristic behavior of the molecule.

Gamma ray Stream of very high-energy photons.

Gas State of matter characterized by the fact that a sample will take both the shape and volume of its container.

Geiger counter Instrument that records the passage of radioactivity by electrically collecting the ions left behind as the rays pass through a gas detection medium.

Geonormal mixture Mixture of isotopes that occurs normally in a sample of an element prepared from natural sources.

Glass Solid in which the particles are not arranged in a repeating pattern.

Glycerol 1,2,3-trihydroxy propane.

Glycogen Sugarlike carbohydrate used by animals for energy storage $(C_6H_{10}O_5)_n$.

Gram One thousandth of a kilogram.

Gram-atomic weight Atomic weight expressed in grams. The mass of 1 mol of a geonormal element.

Gram-molecular weight Molecular weight expressed in grams. The mass of 1 mol of a compound.

Graphite One of the allotropes of carbon.

Half-life Amount of time required for half of any sample of a radioactive isotope to decay.

Halogens Family of elements the atoms of which all have seven valence electrons; group VIIA.

Hard water Water containing undesirable amounts of ions (such as Ca^{2+} and Mg^{2+}) that react with soap to form boiler scale in plumbing.

Heat Transfer of energy between a sample of matter and another sample at a lower temperature.

Heat of fusion Amount of energy required to melt a sample of any given substance. The *molar heat of fusion* is the amount required to melt 1 mol.

Heat of vaporization Amount of energy required to vaporize a sample of any given substance. The *molar heat of vaporization* is the amount required to vaporize 1 mol.

Hemoglobin Respiratory protein that enables blood to carry oxygen and carbon dioxide.

Heterogeneous sample System that has more than one phase.

Homogeneous sample System that has only a single phase.

Hund's rule States that within any subshell, electrons preferentially occupy empty orbitals when such are available.

Hydrate Crystalline solid that contains water molecules in stoichiometric quantities and occupying specific locations in the crystal lattice.

Hydrated ion In aqueous solution, an ion associated with one or more water molecules.

Hydride ion Hydrogen atom with a negative charge, H^-.

Hydrocarbon Compound consisting only of carbon and hydrogen.

Hydrogen bond Unusually strong dipole-dipole interaction between a covalently bonded hydrogen atom and any of the atoms N, O, or F in adjacent molecules.

Hydrogen ion   Hydrogen atom with a positive charge, H^+.

Hydrogenation   Process of adding hydrogen to unsaturated organic compounds, as in the hydrogenation of vegetable oil to make margarine.

Hydrolysis   Reaction of something with water.

Hydronium ion   Singly hydrated hydrogen ion, H_3O^+.

Hydroxide ion   Negative ion containing a hydrogen and an oxygen atom, OH^-.

IUPAC naming system   Internationally accepted system for naming organic molecules in such a way as to describe their structures.

Ideal gas law (combined gas law)   Natural law that combines Boyle's law, Charles's law, the law of Gay-Lussac, and Avogadro's principle. The equation of state for an ideal gas, $PV = nRT$.

Ideal gas (perfect gas)   Imaginary gas that perfectly obeys the ideal gas law. Under ordinary conditions, many gases approach ideal behavior.

Intensive property   Property that is independent of the amount of sample taken. Temperature, for example, is an intensive property, as is density.

Ion   Atom or molecule possessing an electric charge.

Ionic bond   Electric force uniting the ions in an ionic crystal.

Ionic crystal   Crystalline assembly of ions.

Ionization constant   Equilibrium constant describing the extent to which a weak electrolyte is ionized in aqueous solution.

Ionization energy   The energy required to remove an electron from a gaseous ion or molecule.

Ionizing solvent   Solvent, such as water, that causes dissociation in ionic compounds or that reacts with a molecular compound to produce ions.

Isotopes   Two or more atoms that have the same atomic number but different mass numbers.

Isotopic abundance   Relative amounts of the different isotopes in a sample of an element.

Joule   Unit of energy defined in SI as the amount of energy required to accelerate 1 kg of mass from a position of rest to a velocity of 1 m/sec. A calorie is 4.18 J.

Kelvin scale   Temperature scale that defines zero at absolute zero and the freezing temperature of water at 273 kelvins (where the kelvin is the same as a degree Celsius).

Kernel charge   Net charge possessed by the nucleus of an atom and all but the valence electrons.

Ketone   A compound, the molecules of which contain an oxygen atom double-bonded to a carbon atom, which in turn is bonded to two alkyl groups.

Kilogram   Defined unit of mass in SI. Equivalent to 1000 g.

Kilojoule   Commonly used unit of energy in SI. A kilocalorie is the same as 4.18 kJ.

Kilometer   One thousand meters.

Kinetic energy   Energy possessed by a moving body; $K.E. = 1/2\ mv^2$.

Kinetic molecular theory   States that the molecules of a gas are in rapid, random movement and that their collisions with the walls of their vessel produce pressure.

LOX   Liquid oxygen.

Lanthanide, actinide   The f blocks of elements. These elements usually appear separately at the bottom of the periodic table.

Law of definite proportions   Natural law stating that all compounds contain their elements in fixed proportions by mass.

Law of Gay-Lussac   Natural law stating that the pressure of a gas at fixed volume is proportional to its absolute temperature.

Le Chatelier's principle   States that if an external change is made in the intensive properties of a system at equilibrium, the system will respond so as to counteract the effects of the change, thus reaching a new position of equilibrium.

Lewis structure   Schematic diagram of a molecule showing the valence electrons in the covalent bonds.

Ligand   Electron-rich species that forms a coordinate covalent bond with the central ion of a complex ion.

Limiting reactant (limiting reagent)   Reactant first exhausted in a chemical reaction, thus the reactant that limits the extent of the reaction.

Linear molecule Molecule in which a straight line can be drawn through the centers of all the atoms; molecule that has chemical bonds only at 180° angles.

Lipid Substance that can be dissolved in ether or similar solvents, thus an oily or greasy substance. The term is usually applied to biological compounds.

Liquid Substance that will take the shape of the part of the vessel it occupies but will not necessarily occupy the entire volume of the vessel.

Liter One thousandth of a cubic meter, where the cubic meter is the standard SI volume. A liter has very nearly the same volume as that of 1 kg of water.

London force Intermolecular force arising from temporary electron imbalances in molecules and the adjacent dipoles induced by those imbalances. Often called *temporary dipole force* or *induced dipole force*. One of the kinds of van der Waals forces.

Macroscopic Large enough to be visible.

Manometer Device, similar to a barometer, for measuring the pressures of gases.

Mass Amount of matter that an object contains. Mass makes matter subject to gravitational forces and enables it to acquire momentum.

Mass defect Binding energy of a nucleus expressed in terms of mass. The difference in mass between the separate protons and neutrons of a nucleus and the nucleus itself.

Mass number Sum of the number of protons and neutrons in a nucleus. The whole number nearest the atomic mass of an isotope.

Matter Anything that has mass and occupies space.

Melting point Temperature at which a crystalline solid is in equilibrium with its liquid phase.

Metal Any of the elements listed on the left of the periodic table that are shiny, ductile, and malleable and tend to lose electrons in chemical reactions.

Metalloid Class of elements listed between the metals and nonmetals in the periodic table that are neither entirely metallic nor nonmetallic in character.

Meter Defined unit of length in SI and in the metric system.

Metric system System of measurement of which SI is a subset. Subunits in the metric system are all related by powers of ten.

Milligram One thousandth of a gram.

Milliliter One thousandth of a liter.

Millimeter One thousandth of a meter.

Mitochondrion One of the internal parts of a biological cell believed to produce the energy with which cells operate.

Molar gas volume Volume of 1 mol of an ideal gas at STP (standard temperature and pressure).

Molar heat capacity Amount of energy required to raise the temperature of 1 mol of a substance 1 °C.

Molarity Concentration unit of moles of solute per liter of solution.

Mole Avogadro's number of any discrete, countable object such as an atom, molecule, ion, electron, or any other specified object or group.

Molecular compound Compound consisting of individual, discrete molecules.

Molecular crystal Crystalline solid composed of individual molecules held together by van der Waals forces.

Molecular formula Chemical formula that specifies the actual numbers of the atoms of the different elements in the molecules of a molecular compound.

Molecular weight Number that specifies the average mass of the molecules of a compound in terms of the atomic mass scale.

Molecule Discrete group of atoms held together by covalent bonds. The smallest individually stable unit of an element or compound.

Monomer Molecule or molecular fragment that forms the repeating unit of a polymer.

Monosaccharide A single, simple, sugar molecule.

Net ionic equation Reaction equation written for the reactions of ionic compounds in aqueous solution and showing only the ions and molecules that actually participate in the reaction.

Network polymer Polymeric substance that is crosslinked; that is, in which covalent bonds extend throughout the sample in all directions.

Neutral solution Solution that has equal concentrations of H_3O^+ and OH^- ions.

Neutralization The process of making a neutral solution (one that is neither acidic nor basic) from one that is acidic or basic.

Neutron One of the fundamental particles of which nuclei consist. Has approximately the same mass as a proton but no electric charge.

Noble gas One of the family of essentially unreactive gases at the right side of the periodic table (Group 0). Noble gas atoms each have eight valence electrons.

Nonmetal One of the class of elements listed on the right side of the periodic table that tends to gain electrons or to make covalent bonds in chemical reaction.

Normal boiling point Boiling temperature of a liquid at a pressure of 1 atm.

Nuclear fission Process of fragmentation of unstable nuclei; smaller nuclei are formed and energy is released.

Nuclear fusion Process in which two atomic nuclei combine to make a larger nucleus, thereby releasing energy.

Nuclear reactor Array of nuclear fuel units and controls that allow nuclear fission reactions to be run at a desired rate.

Nuclear waste Radioactive by-products of nuclear reactors.

Nucleic acid Class of compounds found in biological cell nuclei and used to store and transmit genetic information.

Octet rule States the tendency of an atom to gain, lose, or share electrons during chemical reaction thereby enabling the valence shell to have the electron configuration ns^2, np^6.

Orbital A kind of energy sub-sublevel that can be occupied by an electron in an atom. An electron in any particular orbital spends most of its time in a particular region of the atom.

Organic chemistry Chemistry of carbon compounds (excluding only a few "inorganic" carbon compounds such as carbon dioxide).

Ostwald oxidation Catalytically encouraged reaction of NH_3 with O_2 to form NO and eventually NO_2.

Oxidation Chemical reaction resulting in the loss of one or more electrons by an atom, ion, or molecule.

Oxidation number Number assigned to each atom in the formula of a compound, used to keep track of electrons lost and gained in oxidation-reduction reactions.

Oxidizing agent Substance that oxidizes another substance in an oxidation-reduction reaction.

Ozone One of the allotropes of oxygen, O_3.

Ozone layer Region high in the atmosphere containing a concentration of ozone that absorbs and prevents certain high-energy ultraviolet components of sunlight from reaching the surface of the earth.

pH Method of designating the acidity or basicity of a solution. The negative logarithm of the molar concentration of the hydronium ion.

Partial pressure Pressure exerted by one of the components of a mixture of gases.

Percent composition A list, in percent terms, of the proportions by mass of the elements in a compound.

Period Horizontal row of elements in the periodic table.

Periodic table of the elements Table of the known elements listed in order of atomic number and arranged according to valence electron configuration.

Phase In a sample of matter, a region with distinct boundaries across which one or more properties change discontinuously. Also used to denote one of the states of matter, as in "liquid phase."

Photosynthesis Biological process by which the chlorophyll in plants uses light energy to convert carbon dioxide and water to glucose.

Polar covalent bond Covalent bond in which the pair of electrons is not equally shared between the two atoms, giving rise to an electric dipole.

Polyamide Polymer, such as nylon or a protein, in which the monomer units are joined by amide linkages.

Polyfunctional molecule A molecule, usually organic, in which more than one group of atoms have chemical or physical significance.

Polymer Large molecule composed of repeating units, which can be alike or assorted.

Potential energy Energy stored in a system by virtue of the relative positions of its parts.

Precision Closeness of agreement between successive measurements of the same quantity.

Pressure Force per unit area.

Primary alcohol Alcohol in which the —OH group is attached to a carbon atom which itself is attached to no more than one other carbon atom, R—CH₂OH.

Proportionality Mathematical relation in which one variable is equal to another variable multiplied by a constant. A linear relation.

Protein Biological polymer, the characteristic monomer units of which all are alpha amino acids from a list of about 20.

Proton One of the fundamental particles from which nuclei are constructed. The proton has a mass of about 1 amu and a single positive charge.

Proton transfer reaction Another term for acid-base reaction. A process in which a proton is transferred from an acid to a base.

Pure substance Substance in which no purification technique can produce further change. Element in which all the atoms have the same atomic number or molecular compound in which all the molecules are alike.

Qualitative analysis Process that yields a list of the compounds present in a sample.

Quantitative analysis Process that yields a list of the compounds present in a sample and gives the relative amounts of each.

Quantum number Designation of the energy level or of one of the classes of sublevel occupied by an electron in an atom.

R group Shorthand designation used by organic chemists to designate any portion of a molecule that is not relevant to the subject at hand.

Radioactive decay Process by which the nuclei of unstable isotopes change to more stable arrangements by emitting high-energy particles or gamma rays.

Radioactivity Emission of high-energy particles or gamma rays during radioactive decay.

Radiocarbon dating Measuring the amounts of radioactive isotopes present in an object and using the known half-lives of those isotopes to make calculations of its age.

Rate constant A constant number that when multiplied appropriately by the concentrations of reactants gives the rate of a chemical reaction.

Reaction equation Statement that describes a chemical reaction by giving the kinds and relative amounts of reactants and products.

Redox reaction Oxidation-reduction or electron transfer reaction.

Reducer Substance that donates electrons in a redox reaction.

Reducing agent Reducer.

Reduction Process of gaining electrons in a redox reaction.

Representative element Any element with a partially filled *s* or *p* subshell. An element of any group designated with a group number ending in A.

Salt Product of the reaction between an acid and a base. An ionic compound containing the cation of the base and the anion of the acid.

Saturated fat A fat, the molecules of which have no (or very few) carbon-carbon multiple bonds.

Saturated hydrocarbon Compound that contains only carbon and hydrogen and that has no carbon-carbon multiple bonds.

Saturated solution Solution that will dissolve no more of the solute with which it is said to be saturated at any particular temperature.

Scale Linear measuring device marked with divisions showing units of length. Also, the lining deposited in pipes and boilers by hard water.

Secondary alcohol An alcohol the molecules of which have the —OH group attached to a carbon atom which is itself attached to two other carbon atoms.

Sickle-cell anemia Genetic disease of the red blood cells caused by a defect in the secondary structure of the hemoglobin molecules.

Significant figures Those digits in a reported value that are known to be accurate, plus one additional digit that is approximate.

Solid State of matter in which a sample has both its own volume and its own shape.

Solubility Amount of a solute dissolved in a particular solvent at saturation at a specified temperature.

Solubility product An equilibrium constant that describes the concentrations of the ions of a compound in its saturated solution. Sometimes called *ion product*.

Solute Substance that is being dissolved in a solution. The component of a solution that is present in the smaller amount.

Solution Homogeneous mixture of two or more separate substances.

Solvated ion In a solution, an ion that is associated with one or more molecules of the solvent.

Solvent Solution component present in the greatest relative amount. The substance in which another is being dissolved.

Space-filling model As distinguished from a "tinkertoy" or stick-and-ball model, a model of a molecule that shows the relative volumes of the atoms or ions in their relative positions.

Specific heat Amount of energy required to raise the temperature of 1 g of a substance 1 °C.

Spectroscopy Study of the energies of the photons emitted by atoms and molecules under certain conditions.

Standard atmosphere Pressure of 760 torr.

Standard temperature and pressure (STP) A pressure of 760 torr and a temperature of 0 °C.

Starch Biological polymer composed of glucose monomers linked in a particular pattern.

Stoichiometric calculation Calculation involving the quantities of elements or compounds in a chemical reaction.

Strong acid Acid that is entirely or almost entirely ionized and dissociated in aqueous solution.

Strong base Base that is entirely or almost entirely ionized and dissociated in aqueous solution.

Strong electrolyte Acid, base, or salt that is entirely or almost entirely dissociated in aqueous solution.

Structural isomers Two molecules with the same empirical formula but different arrangements of which atom is bonded to which.

Sublimation Passage of a solid directly into a vapor without first becoming a liquid.

Subshell One of the allowed energy levels into which the main or principal electron energy levels of atoms are split.

Substitution Reaction in which an atom or group of atoms in a molecule is replaced by an atom or group of another kind.

Supersaturation State of a solution that has dissolved more solute than it can contain when saturated. Supersaturated solutions are not stable.

Suspension Mixture of more than one phase in which there remain suspended particles that are much larger than individual molecules but too small to settle out of the solution.

Systematic error Measurement error that occurs the same way in every measurement and is caused by a characteristic of the measuring system or method.

Système International (SI) The official version of the modern metric system.

Temperature Measure of how hot or cold a substance is. An indication of the average kinetic energy of the individual particles of a substance.

Temporary dipole force See *London force.*

Tertiary alcohol An alcohol the molecules of which contain —OH groups attached to carbon atoms, each of which is itself attached to three other carbon atoms.

Tetrahedral angle An angle of 109.5°. The angle between any two of four lines radiating at equal angles from a single point.

Tinkertoy model Stick-and-ball model of a molecule, designed to show the bond structure of the molecule.

Torr Metric unit of pressure equivalent to the manometer height of 1 mm of mercury or 1/760 standard atm.

Transition elements The three groups of ten elements each in which the d shell is partly filled.

Transmutation Transformation of one element to another.

Triple bond Multiple covalent bond consisting of three pairs of shared electrons.

Universal gas constant The constant R that appears in the universal gas equation (the ideal gas equation). One of the values of R is 0.08206 L atm/mol kelvin.

Unsaturated fat A fat in which the long carbon chains contain some number of double bonds.

Unsaturated hydrocarbon Compound of carbon and hydrogen in which there are one or more multiple carbon-carbon bonds.

Valence electrons Electrons with the highest N (or main shell) value in an atom.

van der Waals forces Dipole or induced dipole forces that hold together the condensed states of molecular substances.

Vapor Gaseous phase of a substance in contact with its condensed state. A vapor is simply a gas in particular circumstances.

Vapor pressure   Pressure exerted by a vapor in equilibrium with its condensed state at a particular temperature.

Vaporization   Transformation of a condensed state (usually taken to be a liquid—see *sublimation*) into the vapor state.

Velocity   Rate of change of position (or speed) in a given direction.

Vinyl monomer   Ethene, or ethene with an R group replacing one hydrogen atom.

Vinyl polymer   Addition polymer made through a free radical reaction of carbon-carbon double bonds in vinyl monomers.

Volatility   Measure of the tendency of a substance to evaporate or vaporize. Volatile substances vaporize easily and have high vapor pressures.

Wave mechanics   Theory that explains the behavior of electrons and other fundamental particles by regarding them as waves instead of moving particles.

Weak acid   Acid that is only partially ionized in aqueous solution.

Weak base   Base that is only partially ionized in aqueous solution.

Weight   Force by which a given mass is attracted to the earth or to some other body.

Answers to Exercises and Selected Problems

CHAPTER 1

Exercises

1.1 $5 \text{ cups} \times \dfrac{4 \text{ teaspoons}}{\text{cup}} \times \dfrac{\text{tablespoon}}{3 \text{ teaspoons}} = 6\frac{2}{3} \text{ tablespoons}$

1.2 $0.242 \text{ mi} \times \dfrac{5{,}280 \text{ ft}}{\text{mi}} \times \dfrac{\text{yd}}{3 \text{ ft}} = 426 \text{ yd}$

1.3 $1.32 \text{ yd} \times 3.55 \text{ yd} \times 0.23 \text{ yd} \times \left(\dfrac{36 \text{ in}}{\text{yd}}\right)^3 \times \dfrac{\text{gal}}{231 \text{ in}^3} = 220 \text{ gal}$

1.4 $10.5 \text{ ft} \times 8.0 \text{ ft} \times \left(\dfrac{\text{yd}}{3 \text{ ft}}\right)^2 \times \dfrac{1 \text{ roll}}{4.0 \text{ yd}^2} = 2.3 \text{ rolls}$

1.5 $1 \text{ ft}^3 \times \left(\dfrac{12 \text{ in}}{\text{ft}}\right)^3 \times \dfrac{4.55 \text{ oz}}{\text{in}^3} \times \dfrac{\text{lb}}{16 \text{ oz}} = 491 \text{ lb}$

1.6 $0.050 \text{ yd}^3 \times \left(\dfrac{3 \text{ ft}}{\text{yd}}\right)^3 \times \dfrac{12 \text{ in}}{\text{ft}} \times \dfrac{1}{0.25 \text{ in}} = 65 \text{ ft}^2$

Problems

1. $5.0 \text{ lb} + 3.0 \text{ lb} + 3.0 \text{ lb} + 8.0 \text{ lb} + 10.0 \text{ lb} = 29.0 \text{ lb}$

     (You can ignore the 0.03 lb and the 0.02 lb, because these quantities are negligible compared to the other quantities.)

7. $\dfrac{\text{day}}{145 \text{ francs}} \times \dfrac{\text{franc}}{0.185 \text{ dollars}} \times \dfrac{0.49 \text{ dollars}}{\text{mark}} \times 550 \text{ marks} = 10 \text{ days}$

     Yes, she has enough money.

11. $4.0 \text{ in} \times \dfrac{\text{ft}}{12 \text{ in}} = 0.33 \text{ ft}$

$$\frac{\text{line}}{2.5 \text{ ft} \times 0.33 \text{ ft}} \times \frac{43 \text{ ft}^2}{\text{qt}} \times \frac{4 \text{ qt}}{\text{gal}} \times 15 \text{ gal} = 3.1 \times 10^3 \text{ lines}$$

19. $$6.0 \times 10^{23} \text{ snowflakes} \times \frac{0.0000223 \text{ in} \times 0.250 \text{ in}^2}{\text{snowflake}} \times \left(\frac{\text{ft}}{12 \text{ in}}\right)^3 \times \left(\frac{\text{mi}}{5280 \text{ ft}}\right)^2$$

$$\times \frac{1}{400 \text{ mi} \times 500 \text{ mi}} = 348 \text{ ft} = 3.5 \times 10^2 \text{ ft}$$

CHAPTER 2

Exercises

2.1 $$290 \text{ mm} \times \frac{\text{cm}}{10 \text{ mm}} \times \frac{\text{in}}{2.54 \text{ cm}} \times \frac{\text{ft}}{12 \text{ in}} = 0.95 \text{ ft}$$

2.2 $$24,000 \text{ mi} \times \frac{5,280 \text{ ft}}{\text{mi}} \times \frac{12 \text{ in}}{\text{ft}} \times \frac{2.54 \text{ cm}}{\text{in}} \times \frac{\text{m}}{100 \text{ cm}} = 3.9 \times 10^7 \text{ m}$$

2.3 $$\frac{\text{trip}}{1 \text{ gal}} \times \frac{\text{gal}}{4 \text{ qt}} \times \frac{\text{qt}}{0.946 \text{ L}} \times \frac{\text{L}}{1000 \text{ cm}^3} \times \left(\frac{100 \text{ cm}}{\text{m}}\right)^3 \times 0.75 \text{ m}^3 = 2 \times 10^2 \text{ trips}$$

2.4 $$12.0 \text{ oz} \times \frac{\text{qt}}{32 \text{ oz}} \times \frac{0.946 \text{ L}}{\text{qt}} \times \frac{10 \text{ dL}}{\text{L}} \times \frac{1.10 \text{ marks}}{5.0 \text{ dL}} = 0.78 \text{ marks}$$

2.5 $$0.500 \text{ lb} \times \frac{453.6 \text{ g}}{\text{lb}} = 227 \text{ g}; \text{ You should buy 200 g.}$$

2.6 $$167.5 \text{ lb} \times \frac{453.6 \text{ g}}{\text{lb}} = 7.598 \times 10^4 \text{ g}$$

$$7.598 \times 10^4 \text{ g} \times \frac{\text{kg}}{1000 \text{ g}} = 75.98 \text{ kg}$$

2.7 $$1 \text{ L} \times \frac{1000 \text{ mL}}{\text{L}} \times \frac{1 \text{ cal}}{\text{mL degree}} \times (100 \text{ degrees} - 25 \text{ degrees}) = 7.5 \times 10^4 \text{ cal}$$

2.8 $$25.0 \text{ g} \times \frac{1 \text{ cal}}{\text{g } °C} \times \frac{\text{kJ}}{239 \text{ cal}} \times 1 \text{ } °C = 0.105 \text{ kJ}$$

$$25.0 \text{ g} \times \frac{1 \text{ cal}}{\text{g } °C} \times \frac{\text{kJ}}{239 \text{ cal}} \times 41 \text{ } °C = 4.29 \text{ kJ}$$

2.9 $$32 \text{ } °C + 273 = 305 \text{ K}; 178 \text{ } °C + 273 = 451 \text{ K}$$

2.10 $$212 \text{ K} - 273 = -61 \text{ } °C; 298 \text{ K} - 273 = 25 \text{ } °C$$

2.11 $$3.31 \text{ kg} \times \frac{1000 \text{ g}}{\text{kg}} \times \frac{\text{cm}^3}{2.16 \text{ g}} = 1.53 \times 10^3 \text{ cm}^3$$

2.12 $$\frac{32.7 \text{ g}}{33.1 \text{ mL}} = 0.988 \text{ g/mL}; \text{ This is less dense then the density of water at 4 } °C.$$

Problems

2.   a. $$2.3 \text{ km} \times \frac{1000 \text{ m}}{\text{km}} \times \frac{100 \text{ cm}}{\text{m}} \times \frac{\text{in}}{2.54 \text{ cm}} \times \frac{\text{ft}}{12 \text{ in}} \times \frac{\text{mi}}{5280 \text{ ft}} = 1.4 \text{ mi}$$

   b. $$4 \text{ ft} \times \frac{12 \text{ in}}{\text{ft}} + 2.2 \text{ in} \times \frac{2.54 \text{ cm}}{\text{in}} = 1 \times 10^2 \text{ cm}$$

c. $223 \text{ Å} \times \dfrac{10^{-8} \text{ cm}}{\text{Å}} \times \dfrac{10 \text{ mm}}{\text{cm}} = 2.23 \times 10^{-5} \text{ mm}$

d. $1.433 \text{ cm} \times \dfrac{\text{Å}}{10^{-8} \text{ cm}} = 1.433 \times 10^{8} \text{ Å}$

4. $1182 \text{ mi} \times \dfrac{5280 \text{ ft}}{\text{mi}} \times \dfrac{12 \text{ in}}{\text{ft}} \times \dfrac{2.54 \text{ cm}}{\text{in}} \times \dfrac{\text{m}}{100 \text{ cm}} \times \dfrac{\text{km}}{1000 \text{ m}} = 1{,}903 \text{ km}$

8. $30 \ \mu\text{m} \times \dfrac{\text{m}}{10^{6} \ \mu\text{m}} = 3 \times 10^{-5} \text{ m}$

$3 \times 10^{-5} \text{ m} \times \dfrac{1000 \text{ mm}}{\text{m}} = 3 \times 10^{-2} \text{ mm}$

15. $5.89 \times 10^{2} \text{ L} \times \dfrac{1000 \text{ cm}^3}{\text{L}} = 5.89 \times 10^{5} \text{ cm}^3$

20. $1555 \text{ students} \times \dfrac{10.0 \text{ cm}^3}{\text{student}} \times \dfrac{\text{L}}{1000 \text{ cm}^3} = 15.6 \text{ L}$

No, the heath center needs 15.6 L of serum; 1.75 L will not be enough serum.

24. a. $155 \text{ lb} \times \dfrac{453.6 \text{ g}}{\text{lb}} \times \dfrac{\text{kg}}{1000 \text{ g}} = 70.3 \text{ kg}$

b. $44 \text{ kg} \times \dfrac{1000 \text{ g}}{\text{kg}} \times \dfrac{\text{lb}}{453.6 \text{ g}} = 97 \text{ lb}$

33. $\dfrac{12 \text{ yen}}{\text{kg}} \times \dfrac{\$1}{185 \text{ yen}} \times \dfrac{\text{kg}}{2.205 \text{ lb}} = \dfrac{\$0.029}{\text{lb}}$

Rice is more expensive in the U.S.

37. $250 \text{ mL} \times \dfrac{1 \text{ cal}}{\text{g deg}} \times (100 \text{ deg} - 25 \text{ deg}) \times \dfrac{\text{kJ}}{239 \text{ cal}} = 78 \text{ kJ}$

54. $\dfrac{8.56 \text{ L soln two}}{3.45 \text{ L soln one}} \times \dfrac{1000 \text{ mL}}{\text{L}} \times 150 \text{ mL} = 372 \text{ mL soln two}$

58. $58.3 \text{ cm} - 55.0 \text{ cm} = 3.3 \text{ cm}$

$\dfrac{5.21 \text{ kg}}{15.2 \text{ cm} \times 22.9 \text{ cm} \times 3.3 \text{ cm}} \times \dfrac{1000 \text{ g}}{\text{kg}} = 4.5 \dfrac{\text{g}}{\text{cm}^3}$

CHAPTER 4

Exercises

4.1 $2.10 \text{ g} - 1.05 \text{ g} = 1.05 \text{ g Mg}; \dfrac{1.05 \text{ g O}}{1.05 \text{ g Mg}} = \dfrac{1.00}{1.00}$

4.2 $14.50 \text{ g} - 14.09 \text{ g} = 0.41 \text{ g O}$

$14.09 \text{ g} - 8.15 \text{ g} = 5.94 \text{ g Ag}$

$\dfrac{5.94 \text{ g Ag}}{0.41 \text{ g O}} = \dfrac{14}{1.0}$

4.3 $\dfrac{9.012 \text{ g Be}}{44.96 \text{ g Sc}} = \dfrac{0.2004}{1.000}$

4.4 $\dfrac{126.90 \text{ amu}}{\text{atom}} \times \dfrac{1.660 \times 10^{-24} \text{ g}}{\text{amu}} = 2.107 \times 10^{-22}$ g/atom

$\dfrac{10.81 \text{ amu}}{\text{atom}} \times \dfrac{1.660 \times 10^{-24} \text{ g}}{\text{amu}} = 1.794 \times 10^{-23}$ g/atom

4.5 ^{24}Mg $\times$ 0.787 $\times$ 24.0 = 18.9

^{25}Mg $\times$ 0.101 $\times$ 25.0 = 2.53

^{26}Mg $\times$ 0.112 $\times$ 26.0 = $\underline{2.91}$

atomic weight of Mg 24.3

Problems

1. 12.11 g − 11.78 g = 0.33 g O

12.11 g − 10.43 g = 1.68 g Zn and O

1.68 g − 0.33 g = 1.35 g Zn

$\dfrac{0.33 \text{ g O}}{1.35 \text{ g Zn}} = 0.24$

6. $\dfrac{\text{mass 1 atom Ag}}{\text{mass 1 atom Cu}} = \dfrac{\text{mass x atoms Ag}}{\text{mass x atoms Cu}} = \dfrac{(5.348/2) \text{ g}}{1.588 \text{ g}} = 1.684$

(where there are x atoms of Cu in 1.588 g)

11. a. $\dfrac{4.48 \times 10^{-23} \text{ g}}{2.66 \times 10^{-23} \text{ g}} = 1.69$

12. a. $\dfrac{10.8 \text{ amu}}{1.01 \text{ amu}} = 10.7$

b. $\dfrac{85.5 \text{ amu}}{14.0 \text{ amu}} = 6.11$

14. a. 0.922 × 27.98 = 25.8

0.047 × 28.98 = 1.3

0.031 × 29.97 = $\underline{0.93}$

atomic weight of Si 28.0

16. 0.4946 × 80.9163 = 40.02

79.904 − 40.02 = 39.88

100 − 49.46 = 50.54

$\dfrac{39.88}{0.5054} = 78.91$

17. $\dfrac{39.96 \text{ amu}}{\text{atom Ca}} \times \dfrac{\text{atom H}}{1.008 \text{ amu}} \times \dfrac{1.000 \text{ amu}}{\text{atom H}} = 39.64$ amu

CHAPTER 5

Exercises

5.1 magnesium, molybdenum, manganese, tantalum, titanium, potassium, sodium, silver, iron

5.2   Ba, F, Si, As

5.3   metal, nonmetal, nonmetal, metal, metal, nonmetal, nonmetal, nonmetal

5.4   solid, solid, gas, gas, liquid, solid, liquid, gas

5.5   1 magnesium atom, 2 oxygen atoms

        2 sodium atoms, 1 sulfur atom, 4 oxygen atoms

        2 aluminum atoms, 3 sulfur atoms, 12 oxygen atoms

5.6 HO, CH_2, C_2H_3O

5.7   a. $14.00307 + 2(1.00797) + 3.0 = 19$

        b. $15.00011 + 1.00797 + 2(3.0) = 23$

        c. $14.00307 + 1.00797 + 2(2.0141) = 19.0392$

5.8 $^{15}N + 2\ ^1H + ^3H$

        $15.00011 + 2(1.00797) + 3.0 = 20$

5.9 $137.34 + 2(18.998) = 175.34$

        $55.847 + 6(51.996) + 21(15.9994) = 703.81$

        $39.098 + 79.904 + 3(15.9994) = 167.00$

5.10   barium fluoride, molybdenum triboride, aluminum hydroxide

5.11 $NaCl + 3\ NaBrO \longrightarrow NaClO_3 + 3\ NaBr$

5.12 $2\ C_7H_8 + 18\ O_2 \longrightarrow 14\ CO_2 + 8\ H_2O$

5.13   30.9738 g/mol, 131.30 g/mol

5.14 NO_2: $14.0067 + 2(15.9994) = 46.0055$; BF_3: $10.81 + 3(18.99840) = 67.81$

Problems

1.   manganese, tantalum, titanium, and potassium

3.   Sr, Rh, P, Ce, Cs

5.   metals: Cr, Ba, K; nonmetals: Te, B, O, Ar, Kr, As, P

6.   solid: Na, Ti, B, C, P, Rh, Cs, I, Si, As; liquid: Br_2; gas: Cl_2

7.   2 hydrogen atoms, 1 sulfur atom, 4 oxygen atoms

        1 sodium atom, 1 nitrogen atom, 3 oxygen atoms

        1 calcium atom, 2 chlorine atoms

        2 chlorine atoms, 1 carbon atom, 1 oxygen atom

        1 aluminum atom, 3 bromine atoms

10.   one ^{12}C atom   12.0000

        one ^{18}O atom   <u>17.999</u>

        mass of molecule 29.999

12.   a.   one ^{14}N atom   14.00307

             two ^{16}O atoms $2 \times 15.995 =$ <u>31.998</u>

             mass of molecule   46.993

14. 1×40.08 = 40.08

        2×35.453 = <u>70.906</u>

        mass of $CaCl_2$ 110.986

17. calcium chloride, nitrogen dioxide, magnesium bromide, potassium fluoride, arsenic diiodide, barium trisulfide, lead oxide

19. a. KCl   b. BeS   c. BN

20. a. BrF_5 b. CCl_4

21. a. $2 H_2 + O_2 \longrightarrow 2 H_2O$

   b. $N_2 + 3 I_2 \longrightarrow 2 NI_3$

   c. $HCl + NaOH \longrightarrow NaCl + H_2O$

   d. $2 KClO_3 \longrightarrow 2 KCl + 3 O_2$

   e. $CuCl_2 + H_2S \longrightarrow CuS + 2 HCl$

   f. $Ca(NO_3)_2 + 2 Pb \longrightarrow Ca(NO_2)_2 + 2 PbO$

   g. $H_2SO_4 + 2 KOH \longrightarrow K_2SO_4 + 2 H_2O$

   h. $2 C_6H_{14}N + 21 O_2 \longrightarrow 12 CO_2 + 2 NO_2 + 14 H_2O$

   i. $BaBr_2 + Cs_2CO_3 \longrightarrow BaCO_3 + 2 CsBr$

   j. $3 CuO + 2 NH_3 \longrightarrow 3 Cu + N_2 + 3 H_2O$

23. $3.0 \text{ mol Ag} \times \dfrac{6.023 \times 10^{23} \text{ atoms}}{\text{mol}} = 1.8 \times 10^{24} \text{ atoms}$

25. $\dfrac{134.9056 \text{ g Ba}}{\text{mol Ba}} \times \dfrac{\text{mol Ba}}{6.023 \times 10^{23} \text{ atoms Ba}} = 2.240 \times 10^{-22} \dfrac{\text{g}}{\text{atom}}$

27. $1 \text{ mol letters} \times \dfrac{6.023 \times 10^{23} \text{ letters}}{\text{mol letters}} \times \dfrac{\text{line}}{75 \text{ letters}} \times \dfrac{\text{page}}{27 \text{ lines}} = 2.974 \times 10^{20} \text{ pages}$

   (If a library has a million books, each with 1000 pages, this is enough letters for about 300 billion libraries)

32. P: 30.9738 g/mol; Ba: 137.34 g/mol; Pd: 106.4 g/mol; Sn: 118.69 g/mol

34.
$$4 \times 30.9738 = 959.376 \text{ g/mol}$$
$$10 \times 15.9994 = \underline{159.994 \text{ g/mol}}$$
g-molecular weight P_4O_{10} 1119.370 g/mol

$$2 \times 14.0067 = 28.0134 \text{ g/mol}$$
$$4 \times 1.00797 = \underline{4.03188 \text{ g/mol}}$$
g-molecular weight N_2H_4 32.0452 g/mol

36. $\dfrac{6.023 \times 10^{23} \text{ atoms}}{\text{mol}} \times \dfrac{\text{mol}}{12.000 \text{ g C}} \times \dfrac{1000 \text{ g}}{\text{kg}} \times \dfrac{15.000 \text{ kg C}}{\text{mol}} = 7.529 \times 10^{26} \dfrac{\text{atoms}}{\text{mol}}$

CHAPTER 6

Exercises

6.1 $1.31 \text{ mol } C_6H_5O_4N \times \dfrac{4 \text{ mol O}}{\text{mol } C_6H_5O_4N} = 5.24 \text{ mol O}$

6.2 $10.5 \text{ g He} \times \dfrac{\text{mol He}}{4.00 \text{ g He}} = 2.62 \text{ mol He}$

6.3 $0.0562 \text{ mol Ag} \times \dfrac{107.87 \text{ g Ag}}{\text{mol Ag}} = 6.06 \text{ g Ag}$

6.4 $176.2 \text{ g K}_2\text{SO}_4 \times \dfrac{\text{mol K}_2\text{SO}_4}{174.26 \text{ g K}_2\text{SO}_4} = 1.011 \text{ mol K}_2\text{SO}_4$

6.5 $1.55 \text{ mol CO} \times \dfrac{28.0 \text{ g CO}}{\text{mol CO}} = 43.4 \text{ g CO}$

6.6 $320.5 \text{ g Al}_2\text{O}_3 \times \dfrac{\text{mol Al}_2\text{O}_3}{101.96 \text{ g Al}_2\text{O}_3} \times \dfrac{3 \text{ mol O}}{\text{mol Al}_2\text{O}_3} = 9.430 \text{ mol O}$

6.7 $156 \text{ g CO}_2 \times \dfrac{\text{mol CO}_2}{44.0 \text{ g CO}_2} \times \dfrac{2 \text{ mol O}}{\text{mol CO}_2} = 7.09 \text{ mol O}$

6.8 $82 \text{ g Ca(NO}_3)_2 \times \dfrac{\text{mol Ca(NO}_3)_2}{164 \text{ g Ca(NO}_3)_2} \times \dfrac{6 \text{ mol O}}{\text{mol Ca(NO}_3)_2} \times \dfrac{16.0 \text{ g O}}{\text{mol O}} = 48 \text{ g O}$

6.9 $46.3 \text{ g H}_2\text{S} \times \dfrac{\text{mol H}_2\text{S}}{34.08 \text{ g H}_2\text{S}} \times \dfrac{2 \text{ mol H}}{\text{mol H}_2\text{S}} \times \dfrac{1.008 \text{ g H}}{\text{mol H}} = 2.74 \text{ g H}$

6.10 $\dfrac{3.20 \text{ g}}{4.40 \text{ g}} \times 100\% = 72.7\%$

6.11 $1.20 \text{ g} + 0.10 \text{ g} + 0.70 \text{ g} = 2.00 \text{ g}$

$\dfrac{1.20 \text{ g}}{2.00 \text{ g}} \times 100\% = 60.0\%; \dfrac{0.10 \text{ g}}{2.00 \text{ g}} \times 100\% = 5.0\%; \dfrac{0.70 \text{ g}}{2.00 \text{ g}} \times 100\% = 35\%$

6.12 $0.403 \text{ g} + 6.412 \text{ g} + 12.800 \text{ g} = 19.615$

$\dfrac{0.403 \text{ g}}{19.615 \text{ g}} \times 100\% = 2.05\%; \dfrac{6.412 \text{ g}}{19.615 \text{ g}} \times 100\% = 32.69\%; \dfrac{12.800 \text{ g}}{19.615 \text{ g}} \times 100\% = 65.256\%$

6.13 $118.9 \text{ g} - (23.0 \text{ g} + 16.0 \text{ g}) = 79.9 \text{ g Br}$

$\dfrac{23.0 \text{ g}}{118.9 \text{ g}} \times 100\% = 19.3\%; \dfrac{16.0 \text{ g}}{118.9 \text{ g}} \times 100\% = 13.5\%; \dfrac{79.9 \text{ g}}{118.9 \text{ g}} \times 100\% = 67.2\%$

6.14 $3.0 \text{ g C} \times \dfrac{\text{mol C}}{12.0 \text{ g C}} = 0.25 \text{ mol C}; 1.0 \text{ g H} \times \dfrac{\text{mol H}}{1.008 \text{ g H}} = 0.99 \text{ mol H};$

$4.0 \text{ g O} \times \dfrac{\text{mol O}}{16.0 \text{ g O}} = 0.25 \text{ mol O}$

$\dfrac{0.25 \text{ mol C}}{0.25} = 1.0 \text{ mol C}; \dfrac{0.99 \text{ mol H}}{0.25} = 4.0 \text{ mol H}$

$\dfrac{0.25 \text{ mol O}}{0.25} = 1.0 \text{ mol O}; \quad \text{CH}_4\text{O}$

6.15 $5.06 \text{ g Na} \times \dfrac{\text{mol Na}}{23.0 \text{ g Na}} = 0.220 \text{ mol Na}; 2.64 \text{ g C} \times \dfrac{\text{mol C}}{12.0 \text{ g C}} = 0.220 \text{ mol C};$

$10.56 \text{ g O} \times \dfrac{\text{mol O}}{16.00 \text{ g O}} = 0.660 \text{ mol O}; 2.22 \times 10^{-1} \text{ g H} \times \dfrac{\text{mol H}}{1.008 \text{ g H}} = 0.220 \text{ mol H}$

$\dfrac{0.220 \text{ mol Na}}{0.220} = 1.00 \text{ mol Na}; \dfrac{0.220 \text{ mol C}}{0.220} = 1.00 \text{ mol C}$

$\dfrac{0.660 \text{ mol O}}{0.220} = 3.00 \text{ mol O}; \dfrac{0.220 \text{ mol H}}{0.220} = 1.00 \text{ mol H}; \quad \text{NaHCO}_3$

6.16 $52 \text{ g Zn} \times \dfrac{\text{mol Zn}}{65.4 \text{ g Zn}} = 0.795 \text{ mol Zn}; 9.6 \text{ g C} \times \dfrac{\text{mol C}}{12.0 \text{ g C}} = 0.800 \text{ mol C};$

$38.4 \text{ g O} \times \dfrac{\text{mol O}}{16.0 \text{ g O}} = 2.40 \text{ mol O}$

$$\frac{0.795 \text{ mol Zn}}{0.795} = 1.00 \text{ mol Zn}; \quad \frac{0.800 \text{ mol C}}{0.795} = 1.01 \text{ mol C};$$

$$\frac{2.40 \text{ mol O}}{0.795} = 3.02 \text{ mol O}; \qquad ZnCO_3$$

6.17 $11.25 \text{ g C} \times \dfrac{\text{mol C}}{12.01 \text{ g C}} = 0.9367 \text{ mol C}; \; 2.5 \text{ g H} \times \dfrac{\text{mol H}}{1.008 \text{ g H}} = 2.5 \text{ mol H}$

$$\frac{0.9367 \text{ mol C}}{0.9367} = 1.000 \text{ mol C}; \frac{2.5 \text{ mol H}}{0.9367} = 2.67 \text{ mol H}$$

$3 \times 1.000 \text{ mol C} = 3.000 \text{ mol C}; \; 3 \times 2.67 \text{ mol H} = 8.0 \text{ mol H}; \qquad C_3H_8$

6.18 $72.36 \text{ g Fe} \times \dfrac{\text{mol Fe}}{55.85 \text{ g Fe}} = 1.296 \text{ mol Fe}; \; 27.64 \text{ g O} \times \dfrac{\text{mol O}}{16.00 \text{ g O}} = 1.728 \text{ mol O}$

$$\frac{1.296 \text{ mol Fe}}{1.296} = 1.000 \text{ mol Fe}; \frac{1.728 \text{ mol O}}{1.296} = 1.333 \text{ mol O}$$

$3 \times 1.000 \text{ mol Fe} = 3.000 \text{ mol Fe}; \; 3 \times 1.333 \text{ mol O} = 3.999 \text{ mol O}; \qquad Fe_3O_4$

6.19 $\dfrac{1 \text{ mol Na}}{\text{mol NaOH}} \times \dfrac{23.0 \text{ g Na}}{\text{mol Na}} \times \dfrac{\text{mol NaOH}}{40.0 \text{ g NaOH}} \times 100\% = 57.5\% \text{ Na}$

$\dfrac{1 \text{ mol O}}{\text{mol NaOH}} \times \dfrac{16.0 \text{ g O}}{\text{mol O}} \times \dfrac{\text{mol NaOH}}{40.0 \text{ g NaOH}} \times 100\% = 40.0\% \text{ O}$

$\dfrac{1 \text{ mol H}}{\text{mol NaOH}} \times \dfrac{1.01 \text{ g H}}{\text{mol H}} \times \dfrac{\text{mol NaOH}}{40.0 \text{ g NaOH}} \times 100\% = 2.53\% \text{ H}$

6.20 $\dfrac{1 \text{ mol Ca}}{\text{mol Ca(ClO}_2)_2} \times \dfrac{40.08 \text{ g Ca}}{\text{mol Ca}} \times \dfrac{\text{mol Ca(ClO}_2)_2}{175.0 \text{ g Ca(ClO}_2)_2} \times 100\% = 22.9\% \text{ Ca}$

$\dfrac{2 \text{ mol Cl}}{\text{mol Ca(ClO}_2)_2} \times \dfrac{35.45 \text{ g Cl}}{\text{mol Cl}} \times \dfrac{\text{mol Ca(ClO}_2)_2}{175.0 \text{ g Ca(ClO}_2)_2} \times 100\% = 40.5\% \text{ Cl}$

$\dfrac{4 \text{ mol O}}{\text{mol Ca(ClO}_2)_2} \times \dfrac{16.00 \text{ g O}}{\text{mol O}} \times \dfrac{\text{mol Ca(ClO}_2)_2}{175.0 \text{ g Ca(ClO}_2)_2} \times 100\% = 36.6\% \text{ O}$

6.21 $3 \text{ mol Cs} \times \dfrac{132.9 \text{ g Cs}}{\text{mol Cs}} = 398.7 \text{ g Cs}; \; 1 \text{ mol N} \times \dfrac{14.01 \text{ g N}}{\text{mol N}} = 14.01 \text{ g N}$

28.46 g Cs to 1.000 g N

6.22 $2.63 \text{ g F} \times \dfrac{\text{mol F}}{19.0 \text{ g F}} \times \dfrac{1 \text{ mol M}}{3 \text{ mol F}} = 0.0461 \text{ mol M}$

$$\frac{9.63 \text{ g M}}{0.0461 \text{ mol M}} = 209 \text{ g/mol}$$

6.23 $28.40 \text{ g O} \times \dfrac{\text{mol O}}{16.00 \text{ g O}} = 1.775 \text{ mol O}$

$60.94 \text{ g X} \times \dfrac{4 \text{ mol O}}{1 \text{ mol X}} \times \dfrac{1}{1.775 \text{ mol O}} = 137.3 \text{ g/mol}$

$10.66 \text{ g Y} \times \dfrac{4 \text{ mol O}}{2 \text{ mol Y}} \times \dfrac{1}{1.775 \text{ mol O}} = 12.01 \text{ g/mol}$

6.24 $\dfrac{60}{31} \approx 2; \quad C_2H_6O_2; \quad 2(12.01) + 6(1.008) + 2(16.00) = 62.07 \text{ g/mol}$

6.25 $\dfrac{200}{102} \approx 2; \quad C_8H_6O_2Cl_2; \quad 8(12.01) + 6(1.008) + 2(16.00) + 2(35.45) = 205 \text{ g/mol}$

6.26 $37.5 \text{ g C} \times \dfrac{\text{mol C}}{12.0 \text{ g C}} = 3.13 \text{ mol C}; 4.20 \text{ g H} \times \dfrac{\text{mol H}}{1.008 \text{ g H}} = 4.17 \text{ mol H};$

$58.37 \text{ g O} \times \dfrac{\text{mol O}}{16.00 \text{ g O}} = 3.648 \text{ mol O}$

$\dfrac{3.13 \text{ mol C}}{3.13} = 1.00 \text{ mol C}; \dfrac{4.17 \text{ mol H}}{3.13} = 1.33 \text{ mol H}; \dfrac{3.648 \text{ mol O}}{3.13} = 1.17 \text{ mol O}$

$6 \times 1.00 \text{ mol C} = 6.00 \text{ mol C}; 6 \times 1.33 \text{ mol H} = 8.00 \text{ mol H};$
$6 \times 1.17 \text{ mol O} = 7.02 \text{ mol O}; \quad C_6H_8O_7$

$6(12.01) + 8(1.008) + 7(16.00) = 192; \quad \dfrac{190}{192} \approx 1$

Problems

2. $4.8 \text{ mol CH}_4 \times \dfrac{4 \text{ mol H}}{\text{mol CH}_4} = 19 \text{ mol H}$

5. $6 \text{ mol H}_2\text{O} \times \dfrac{2 \text{ mol H}}{\text{mol H}_2\text{O}} \times \dfrac{1 \text{ mol C}_6\text{H}_{12}\text{F}_2}{12 \text{ mol H}} = 1 \text{ mol C}_6\text{H}_{12}\text{F}_2$

7.   a. $72.8 \text{ g Al} \times \dfrac{\text{mol Al atoms}}{26.98 \text{ g Al}} = 2.70 \text{ mol Al atoms}$

9.   a. $53.2 \text{ g Fe}_2\text{O}_3 \times \dfrac{\text{mol Fe}_2\text{O}_3}{159.69 \text{ g Fe}_2\text{O}_3} = 0.333 \text{ mol Fe}_2\text{O}_3$

11.   a. $6.30 \text{ mol N}_2 \times \dfrac{28.01 \text{ g N}_2}{\text{mol N}_2} = 1.76 \text{ g N}_2$

15.   a. $66.6 \text{ g NaOH} \times \dfrac{\text{mol NaOH}}{40.00 \text{ g NaOH}} \times \dfrac{\text{mol O}}{\text{mol NaOH}} = 1.67 \text{ mol O}$

17. $0.532 \text{ mol Na}_2\text{S}_2\text{O}_3 \times \dfrac{2 \text{ mol S}}{\text{mol Na}_2\text{S}_2\text{O}_3} \times \dfrac{32.064 \text{ g S}}{\text{mol S}} = 34.1 \text{ g S}$

23. $\dfrac{10.6 \text{ g Na}}{24.4 \text{ g Na}_2\text{CO}_3} \times 100\% = 43.3\% \text{ Na}$

$\dfrac{2.8 \text{ g C}}{24.4 \text{ g Na}_2\text{CO}_3} \times 100\% = 11\% \text{ C}$

$\dfrac{11.1 \text{ g O}}{24.4 \text{ g Na}_2\text{CO}_3} \times 100\% = 45.5\% \text{ O}$

27. $17.6 \text{ g} - (9.60 \text{ g} + 7.31 \text{ g}) = 0.7 \text{ g H}$

$\dfrac{9.60 \text{ g C}}{17.6 \text{ g C}_7\text{H}_6\text{O}_4} \times 100\% = 54.6\% \text{ C}$

$\dfrac{7.31 \text{ g O}}{17.6 \text{ g C}_7\text{H}_6\text{O}_4} \times 100\% = 41.5\% \text{ O}$

$\dfrac{0.7 \text{ g H}}{17.6 \text{ g C}_7\text{H}_6\text{O}_4} \times 100\% = 4\% \text{ H}$

32. $13.21 \text{ g Mg} \times \dfrac{\text{mol Mg}}{24.312 \text{ g Mg}} = 0.5434 \text{ mol Mg}$

$40.71 \text{ g As} \times \dfrac{\text{mol As}}{74.922 \text{ g As}} = 0.5434 \text{ mol As}$

MgAs

35. $23.950 \text{ g} - (17.294 \text{ g} + 1.697 \text{ g} + 1.121 \text{ g}) = 3.838 \text{ g O}$

$17.294 \text{ g C} \times \dfrac{\text{mol C}}{12.011 \text{ g C}} = 1.440 \text{ mol C}$

$1.697 \text{ g H} \times \dfrac{\text{mol H}}{1.00797 \text{ g H}} = 1.684 \text{ mol H}$

$1.121 \text{ g N} \times \dfrac{\text{mol N}}{14.007 \text{ g N}} = 0.2399 \text{ mol N}$

$3.838 \text{ g O} \times \dfrac{\text{mol O}}{15.999 \text{ g O}} = 0.2399 \text{ mol O}$

$\dfrac{1.440}{0.8003} \qquad \dfrac{1.684}{0.8003} \qquad \dfrac{0.8003}{0.8003} \qquad \dfrac{0.2399}{0.8003}$

$\quad 18 \qquad\qquad 21 \qquad\qquad 1 \qquad\qquad 3 \qquad\quad C_{18}H_{21}NO_3$

37. $100.00 - 23.5 = 76.5$

$23.5 \text{ g O} \times \dfrac{\text{mol O}}{15.999 \text{ g O}} = 1.47 \text{ mol O}$

$76.5 \text{ g Cr} \times \dfrac{\text{mol Cr}}{52.00 \text{ g Cr}} = 1.47 \text{ mol Cr}$

CrO

40. a. $\dfrac{2 \text{ mol Li}}{\text{mol Li}_2\text{B}_4\text{O}_7} \times \dfrac{\text{mol LiB}_4\text{O}_7}{169.12 \text{ g LiB}_4\text{O}_7} \times \dfrac{6.939 \text{ g Li}}{\text{mol Li}} \times 100\% = 8.206\% \text{ Li}$

41. $\dfrac{4 \text{ mol S}}{\text{mol Co}_3\text{S}_4} \times \dfrac{\text{mol Co}_3\text{S}_4}{305.06 \text{ g Co}_3\text{S}_4} \times \dfrac{32.064 \text{ g S}}{\text{mol S}} \times 100\% = 42.043\% \text{ S}$

$\dfrac{2 \text{ mol S}}{\text{mol Cs}_2\text{S}_2} \times \dfrac{\text{mol Cs}_2\text{S}_2}{329.94 \text{ g Cs}_2\text{S}_2} \times \dfrac{32.064 \text{ g S}}{\text{mol S}} \times 100\% = 19.436\% \text{ S}$

$\dfrac{3 \text{ mol S}}{\text{mol Al}_2\text{S}_3} \times \dfrac{\text{mol Al}_2\text{S}_3}{150.155 \text{ g Al}_2\text{S}_3} \times \dfrac{32.064 \text{ g S}}{\text{mol S}} \times 100\% = 64.062\% \text{ S}$

$\dfrac{2 \text{ mol S}}{\text{mol K}_2\text{S}_2} \times \dfrac{\text{mol K}_2\text{S}_2}{142.33 \text{ g K}_2\text{S}_2} \times \dfrac{32.064 \text{ g S}}{\text{mol S}} \times 100\% = 45.056\% \text{ S}$

$\dfrac{2 \text{ mol S}}{\text{mol SiS}_2} \times \dfrac{\text{mol SiS}_2}{92.214 \text{ g SiS}_2} \times \dfrac{32.064 \text{ g S}}{\text{mol S}} \times 100\% = 69.543\% \text{ S}$

Listed in order: Cs_2S_2, Co_3S_4, K_2S_2, Al_2S_3, SiS_2

43. a. $\dfrac{\text{mol Cl}}{\text{mol C}_{12}\text{H}_9\text{OCl}} \times \dfrac{\text{mol C}_{12}\text{H}_9\text{OCl}}{204.66 \text{ g C}_{12}\text{H}_9\text{OCl}} \times \dfrac{35.453 \text{ g Cl}}{\text{mol Cl}} \times 100\% = 17.323\% \text{ Cl}$

b. $\dfrac{12 \text{ C atoms}}{23 \text{ atoms}} = 0.523$

c. $\dfrac{204.66 \text{ g C}_{12}\text{H}_9\text{OCl}}{\text{mol C}_{12}\text{H}_9\text{OCl}} \times \dfrac{\text{mol C}_{12}\text{H}_9\text{OCl}}{6.023 \times 10^{23} \text{ molecules}} = 3.398 \times 10^{-22} \dfrac{\text{g}}{\text{molecule}}$

d. $250 \text{ mL} \times \dfrac{1.25 \text{ g}}{\text{mL}} \times \dfrac{\text{mol C}_{12}\text{H}_9\text{OCl}}{204.66 \text{ g C}_{12}\text{H}_9\text{OCl}} = 1.5 \text{ mol C}_{12}\text{H}_9\text{OCl}$

44. a. $\dfrac{6 \text{ mol B}}{\text{mol MgB}_6} \times \dfrac{\text{mol MgB}_6}{89.178 \text{ g MgB}_6} \times \dfrac{10.811 \text{ g B}}{\text{mol B}} \times 100\% = 72.738\% \text{ B}$

$\dfrac{\text{mol Mg}}{\text{mol MgB}_6} \times \dfrac{\text{mol MgB}_6}{89.178 \text{ g MgB}_6} \times \dfrac{24.312 \text{ g Mg}}{\text{mol Mg}} \times 100\% = 27.262\% \text{ Mg}$

b. $\dfrac{\text{mol Na}}{\text{mol NaHSO}_4} \times \dfrac{\text{mol NaHSO}_4}{120.06 \text{ g NaHSO}_4} \times \dfrac{22.989 \text{ g Na}}{\text{mol Na}} \times 100\% = 19.149\% \text{ Na}$

$$\frac{\text{mol H}}{\text{mol NaHSO}_4} \times \frac{\text{mol NaHSO}_4}{120.06 \text{ g NaHSO}_4} \times \frac{1.00797 \text{ g H}}{\text{mol H}} \times 100\% = 0.83956\% \text{ H}$$

$$\frac{\text{mol S}}{\text{mol NaHSO}_4} \times \frac{\text{mol NaHSO}_4}{120.06 \text{ g NaHSO}_4} \times \frac{32.064 \text{ g S}}{\text{mol S}} \times 100\% = 26.707\% \text{ S}$$

$$\frac{4 \text{ mol O}}{\text{mol NaHSO}_4} \times \frac{\text{mol NaHSO}_4}{120.06 \text{ g NaHSO}_4} \times \frac{15.999 \text{ g O}}{\text{mol O}} \times 100\% = 53.303\% \text{ O}$$

46.　$6 \text{ mol B} \times \dfrac{10.811 \text{ g B}}{\text{mol B}} = 64.87 \text{ g B}$

$\text{mol Ca} \times \dfrac{40.08 \text{ g Ca}}{\text{mol Ca}} = 40.08 \text{ g Ca}$

1.000 g Ca to 1.619 g B

50.　$\dfrac{1.10 \text{ g K}}{2.10 \text{ g compound}} \times 100\% = 52.4\% \text{ K}$

$\dfrac{1.00 \text{ g Cl}}{2.10 \text{ g compound}} \times 100\% = 47.6\% \text{ Cl}$

53.　$65.82 \text{ g X} \times \dfrac{1 \text{ mol N}}{1 \text{ mol X}} \times \dfrac{14.007 \text{ g N}}{\text{mol N}} \times \dfrac{1}{34.18 \text{ g N}} = 26.97 \text{ g/mol};$　　X is Al.

62.　$\dfrac{650}{328} \approx 2;$　　Hg_2I_2

65.　$12.0 \text{ g H}_2\text{O} \times \dfrac{\text{mol H}_2\text{O}}{18.0 \text{ g H}_2\text{O}} \times \dfrac{2 \text{ mol H}}{\text{mol H}_2\text{O}} = 1.33 \text{ mol H}$

$1.33 \text{ mol H} \times \dfrac{1.008 \text{ g H}}{\text{mol H}} = 1.34 \text{ g H}$

$8.00 \text{ g compound} - 1.34 \text{ g H} = 6.66 \text{ g C}$

$6.66 \text{ g C} \times \dfrac{\text{mol C}}{12.01 \text{ g C}} = 0.555 \text{ mol C}$

$\dfrac{0.555}{0.555} = 1.00 \text{ mol C};$　　$\dfrac{1.33}{0.555} = 2.40 \text{ mol H}$

$5 \times 1.00 = 5.00 \text{ mol C};$　　$5 \times 2.40 = 12.0 \text{ mol H}$

$\dfrac{144}{72} = 2;$　　$C_{10}H_{24}$

67.　$5.96 \text{ g C} \times \dfrac{\text{mol C}}{12.01 \text{ g C}} = 0.496 \text{ mol C}$

$1.00 \text{ g H} \times \dfrac{\text{mol H}}{1.008 \text{ g H}} = 0.992 \text{ mol H}$

$\dfrac{0.496}{0.496} = 1.00 \text{ mol C};$　　$\dfrac{0.992}{0.496} = 2.00 \text{ mol H}$

$\dfrac{112}{14} = 8;$　　C_8H_{16}

68.　$2.50 \text{ g CO}_2 \times \dfrac{\text{mol CO}_2}{44.0 \text{ g CO}_2} \times \dfrac{1 \text{ mol C}}{\text{mol CO}_2} = 0.0568 \text{ mol C}$

$6.5 \text{ g C}_4\text{H}_9 \times \dfrac{\text{mol C}_4\text{H}_9}{57.12 \text{ g C}_4\text{H}_9} \times \dfrac{4 \text{ mol C}}{\text{mol C}_4\text{H}_9} = 0.46 \text{ mol C}$

$\dfrac{0.46}{0.0568} \approx 8;$　　C_8H_{18}

CHAPTER 7

Exercises

7.1 $1.60 \text{ mol CO} \times \dfrac{1 \text{ mol H}_2}{1 \text{ mol CO}} = 1.60 \text{ mol H}_2$

7.2 $5.45 \text{ mol NaOH} \times \dfrac{1 \text{ mol H}_2}{2 \text{ mol NaOH}} = 2.73 \text{ mol H}_2$

7.3 $8.72 \text{ mol NaCl} \times \dfrac{1 \text{ mol Na}_2\text{O}}{2 \text{ mol NaCl}} = 4.36 \text{ mol Na}_2\text{O}$

7.4 $3.40 \text{ mol H}_2\text{O} \times \dfrac{2 \text{ mol HCl}}{1 \text{ mol H}_2\text{O}} = 6.80 \text{ mol HCl}$

7.5 $3.42 \text{ mol O}_2 \times \dfrac{2 \text{ mol C}_2\text{H}_6}{7 \text{ mol O}_2} = 0.977 \text{ mol C}_2\text{H}_6$

7.6 $2.5 \text{ mol Ag} \times \dfrac{3 \text{ mol AgNO}_3}{3 \text{ mol Ag}} = 2.5 \text{ mol AgNO}_3$; $2.0 \text{ mol HNO}_3 \times \dfrac{3 \text{ mol AgNO}_3}{4 \text{ mol HNO}_3}$

$= 1.5 \text{ mol AgNO}_3$

HNO_3 is the limiting reactant, thus 1.5 mol $AgNO_3$ can be produced.

7.7 $0.245 \text{ mol Ag} \times \dfrac{1 \text{ mol NO}}{3 \text{ mol Ag}} = 0.0817 \text{ mol NO}$; $0.320 \text{ mol HNO}_3 \times \dfrac{1 \text{ mol NO}}{4 \text{ mol HNO}_3}$

$= 0.0800 \text{ mol NO}$

HNO_3 is the limiting reactant, therefore 0.0800 mol NO can be produced.

7.8 $3.38 \text{ mol C} \times \dfrac{1 \text{ mol C}_2\text{H}_5\text{OH}}{2 \text{ mol C}} = 1.69 \text{ mol C}_2\text{H}_5\text{OH}$

$3.47 \text{ mol H}_2 \times \dfrac{1 \text{ mol C}_2\text{H}_5\text{OH}}{2 \text{ mol H}_2} = 1.74 \text{ mol C}_2\text{H}_5\text{OH}$

$1.79 \text{ mol H}_2\text{O} \times \dfrac{1 \text{ mol C}_2\text{H}_5\text{OH}}{1 \text{ mol H}_2\text{O}} = 1.79 \text{ mol C}_2\text{H}_5\text{OH}$

Carbon is the limiting reactant, thus 1.69 mol C_2H_5OH can be produced.

7.9 $0.72 \text{ mol K}_2\text{Cr}_2\text{O}_7 \times \dfrac{1 \text{ mol Cr}_2(\text{SO}_4)_3}{1 \text{ mol K}_2\text{Cr}_2\text{O}_7} = 0.72 \text{ mol Cr}_2(\text{SO}_4)_3$

$2.0 \text{ mol H}_2\text{S} \times \dfrac{1 \text{ mol Cr}_2(\text{SO}_4)_3}{3 \text{ mol H}_2\text{S}} = 0.67 \text{ mol Cr}_2(\text{SO}_4)_3$

$3.0 \text{ mol H}_2\text{SO}_4 \times \dfrac{1 \text{ mol Cr}_2(\text{SO}_4)_3}{4 \text{ mol H}_2\text{SO}_4} = 0.75 \text{ mol Cr}_2(\text{SO}_4)_3$

H_2S is the limiting reactant, therefore 0.67 mol $Cr_2(SO_4)_3$ can be produced. Some $K_2Cr_2O_7$ and H_2SO_4 will be left over.

7.10 $16.4 \text{ g KClO}_3 \times \dfrac{\text{mol KClO}_3}{122.6 \text{ g KClO}_3} \times \dfrac{3 \text{ mol O}_2}{2 \text{ mol KClO}_3} = 0.201 \text{ mol O}_2$

Did you remember to balance the reaction equation?

7.11 $24.3 \text{ g HgO} \times \dfrac{\text{mol HgO}}{216.6 \text{ g HgO}} \times \dfrac{2 \text{ mol Hg}}{2 \text{ mol HgO}} = 0.112 \text{ mol Hg}$

7.12 $10.3 \text{ mol AgNO}_3 \times \dfrac{3 \text{ mol Ag}}{3 \text{ mol AgNO}_3} \times \dfrac{107.9 \text{ g Ag}}{\text{mol Ag}} = 1.11 \times 10^3 \text{ g Ag}$

7.13 $8.50 \text{ mol Hg} \times \dfrac{2 \text{ mol HgO}}{2 \text{ mol Hg}} \times \dfrac{216.6 \text{ g HgO}}{\text{mol HgO}} = 1.84 \times 10^3 \text{ g HgO}$

7.14 $76.8 \text{ g NaOH} \times \dfrac{\text{mol NaOH}}{40.00 \text{ g NaOH}} \times \dfrac{1 \text{ mol Na}_2\text{ZnO}_2}{2 \text{ mol NaOH}} \times \dfrac{143.4 \text{ g Na}_2\text{ZnO}_2}{\text{mol Na}_2\text{ZnO}_2} = 138 \text{ g Na}_2\text{ZnO}_2$

7.15 $18.6 \text{ g Al} \times \dfrac{\text{mol Al}}{26.98 \text{ g Al}} \times \dfrac{2 \text{ mol AlBr}_3}{2 \text{ mol Al}} \times \dfrac{266.7 \text{ g AlBr}_3}{\text{mol AlBr}_3} = 184 \text{ g AlBr}_3$

7.16 $22.6 \text{ g Al} \times \dfrac{\text{mol Al}}{26.98 \text{ g Al}} \times \dfrac{2 \text{ mol AlBr}_3}{2 \text{ mol Al}} = 0.838 \text{ mol AlBr}_3$

       $38.5 \text{ g Br}_2 \times \dfrac{\text{mol Br}_2}{159.8 \text{ g Br}_2} \times \dfrac{2 \text{ mol AlBr}_3}{3 \text{ mol Br}_2} = 0.161 \text{ mol AlBr}_3$

       $0.161 \text{ mol AlBr}_3 \times \dfrac{266.7 \text{ g AlBr}_3}{\text{mol AlBr}_3} = 42.9 \text{ g AlBr}_3$

7.17 $76.5 \text{ g Ca(OH)}_2 \times \dfrac{\text{mol Ca(OH)}_2}{74.1 \text{ g Ca(OH)}_2} \times \dfrac{1 \text{ mol Ca}_3(\text{PO}_4)_2}{3 \text{ mol Ca(OH)}_2} = 0.344 \text{ mol Ca}_3(\text{PO}_4)_2$

       $100.0 \text{ g H}_3\text{PO}_4 \times \dfrac{\text{mol H}_3\text{PO}_4}{98.00 \text{ g H}_3\text{PO}_4} \times \dfrac{1 \text{ mol Ca}_3(\text{PO}_4)_2}{2 \text{ mol H}_3\text{PO}_4} = 0.5102 \text{ mol Ca}_3(\text{PO}_4)_2$

       $0.344 \text{ mol Ca}_3(\text{PO}_4)_2 \times \dfrac{310.2 \text{ g Ca}_3(\text{PO}_4)_2}{\text{mol Ca}_3(\text{PO}_4)_2} = 107 \text{ g Ca}_3(\text{PO}_4)_2$

       Did you remember to balance the reaction equation?

7.18 $12.8 \text{ mL Br}_2 \times \dfrac{3.12 \text{ g Br}_2}{\text{mL Br}_2} \times \dfrac{\text{mol Br}_2}{159.8 \text{ g Br}_2} \times \dfrac{2 \text{ mol AlBr}_3}{3 \text{ mol Br}_2}$

       $\times \dfrac{266.7 \text{ g AlBr}_3}{\text{mol AlBr}_3} = 44.4 \text{ g AlBr}_3$

7.19 $250 \text{ mL HCl} \times \dfrac{\text{L}}{1000 \text{ mL}} \times \dfrac{1.097 \text{ g HCl}}{\text{L}} \times \dfrac{\text{mol HCl}}{36.47 \text{ g HCl}}$

       $\times \dfrac{2 \text{ mol FeCl}_3}{6 \text{ mol HCl}} = 2.51 \times 10^{-3} \text{ mol FeCl}_3$

       $43.2 \text{ g Fe}_2\text{O}_3 \times \dfrac{\text{mol Fe}_2\text{O}_3}{159.7 \text{ g Fe}_2\text{O}_3} \times \dfrac{2 \text{ mol FeCl}_3}{1 \text{ mol Fe}_2\text{O}_3} = 0.541 \text{ mol FeCl}_3$

       $2.51 \times 10^{-3} \text{ mol FeCl}_3 \times \dfrac{162.2 \text{ g FeCl}_3}{\text{mol FeCl}_3} = 0.407 \text{ g FeCl}_3$

7.20 $2210 \text{ g CaO} \times \dfrac{\text{mol CaO}}{56.08 \text{ g CaO}} \times \dfrac{-65.2 \text{ kJ}}{\text{mol CaO}} = -2.57 \times 10^3 \text{ kJ}$

       Since $\triangle H = -2.57 \times 10^3 \text{ kJ}$, the reaction *produced* $2.57 \times 10^3 \text{ kJ}$

Problems

2.   a. $2.80 \text{ mol Al} \times \dfrac{2 \text{ mol Al}_2\text{O}_3}{4 \text{ mol Al}} = 1.40 \text{ mol Al}_2\text{O}_3$

4. $2.74 \text{ mol H}_2 \times \dfrac{3 \text{ mol Fe}}{4 \text{ mol H}_2} = 2.06 \text{ mol Fe}$

7.   a. $7.22 \text{ mol CO} \times \dfrac{1 \text{ mol FeO}}{1 \text{ mol CO}} = 7.22 \text{ mol FeO}$

9. $4.3 \text{ mol O}_2 \times \dfrac{2 \text{ mol CO}}{1 \text{ mol O}_2} = 8.6 \text{ mol CO}$

12. $2.40 \text{ mol N}_2 \times \dfrac{2 \text{ mol NH}_3}{\text{mol N}_2} = 4.80 \text{ mol NH}_3$

$$10.0 \text{ mol } H_2 \times \frac{2 \text{ mol } NH_3}{3 \text{ mol } H_2} = 6.67 \text{ mol } NH_3$$

N_2 is the limiting reactant, thus 4.80 mol NH_3 can be produced.

21. $$10.7 \text{ g } NH_3 \times \frac{\text{mol } NH_3}{17.03 \text{ g } NH_3} \times \frac{2 \text{ mol } N_2}{4 \text{ mol } NH_3} = 0.314 \text{ mol } N_2$$

23. $$9.00 \text{ g Fe} \times \frac{\text{mol Fe}}{55.85 \text{ g Fe}} \times \frac{4 \text{ mol } H_2}{3 \text{ mol Fe}} = 0.215 \text{ mol } H_2$$

29. $$37.0 \text{ g PbS} \times \frac{\text{mol PbS}}{239.3 \text{ g PbS}} \times \frac{2 \text{ mol PbO}}{2 \text{ mol PbS}} \times \frac{223.2 \text{ g PbO}}{\text{mol PbO}} = 34.5 \text{ g PbO}$$

34.   a. $$50.0 \text{ g CaO} \times \frac{\text{mol CaO}}{56.08 \text{ g CaO}} \times \frac{\text{mol CO}}{\text{mol CaO}} = 0.892 \text{ mol CO}$$

$$30.5 \text{ g C} \times \frac{\text{mol C}}{12.01 \text{ g C}} \times \frac{\text{mol CO}}{3 \text{ mol C}} = 0.847 \text{ mol CO}$$

C is the limiting reactant.

35. $$75.6 \text{ g KMnO}_4 \times \frac{\text{mol KMnO}_4}{158.04 \text{ g KMnO}_4} \times \frac{1 \text{ mol MnO}_2}{1 \text{ mol KMnO}_4} = 0.478 \text{ mol MnO}_2$$

$$0.68 \text{ g NO} \times \frac{\text{mol NO}}{30.01 \text{ g NO}} \times \frac{1 \text{ mol MnO}_2}{1 \text{ mol NO}} = 0.023 \text{ mol MnO}_2$$

NO is the limiting reactant, hence 0.023 mol MnO_2 can be produced.

$$0.023 \text{ mol MnO}_2 \times \frac{1 \text{ mol KNO}_3}{1 \text{ mol MnO}_2} \times \frac{101.1 \text{ g KNO}_3}{\text{mol KNO}_3} = 2.3 \text{ g KNO}_3$$

41. $$750 \text{ g C}_6H_{12}O_6 \times \frac{\text{mol C}_6H_{12}O_6}{180.2 \text{ g C}_6H_{12}O_6} \times \frac{2 \text{ mol C}_2H_5OH}{1 \text{ mol C}_6H_{12}O_6} \times \frac{46.1 \text{ g C}_2H_5OH}{\text{mol C}_2H_5OH} = 384 \text{ g C}_2H_5OH$$

$$750 \text{ g C}_6H_{12}O_6 \times \frac{\text{mol C}_6H_{12}O_6}{180.2 \text{ g C}_6H_{12}O_6} \times \frac{2 \text{ mol CO}_2}{1 \text{ mol C}_6H_{12}O_6} \times \frac{44.0 \text{ g CO}_2}{\text{mol CO}_2} = 366 \text{ g CO}_2$$

$$384 \text{ g C}_2H_5OH \times \frac{\text{mL}}{0.789 \text{ g}} = 487 \text{ mL}$$

47. released

$$3.5 \text{ mol } H_2 \times \frac{539 \text{ kJ}}{\text{mol } H_2} = 1.9 \times 10^3 \text{ kJ}$$

51. $$1.00 \text{ g C}_6H_{12}O_6 \times \frac{\text{mol C}_6H_{12}O_6}{180.2 \text{ g C}_6H_{12}O_6} \times \frac{2816 \text{ kJ}}{\text{mol C}_6H_{12}O_6} = 15.6 \text{ kJ}$$

53. if $x = 1$, $MBr_x = MBr$

$$135 \text{ g MBr} \times \frac{1 \text{ mol } H_2}{2 \text{ mol MBr}} \times \frac{2.016 \text{ g } H_2}{\text{mol } H_2} \times \frac{1}{1.00 \text{ g } H_2} = 136 \text{ g/mol}$$

136 g/mol − 79.9 g/mol = 56 g/mol; M is Fe

if $x = 2$, $MBr_x = MBr_2$

$$135 \text{ g MBr}_2 \times \frac{1 \text{ mol } H_2}{1 \text{ mol MBr}_2} \times \frac{2.016 \text{ g } H_2}{\text{mol } H_2} \times \frac{1}{1.00 \text{ g } H_2} = 272 \text{ g/mol}$$

272 g/mol − 2(79.9 g/mol) = 112 g/mol; M is Cd

if $x = 3$, $MBr_x = MBr_3$

$$135 \text{ g MBr}_3 \times \frac{3 \text{ mol } H_2}{2 \text{ mol MBr}_3} \times \frac{2.016 \text{ g } H_2}{\text{mol } H_2} \times \frac{1}{1.00 \text{ g } H_2} = 408 \text{ g/mol}$$

408 g/mol − 3(79.9 g/mol) = 168 g/mol; M is Tm

CHAPTER 8

Exercises

8.1   36, 40

8.2 ^{36}Ar, ^{40}Ar

8.3 $35 - 17 = 18$ neutrons; $14 - 6 = 8$ neutrons

8.4 $1 - 1 = 0$ neutrons; $23 - 11 = 12$ neutrons

Problems

1. $(9.1 \times 10^{-28} \text{ g}) \times \dfrac{\text{amu}}{1.66 \times 10^{-24} \text{ g}} = 5.5 \times 10^{-4}$ amu

2.   Hg: 199; Cs: 133

4. $^{20}_{10}$Ne, $^{21}_{10}$Ne, $^{22}_{10}$Ne. Probably ^{20}Ne is the most abundant while ^{21}Ne and ^{22}Ne are present in very small amounts.

7. $^{12}_{6}$C has 6 neutrons; $12 - 6 = 6$; $^{13}_{6}$C has 7 neutrons; $13 - 6 = 7$; $^{14}_{6}$C has 8 neutrons; $14 - 6 = 8$

8.   a. $27 - 13 = 14$ neutrons; b. $3 - 2 = 1$ neutron

CHAPTER 9

Exercises

9.1 $2 \times 4^2 = 32$ s, p, d, f

9.2 $N = 3$

9.3 $4d^3$

9.4 $N = 3, d, 6$

9.5 $1s^2 2s^2 2p^6 3s^2 3p^6 3d^{10} 4s^2 4p^6 4d^2 5s^2$

9.6 $1s^2 2s^2 2p^6 3s^2 3p^6 3d^3 4s^2$

       $1s^2 2s^2 2p^6 3s^2 3p^6 3d^{10} 4s^2 4p^5$

9.7 $\cdot \ddot{\text{P}} \cdot$

9.8   N, P, As

9.9
1↑↓	1↑↓	1↑↓ 1↑↓ 1↑↓	1↑↓	1↑↓ 1↑ 1↑
1s	2s	2p	3s	3p

9.10   Na

Problems

1. $2N^2 = 50$; $N = 5$

3. $2(1)^2 = 1$; s $2(2)^2 = 8$; s and p

6. $N = 2$, subshell $= s$, number of electrons $= 2$

8.   a. $4p^5$

9. $2p^2$

12.   18 electrons

24.   B, Al, Ga, In, Tl

28.   Fe

CHAPTER 10

Exercises

10.1   Si, 14, 28.09, 3, IVA

10.2   3, IIA, S, $1s^22s^22p^63s^2$

10.3   3, 6, 8, 7, 2, 4; He has only two valence electrons because the N = 1 main shell which is the outer shell has a maximum capacity of two electrons.

10.4   Helium has the smallest atoms. It is a nonmetal and a gas. Helium is practically chemically unreactive. Aluminum is a metal with relatively high melting and boiling points.

10.5   RbI

Problems

2.   a. F   b. Cr   c. H, He   d. Ba

3.   a. period = 3, group = noble gases

6.   a. VIIB, the configuration is {Rn} $7s^25f^{14}6d^6$

7.   carbon, $1s^22s^22p^2$, nonmetal

8.   a. Rb   b. Se   c. S   d. I

9.   a. block = s; group = IIA; period = 3; $3s^2$

11.   a. F, As, Ga, K

CHAPTER 11

Exercises

11.1   13 − 10 = 3; 15 − 10 = 5

11.2   C, Si, Ge, Sn, Pb

11.3   1, −1, 2

11.4   3, $\dot{A}l:$, Al^{3+}, 3, 11

11.5 Cl^-, I^-, S^{2-}, Ba^{2+}, Al^{3+}

11.6   CsF   132.90 + 19.00 = 151.90

Al_2O_3 $2(26.98) = 53.96$
$3(16.00) = \underline{48.00}$
101.96

HgI$_2$ $2(126.90) = 253.80$
$\underline{200.59}$
454.39

$Ba_3(PO_4)_2$ $3(137.34) = 412.02$
$2(30.97) = 61.94$
$8(16.00) = \underline{128.00}$
601.96

11.7

		Substance	Substance	Ions
Oxidizer	Reducer	Reduced	Oxidized	Produced
Br	Ba	Br	Ba	—
O	Fe	O	Fe	O^{2-}, Fe^{2+}
Te	Li	Te	Li	Te^{2-}, Li^+

11.8 O^{2-}, Sr^{2+}, Rb^+, Ar does not ordinarily form ions

-2, $+2$, $+1$, 0

11.9 Li^+, F^-, LiF; Sr^{2+}, Cl^-, $SrCl_2$; Al^{3+}, O^{2-}, Al_2O_3

11.10
$$H : \overset{\cdot\cdot}{\underset{\cdot\cdot}{F}} : \qquad H{-}\overline{\underline{F}}| \qquad\qquad :\overset{\cdot\cdot}{\underset{\cdot\cdot}{I}} : \overset{\cdot\cdot}{\underset{\cdot\cdot}{I}} : \qquad |\overline{\underline{I}}{-}\overline{\underline{I}}| \qquad\qquad :\overset{\cdot\cdot}{\underset{\cdot\cdot}{I}} : \overset{\cdot\cdot}{\underset{\cdot\cdot}{Br}} : \qquad |\overline{\underline{I}}{-}\overline{\underline{Br}}|$$

11.11
$$:\overset{\cdot\cdot}{\underset{\cdot\cdot}{Cl}} : \overset{\cdot\cdot}{P} : \overset{\cdot\cdot}{\underset{\cdot\cdot}{Cl}} : \qquad |\overline{\underline{Cl}}{-}P{-}\overline{\underline{Cl}}|$$
$$:\overset{\cdot\cdot}{\underset{\cdot\cdot}{Cl}} : \qquad\qquad\quad |\underline{Cl}|$$

11.12
$$\begin{array}{c} H \\ \cdot\cdot \\ H : C : H \\ \cdot\cdot \\ H \end{array} \qquad \begin{array}{c} H \\ | \\ H{-}C{-}H \\ | \\ H \end{array}$$

11.13
$$\begin{array}{c} H \\ \cdot\cdot \\ :O :: O: \\ \cdot\cdot \\ H \end{array}$$

11.14
$$\begin{array}{c} O \\ \| \\ H{-}O{-}P{-}O{-}H \\ | \\ O{-}H \end{array}$$

11.15
$$\left[\begin{array}{c} H \\ \cdot\cdot \\ H : N : H \\ \cdot\cdot \\ H \end{array} \right]^+ \qquad \left[\begin{array}{c} :\overset{\cdot\cdot}{\underset{\cdot\cdot}{O}} : N :: \overset{\cdot\cdot}{\underset{\cdot\cdot}{O}} \\ \cdot\cdot \\ :\overset{\cdot\cdot}{\underset{\cdot\cdot}{O}}: \end{array} \right]^-$$

11.16
$$\begin{array}{c} \diagdown \quad \diagup \\ C{=}C \\ \diagup \quad \diagdown \end{array} \qquad \begin{array}{c} Cl \\ | \\ H{-}C{-}Cl \\ | \\ Cl \end{array}$$

11.17
$$\begin{array}{c} | \quad | \quad | \quad | \\ {-}C{-}C{-}C{-}C{-} \\ | \quad | \quad | \quad | \end{array} \qquad \begin{array}{c} | \quad | \quad | \\ {-}C{-}C{-}C{-} \\ \quad | \\ {-}C{-} \\ \quad | \end{array}$$

$$\begin{array}{c} | \quad\quad | \quad | \\ {-}C{-}O{-}C{-}C{-} \\ | \quad\quad | \quad | \end{array} \qquad \begin{array}{c} | \quad | \quad | \\ {-}C{-}C{-}C{-} \\ | \quad | \quad | \\ \quad O \\ \quad | \\ \quad H \end{array} \qquad \begin{array}{c} | \quad | \quad | \\ {-}C{-}C{-}C{-}O{-}H \\ | \quad | \quad | \end{array}$$

Problems

1.   a. $12 - 10 = 2$;   b. $14 - 10 = 4$

5.   a.   Be

7.   a.   Mg

9.   3, 1, 1, 2, 2, 3

12. N^{3-}, Cl^-, Br^-, O^{2-}, S^{2-}, P^{3-}

14.   a. $1s^2 2s^2 2p^6$

15.   a. Na^+, Cl, P, S^{2-}, Na

17.

Oxidizer	Reducer	Substance Reduced	Substance Oxidized
O	Mg	O	Mg
P	V	P	V
Cl	Co	Cl	Co
O	Ni	O	Ni
I	Mn	I	Mn

19.   a.   Li ion is $+1$

21.   a.   MgO

23.   a. X_3Y_2 (example Al_2O_3)

24.   a. Ba^{2+}, Br^-

25.

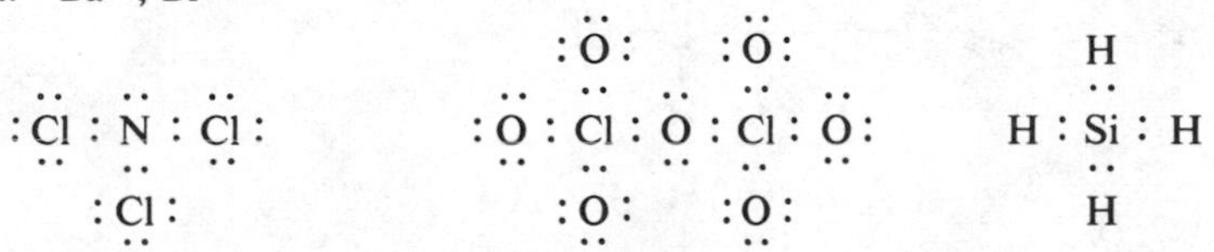

27.

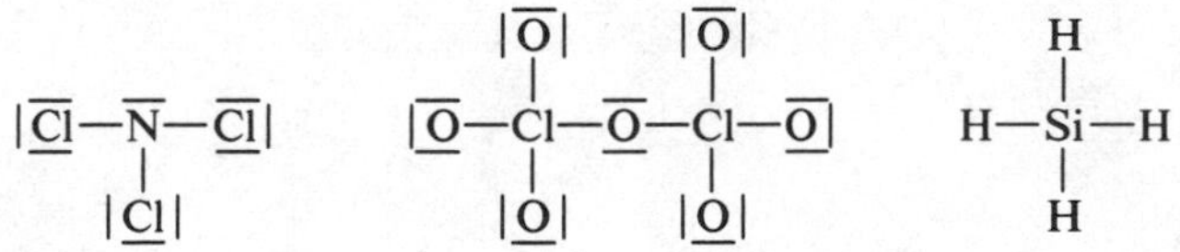

29.

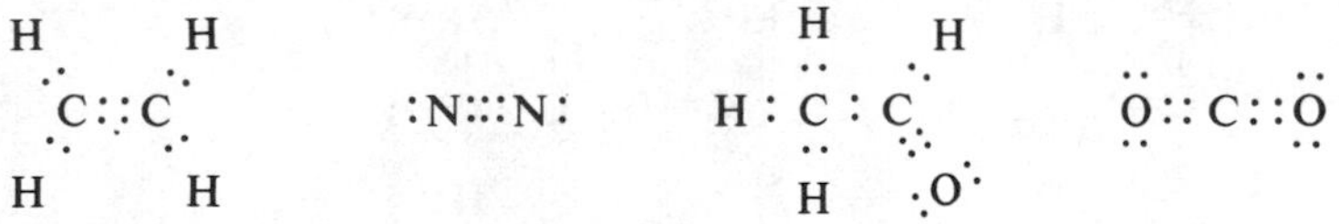

31.

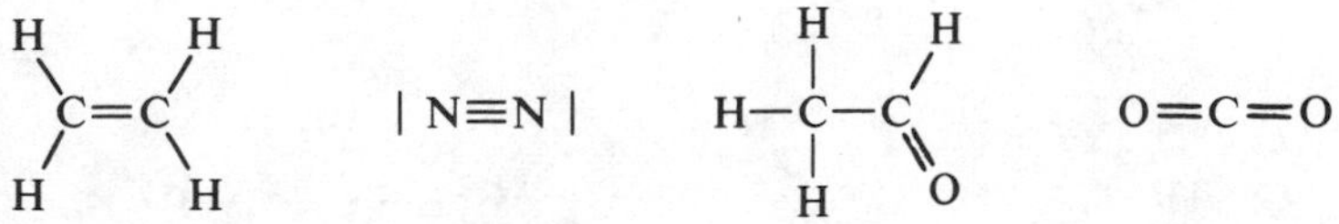

35.

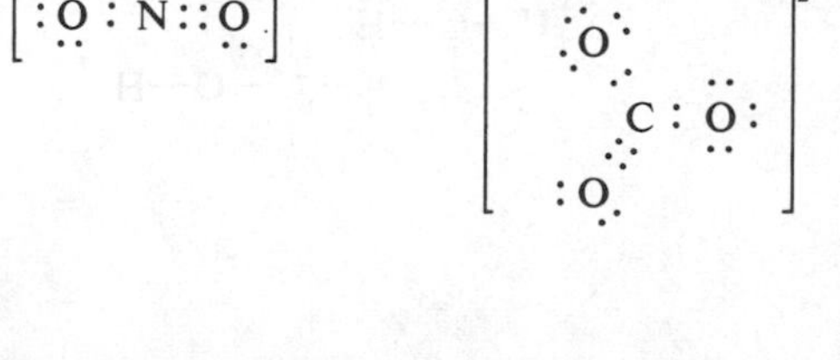

$$\left[\begin{array}{c} :\overset{\cdot\cdot}{O}: \\ :\overset{\cdot\cdot}{O}:\overset{\cdot\cdot}{Cl}:\overset{\cdot\cdot}{O}: \\ :\overset{\cdot\cdot}{O}: \end{array}\right]^{-} \qquad \left[\begin{array}{c} :\overset{\cdot\cdot}{O}: \\ :\overset{\cdot\cdot}{O}:P:\overset{\cdot\cdot}{O}:H \\ :\overset{\cdot\cdot}{O}: \end{array}\right]^{2-}$$

37.

$$H-C\equiv C-H \qquad \begin{array}{c} H \\ | \\ H-C-O-H \\ | \\ H \end{array} \qquad \begin{array}{c} H \\ | \\ C=O \\ | \\ H \end{array}$$

39.

$$\begin{array}{c} H\;\;H\;\;\;\;H \\ |\;\;\;|\;\;\;\;| \\ H-C-C-N \\ |\;\;\;|\;\;\;\;| \\ H\;\;H\;\;\;\;H \end{array} \qquad \begin{array}{c} H\;\;\;H\;\;\;H \\ |\;\;\;\;|\;\;\;| \\ H-C-N-C-H \\ |\;\;\;\;|\;\;\;| \\ H\;\;\;\;\;\;\;H \end{array}$$

$$\begin{array}{c} H\;\;\;\;\;H\;\;H \\ \backslash\;\;\;\;\;\;|\;\;\;| \\ C=C-C-H \\ /\;\;\;\;\;\;\;\;\;| \\ H\;\;\;\;\;\;\;\;\;Cl \end{array} \qquad \begin{array}{c} H\;\;\;\;\;H\;\;H \\ \backslash\;\;\;\;\;\;|\;\;\;| \\ C=C-C-H \\ /\;\;\;\;\;\;\;\;\;| \\ Cl\;\;\;\;\;\;\;\;H \end{array}$$

$$\begin{array}{c} H\;\;\;\;Cl\;\;H \\ \backslash\;\;\;\;|\;\;\;\;| \\ C=C-C-H \\ /\;\;\;\;\;\;\;\;\;| \\ H\;\;\;\;\;\;\;\;\;H \end{array} \qquad \begin{array}{c} H\;\;\;\;\;H \\ \backslash\;\;/ \\ C \\ H-C\!-\!-\!C-Cl \\ |\;\;\;\;\;\;| \\ H\;\;\;\;\;\;H \end{array}$$

$$\begin{array}{c} H\;\;\;\;H \\ H\;\;\backslash\;/ \\ \backslash\;\;C \\ C\;\;\;\;\;\;C-H \\ H\;/\;\;\;\;\;\;\| \\ H-C\;\;\;\;\;C \\ |\;\;\;\;\;\;\backslash \\ H\;\;\;\;\;\;H \end{array} \qquad \begin{array}{c} H\;\;\;H\;H\;H\;H \\ \backslash\;\;\;|\;\;|\;\;|\;\;| \\ C=C-C=C-C-H \\ /\;\;\;\;\;\;\;\;\;\;\;\;\;\;| \\ H\;\;\;\;\;\;\;\;\;\;\;\;\;\;H \end{array}$$

$$\begin{array}{c} H\;\;\;H\;H\;H\;\;\;H \\ \backslash\;\;\;|\;\;|\;\;|\;\;/ \\ C=C-C-C=C \\ /\;\;\;\;\;\;|\;\;\;\;\;\backslash \\ H\;\;\;\;\;\;H\;\;\;\;\;H \end{array} \qquad \begin{array}{c} H\;\;\;\;\;H\;H\;H \\ \backslash\;\;\;\;\;|\;\;|\;\;| \\ C=C=C-C-C-H \\ /\;\;\;\;\;\;\;\;\;\;|\;\;| \\ H\;\;\;\;\;\;\;\;\;\;H\;H \end{array}$$

$$\begin{array}{c} H \\ H\;\;\;\;\;\;| \\ \backslash\;\;\;\;\;C \\ H-C-C=C \\ |\;\;\;\;\;\;\;\;\backslash \\ H-C-C-H\;\;\;H \\ |\;\;\;| \\ H\;\;H \end{array} \qquad \begin{array}{c} H \\ | \\ H-C-H\;\;H \\ |\;\;\;\;\;\;\;| \\ H-C-C \\ |\;\;\;\backslash\;\; \\ C=C \\ /\;\;\;\;\backslash \\ H\;\;\;\;\;\;H \end{array} \qquad \begin{array}{c} H\;\;\;H \\ |\;\;\;| \\ H-C-C-H \\ |\;\;\;\| \\ C=C \\ /\;\;\;\;\;\backslash \\ H-C-H\;\;\;\;H \\ | \\ H \end{array}$$

CHAPTER 12

Exercises

12.1 $720 \text{ torr} \times \dfrac{1 \text{ atm}}{760 \text{ torr}} = 0.947 \text{ atm}$

12.2 $2.00 \text{ L} \times \dfrac{1.00 \text{ atm}}{0.750 \text{ atm}} = 2.67 \text{ L}$

12.3 $900 \text{ torr} \times \dfrac{6.50 \text{ L}}{10.00 \text{ L}} = 585 \text{ torr}$

12.4 $695 \text{ torr} \times \dfrac{3.1 \text{ L}}{5.7 \text{ L}} \times \dfrac{1 \text{ atm}}{760 \text{ torr}} = 0.50 \text{ atm}$

12.5 $8.74 \text{ L} \times \dfrac{1.25 \text{ mol}}{2.50 \text{ mol}} = 4.37 \text{ L}$

12.6 $5.32 \text{ mol} \times \dfrac{30.4 \text{ L}}{18.6 \text{ L}} = 8.70 \text{ mol}$

12.7 $5.25 \text{ L} \times \dfrac{500 \text{ K}}{298 \text{ K}} = 8.81 \text{ L}$

12.8 $273 \text{ K} \times \dfrac{2.74 \text{ L}}{4.28 \text{ L}} = 175 \text{ K}; \; 175 \text{ K} - 273 = -98 \; ^\circ\text{C}$

12.9 $298 \text{ K} \times \dfrac{5.00 \text{ atm}}{1.50 \text{ atm}} = 993 \text{ K}$

12.10 $790 \text{ torr} \times \dfrac{273 \text{ K}}{453 \text{ K}} = 476 \text{ torr}$

12.11 $\dfrac{1.25 \text{ atm} \times 0.250 \text{ L}}{0.0130 \text{ mol} \times 0.0821 \text{ L atm/mol K}} = 293 \text{ K}; \; 293 \text{ K} - 273 = 20 \; ^\circ\text{C}$

12.12 $55.5 \text{ g N}_2 \times \dfrac{\text{mol N}_2}{28.0 \text{ g N}_2} = 1.98 \text{ mol N}_2$

$\dfrac{1.98 \text{ mol} \times 62.4 \text{ L torr/mol K} \times 273 \text{ K}}{850 \text{ torr}} = 39.7 \text{ L}$

12.13 $\dfrac{1.75 \text{ atm} \times 4.00 \text{ L}}{0.0821 \text{ L atm/mol K} \times 323 \text{ K}} = 0.264 \text{ mol}$

12.14 $2.50 \text{ mol} - 1.35 \text{ mol} = 1.15 \text{ mol remains}$

$733 \text{ torr} \times \dfrac{1.30 \text{ L}}{2.00 \text{ L}} \times \dfrac{1.15 \text{ mol}}{2.50 \text{ mol}} \times \dfrac{T_1/3}{T_1} = 73.1 \text{ torr}$

12.15 $0.657 \text{ mol} \times \dfrac{P_1}{2 P_1} \times \dfrac{1.14 \text{ L}}{2.38 \text{ L}} \times \dfrac{298 \text{ K}}{273 \text{ K}} = 0.172 \text{ mol}$

$0.657 \text{ mol} - 0.172 \text{ mol} = 0.485 \text{ mol}$

12.16 $962 \text{ torr} \times \dfrac{2.90 \text{ L}}{7.08 \text{ L}} \times \dfrac{2 \, n_1}{n_1} = 788 \text{ torr}$

12.17 $6.72 \text{ mol} \times \dfrac{2.90 \text{ atm}}{4.68 \text{ atm}} \times \dfrac{V_1/2}{V_1} = 2.08 \text{ mol remains}$

$6.72 \text{ mol} - 2.08 \text{ mol} = 4.64 \text{ mol escaped}$

12.18 $\dfrac{0.500 \text{ atm} \times 1.23 \text{ L}}{0.0821 \text{ L atm/mol K} \times 300 \text{ K}} = 0.0250 \text{ mol}$

$\dfrac{0.700 \text{ g}}{0.0250 \text{ mol}} = 28.0 \text{ g/mol}$

12.19 $\dfrac{17.0 \text{ g NH}_3}{\text{mol NH}_3} \times \dfrac{1 \text{ mol NH}_3}{22.4 \text{ L}} = 0.759 \text{ g/L}$

12.20 $\dfrac{62.4 \text{ L torr/mol K} \times 303 \text{ K}}{800 \text{ torr}} = 23.6 \text{ L/mol}; \dfrac{44.0 \text{ g/mol}}{23.6 \text{ L/mol}} = 1.86 \text{ g/L}$

12.21 $\dfrac{0.0821 \text{ L atm/mol K} \times 298 \text{ K}}{0.0085 \text{ atm}} = 2.9 \times 10^3 \text{ L/mol}$

$\dfrac{108.0 \text{ g/mol}}{2.9 \times 10^3 \text{ L/mol}} = 0.037 \text{ g/L}$

12.22 $\dfrac{5.28 \text{ g}}{\text{L}} \times \dfrac{22.4 \text{ L}}{\text{mol}} = 118 \text{ g/mol}$

12.23 $15.5 \text{ g N}_2 \times \dfrac{\text{mol N}_2}{28.0 \text{ g N}_2} = 0.554 \text{ mol N}_2$

$\dfrac{0.554 \text{ mol} \times 0.0821 \text{ L atm/mol K} \times 273 \text{ K}}{7.00 \text{ L}} = 1.77 \text{ atm}$

$1.77 \text{ atm} + 0.918 \text{ atm} = 2.69 \text{ atm}$

12.24 $32 \text{ torr} \times \dfrac{1 \text{ atm}}{760 \text{ torr}} = 0.042 \text{ atm}$

$0.750 \text{ atm} - 0.042 \text{ atm} = 0.708 \text{ atm}$

12.25 $2.96 \text{ L CO} \times \dfrac{1 \text{ L H}_2}{1 \text{ L CO}} = 2.96 \text{ L H}_2$

12.26 $255 \text{ L H}_2\text{O} \times \dfrac{1 \text{ L H}_2}{1 \text{ L H}_2\text{O}} = 255 \text{ L H}_2$

12.27 $CaCO_3 \longrightarrow CaO + CO_2$ (continued on next page)

$$5.20 \text{ g CaCO}_3 \times \frac{\text{mol CaCO}_3}{100.1 \text{ g CaCO}_3} \times \frac{1 \text{ mol CO}_2}{1 \text{ mol CaCO}_3} = 0.0519 \text{ mol CO}_2$$

$$\frac{0.0519 \text{ mol} \times 62.4 \text{ L torr/mol K} \times 300 \text{ K}}{730 \text{ torr}} = 1.33 \text{ L}$$

12.28 $C_5H_{12} + 8 \text{ O}_2 \longrightarrow 5 \text{ CO}_2 + 6 \text{ H}_2\text{O}$

$$0.265 \text{ L C}_5\text{H}_{12} \times \frac{1000 \text{ mL}}{\text{L}} \times \frac{0.626 \text{ g C}_5\text{H}_{12}}{\text{mL C}_5\text{H}_{12}} \times \frac{\text{mol C}_5\text{H}_{12}}{72.2 \text{ g C}_5\text{H}_{12}} \times \frac{5 \text{ mol CO}_2}{1 \text{ mol C}_5\text{H}_{12}} = 11.5 \text{ mol CO}_2$$

$$\frac{11.5 \text{ mol} \times 0.0821 \text{ L atm/mol K} \times 298 \text{ K}}{0.950 \text{ atm}} = 296 \text{ L}$$

Problems

1.   a. $700 \text{ torr} \times \dfrac{760 \text{ mm Hg}}{760 \text{ torr}} = 700 \text{ mm Hg}$

3.   a. $450 \text{ mL} \times \dfrac{850 \text{ torr}}{524 \text{ torr}} = 730 \text{ mL}$

4. $22.86 \text{ atm} \times \dfrac{9.25 \text{ L}}{5.42 \text{ L}} = 39.0 \text{ atm}$

10. $148 \text{ L} \times \dfrac{5.00 \text{ mol}}{3.40 \text{ mol}} = 218 \text{ L}$

13.   The question cannot be answered because the original number of moles is not given.

14.   a. $200 \text{ K} - 273 = -73 \text{ °C}$

15. $6.50 \text{ L} \times \dfrac{555 \text{ K}}{298 \text{ K}} = 12.1 \text{ L}$

16. $273 + 20 \text{ °C} = 293 \text{ K}$

$$293 \text{ K} \times \frac{2.55 \text{ L}}{4.75 \text{ L}} = 157 \text{ K}$$

$$157 \text{ K} - 273 = -116 \text{ °C}$$

20. $26 \text{ °C} + 273 = 299 \text{ K}$

$534 \text{ °C} + 273 = 807 \text{ K}$

$$760 \text{ torr} \times \frac{807 \text{ K}}{299 \text{ K}} = 2.05 \times 10^3 \text{ torr}$$

$$-120 \text{ °C} + 273 = 153 \text{ K}$$

$$760 \text{ torr} \times \frac{153 \text{ K}}{299 \text{ K}} = 389 \text{ torr}$$

23. $200 \text{ °C} + 273 = 473 \text{ K}$

$$473 \text{ K} \times \frac{733 \text{ torr}}{4.00 \text{ atm}} \times \frac{1 \text{ atm}}{760 \text{ torr}} = 114 \text{ K}$$

$$114 \text{ K} - 273 = -159 \text{ °C}$$

26. $\dfrac{760 \text{ torr} \times 0.250 \text{ L}}{62.4 \text{ L torr/mol K} \times 303 \text{ K}} = 0.0100 \text{ mol}$

28. $\dfrac{1.40 \text{ mol} \times 0.0821 \text{ L atm/mol K} \times 288 \text{ K}}{0.555 \text{ L}} = 59.6 \text{ atm}$

30. $11.3 \text{ g O}_2 \times \dfrac{\text{mol O}_2}{32.0 \text{ g O}_2} = 0.353 \text{ mol O}_2;$ (continued on next page)

$$\frac{0.353 \text{ mol} \times 0.0821 \text{ L atm/mol K} \times 295 \text{ K}}{1.275 \text{ L}} = 6.71 \text{ atm}$$

32. $2.16 \text{ g} \times \dfrac{V/2}{V} \times \dfrac{2.00 \text{ atm}}{5.50 \text{ atm}} \times \dfrac{293}{323} = 0.356 \text{ g}$

(The problem can be worked this way because the mass of the sample is directly proportional to n.)

37. $45.0 \text{ g He} \times \dfrac{\text{mol He}}{4.003 \text{ g He}} = 11.2 \text{ mol He}$

$11.2 \text{ mol} \times \dfrac{950 \text{ torr}}{2.00 \text{ atm}} \times \dfrac{1 \text{ atm}}{760 \text{ torr}} = 7.03 \text{ mol}$

$7.03 \text{ mol He} \times \dfrac{4.003 \text{ g He}}{\text{mol He}} = 28.1 \text{ g He}$

(Compare this solution to that of Problem 32.)

39. $2.24 \text{ L} \times \dfrac{1.00 \text{ mol}}{22.4 \text{ L}} = 0.100 \text{ mol}$

$\dfrac{5.40 \text{ g}}{0.100 \text{ mol}} = 54.0 \text{ g/mol}$

43. $\dfrac{20.2 \text{ g Ne}}{\text{mol Ne}} \times \dfrac{\text{mol Ne}}{22.4 \text{ L}} = 0.908 \text{ g/L}$

45. $\dfrac{62.4 \text{ L torr/mol K} \times 263 \text{ K}}{725 \text{ torr}} = 22.6 \text{ L/mol}$

$\dfrac{44.0 \text{ g CO}_2}{\text{mol CO}_2} \times \dfrac{\text{mol}}{22.6 \text{ L}} = 1.95 \text{ g/L}$

47. $\dfrac{3.92 \text{ g}}{\text{L}} \times \dfrac{22.4 \text{ L}}{\text{mol}} = 87.8 \text{ g/mol}$

50. $1.02 \text{ atm} + 2.50 \text{ atm} + 0.505 \text{ atm} = 4.02 \text{ atm}$

51. $3.05 \text{ atm} \times \dfrac{760 \text{ torr}}{1 \text{ atm}} = 2318 \text{ torr}$

$2318 \text{ torr} - 833 \text{ torr} - 557 \text{ torr} = 928 \text{ torr}$

52. $100.0\% - 35.4\% - 50.8\% = 13.8\% \text{ O}$

$\dfrac{100.0 \text{ atm}}{13.8 \text{ atm O}} \times 0.550 \text{ atm O} = 3.99 \text{ atm}$

57. $0.400 \text{ g CO} \times \dfrac{\text{mol CO}}{28.0 \text{ g CO}} = 0.0143 \text{ mol CO}$

$\dfrac{0.0143 \text{ mol} \times 0.0821 \text{ L atm/mol K} \times 283 \text{ K}}{1.00 \text{ L}} = 0.332 \text{ atm}$

$2.00 \text{ atm} - 0.332 \text{ atm} = 1.67 \text{ atm}$

59. $24 \text{ torr} \times \dfrac{1 \text{ atm}}{760 \text{ torr}} = 0.0316 \text{ atm}$

$0.748 \text{ atm} - 0.0316 \text{ atm} = 0.716 \text{ atm}$

61. $5.75 \text{ L CH}_4 \times \dfrac{1 \text{ L CO}_2}{1 \text{ L CH}_4} = 5.75 \text{ L CO}_2$

$5.75 \text{ L CH}_4 \times \dfrac{2 \text{ L H}_2\text{O}}{1 \text{ L CH}_4} = 11.5 \text{ L H}_2\text{O}$ (continued on next page)

11.5 L + 5.75 L = 17.2 L total

64. $2.40 \text{ L Cl}_2 \times \dfrac{3 \text{ L HCl}}{3 \text{ L Cl}_2} = 2.40 \text{ L HCl}$

$5.75 \text{ L CH}_4 \times \dfrac{3 \text{ L HCl}}{1 \text{ L CH}_4} = 17.2 \text{ L HCl}$

Cl_2 is the limiting reactant; 2.40 L HCl will be produced.

$2.40 \text{ L Cl}_2 \times \dfrac{1 \text{ L CHCl}_3}{3 \text{ L Cl}_2} = 0.800 \text{ L CHCl}_3$

$\dfrac{1 \text{ atm} \times 0.800 \text{ L}}{0.0821 \text{ L atm/mol K} \times 298 \text{ K}} = 0.03 \text{ mol CHCl}_3$

$0.03 \text{ mol CHCl}_3 \times \dfrac{119 \text{ g CHCl}_3}{\text{mol CHCl}_3} = 6 \text{ g CHCl}_3$

65. $63.2 \text{ g KNO}_3 \times \dfrac{\text{mol KNO}_3}{101.1 \text{ g KNO}_3} \times \dfrac{1 \text{ mol N}_2}{2 \text{ mol KNO}_3} \times \dfrac{22.4 \text{ L}}{\text{mol}} = 7.00 \text{ L N}_2$

71. $54.0 \text{ g SO}_2 \times \dfrac{\text{mol SO}_2}{64.06 \text{ g SO}_2} \times \dfrac{2 \text{ mol SO}_3}{2 \text{ mol SO}_2} = 0.843 \text{ mol SO}_3$

$28.0 \text{ g O}_2 \times \dfrac{\text{mol O}_2}{32.0 \text{ g O}_2} \times \dfrac{2 \text{ mol SO}_3}{1 \text{ mol O}_2} = 1.75 \text{ mol SO}_3$

SO_2 is the limiting reactant, thus 0.843 mol of SO_3 will be formed.

$28.0 \text{ g O}_2 \times \dfrac{\text{mol O}_2}{32.0 \text{ g O}_2} = 0.875 \text{ mol O}_2 \text{ initially}$

$0.843 \text{ mol SO}_3 \times \dfrac{1 \text{ mol O}_2}{2 \text{ mol SO}_3} = 0.422 \text{ mol O}_2 \text{ used}$

0.875 mol − 0.422 mol = 0.453 mol O_2 left

$0.843 \text{ mol SO}_3 + 0.453 \text{ mol O}_2 = 1.296 \text{ mol}$

$\dfrac{1.296 \text{ mol} \times 0.0821 \text{ L atm/mol K} \times 773 \text{ K}}{5.00 \text{ L}} = 16.4 \text{ atm}$

73. $0.26 \text{ mol} \times \dfrac{0.78 \text{ L}}{1.55 \text{ L}} = 0.13 \text{ mol}$

0.26 mol − 0.13 mol = 0.13 mol escaped

76. $\dfrac{770 \text{ torr} \times 0.250 \text{ L}}{62.4 \text{ L torr/mol K} \times 298 \text{ K}} = 0.0104 \text{ mol}$

100% − 46% = 54% N_2O_4

$0.0104 \text{ mol} \times \dfrac{54 \text{ mol N}_2\text{O}_4}{100 \text{ mol}} \times \dfrac{92.0 \text{ g N}_2\text{O}_4}{\text{mol N}_2\text{O}_4} = 0.52 \text{ g N}_2\text{O}_4$

CHAPTER 13

Exercises

13.1 $\dfrac{3.08 \times 10^{-2} \text{ kJ}}{(4.8 \times 10^{-5} \text{ kJ/g }^\circ\text{C}) \times 13.1 \text{ g}} = 49 \text{ }^\circ\text{C}$

The specific heat of Al is much greater than the specific heat of H_2O.

13.2 $26.10 \text{ }^\circ\text{C} - 18.51 \text{ }^\circ\text{C} = 7.59 \text{ deg}$ (continued on next page)

$$\frac{4 \text{ kJ}}{(1.75 \times 10^3 \text{ g}) \times 7.59 \text{ deg}} = 3 \times 10^{-4} \text{ kJ/g deg}$$

13.3   42 °C − 25 °C = 17 deg

$$\frac{(2.6 \times 10^{-2} \text{ kJ/mol deg}) \times 500 \text{ g} \times 17 \text{ deg}}{2.15 \text{ kJ}} = 1.0 \times 10^2 \text{ g/mol}$$

13.4 $(1.35 \times 10^3 \text{ g}) \times 0.520 \text{ kJ/g} = 702 \text{ kJ}$

13.5 $21.0 \text{ g} \times 4.27 \text{ kJ/g} = 89.7 \text{ kJ}$

13.6   O, F

13.7

Electronegativities of the Elements	Difference in Electronegativities	Kind of Bond Formed
O, 3.5; Na, 0.9	2.6	ionic
C, 2.5; H, 2.1	0.4	polar covalent
Si, 1.8; Cl, 3.0	1.2	polar covalent
Br, 2.8; K, 0.8	2.0	ionic

Problems

1. $\dfrac{7.94 \times 10^{-4} \text{ kJ}}{\text{g °C}} \times 3.75 \text{ g} \times (29 \text{ °C} - 16 \text{ °C}) = 3.9 \times 10^{-2} \text{ kJ}$

3. $\dfrac{\text{g °C}}{1.8 \times 10^{-5} \text{ kJ}} \times \dfrac{7.15 \times 10^{-2} \text{ kJ}}{257 \text{ g}} = 15 \text{ °C}$

      25 °C + 15 °C = 40 °C

6. $1.35 \text{ g} \times \dfrac{2.26 \text{ kJ}}{\text{g}} = 3.05 \text{ kJ}$

8.   a. equal attraction   b. N   c. O

10.   a. H—C, C; C—N, N   b. C—H, C; N—H, N; C—N, N

11.   a. polar covalent   b. polar covalent

12.   a. no dipole   b. dipole

CHAPTER 15

Exercises

15.1 K_2SO_4

15.2   reaction, $Sr_3(PO_4)_2$ is insoluble
no reaction
reaction, H_2O is formed
reaction, CuS is insoluble
reaction, Ag_2SO_4 is only moderately soluble

15.3 $\dfrac{21.3 \text{ g}}{450 \text{ g} + 21.3 \text{ g}} \times 100\% = 4.52\%$

15.4 $75 \text{ g soln} \times \dfrac{8.0 \text{ g } SnCl_2}{100 \text{ g soln}} \times \dfrac{\text{mol } SnCl_2}{189.6 \text{ g } SnCl_2} \times$

$$\dfrac{1 \text{ mol } SnCl_2 \cdot 2 \text{ } H_2O}{1 \text{ mol } SnCl_2} \times \dfrac{225.6 \text{ g } SnCl_2 \cdot 2 \text{ } H_2O}{\text{mol } SnCl_2 \cdot 2 \text{ } H_2O} = 7.1 \text{ g } SnCl_2 \cdot 2 \text{ } H_2O$$

15.5 $\dfrac{40.3 \text{ g Fe(ClO}_4)_2}{750 \text{ mL soln}} \times \dfrac{1000 \text{ mL}}{\text{L}} \times \dfrac{\text{mol Fe(ClO}_4)_2}{254.7 \text{ g Fe(ClO}_4)_2} = 0.211 \text{ M}$

15.6 $0.137 \text{ L soln} \times \dfrac{0.25 \text{ mol Cl}^-}{\text{L soln}} \times \dfrac{1 \text{ mol CoCl}_3}{3 \text{ mol Cl}^-} \times \dfrac{165.3 \text{ g CoCl}_3}{\text{mol CoCl}_3} = 1.9 \text{ g CoCl}_3$

15.7 $\dfrac{2.0 \text{ mol Na}_3\text{PO}_4}{\text{L soln}} \times \dfrac{3 \text{ mol Na}^+}{\text{mol Na}_3\text{PO}_4} = 6.0 \text{ M Na}^+$

15.8 $\dfrac{119.3 \text{ g Be}_3(\text{PO}_4)_2}{1.73 \text{ L soln}} \times \dfrac{\text{mol Be}_3(\text{PO}_4)_2}{217.0 \text{ g Be}_3(\text{PO}_4)_2} \times \dfrac{3 \text{ mol Be}^{2+}}{\text{mol Be}_3(\text{PO}_4)_2} = 0.953 \text{ M Be}^{2+}$

$\dfrac{0.953 \text{ mol Be}^{2+}}{\text{L soln}} \times \dfrac{2 \text{ mol PO}_4^{3-}}{3 \text{ mol Be}^{2+}} = 0.635 \text{ M PO}_4^{-3}$

15.9 $2.00 \text{ M} \times \dfrac{43 \text{ mL}}{155 \text{ mL}} = 0.55 \text{ M}$

15.10 $2.83 \text{ g Cu} \times \dfrac{\text{mol Cu}}{63.55 \text{ g Cu}} \times \dfrac{8 \text{ mol HNO}_3}{3 \text{ mol Cu}} \times \dfrac{\text{L soln}}{3.50 \text{ mol HNO}_3} \times \dfrac{1000 \text{ mL}}{\text{L}} = 33.9 \text{ mL}$

15.11 $\text{Ba(OH)}_2 + \text{H}_2\text{SO}_4 \longrightarrow \text{BaSO}_4 + 2 \text{ H}_2\text{O}$

$550 \text{ mL} \times \dfrac{\text{L}}{1000 \text{ mL}} \times \dfrac{1.55 \text{ mol Ba(OH)}_2}{\text{L soln}} \times \dfrac{1 \text{ mol H}_2\text{SO}_4}{1 \text{ mol Ba(OH)}_2} \times \dfrac{98.1 \text{ g H}_2\text{SO}_4}{\text{mol H}_2\text{SO}_4} = 83.6 \text{ g H}_2\text{SO}_4$

15.12 $2 \text{ HClO}_4 + \text{Ba(OH)}_2 \longrightarrow \text{Ba(ClO}_4)_2 + 2 \text{ H}_2\text{O}$

$500 \text{ mL} \times \dfrac{\text{L}}{1000 \text{ mL}} \times \dfrac{1.00 \text{ mol Ba(OH)}_2}{\text{L soln}} \times \dfrac{2 \text{ mol HClO}_4}{1 \text{ mol Ba(OH)}_2}$

$\times \dfrac{\text{L soln}}{0.500 \text{ mol HClO}_4} \times \dfrac{1000 \text{ mL}}{\text{L}} = 2.00 \times 10^3 \text{ mL}$

Notice that the term (L/1000 mL) cancels the term (1000 mL/L).

15.13 $\text{pH} = -\log (4.0 \times 10^{-2}) = 1.40$

15.14 $\text{pH} = -\log\left(\dfrac{10^{-14}}{3.00}\right) = 14.48$

15.15 $\text{pH} = -\log\left(\dfrac{10^{-14}}{3.3 \times 10^{-9}}\right) = 5.52$

15.16 $[\text{H}_3\text{O}^+] = 10^{-9.5} \text{ mol/L} = 3.2 \times 10^{-10} \text{ mol/L}$

15.17 $[\text{H}_3\text{O}^+] = 10^{-12.84} \text{ mol/L} = 1.45 \times 10^{-13} \text{ mol/L}$

$[\text{OH}^-] = \dfrac{10^{-14}}{1.45 \times 10^{-13}} \text{ mol/L} = 6.92 \times 10^{-2} \text{ mol/L}$

Problems

8. a. $3 \text{ KOH} + \text{H}_3\text{PO}_4 \longrightarrow \text{K}_3\text{PO}_4 + 3 \text{ H}_2\text{O}$
    b. $\text{HNO}_3 + \text{NaOH} \longrightarrow \text{NaNO}_3 + \text{H}_2\text{O}$
    c. $\text{Ca(OH)}_2 + 2 \text{ HCl} \longrightarrow \text{CaCl}_2 + 2 \text{ H}_2\text{O}$

9. a. Fe(OH)_3 b. AgCl

10. $\dfrac{65 \text{ g}}{110 \text{ g} + 65 \text{ g}} \times 100\% = 37\%$

15. $100.0 \text{ g soln} - 25.0 \text{ g LiNO}_3 = 75.0 \text{ g H}_2\text{O}$

$\dfrac{75.0 \text{ g H}_2\text{O}}{25.0 \text{ g LiNO}_3} \times \dfrac{68.95 \text{ g LiNO}_3}{\text{mol LiNO}_3} \times 0.71 \text{ mol LiNO}_3 = 147 \text{ g H}_2\text{O}$

16. $\dfrac{3.25 \text{ g NaCl}}{19.1 \text{ g soln}} \times 100\% = 17.0\%$

18. $\dfrac{5.0 \text{ g NiCl}_2}{100 \text{ g soln}} \times \dfrac{\text{mol NiCl}_2}{129.6 \text{ g NiCl}_2} \times \dfrac{1 \text{ mol NiCl}_2 \cdot 6 \text{ H}_2\text{O}}{1 \text{ mol NiCl}_2}$

$\times \dfrac{237.7 \text{ g NiCl}_2 \cdot 6 \text{ H}_2\text{O}}{\text{mol NiCl}_2 \cdot 6 \text{ H}_2\text{O}} \times 150 \text{ g soln} = 14 \text{ g NiCl}_2 \cdot 6 \text{ H}_2\text{O}$

19. $\dfrac{15 \text{ g C}_5\text{H}_{12}}{250 \text{ mL soln}} \times \dfrac{1000 \text{ mL}}{\text{L}} \times \dfrac{\text{mol C}_5\text{H}_{12}}{72.15 \text{ g C}_5\text{H}_{12}} = 0.83 \text{ M}$

20. $\dfrac{1.50 \text{ mol Co}^{2+}}{\text{L soln}} \times \dfrac{1 \text{ mol CoCl}_2}{1 \text{ mol Co}^{2+}} \times \dfrac{129.8 \text{ g CoCl}_2}{\text{mol CoCl}_2} \times \dfrac{\text{L}}{1000 \text{ mL}} \times 550 \text{ mL} = 107 \text{ g CoCl}_2$

26. $\dfrac{4.00 \text{ mol HBr}}{\text{L soln}} \times \dfrac{\text{L}}{1000 \text{ mL}} \times 300 \text{ mL} = 1.20 \text{ mol HBr}$

$\dfrac{1.20 \text{ mol} \times 62.4 \text{ L torr/mol K} \times 293 \text{ K}}{750 \text{ torr}} = 29.3 \text{ L}$

27. $\dfrac{1.43 \text{ g H}_3\text{PO}_4}{2.50 \text{ L soln}} \times \dfrac{\text{mol H}_3\text{PO}_4}{98.0 \text{ g H}_3\text{PO}_4} \times \dfrac{2 \text{ mol H}_3\text{O}^+}{\text{mol H}_3\text{PO}_4} = 0.0117 \text{ M}$

30. $2.5 \text{ mL} \times \dfrac{\text{L}}{1000 \text{ mL}} \times \dfrac{1.50 \text{ mol NaOH}}{\text{L}} \times \dfrac{1}{1.00 \text{ L}} = 0.0038 \text{ M}$

34. $250 \text{ mL} \times \dfrac{\text{L}}{1000 \text{ mL}} \times \dfrac{0.44 \text{ mol HCl}}{\text{L soln}} \times \dfrac{\text{mol Cr}}{2 \text{ mol HCl}} \times \dfrac{52.0 \text{ g Cr}}{\text{mol Cr}} = 2.9 \text{ g Cr}$

$2.9 \text{ g Cr} \times \dfrac{\text{mol Cr}}{52.0 \text{ g Cr}} \times \dfrac{1 \text{ mol H}_2}{1 \text{ mol Cr}} = 0.056 \text{ mol H}_2$

$\dfrac{0.056 \text{ mol} \times 62.4 \text{ L torr/mol K} \times 293 \text{ K}}{760 \text{ torr}} = 1.3 \text{ L}$

36. $50 \text{ mL} \times \dfrac{\text{L}}{1000 \text{ mL}} \times \dfrac{0.00211 \text{ mol Ca(NO}_3)_2}{\text{L}} \times \dfrac{\text{mol Ca}^{2+}}{\text{mol Ca(NO}_3)_2}$

$= 1.1 \times 10^{-4} \text{ mol Ca}^{2+} \text{ (limiting reagent)}$

$25 \text{ mL} \times \dfrac{\text{L}}{1000 \text{ mL}} \times \dfrac{1.10 \text{ mol H}_2\text{SO}_4}{\text{L}} \times \dfrac{\text{mol SO}_4^{2-}}{\text{mol H}_2\text{SO}_4} = 0.028 \text{ mol SO}_4^{2-}$

$1.1 \times 10^{-4} \text{ mol CaSO}_4 \times \dfrac{136 \text{ g CaSO}_4}{\text{mol CaSO}_4} = 0.015 \text{ g CaSO}_4$

39. $100.0 \text{ mL} \times \dfrac{\text{L}}{1000 \text{ mL}} \times \dfrac{2.05 \text{ mol KOH}}{\text{L}} \times \dfrac{1 \text{ mol HNO}_3}{1 \text{ mol KOH}} \times \dfrac{1}{53.2 \text{ mL}} \times \dfrac{1000 \text{ mL}}{\text{L}} = 3.85 \text{ M}$

44. $\dfrac{10^{-14} \text{ M}^2}{5.5 \times 10^{-3} \text{ M}} = 1.8 \times 10^{-12} \text{ M}$

45. a. $\text{pH} = -\log(8.4 \times 10^{-2}) = 1.08$

b. $\dfrac{10^{-14} \text{ M}^2}{1 \times 10^{-4} \text{ M}} = 1 \times 10^{-10} \text{ M}; \text{pH} = -\log(1 \times 10^{-10}) = 10.0$

47. a. $10^{-3.82} = 1.5 \times 10^{-4} \text{ M}$

48. a. $10^{-12.7} = 2.0 \times 10^{-13} \text{ M}; \dfrac{10^{-14} \text{ M}^2}{2.0 \times 10^{-13} \text{ M}} = 0.050 \text{ M}$

50. $10^{-9.40} = 3.98 \times 10^{-10} \text{ M}; \dfrac{10^{-14} \text{ M}^2}{3.98 \times 10^{-10} \text{ M}} = 2.51 \times 10^{-5} \text{ M}$

$$250 \text{ mL} \times \frac{\text{L}}{1000 \text{ mL}} \times \frac{2.51 \times 10^{-5} \text{ mol OH}^-}{\text{L}} \times \frac{1 \text{ mol Li}_2\text{O}}{2 \text{ mol OH}^-}$$

$$\times \frac{29.9 \text{ g Li}_2\text{O}}{\text{mol Li}_2\text{O}} = 9.38 \times 10^{-5} \text{ g Li}_2\text{O}$$

CHAPTER 16

Exercises

16.1 $\dfrac{[\text{NOCl}]^2}{[\text{NO}]^2 \, [\text{Cl}_2]}$, L/mol

16.2 $\dfrac{[\text{O}_3]^2}{[\text{O}_2]^3}$, L/mol

16.3 $\dfrac{(0.936)^2}{(0.053)^2 \, (0.110)} = 2.8 \times 10^3$ L/mol

16.4 $[\text{O}_3] = \left[(1 \times 10^{-48}) \left(\dfrac{1.00 \text{ mol}}{2.00 \text{ L}} \right)^3 \right]^{\frac{1}{2}} = 4 \times 10^{-25}$ mol/L

$[\text{O}_2] = 0.50$ mol/L   (essentially unchanged)

16.5   After the addition, $[\text{CO}_2]$ and $[\text{H}_2]$ will increase while $[\text{H}_2\text{O}]$ and $[\text{CO}]$ decrease; however, at the end, $[\text{H}_2\text{O}]$ will still be greater than before the addition was made.

16.6 $[\text{HBr}]$ will decrease.

16.7 $\dfrac{(10^{-7.68})^2}{0.001} = 4 \times 10^{-13}$

16.8 $[(1.8 \times 10^{-5})(0.55)]^{\frac{1}{2}} = 3.1 \times 10^{-3}$ M

16.9 $\dfrac{(4.5 \times 10^{-4})(0.10)}{0.10} = 4.5 \times 10^{-4}$ M H_3O^+; pH $= -\log(4.5 \times 10^{-4}) = 3.35$

16.10 $\dfrac{(4.5 \times 10^{-4})(3.2)}{0.5} = 3 \times 10^{-3}$ M H_3O^+; pH $= -\log(3 \times 10^{-3}) = 2.52$

16.11 $\dfrac{(5.0 \times 10^{-6})(0.5)}{1.5} = 2 \times 10^{-6}$ M OH^-; pH $= -\log\left(\dfrac{10^{-14}}{2 \times 10^{-6}} \right) = 8.30$

16.12 $(2.12 \times 10^{-3})^2 = 4.49 \times 10^{-6}$

16.13 $(2.145 \times 10^{-4})[(2)(2.145 \times 10^{-4})]^2 = 3.95 \times 10^{-11}$ M³

16.14 $(4 \times 10^{-5})^{\frac{1}{2}} = 6 \times 10^{-3}$ M

Problems

2.   a. $\dfrac{[\text{NOBr}]^2}{[\text{NO}]^2 \, [\text{Br}_2]}$

4. $\dfrac{1.15 \text{ atm}}{0.0821 \text{ L atm/mol K} \times 300 \text{ K}} = 0.0467$ mol/L

7. $\dfrac{(1.56 \times 10^{-3})^2}{(4.53)^2 \, (5.17)} = 2.29 \times 10^{-8}$ L/mol

11. $\dfrac{0.0830 \text{ atm}}{0.0821 \text{ L atm/mol K} \times 300 \text{ K}} = 3.37 \times 10^{-3}$ mol/L   (continued on next page)

$$\frac{1.91 \text{ atm}}{0.0821 \text{ L atm/mol K} \times 300 \text{ K}} = 7.75 \times 10^{-2} \text{ mol/L}$$

$$\frac{0.0664 \text{ atm}}{0.0821 \text{ L atm/mol K} \times 300 \text{ K}} = 2.70 \times 10^{-3} \text{ mol/L}$$

$$\frac{3.37 \times 10^{-3}}{(2.70 \times 10^{-3})(7.75 \times 10^{-2})} = 16.1 \text{ L/mol}$$

15.　a.　$\dfrac{[H_3O^+] \, [HCO_3^-]}{[H_2CO_3]}$

16.　a.　$[(4.4 \times 10^{-7})(1.00)]^{\frac{1}{2}} = 6.6 \times 10^{-4} \text{ M}$

17.　a.　$[(9.1 \times 10^{-8})(2.50)]^{\frac{1}{2}} = 4.8 \times 10^{-4} \text{ M}$

19.　$[(2 \times 10^{-9})(0.0250)]^{\frac{1}{2}} = 7 \times 10^{-6} \text{ M}; \text{ pH} = -\log(7 \times 10^{-6}) = 5.15$

20.　$10^{-5.87} = 1.35 \times 10^{-6} \text{ M}; \dfrac{(1.35 \times 10^{-6})^2}{0.875} = 2.08 \times 10^{-12}$

25.　$10^{-4.2} = 6.3 \times 10^{-5} \text{ M}; \dfrac{(6.3 \times 10^{-5})^2}{0.010} = 4.0 \times 10^{-7}$

27.　a.　$\dfrac{(1.7 \times 10^{-4})(1.0)}{1.0} = 1.7 \times 10^{-4} \text{ M}; \text{ pH} = -\log(1.7 \times 10^{-4}) = 3.77$

30.　$\dfrac{(1.7 \times 10^{-6})(0.100)}{0.300} = 5.7 \times 10^{-7} \text{ M}; \text{ pH} = -\log(5.7 \times 10^{-7}) = 6.24$

32.　$10^{-7.50} = 3.16 \times 10^{-8} \text{ M}$

$$\frac{10^{-14} \text{ M}^2}{3.16 \times 10^{-8} \text{ M}} = 3.16 \times 10^{-7} \text{ M}$$

$$\frac{(3.16 \times 10^{-7})(0.085)}{2.50} = 1.1 \times 10^{-8}$$

33.　$10^{-6.5} = 3.2 \times 10^{-7} \text{ M}; \dfrac{(2.95 \times 10^{-8})(1.25)}{3.2 \times 10^{-7}} = 0.12 \text{ M}$

35.　a.　$10^{-3.2} = 6.3 \times 10^{-4} \text{ M}; \text{ Use } H_3COOH \text{ and } H_3COO^-$

$$\frac{1.7 \times 10^{-4}}{6.3 \times 10^{-4}} = \frac{0.27}{1.0} = \frac{[H_3COO^-]}{[H_3COOH]}$$

37.　$(2.37 \times 10^{-3})(2.37 \times 10^{-3}) = 5.62 \times 10^{-6}$

40.　a.　$(2 \times 10^{-5})^{\frac{1}{2}} = 4 \times 10^{-3} \text{ M}$

CHAPTER 17

Exercises

17.1　$+3, -1; +5, -1$

17.2　C, $+2$; H, $+1$; Cl, -1

17.3　N, $+5$; O, -2

17.4　$+3$

17.5　$2(Cu^{2+} + e^- \longrightarrow Cu^{1+})$

$$\frac{2 \, CN^- \longrightarrow 2 \, e^- + (CN)_2}{2 \, Cu^{2+} + 2 \, CN^- \longrightarrow 2 \, Cu^{1+} + (CN)_2}$$

17.6 $3(Cu^{2+} + 2\ e^- \longrightarrow Cu)$

$\underline{2\ N^{3-} \longrightarrow 6\ e^- + N_2}$

$3\ CuO + 2\ NH_3 \longrightarrow 3\ Cu + N_2 + 3\ H_2O$

17.7 $3(Sn^{2+} \longrightarrow Sn^{4+} + 2\ e^-)$

$\underline{I^{5+} + 6\ e^- \longrightarrow I^-}$

$6\ H_3O^+ + IO_3^- + 3\ Sn^{2+} \longrightarrow 3\ Sn^{4+} + I^- + 9\ H_2O$

17.8   cobalt (III) chloride; chromium (IV) oxide; mercury (I) chloride

17.9   yes. $2\ Ag^+ + Fe \longrightarrow Fe^{2+} + 2\ Ag$

17.10   no

17.11   no

17.12 $NO_3^- + Zn^{2+}$ no reaction; both in most oxidized state.

$NO_3 + Mn^{2+}$ no reaction; Mn^{2+} can be further oxidized, but NO_3^- is not strong enough to do it.

$Mn^{2+} + Zn^{2+}$ no reaction

Problems

1. a. HCl, -1 b. HClO, $+1$ c. $NaClO_4$, $+7$

3. a. Ba, $+2$; C, $+3$; O, -2 b. H, $+1$; B, $+3$; O, -2

5. a. $Cr^{2+} + 2\ H_3O^+ + NO_3^- \longrightarrow Cr^{3+} + NO_2 + 3\ H_2O$

   b. $SO_4^{2-} + I^- + 4\ H_3O^+ \longrightarrow SO_2 + I_2 + 6\ H_2O$

7. a. oxidizing agent: H_2O_2; reducing agent: CH_3OH

   b. oxidizing agent: CrO_4^{2-}; reducing agent: Fe^{3+}

10. a. chromium (III) fluoride;   b. cobalt (III) oxide

12. a. Zn/Zn^{2+} b. Na/Na^+ c. Li/Li^+ d. Mg/Mg^{2+}

14. no

CHAPTER 19

Exercises

19.1   C—C—C—C—C—C

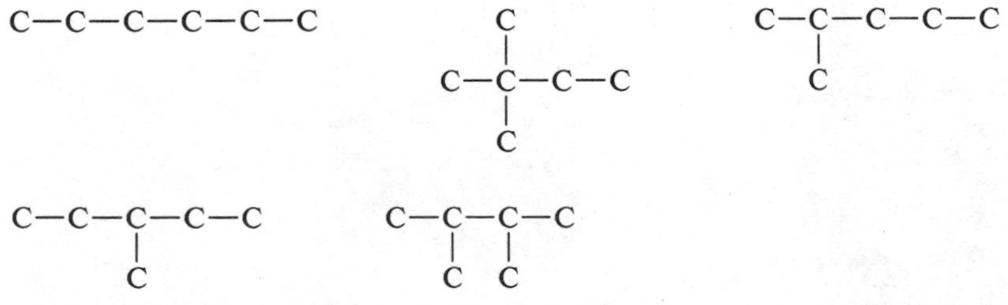

19.2   3-bromopropanal

19.3   cyclopentanone

19.4   3-bromobutanoic acid

19.5 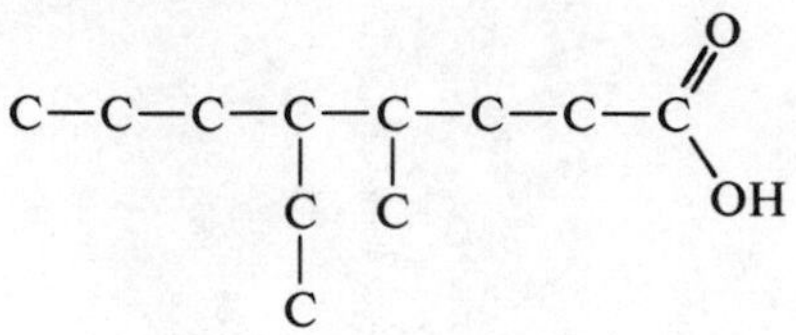

19.6 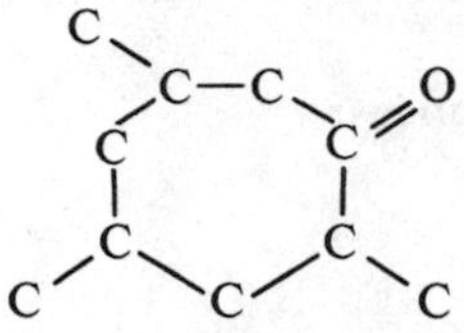

Problems

2. 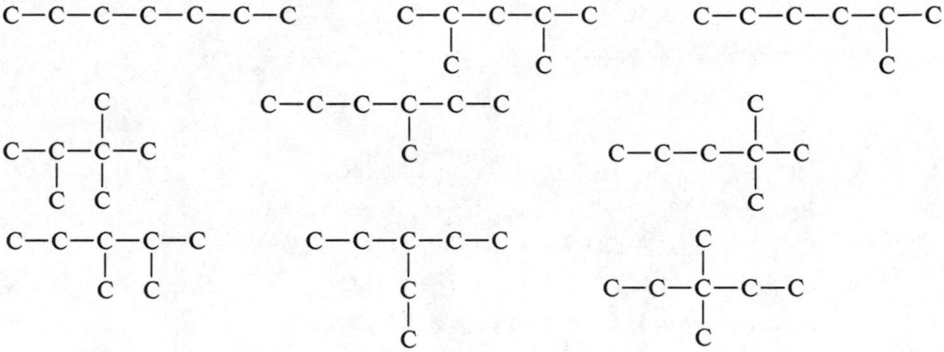

6. a. 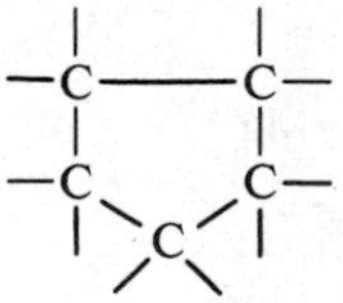

b. 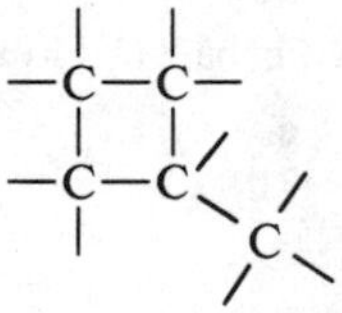

c. 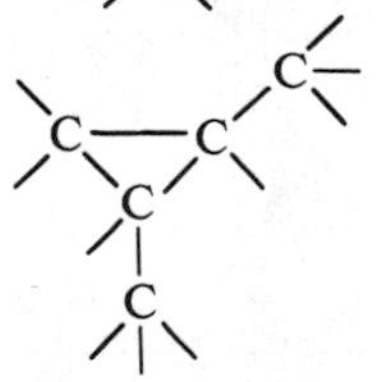

d. 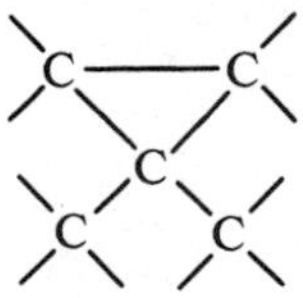

8. a. 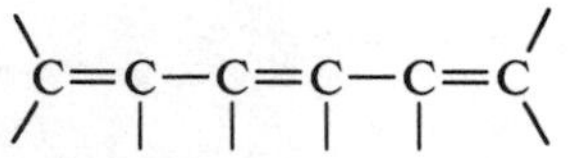

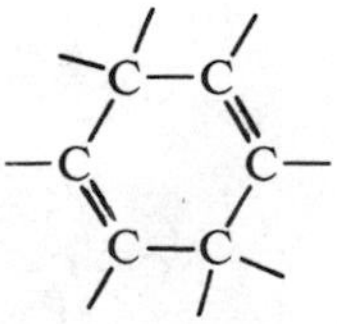

9. a. 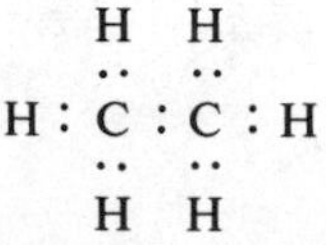

10. a. none

11. a.

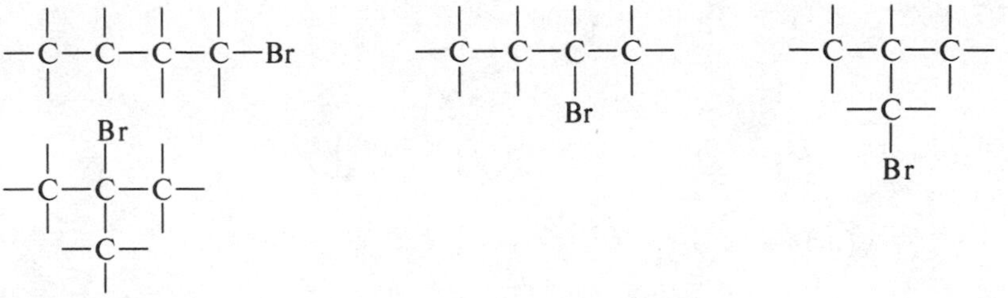

13. a. cyclopentane b. methylcyclobutane
c. 1,2-dimethylcyclopropane d. 1,1-dimethylpropane

14. a. 2-methylpropanol

16. a. 3-methylhexane

17. a.

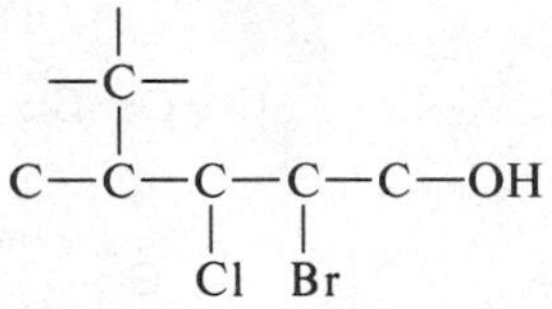

19. a. $CH_3CH_2Br + CN^- \longrightarrow CH_3CH_2CN + Br^-$

20. a. $CH_3CH_2CH_2Cl + F^- \longrightarrow CH_3CH_2CH_2F$

21. a. —C=C—

22. a. $CH_3C{\equiv}CCH_3$

24. a. $CH_3\overset{O}{\overset{\|}{C}}H \quad CH_3\overset{O}{\overset{\|}{C}}OH$

26. a. $CH_3\underset{CH_3}{\underset{|}{C}}HCH_2\overset{O}{\overset{\|}{C}}OH$

27. a. $CH_3\underset{OH}{\underset{|}{C}}HCH_2CH_3$

28. a. $CH_3CH_2\overset{O}{\overset{\|}{C}}OCH_2CH_3$

29. a. $HOCH_2CH_2\overset{O}{\overset{\|}{C}}OH$

30. a.

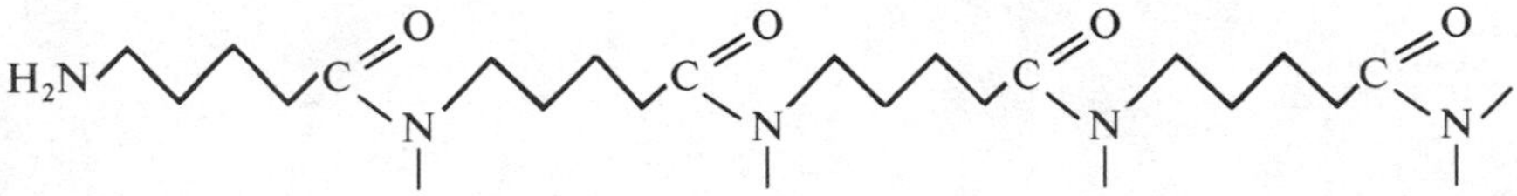

CHAPTER 21

Exercises

21.1 $^{199}_{83}Bi \longrightarrow {}^{4}_{2}He + {}^{195}_{81}Ti$

21.2 $^{14}_{7}N + ^{1}_{0}n \longrightarrow ^{14}_{6}C + ^{1}_{1}H$; a proton is produced

21.3 $^{97}_{43}Tc$

21.4 $(7.01601 + 1.00797) - 2(4.00260) = 0.01878$

      0.01878 amu $\times (1.5 \times 10^{-13}$ kJ/amu$) = 2.8 \times 10^{-15}$ kJ

21.5 $^{103}_{43}Tc$

21.6 $2\ ^{1}_{1}H$ (There is no $^{2}_{2}He$.)

Problems

1. 36 hr $\times \dfrac{\text{half-life}}{7.2 \text{ hr}} = 5$ half-lives

      $\dfrac{2.5 \times 10^7 \text{ atoms}}{2^5} = 7.81 \times 10^5$ atoms

2. $\dfrac{2.5 \text{ particles}}{0.311 \text{ g min}} \times \dfrac{\text{g min}}{16 \text{ particles}} = 0.50$; one half-life $= 5730$ years

3.   a. $^{23}_{10}Ne \longrightarrow {}^{0}_{-1}e + ^{23}_{11}Na$

5.   a. $^{13}_{6}C + ^{1}_{0}n \longrightarrow ^{14}_{6}C$

8. $^{241}_{94}Pu \longrightarrow ^{209}_{83}Bi + 8\ ^{4}_{2}He + 5\ _{-1}^{0}e$

11. $^{90}_{35}Br \longrightarrow ^{90}_{38}Sr + 3\ _{-1}^{0}e;\ ^{131}_{50}Sn$

APPENDIX A

Problems

1.   a. $3^4 = 3 \times 3 \times 3 \times 3 = 81$

      b. $5^3 = 5 \times 5 \times 5 = 125$

      c. $4^{\frac{1}{2}} = \sqrt{4} = 2$

      d. $8^{\frac{1}{5}} = \sqrt[5]{8} = 1.5157$

      e. $27^{\frac{2}{3}} = \sqrt[3]{(27)^2} = 9$

      f. $(64^{\frac{1}{2}})^{\frac{1}{3}} = (8)^{\frac{1}{3}} = \sqrt[3]{8} = 2$

      g. $5^{-2} = \dfrac{1}{5^2} = \dfrac{1}{25} = 0.04$

      h. $2^{-3} = \dfrac{1}{2^3} = \dfrac{1}{8} = 0.125$

3.   a. $2^2 \times 2^3 = 2^{2+3} = 2^5 = 32$

      b. $7^5 \times 7^{-3} = 7^{5-3} = 7^2 = 49$

      c. $8^2 \times 8^{\frac{1}{3}} = 8^{2+\frac{1}{3}} = 8^{\frac{7}{3}} = \sqrt[3]{8^7} = 128$

      d. $3^3 \cdot 3^2 = 3^{3+2} = 3^5 = 15552$

      e. $5^3 \times 5^{-2} = 5^{3+(-2)} = 5^1 = 5$

      f. $2^3/2^{-2} = 2^3 \times 2^2 = 2^5 = 32$

5.   a. $\dfrac{1}{2} \times \dfrac{1}{3} = \dfrac{1}{2 \times 3} = \dfrac{1}{6}$

      b. $\dfrac{1}{3} \times \dfrac{1}{4} \times \dfrac{1}{5} = \dfrac{1}{3 \times 4 \times 5} = \dfrac{1}{60}$

      c. $\dfrac{2}{3} \times \dfrac{3}{2} = \dfrac{2 \times 3}{3 \times 2} = \dfrac{6}{6} = 1$

7.   a. $\dfrac{\frac{1}{4}}{\frac{1}{2}} = \dfrac{1}{4} \times \dfrac{2}{1} = \dfrac{1 \times 2}{4 \times 1} = \dfrac{2}{4} = \dfrac{1}{2}$

b.

$$\frac{\frac{1}{4}}{\frac{1}{7}} = \frac{1}{4} \times \frac{7}{1} = \frac{1 \times 7}{4 \times 1} = \frac{7}{4} = 1\frac{3}{4}$$

9. a. $\frac{1}{2} \times \frac{4}{4} = \frac{4}{8}$ and $\frac{1}{8} \times \frac{1}{1} = \frac{1}{8}$ so 8 is a common denominator.

b. $\frac{1}{4} \times \frac{3}{3} = \frac{3}{12}$; $\frac{1}{12} \times \frac{1}{1} = \frac{1}{12}$; $\frac{1}{6} \times \frac{2}{2} = \frac{2}{12}$ so 12 is a common denominator.

11. a. $\frac{1}{3} + \frac{1}{5} = \frac{5}{15} + \frac{3}{15} = \frac{8}{15}$

b. $\frac{2}{5} + \frac{4}{3} + \frac{1}{2} = \frac{12}{30} + \frac{40}{30} + \frac{15}{30} = \frac{67}{30} = 2\frac{7}{30}$

c. $\frac{4}{5} - \frac{1}{10} - \frac{1}{6} = \frac{24}{30} - \frac{3}{30} - \frac{5}{30} = \frac{16}{30} = \frac{8}{15}$

13. a. $\frac{1}{4} = \frac{1 \times 25}{4 \times 25} = \frac{25}{100} = 0.25$

b. $\frac{2}{6} = \frac{1}{3} = \frac{1 \times 33.333}{3 \times 33.333} = \frac{33.333}{100} = 0.33333$

c. $\frac{3}{10} = \frac{3 \times 10}{10 \times 10} = \frac{30}{100} = 0.30$

15. a. $0.311 \times 1.82 = 0.56602$

b. $0.0033 \times 0.14 = 0.000462$

c. $0.08206/22.414 = 0.003661104$

d. $\frac{760 \times 22.4}{1.97 \times 274} = \frac{17024.}{539.78} = 31.53877506$

17. a. $3.175 + 221.01 + 15.33 + 0.002 = 239.517$

b. $431.25 - 0.07 = 431.18$

19. a. $\frac{1}{3} = \frac{1 \times 33.333}{3 \times 33.333} = \frac{33.333}{100} = 33.333\%$

b. $\frac{1}{6} = \frac{1 \times 16.667}{6 \times 16.667} = \frac{16.667}{100} = 16.667\%$

c. $\frac{1}{9} = \frac{1 \times 11.111}{9 \times 11.111} = \frac{11.111}{100} = 11.111\%$

21. a. $0.067 \times 100\% = 6.7\%$

b. $0.121 \times 100\% = 12.1\%$

c. $0.000335 \times 100\% = 0.0335\%$

d. $9.3/11.5 = 0.80869; 0.80869 \times 100\% = 80.869\%$

23. a. $12 - 7 = 5;$ b. $(-12) - 7 = -19$

26. a. $2(4x - 5) = 5x - 1$

$8x - 10 = 5x - 1$

$-5x + 8x = -1 + 10$

$3x = 9$

$x = 9/3 = 3$

b.
$$4y = y + \frac{(10y - 7)}{3}$$

$$(4y - y) \times 3 = 10y - 7$$

$$12y - 3y - 10y = -7$$

$$-y = -7 \qquad \text{so} \qquad y = 7$$

28. a. $\dfrac{b}{1} \cdot \dfrac{(ax)}{b} = [3a^2b + y/2b] \cdot b$

$$ax = 3a^2b^2 + \frac{y}{2}$$

so $\quad x = 3ab^2 + \dfrac{y}{2a}$

29. a. $.225 = 2.25 \times 10^{-1}$ d. $2,000,019 = 2.000019 \times 10^6$

b. $253 = 2.53 \times 10^2$ e. $0.00000321 = 3.21 \times 10^{-6}$

c. $44,163 = 4.4163 \times 10^4$

31. a. $\dfrac{12}{335} = .03582 = 3.582 \times 10^{-2}$

b. $\dfrac{441}{2.3} = 191.7391 = 1.917391 \times 10^2$

33. a. $(1.15 \times 10^{-3})(2.4 \times 10^5) = 12.76 \times 10^{-3+5} = 276 \times 10^2$

b. $(3.5 \times 10^{-7})(2.55 \times 10^{-3}) = 8.925 \times 10^{-7-3} = 8.925 \times 10^{-10}$

c. $\dfrac{2.15 \times 10^3}{9.93 \times 10^5} = .2165 \times 10^{3-5} = 2.165 \times 10^{-3}$

35. a. $\log 10 = \log 10^1 = 1$ e. $\log 1,000 = \log 10^3 = 3$

b. $\log 10^3 = 3$ f. $\log 0.0001 = \log 10^{-4} = -4$

c. $\log 10^{a3} = a3$

d. $\log 10^{0.221} = 0.221$ g. $\log 3.1623 = \log 10^{\frac{1}{2}} = \dfrac{1}{2}$

h. $\log 316.23 = \log 10^{2.5} = 2.5$

37. a. $\log 3.16 \times 10^4 = \log 3.16 + \log 10^4 = .499 + 4.0 = 4.499$

b. $\log 7.7 \times 10^{-6} = \log 7.7 + \log 10^{-6} = .886 + (-6) = -5.113$

c. $\log 0.00012 = \log 1.2 + \log 10^{-4} = .079 + (-4) = -3.920$

39. a. antilog $3.5 = 10^{3.5} = 3162.277 = 3.162 \times 10^3$

b. antilog $-6.7 = 10^{-6.7} = 10^{-7} \times 10^{.3} = 1.995 \times 10^{-7}$

c. antilog $2.02 = 10^{2.02} = 104.712 = 1.047 \times 10^2$

d. antilog $-1.2 = 10^{-2} \times 10^{.8} = 0.0639 = 6.309 \times 10^{-2}$

41. a. 225.0 has 4 significant figures.

b. 1000 as 1 significant figure.

c. 0.0003210 has 4 significant figures.

d. 0.000312 has 3 significant figures.

e. 1,000,000. has 7 significant figures.

f. 0.121000 has 6 significant figures.

g. 2.00001 has 6 significant figures.

43.   a. $\dfrac{1}{4}$ = .2;   b. $\dfrac{2}{6}$ = .3;   c. $\dfrac{3}{10}$ = .3

47.   a. 239.517 becomes 239.52;   b. 431.18 stays 431.18

48.   a. $77.981 \times 2.33 = 181.6957 = 1.82 \times 10^2$

       b. $4 \times 0.0665 = 0.266 = 3 \times 10^{-1}$

       c. $88.7/32 = 2.771 = 2.8$

APPENDIX B

Problems

2.   a. average is 437.5 = 438;   b. average is 1.180;   c. average if 77.0

3.   a. −66, −217, −351, 895, 162, −426; average deviation is 353

       b. −1.078, 0.12, 0.371, 0.864, −0.277; average deviation is 0.542

       c. −3.9, 0.8, 2.9, .2; average deviation is 2.0

4.   a. relative average deviation is $\dfrac{353}{438}$ = 0.806 = 80.6%

       b. relative average deviation is $\dfrac{0.542}{1.180}$ = 0.459 = 45.9%

       c. relative average deviation is $\dfrac{2.0}{77.0}$ = .026 = 2.6%

5.   a. 183 miles is average distance

       b. $1 + 3 + 14 + 5 + 6 + 2 + 1 + 4 = 36$ so $\dfrac{36}{8}$ = 4.5 miles is average deviation.

       c. $\dfrac{4.5 \text{ miles}}{183 \text{ miles}}$ = 0.025 = 2.5% is the percent relative average deviation.

7.   a. 35.04 mL is average volume

       b. 0.08 mL, −0.14 mL, −0.29 mL, 0.37 mL, −0.04 mL

       c. the average deviation is 0.18 mL

       d. $\dfrac{0.18}{35.04}$ = 0.005 = .5% is the percent relative deviation.

APPENDIX C

Problems

2.   a. 331 °C = 604 K   d. 95 °F = 35 °C

       b. 285 K = 12 °C   e. 444 °F = 501.89 K

       c. 23 °C = 73.4 °F

3.   a. 44 °F = 6.67 °C; 6.0 °C; 277 K = 4 °C so 44 °F is the highest.

       b. 12 °C; 20 °F = −6.66 °C; 200 K = −73 °C so 12 °C is the highest.

7.   solve for x : $x = (x - 32)\dfrac{5}{9}$

$9x = 5x - 160$

$4x = -160$

so $x = -40$ $-40\ °F = -40\ °C$

APPENDIX D

Problems

1.

	Species	Bonded Atoms	Unshared Pairs	Group Number	Pair Geometry	Molecular Geometry	Bond Angle
a.	CO_2	2	0	2	Linear	Linear	180°
b.	BH_3	3	0	3	Triangular	Triangular	120°
c.	NO_3^-	3	0	3	Triangular	Triangular	120°
d.	SO_3	3	0	3	Triangular	Triangular	120°
e.	H_2SO_4	4	0	4	Tetrahedral	Tetrahedral	109°

Index

Page numbers on which a definition appears are printed in boldface. An *f* following a page number indicates figure; *t* indicates table; and *e* indicates exercise.

Abbreviations, 13
Absolute temperature scale. *See* Kelvin scale
Absolute zero, **263**–264, 281
Acceleration, **46**, 47
Accident, nuclear, 558
Accuracy, **6**
Acetate ion, 343
Acetic acid, 492
  equilibrium in solutions of, 373
  ionization of in water, 343
  properties of, 434
Acetone, 490–491
Acetylene, **467**–468, 473
Acidic, **343**
Acids, **342**
  carboxylic, reactions of, 493
  equilibrium in weak, 391–395
  fatty. *See* Fatty acids
  list of strong, 344*t*
  nucleic, **521**
  organic, 492–494
  from oxides, 344
  oxyacids, 87
  reaction with base, 342–343
  strength vs structure, 434, 492*f*, 493
  strengths of oxyacids, 440, 440*t*
  strong, **343**
  weak, **343**
Activation energy, **376**
Active metals, 449
  as reducers, 418
Activity series, **426**–427, 427*t*
ADP-ATP, 512, 514
Addition reactions, **484**
  with bromine, 488

Adenine, 522
Adenosine, *See* ADP-ATP
Aerosol, 335
  liquid, 335
  solid, 335
Aerosol cans, 326
Adipic acid, **494**
Albumin, 520
Alchemy, **59**, 546
Alcohol, 478, 489
  primary, 490
  secondary, 490
Alcohol, ethyl
  formula of, **69**
  isomer of C_2H_6O, 236
  purification, 53
  synthesis of, 133
Alcohols, reactions of, 489
Aldehyde, **490**
Algebra, 574
Alkali metals, **191**, 199
  melting temperature, hardness, other properties of, 200
  properties of, 448–449, 449*t*
Alkali
  derivation of name, 448
Alkaline earths, 449–450
  properties of, 450*t*
  reactivities of, 450
  second level ionization in, 450
Alkanes, **471**
  general formulas of, 471
  straight chain, properties of, 472*t*
Alkene, **473**
Alkyl group, **471**

Alkylation of petroleum, **499**
Alkyne, **473**
Allotrope, **325**
Alloys, of steel, 460
Alpha amino acid, **515**
Alpha helix, 519, 519*f*
Alpha particle. *See* Alpha ray
Alpha ray, **153**, 540
   penetrating power of, 541*ft*
Alum, 105
Aluminum
   history, 451
   chemistry of, 451
   metallurgy, 452, 453*f*
AMU, 71
Amide, **493**–494
   linkages in nylon, 495
   linkages in proteins, 515
Amine, **491**–492
Amino acid, **515**
   twenty found in proteins, 517*t*
Ammonia, 85*e*
   complexes of, 456
   as fertilizer, 444
   formation of (Haber process), 90
   oxidation of (Ostwald oxidation), 126
   properties of, 444
   reaction with boron trifluoride, 226*f*
   structure of, 233*f*, 602*e*
   substituted with alkyl groups, 492–493
   as weak base, 344
Ammonia-ammonium buffer, 395–396
Amphoterism, 451
   of aluminum, 451
Analysis, chemical, **60**–62
   and atomic theory, 66–68
   qualitative, **60**
   quantitative, **60**–62
Analysis, dimensional, 13
Analysis, unit, 13
Anaerobic bacteria, 331
Angstrom, **23**
Anhydrous, **457**
Anion, **214**, 421
Anode, 421
Anthracene, 489
Antibodies, **521**
Antiseptic, use of H_2O_2 as, 328
Approximations, 9
   in equilibrium calculations, 394
Aqueous solution, **336**
Argon, 435
Arithmetic, 563–574
Aromatic compounds, **476**
   reactions of, 488
   structures of, 476
Asbestos fibers, as water contaminants, 329
Assumptions in calculations, 12
Atmosphere, 324
   as a unit of pressure. *See* Standard atmosphere

pressure of, 251
Atom, 104
   and Dalton's theory, 63–68
   early Greek view, 59
   kernel of, **177**
   in modern atomic theory, 68–74
   pudding model, 153
   Rutherford model, 154
   sizes, and ionization, 218*f*, 219*f*
Atomic bomb, 555
Atomic mass, 70–73, **98**, 156
Atomic mass scale, 71–73
Atomic mass unit (AMU), 71
Atomic number, 155–157
   and periodic table, 190
Atomic size
   diameters of several kinds, 197*f*, 199*f*
   effect on valence electrons, 211–212
Atomic weight, **72**, 98
   how to calculate, 116–117
   used as *gram atomic weight,* 96
Atomic weight scale, **72**, 104
   and isotopes, 73
   versus atomic mass, 72
Avogadro, Amadeo, 94
Avogadro's hypothesis, 260–**262**
Avogadro's number, **94**, 95, 104

Backside attack, 485*f*
Bakelite
   manufacture of, 496
   structure of, 497
Balance, **27**, 35
Balancing reaction equations, 88–92
   oxidation-reduction equations, 413
Barium, 450
Barometer, **251**, 251*f*
Bases, **342**
   list of strong, 344*t*
   strong, **343**
   weak, **344**
Basic, **343**
Battery, 421, **423**–425
   automotive (storage), 425–426*f*
   mechanism of operation, 424*f*
Bauxite, 452
Becquerel, Henry, 539
Benzene, **475**
   source of name, 493
   symbol for, 476
Bessemer converter, 325, 458
Beta ray, **540**
   penetrating power of, 541*ft*
Betatron, 550
Bifunctional molecule, **494**
Binary compounds, 86
Binding energy, **553**
Black water, 331
Blast furnace, 458, 458*f*, 459*f*
Bleach, chlorine, 438

Blocks, of periodic table, 196*f*
Blood
  acid balance in, 446
  buffer systems in, 395
  carbon monoxide poisoning, 446
  hemoglobin in, 515–519
Bohr atom, **165**
Bohr, Neils, 165
Boiling, 51
Boiling temperature (boiling point), **303**
  effect of dissolved substances, 52*f*
Bond angles, 232
  in organic molecules, 469
Bond line drawing, **224**
Bond structures, 232–236
  diagrams of in organic molecules, 470
Bonds, chemical, **207**
  covalent, **222–223**
  coordinate-covalent. *See* Coordinate covalent
      bonds
  dative. *See* Coordinate-covalent bonds
  double, 467
  electron densities in double, 468
  hydrogen, **310**
  hydrogen, influence on boiling point, 311*t*
  ionic, **216**, 217
  multiple, 225, 227–234. *See also* Multiple bonds
  polar covalent, **304**
  triple, 467
  triple in N₂, 225
Boron trifluoride, 226
Boyle, Robert, 256
Boyle's law, **256–257**
  calculations using, 257–260
Branched chain molecules, **468**
Breeder reactor, **558**
Bromine, 82, 438
  addition to double bond, 488
Buffer systems, **395–397**
Butadiene, **473**, 487
Butane, **471**
  structure of, 235
Butyric acid, name of, 493

Cadmium, as control in reactor, 557
calorie, **29**
Cancer, radioactive treatment of, 550
Caramel, 315
Carbohydrates, **531–533**
Carbon
  isotope 14 of, 543, 556*t*
  nonorganic, 445–447
  properties of, 466
  tetrahedral, 466
Carbon black, 445
Carbon compounds. *See also* Organic listings
  number and variety of, 468–469
Carbon dioxide
  dipole cancellation in, 307
  preparation and properties, 446

Carbon monoxide, preparation and properties, 446
Carbonates, **447**
Carbonic acid, **446**
Carbonyl, **490**
Carboxylic acids
  reactions of, 493
  salts of, 493
Catalyst, **380**
  and activation energy, 380, 381*f*
  biological. *See* Enzyme
Cathode, 421
Cation, **215**, 421
Cell, **510–511***f*
Cellulose, **532–533**
  digestion by bacteria, 533
Celsius scale, **31**
Centimeter, **23**
Central ion of complex ion, **455**
Chadwick, James, 156
Chain length, names for, 478*t*
Chain reaction (nuclear), **555**, 555*t*
Chalcocite, 460
Changes of state, 295–298
  in separations, 51
  particle energy and, 295–296
  volume changes during, 291, 291*f*
Charcoal, 445
Charge, electric, **149**
Charles's law, **262–265**
Chemical attack, on C atoms, 466
Chemical bond. *See* Bond, chemical
Chemistry, definitions and goals, **4, 45**
Chile saltpeter, 438
Chlorine dioxide, 438
Chlorine, 437–438
  covalent bond of, 222, 222*f*
  electrolytic preparation of, 422
  oxyacids of, 438
  uses, 437
  in water supplies, 438
Chlorophyll, **513**
Chromatography, **54**, 55*f. See also* cover and color
      insert
Chromium, electron population of, 174
Chromosome, **511**
Citric acid, name derivation, 493
Coal, 445
Coefficients, **88**
Coinage metals, **460**
Coke, 445
Collision theory, 376
Collisions
  effect of temperature, 377
  energy of, 377
  of gas molecules, 247
  molecular in reaction, 376
Colloids, 334
  light scattering by, 336*f*
Color, in complex ions, 456, 456*t*
Combined gas equation. *See* Ideal gas law

Combustion, **90,** 90*e,* 419
Common substances, chemical names and formulas
    for, 605*t*
Complex ions, **455**
    formation of from ions, 346
Compound, **50,** 62
Compound
    binary, 86
    nonstoichiometric, 207
Compressibility, 292
Concentration, **335**
    units of, 347–352
Concrete, 447
Condensation. *See* Vaporization
Condensed ring structures, **476**
Condensed state, **44,** 291–317
Conduction, in solution, 421
Conservation of energy, law of, **48**
Conservation of mass, law of, **48**
    and chemical analysis, 60
Constant composition, law of, 62
Controlled experiment, 252–253
Conversion factor, **13**
Conversions, temperature, 31, 595–598
Coordinate covalent bonds, **226**–227
    formation of, 226, 226*f*
    in transition metals, 455
Copper, 61–62
    electrolytic refining of, 423
    oxidation by silver ion, 424*f*
    smelting of, 460
    uses, 460
Copper-ammonia complex, 457
Copper sulfate, structure of hydride, 457
Cortisone, 530*f*
Corundum, 313
Covalency. *See also* Bonds, covalent
    in metals, 448
    in transition metals, 454–455
Covalent bond. *See* Bonds, covalent
Critical mass, **556**
Crude oil, 485
Cryolite, 452
Crystal, **45**
    conduction of, 315
    growth of from solution, 338*f*
    ionic, **311**
    metallic, change of shape of, 315
    molecular, properties of, 315
    properties of various kinds, 317*t*
    structures of, 311
Cupric oxide, 60–61
Curie, 546
Curie, Marija Sklodowska, 539
Current, electric, **149**
Cyanide ion, 217
    as a ligand, 455–456
Cyclobutane, **474**
Cyclohexane, 473–**474**
    models of, 475*f*
Cyclopropane, **474**
Cyclotron, **550**

Cytoplasm, **510**
Cytosine, 522

d block of periodic table, 196
d levels, splitting of in complex ions, 456
Dalton, John, 62–70, 547
Dalton's law of partial pressures, **275**–276
Daniell cell, **425**
de Broglie, Louis, 166
Decimal system, 566
Definite proprtions, law of, 62–63
    and Dalton, 63–66
Degree, **31**
Delocalized electrons, **475**
Denaturation of proteins, **520**
Density, **33**
    of nuclear matter, 154–155
    of some substances, 34*t*
Detergent, 493
Diamond, 313, 445*f*
    structure of, 313*f*
Diatomic elements, 83*t*
Dibromoethane, in gasoline, 438
Dichromate ion, 408, 418
    use in oxidizing alcohols, 490
Diffusion, **243**–244
Diluent, **335**
Dilute, **335**
Dilution, calculations involving, 352
Dimensional analysis, 13
Dimensions, **6**
Dipeptide, **515**
Dipole, electric, **305**
    cancellation of, 307
    of several molecules, 307*f*
Dipole force, **304**
    temporary, **304**
    in I_2, 308
Disaccharides, **532**
Disproportionation, **419**
Dissociation, **337**
    of ionic compounds in solution, **217**
Dissolving, mechanism of, 338, 339*f*
Distance, interparticle, 292
Distillation, **53,** 53*f*
    apparatus for, 53*f,* 54*f*
    of petroleum, 499
    of seawater, 53
    by sunlight, 53
DNA (deoxyribonucleic acid), 511, **522**
Double helix, **522,** 525*f*
Driving reactions, 387
Dry ice, 446
Dulong and Petit relation, 295

Edible oils, **533**
Einstein equation, 552
Electric charge, **149**
Electric circuit, 421
Electric power, **149,** 421
    in making aluminum, 453
    use in making NaOH, 364, 366*f*

Electrochemical series. *See* Activity series
Electrode potential, 427
Electrolyte, **340,** 423
    strong, 340
    weak, 341
Electrolytic reactions. *See* Reactions, electrolytic
Electron affinity, **211**–214
    of C atoms, 466
    of nonmetals, 433
"Electron sea, electron glue," 315, 448
Electron sharing, 222–223. *See also* Bonds, covalent
Electron transfer reaction, **213**–214
Electronegativity, **305,** 310
    in complex ions, 456
    differences in polar molecules, 305, 305*f*
    and oxidation numbers, 411
    of some elements, 306*t*
Electrons, **149,** 152, 156
    configurations in atoms, **173**–176, 196
    deflection experiments, 150*f*
    delocalized, 475
    dot diagrams for atoms, **175**–178
    dot diagrams for molecules, **223**–225
    energy levels, **163,** 166–173
    locations of in atoms, 181, 182
    mass of, 151
    shared pairs, 222
    shell structure, 163, 164, **168**–180
    unshared pairs of, 223
    valence, **175,** 196, 208–215, 223–224, 227–234
    wave mechanics of, **166**–168
Electroplating, 423
Elemental abundance, 79*t*
Elements, **50,** 79–82
    actinide, **191,** 201
    defined by atomic theory, **69**
    families of, 189
    lanthanide, **191,** 201
    representative, **191**
    synthetic, 551
    transition, **191,** 201
Elimination reactions, **484**
Empirical formula, **84,** 106, 117–119
    how to calculate, 111–114
Emulsifier, operation of, 494*f*
Emulsion, 335
Endocrine glands, 530
Endoenergetic. *See* Endothermic reactions
Endothermic reactions, **50**
    calculations with, 140
    reaction equations for, 92–93
Energy levels
    called shells and subshells, 192
    energies of, 171–173
    filling sequence, 164
    lowest, 164
    main, 170
    quantization of, 165
Energy, **28,** 45
    activation, **376**
    barrier, 376*f*
    biological, 511

change during reaction, 93*f*
chemical, 47
distribution in gas molecules, 248
effect of kernel charge on, 209–210
of gas molecules, temperature effects on, 248*f*
ionization, **211**–214
    effect of atom size, 212
    effect of subshell energy, 211
    of C atoms, 466
    of nonmetals, 433
kinetic, **28,** 29, 47
and mass defect, 552–553
molecular distribution, 377*f*
nuclear, 552–559
potential, 47
and reaction rate, 377, 376*f*
temperature change and, 296
Enriched nuclear fuel, 557
Enthalpy, **92**
    notation in reaction equations, 92–93
    reaction, calculations with, 140
Enzymes, 380, 510, 521, **527**
    characteristics of, 527
    mechanism of operation, 528*f*
    names of, 527
    specificity of action, 529*f*
Equations, reaction. *See* Reaction equations
Equations, nuclear. *See* Nuclear equations
Equilibrium
    chemical, 375
    dynamic, **300**
    phase, 300
Equilibrium constant, **382**
    acid-base, 391–395
    finding values of, 384–386
    interpreting, 386–387
    and redox reactions, 426
    solubility, 397–398
Equilibrium equation, **382**
Error, **6**
    experimental, 7
    assessing, 255
    random, **7**
    systematic, **6**
Essential amino acid, **516**
Essential fatty acids, **535**
Esters, **493**
    names of, 478
Estradiol, 530*f*
Ethane, **467,** 470
    properties of, 470
    structure of, 471*f*
Ethanol. *See* Ethyl alcohol
Ethyl alcohol
    geometry of, 602*e*
    from glucose oxidation, 419
    properties of, 434
    oxidation by dichromate, 408
    synthesis, 499
Ethyl group, **471**
Ethylene. *See* Ethane
Ethylene glycol, 495

Evaporation, **298**

Exoenergetic. *See* Exothermic reaction

Exothermic reaction, **50**
  calculations with, 140
  reaction equations for, 92–93

Experiment, 3
  controlled, 252–253
  for measuring pressure, 253–254

Explosives, nitrates in, 445

Exponential arithmetic, 576

Extensive quantities (properties), **29**

Extrapolation, of temperature-volume curve, 263

f block of periodic table, 196

Fahrenheit scale, **31**

Fallout, nuclear, 556

Faraday, **149**

Faraday, Michael, 149

Fats, **533**
  compositions of, 534*t*
  hydrogenation of, 535
  melting temperatures of, 533–535
  saturated, structures of, 535
  structures of, 533–534*t*
  unsaturated, structures of, 535

Fatty acids, **534**
  essential, 535

Ferrocyanide ion, 455

Fire extinguishers, CO_2, 446

Fission, nuclear, **554–559**

Flask, volumetric, use of, 350*f*

Flow, of liquid, 292, 296

Fluorides, activity of, 427

Fluorine, 437

Foam, 335

Force, **45, 46**
  and pressure, 245
  intermolecular and boiling point, 303
  intermolecular vs chemical bonds, 304
  interionic, 311–312
  nuclear, 552
  unbalanced, 46
  van der Waals, **304**, 318

Formaldehyde. *See also* Methanal
  in preparation of Bakelite, 497

Formula weight, **97**

Formula, 105
  chemical, **82–83**
  empirical. *See* Empirical formula
  of ionic compounds, 220–221
  molecular. *See* Molecular formula
  simplest, **84**

Fossil dating, 544

Fractional crystallization, **54**

Fractions, working with, 564

Frasch process for mining sulfur, 441*f*

Free radicals, 484, **486**
  reactions of, 486–487

Freezing, 51. *See also* Melting

Fructose, 531

Fuel element, nuclear, 557

Fuel oil, 471, 499

Functional groups in organic molecules, **476–478**
  names and structures, 477*t*
  name endings for, 478

Fusion, nuclear, 556–559

Gamma ray, **540**
  penetrating power of, 541*ft*

Gas, **44,** 291
  collection in laboratory, 276*f*
  and density, 273
  formation of from ions, 346
  mass of, 245
  molecular weight of, 268
  pressure of, **245**
  properties of, 243–246
  relation between volume and mass, 261*f*
  stoichiometric calculations involving, 277–280
  temperature of, 245
  volume of, **243**

Gaseous elements, 83*t*

Gasoline, 318, 471, 499

Gauge, pressure, 249*f*

Gay-Lussac's law, **265**

Geiger counter, 544–545

Gene, **511**

Geneva naming system, 477

Geometry, molecular, of some compounds, 601*t*

Geonormal samples, 70–73

Germanium, Mendeleev's prediction of, 191

Glass, **45**
  silica, 314*f*

Glucose, **531**
  oxidation of, 329, 513–514
  structure of, 531
  synthesis of, 512

Glutamic acid, 517

Glycerol (glycerine), **533**

Glycine, **515**

Glycogen, **532–533**

Gram-atomic weight, **96–98**

Gram-molecular weight, **97–98**, 104

Graphite, 313, 445
  structure of, 314*f*

Gravity, 26, 47

Group, in periodic table, **191**

Guanine, 522

ΔH (enthalpy change), **92**
  calculations involving, 140–141

Haber process for ammonia, 419, 443

Half-life, **543**

Half-reaction
  in balancing equations, 414
  in electrolytic reactions, 422–423

Hall, Charles, 452

Hall process for smelting Al, 452

Halogens, **197–199, 435–440**
  electron affinities and ionization energies of, 436*t*

liquification of, 309
oxyacids of, 440, 440*t*
properties of, 437*t*
relative atomic sizes, 436*f*
temporary dipoles in, 309
Hardness, of solids, 311–317, 317*t*
Heat, **28**–29
of fusion (melting), **297**
of fusion, and interparticle forces, 298
of phase change, values, of, 297*t*
of vaporization (condensation), **297**
and interparticle forces, 298
Heat content. *See* Enthalpy
Heisenberg uncertainty principle. *See* Uncertainty
principle
Helium, 69, 435
Helix
alpha, 519
double in nucleic acids, 522–526
Hemoglobin, **516**–519
beta chain of, 516–518, 518*f*
and sickle cell anemia, 517
Heterogeneous samples, **49**
Hexose, **531**
"Holes," in liquids, 292, 296
Homogeneous samples, **49**
Hormone, 510, 521, **529**–530
Hund's rule, **178**–180
Hydrates, **341**
Hydration, **341**
Hydride ions, 328
Hydrides
metal, as hydrogen storage medium, 331
structures and properties, 447–448
Hydrocarbons, **469**
ring, **473**
unsaturated, **473**
Hydrogen, 61–62, 70, 152
in balloons, 327
covalent bond of, 222–223*f*
as a fuel, 331
generation in laboratory, 276*f*
liquification of, 331
in making ammonia, 327
portability of, 331
preparation and uses, 326
special nature of, 201
structure of atom, 151
in torches, 327
unique properties of, 327
visualization of atom, 154
Hydrogen bonds. *See* Bonds, hydrogen
Hydrogen chloride, 439
dipole in, 308*f*
properties of, 307
reaction with water, 341, 434, 467
Hydrogen fluoride, 63
Hydrogen halides, relative sizes of, 439*t*
Hydrogen iodide, 439
equilibrium system, 374, 375*f*

Hydrogen ions, 151
as acids, **342**
Hydrogen peroxide, 328
as rocket fuel, 328
Hydrogen sulfide, in water supplies, 331
Hydrogenation, 485
of fats, 535
Hydrolysis, **343**
Hydronium ions (as acids), **342**
Hydroxide ions (as bases), **342**

Ice, structure of, 315, 316*f*
Ideal gas, **281**
Ideal gas constant, R, **266**–267
Ideal gas law, **266**–267
calculations with, 267–280
Ideal metals, 448
Indicator, acid-base, 357
Inertia, 26
Insoluble solid, formation of from ions, 346
Insulin, 530
Intensive quantities (properties), **29**, 33
Intermolecular force
and phase change, 303
International System of Units, **22**
Invert sugar, 532
Iodine, 438–439
intermolecular forces in solid, 308
sublimation of, 55*f*
solid, properties of, 315
tincture of, 439
Iodine-131, 556
Iodized salt, 439
Ion, 104, **151,** 156, **213**–221
creation by electron transfer, 213–214
creation of by hydrolysis, 341
polyatomic, 230–232
reactions in solution, 345
Ionic bond. *See* Bonds, ionic
Ionic solid, **216**
Ionic compounds, names of, 86–87, 603–605
Ionization constant, **392**
values for weak acids, 393*t*
Ionization
caused by radioactivity, 544–546
Ionization, successive and atomic number, 155
Ionization energy. *See* Energy, ionization
Ionizing solvent, 339
Iron
metallurgy of, 457–460
properties of, 52*t*
Isolated system, **252**
Isomers, **236**
examples of in organic compounds, 472
of organic molecules, 469
structural sketches of, 472
Isopropyl alcohol, 490
Isotopes, 68–73, 155
abundance of, 70–73
artificial, 550

and atomic number, 155
daughter, **548**
fissionable, 554
and law of definite proportions, 71
parent, **548**
radioactive, 542
radioactive, uses of, 550
stability of, 543
unstable, 542
IUPAC naming system, 477

K_a, **392**
K_b, **392**
Kelvin, 22, **31**
Kelvin scale, **31,** 263
Keratin, 519
Kernel, atomic, **175,** 208–209
Kernel charge, **208,** 209–214, 440
    effect on valence electrons, 209
    in nonmetals, 447
Kerosene, 471, 499
Ketone, **490**
Kilocalorie, **29**
Kilogram, **22**
Kilojoule, **30**
Kinetic energy. *See also* Energy, kinetic
    of molecules during heating, 296–300
    and temperature, 296
Kinetic molecular theory, **246**
Krebs cycle, 513–**514**
Krebs, Sir Hans, 513

Lactose, 532
Lavoister, Antoine, 48, 60
    and air composition, 324
Lawrence, E. O., 550, 551*f*
Lead, in Lavoisier experiment, 48
Le Chatelier, Henri, 390
Le Chatelier's principle, 390–391
    and buffer systems, 396
Levels, energy. *See* Energy levels
Lewis, G. N., 224
Lewis structures (Lewis diagrams), **224,** 226–232
Ligand, **455**
Limestone, 447
    in blast furnace, 457
Limiting reactant, **130**–133, 136–138
Linear accelerator, 550, 551*f*
Linear graph, 257*f*
Linear molecules, 468
Lipase, 527
Lipids, **533**
Liquid air, fractionation of, 325
Liquids, **44,** 291
    particle motion in, 292
    properties of, 318
Liquid-vapor equilibrium, 300
Liquification, 291
Liter, **25**
Lithosphere, 324

LOX, 324
Logarithms
    in pH calculation, 359
    working with, 580–582
London, Fritz, 309
London forces (dispersion forces), **309**
    how developed, 309*f*
Long chain molecules, 468
LPG (liquified petroleum gas), **471,** 499
Lubricating oil, 499

Macroscopic, **45**
Magnesium, 450
Main chain, 478
Main shell, **168**
Maltose, 532
Manganese, in steel, 460
Manometer, **251**
    operation of, 252*f*
Margarine, 535
Mass, **26,** 44
    atomic. *See* Atomic mass
    molecular. *See* Molecular mass
Mass defect, **553**
Mass-energy conversion, 552–559
Mass number, **156**–157
Mass spectrometer, **70, 151**
Mass spectrograph, 152*f*
Matter, **44**
    classifications of, 49
Meltdown, 558
Melting, 291, 312
    energetics of, 297–300
    work and interparticle distance in, 312
Melting point. *See* Melting temperature
Melting temperature, 45, **302**
Membrane, cell, **510**
Memory device for shell filling sequence, 173
Mendeleev, D. I., 189–191
Mercuric oxide, in Lavoisier experiment, 48
Mercury, 82
    in Lavoisier experiment, 48
    in Torricelli tube, 250, 251*f*
Meta, **488**
Metalloids, **82**
Metals, **80,** 199, 448–460
    melting and boiling temperatures of, 316
Meter, 22–**23**
Methanal, **491**
    in making Bakelite, 496
    in making Melmac, 497
Methane
    dipole cancellation in, 307
    reactions of, 470
    stability of, 467
    structure of, 235–236
Methyl alcohol, 489
    preparation, 327
Methyl group, **471**
Methylethylketone (MEK), **491**

Metric system, **6,** 22
  prefixes of, 23*t*
  units, 23–25
Metric-English conversions, 23–24, 28, 31
Meyer, Lothar, 189
Microscopic, **45**
Milli, **23**
Milliliter, **25**
Millimeter, **23**
Millimeter of mercury. *See* Torr
Millimole, **357**
Mineral oil, 471
Mitochondrion, **511,** 513
Mixtures, 50
Models
  "Tinkertoy," **233**–236, 475
  "pudding," of atom, 153
  space-filling, **233**–236
Molar gas volume, **271**–273
Molar heat capacity, **293**
Molar heat of fusion, **297**
Molar heat of vaporization, **297**
Molarity, **349**
  calculating, 350
Mole, 22, **94, 98,** 104–105
  and atomic weight scale, 94–96
  and mass calculations, 107–109
Mole ratios, 126–130
  as conversion factors, 128–130, 133–136
Molecular crystals. *See* Crystal, molecular
Molecular formula, 84
  calculating from empirical formula, 117–119
Molecular geometry, of some molecules, 601*t*
Molecular mass, 84
Molecular structures
  effect of on properties, 236
  sketches of organic, 471
Molecular weight, **85,** 97–98, 104
  of ionic compounds, 217
Molecule, **69,** 104
  and Dalton theory, 63–68
  defined by atomic theory, **68–69**
Monomer, **486**
Monosaccharides, **531**
Motion, 45
Multiple bonds
  geometry of, 468
  in organic molecules, 467
  structure of, 467
Multiple proportions, law of, **66**
Muriatic acid, 437

Names
  of compounds, 86, 87*t*
  indicating oxidation state, 419–421, 420*t*
  of ionic compounds, 603–605
  system for organic compounds, 476–483
Naphtha, 499
Naphthalene, 489
Natural law, 3

Negative numbers, working with, 572
Net ionic equation, **339**
Network polymers, **496**
Neutrality, electric, **151**
Neutralization, acid-base, **345**
Neutrons, **156**–157, 552
  in nuclear fission, 554–556
  numbers in nuclei, 157
  penetrating power of, 541*t*
  Ra-Be source of, 541*t*, **550**
Newlands, John, 189
Nitrate ion, as oxidizer, 418
Nitric acid
  Lewis diagram for, 229*e*
  as nitrating agent, 445
  preparation of, 126, 132
  preparation and uses, 444–445
NItric oxide, 126
Nitrogen, 442–445
  oxides of, atom ratios in, 67*f*
  triple bond in, 225, 443
Noble gases, **82,** 191, 197–198
  compounds of, 435
Noble metals, **460**
Nomenclature, for organic structures, 479
Nonideality of gases, 280
Nonmetals, **80**–82, 199
  characteristics of, 433–434
Normal boiling point, **303**
  of several liquids, 303*t*
Nuclear accident, 558
Nuclear atom, 154
Nuclear equations, **546**
  balancing of, 547–548
Nuclear power, 557–559
Nuclear stability curve, 542*f*
Nuclear waste, 558
Nucleic acid, **521,** 524*f*
Nucleoproteins, **521**
Nucleotide, 522, 523*f*
Nucleus, 154, 156
  of cell, 510–511
Number line, 572
Nylon, 495, 495*f*, 519

Observations, 3
Octet rule, 175, 214, 220, 223, 227
  and complex ions, 455
Open hearth, 458, 459*f*
Orbitals, **168**–180
  patterns in space, 182
Orbital distribution of electrons
  pictorial representation for first 11 elements, 180*f*
  table for all elements, 175–176
Organic chemistry, derivation of name, **469**
Organic compounds
  naming, 477
  properties and reactions of, 483–497
Organic molecules
  sizes of, 469

Ortho, **488**
Ostwald oxidation of ammonia, 126, 132, 418, 444, 444*f*
Overlap, of energy levels, 172, 177
Oxidation, **219**
Oxidation number, **220,** 408
 of covalently bonded atoms, 410–412
 invariant, 410, 410*t*
 rules for assigning, 409
 for several elements, 413*t*
Oxidation-reduction, **213,** 218–220, 408
 some important reactions of, 417–419
Oxidation-reduction equations
 balancing, 413–417
Oxides, 324
 acidic, 344
 basic, 344
 of nitrogen, and law of definite proportions, 66–67
 periodic repetition of, 200*t*
 of sulfur, 434
Oxidizing agent, **219**
Oxyacetylene welding, 325
Oxyacids, 87, 226–234
 common, names and formulas, 604*t*
 ions of, formulas and names, 605*t*
 names and formulas of, 604*t*
Oxygen, 60–62
 for breathing and anaesthesia, 325
 combustion in, 90*e*
 density at STP, 273*e*
 depletion, in water supplies, 331
 in Lavoisier experiment, 48
 particles in nucleus of O-18, 157*f*
 preparation from KCIO₃, 267*f*
 properties and history, 325
Ozone, **325**
 layer, 325

p block of periodic table, 195–196
p orbitals, sketches of, 183
Pancreas, 512
Para, **488**
Paraffins, **471.** *See also* Alkanes
Partial pressure of gas, **275–276**
Pauling, Linus, 519
Penetration power of radiation, 540, 541*f*
Pentane, isomers of, 472
Pentose, **531**
Percent, 570
Percent composition, **109**
 how to calculate, 114–116
Periodic table, **80,** 189–201
 blocks in, 196*f*
 periods in, 191
Permanganate ion, 455
 as oxidizer, 418
 in oxidizing alcohols, 490
Perspiration, cooling by, 299
Petroleum, **497**–499
 fractionation of, 499

pH, 359
 meter, 363*f*
 noninteger values of, 360
 tabular version of scale, 360*t*
Phases, **49**
 coexistence of, 298–302
 change of, 53, 291, 295–300
Phenanthrene, 489
Phenol (carbolic acid), 488
 in making Bakelite, 496
Phlogiston, 324
Phosphate group, in DNA, 522
Phosphorus, 442
Photosynthesis, **512,** 526
Physical changes, **50**
Physical process, 51
Pipet, 35
Planar molecules, 468
Plant nutrients, 444
Plutonium, 79, 554–559
Polar covalent bond, **304**
Polarity, molecular, **305**
Pollutants, water, 329
Polonium, 442, 539
Polyamide, **495,** 515
Polyethylene, **486**
Polyfunctional molecule, **494**
Polymer, **486**
Polymerization, **486**
Polypeptide, **516**
Polysaccharides, **532**
Polystyrene, **486**
Polyvinylchloride, **486**
Positrons, **541**
 penetrating power of, 541*t*
Potassium chlorate, in preparation of O₂, 267*f*
Potential energy, of substances during heating, 296–300
Precision, **6**
 relative, calculating, 589–594
Prefixes, metric, 23*t*
Pressure, **245**
 atmospheric, 251
 and force, 245
 forces causing, 246
 and kinetic molecular theory, 247
 standard, 271
 visible effects of, 246
Pressure-volume relations in gases, 253–257
Principal quantum number, **168**
Probability density plots for electrons, **181**
Problem solving, 9
Product, **50, 88**
Progesterone, 530*f*
Propane, **471**
Properties
 chemical, 51–52
 examples of chemical, 52*t*
 examples of physical, 52*t*
 physical, 51–52

Proportionality, **256**
 inverse, 257
Protein, 497, 514, **516,** 530
 globular, 521
 muscular, 521
 structure of, 516–521
 structural, 521
 synthesis of, 526
Proton, **151,** 152–157
 mass of, 151
Proton-neutron ratio in nuclei, 542
Proton transfer reactions, **343**
Pudding model, 153
Pure substances, 50
Purification, **51**

Quantization, **163**
Quantum mechanics, **166**
Quantum number, **168**–180

R shorthand, **483**
Radioactive emission, 539
Radioactive intensity, 543
Radioactivity, 542
Radium, 450, 539
 Ra/Be neutron source, 541*t*, **550**
Radii, of ions and atoms, 218
Rate constant, **380**
Rate of reaction. *See* Reaction rate
Reactant, **50,** 88
Reactant-product
 in relation to equilibrium, 374
Reaction equations, **88**
 balancing, 88
 balancing redox, 413–417
 reading directly in gas volume, 277
 use in calculations, 126
Reaction rate, 375–376
 effect of concentration, 379–380
 external factors affecting, 376–380
Reaction, chemical, **50**
 completion in, 373
 effect of temperature on exothermic and
  endothermic, 379
 electrolytic, 421–**423,** 422*f*
 energy change during, 93*f*
 of ions in solution, 345
 organic, mechanisms of, 484
 proton transfer, **343**
 reverse, 374
 in solution, 354–356
 typical or organic compounds, 500*t*
Reactor, nuclear, 557–558
Redox. *See* Oxidation-reduction
Reducing agent, **219**
Reduction, **219**
Relative scale, 71
Relative precision, 589–594
Rem, millirem, 546
Replication of DNA, 523*f*

Respiration, **513,** 526
Respiratory proteins, **521**
Reversibility, 378
Ribosome, **511**–512, 511*f*
Ring structures, **469**
RNA (ribonucleic acid), 523
Rockets, use of H_2O_2 in 328
Rocks, radioactive analysis of, 548*e*
Roentgen (unit of radioactivity), 546
Roentgen, Wilhelm, 539
Rounding, **8**
Rubbing alcohol, 490
Rutherford, Ernest, 153–154, 163, 540, 550

s block of periodic table, 195
Salt, **345**
 formation of, 345
Saturated hydrocarbons, **470**
Saturated solutions, 337
Scale, **7**
Scale, boiler, 330, 330*f*
Scattering of alpha particles, 153–154
Scientific method, 3
Scientific notation, 577–580
Scintillation, **540**
 counter, 545
Second, **22**
Selenium, use in copiers, 442
Sewage disposal, and water supplies, 330
Shared pair of electrons, 222–223
SI, **22**
Sickle cell anemia, 517
Significant figures, **8**
 working with, 582–584
Silicon, properties and uses, 447
Silicon dioxide (SiO_2), model of, 314*f*
Silicone rubber, 447
Silver, 460
 and photography, 460
Silver iodide, and law of definite proportions, 62
Sizes of atoms and ions, 218, 219*t*
Sklodowska, Marija. *See* Curie, Marija Sklodowska
Soap, 493
Sodium, preparation of, 422
Sodium chloride, 69
 conduction in solutions, 340
 crystal structure, 216, 216*f*, 311
 dissociation of, 218
 electrolysis of, 421–422
 mechanism of solution of, 338, 339*f*
 properties of, 52*t*
 structure of, 311
 as table salt, 217
Sodium hydroxide, 343
 industrial preparation of, 364, 366*f*
Sodium oxide, reaction with water, 344
Solids, **44,** 291–292, 317
Solubility, **337,** 397
 of some ionic compounds in water, 342*t*
 particle motion in, 338

terms describing, 341
Solubility product, **398**
  table of, 398*t*
Solute, **335**
Solutions, **334**
  acidic, **343**
  basic, **343**
  gaseous, 334
  molarity of, 349
  neutral, **345**, 359
  of nonpolar molecules, 336
  of polar molecules, 336, 338
  solid, 334
Solution map, **9**–10
Solvation, **341**
Solvent, **335**
Space-filling models, **233**–236
  examples of, 69*f*
Specific heat, **293**
  values of for some substances, 293*t*
Spectroscopy, **163**, 165
Spray cans, 326
Standard atmosphere, **251**
Standard temperature and pressure (STP), **271**
Standing waves, **167**
Starch, **532**–533
States of matter, 44
  characteristics of, 291–292
Steel, making of, 460
Steroids, 530
Stock system, **420**
Stoichiometry, **126**
  and density, 138
  and equilibrium systems, 373
STP. *See* Standard temperature and pressure
Straight chain molecules, **468**
Strontium-90, 556*t*
Structural isomers, **236**
Sublimation, **54**, 298
Subshell, **168**–180
  filling sequence, 171–173
  filling sequence, memory device for, 173
  Hund's rule, **179**–180
Subshell energy
  effect on ionization energy, 211
Substances, 50
Substitution reactions, **484**
Substrate, **527**
Sucrose. *See* Sugar
Sugar, 50, **532**
  purification of, 54
  solution of, 337
Sulfate ion, Lewis diagram for, 231*e*
Sulfur, 441
  mining of, 441*f*
  oxidation states of, 441
  properties of, 52*t*
Sulfur dioxide
  as fumigant, 442
  geometry of, 601*e*

Sulfur trioxide, SO₃
  equilibrium reaction system, 378, 379*f*
  reaction with water, 344
Sulfuric acid
  importance of, 363
  industrial preparation of, 364, 365*f*
Sulfuric-sulfurous acids, comparisons of, 434
Supersaturation, 337, 338*f*
Suspension, 334, 335*f*
Symbol, of element, **80**
Synchrotron, 550
Système International, **22**
Systems, **49**

Tantalum, 455
Teflon, **487**, 487*f*
Tellurium, 442
Temperature, **28**–29
  and reaction rate, 377
  of gas and kinetic energy of molecules, 247–248,
    248*f*
  standard, 271
Temperature scales, 31
  conversions among, 31, 595–598
Temporary dipole force. *See* London forces
Terephthalic acid, 495
  in making mylar, 496
Testosterone, 530*f*
Tetraene, **473**
Tetrahedral angle, **474**
Tetrahedral carbon, 466
Theory, 3, 62
  atomic. *See* Atomic theory
Thomson, J. J., 149, 540
Thymine, 522
Thyroid
  and iodine, 439
  radioactive treatment of, 550
Tin, in Lavoisier experiment, 48
Titanium, 455
Titration, **356**, 358*f*
Torr, **251**
Torricelli, Evangelista, 249
Torricelli tube, **250**–252
Transition metals, 452–460
  inner shell electrons in, 454
  oxidation states, 454*t*
  polyvalency in, 454
  structures of, 454
Transmutation of elements, 546–549
  induced, 550–551
Trichloroacetic acid, 493
Triene, **473**
Trigonal planar molecules, 468
Triple bond, 467
  in nitrogen, 225, 443
  in acetylene, 467–468
Tungsten, 455
Tyndall effect of light scattering, 336*f*

Uncertainty, 7
Uncertainty principle, **163,** 165
Unequally shared electrons, 304
Un-ionized substance, formation of from ions, 346
Unit factor. *See* Conversion factor
Unit of electric charge, 151
Units, 6
Universal gas constant. *See* Ideal gas constant
Unsaturated hydrocarbons, **473**
Unsaturated ring compounds, 474
Unshared pairs of electrons, 223
Unusual organic structures, 498
Uranium, 539
  isotope 238, decay series of, 549*f*
Urease, 527

Valence electron, **175**–180
Valence shell electron pair repulsion method, 234,
    599–602
Valine, 517
Value, **6**
Vanadium, in steel, 460
van der Waals, Johannes, 304
van der Waals forces. *See* Forces, van der Waals
Vapor, **53,** 276, **299**
Vapor pressure, **300**
  dependence on temperature, 301
  establishment of, 299*f*, 298–300, 303*f*
  of several liquids, as affected by temperature, 301*f*
  of water, 276
  of water at several temperatures, 606*t*
Vaporization, **53,** 291, 300, 311
  energetics of, 297–300
  temperature of, 302
  work and interparticle distance in, 311
Velocity, **45**
Vibration, of particles in condensed phases, 291,
    295–296
Vinyl, **486**
Vital force, 509
Volatility, **300**
Volumetric glassware, 35
VSEPR, 234, 599–602

Water gas reaction, 326, 419
Water, 318
  boiling at room temperature, 302*f*
  bond angle in, 233*f*
  dipole in, 307, 307*f*
  electric decomposition of, 325
  electrolysis of, 331
  energy of decomposition and formation of, 329
  fresh, source of, 329–330
  hard, 330
  heats of fusion and vaporization of, 297
  hydrogen bonds in, 310
  ionization of, 357–359
  as a ligand, 456
  in oceans, 329
  polarity of, 318
  properties of, 328
  reactions forming, 329
  reaction with metals, 343
  softening, 330
  structure of molecule, 233*f*
  vapor pressure of at several temperatures, 301*f*,
      606*t*
  waste, 330
Water supplies, 329
  "black," 329
  pollution in, 329
  purification, 330
Wavelength, **167**
  allowed and forbidden, 167*f*
Waves, 163
  electron, **166**
Weight, 26–**27,** 35
Weight fraction, of solutions, 347
Weight percent, of solutions, 347
Wilson cloud chamber, **545**–546
Work, **45,** 297

X rays, 539, 546
Xenon, 435

Student Survey

E. Russell Hardwick, *Principles of College Chemistry*

Students, send us your ideas!

The author and the publisher want to know how well this book served you and what can be done to improve it for those who will use it in the future. By completing and returning this questionnaire, you can help us develop better textbooks. We value your opinion and want to hear your comments. Thank you.

Your name (optional) _____ School _____

Your mailing address _____

City _____ State _____ ZIP _____

Instructor's name (optional) _____ Course title _____

1. How does this book compare with other texts you have used? (Check one.)

    ☐ Better than any other   ☐ Better than most

    ☐ About the same as the rest   ☐ Not as good as most

2. Circle those chapters you especially liked:

    Chapters: 1   2   3   4   5   6   7   8   9   10   11   12   13   14   15
                16   17   18   19   20   21

    Comments:

3. Circle those chapters you think could be improved:

    Chapters: 1   2   3   4   5   6   7   8   9   10   11   12   13   14   15
                16   17   18   19   20   21

    Comments:

4. Please give us your impressions of the text. (Check your rating below.)

	Excellent	Good	Average	Poor
Logical organization	( )	( )	( )	( )
Readability of text material	( )	( )	( )	( )
General layout and design	( )	( )	( )	( )
Match with instructor's course organization	( )	( )	( )	( )
Illustrations that clarify the text	( )	( )	( )	( )
Up-to-date treatment of subject	( )	( )	( )	( )
Explanation of difficult concepts	( )	( )	( )	( )
Selection of topics in the text	( )	( )	( )	( )

OVER, PLEASE

5. Please list any chapters that your instructor did not assign. _____

   _____

6. What additional topics did your instructor discuss that were not covered in the text? __

   _____

7. Did you buy this book new or used?　　　　　□ New　　　□ Used

   Do you plan to keep the book or sell it?　　□ Keep it　□ Sell it

   Do you think your instructor should continue to
   assign this book?　　　　　　　　　　　　□ Yes　　　□ No

   Did you purchase the study guide for the text?　□ Yes　　　□ No

8. After taking the course, are you interested in taking
   more courses in this field?　□ Yes　　　□ No

   Are you a major in chemistry?　　□ Yes　　□ No

9. Is a lab offered in conjunction with the course?　　□ Yes　　□ No

   If so, author and title of the lab book used: _____

   Kindly rate the coordination of the lab book with the text.

   　□ Excellent　　□ Good　　□ Average　　□ Poor　　□ Very poor

10. GENERAL COMMENTS:

May we quote you in our advertising?　□ Yes　　□ No

To mail, remove this page and mail to:　　Mary L. Paulson
　　　　　　　　　　　　　　　　　　　Burgess Publishing Company
　　　　　　　　　　　　　　　　　　　7108 Ohms Lane
　　　　　　　　　　　　　　　　　　　Minneapolis, MN 55435

THANK YOU!